Together with®

Mathematics with Solution

CLASS - 10 | TERM - 1

BASED ON THE LATEST SYLLABUS
ISSUED BY CBSE AND NCERT TEXTBOOKS

Author
S.K. Batra

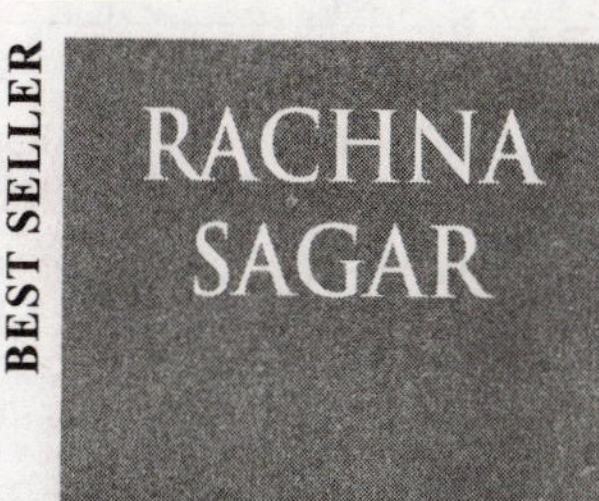

Mathematics

CLASS - 10 | TERM - 1

ISBN 978-81-8137-695-4

First Edition 1999
Twenty-fifth Latest Revised Edition

PUBLISHED BY

RACHNA SAGAR PVT. LTD.
– EDUCATIONAL PUBLISHERS –
where quality speaks for itself...

4583 / 15, Daryaganj,
New Delhi - 110 002
PO Box 7226
Phone 011 - 4358 5858, 2328 5568
Fax 011 - 2324 3519, 4311 5858
Email info@rachnasagar.in,
rachnasagar@hotmail.com,
editorial@rachnasagar.in
Web www.rachnasagar.in
IE License No. 0501009426

Offices

AHMEDABAD	Shop No.-18, Chanakya Plaza, Opp. Swagat Bunglows-1, New C.G. Road, Chandkheda, 382 424, Phone 0 99246 45576
BENGALURU	90 / 7 & 90 / 8, 1st Floor, 1st Cross, Vittal Nagar, Mysore Road, 560 026 Phone 0 90085 57707, (080) 2674 7475, 2674 7476
BHOPAL	E 6 / 127, Ground Floor, Arera Colony, 462 016 Phone 0 97525 93355, (0755) 422 3838
CHANDIGARH	S.C.O. 31, Second Floor, Sector 31D, 160 031 Phone (0172) 262 4882, Fax (0172) 508 6882
CHENNAI	Old No. 18, New No. 80, Ramar Koil Street (Opp. Chennai Trade Centre), Nandambakkam, 600 089 Phone 0 87545 80791
COCHIN	Building No. 52 / 1000, Pravada Nivas, Muttathil Lane, Kadavanthra, 682 020 Phone 0 96332 76365
DEHRADUN	I - 15, Nehru Colony, 248 001, Phone 0 73889 33938
HYDERABAD	H. No. 1 - 8 - 607 / 8, Achaiah Nagar Nallakunta, 500 020, Phone 0 90104 44006
JAIPUR	Plot No. 32 - 33, IInd Floor, Garg Complex, Prince Road (Opp. Shubh Hospital), Vidyut Nagar A, Ajmer Road, 302 021, Phone 0 97999 99123
KOLKATA	139, Hemanta Basu Sarani, (Jawpur Road), 700 074, Phone 0 93301 02176
LUCKNOW	C 1649, Indira Nagar, 226 016, Phone (0522) 400 4909
MUMBAI	VAASTU SIDDHI, 003 / 004 A Wing, Ground Floor, Shree Vaastu Enclave, Behind Manish Park, Jijamata Road, Pump House, Andheri (East), 400 093 Phone 0 81084 48884, 0 84258 69445
PATNA	1st Floor, Annpurna Market, K 54, Hanuman Nagar, Kankarbagh, 800 020 Phone 0 97714 41611, (0612) 235 1127
RANCHI	D 17 MIG, Harmu Housing Colony, 834 012, Phone 0 97714 41620

Editorial

Chief Editor	Ranvir Singh
Team In-charge	Nidhi Gupta
Subject Editor	Avnish Tyagi

Designing & Typesetting Team

Heads	Deepak Ganguly, Dinesh Vashistha
Team In-charge	Shiv Lal Koli
Subject Operator	B.S. Rawat

PREFACE

The Central Board of Secondary Education has issued new guidelines in Continuous and Comprehensive Evaluation (C.C.E.). As a result, the technique of assessing (Examination) students has undergone some changes during the last few years. To overcome these changes, a good practice material is required.

Together with® Mathematics - 10 with solution (Term - 1) is strictly designed in accordance with the syllabus and guidelines prescribed by CBSE and NCERT for C.C.E.

Term - 1 consists of six Chapters. The book is divided in two parts. Part I contains practice questions with solutions for all the six chapters. Part II consists of practice papers for Term - 1.

Practice material for each chapter consists of

- Basic concepts, supported by a number of solved examples.
- Very short answer type questions for better understanding of the topic.
- Short answer type questions - I, Short answer type questions - II and Long answer type questions for thorough practice and time management.
- Questions asked in board exams have been included under the Previous Years' Questions in each category.
- NCERT Questions contains representative problems from the NCERT textbook.
- Questions from NCERT Exemplar Problems to provide the students with an insight into and proficiency of the topic.
- HOTS (Higher Order Thinking Skills): These questions are meant for concentrating on the application part of the topics.
- Multiple Choice Questions: At the end of each chapter, MCQs have been given to help students revise all the concepts at one go.
- Value Based Questions: As per the guidelines issued by CBSE, Value-based questions will be a part of the examination. To familiarise the students with this concept, value-based questions have been added at the end of each chapter.
- Solutions: Detailed solutions of each and every problem have been given at the end of each chapter. These solutions are complete and easy to understand. Students are advised to attempt the questions given in the problem sets independently. If they fail to do so, only then they should look into the solutions.
- Formative Assessments: These include formative activities with class work, homework, oral questions and quiz questions.
- Practice Papers: 5 Practice papers (2 solved, 3 unsolved) have been prepared strictly in accordance with the guidelines issues by the CBSE. These practice papers are fully solved for students which will enable them to have a view of the standard solutions.

The interesting feature of these Practice Papers is that we have given a chart explaining the time required to attempt a particular question. Our purpose is to give an idea so as to master the subject. However, it should not be used as a tool to create panic and havoc among the students.

Answers to 3 Unsolved Practice Papers will be available on **gowebrachnasagar.com**

I hope that with all these unique features, the book will meet the requirements of all those for whom it is meant.

I am grateful to **Mr. S P Gupta and Mr. A K Jha for** constructive feedback and correcting the mistakes from time to time.

In spite of all the efforts, some errors might have crept in. I shall be thankful to the readers if they send these errors/suggestions at editorial@rachnasagar.in

I understand the scope for further improvement is always there, and hence, suggestions from all are welcome.

—Author

SYLLABUS

TERM - 1

General Instructions

- As per CCE guidelines, the syllabus of Mathematics for classes X has been divided termwise.
- The units specified for each term shall be assessed through both formative and summative assessment.
- In each term, there will be two formative assessments, each carrying 10% weightage.
- The summative assessment in term I will carry 30% weightage and the summative assessment in the II term will carry 30% weightage.
- Listed laboratory activities and projects will necessarily be assessed through formative assessments.

Course Structure

UNITS		MARKS
I	NUMBER SYSTEMS	11
II	ALGEBRA	23
III	GEOMETRY	17
IV	TRIGONOMETRY	22
V	STATISTICS	17
	TOTAL	**90**

: GENERAL GUIDELINES :

(*i*) All concepts/identities must be illustrated by situational examples.

(*ii*) The language of 'word problems' must be clear, simple and unambiguous.

(*iii*) All proofs to be produced in a non-didactic manner, allowing the learner to see flow of reason. Wherever possible, give more than one proof.

(*iv*) Motivate most results. Prove explicitly those where a short and clear argument reinforces mathematical thinking and reasoning. There must be emphasis on correct way of expressing their arguments.

(*v*) The reason for doing ruler and compass construction is to motivate and illustrate logical argument and reasoning. All constructions must include an analysis of the construction, and proof for the steps taken to do the required construction must be given.

UNIT I : NUMBER SYSTEMS

1. REAL NUMBERS ....(15) Periods

Euclid's division lemma, Fundamental Theorem of Arithmetic - statements after reviewing work done earlier and after illustrating and motivating through examples. Proofs of results – irrationality of $\sqrt{2}$, $\sqrt{3}$, $\sqrt{5}$ decimal expansions of rational numbers in terms of terminating/non-terminating recurring decimals.

UNIT II : ALGEBRA

1. POLYNOMIALS ....(7) Periods

Zeroes of a polynomial. Relationship between zeroes and coefficients of quadratic polynomials. Statement and simple problems on division algorithm for polynomials with real coefficients.

2. PAIR OF LINEAR EQUATIONS IN TWO VARIABLES ....(15) Periods

Pair of linear equations in two variables and their graphical soluion. Geometric representation of different possibilities of solutions/inconsistency.

Algebraic conditions for number of solutions. Solution of pair of linear equations in two variables algebraically – by substitution, by elimination and by cross multiplication. Simple situational problems must be included. Simple problems on equations reducible to linear equations may be included.

UNIT III : GEOMETRY

1. TRIANGLES ....(15) Periods

Definitions, examples, counter examples of similar triangles.

1. (Prove) If a line is drawn parallel to one side of a triangle to intersect the other two sides in distinct points, the other two sides are divided in the same ratio.
2. (Motivate) If a line divides two sides of a triangle in the same ratio, the line is parallel to the third side.
3. (Motivate) If in two triangles, the corresponding angles are equal, their corresponding sides are proportional and the triangles are similar.
4. (Motivate) If the corresponding sides of two triangles are proportional, their corresponding angles are equal and the two triangles are similar.
5. (Motivate) If one angle of a triangle is equal to one angle of another triangle and the sides including these angles are proportional, the two triangles are similar.
6. (Motivate) If a perpendicular is drawn from the vertex of the right angle of a right triangle to the hypotenuse, the triangles on each side of the perpendicular are similar to the whole triangle and to each other.
7. (Prove) The ratio of the areas of two similar triangles is equal to the ratio of the squares on their corresponding sides.
8. (Prove) In a right triangle, the square on the hypotenuse is equal to the sum of the squares on the other two sides.
9. (Prove) In a triangle, if the square on one side is equal to sum of the squares on the other two sides, the angles opposite to the first side is a right triangle.

UNIT IV : TRIGONOMETRY

1. INTRODUCTION TO TRIGONOMETRY ....(10) Periods

Trigonometric ratios of an acute angle of a right-angled triangle. Proof of their existence (well defined); motivate the ratios, whichever are defined at 0° and 90°. Values (with proofs) of the trigonometric ratios of 30°, 45° and 60°. Relationships between the ratios.

2. TRIGONOMETRIC IDENTITIES ....(15) Periods

Proof and applications of the identity $\sin^2 A + \cos^2 A = 1$. Only simple identities to be given. Trigonometric ratios of complementary angles.

UNIT V : STATISTICS AND PROBABILITY

1. STATISTICS ....(18) Periods

Mean, median and mode of grouped data (bimodal situation to be avoided). Cumulative frequency graph.

LIST OF LAB ACTIVITIES

ACTIVITIES

1. To obtain the conditions for consistency of a system of linear equations in two variables by graphical method.
2. To verify the Basic Proportionality Theorem using parallel line board and triangle cut-outs.
3. To verify the Pythagoras Theorem by the method of paper folding, cutting and pasting.
4. To illustrate that the medians of a triangle concurred at a point (centroid) which always lies inside triangle the circle.

PROJECTS

Project 1 : Frequency of letters/words in a language

Analysis of a language text using graphical and pie chart techniques.

GROUP ACTIVITIES

Group activity 1 : Fourth order Magic Dance

The interplay of mathematics and art can be very appealing. This activity makes an attempt to present a versatile form of the fourth order magic square through a dance.

Group activity 2 : Live Lattice

Live lattice is a lattice formed by students placed in square or rectangular formation.

SUGGESTED PROJECTS

Project 1

Mathematical designs and patterns using arithmetic progression

Project 2

Early history of Mathematics

Project 3

Analysis of test results and interpretation

CONTENTS

PART-I

PART-II

PART - I

• SUMMATIVE ASSESSMENTS

Multiple Choice Questions have been designed keeping in view the following Points:

- The stem of the question does not include any irrelevant material, question is in positive form, and options given are clear, concise and the best plausible answers.
- NCERT Exempler Problem (MCQs)

Subjective Questions

- Very Short Answer Type Questions, Short Answer Type Questions-I, Short Answer Type Questions - II, Long Answer Type Questions, NCERT Exempler Problem, HOTS Questions and Value-based Questions.
- Above type,includes Previous Years' Questions, Practice Questions and NCERT Questions

• FORMATIVE ASSESSMENTS

It is seen that schools conduct a series of 'tests'. The argument put forth is that only by conducting frequent tests continuous assessment can be ensured. However, such practices can hardly be called formative assessments because they are not integrated with the teaching-learning process.

The most important aspect is that the tasks are meant to be integrated with the teaching-learning process, i.e., while teaching a unit/lesson and not after.

Formative Assessment will include the following :

- Test Your Knowledge-I and Test Your Knowledge-II in each chapter should be assessed under formative assessment.

  (*i*) Unit tests based on the content taught during the respective periods.

  (*ii*) Written tests / oral tests

  (*iii*) H.W. / C.W.

  (*iv*) Worksheets / assignments

  (*v*) Quizzes

  (*vi*) Group activities / discussions

  (*vii*) Mathematics projects in groups of 3 to 4 students. It may be presented in any of the following forms :

  (*a*) Written project reports,

  (*b*) Charts / Models,

  (*c*) Power point presentations

  (*d*) Survey analysis, etc.

  (*viii*) Mathematics activities (Hands-on) listed in CBSE Lab Activity Manual.

  The student may be evaluated on the following :

  (*a*) Performance of activity,

  (*b*) File record of the activities performed,

  (*c*) Viva-Voce

Real Numbers

CHAPTER 1

Basic Concepts with Examples

- ***Algorithm :*** Algorithm is a series of well defined steps which gives a procedure for solving a type of problem.
- ***Lemma :*** A lemma is a proven statement used for proving another statement.
- ***Euclid's Division Lemma :*** Given positive integers a and b there exists unique integers q and r satisfying $a = bq + r$, $0 \le r < b$.
- ***Euclid's Division Algorithm :*** It is a technique to compute the Highest Common Factor (HCF) of two given positive integers.

  Euclid's Division algorithm to obtain the HCF of two positive integers say m and n with $m > n$ is :
- ***Step I :*** We find whole numbers, q and r such that $m = nq + r, \quad 0 \le r < n$
- ***Step II :*** If $r = 0$ then n is the HCF of m and n. If $r \ne 0$ apply the division lemma to n and r.
- ***Step III :*** Continue this process till the remainder is zero. The divisor at this stage will be the required HCF.

  Note : Let a and b be positive integers. If $a = bq + r$, $0 \le r < b$, then HCF (a, b) = HCF (b, r)

1. Use Euclid's Division algorithm to find the HCF of 504 and 980.

Sol. Start with the larger integer, that is 980. Using Euclid's division algorithm, we get

$$980 = 504 \times 1 + 476$$

Now considering the divisor 504 and the remainder 476, and applying the division algorithm, we get

$$504 = 476 \times 1 + 28$$

Now considering the divisor 476 and remainder 28, and applying the division algorithm, we get

$$476 = 28 \times 17 + 0$$

Notice that the remainder has become zero and we cannot proceed further. Hence the HCF of 504 and 980 is 28.

Note : 28 = HCF (476, 28) = HCF (504, 476) = HCF (980, 504)

2. Show that any odd positive integer is of the form $6q + 1$ or $6q + 3$ or $6q + 5$, where q is some integer. [NCERT]

Sol. Let a is a odd positive integer and apply division algorithm with a,

$$a = bq + r$$

Here $b = 6$, $\therefore 0 \le r < 6$

The possible remainders are 0, 1, 2, 3, 4, 5

i.e., a can be $6q$, $6q + 1$, $6q + 2$, $6q + 3$, $6q + 4$, $6q + 5$, where q is the quotient.

Since a is odd, neglect $6q$, $6q + 2$, $6q + 4$ as they are divisible by 2.

So any odd positive integer is of the form $6q + 1$, $6q + 3$ or $6q + 5$.

- ***Fundamental Theorem of Arithmetic***

  Every composite number can be expressed as product of primes and this decomposition is unique, apart from the order in which prime factors occur.

  Let p be a prime number and a be a positive integer. If p divides a^2, then p divides a.

  For two positive numbers a and b.

  HCF × LCM = $a \times b$.

3. Decompose 32760 into prime factors.

Sol. $32760 = 2 \times 2 \times 2 \times 3 \times 3 \times 5 \times 7 \times 13$
$= 2^3 \times 3^2 \times 5 \times 7 \times 13$

4. Find LCM of 26 and 91.

Sol. $26 = 2 \times 13, \quad 91 = 13 \times 7$

So, HCF of 26 and 91 = 13

$$\text{LCM} = \frac{\text{First Number} \times \text{Second Number}}{\text{HCF}}$$

$\therefore$ LCM of 26 and 91

$$= \frac{26 \times 91}{13} = 26 \times 7 = 182$$

- ***Irrational Number***

  A number is said to be irrational if it cannot be written in the form $\frac{p}{q}$, where p and q are integers and $q \ne 0$.

5. Prove that $\sqrt{2}$ is irrational. **[NCERT]** **[AI, Foreign 2010, Delhi 2009, AI 2008]**

Sol. Let $\sqrt{2} = \frac{p}{q}$ where p and q are coprime integers.

$\Rightarrow \quad q\sqrt{2} = p$

$\Rightarrow \quad q^2.2 = p^2 \quad ... (i)$

$\Rightarrow$ 2 divides p^2 as p^2 is even. If p^2 is even then p should also be even.

Let $p = 2c$ for some integer, then equation (*i*) becomes

$\Rightarrow q^2.2 = 4c^2$

$\Rightarrow q^2 = 2c^2$

$\Rightarrow q^2$ is even

$\Rightarrow q$ is also even.

This contradicts our assumption that p and q are coprime as they both have a common factor 2.

Hence $\sqrt{2}$ is irrational.

- For any rational number $\frac{p}{q}$ with terminating decimal representation, the prime factorisation of q is of the form $2^n 5^m$, where n and m are non-negative integers.
- For any rational number $\frac{p}{q}$ where prime factorisation of q is of the form $2^n\ 5^m$, where n and m are non-negative, the decimal representation is terminating.
- For any rational number $\frac{p}{q}$ where prime factorisation of q is not of the form $2^n\ 5^m$, where n and m are non-negative integers, the decimal representation is non-terminating and repeating.

Summative Assessment

Subjective Questions

VERY SHORT ANSWER TYPE QUESTIONS

This category is given for better understanding of the topic and for EXTRA PRACTICE.

Previous Years' Questions

1. If $\frac{p}{q}$ is a rational number ($q \neq 0$), what is condition of q so that the decimal representation of $\frac{p}{q}$ is terminating ? [Delhi 2008]
2. Write a rational number between $\sqrt{2}$ and $\sqrt{3}$. [AI 2008]
3. Complete the missing entries in the following factor tree : [Foreign 2008]

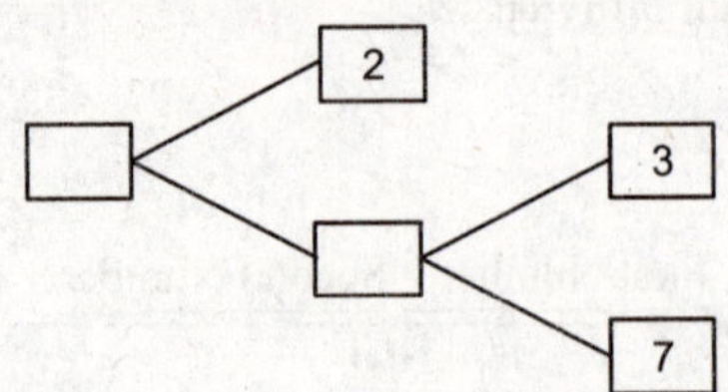

4. Write the HCF of smallest composite number and smallest prime number. [Delhi 2008 C]
5. Find the HCF of 960 and 432. [Delhi 2008 C]
6. The decimal expansion of the rational number $\frac{43}{2^4 . 5^3}$, will terminate after how many places of decimal. [Delhi 2009]
7. Find the [HCF × LCM] for the numbers 100 and 190. [AI 2009]
8. Write whether the rational number $\frac{51}{1500}$ will have a terminating decimal expansion or a non-terminating repeating decimal expansion. [Foreign 2009]
9. (*i*) The HCF and LCM of two numbers are 9 and 360 respectively. If one number is 45, write the other number. [Foreign 2009]

   (*ii*) Has the rational number $\frac{441}{2^2.5^7.7^2}$, a terminating or a non-terminating decimal representation ? [Delhi 2010]

   (*iii*) Write whether $\frac{2\sqrt{45}+3\sqrt{20}}{2\sqrt{5}}$ on simplification gives a rational or an irrational number. [All India 2010]

   (*iv*) The HCF of 45 and 105 is 15. Write their LCM. [Foreign 2010]

PRACTICE QUESTIONS

10. Express the following as product of its prime factors
    (*i*) 5005 (*ii*) 140 (*iii*) 3825
11. There is a circular path around a sports field. Priya takes 18 minutes to drive one round of the field, while Ravish takes 12 minutes for the same. Suppose they both start at the same point and at the same time, and go in the same direction. After how many minutes will they meet again at the starting point ?
12. Given that LCM (77, 99) = 693, find the HCF (77, 99).

Without actually performing the long division, state whether the following rational numbers will have terminating or non-terminating decimal expansion :

13. $\frac{13}{3125}$ 14. $\frac{129}{2^2 5^7 7^5}$ 15. $\frac{23}{2^3 5^2}$ 16. $\frac{64}{455}$

17. $\frac{63}{90}$ 18. $\frac{15}{1600}$ 19. $\frac{125}{441}$

What can you say about the prime factorisation of the denominators of the following rationals (Q.20, 21) ?

20. 34.12345 21. $34.\overline{5678}$

22. If the prime factorisation of a natural number N is $2^4 \times 3^4 \times 5^3 \times 7$, write the number of consecutive zeroes in N.
23. Express $\left(\frac{17}{8}+\frac{7}{80}\right)$ as a decimal representation.
24. Explain why $7 \times 11 \times 13 + 11$ is a composite number ?
25. Write the exponent of 3 in the prime factorisation of 162.
26. If the product of two numbers is 2500 and their HCF is 50, find their LCM.
27. If product of two numbers is 3691 and their LCM is 3691, find their HCF.
28. If $\frac{278}{2^3 m}$ has a terminating decimal expansion and m is a positive integer such that $2 < m < 9$, then find the value of m.
29. Write the sum of exponents of prime factors in the prime factorisation of 250.
30. The HCF of two numbers is 145 and their LCM is 2175. If one number is 725, find the other number.

SHORT ANSWER TYPE QUESTIONS-I

PREVIOUS YEARS' QUESTIONS

1. Prove that $15 + 17\sqrt{3}$ is an irrational number. [CBSE (CCE) 2011]
2. (*i*) Find the LCM and HCF of 120 and 144 by using Fundamental Theorem of Arithmetic. [CBSE (CCE) 2011]

   (*ii*) If $\frac{241}{4000} = \frac{241}{2^m 5^n}$, find the values of m and n where m and n are non-negative integers. Hence write its decimal expansion without actual division. [CBSE (CCE) 2012]

PRACTICE QUESTIONS

Use Euclid's division algorithm to find the HCF of (Q. 3 to Q. 10) :

3. (*a*) 426 and 576 (*b*) 56 and 814
4. 135 and 714, 38540 and 48250
5. 6265 and 76254
6. 2825 and 70625
7. 8262 and 101592
8. 26676 and 337554
9. 273070 and 434330
10. 81445 and 687897

Find HCF and LCM of following using Fundamental Theorem of Arithmetic method (Q. 11 to Q. 18) :

11. 426 and 576
12. 26676 and 337554
13. 1376 and 15428
14. 11008 and 7344
15. 625, 1125 and 2125
16. 448, 1008 and 168
17. 270, 405 and 315
18. 377, 435 and 667

NCERT QUESTIONS

Use Euclid's division algorithm to find the HCF of (Q. 19 to Q. 22) :

19. 12576 and 4052
20. 867 and 255
21. 196 and 38220
22. 135 and 225

Find HCF and LCM of following using Fundamental Theorem of Arithmetic method (Prime Factorisation) (Q. 23 to Q. 28) :

23. 96 and 404
24. 6, 72 and 120
25. 26 and 91
26. 6 and 20
27. 17, 23 and 29
28. 8, 9 and 25

Without actually performing the long division, state whether the following rational numbers will have a terminating decimal expansion or non-terminating repeating decimal expansion (Q. 29 to Q. 34) :

29. $\frac{13}{3125}$

30. $\frac{129}{2^2 5^7 7^5}$

31. $\frac{64}{455}$

32. $\frac{15}{1600}$

33. $\frac{23}{2^3 5^2}$

34. $\frac{77}{210}$

SHORT ANSWER TYPE QUESTIONS-II

Previous Years' Questions

1. Show that $5 - 2\sqrt{3}$ is an irrational number. [Delhi 2008, AI 2009]

2. (*a*) Show that $5 + 3\sqrt{2}$ is an irrational number. [Delhi 2008, Foreign 2010]

   (*b*) Prove that $\sqrt{3}$ is irrational. [AI 2008, Delhi 2009, AI 2010]

   (*c*) Use Euclid's Division Algorithm to show that the square of any positive integer is either of the form $3m$ or $3m + 1$ for some integer m. [Foreign 2008, AI 2008C]

3. Show that only one of the numbers n, $n + 2$ and $n + 4$ is divisible by 3. [Delhi 2008C]

4. Show that $3 + 5\sqrt{2}$ is an irrational number. [AI 2009]

5. (*i*) Show that for odd positive integer to be a perfect square, it should be of the form $8k + 1$.

   Or

   Show that the square of any positive odd integer is of the form $8m + 1$, for some integer m. [Foreign 2009]

   (*ii*) Prove that $2\sqrt{3} - 1$ is an irrational number. [Delhi 2010]

   (*iii*) Prove that $7 - 2\sqrt{3}$ is an irrational number. [Foreign 2010]

   (*iv*) An army contingent of 1000 members is to march behind an army band of 56 members in a parade. The two groups are to march in the same number of columns. What is the maximum number of columns in which they can march ? [CBSE (CCE) 2011]

   (*v*) Show that any positive odd integer is of the form $4q + 1$ or $4q + 3$ where q is a positive integer. [CBSE (CCE) 2011]

   (*vi*) Prove that $\frac{2\sqrt{3}}{5}$ is irrational. [CBSE (CCE) 2011]

   (*vii*) Express the number $0.3\overline{178}$ in the form of rational number $\frac{a}{b}$. [CBSE (CCE) 2012]

   (*viii*) Using Euclid's division algorithm, find whether the pair of numbers 847, 2160 are coprimes or not. [CBSE (CCE) 2012]

   (*ix*) The LCM of 2 numbers is 14 times their HCF. The sum of LCM and HCF is 600. If one number is 280, then find the other number. [CBSE (CCE) 2012]

Practice Questions

6. Prove that $3\sqrt{5}$ is irrational.
7. Prove that $\frac{1}{\sqrt{3}}$ is irrational.
8. Show that $\sqrt[3]{6}$ is an irrational number.
9. Can we have any $n \in$ N, where 4^n ends with the digit zero ?
10. Can we have any $n \in$ N, where 6^n ends with the digit zero ?
11. Can we have any $n \in$ N, where 7^n ends with the digit zero ?
12. Show that square of any positive integer is of the form $3q$ or $3q + 1$ for some integer q.
13. Show that p^2 will leave a remainder 1 when divided by 8, if p is an odd positive integer.
14. Prove that the product of two consecutive positive integers is divisible by 2.
15. Prove that square of any positive integer is of the form $4m$ or $4m + 1$ for some integer m.
16. Prove that square of any positive integer is of the form $5m + 1$ will leave a remainder 1 when divided by 5 for some integer m.
17. Find the largest number which divides 615 and 963 leaving remainder 6 in each case.
18. Find the largest number that divides 2053 and 967 and leaves a remainder of 5 and 7 respectively.
19. Find the greatest number that will divide 445, 572 and 699 leaving remainders 4, 5 and 6 respectively.
20. In a seminar the number of participants in Mathematics, Physics and Biology are 336, 240 and 96. Find the minimum number of rooms required if in each room same

number of participants is to be seated and all of them being in the same subject.

21. The length, breadth and height of a room are 8m 25cm, 6m 75cm and 4m 50cm respectively. Determine the length of the longest rod which can measure the three dimensions of the room exactly.

22. Find the smallest number which when increased by 17 is exactly divisible by 520 and 468.

NCERT Questions

23. Show that any positive odd integer is of the form $4q + 1$ or $4q + 3$, where q is some integer.

24. Use Euclid's Division Algorithm to show that the cube of any positive integer is of the form $9m$, $9m + 1$ or $9m + 8$.

25. An army contingent of 616 members is to march behind an army band of 32 members in a parade. The two groups are to march in the same number of columns. What is the maximum number of columns in which they can march ?

NCERT EXEMPLAR PROBLEMS

Short Answer Questions with Reasoning

1. Write whether every positive integer can be of the form $4q + 2$, where q is an integer. Justify your answer.

2. "The product of three consecutive positive integers is divisible by 6". Is this statement true or false ? Justify your answer.

3. Write whether the square of any positive integer can be of the form $3m + 2$, where m is a natural number. Justify your answer.

4. Can two numbers have 18 as their HCF and 380 as their LCM ? Give reasons.

5. Without actually performing the long division, find if $\frac{987}{10500}$ will have terminating or non-terminating (repeating) decimal expansion. Give reasons for your answer.

6. A rational number in its decimal expansion is 327.7081. What can you say about the prime factors of q, when this number is expressed in the form $\frac{p}{q}$? Give reasons.

Short Answer Questions

7. On a morning walk, three persons step off together and their steps measure 40 cm, 42 cm and 45 cm, respectively. What is the minimum distance each should walk so that each can cover the same distance in complete steps ?

8. Prove that $\sqrt{p} + \sqrt{q}$ is irrational, where p, q are primes.

Long Answer Questions

9. For any positive integer n, prove that $n^3 - n$ is divisible by 6.

HIGHER ORDER THINKING SKILLS (HOTS) QUESTIONS

1. Find the HCF of 52 and 117 and express it in form $52x + 117y$.

2. Prove that $x^2 - x$ is divisible by 2 for all positive integer x.

3. If m and n are odd positive integers, then $m^2 + n^2$ is even, but not divisible by 4. Justify.

4. If HCF $(6, a) = 2$ and LCM $(6, a) = 60$, then find a.

5. If remainder of $\frac{(5m+1)(5m+3)(5m+4)}{5}$ is a natural number, then find it.

6. If n is any prime number and a^2 is divisible by n, then n will also divide a. Justify.

7. Find the missing numbers in prime factors tree.

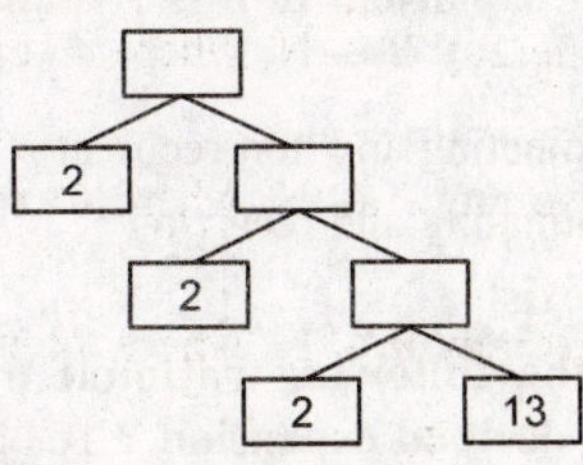

8. Find the greatest number of 5 digits exactly divisible by 12, 15 and 36.

9. Find the smallest number which when increased by 20 is exactly divisible by 90 and 144.

10. Find the smallest number which leaves remainder 8 and 12 when divided by 28 and 32 respectively.

11. Floor of a room is to be fitted with square marble tiles of the largest possible size. The size of the room is 10 m × 7 m. What should be the size of tiles required that has to be cut and how many such tiles are required ?

12. If the HCF of 408 and 1032 is expressible in the form $1032p - 408 \times 5$ find p.

VALUE BASED QUESTIONS

1. 96 defective pens are accidentally mixed with 105 good pens
   (*a*) what is LCM of 96 and 105 ?
   (*b*) the shopkeeper draws one pen and finds it to be defective. The shopkeeper did not sell and kept the pen aside. Which value is shown by the shopkeeper ?

2. 50 people work in a cooperative society. They all use their own conveyance. 20 people use their scooters, 12 go by their cars, 16 go by public transport and 2 use bicycles.
   (*a*) Find HCF of 20, 12, 16 and 2.
   (*b*) One day they all decided to go by public transport which value is shown by them ?

3. Priya and Ravish planned to participate in a cycle race to be organised for national integration. They decided to practice on a circular path around a sports field. Priya takes 18 minutes, to complete one round, while Ravish takes 12 minutes for the same. Suppose they both start at the same time and go in the same direction.
   (*a*) After how many minutes will they meet again at the starting point.
   (*b*) What is the value depicted by Priya and Ravish ?

4. Atul, Ravi and Tarun go for a morning walk. They step off together and their steps measure 40 cm, 42 cm and 45 cm, respectively.
   (*a*) What is the minimum distance each should walk so that each can cover the same distance in complete steps ?
   (*b*) How is morning walk useful ?

Multiple Choice Questions

Previous Years' Questions

1. The decimal expansion of $\frac{17}{8}$ will terminate after how many places of decimals ? **[CBSE (CCE) 2011]**

(*a*) 1 (*b*) 2
(*c*) 3 (*d*) will not terminate

2. From the following, the rational number whose decimal expansion is terminating is : **[CBSE (CCE) 2011]**

(*a*) $\frac{2}{15}$ (*b*) $\frac{11}{160}$
(*c*) $\frac{17}{60}$ (*d*) $\frac{6}{35}$

3. The decimal expansion of π is : **[CBSE (CCE) 2011]**

(*a*) terminating
(*b*) non-terminating and non-recurring
(*c*) non-terminating and recurring
(*d*) doesn't exist

4. Which of the following rational numbers have a terminating decimal expansion ? **[CBSE (CCE) 2011]**

(*a*) $\frac{125}{441}$ (*b*) $\frac{77}{210}$
(*c*) $\frac{15}{1600}$ (*d*) $\frac{129}{2^2 \times 5^2 \times 7^2}$

5. If the HCF of 55 and 99 is expressible in the form $55m - 99$, then the value of m is : **[CBSE (CCE) 2011]**

(*a*) 4 (*b*) 2
(*c*) 1 (*d*) 3

6. Given that LCM (91, 26) = 182, then HCF (91, 26) is : **[CBSE (CCE) 2011]**

(*a*) 13 (*b*) 26
(*c*) 7 (*d*) 9

7. The decimal expansion of number $\frac{441}{2^2 \times 5^3 \times 7}$ has : **[CBSE (CCE) 2012]**

(*a*) a terminating decimal
(*b*) non-terminating but repeating
(*c*) non-terminating non repeating
(*d*) terminating after two places of decimal

8. The values of x and y in the given figure are : **[CBSE (CCE) 2012]**

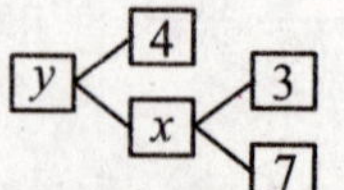

(*a*) $x = 10; y = 14$ (*b*) $x = 21; y = 84$
(*c*) $x = 21; y = 25$ (*d*) $x = 10; y = 40$

9. **For any positive integer a and 3, there exist unique integers q and r such that $a = 3q + r$, where r must satisfy :** **[CBSE (CCE) 2012]**

(a) $0 \le r < 3$

(b) $1 < r < 3$

(c) $0 < r < 3$

(d) $0 < r \le 3$

10. **$\pi - \frac{22}{7}$ is :** **[CBSE (CCE) 2012]**

(a) a rational number (b) an irrational number

(c) a prime number (d) an even number

11. **L.C.M. of $2^3 \times 3^2$ and $2^2 \times 3^3$ is :** **[CBSE (CCE) 2012]**

(a) 2^3 (b) 3^3

(c) $2^3 \times 3^3$ (d) $2^2 \times 3^2$

Practice Questions

12. **Two alarm clocks ring their alarms at regular intervals of 50 seconds and 48 seconds. If they first beep together at 12 noon, at what time will they beep again for the first time ?**

(a) 12.20 pm (b) 12.12 pm

(c) 12.11 pm (d) none of these

13. **On MG road, three consecutive traffic lights change after 36, 42 and 72 seconds. If the lights are first switched on at 9.00 am, at what time will they change simultaneously ?**

(a) 9:08:04 (b) 9:08:24

(c) 9:08:44 (d) none of these

14. **Two equilateral triangles have the sides of lengths 34 cm and 85 cm respectively. The greatest length of tape that can measure the sides of both of them exactly is :**

(a) 34 cm (b) 17 cm

(c) 51 cm (d) none of these

15. **The HCF and LCM of two numbers are 33 and 264 respectively. When the first number is completely divided by 2 the quotient is 33. The other number is :**

(a) 66 (b) 130

(c) 132 (d) 196

16. **Find the greatest number of 5 digits, that will give us remainder of 5, when divided by 8 and 9 respectively.**

(a) 99921 (b) 99931

(c) 99941 (d) 99951

17. **What will be the least possible number of the planks, if three pieces of timber 42 m, 49 m and 63 m long have to be divided into planks of the same length ?**

(a) 5 (b) 6

(c) 7 (d) none of these

18. **The LCM of 2.5, 0.5 and 0.175 is**

(a) 2.5 (b) 5

(c) 7.5 (d) 17.5

19. **What is the smallest number which when increased by 6 becomes divisible by 36, 63 and 108 ?**

(a) 750 (b) 752

(c) 754 (d) 756

20. **What is the greatest possible speed at which a man can walk 52 km and 91 km in an exact number of minutes ?**

(a) 17 m/min (b) 7 m/min

(c) 13 m/min (d) 26 m/min

21. **LCM of 5, 8, 12, 20 will not be a multiple of :**

(a) 5 (b) 3

(c) 9 (d) 2

22. **If $A = 2n + 13$, $B = n + 7$, where n is a natural number then HCF of A and B is :**

(a) 2 (b) 1

(c) 3 (d) 4

23. **Three runners running around a circular track, can complete one revolution in 2, 3 and 4 hrs respectively. They will meet again at the starting point after :**

(a) 8 hrs (b) 6 hrs

(c) 12 hrs (d) 18 hrs

24. **Two natural numbers whose difference is 66 and the least common multiple is 360, are :**

(a) 120 and 54 (b) 90 and 24

(c) 180 and 114 (d) 130 and 64

25. **Pairs of natural numbers whose least common multiple is 78 and the greatest common divisor is 13 are :**

(a) 58 and 13 or 16 and 29 (b) 68 and 23 or 36 and 49

(c) 18 and 73 or 56 and 93 (d) 78 and 13 or 26 and 39

26. **Two natural numbers whose sum is 85 and the least common multiple is 102 are :**

(a) 30 and 55 (b) 17 and 68

(c) 35 and 55 (d) 51 and 34

27. **4 Bells toll together at 9.00 am. They toll after 7, 8, 11 and 12 seconds respectively. How many times will they toll together again in the next 3 hours ?**

(a) 3 (b) 4

(c) 5 (d) 6

28. **The HCF of 2472, 1284 and a third number N is 12. If their LCM is $2^3.3^2.5 \times 103 \times 107$, then the number N is :**

(a) $2^2 \times 3^2 \times 7$ (b) $2^2 \times 3^3 \times 103$

(c) $2^2 \times 3^2 \times 5$ (d) $2^4 \times 3^2 \times 11$

29. **A forester wants to plant 66 apple trees, 88 banana trees and 110 mango trees in equal rows (in terms of number of trees). Also he wants to make distinct rows of trees (*i.e.*, only one type of trees in one row). The number of minimum rows required are**

(a) 2 (b) 3

(c) 10 (d) 12

30. **Which of the following represents the largest 4 digit number which can be added to 7249 in order to make the derived number divisible by each of 12, 14, 21, 33 and 54 ?**

(*a*) 9123 (*b*) 9383
(*c*) 8727 (*d*) none of these

31. **The ratio between the LCM and HCF of 5, 15 and 20 is :**

(*a*) 9 : 1 (*b*) 4 : 3
(*c*) 11 : 1 (*d*) 12 : 1

32. **There are 576 boys and 448 girls in a school that are to be divided into equal sections of either boys or girls alone. The total number of sections thus formed are :**

(*a*) 22 (*b*) 16
(*c*) 36 (*d*) 21

33. **Find the least number that when divided by 16, 18 and 20 leaves a remainder 4 in each case, but is completely divisible by 7.**

(*a*) 365 (*b*) 2884
(*c*) 2774 (*d*) 2974

34. **A number $10x + y$ is multiplied by another number $10a + b$ and the result comes as $100p + 10q + r$, where $r = 2y$, $q = 2(x + y)$ and $p = 2x$; $x, y < 5$, $q \neq 0$.**
The value of $10a + b$ may be :

(*a*) 11 (*b*) 13
(*c*) 31 (*d*) 22

35. **Two friends were discussing their marks in an examination. While doing so, they realised that both the numbers had the same prime factors, although one got a score which had two more factors than other. If their marks are represented by one of the options as given below, which of the following options would correctly represent the number of marks they got ?**

(*a*) 30, 60 (*b*) 20, 80
(*c*) 40, 80 (*d*) 20, 60

NCERT EXEMPLAR PROBLEMS

Multiple Choice Questions

1. If the HCF of 65 and 117 is expressible in the form $65m - 117$, then the value of m is

(*a*) 4 (*b*) 2
(*c*) 1 (*d*) 3

2. The largest number which divides 70 and 125, leaving remainders 5 and 8 respectively, is

(*a*) 13 (*b*) 65
(*c*) 875 (*d*) 1750

3. If two positive integers a and b are written as $a = x^3y^2$ and $b = xy^3$; x, y are prime numbers, then HCF (a, b) is

(*a*) xy (*b*) xy^2
(*c*) x^3y^3 (*d*) x^2y^2

4. If two positive integers p and q can be expressed as $p = ab^2$ and $q = a^3b$; a, b being prime numbers, then LCM (p, q) is

(*a*) ab (*b*) a^2b^2
(*c*) a^3b^2 (*d*) a^3b^3

5. The least number that is divisible by all the numbers from 1 to 10 (both inclusive) is

(*a*) 10 (*b*) 100
(*c*) 504 (*d*) 2520

6. The decimal expansion of the rational number $\frac{14587}{1250}$ will terminate after

(*a*) one decimal place (*b*) two decimal places
(*c*) three decimal places (*d*) four decimal places

ANSWERS

Very Short Answer Type Questions

1. $q = 2^n \times 5^m$, where n and m are whole numbers. **2.** 1.5 **3.** 42, 21 **4.** 2 **5.** 48
6. 4 **7.** 19000 **8.** Terminating **9.** (*i*) 72 (*ii*) non-terminating (*iii*) rational number (*iv*) 315
10. (*i*) $5 \times 7 \times 11 \times 13$ (*ii*) $2^2 \times 5 \times 7$ (*iii*) $3^2 \times 5^2 \times 17$ **11.** 36 minutes **12.** 11
13. Terminating **14.** Non-terminating **15.** Terminating **16.** Non-terminating
17. terminating **18.** Terminating **19.** Non-terminating
20. Prime factor of the denominator is of the form $2^m \times 5^n$, where m, n are non-negative integers.
21. Prime factors of the denominator contains factors other than 2 or 5. **22.** 3 **23.** 2.2125 **25.** 4
26. 50 **27.** 1 **28.** 5 **29.** 4 **30.** 435

Short Answer Type Questions-I

2. (*i*) 720 (*ii*) $m = 5, n = 3$, 0.06025 terminating **3.** (*a*) 6 (*b*) 2 **4.** 3, 10 **5.** 179 **6.** 2825 **7.** 306
8. 1026 **9.** 10 **10.** 1253 **11.** 6, 40896 **12.** 1026, 8776404 **13.** 4, 5307232
14. 16, 5052672 **15.** 125, 95625 **16.** 56, 4032 **17.** 45, 5670 **18.** 29, 130065 **19.** 4
20. 51 **21.** 196 **22.** 45 **23.** 4, 9696 **24.** 6, 360 **25.** 13, 182
26. 2, 60 **27.** 1, 11339 **28.** 1, 1800 **29.** Terminating **30.** Non-terminating
31. Non-terminating repeating **32.** Terminating **33.** Terminating **34.** Non-terminating repeating

Short Answer Type Questions-II

5. (*iv*) 8 (*vii*) $\frac{635}{1998}$ (*viii*) co-prime (*ix*) 80 **9.** No **10.** No **11.** No **17.** 87
18. 64 **19.** 63 **20.** 14 **21.** 75 cm **22.** 4663 **25.** 8

NCERT Exemplar Problems

1. No **2.** True **3.** No **4.** No **5.** Yes **6.** Prime factors of q are $2^4.2^5$
7. 2520 cm

Higher Order Thinking Skills (Hots) Questions

1. 13, $13 = 52(-2) + 117 \times 1$; $x = -2, y = 1$ **4.** 20 **5.** 2 **7.** 104, 52, 26
8. 99900 **9.** 700 **10.** 204 **11.** 1 m × 1 m and 70 **12.** $p = 2$

Value Based Questions

1. (*a*) 3360 (*b*) Honesty **2.** (*a*) 2 (*b*) Environment awareness **3.** (*a*) 36 minutes (*b*) Love for their country
4. (*a*) 2520 cm, (*b*) To stay fit and healthy

Multiple Choice Questions

1. (*c*) **2.** (*b*) **3.** (*b*) **4.** (*c*) **5.** (*b*) **6.** (*a*) **7.** (*a*)
8. (*b*) **9.** (*a*) **10.** (*b*) **11.** (*c*) **12.** (*a*) **13.** (*b*) **14.** (*b*)
15. (*c*) **16.** (*c*) **17.** (*d*) **18.** (*d*) **19.** (*a*) **20.** (*c*) **21.** (*c*)
22. (*b*) **23.** (*c*) **24.** (*b*) **25.** (*d*) **26.** (*d*) **27.** (*c*) **28.** (*c*)
29. (*d*) **30.** (*b*) **31.** (*d*) **32.** (*b*) **33.** (*b*) **34.** (*d*) **35.** (*b*)

NCERT Exemplar Problems

1. (*b*) **2.** (*a*) **3.** (*b*) **4.** (*c*) **5.** (*d*) **6.** (*d*)

SOLUTIONS

VERY SHORT ANSWER TYPE QUESTIONS

1. For any rational number $\frac{p}{q}$ with terminating decimal representation, the prime factorisation of q is of the form $2^n.5^m$ where n and m are non-negative integers.

2. $(\sqrt{2})^2 = 2$ and $(\sqrt{3})^2 = 3$

Let Q is any rational number between $\sqrt{2}$ and $\sqrt{3}$.

$\Rightarrow (\sqrt{2})^2 < (Q)^2 < (\sqrt{3})^2$

$\Rightarrow 2 < (Q)^2 < 3$

One possible value of $(Q)^2 = 2.25$

$\Rightarrow Q = 1.5$

3.
42 → 2, 21
21 → 3, 7

4. Smallest composite number = 4 and

Smallest prime number = 2

$\therefore$ HCF = 2

5. $960 = 2^6 \times 3 \times 5$

$432 = 2^4 \times 3^3$

$\therefore$ HCF $= 2^4 \times 3 = 48$

6. The decimal expansion of the rational numbers $\frac{43}{2^4.5^3}$, will terminate after 4 place of decimal.

$$\frac{43}{2^4.5^3} = \frac{43}{16\times125} = \frac{43}{2000} = 0.0215$$

7. HCF × LCM = one number × another number

$= 100 \times 190 = 19000$

8. $\frac{51}{1500} = \frac{17}{500}$

Prime factor of $500 = 2 \times 2 \times 5 \times 5 \times 5 = 2^2 \times 5^3$

∵ Denominator has prime factors of form $2^m.5^n$,

So, it has terminating decimal expansion.

9. (*i*) Let another number = x

HCF $(45, x) = 9$

LCM $(45, x) = 360$

HCF × LCM $= 45 \times x$

$9 \times 360 = 45 \times x$

$x = \frac{360}{5} = 72$

(*ii*) $\frac{441}{2^2.5^7.7^2}$ non-terminating decimal.

Since $q = 2^2 \times 5^7 \times 7^2$ is not of the form $2^m \times 5^n$.

(*iii*) $\frac{2\sqrt{45}+3\sqrt{20}}{2\sqrt{5}} = \frac{2\sqrt{9\times5}+3\sqrt{4\times5}}{2\sqrt{5}}$

$= \frac{6\sqrt{5}+6\sqrt{5}}{2\sqrt{5}} = \frac{12\sqrt{5}}{2\sqrt{5}}$

$= 6$ which is rational number.

(*iv*) HCF (45, 105) = 15

∴ LCM $= \frac{45\times105}{15} = 315$

10. (*i*) $5005 = 5 \times 7 \times 11 \times 13$

(*ii*) $140 = 2^2 \times 5 \times 7$

(*iii*) $3825 = 3^2 \times 5^2 \times 17$

11. They will meet again at the starting point after the time, which is equal to LCM of 18 and 12.

$18 = 2 \times 3^2, 12 = 2^2 \times 3$

∴ LCM $= 2^2 \times 3^2 = 36$

They will meet again at the starting point after 36 minutes.

12. LCM × HCF $= 77 \times 99$

⇒ $693 \times$ HCF $= 77 \times 99$

⇒ HCF $= \frac{77\times99}{693} = 11$

13. Prime factorisation of $3125 = 5^5 = 2^0 \times 5^5$ which is of the type $2^n \times 5^m$.

∴ Rational number will have a terminating decimal expansion.

14. Prime factorisation of denominator $= 2^2.5^7.7^5$

which is not of the type $2^n \times 5^m$.

∴ Number will have non-terminating repeating decimal expansion.

15. $\frac{23}{2^3 \cdot 5^2}$ denominator is of the type $2^m.5^n$

∴ Rational number will have terminating decimal expansion.

16. $\frac{64}{455} = \frac{64}{5\times7\times13}$ Denominator is not of type $2^m \times 5^n$.

∴ Rational number will have non-terminating decimal expansion.

17. $\frac{63}{90} = \frac{63}{2\times3^2\times5} = \frac{7}{2\times5}$ denominator is of the type $2^m.5^n$

∴ Rational number will have terminating decimal expansion.

18. $\frac{15}{1600} = \frac{15}{2^6\times5^2} = \frac{3}{2^6\times5}$ denominator is of the type $2^m.5^n$

∴ Rational number will have terminating decimal expansion.

19. $\frac{125}{441} = \frac{125}{3^2\times7^2}$ denominator is not of the type $2^m.5^n$

∴ Rational number will have non-terminating decimal expansion.

20. ∵ Rational number has a terminating decimal expansion.

∴ Prime factorisation of denominator will be of the type $2^n.5^m$ where n and m are non-negative integers.

21. ∵ Decimal expansion of the rational number is non-terminating repeating,

∴ Prime factorisation of denominator is not of the type $2^n \times 5^m$.

22. Number of consecutive zeroes = zeroes in $2^3 \times 5^3$

= zeroes in $(10)^3 = 3$

23. $\frac{17}{8} + \frac{7}{80} = \frac{170+7}{80} = \frac{177}{80} = 2.2125$

24. ∵ $7 \times 11 \times 13 + 11 = 11[7 \times 13 + 1]$

$= 11 \times 92 = 11 \times 2^2 \times 23$

∴ Prime factorisation has more than 1 prime number.

∴ Number is composite.

25. $162 = 2 \times 3^4$

∴ Exponent of 3 = 4

26. LCM $= \frac{\text{Product of two numbers}}{\text{HCF}} = \frac{2500}{50} = 50$

27. HCF $= \frac{\text{Product of two numbers}}{\text{LCM}} = \frac{3691}{3691} = 1$

28. $m = 5$

29. $250 = 2 \times 5^3$

∴ Sum of exponents = 1 + 3 = 4

30. HCF × LCM = Product of two numbers.

⇒ $145 \times 2175 = 725 \times$ other number

∴ Other number $= \frac{145\times2175}{725} = 435$

SHORT ANSWER TYPE QUESTIONS-I

1. Let $15 + 17\sqrt{3}$ is a rational number.

∴ $15 + 17\sqrt{3} = \frac{a}{b}$, where a and b are coprime, $b \neq 0, b \neq 1$

⇒ $17\sqrt{3} = \frac{a}{b} - 15$

$\sqrt{3} = \frac{\frac{a}{b}-15}{17}$

$\frac{a}{b} - 15$ is rational number.

$\Rightarrow \quad \dfrac{\frac{a}{b}-15}{17}$ is also a rational number.

$\because$ Rational number divided by a rational number is also rational.

But $\sqrt{3}$ is irrational.

$\therefore \quad \sqrt{3} \neq \dfrac{\frac{a}{b}-15}{17}$

$\therefore$ Our supposition is wrong.

$\Rightarrow$ $15 + 17\sqrt{3}$ is irrational.

2. (*i*)

$120 = 2^3 \times 3 \times 5$

$144 = 2^4 \times 3^2$

$\therefore \quad HCF = 2^3 \times 3 = 24$

$LCM = 2^4 \times 5 \times 3^2 = 720$

(*ii*) $\dfrac{241}{4000} = \dfrac{241}{2^m \cdot 5^n}$

$\Rightarrow \quad 4000 = 2^m.5^n$

$\Rightarrow \quad 2^5 \times 5^3 = 2^m.5^n$

$\Rightarrow \quad m = 5, n = 3$

$\dfrac{241}{4 \times 1000} = \dfrac{1}{4} \times 0.241 = 0.06025$. terminating

3. (*a*) $576 = 426 \times 1 + 150$

$\Rightarrow 426 = 150 \times 2 + 126$

$\Rightarrow 150 = 126 \times 1 + 24$

$\Rightarrow 126 = 24 \times 5 + 6$

$\Rightarrow 24 = 6 \times 4 + 0$

Hence, the HCF of 426 and 576 is 6.

Note : The remainder has become zero, so we cannot proceed further.

(*b*) $814 = 56 \times 14 + 30$

$56 = 30 \times 1 + 26$

$30 = 26 \times 1 + 4$

$26 = 4 \times 6 + 2$

$4 = 2 \times 2 + 0$

$\therefore$ HCF = 2

4. $714 = 135 \times 5 + 39$

$135 = 39 \times 3 + 18$

$39 = 18 \times 2 + 3$

$18 = 3 \times 6 + 0$

$\therefore$ H.C.F. = 3

$48250 = 38540 \times 1 + 9710$

$38540 = 9710 \times 3 + 9410$

$9710 = 9410 \times 1 + 300$

$9410 = 300 \times 31 + 110$

$300 = 110 \times 2 + 80$

$110 = 80 \times 1 + 30$

$80 = 30 \times 2 + 20$

$30 = 20 \times 1 + 10$

$20 = 10 \times 2 + 0$

$\therefore$ H.C.F. = 10

5. $76254 = 6265 \times 12 + 1074$

$6265 = 1074 \times 5 + 895$

$1074 = 895 \times 1 + 179$

$895 = 179 \times 5 + 0$

$\therefore$ H.C.F = 179

6. $70625 = 2825 \times 25 + 0$

$\therefore$ H.C.F. = 2825

7. $101592 = 8262 \times 12 + 2448$

$8262 = 2448 \times 3 + 918$

$2448 = 918 \times 2 + 612$

$918 = 612 \times 1 + 306$

$612 = 306 \times 2 + 0$

$\therefore$ H.C.F. = 306

8. $337554 = 26676 \times 12 + 17442$

$26676 = 17442 \times 1 + 9234$

$17442 = 9234 \times 1 + 8208$

$9234 = 8208 \times 1 + 1026$

$8208 = 1026 \times 8 + 0$

H.C.F = 1026

9. $434330 = 273070 \times 1 + 161260$

$273070 = 161260 \times 1 + 111810$

$161260 = 111810 \times 1 + 49450$

$111810 = 49450 \times 2 + 12910$

$49450 = 12910 \times 3 + 10720$

$12910 = 10720 \times 1 + 2190$

$10720 = 2190 \times 4 + 1960$

$2190 = 1960 + 230$

$1960 = 230 \times 8 + 120$

$230 = 120 \times 1 + 110$

$120 = 110 \times 1 + 10$

$110 = 10 \times 11 + 0$

H.C.F. = 10

10. $687897 = 81445 \times 8 + 36337$

$81445 = 36337 \times 2 + 8771$

$36337 = 8771 \times 4 + 1253$

$8771 = 1253 \times 7 + 0$

$\therefore$ H.C.F. = 1253

11. Using prime factorisation method.

$426 = 2 \times 3 \times 71$

$576 = 2 \times 3 \times 3 \times 2 \times 2 \times 2 \times 2 \times 2$

Common factors of 426 and 576 are 2 and 3.

$HCF = 2 \times 3 = 6$

LCM × HCF = First No. × Second No.

$LCM = \dfrac{\text{First No.} \times \text{Second No.}}{\text{HCF}}$

$LCM = \dfrac{426 \times 576}{6}$

$\therefore$ LCM (426, 576) = 40896.

12. $26676 = 2 \times 2 \times 3 \times 3 \times 3 \times 13 \times 19$

$337554 = 2 \times 3 \times 3 \times 3 \times 7 \times 19 \times 47$

$\therefore$ H.C.F. $= 2 \times 3 \times 3 \times 3 \times 19 = 1026$

L.C.M. $= 2^2 \times 3^3 \times 13 \times 19 \times 7 \times 47 = 8776404$

13. $1376 = 2 \times 2 \times 2 \times 2 \times 2 \times 43 = 2^5 \times 43$

$15428 = 2 \times 2 \times 7 \times 19 \times 29 = 2^2 \times 7 \times 19 \times 29$

$\therefore$ HCF $= 2^2 = 4$

LCM $= 2^5 \times 43 \times 7 \times 19 \times 29 = 5307232$

14. $11008 = 2 \times 2 \times 2 \times 2 \times 2 \times 2 \times 2 \times 2 \times 43 = 2^8 \times 43$

$7344 = 2 \times 2 \times 2 \times 2 \times 3 \times 3 \times 3 \times 17$

$= 2^4 \times 3^3 \times 17$

$\therefore$ HCF $= 2^4 = 16$

LCM $= 2^8 \times 43 \times 3^3 \times 17 = 5052672$

15. $625 = 5^4$

$1125 = 3^2 \times 5^3$

$2125 = 5^3 \times 17$

$\therefore$ H.C.F $= 5^3 = 125$

L.C.M. $= 5^4 \times 3^2 \times 17 = 95625$

16.
$$448 = 2\times2\times2\times2\times2\times2\times7 = 2^6\times7$$
$$1008 = 2\times2\times2\times2\times3\times3\times7 = 2^4\times3^2\times7$$
$$168 = 2\times2\times2\times3\times7 = 2^3\times3\times7$$
$\therefore$ HCF $= 2^3\times7 = 56$

LCM $= 2^6\times7\times3^2 = 4032$

17.
$$270 = 2\times3^3\times5$$
$$405 = 3^4\times5$$
$$315 = 3^2\times5\times7$$
$\therefore$ H.C.F. $= 3^2\times5 = 45$

L.C.M. $= 2\times3^4\times5\times7 = 5670.$

18.
$$377 = 13\times29$$
$$435 = 3\times5\times29$$
$$667 = 23\times29$$
$\therefore$ HCF $= 29$

LCM $= 29\times13\times3\times5\times23 = 130065$

19.
$$12576 = 4052\times3+420$$
$$4052 = 420\times9+272$$
$$420 = 272\times1+148$$
$$272 = 148\times1+124$$
$$148 = 124\times1+24$$
$$124 = 24\times5+4$$
$$24 = 4\times6+0$$
$\therefore$ HCF $= 4$

20.
$$867 = 255\times3+102$$
$$255 = 102\times2+51$$
$$102 = 51\times2+0$$
$\therefore$ H.C.F $= 51$

21.
$$38220 = 196\times195+0$$
$\therefore$ H.C.F. $= 196$

22.
$$225 = 135\times1+90$$
$$135 = 90\times1+45$$
$$90 = 45\times2+0$$
$\therefore$ H.C.F. $= 45$

23.
$$96 = 2^5\times3$$
$$404 = 2^2\times101$$
$\therefore$ H.C.F. $= 2^2 = 4$

L.C.M. $= 2^5\times3\times101 = 9696$

24.
$$6 = 2\times3$$
$$72 = 2\times2\times2\times3\times3$$
$$120 = 2\times2\times2\times3\times5$$
Common factors of 6, 72 and 120 are 2 and 3.

HCF $= 2\times3 = 6$

LCM $= 2\times2\times2\times3\times3\times5$

$\therefore$ LCM $= 360$

25.
$$26 = 2\times13$$
$$91 = 7\times13$$
$\therefore$ H.C.F. $= 13$, L.C.M. $= 2\times13\times7 = 182$

26.
$$6 = 2\times3$$
$$20 = 2^2\times5$$
$\therefore$ H.C.F. $= 2$

L.C.M. $= 2^2\times3\times5 = 60$

27.
$$17 = 17\times1$$
$$23 = 23\times1$$
$$29 = 29\times1$$
$\therefore$ H.C.F. $= 1$

L.C.M. $= 17\times23\times29 = 11339$

28.
$$8 = 2^3$$
$$9 = 3^2$$
$$25 = 5^2$$
$\therefore$ H.C.F. $= 1$

L.C.M. $= 8\times9\times25 = 1800$

29. $$\frac{13}{3125} = \frac{13}{5^5} = \frac{13}{2^0\times5^5}$$

$\because$ denominator is of the type $2^m\times5^n$

$\therefore$ Rational number has a terminating decimal expansion.

30. $$\frac{129}{2^2\times5^7\times7^5}$$

$\because$ denominator is not of the type $2^m\times5^n$

$\therefore$ Rational number has a non-terminating decimal expansion.

31. $$\frac{64}{455} = \frac{64}{5\times7\times13}$$

$\because$ denominator is not of the type $2^m\times5^n$

$\therefore$ Rational number has a non-terminating decimal expansion.

32. $$\frac{15}{1600} = \frac{15}{2^6\times5^2} = \frac{3}{2^6\times5}$$

$\because$ denominator is of the type $2^m\times5^n$

$\therefore$ Rational number has a terminating decimal expansion.

33. $$\frac{23}{2^3\cdot5^2}$$

$\because$ denominator is of the type $2^m\times5^n$

$\therefore$ Rational number has a terminating decimal expansion.

34. $$\frac{77}{210} = \frac{77}{2\times3\times5\times7} = \frac{11}{2\times3\times5}$$

$\because$ denominator is not of the type $2^m\times5^n$

$\therefore$ Rational number has a non-terminating decimal expansion.

SHORT ANSWER TYPE QUESTIONS-II

1. Let $5-2\sqrt{3} = \frac{p}{q}$ be a rational number.

$\therefore$ $5-\frac{p}{q} = 2\sqrt{3}$ *or* $\frac{5q-p}{2q} = \sqrt{3}$,

which is contradiction because $\sqrt{3}$ is an irrational number and $\frac{5q-p}{2q}$ is a rational.

$\therefore$ Our supposition is wrong and hence, $5-2\sqrt{3}$ is irrational.

2. (*a*) Let $x = 5+3\sqrt{2}$, where x is a rational number.

$\therefore$ $3\sqrt{2} = x-5$

$\Rightarrow$ $\sqrt{2} = \frac{x-5}{3}$ ...(*i*)

As x is rational, $\frac{x-5}{3}$ is also rational.

But $\sqrt{2}$ is irrational.

$\therefore$ (*i*) presents a contradiction

$\therefore$ x is not rational.

$\therefore$ $5+3\sqrt{2}$ is irrational.

(b) Let $\sqrt{3} = \frac{a}{b}$, where a and b are coprime integers, $b \neq 0$.

Squaring both sides, we get $3 = \frac{a^2}{b^2}$.

Multiplying with b on both sides, we get

$$3b = \frac{a^2}{b}$$

$$\text{LHS} = 3 \times b = \text{Integer}$$

$$\text{RHS} = \frac{a^2}{b} = \frac{\text{Integer}}{\text{Integer}}$$

$$= \text{Rational no.}$$

$\therefore$ LHS $\neq$ RHS

$\therefore$ Our supposition is wrong.

$\Rightarrow \sqrt{3}$ is irrational.

(c) Let N is a positive integer.

$\therefore$ N = $3k$, $3k + 1$ or $3k + 2$, where k is any integer.

when N = $3k$

$$N^2 = (3k)^2$$

$$\Rightarrow N^2 = 9k^2$$

$$= 3.3k^2 = 3.\ m. \quad (\text{where } m = 3k^2)$$

when N = $3k + 1$

$$N^2 = (3k + 1)^2$$

$$\Rightarrow N^2 = 9k^2 + 6k + 1$$

$$\Rightarrow N^2 = 3(3k^2 + 2k) + 1$$

$$= 3m + 1 \quad (\text{where } m = 3k^2 + 2k)$$

when N = $3k + 2$

$$N^2 = (3k + 2)^2$$

$$\Rightarrow N^2 = 9k^2 + 12k + 4$$

$$= 3(3k^2 + 4k + 1) + 1$$

$$= 3m + 1 \quad (\text{Where } m = 3k^2 + 4k + 1)$$

$\therefore$ Square of any positive integer is either of the form $3m$ or $3m + 1$.

3. Let $n = 3k$; $3k + 1$ or $3k + 2$

(i) When $n = 3k$ $\Rightarrow$ n is divisible by 3.

$n + 2 = 3k + 2$ $\therefore$ $n + 2$ is not divisible by 3.

and $n + 4 = 3k + 4$

$= 3k + 3 + 1 = 3(k + 1) + 1$

$\therefore$ $n + 4$ is not divisible by 3.

(ii) When $n = 3k + 1$, $\therefore$ n is not divisible by 3

$$n + 2 = (3k + 1) + 2 = 3k + 3$$

$$= 3(k + 1)$$

$\Rightarrow$ $n + 2$ is divisible by 3.

$$n + 4 = 3k + 1 + 4 = 3k + 5$$

$$= 3k + 3 + 2 = 3(k + 1) + 2$$

$\therefore$ $n + 4$ is not divisible by 3.

(iii) When $n = 3k + 2$, n is not divisible by 3.

$$n + 2 = (3k + 2) + 2 = 3k + 4$$

$$= 3(k + 1) + 1$$

$\therefore$ $n + 2$ is not divisible by 3.

$$n + 4 = (3k + 2) + 4 = 3k + 6 = 3(k + 2)$$

$\therefore$ $n + 4$ is divisible by 3.

$\therefore$ Only one of the numbers n, $n + 2$ and $n + 4$ is divisible by 3.

4. Let, $x = 3 + 5\sqrt{2}$ is a rational number

$$\Rightarrow 5\sqrt{2} = x - 3 \Rightarrow \sqrt{2} = \frac{x - 3}{5} \quad \text{...}(i)$$

As x is rational $\therefore$ $\frac{x-3}{5}$ is also rational.

But $\sqrt{2}$ is irrational.

$\therefore$ (i) represents a contradiction.

$\Rightarrow x$ is not rational.

$\Rightarrow 3 + 5\sqrt{2}$ is an irrational.

5. (i) Let n be a positive integer.

We can express it as $8q + 1$.

If $n = 8q + 1$, we have

$$n^2 = (8q + 1)^2 = 64q^2 + 1 + 16q$$

$$= 8(8q^2 + 2q) + 1$$

$$= 8k + 1 \text{ where } k = 8q^2 + 2q$$

Hence, we can say that n^2 is in the form of $8k + 1$ for some integer k.

$\therefore$ Square of any positive odd integer is of the form $8k + 1$, for some integer k.

(ii) Let, if possible, $2\sqrt{3} - 1$ is a rational number.

$$\therefore \quad 2\sqrt{3} - 1 = \frac{a}{b},\ b \neq 0,\ a, b \in Z$$

$$\Rightarrow \quad 2\sqrt{3} = \frac{a}{b} + 1$$

$$\Rightarrow \quad 2\sqrt{3} = \frac{a + b}{b}$$

$$\Rightarrow \quad \sqrt{3} = \frac{a + b}{2b} \quad \text{...}(i)$$

From (i), we notice

LHS is an irrational number and RHS is rational number, which is not possible. Hence, our supposition is wrong.

Hence, $2\sqrt{3} - 1$ is an irrational number.

(iii) Let, if possible, $7 - 2\sqrt{3}$ is a rational number.

Let $7 - 2\sqrt{3} = \frac{a}{b},\ b \neq 0,\ a, b \in Z$

$$\Rightarrow \quad 7 - \frac{a}{b} = 2\sqrt{3}$$

$$\Rightarrow \quad \frac{7b - a}{b} = 2\sqrt{3}$$

$$\Rightarrow \quad \frac{7b - a}{2b} = \sqrt{3}$$

We notice LHS is a rational number, whereas RHS is an irrational number, which is a contradiction.

Hence, our supposition is wrong.

Hence, $7 - 2\sqrt{3}$ is an irrational number.

(iv)

$$1000 = 2 \times 2 \times 2 \times 5 \times 5 \times 5$$

$$56 = 2 \times 2 \times 2 \times 7$$

HCF of 1000 and 56 = 8

$\therefore$ Maximum number of columns = 8

(v) Let N is any positive integer

$\therefore N = 4q$ or $4q + 1$ or $4q + 2$ or $4q + 3$

(i) when $N = 4q = 2(2q)$ = even

(ii) when $N = 4q + 1 = 2(2q) + 1$ = even + 1 = odd

(iii) when $N = 4q + 2 = 2(2q + 1)$ = even

(iv) when $N = 4q + 3 = 4q + 2 + 1$

$= 2(2q + 1) + 1$ = Even + 1 = odd

$\therefore$ When $N = 4q + 1$ or $4q + 3$, then it is odd

$\Rightarrow$ any positive odd integer is of the form $4q + 1$ or $4q + 3$

(vi) Let $\frac{2\sqrt{3}}{5}$ is rational.

$\therefore \quad \frac{2\sqrt{3}}{5} = \frac{a}{b}$,

where a and b are coprime, $b \neq 0$, $b \neq 1$.

$\Rightarrow \quad \frac{2\sqrt{3}}{5} = \frac{a}{b}$

$\Rightarrow \quad \sqrt{3} = \frac{5a}{2b}$

where $5a$ and $2b$ are integer and $\sqrt{3}$ is irrational. But $\frac{5a}{2b}$ is rational. $\therefore \sqrt{3} \neq \frac{5a}{2b}$

$\therefore$ Our supposition is wrong.

$\Rightarrow \frac{2\sqrt{3}}{5}$ is irrational.

(vii) Let $x = 0.3\overline{178}$

$\Rightarrow \quad 10x = 3.\overline{178}$ ...(i)

$\Rightarrow \quad 10000x = 3178.\overline{178}$ ...(ii)

Subtracting (i) from (ii)

$\Rightarrow \quad 9990x = 3175$

$x = \frac{635}{1998}$

(viii) Co-prime $\because$ HCF = 1

(ix) Let HCF = x

$\therefore \quad$ LCM = $14x$

A.T.Q.

$x + 14x = 600 \Rightarrow x = 40$

Now 280 × other number = HCF × LCM = 40 × 560

Other number = 80

6. Let $3\sqrt{5}$ be rational.

$3\sqrt{5} = \frac{a}{b} \quad \Rightarrow \sqrt{5} = \frac{a}{3b}$

$\Rightarrow \sqrt{5}$ is rational, this is contradiction because $\frac{a}{3b}$ is rational.

$\therefore \quad 3\sqrt{5}$ is irrational.

7. Let $\frac{1}{\sqrt{3}}$ is rational.

$\frac{1}{\sqrt{3}} = \frac{a}{b}$, where a and b are positive co-primes.

$(\sqrt{3}a)^2 = b^2 \Rightarrow 3a^2 = b^2$

$\therefore \quad b^2$ is divisible by 3 $\Rightarrow b$ is divisible by 3

let $b = 3c$ for some positive integer c.

$\therefore \quad 3a^2 = b^2 \Rightarrow 3a^2 = 9c^2 \Rightarrow a^2 = 3c^2$

$\therefore a^2$ is divisible by 3 $\Rightarrow a$ is divisible by 3

$\Rightarrow \quad a$ and b have a common factor

This is contradiction to the fact that a, b are co-primes.

Hence, $\frac{1}{\sqrt{3}}$ is irrational.

8. Let $\sqrt[3]{6}$ is rational number and $\sqrt[3]{6} = \frac{a}{b}$, where a and b are coprime and $b \neq 0$.

$\Rightarrow \quad (\sqrt[3]{6})^3 = \frac{a^3}{b^3} \Rightarrow 6 = \frac{a^3}{b^3}$

$\Rightarrow \quad 6.b^3 = a^3$ ...(i)

$\Rightarrow a^3$ is divisible by 6

$\Rightarrow a$ is divisible by 6

Let $a = 6c \quad \therefore$ (i) becomes

$6b^3 = (6c)^3$

$\Rightarrow \quad b^3 = 36c^3$

$\Rightarrow b^3$ is divisible by 6

$\Rightarrow b$ is divisible by 6

$\Rightarrow a$ and b have a common factor *i.e.* 6.

$\Rightarrow a$ and b are not coprime which is a contradiction.

$\therefore \sqrt[3]{6}$ is an irrational.

9. No, 4^n, units digit would be 4 or 6 only.

10. No

11. No

12. Let N is any positive integer.

$\therefore \quad N = 3m$ or $3m + 1$ or $3m + 2$

When $N = 3m$

$N^2 = 9m^2 = 3 \times 3m^2 = 3q$ form

When $N = 3m + 1$

$\Rightarrow \quad N^2 = (3m + 1)^2 = 9m^2 + 6m + 1$

$= 3(3m^2 + 2m) + 1$

$= 3q + 1$ [where $3m^2 + 2m = q$]

When $N = 3m + 2$

$N^2 = (3m + 2)^2$

$N^2 = 9m^2 + 12m + 4$

$N^2 = 9m^2 + 12m + 3 + 1$

$= 3(3m^2 + 4m + 1) + 1$

$= 3q + 1$ [where $q = 3m^2 + 4m + 1$]

$\because$ Square of any positive integer is of the form $3q$ or $3q + 1$ for some integer.

13. $\because \quad p$ is an odd positive integer.

$\therefore \quad p = 4q + 1$ or $4q + 3$ for some integer q.

When $p = 4q + 1$

When $p^2 = 16q^2 + 8q + 1$

$= 8(2q^2 + q) + 1$

$p^2 = 8m + 1$

$\Rightarrow p^2$ leaves remainder 1 when divided by 8.

When $p = 4q + 3$

$p^2 = 16q^2 + 24q + 9$

$p^2 = 16q^2 + 24q + 8 + 1$

$p^2 = 16q^2 + 24q + 8 + 1$

$= 8(2q^2 + 3q + 1) + 1$

$= 8m + 1$

$\therefore \quad p^2$ leaves a remainder 1 when divided by 8.

14. Let two consecutive positive integers are $2n$ and $2n + 1$

$\therefore \quad (2n)(2n + 1) = 4n^2 + 2n = 2(2n^2 + n)$

$\Rightarrow$ Product is divisible by 2.

15. Let N is any positive integer.

$\therefore N = 2q$ on $N = 2q + 1$ for some non negative integer q.

When $N = 2q \Rightarrow N^2 = 4q^2 = 4m$ (where $m = q^2$)

When $N = 2q + 1$ ...(i)

$\Rightarrow N^2 = (2q+1)^2 \Rightarrow N^2 = 4q^2 + 4q + 1$

$\Rightarrow N^2 = 4(q^2 + q) + 1 = 4m + 1$ ...(ii) (where $m = q^2 + q$)

$\therefore$ From (i) and (ii) square of any positive integer is of the type $4m$ or $4m + 1$

16. Let $N = 5m + 1$

$\Rightarrow N^2 = 25m^2 + 10m + 1$

$N^2 = 5(5m^2 + 2m) + 1$

$= 5q + 1$

17. A.T.Q.

Number divides $615 - 6$ and $963 - 6$

or 609 and 957

$\therefore$ Number = HCF of 609 and 957 = 87

18. Required number is HCF of

$2053 - 5$ and $967 - 7$

= HCF of 2048 and 960 = 64

19. A.T.Q. greatest number will divide $445 - 4$, $572 - 5$ and $699 - 6$

$\Rightarrow$ greatest number will be H.C.F. of 441, 567 and 693

Now $441 = 3^2 \times 7^2$

$567 = 3^4 \times 7$

$693 = 3^2 \times 7 \times 11$

$\therefore$ H.C.F. $= 3^2 \times 7 = 63$

$\therefore$ 63 is the required number.

20. HCF of 336, 240, 96 = 48

Now number of rooms for participants in Mathematics

$$= \frac{336}{48} = 7$$

Number of rooms for participants in Physics

$$= \frac{240}{48} = 5$$

Number of rooms for participants in Biology

$$= \frac{96}{48} = 2$$

$\therefore$ Total no. of rooms = 14

21. Length of the longest rod = HCF of 825 cm, 675 cm and 450 cm.

$\Rightarrow$ HCF = 2520 cm

22. Smallest number exactly divisible by 520 and 468

= LCM. of 520 and 468 = 4680

$\therefore$ Required number = 4680 – 17 = 4663

23. Let $N = 4q, 4q + 1, 4q + 2, 4q + 3$.

When $N = 4q = 2(2q)$ = Even

When $N = 4q + 1 = 2 \times (2q) + 1$ = Even + 1 = Odd

When $N = 4q + 2 = 2(2q + 1)$ = Even

When $N = 4q + 3 = 4q + 2 + 1 = 2(2q + 1) + 1$

= Even + 1 = Odd

$\therefore$ Any positive odd integer is of the form $4q + 1$ or $4q + 3$.

24. Let x be any positive integer. Then, it is of the form $3m$, $3m + 1$ or $3m + 2$.

Take $x = 3m$

$x^3 = (3m)^3 = 27m^3$

$= 9(3m^2) = 9q$

where $q = 3m^2$

Similarly, we can take

$x = 3m + 1$ and $x = 3m + 2$

and solve above process and procced.

25. Maximum number of columns = H.C.F. of 616 and 32

$616 = 2^3 \times 7 \times 11$

$32 = 2^5$

$\therefore$ H.C.F. $= 2^3 = 8$

$\therefore$ Maximum number of columns = 8

NCERT EXEMPLAR PROBLEMS

1. No, by Euclid's division lemma,

$b = aq + r, 0 \le r < a$

Here b is any positive integer

$a = 4$

$b = 4q + r$ for $0 \le r < 4$

So, this must be in the form $4q$, $4q + 1$, $4q + 2$ or $4q + 3$.

2. Yes, three positive consecutive integers can be

$n, (n + 1), (n + 2)$

So, one number of these three must be divisible by 2 and another one must be divisible by 3 for any positive integeral value of n.

So, product of numbers is divisible by 6.

3. No, any positive integer is of the form $3k$, $3k + 1$ or $3k + 2$

Now $(3k)^2 = 9k^2 = 3m$ (Where $m = 3k^2$)

and $(3k + 1)^2 = 9k^2 + 6k + 1$

$= 3p + 1$ (Where $p = 3k^2 + 2k$)

also $(3k + 2)^2 = 9k^2 + 12k + 4$

$= 9k^2 + 12k + 3 + 1$

$= 3q + 1$ (where $q = 3k^2 + 4k + 1$)

4. No, because HCF is always a factor of LCM.

But 18 is not a factor of 380.

5. Yes, after simplification denominator has factors $5^3 2^2$ and this is of the type $2^m \cdot 5^n$ so, this is terminating decimal.

$$\left(\because \frac{987}{10500} = \frac{47}{500} = \frac{47}{5^3 2^2}\right)$$

6. The prime factors of $\frac{p}{q}$ is $\frac{3277081}{10000}$ and q has factors $2^4 5^4$.

7. LCM of 40 cm, 42 cm, 45 cm is 2520 cm.

Minimum distance each should walk in 2520 cm so that each can cover the same distance in complete steps.

8. Let $\sqrt{p}$ is rational so that it can be written in the form of $\frac{a}{b}$.

$\sqrt{p} = \frac{a}{b}$ (where a and b are coprimes)

Squaring both sides, $p = \frac{a^2}{b^2}$

a^2 has a factor p.

$pb^2 = a^2$ ...(i)

so, a also has a factor p.

so, $a = pc$

$a^2 = p^2c^2$

Put the value of a^2 in equation (*i*).

$pb^2 = p^2c^2$

$b^2 = pc^2$

b^2 has a factor p, $\therefore$ b has a factor p

But a and b have common factor p.

But as stated earlier a, b are coprimes.

So, our supposition is wrong.

$\sqrt{p}$ must be irrational number. (where p is prime number)

we can prove $\sqrt{q}$ is also an irrational number. (where q is prime number)

Sum of two irrational numbers is irrational if both are prime numbers.

So, $\sqrt{p}+\sqrt{q}$ is irrational number.

9. $n^3 - n$

$n(n^2 - 1)$

$n(n - 1)(n + 1)$

$(n - 1)$, n, $(n + 1)$ are three consecutive positive integers. One number must be divisible by 3 and one must be even number so it must be divisible by 2 and the product of these numbers must have factor product of 2 and 3 *i.e.*, $2 \times 3 = 6$.

HOTS QUESTIONS

1. $a = bq + r, 0 \le r < b$

Since $52 < 117$, then apply division algorithm

$$52 \overline{)117}\ (2,\quad 104,\quad \text{remainder } 13$$

$117 = 52 \times 2 + 13$

Since remainder $13 \neq 0$ then

$$13 \overline{)52}\ (4,\quad 52,\quad \text{remainder } 0$$

As remainder is 0, then HCF = 13.

now in order to write HCF inform of 52, 117 we have

HCF $= 13 = 52 \times (-2) + 117 \times (1)$

i.e., $x = -2,\ y = 1$

2. Let us suppose that two integers are $2q$ & $2q + 1$ for some integer q, then at $x = 2q$

$x^2 - x = (2q)^2 - 2q = 4q^2 - 2q = 2q(2q - 1) = 2R$

then $x^2 - x$ is divisible by 2.

again when $x = 2q + 1$ then

$x^2 - x = (2q + 1)^2 - (2q + 1)$

$= 4q^2 + 1 + 4q - 2q - 1$

$= 4q^2 + 2q = 2q(2q + 1) = 2R.$

then $x^2 - x$ is divisible by 2

Hence, $x^2 - x$ is divisible by 2 $\forall\, x \in$ Integer.

3. Let $m = 2x + 1$ and $n = 2y + 1$ for some positive integers x, y.

then $m^2 + n^2 = (2x + 1)^2 + (2y + 1)^2$

$= 4x^2 + 1 + 4x + 4y^2 + 1 + 4y$

$= 4x^2 + 4y^2 + 4x + 4y + 2$

$= 2[2x^2 + 2y^2 + 2x + 2y + 1]$

= even integer.

again $m^2 + n^2 = 4[(x^2 + y^2) + (x + y)] + 2$

Let $q = (x^2 + y^2) + (x + y)$

then $m^2 + n^2 = 4q + 2$

Since $m^2 + n^2$ is even and leaves remainder 2 on dividing by 4.

$\therefore$ $m^2 + n^2$ is not divisible by 4.

4. HCF of 6 and a is 2 and LCM of 6 & $a = 60$

but we know that LCM × HCF = $6 \times a$.

$6a = 60 \times 2 \ \Rightarrow\ a = 20$

5. Since remainder of $\frac{(5m+1)(5m+3)(5m+4)}{5}$ is a natural number then at $m = 1, 2, \ldots$ we get the remainder

i.e., $m = 1$

$$\frac{(5+1)(5+3)(5+4)}{5} = \frac{432}{5} = 86\frac{2}{5}$$

Hence, remainder is always 2.

6. Let $a = n_1 \times n_2 \ldots n_n$

p be the prime factorisation of a.

where $n_1, n_2, n_3, \ldots$ are prime numbers not necessarily distinct

Now $a = n_1 \times n_2 \ldots n_n$

when $a^2 = (n_1 \times n_2 \ldots n_n) \times (n_1 \times n_2 \ldots n_n)$

$a^2 = n_1^2 \times n_2^2 \ldots n_n^2$

It is given that n is prime and it divides a^2.

$\therefore$ n is a prime factor of a^2.

$\therefore$ n is one of the $n_1 \times n_2 \times n_3 \ldots n_n$

$\Rightarrow$ n divides by $n_1 \times n_2 \times n_3 \ldots n_n = n$ divides a.

7.

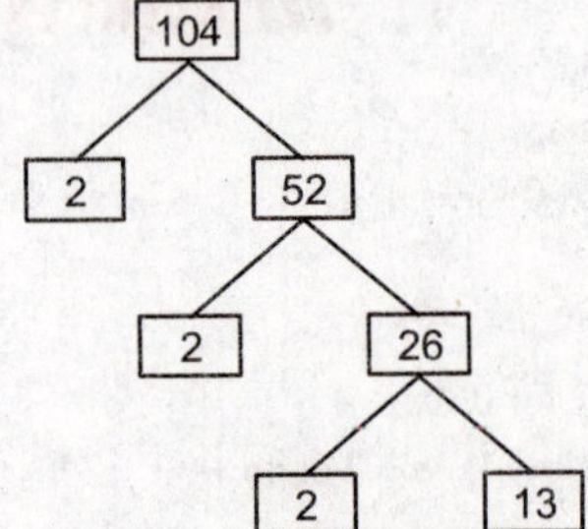

104, 52, 26 are answers.

8. The greatest number of 5 digits divisible by 12, 15 and 36 will be divisible by LCM of 12, 15 and 36.

$12 = 2^2 \times 3,\ 15 = 3 \times 5,\ 36 = 2^2 \times 3^2$

$\therefore$ LCM $= 2^2 \times 3^2 \times 5 = 180$

The greatest number of 5 digits = 99999

$99999 = 180 \times 555 + 99$

$\therefore$ The greatest number divisible by

$180 = 99999 - 99 = 99900$

9. The smallest number exactly divisible by 90 and 144

= LCM of 90 and 144 = 720

$\therefore$ The smallest number when increased by 20 and divisible by 90 and 144 = 720 – 20 = 700

10. Let the smallest number = N

LCM of 28 and 32

$28 = 2^2 \times 7, 32 = 2^5$

$\therefore$ LCM $= 2^5 \times 7 = 224$

Now number leaves remainder 8

When divided by 28 = 224 – 20 = 224 – (28 – 8) = 204

Number leaves remainder 12

When divided by 32 = 224 – (32 – 12) = 204

The smallest number = 204

11. The size of tile should be equal to

HCF of 10 and 7 = 1

$\therefore$ Size of square tiles = 1 m × 1 m

Area of floor = 10 × 7 = 70 m^2

Area of 1 tile = 1 × 1 = 1 m^2

$\therefore$ Number of tiles $= \frac{70}{1} = 70$

12. HCF of 408 and 1032

$408 = 2^3 \times 3 \times 17$

$1032 = 2^3 \times 3 \times 43$

$\therefore$ HCF $= 2^3 \times 3 = 24$

$24 = 1032 \times p - 408 \times 5$

$\Rightarrow 24 + 2040 = 1032 \times p$

$\Rightarrow 2064 = 1032 \times p \Rightarrow p = 2$

/// VALUE BASED QUESTIONS ///

1. (*a*) L.C.M. of 96 and 105

$96 = 2^5 \times 3$

$105 = 3 \times 7 \times 5$

L.C.M. $= 2^5 \times 3 \times 7 \times 5 = 3360$

(*b*) Honesty.

2. (*a*)

$20 = 2^2 \times 5$

$12 = 2^2 \times 3$

$16 = 2^4$

$2 = 2^1$

$\therefore$ H.C.F. = 2

(*b*) Environment awareness.

3. (*a*) L.C.M. of 12 and 18 = 36

$\therefore$ After 36 minutes they will meet again at the starting point.

(*b*) Love for their country.

4. (*a*) L.C.M. of 40, 42 and 45 = 2520

So, after walk of 2520 cm each can cover the same distance in complete steps.

(*b*) To stay fit and healthy.

/// MULTIPLE CHOICE QUESTIONS ///

1. (*c*) Will terminate after 3 places of decimal $\frac{17}{8} = 2.125$

2. (*b*) $\because \frac{11}{160} = \frac{11}{2^5 \cdot 5}$

Denominator is of the form $2^m.5^n$

$\therefore$ its decimal expansion is terminating.

3. (*b*) $\because$ π is irrational number $\therefore$ its decimal expansion is non terminating and non recurring.

4. (*c*) $\frac{15}{1600} = \frac{15}{2^6 \times 5^2}$. Denominator is of the type $2^m.5^n$ $\therefore$ its decimal expansion is terminating.

5. (*b*) $55 = 5 \times 11, 99 = 3^2 \times 11$

$\therefore$ H.C.F. = 11

A.T.Q. $11 = 55m - 99 \Rightarrow 110 = 55m$

$\Rightarrow m = 2$

6. (*a*) HCF $= \frac{91 \times 26}{182} = 13$

7. (*a*) *a* terminating decimal

8. (*b*) $x = 21, y = 84$

9. (*a*) $0 \le r < 3$

10. (*b*) an irrational number

11. (*c*) $2^3 \times 3^3$

12. (*a*) LCM of 50 and 48 = 1200

$\therefore$ 1200 sec = 20 min

Hence at 12.20 pm they will beep again for the first time.

13. (*b*) LCM of 36, 42 and 72 = 504

504 sec = 8 min 24 sec

Hence, at 9:08:24 they will change simultaneously

14. (*b*) Length of tape = HCF of 34 and 85 = 17

Hence length is 17.

15. (*c*) First number = 2 × 33 = 66

$\therefore$ other number $= \frac{\text{HCF} \times \text{LCM}}{\text{1st number}}$

$= \frac{33 \times 264}{66} = 132$

16. (*c*) The greatest number will be (multiple of LCM of 8 and 9) + 5

LCM of 8 and 9 = 72

On verification we find that 99941 when divided by 72 leaves remainder 5.

17. (*d*) Length of plank = HCF of 42, 49 and 63 = 7

$\therefore$ Number of planks $= \frac{42}{7} + \frac{49}{7} + \frac{63}{7}$

$= 6 + 7 + 9 = 22$

18. (*d*) LCM of rational number $= \frac{\text{LCM of numerators}}{\text{GCD of denominators}}$

Numbers are $\frac{25}{10}, \frac{5}{10}, \frac{175}{1000}$

$\therefore$ LCM $= \frac{175}{10} = 17.5$

19. (*a*) Number =(LCM of 36, 63 and 108) – 6

= 756 – 6 = 750

20. (*c*) Speed = HCF of 52 and 91 = 13 m/min

21. **(c)** LCM of 5, 8, 12 and 20 = $2^3 \times 5 \times 3 = 120$

which is not a multiple of 9

22. **(b)** Taking different values of n we find that A and B are coprime.

$\therefore$ HCF = 1

23. **(c)** They will meet after LCM of 2, 3 and 4 *i.e.*, they will meet after 12 hrs.

24. **(b)** Difference of 90 and 24 = 66

and LCM of 90 and 24 = 360

$\therefore$ Numbers are 90 and 24.

25. **(d)** Let numbers are $13x$ and $13y$ where x and y are coprime.

$\therefore$ LCM $= 13 \times x \times y = 78$

$x \times y = 6$

$\therefore$ Possible values of x and y

$x = 1, y = 6, x = 2, y = 3$

$\therefore$ Pairs of natural numbers are 78 and 13 or 26 and 39.

26. **(d)** In the given options, we find that sum of

51 and 34 = 85

and LCM = 102

$\therefore$ Numbers are 51 and 34.

27. **(c)** LCM of 7, 8, 11, 12 = 1848

$\therefore$ Bells will toll together after every 1848 sec.

$\therefore$ In next 3 hrs, number of times the bells will toll together

$$= \frac{3 \times 3600}{1848} = 5.84$$

$\Rightarrow$ 5 times.

28. **(c)**

$2472 = 2^3 \times 3 \times 103$

$1284 = 2^2 \times 3 \times 107$

$\because$ LCM $= 2^3 \times 3^2 \times 5 \times 103 \times 107$

$\therefore$ N $= 2^2 \times 3^2 \times 5 = 180$

29. **(d)** HCF of 66, 88 and 110 = 22

$\therefore$ Number of rows $= \frac{66}{22} + \frac{88}{22} + \frac{110}{22} = 3 + 4 + 5 = 12$

30. **(b)** Number when added to 7249 gives a number which is divisible by LCM of 12, 14, 21, 33 and 54.

$\therefore$ 9383 satisfy the conditions.

31. **(d)** LCM of 5, 15, 20 = 60 and

HCF = 5

$\therefore$ Ratio = 60 : 5 = 12 : 1

32. **(b)** HCF of 576 and 448 = 64

$\therefore$ Number of sections $= \frac{576}{64} + \frac{448}{64}$

$= 9 + 7 = 16$

33. **(b)** On checking the given options, we find that 2884 satisfies the condition.

34. **(d)** $(10x + y)\,(10a + b) = 100\,p + 10q + r$

$\Rightarrow (10x + y)\,(10a + b) = 100 \times 2x + 20(x + y) + 2y$

$\Rightarrow (10x + y)\,(10a + b) = 220x + 22y$

$\Rightarrow (10x + y)\,(10a + b) = 22(10x + y)$

$\Rightarrow \quad 10a + b = 22$

35. **(b)** On checking the options, 20,80 satisfy the given conditions.

NCERT EXEMPLAR PROBLEMS

1. **(b)** HCF of 65 and 117 = 13

$\therefore$ $65m - 117 = 13;\ m = 2$

$= 65(2) - 117$

$= 130 - 117 = 13$

2. **(a)** Numbers are 70 and 125.

Leaving remainder numbers becomes = 65, 117

HCF of 65, 117 is 13

3. **(b)** HCF of $a = x^3y^2$ and $b = xy^3$ is xy^2.

4. **(c)** LCM $(p, q) = a^3b^2$

5. **(d)** LCM of numbers from 1 to 10 = 2520.

6. **(d)** Four decimal places.

Formative Assessment

Activity/Project 1

- **Topic Covered :** Understanding of rational numbers.
- **Activity Over View**

Students learn to identify real numbers as rational numbers and irrational numbers. They review rational numbers, a non terminating repeating decimals or terminating decimals.

- **Pre-requisite Knowledge :** A rational number has terminating decimal representation if the prime factorization of denominator is of the type $2^n \times 5^m$ (where m & n are non-negative integers).
- **Materials Required**

Paper sheets as shown below.

Topic : Understanding of rational numbers.			**Name : ....** **Date : ....**	
S.No.	**Number**	**Terminating Decimal**	**Non terminating but repeating decimals**	**Score**
1.				
2.				
3.				
4.				
5.				
			Total	

- **Procedure**
  - Teacher writes a rational number on the board, students quickly determine if the number is terminating decimal or non terminating but repeating decimals. They will write in the column in the sheet shown above.
  - Process is repeated at fixed interval of time say 20 seconds.
  - Students play until a certain amount of time has expired.
  - Teacher will give score according to their performance in sheet. The student with highest score wins.
  - Activity is repeated over a period of time (day, weeks etc.) for tracking improvement of high scores.

Now answer the following questions :

1. What was your score ?
2. How many incorrect answers did you have ?
3. Write a paragraph describing the activity. Describe your strategy for playing.

Activity/Project 2

- **Topic Covered :** Real Numbers
- **Objective :** Approximation of irrational number.
- **Skills Developed :** Analysis, Problem solving, Creativity.
- **Word Problem**

Ruchil wants to build a small fence around the garden in his backyard. The garden is in the shape of a right triangle. One leg is 2 meters and the other leg is 1 meter. He goes to a store for purchasing fencing. The store sells fencing in tenths of a metre.

1. What is perimeter of Ruchil's garden ? Show all your work.
2. What length of fencing should Ruchil buy if the store only sells the fencing in tenth of a metre ?

Class Work

1. Find the hypotenuse of a right triangle with sides 3 m and 5 m.
2. Find the sum $3 + \sqrt{3}$ up to 2 places of decimals.

Homework

Construct a right triangle with side 2 cm and 4 cm. Measure its hypotenuse. Also use pythagoras theorem to find the hypotenuse. Compare your result.

Oral Questions

1. What is rational number ?
2. $\sqrt{5}$ is .... number.

Quiz Questions

1. Sum of two irrational numbers is always irrational. State True/False.
2. Every real number is rational number. State True/False.

Suggested Activity/Assignment

Make a list of ten irrational numbers of the form $\sqrt{n}$, where n is a positive integer and find their values upto three places of decimal.

Activity/Project 3

- **Topic Covered :** Irrational Numbers
- **Objective :** To find irrational number between two consecutive positive integers.
- **Skills Developed :** Creativity, Analysis, Observations.
- **Method**

Writing 10 irrational numbers between 1 and 2:

1. Take $x = \sqrt{3}$
2. Use formula $N_i = \frac{Nx}{N+1}$, where N is natural number ≥ 2, $i = 1, 2, 3, ...,$
3. $N_1 = \frac{2\sqrt{3}}{3}, N_2 = \frac{3\sqrt{3}}{4}, N_3 = \frac{4\sqrt{3}}{5}, N_4 = \frac{5\sqrt{3}}{6}, N_5 = \frac{6\sqrt{3}}{7},$

$N_6 = \frac{7\sqrt{3}}{8}, N_7 = \frac{8\sqrt{3}}{9}, N_8 = \frac{9\sqrt{3}}{10}, N_9 = \frac{10\sqrt{3}}{11}, N_{10} = \frac{11\sqrt{3}}{12}$

- **Observation**

$(N_1)^2 = \frac{12}{9} \Rightarrow (1)^2 < (N_1)^2 < (2)^2 \therefore 1 < N_1 < 2$

Similarly $(N_{10})^2 = \frac{363}{144} \Rightarrow (1)^2 < (N_{10})^2 < (2)^2 \therefore 1 < N_{10} < 2$

$\therefore N_1, N_2, ... N_{10}$ are 10 irrational numbers between 1 and 2.

Class Work

1. Find one irrational number between 4 and 5.
2. Find two irrational numbers between 6 and 7.

Homework

Write 10 rational numbers between 2 and 3.

Quiz Questions

1. Is $\sqrt{3} > 1$? State True/False.
2. Is $\sqrt{5} < 2$? State True/False.

Oral Questions

1. π is .... number.
2. Find LCM of $\sqrt{2}$ and $\sqrt{3}$.

Suggested Activity/Assignment

Write 15 irrational numbers in between 6 and 7.

Polynomials

CHAPTER 2

Basic Concepts with Examples

- **Zeroes of a polynomial.** k is said to be zero of a polynomial $p(x)$ if $p(k) = 0$
- **Graph of polynomial.**
  - (i) Graph of a linear polynomial $ax + b$ is a straight line.
  - (ii) Graph of a quadratic polynomial $p(x) = ax^2 + bx + c$ is a parabola open upwards like $\cup$, if $a > 0$.
  - (iii) Graph of a quadratic polynomial $p(x) = ax^2 + bx + c$ is a parabola open downwards like $\cap$, if $a < 0$.
  - (iv) In general a polynomial $p(x)$ of degree n crosses the x-axis at atmost n points.

y-axis
x-axis
O
y-axis
x-axis
O

- **Relationship between the zeroes and the coefficients of a Polynomial.**
  - (i) If α, β are zeroes / roots of $p(x) = ax^2 + bx + c$, then

$$\text{Sum of roots} = \alpha + \beta = \frac{-b}{a} \Rightarrow \alpha + \beta = \frac{-(\text{coefficient of } x)}{\text{coefficient of } x^2}$$

$$\text{Product of roots} = \alpha\beta = \frac{c}{a} \Rightarrow \alpha\beta = \frac{\text{constant term}}{\text{coefficient of } x^2}$$

  - (ii) If α, β and γ are zeroes / roots of $p(x) = ax^3 + bx^2 + cx + d$

$$\text{Then, sum of roots} = \alpha + \beta + \gamma = \frac{-b}{a} = \frac{-(\text{coefficient of } x^2)}{\text{coefficient of } x^3}$$

Sum of product of roots taken, two at a time

$$\alpha\beta + \beta\gamma + \alpha\gamma = \frac{c}{a} = \frac{\text{coefficient of } x}{\text{coefficient of } x^3}$$

$$\text{Product of roots} = \frac{-d}{a} \Rightarrow \alpha\beta\gamma = \frac{-(\text{constant term})}{\text{coefficient of } x^3}$$

  - (iii) If α, β are roots of a quadratic polynomial $p(x)$, then $p(x) = x^2 - (\alpha + \beta)\, x + \alpha\beta$

$\Rightarrow$ $p(x) = x^2 - (\text{sum of roots})\, x + \text{product of roots}$

  - (iv) If α, β and γ are zeroes of a cubic polynomial $p(x)$,

Then, $p(x) = x^3 - (\alpha + \beta + \gamma)\, x^2 + (\alpha\beta + \beta\gamma + \alpha\gamma)\, x - (\alpha\beta\gamma)$

$\Rightarrow$ $p(x) = x^3 - (\text{sum of zeroes})\, x^2 + (\text{sum of product of zeroes / roots taken two at a time})\, x - (\text{product of zeroes})$

1. Find the zeroes of the quadratic polynomial and verify the relationship between the zeroes and coefficient of polynomial $p(x) = x^2 + 7x + 12$.

Sol. $p(x) = x^2 + 7x + 12$

$\Rightarrow$ $p(x) = (x + 3)(x + 4)$

$\therefore$ $p(x) = 0$ if $x + 3 = 0$ or $x + 4 = 0$

$\Rightarrow$ $x = -3$ or $x = -4$

$\therefore$ -3 and -4 are zeroes of the $p(x)$.

Now,

Sum of the zeroes $= -3 + (-4) = -7$

$$= \frac{-7}{1} = \frac{-(\text{coefficient of } x)}{\text{coefficient of } x^2}$$

$$\text{Product of the zeroes} = (-3) \times (-4) = 12 = \frac{12}{1}$$

$$= \frac{\text{constant term}}{\text{coefficient of } x^2}$$

2. Find the zeroes of $4x^2 - 7$ and verify the relationship between the zeroes and its coefficients.

Sol. Let $p(x) = 4x^2 - 7$

Here coefficient of $x^2 = 4$,

Coefficient of $x = 0$ and constant term $= -7$.

Now $p(x) = 4x^2 - 7 = (2x - \sqrt{7})(2x + \sqrt{7})$

$\therefore$ $p(x) = 0$, if $2x - \sqrt{7} = 0$ or $2x + \sqrt{7} = 0$

$\Rightarrow \quad x = \dfrac{\sqrt{7}}{2}$ or $x = \dfrac{-\sqrt{7}}{2}$

$\therefore \quad \dfrac{\sqrt{7}}{2}$ and $\dfrac{-\sqrt{7}}{2}$ are zeroes of $p(x)$.

Now,

$$\text{Sum of zeroes} = \frac{\sqrt{7}}{2} + \left(\frac{-\sqrt{7}}{2}\right) = 0$$

$$= \frac{0}{4} = \frac{-(\text{coefficient of } x)}{\text{coefficient of } x^2}$$

$$\text{Product of zeroes} = \frac{\sqrt{7}}{2} \times \frac{-\sqrt{7}}{2} = \frac{-7}{4}$$

$$= \frac{\text{constant term}}{\text{coefficient of } x^2}$$

3. Find a quadratic polynomial whose zeroes are $5 + \sqrt{2}$ and $5 - \sqrt{2}$.

Sol. Let α, β are zeroes of quadratic polynomial $p(x)$.

$\therefore \quad p(x) = x^2 - (\alpha + \beta)\, x + \alpha\beta$

Here, $\quad \alpha = 5 + \sqrt{2}, \beta = 5 - \sqrt{2}$

$\therefore \quad \alpha + \beta = 5 + \sqrt{2} + 5 - \sqrt{2} = 10$

and $\quad \alpha\beta = (5 + \sqrt{2})(5 - \sqrt{2})$

$= 25 - 4 = 21$

$\therefore \quad p(x) = x^2 - 10x + 21$

4. Find a quadratic polynomial, the sum of whose zeroes is 0 and one zero is 5.

Sol. Let zeroes are α and β.

$\Rightarrow \quad \alpha + \beta =$ Sum of zeroes

$\Rightarrow \quad \alpha + \beta = 0 \Rightarrow 5 + \beta = 0 \Rightarrow \beta = -5$

Now product of zeroes $= \alpha\beta = 5 \times (-5) = -25$

Let polynomial $p(x) = ax^2 + bx + c$

$\therefore \quad \alpha + \beta = 0 = \dfrac{-b}{a}$; $\alpha\beta = \dfrac{c}{a} = -25$

$\therefore \quad a = 1, b = 0, c = -25$

Hence, $p(x) = x^2 - 25$

5. Find a cubic polynomial with the sum, sum of the product of its zeroes taken two at a time, and product of its zeroes are 5, –6 and –20 respectively.

Sol. Let $p(x) = ax^3 + bx^2 + cx + d$

and α, β, γ are its zeroes.

$\therefore \quad \alpha + \beta + \gamma =$ Sum of zeroes $= 5 = \dfrac{-b}{a}$

$\alpha\beta + \alpha\gamma + \beta\gamma =$ Sum of the products of zeroes taken two at a time

$= -6 = \dfrac{c}{a}$

$\alpha\beta\gamma =$ Product of the zeroes $= -20 = \dfrac{-d}{a}$

If $a = 1$, then $b = -5$, $c = -6$ and $d = 20$

$\therefore \quad$ Polynomial, $p(x) = x^3 - 5x^2 - 6x + 20$.

- **Division Algorithm for polynomials.**

If $p(x)$ and $g(x)$ are any two polynomials with $g(x) \neq 0$, then we can find polynomials $q(x)$ and $r(x)$ such that $p(x) = q(x) \times g(x) + r(x)$, where $r(x) = 0$ or degree of $r(x) <$ degree of $g(x)$.

Step-1. Divide the highest degree term of the dividend by the highest degree term of the divisor and obtain the remainder.

Step-2. If the remainder is 0 or degree of remainder is less than divisor, then we cannot continue the division any further. If degree of remainder is equal to or more than divisor repeat step-1.

6. Divide $4x^3 + 2x^2 + 5x - 6$ by $2x^2 + 3x + 1$.

Sol.

$$\begin{array}{r|l} & 2x - 2 \\ 2x^2 + 3x + 1 & 4x^3 + 2x^2 + 5x - 6 \\ & 4x^3 + 6x^2 + 2x \\ & - \quad - \quad - \\ \hline & -4x^2 + 3x - 6 \\ & -4x^2 - 6x - 2 \\ & + \quad + \quad + \\ \hline & 9x - 4 \end{array}$$

$\therefore \quad$ Quotient is $2x - 2$ and remainder is $9x - 4$.

7. Find all zeroes of polynomial $4x^4 - 20x^3 + 23x^2 + 5x - 6$ if two of its zeroes are 2 and 3.

Sol. Since two zeroes are 2 and 3.

$\therefore (x - 2)(x - 3)$ is a factor of given polynomial.

$\Rightarrow x^2 - 5x + 6$ is a factor of given polynomial.

Now

$$\begin{array}{r|l} & 4x^2 - 1 \\ x^2 - 5x + 6 & 4x^4 - 20x^3 + 23x^2 + 5x - 6 \\ & 4x^4 - 20x^3 + 24x^2 \\ & - \quad + \quad - \\ \hline & -x^2 + 5x - 6 \\ & -x^2 + 5x - 6 \\ & + \quad - \quad + \\ \hline & 0 \end{array}$$

$\therefore \quad 4x^4 - 20x^3 + 23x^2 + 5x - 6$

$= (x^2 - 5x + 6)(4x^2 - 1)$

$= (x - 2)(x - 3)(2x - 1)(2x + 1)$

$\therefore$ Zeroes of the given polynomial are

$2, 3, \dfrac{1}{2}, \dfrac{-1}{2}$.

Summative Assessment

Subjective Questions

VERY SHORT ANSWER TYPE QUESTIONS

This category is given for better understanding of the topic and for EXTRA PRACTICE.

Previous Years' Questions

1. Write the zeroes of polynomial $x^2 - x - 6$. [Delhi 2008]
2. Write a quadratic polynomial, the sum and product of whose zeroes are 3 and -2. [Delhi 2008]
3. Is $x = -3$, solution of the equation $2x^2 + 5x + 3 = 0$? [AI 2008]
4. If $x + a$ is a factor of $2x^2 + 2ax + 5x + 10$, find a. [Foreign 2008]
5. Show that $x = -2$ is a solution of $3x^2 + 13x + 14 = 0$ [Foreign 2008]
6. The graph of $y = f(x)$ is given, how many zeroes are there of $f(x)$? [AI 2008 C]

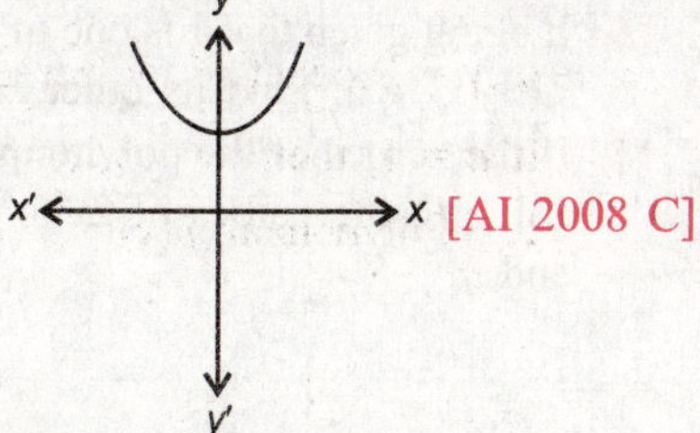

7. For what value of k, (-4) is a zero of the polynomial $x^2 - x - (2k + 2)$? [Delhi 2009]
8. If 1 is a zero of polynomial $p(x) = ax^2 - 3(a - 1) - 1$, then find the value of a. [AI 2009]
9. Write the polynomial, the product and sum of whose zeroes are $-\frac{9}{2}$ and $-\frac{3}{2}$ respectively. [Foreign 2009]
10. (*i*) If α, β are the zeroes of a polynomial, such that $\alpha + \beta = 6$ and $\alpha\beta = 4$, then write the polynomial. [Delhi 2010]

    (*ii*) If α, β are the zeroes of the polynomial $2y^2 + 7y + 5$, write the value of $\alpha + \beta + \alpha\beta$. [AI 2010]

    (*iii*) If one zero of the polynomial $x^2 - 4x + 1$ is $2 + \sqrt{3}$, write the other zero. [Foreign 2010]

Practice Questions

11. The graph of $y = f(x)$ is given below, for some polynomial $f(x)$. Find the number of zeroes of $f(x)$.

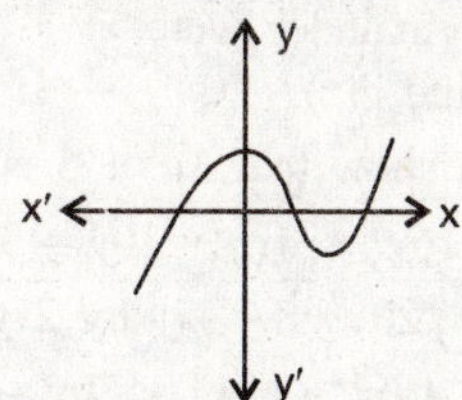

12. The graph of $x = p(y)$ is given below, for some polynomial $p(y)$. Find the number of zeroes of $p(y)$.

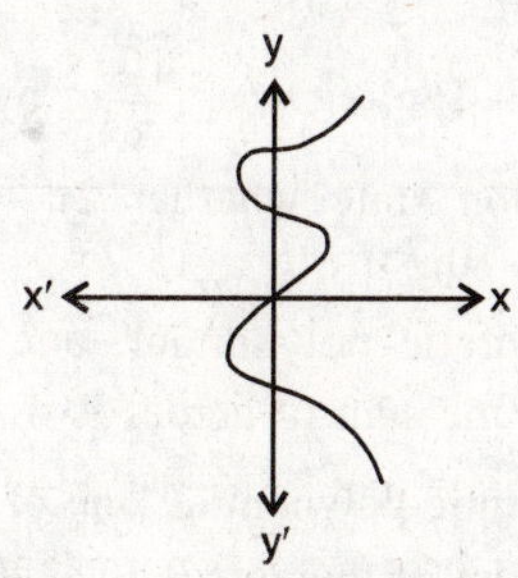

13. A polynomial of degree five is divided by a quadratic polynomial. If it leaves a remainder, then find the degree of remainder.
14. If $x^3 + x^2 - ax + b$ is divisible by $x^2 - x$, write the values of a and b.
15. Find the sum of the zeroes of quadratic polynomial $x^2 + 7x + 10$.
16. Find the product of the zeroes of quadratic polynomial $x^2 - 3$.
17. If α and β are the zeroes of the quadratic polynomial $f(x) = x^2 - px + q$, then find the value of $\alpha^2 + \beta^2$.
18. If α and β are the zeroes of the quadratic polynomial $f(x) = ax^2 + bx + c$, find the value of $\frac{1}{\alpha} + \frac{1}{\beta}$.
19. Show that the polynomial $x^2 - 2x + 5$ has no real zeroes.
20. Check whether $3x - 7$ is a factor of polynomial $6x^3 + x^2 - 26x - 25$?
21. If 1 is one of the zeroes of polynomial $x^2 - x + k$, then find value of k.
22. If α, β are the zeroes of the polynomial $p(x) = x^2 - p(x + 1) - c$ such that $(\alpha + 1)(\beta + 1) = 0$. What is the value of c ?
23. Find the product of the zeroes of $-2x^2 + kx + 6$.
24. Find the sum of the zeroes of the given quadratic polynomial $-3x^2 + k$.

SHORT ANSWER TYPE QUESTIONS-I

PREVIOUS YEARS' QUESTIONS

1. Find the zeroes of the quadratic polynomial $6x^2 - 3 - 7x$ and verify the relationship between the zeroes and the coefficients of the polynomial. [Delhi 2008]
2. Find the quadratic polynomial sum of whose zeroes is 8 and their product is 12. Hence, find the zeroes of polynomial. [AI 2008]
3. If one zero of the polynomial $(a^2 + 9)x^2 + 13x + 6a$ is reciprocal of the other. Find the value of a. [AI 2008]
4. If the product of zeroes of the polynomial $ax^2 - 6x - 6$ is 4. Find the value of a. [AI 2008]
5. Find all the zeroes of the polynomial $x^4 + x^3 - 34x^2 - 4x + 120$, if two of its zeroes are 2 and –2. [Foreign 2008]
6. Find all the zeroes of the polynomial $2x^4 + 7x^3 - 19x^2 - 14x + 30$, if two of its zeroes are $\sqrt{2}$ and $-\sqrt{2}$. [Foreign 2008]
7. Find a quadratic polynomial whose zeroes are 1 and –3. Verify the relation between the coefficients and zeroes of polynomial. [Delhi 2008 C]
8. Find the zeroes of the quadratic polynomial $\sqrt{3}x^2 - 8x + 4\sqrt{3}$. [AI 2008 C]
9. Using division algorithm, find the quotient and remainder on dividing $f(x)$ by $g(x)$ where $f(x) = 6x^3 + 13x^2 + x - 2$ and $g(x) = 2x + 1$ [AI 2008 C]
10. If the polynomial $6x^4 + 8x^3 + 17x^2 + 21x + 7$ is divided by another polynomial $3x^2 + 4x + 1$, the remainder comes out to be $(ax + b)$, find a and b. [Delhi 2009]
11. Find all the zeroes of the polynomial $2x^3 + x^2 - 6x - 3$, if two of its zeroes are $-\sqrt{3}$ and $\sqrt{3}$. [AI 2009]
12. For what value of k, is 3 a zero of the polynomial $2x^2 + x + k$? [AI 2010]
13. If –1 and 2 are two zeroes of the polynomial $2x^3 - x^2 - 5x - 2$, find its third zero. [Foreign 2010]
14. Divide $x^4 - 3x^2 + 4x + 5$ by $x^2 - x + 1$, find quotient and remainder. [CBSE (CCE) 2011]
15. If 2 and –3 are the zeroes of the quadratic polynomial $x^2 + (a + 1)x + b$; then find the values of a. [CBSE (CCE) 2011]
16. It being given that 1 is one of the zeroes of the polynomial $7x - x^3 - 6$. Find its other zeroes. [CBSE (CCE) 2011]
17. If the zeroes of the polynomial $x^2 + px + q$ are double in value to the zeroes of $2x^2 - 5x - 3$, find the value of p and q. [CBSE (CCE) 2012]

PRACTICE QUESTIONS

Find the zeroes of following quadratic polynomials and verify the relationship between the zeroes and their coefficients. (Q 18 to 22)

18. $p(x) = x^2 - 7x + 12$
19. $p(x) = 4x^2 + 24x + 36$
20. $p(x) = 100x^2 - 81$
21. $p(x) = 25x^2 + 5x$
22. $p(y) = 4\sqrt{3}y^2 + 5y - 2\sqrt{3}$
23. Find a quadratic polynomial whose zeroes are 3 and –5.
24. Find a quadratic polynomial whose zeroes are –9 and $-\frac{1}{9}$.
25. Sum and product of zeroes of quadratic polynomial are 5 and 17 respectively. Find the polynomial.
26. Sum and product of zeroes of a quadratic polynomial are 0 and $\sqrt{15}$ respectively. Find the quadratic polynomial.
27. Find a quadratic polynomial whose one zero is 5 and product of zeroes is 30.
28. Find a quadratic polynomial whose one zero is 7 and sum of zeroes is –18.
29. Find a quadratic polynomial whose zeroes are $3 + \sqrt{5}$ and $3 - \sqrt{5}$.
30. Find a quadratic polynomial whose one zero is – 8 and sum of zeroes is 0.
31. Find a quadratic polynomial whose one zero is –5 and product of zeroes is 0.
32. Verify that $\frac{1}{2}$, 3 are zeroes of cubic polynomial $2x^3 - 17x^2 + 38x - 15$.
33. On dividing $x^3 + x^2 + x - 2$ by a polynomial $g(x)$, the quotient and remainder were $x^2 + 2x + 1$ and $2x - 1$ respectively. Find $g(x)$.
34. Using division show that $3y^2 + 5$ is a factor of $6y^5 + 15y^4 + 16y^3 + 4y^2 + 10y - 35$.
35. Divide $24x^3y + 20x^2y^2 - 4xy$ by $2xy$.
36. Divide $10x^4 + 17x^3 - 62x^2 + 30x - 3$ by $2x^2 + 7x - 1$.
37. Divide $ax^2 + (b + ac)x + bc$ by $x + c$.
38. Divide $64y^3 - 1000$ by $8y - 20$.
39. Divide $15y^4 - 16y^3 + 9y^2 - \frac{10}{3}y$ by $3y - 2$.
40. Using division state whether $2y - 5$ is a factor of $4y^4 - 10y^3 - 10y^2 + 30y - 15$?
41. Form a quadratic polynomial, one of whose zero is $2 + \sqrt{5}$ and the sum of zeroes is 4.
42. Form a quadratic polynomial, one of whose zero is $\sqrt{5}$ and the product of the zeroes is $-2\sqrt{5}$.
43. Determine is 3 a zero of the polynomial
$$p(x) = \sqrt{x^2 - 4x + 3} + \sqrt{x^2 - 9} - \sqrt{4x^2 - 14x + 6}$$

44. α, β are zeroes of the quadratic polynomial $x^2 - (k + 6)x + 2(2k - 1)$. Find the value of k if $\alpha + \beta = \frac{1}{2}\alpha\beta$.

45. m, n are zeroes of $ax^2 - 5x + c$. Find the value of a and c if $m + n = m.n = 10$.

46. Form a quadratic polynomial whose zeroes are $\frac{3-\sqrt{3}}{5}$ and $\frac{3+\sqrt{3}}{5}$.

47. α and $\frac{1}{\alpha}$ are zeroes of polynomial $4x^2 - 2x + (k - 4)$. Find the value of k.

48. If α, β are zeroes of $x^2 + 5x + 5$, find the value of $\alpha^{-1} + \beta^{-1}$.

49. If α, β are zeroes of the $x^2 + 7x + 7$, find the value of $\frac{1}{\alpha} + \frac{1}{\beta} - 2\alpha\beta$.

50. Find the zeroes of $\sqrt{3}x^2 + 10x + 7\sqrt{3}$.

51. Write a quadratic polynomial whose one zero is $3 - \sqrt{5}$ and product of zeroes is 4.

52. Check whether the $g(x)$ is a factor of the $p(x)$ by applying the division algorithm.

$p(x) = 2x^5 - 4x^3 + 2x^2 + 5x + 1$

$g(x) = x^3 - 4x + 1$

NCERT Questions

Find the zeroes of following quadratic polynomials and verify the relationship between the zeroes and their coefficients. (Q. 53 to Q. 55)

53. $p(x) = x^2 - 2x - 8$ **54.** $p(s) = 4s^2 - 4s + 1$

55. $p(t) = t^2 - 15$

56. Verify that 2, 1, 1 are zeroes of polynomial $x^3 - 4x^2 + 5x - 2$. Also verify the relationship between the zeroes and the coefficients.

57. Find a cubic polynomial with the sum, sum of the product of its zeroes taken two at a time and product of its zeroes are 2, –7, –14 respectively.

58. Apply the division algorithm to find the quotient and remainder on dividing $p(x)$ by $g(x)$ where

(i) $p(x) = x^3 - 3x^2 + 5x - 3$, $g(x) = x^2 - 2$

(ii) $p(x) = x^4 - 3x^2 + 4x + 5$, $g(x) = x^2 + 1 - x$.

(iii) $p(x) = x^4 - 5x + 6$, $g(x) = 2 - x^2$

59. Check whether the $g(x)/g(t)$ is a factor of the $p(x)/p(t)$ by applying the division algorithm.

(i) $p(x) = x^5 - 4x^3 + x^2 + 3x + 1$, $g(x) = x^3 - 3x + 1$

(ii) $p(t) = 2t^4 + 3t^3 - 2t^2 - 9t - 12$, $g(t) = t^2 + 3t + 1$

60. On dividing $x^3 - 3x^2 + x + 2$ by a polynomial $g(x)$, the quotient and remainder were $x - 2$ and $-2x + 4$, respectively. Find $g(x)$.

SHORT ANSWER TYPE QUESTIONS-II

Previous Years' Questions

1. Find others zeroes of $2x^4 - 9x^3 + 5x^2 + 3x - 1$, if two of its zeroes are $2+\sqrt{3}$ and $2-\sqrt{3}$ [Delhi 2008 C]

2. If the polynomial $6x^4 + 8x^3 - 5x^2 + ax + b$ is exactly divisible by the polynomial $2x^2 - 5$, then find the values of a and b. [Foreign 2009]

3. If the polynomial $6x^4 + 8x^3 + 17x^2 + 21x + 7$ is divided by another polynomial $3x^2 + 4x + 1$, then what will be the quotient and remainder ? [CBSE (CCE) 2011]

4. On dividing the polynomial $4x^4 - 5x^3 - 39x^2 - 46x - 2$ by the polynomial $g(x)$, the quotient and remainder were $x^2 - 3x - 5$ and $-5x + 8$ respectively. Find $g(x)$. [CBSE (CCE) 2011]

5. Show that $\frac{1}{2}$ and $\frac{-3}{2}$ are the zeroes of the polynomial $4x^2 + 4x - 3$ and verify the relationship between zeroes and co-efficients of polynomial. [CBSE (CCE) 2012]

6. Find the value of b for which $(2x + 3)$ is a factor of $2x^3 + 9x^2 - x - b$. [CBSE (CCE) 2012]

Practice Questions

7. If one zero of the quadratic polynomial $2x^2 + px + 4$ is 2, find the other zero. Also, find the value of p.

8. If α, β are zeroes/roots of polynomial $x^2 - 6x + k$, find the value of k such that $(\alpha + \beta)^2 - 2\alpha\beta = 40$.

9. If α, β are zeroes of quadratic polynomial $2x^2 + 5x + k$, find the value of k such that $(\alpha + \beta)^2 - \alpha\beta = 24$.

10. If one root of the quadratic polynomial $2x^2 - 3x + p$ is 3, find the other root. Also, find the value of p.

11. If 2 and 3 are zeroes of polynomial $3x^2 - 2kx + 2m$, find the values of k and m.

12. α, β are zeroes of the polynomial $x^2 - 6x + a$. Find the value of a, if $3\alpha + 2\beta = 20$.

13. If one zero of polynomial $3x^2 = 8x + 2k + 1$ is seven times the other, then find the zeroes and value of k.

14. Divide $p(x)$ by $q(x)$ and check your answer using division algorithm.
(Dividend = Divisor × Quotient + Remainder)
(i) $p(x) = 3x^5 - 2x^4 + x^2 - 2$, $q(x) = x^2 + x + 1$
(ii) $p(x) = 6x^5 + 4x^4 - 27x^3 - 7x^2 - 27x - 6$, $q(x) = 2x^2 - 3$

15. $p(x)$ is a polynomial of degree more than 2. When $p(x)$ is divided by $x - 2$, it leaves remainder 1 and when it is divided by $x - 3$ it leaves a remainder 3. Find the remainder when $p(x)$ is divided by $(x - 2)(x - 3)$.

16. Obtain all the zeroes of $3x^4 + 6x^3 - 2x^2 - 10x - 5$, if two of its zeroes are $\sqrt{\frac{5}{3}}$ and $-\sqrt{\frac{5}{3}}$.

17. Obtain all the zeroes of $x^4 - 7x^3 + 17x^2 - 17x + 6$, if two of its zeroes are 3 and 1.

LONG ANSWER TYPE QUESTIONS

Previous Years' Questions

1. Find other zeroes of the polynomial $x^4 - 7x^2 + 12$ if it is given that two of its zeroes are $\sqrt{3}$ and $-\sqrt{3}$. [CBSE (CCE) 2011]

2. Obtain all other zeroes of the polynomial $x^4 - 3x^3 - x^2 + 9x - 6$, if two of its zeroes are $\sqrt{3}$ and $-\sqrt{3}$. [CBSE (CCE) 2011]

3. Divide $2x^4 - 9x^3 + 5x^2 + 3x - 8$ by $x^2 - 4x + 1$ and verify the division algorithm. [CBSE (CCE) 2011]

4. Given that $x - \sqrt{5}$ is factor of the polynomial $x^3 - 3\sqrt{5}x^2 - 5x + 15\sqrt{5}$, find, all the zeroes of the polynomial. [CBSE (CCE) 2012]

5. If the polynomial $x^4 - 6x^3 + 16x^2 - 25x + 10$ is divided by $(x^2 - 2x + k)$ the remainder comes out to be $x + a$, find k and a. [CBSE (CCE) 2012]

6. What must be subtracted or added to $p(x) = 8x^4 + 14x^3 - 2x^2 + 8x - 12$ so that $4x^2 + 3x - 2$ is a factor of $p(x)$? [CBSE (CCE) 2012]

NCERT EXEMPLAR PROBLEMS

Short Answer Questions with Reasoning

1. Answer the following and justify :
(i) Can $x^2 - 1$ be the quotient on division of $x^6 + 2x^3 + x - 1$ by a polynomial in x of degree 5 ?
(ii) If on division of a polynomial $p(x)$ by a polynomial $g(x)$, the quotient is zero, what is the relation between the degrees of $p(x)$ and $g(x)$?
(iii) If on division of a non-zero polynomial $p(x)$ by a polynomial $g(x)$, the remainder is zero, what is the relation between the degrees of $p(x)$ and $g(x)$?

2. Are the following statements 'True' or 'False' ? Justify your answers.
(i) If the zeroes of a quadratic polynomial $ax^2 + bx + c$ are both positive, then a, b and c all have the same sign.
(ii) If the graph of a polynomial intersects the x-axis at only one point, it cannot be a quadratic polynomial.
(iii) If two of the zeroes of a cubic polynomial are zero, then it does not have linear and constant terms.

Short Answer Questions

Find the zeroes of the following polynomials by factorisation method and verify the relations between the zeroes and the coefficients of the polynomials :

3. $4x^2 + 5\sqrt{2}x - 3$

4. $2s^2 - (1 + 2\sqrt{2})s + \sqrt{2}$

5. $v^2 + 4\sqrt{3}v - 15$

6. $y^2 + \frac{3}{2}\sqrt{5}y - 5$

7. $7y^2 - \frac{11}{3}y - \frac{2}{3}$

Long Answer Questions

8. Given that the zeroes of the cubic polynomial $x^3 - 6x^2 + 3x + 10$ are of the form a, $a + b$, $a + 2b$ for some real numbers a and b, find the values of a and b as well as the zeroes of the given polynomial.

9. Find k so that $x^2 + 2x + k$ is a factor of $2x^4 + x^3 - 14x^2 + 5x + 6$. Also find all the zeroes of the two polynomials.

10. For which values of a and b, are the zeroes of $q(x) = x^3 + 2x^2 + a$ also the zeroes of the polynomial $p(x) = x^5 - x^4 - 4x^3 + 3x^2 + 3x + b$? Which zeroes of $p(x)$ are not the zeroes of $q(x)$?

HIGHER ORDER THINKING SKILLS (HOTS) QUESTIONS

1. If one zero of the polynomial $5z^2 + 13z - p$ is reciprocal of the other, then find p.
2. If the product of two zeroes of polynomial $2x^3 + 3x^2 - 5x - 6$ is 3, then find its third zero.
3. Find the polynomial of least degree which should be subtracted from the polynomial $x^4 + 2x^3 - 4x^2 + 6x - 3$ so that it is exactly divisible by $x^2 - x + 1$.
4. Is polynomial $y^4 + 4y^2 + 5$ have zeroes or not ?
5. Write a quadratic polynomial, sum of whose zeroes is $2\sqrt{3}$ and product is 5.
6. Write the zeroes of the polynomial $x^2 + 2x + 1$. [Delhi 2008]
7. If the zeroes of the polynomial $f(x) = x^3 - 12x^2 + 39x + a$ are in AP, find the value of a.
8. A polynomial $g(x)$ of degree zero is added to the polynomial $2x^3 + 5x^2 - 14x + 10$ so that it becomes exactly divisible by $2x - 3$. Find the $g(x)$.
9. Find the zeroes of the quadratic polynomial $x^2 + 5x + 6$ and verify the relationship between the zeroes and the coefficients.
10. If the zeroes of polynomial $x^3 - ax^2 + bx - c$ are in AP then show that $2a^3 - 9ab + 27c = 0$
11. If 1 and –1 are zeroes of polynomial $Lx^4 + Mx^3 + Nx^2 + Rx + P$, show that $L + N + P = M + R = 0$
12. Draw graph of the function $f(x) = -2x^2 + 4x$.
13. If $x + a$ is a factor of the polynomial $x^2 + px + q$ and $x^2 + mx + n$ prove that $a = \dfrac{n-q}{m-p}$.
14. Find a cubic polynomial with the sum, sum of the product of its zeroes taken two at a time and product of its zeroes are 3, $\dfrac{-1}{2}$, $\dfrac{5}{4}$ respectively.
15. Write cubic polynomial whose zeroes are $\dfrac{2+\sqrt{5}}{2}, \dfrac{2-\sqrt{5}}{2}$, 4.
16. α, β, γ are zeroes of cubic polynomial $kx^3 - 5x + 9$. If $\alpha^3 + \beta^3 + \gamma^3 = 27$, find the value of k.
17. α, β, γ are zeroes of cubic polynomial $x^3 - 12x^2 + 44x + c$. If α, β, γ are in AP, find the value of c.
18. Two zeroes of cubic polynomial $ax^3 + 3x^2 - bx - 6$ are –1 and –2. Find the third zero and value of a and b.
19. α, β, γ are zeroes of cubic polynomial $x^3 - 2x^2 + qx - r$. If $\alpha + \beta = 0$ then show that $2q = r$.
20. α, β and γ are zeroes of polynomial $x^3 + px^2 + qx + 2$ such that $\alpha . \beta + 1 = 0$. Find the value of $2p + q + 5$.

VALUE BASED QUESTIONS

1. A group consists of 12 honest people and 8 dishonest people. Write a quadratic polynomial whose roots are equal to number of honest people and number of dishonest people. Which value do you prefer ?
2. Manish engages a labour to get some repair work. Charges to be paid for this work are zeroes of the polynomial $x^2 - 300x + 22500$.
   (*a*) Find zeroes of this polynomial.
   (*b*) Labour claims ₹ 125 for the whole work.
   Manish paid the actual amount. What value is depicted by Manish ?
3. Sum of zeroes of the polynomial $2x^2 - 4x + 5$ is 4. Navdeep at once said "it is false"
   (*a*) Do you agree with Navdeep ? Justify.
   (*b*) What value are depicted by Navdeep ?

Multiple Choice Questions

PREVIOUS YEARS' QUESTIONS

1. The quadratic polynomial whose sum of zeroes is 3 and product of zeroes is –2 is : [CBSE (CCE) 2011]

(*a*) $x^2 + 3x - 2$ (*b*) $x^2 - 2x + 3$
(*c*) $x^2 - 3x + 2$ (*d*) $x^2 - 3x - 2$

2. If $(x + 1)$ is a factor of $2x^3 + ax^2 + 2bx + 1$, then find the values of a and b given that $2a - 3b = 4$ [CBSE (CCE) 2011]

(*a*) $a = -1, b = -2$ (*b*) $a = 2, b = 5$
(*c*) $a = 5, b = 2$ (*d*) $a = 2, b = 0$

3. The number of zeroes that polynomial $f(x) = (x - 2)^2 + 4$ can have is : [CBSE (CCE) 2012]

(*a*) 1 (*b*) 2
(*c*) 0 (*d*) 3

4. The zeroes of the polynomial $f(x) = 4x^2 - 12x + 9$ are : [CBSE (CCE) 2012]

(*a*) $\dfrac{3}{2}, \dfrac{3}{2}$ (*b*) $-\dfrac{3}{2}, -\dfrac{3}{2}$
(*c*) 3, 4 (*d*) – 3, – 4

5. **If $p(x)$ is a polynomial of at least degree one and $p(k) = 0$, then k is known as**
   (a) value of $p(x)$ (b) zero of $p(x)$
   (c) constant term of $p(x)$ (d) none of these

6. **If $p(x) = ax + b$, then zero of $p(x)$**
   (a) a (b) b
   (c) $\frac{-a}{b}$ (d) $\frac{-b}{a}$

7. **Graph of a quadratic polynomial is a**
   (a) straight line (b) circle
   (c) parabola (d) ellipse

8. **Zeroes of a polynomial can be determined graphically. No. of zeroes of a polynomial is equal to no. of points where the graph of polynomial**
   (a) intersects y-axis
   (b) intersects x-axis
   (c) intersects y-axis or intersects x-axis
   (d) none of these

9. **If graph of a polynomial does not intersects the x-axis but intersects y-axis in one point, then no. of zeroes of the polynomial is equal to**
   (a) 0 (b) 1
   (c) 0 or 1 (d) none of these

10. **A polynomial of degree n has**
    (a) only 1 zero (b) exactly n zeroes
    (c) atmost n zeroes (d) more than n zeroes

11. **If $p(x) = ax^2 + bx + c$, then $\frac{c}{a}$ is equal to**
    (a) 0 (b) 1
    (c) sum of zeroes (d) product of zeroes

12. **If $p(x) = ax^2 + bx + c$, then $-\frac{b}{a}$ is equal to**
    (a) 0 (b) 1
    (c) product of zeroes (d) sum of zeroes

13. **If $p(x) = ax^2 + bx + c$ and $a + b + c = 0$, then one zero is**
    (a) $\frac{-b}{a}$ (b) $\frac{c}{a}$
    (c) $\frac{b}{c}$ (d) none of these

14. **If $p(x) = ax^2 + bx + c$ and $a + c = b$, then one of the zeroes is**
    (a) $\frac{b}{a}$ (b) $\frac{c}{a}$
    (c) $\frac{-c}{a}$ (d) $\frac{-b}{a}$

15. **A quadratic polynomial whose one zero is 6 and sum of the zeroes is 0, is**
    (a) $x^2 - 6x + 2$ (b) $x^2 - 36$
    (c) $x^2 - 6$ (d) $x^2 - 3$

16. **A quadratic polynomial whose one zero is 5 and product of zeroes is 0, is**
    (a) $x^2 - 5$ (b) $x^2 - 5x$
    (c) $5x^2 + 1$ (d) $x^2 + 5x$

17. **If $p(x)$ and $g(x)$ are any two polynomials with $g(x) \neq 0$, then we can find polynomials $q(x)$ and $r(x)$ such that $p(x) = g(x) \cdot q(x) + r(x)$ where $r(x) = 0$ or degree of $r(x) <$ degree of $g(x)$. This result is known as**
    (a) Euclid's algorithm for division
    (b) Division algorithm for polynomials
    (c) Multiplication algorithm for polynomials
    (d) None of these

18. **If $p(x) = g(x) \cdot q(x) + r(x)$, then degree of $p(x)$ is equal to**
    (a) product of degrees of $g(x)$ and $q(x)$
    (b) product of degrees of $g(x)$ and $r(x)$
    (c) sum of degrees of $g(x)$ and $q(x)$
    (d) sum of degrees of $g(x)$ and $r(x)$

19. **If $p(x) = g(x) \cdot q(x) + r(x)$, then degree of $q(x)$ is always less than**
    (a) the degree of $g(x)$ (b) the degree of $p(x)$
    (c) equal to degree of $p(x)$ (d) the degree of $r(x)$

20. **If $p(x) = g(x) \cdot q(x) + r(x)$ degree of $p(x) = 6$, degree of $g(x) = 3$, then degree of $q(x)$ is**
    (a) 2 (b) 3
    (c) 1 (d) 4

21. **The value of m so that $4x^2 - 6x - m$ is divisible by $x - 3$, is exact divisor of**
    (a) 9 (b) 45
    (c) 20 (d) 36

22. **For a given value of k, the product of the zeroes of $x^2 - 3kx + 2k^2 - 1$ is 7, then zeroes are**
    (a) rational numbers
    (b) irrational numbers
    (c) one rational, other irrational
    (d) none of these

23. **For $x^2 + 2x + 5$ to be a factor of $x^4 + px^2 + q$, the value of p and q must be respectively**
    (a) $-2, 5$ (b) 5, 25
    (c) 10, 20 (d) 6, 25

24. **Which one of the following statements is correct**
    (a) if $x^6 + 1$ is divided by $x + 1$, then the remainder is -2.
    (b) if $x^6 + 1$ is divided by $x - 1$, then the remainder is 2.
    (c) if $x^6 + 1$ is divided by $x + 1$, then the remainder is 1.
    (d) if $x^6 + 1$ is divided by $x - 1$, then the remainder is -1.

25. **Consider the following statements**

(*i*) $x - 2$ **is a factor of** $x^3 - 3x^2 + 4x - 4$.

(*ii*) $x + 1$ **is a factor of** $2x^3 + 4x + 6$.

(*iii*) $x - 1$ **is a factor of** $x^5 + x^4 - x^3 + x^2 - x + 1$.

In these statements

(*a*) 1 and 2 are correct (*b*) 1, 2 and 3 are correct

(*c*) 2 and 3 are correct (*d*) 1 and 3 are correct

26. **The value of** m**, in order that** $x^2 - mx - 2$ **is the quotient where** $x^3 + 3x^2 - 4$ **is divided by** $x + 2$ **is**

(*a*) –1 (*b*) 1

(*c*) 0 (*d*) –2

27. **If** $f(x) = 5x - 10$ **is divided by** $x - \sqrt{2}$**, then the remainder will be**

(*a*) non zero rational number (*b*) an irrational number

(*c*) 0 (*d*) $f\left(\frac{1}{\sqrt{2}}\right)$

28. **If one factor of** $x^3 + 7kx^2 - 4kx + 12$ **is** $(x + 3)$**, then the value of** k **is**

(*a*) 5 (*b*) $\frac{1}{5}$

(*c*) $\frac{-13}{17}$ (*d*) $\frac{-17}{13}$

29. **If zeroes of** $p(x) = 2x^2 - 7x + k$ **are reciprocal of each other, then value of** k **is**

(*a*) 1 (*b*) 2

(*c*) 3 (*d*) –7

30. **Zeroes of** $p(z) = z^2 - 27$ **are**

(*a*) $2\sqrt{3}, 3\sqrt{3}$ (*b*) $3\sqrt{3}, -3\sqrt{3}$

(*c*) $\sqrt{3}, -\sqrt{3}$ (*d*) $2\sqrt{2}, -2\sqrt{2}$

NCERT EXEMPLAR PROBLEMS

Multiple Choice Questions

1. If one of the zeroes of the quadratic polynomial $(k - 1)x^2 + kx + 1$ is –3, then the value of k is

(*a*) $\frac{4}{3}$ (*b*) $\frac{-4}{3}$

(*c*) $\frac{2}{3}$ (*d*) $\frac{-2}{3}$

2. If the zeroes of the quadratic polynomial $x^2 + (a + 1)x + b$ are 2 and –3, then

(*a*) $a = -7, b = -1$ (*b*) $a = 5, b = -1$

(*c*) $a = 2, b = -6$ (*d*) $a = 0, b = -6$

3. The number of polynomials having zeroes as –2 and 5 is

(*a*) 1 (*b*) 2

(*c*) 3 (*d*) more than 3

4. Given that one of the zeroes of the cubic polynomial $ax^3 + bx^2 + cx + d$ is zero, the product of the other two zeroes is

(*a*) $-\frac{c}{a}$ (*b*) $\frac{c}{a}$

(*c*) 0 (*d*) $-\frac{b}{a}$

5. The zeroes of the quadratic polynomial $x^2 + kx + k$, $k \neq 0$,

(*a*) cannot both be positive (*b*) cannot both be negative

(*c*) are always unequal (*d*) are always equal

6. If the zeroes of the quadratic polynomial $ax^2 + bx + c$, $c \neq 0$ are equal, then

(*a*) c and a have opposite signs

(*b*) c and b have opposite signs

(*c*) c and a have the same sign

(*d*) c and b have the same sign

7. If one of the zeroes of a quadratic polynomial of the form $x^2 + ax + b$ is the negative of the other, then it

(*a*) has no linear term and the constant term is negative.

(*b*) has no linear term and the constant term is positive.

(*c*) can have a linear term but the constant term is negative.

(*d*) can have a linear term but the constant term is positive.

8. Which of the following is not the graph of a quadratic polynomial ?

(*a*)
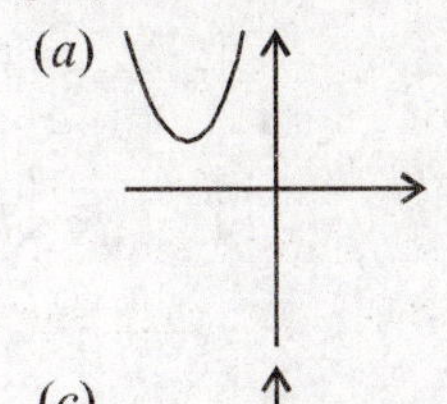

(*b*)
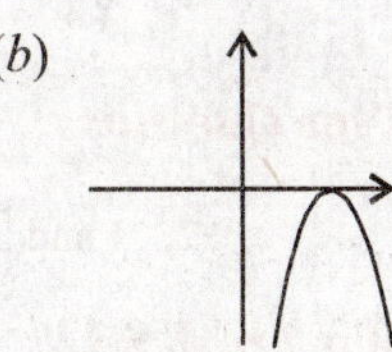

(*c*)
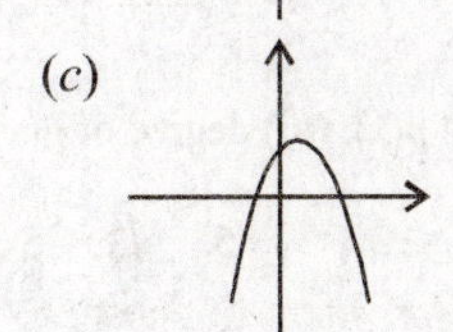

(*d*)
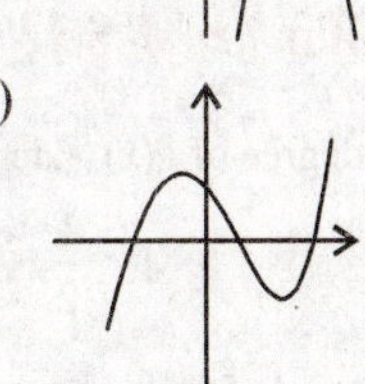

ANSWERS

Very Short Answer Type Questions

1. 3, – 2 **2.** $x^2 - 3x - 2$ **3.** No **4.** $a = 2$ **6.** No zero **7.** 9 **8.** 1

9. $k(2x^2 + 3x - 9)$ **10.** (*i*) $k(x^2 - 6x + 4)$ (*ii*) – 1 (*iii*) $2 - \sqrt{3}$ **11.** 3 **12.** 4 **13.** 1 or 0 **14.** $a = 2, b = 0$

15. – 7 **16.** – 3 **17.** $p^2 - 2q$ **18.** $\frac{-b}{c}$ **20.** No **21.** 0 **22.** 1

23. – 3 **24.** 0

Short Answer Type Questions-I

1. $\frac{-1}{3}, \frac{3}{2}$ **2.** $x^2 - 8x + 12, 6, 2$ **3.** $a = 3$ **4.** $a = \frac{-3}{2}$ **5.** $2, -2, -6, 5$ **6.** $-\sqrt{2}, \sqrt{2}, -5, \frac{3}{2}$

7. $x^2 + 2x - 3$ **8.** $\frac{2}{\sqrt{3}}, 2\sqrt{3}$ **9.** Quotient $= 3x^2 + 5x - 2$, remainder $= 0$ **10.** $a = 1, b = 2$

11. $-\sqrt{3}, \sqrt{3}$ and $\frac{-1}{2}$ **12.** -21 **13.** $-\frac{1}{2}$ **14.** $Q = x^2 + x - 3, R = 8$ **15.** $a = 0, b = -6$

16. -3 and 2 **17.** $p = -5, q = -6$ **18.** 3, 4 **19.** $-3, -3$ **20.** $\frac{9}{10}, \frac{-9}{10}$ **21.** $0, -\frac{1}{5}$

22. $-\frac{2}{\sqrt{3}}, \frac{\sqrt{3}}{4}$ **23.** $x^2 + 2x - 15$ **24.** $9x^2 + 82x + 9$ **25.** $x^2 - 5x + 17$ **26.** $x^2 + \sqrt{15}$ **27.** $x^2 - 11x + 30$

28. $x^2 + 18x - 175$ **29.** $x^2 - 6x + 4$ **30.** $x^2 - 64$ **31.** $x^2 + 5x$ **33.** $x - 1$

35. $Q = 12x^2 + 10xy - 2, R = 0$ **36.** $Q = 5x^2 - 9x + 2, R = 0$ **37.** $Q = ax + b, R = 0$ **38.** $Q = 8y^2 + 20y + 50, R = 0$

39. $Q = 5y^3 - 2y^2 + \frac{5}{3}y, R = 0$ **40.** No **41.** $x^2 - 4x - 1$ **42.** $x^2 + (2 - \sqrt{5})x - 2\sqrt{5}$

43. Yes **44.** $k = 7$ **45.** $a = \frac{1}{2}, c = 5$ **46.** $k(25x^2 - 30x + 6)$ **47.** $k = 8$ **48.** -1

49. -15 **50.** $-\sqrt{3}, \frac{-7}{\sqrt{3}}$ **51.** $x^2 - 6x + 4$ **52.** No **53.** $-2, 4$ **54.** $\frac{1}{2}, \frac{1}{2}$

55. $\sqrt{15}, -\sqrt{15}$ **57.** $x^3 - 2x^2 - 7x + 14$

58. (*i*) Quotient $= x - 3$; remainder $= 7x - 9$ (*ii*) Quotient $= x^2 + x - 3$; remainder $= 8$ (*iii*) Quotient $= -x^2 - 2$; remainder $= -5x + 10$

59. (*i*) No, (*ii*) No **60.** $x^2 - x + 1$

Short Answer Type Questions-II

1. $1, -\frac{1}{2}$ **2.** $a = -20, b = -25$ **3.** $Q = 2x^2 + 5, R = x + 2$ **4.** $4x^2 + 7x + 2$ **6.** $b = 15$ **7.** $1, p = -6$

8. $k = -2$ **9.** $k = \frac{-71}{2}$ **10.** $\frac{-3}{2}, p = -9$ **11.** $k = \frac{15}{2}, m = 9$ **12.** $a = -16$ **13.** $\frac{1}{3}, \frac{7}{3}, k = \frac{-5}{3}$

14. (*i*) Quotient $= 3x^3 - 5x^2 + 2x + 4$; remainder $= -6x - 6$ (*ii*) Quotient $= 3x^3 + 2x^2 - 9x - \frac{1}{2}$; remainder $= -54x - \frac{15}{2}$

15. $2x - 3$ **16.** $\sqrt{\frac{5}{3}}, -\sqrt{\frac{5}{3}}, -1, -1$ **17.** 1, 2, 3, 1

Long Answer Type Questions

1. 2 and -2 **2.** 1 and 2 **4.** $\pm\sqrt{5}, 3\sqrt{5}$ **5.** $k = 5, a = -5$ **6.** Subtract $15x - 14$, or add $-15x + 14$

NCERT Exemplar Problems

1. (*i*) No (*ii*) degree of $g(x) >$ degree of $p(x)$. (*iii*) degree of $g(x) \le$ degree of $p(x)$ **2.** (*i*) False (*ii*) False (*iii*) True

3. $\frac{1}{2\sqrt{2}}, \frac{-3}{\sqrt{2}}$ **4.** $\frac{1}{2}, \sqrt{2}$ **5.** $\sqrt{3}, -5\sqrt{3}$ **6.** $-2\sqrt{5}, \frac{\sqrt{5}}{2}$ **7.** $\frac{2}{3}, \frac{-1}{7}$

8. $b = \pm 3, a = -1, 5$, zeroes are $-1, 2, 5$, or $5, 2, -1$ **9.** $1, 2, -\frac{1}{2}, -3$

10. $a = -1, b = -2$, $q(x)$ has all zeroes of $p(x)$; $x = 1, 2$ are zeroes of $p(x)$ but not of $q(x)$.

Higher Order Thinking Skills (Hots) Questions

1. -5 **2.** 1 **3.** $x - 1$ **4.** No **5.** $x^2 - 2\sqrt{3}x + 5$

6. $-1, -1$ **7.** -28 **8.** $g(x) = -7$ **9.** $-3, -2$ **14.** $k(4x^3 - 12x^2 - 2x - 5)$

15. $k(4x^3 - 24x^2 + 31x + 4)$ **16.** $k = -1,$ **17.** $c = -48$ **18.** $a = 2, b = 5$, third zero $= \frac{3}{2}$

20. 0

Value Based Questions

1. $k(x^2 - 20x + 96)$, Honesty 2. (*a*) 150, 150 (*b*) Honesty 3. (*a*) Yes (*b*) Sincerity, Brilliance

Multiple Choice Questions

1. (*d*)	2. (*c*)	3. (*c*)	4. (*a*)	5. (*b*)	6. (*d*)	7. (*c*)	8. (*b*)
9. (*a*)	10. (*c*)	11. (*d*)	12. (*d*)	13. (*b*)	14. (*c*)	15. (*b*)	16. (*b*)
17. (*c*)	18. (*c*)	19. (*c*)	20. (*b*)	21. (*a*)	22. (*a*)	23. (*d*)	24. (*b*)
25. (*a*)	26. (*a*)	27. (*b*)	28. (*b*)	29. (*b*)	30. (*b*)		

NCERT Exemplar Problems

1. (*a*)	2. (*d*)	3. (*d*)	4. (*b*)	5. (*b*)	6. (*c*)	7. (*a*)	8. (*d*)

SOLUTIONS

VERY SHORT ANSWER TYPE QUESTIONS

1. $p(x) = x^2 - x - 6$

$\Rightarrow \quad p(x) = (x-3)(x+2)$

If $\quad p(x) = 0$

$\Rightarrow \quad x - 3 = 0$ or $x + 2 = 0$

$\Rightarrow \quad x = 3, x = -2$

2. Quadratic polynomial

$= x^2 -$ (sum of zeroes) x + product of zeroes

$= x^2 - 3x - 2$

3. When $\quad x = -3$

LHS $= 2x^2 + 5x + 3 = 2(-3)^2 + 5(-3) + 3$

$= 18 - 15 + 3 = 6 \neq 0$

$\therefore$ $x = -3$ is not a solution of given equation.

4. $p(x) = 2x^2 + 2ax + 5x + 10$

$\because$ $x + a$ is a factor of $p(x)$

$\therefore$ $p(-a) = 0$

$\Rightarrow \quad 2(-a)^2 + 2a(-a) + 5(-a) + 10 = 0$

$\Rightarrow \quad -5a + 10 = 0$

$\Rightarrow \quad a = 2$

5. Given equation is $3x^2 + 13x + 14 = 0$

When $x = -2$

LHS $= 3(-2)^2 + 13 \times (-2) + 14$

$= 12 - 26 + 14 = 0 =$ RHS

$\therefore$ $x = -2$ is a solution of the given equation.

6. $\because$ Graph $y = f(x)$ does not intersect x-axis.

$\therefore$ $f(x)$ has no zeroes.

7. By putting $x = -4$ in the polynomial

$x^2 - x - (2k + 2) = 0$

$\Rightarrow \quad 16 + 4 - 2k - 2 = 0$

$\Rightarrow \quad 18 = 2k \Rightarrow k = 9$

8. $\because$ 1 is a zero of $p(x)$

$\Rightarrow \quad p(1) = 0$

$\Rightarrow \quad a(1)^2 - 3(a - 1) - 1 = 0$

$\Rightarrow \quad a - 3a + 3 - 1 = 0$

$\Rightarrow \quad a = 1$

9. Product of zeroes $= -\dfrac{9}{2} = \dfrac{c}{a}$

Sum of zeroes $= -\dfrac{3}{2} = -\dfrac{b}{a}$

Polynomial is $k\left(x^2 - \dfrac{b}{a}x + \dfrac{c}{a}\right)$

i.e., $k\left(x^2 + \dfrac{3}{2}x - \dfrac{9}{2}\right)$ or $k(2x^2 + 3x - 9)$

10. (*i*) α, β are the zeroes of a polynomial.

Sum of zeroes, $S = \alpha + \beta = 6$, product of zeroes, $P = \alpha\beta = 4$

The required polynomial $g(x)$ is given by

$g(x) = k(x^2 - sx + p)$

$g(x) = k(x^2 - 6x + 4)$

where k is any non zero real number.

(*ii*) $p(y) = 2y^2 + 7y + 5$

$a = 2, b = 7, c = 5$

$\because$ α, β are zeroes of $p(y)$

$\therefore \quad \alpha + \beta = \dfrac{-b}{a} = \dfrac{-7}{2}$

$\alpha\beta = \dfrac{c}{a} = \dfrac{5}{2}$

$\alpha + \beta + \alpha\beta = \dfrac{-7}{2} + \dfrac{5}{2} = \dfrac{-2}{2} = -1$

(*iii*) Let other zero be α,

$\therefore \quad (2 + \sqrt{3}) + \alpha = -\left(\dfrac{-4}{1}\right)$

$\Rightarrow \alpha = 4 - 2 - \sqrt{3} = 2 - \sqrt{3}$

11. $\because$ Graph of $f(x)$ intersects x-axis at three different points.

$\therefore$ Number of zeroes of $f(x) = 3$.

12. $\because$ Graph of $p(y)$ intersects y-axis in four different points.

$\therefore$ Numbers of zeroes $= 4$

13. Degree of remainder is always less than divisor.

$\therefore$ Degree of remainder will be less than 2.

Hence, degree is 1 or 0.

14. $\because$ $p(x) = x^3 + x^2 - ax + b$ is divisible by $x^2 - x$.

$\therefore$ $x(x-1)$ is a factor of $p(x)$.

$\Rightarrow$ $x = 0$ and $x = 1$ are zeroes of $p(x)$.

$\Rightarrow$ $p(0) = 0 \Rightarrow (0)^3 + (0)^2 - a \times 0 + b = 0 \Rightarrow b = 0$

and $p(1) = 0 \Rightarrow 1^3 + (1)^2 - a \times 1 + b = 0$

$\Rightarrow$ $2 - a + 0 = 0 \Rightarrow a = 2$

15. Here, $a = 1, b = 7, c = 10$

Sum of zeroes $= -\dfrac{b}{a} = -7$

16. Here, $a = 1, b = 0, c = -3$.

$\Rightarrow$ Product of zeroes $= \dfrac{c}{a} = -3$

17. Here, $a = 1, b = -p, c = q$

$\alpha + \beta = -\dfrac{b}{a} = \dfrac{-(-p)}{1} = p, \alpha\beta = \dfrac{c}{a} = q$

Now $\alpha^2 + \beta^2 = (\alpha + \beta)^2 - 2\alpha\beta = p^2 - 2q$

18. Here, $a = a, b = b, c = c$

$\alpha + \beta = -\dfrac{b}{a}, \alpha\beta = \dfrac{c}{a}$

Now $\dfrac{1}{\alpha} + \dfrac{1}{\beta} = \dfrac{\alpha + \beta}{\alpha\beta} = -\dfrac{b/a}{c/a} = -\dfrac{b}{c}$

19. $p(x) = x^2 - 2x + 5 = x^2 - 2x + 1 + 4 = (x-1)^2 + 4$

If $p(x) = 0$

$\Rightarrow$ $(x-1)^2 + 4 = 0 \Rightarrow (x-1)^2 = -4$

which is not true for any real value of x.

$\therefore$ $p(x)$ has no real zeroes.

20. If $3x - 7$ is a factor of $p(x) = 6x^3 + x^2 - 26x - 25$ then $p\left(\dfrac{7}{3}\right)$ should be equal to 0.

Now, $p\left(\dfrac{7}{3}\right) = 6 \times \left(\dfrac{7}{3}\right)^3 + \left(\dfrac{7}{3}\right)^2 + 26 \times \dfrac{7}{3} - 25 \neq 0$

$\therefore$ $3x - 7$ is not a factor of $p(x)$.

21. $p(x) = x^2 - x + k$

$\because$ 1 is one of the zeroes.

$\therefore$ $p(1) = 0$

$\Rightarrow$ $1^2 - 1 + k = 0 \Rightarrow k = 0$

22. $p(x) = x^2 - px - p - c$

$\therefore$ $\alpha + \beta = p, \alpha\beta = (-p - c)$

$\because$ $(\alpha + 1)(\beta + 1) = 0 \Rightarrow \alpha\beta + \alpha + \beta + 1 = 0$

$\Rightarrow$ $-p - c + p + 1 = 0$

$c = 1$

23. Product of zeroes $= \dfrac{c}{a}$

i.e., $\alpha \times \beta = \dfrac{6}{-2} = -3$

24. Since equation is $-3x^2 + 0x + k$

and sum of zeroes $= \dfrac{-b}{a}$

i.e., $\alpha + \beta = \dfrac{-b}{a} \Rightarrow \alpha + \beta = \dfrac{0}{-3} = 0$

SHORT ANSWER TYPE QUESTIONS-I

1. $p(x) = 6x^2 - 7x - 3 = 6x^2 - 9x + 2x - 3$

$= 3x(2x - 3) + 1(2x - 3) = (3x + 1)(2x - 3)$

For $p(x) = 0$, $3x + 1 = 0$ or $2x - 3 = 0$

$\Rightarrow$ $x = \dfrac{-1}{3}$ or $x = \dfrac{3}{2}$

$a = 6, b = -7, c = -3$

Sum of zeroes $= \dfrac{-1}{3} + \dfrac{3}{2} = \dfrac{7}{6}$

Also, $\dfrac{-b}{a} = \dfrac{-(-7)}{6} = \dfrac{7}{6}$

$\Rightarrow$ Sum of zeroes $= \dfrac{-b}{a}$

Now, product of zeroes $= \left(-\dfrac{1}{3}\right) \times \dfrac{3}{2} = \dfrac{-1}{2}$

Also, $\dfrac{c}{a} = \dfrac{-3}{6} = \dfrac{-1}{2}$.

$\Rightarrow$ Product of zeroes $= \dfrac{c}{a}$.

2. Quadratic polynomial

$= x^2 - (\text{sum of zeroes})\,x + (\text{product of zeroes})$

$= x^2 - 8x + 12$

Now $x^2 - 8x + 12 = 0$

$\Rightarrow$ $(x - 6)(x - 2) = 0 \Rightarrow x = 6, x = 2$

$\therefore$ Zeroes are 6 and 2.

3. Let one zero $= \alpha$

$\therefore$ Other zero $= \dfrac{1}{\alpha}$

Product of zeroes $= \dfrac{c}{a}$

$\Rightarrow$ $\alpha \times \dfrac{1}{\alpha} = \dfrac{6a}{a^2 + 9}$

$\Rightarrow$ $a^2 + 9 = 6a \Rightarrow a^2 - 6a + 9 = 0$

$\Rightarrow$ $(a - 3)^2 = 0 \Rightarrow a = 3$

4. $p(x) = ax^2 - 6x - 6$

Product of zeroes $= \dfrac{c}{a}$ $\quad (c = -6, a = a)$

$\Rightarrow$ $4 = \dfrac{-6}{a} \Rightarrow 4a = -6$

$a = \dfrac{-6}{4} = \dfrac{-3}{2}$

$a = \dfrac{-3}{2}$

5. $\because$ Two zeroes are 2 and -2.

$\therefore$ $(x + 2)(x - 2)$ is a factor of $x^4 + x^3 - 34x^2 - 4x + 120$

$$
\begin{array}{r|l}
 & x^2 + x - 30 \\
x^2 - 4 & x^4 + x^3 - 34x^2 - 4x + 120 \\
 & x^4 \qquad - 4x^2 \\
 & - \qquad\quad + \\
\hline
 & x^3 - 30x^2 - 4x + 120 \\
 & x^3 \qquad\quad - 4x \\
 & - \qquad\qquad + \\
\hline
 & -30x^2 + 120 \\
 & -30x^2 + 120 \\
 & + \qquad\quad - \\
\hline
 & 0
\end{array}
$$

$\therefore$ Other factor is $x^2 + x - 30$.

For other zeroes,

$x^2 + x - 30 = 0$

$\Rightarrow$ $(x + 6)(x - 5) = 0$

$\Rightarrow$ $x = -6, x = 5$

$\therefore$ $x = 2, -2, -6$ and 5 are zeroes of given polynomial.

6. $\because$ $\sqrt{2}$ and $-\sqrt{2}$ are zeroes.

$\therefore$ $(x-\sqrt{2})(x+\sqrt{2}) = x^2-2$ is a factor of given polynomial.

$$\begin{array}{r|l} & 2x^2+7x-15 \\ \hline x^2-2 & 2x^4+7x^3-19x^2-14x+30 \\ & \underline{\overset{-}{2x^4} \qquad \overset{+}{-4x^2}} \\ & 7x^3-15x^2-14x+30 \\ & \underline{\overset{-}{7x^3} \qquad \overset{+}{-14x}} \\ & -15x^2 \quad +30 \\ & \underline{\overset{+}{-15x^2} \quad \overset{-}{+30}} \\ & 0 \end{array}$$

$\therefore$ Other factor is $2x^2+7x-15$

Now $2x^2+7x-15 = 0$

$\Rightarrow$ $2x^2+10x-3x-15 = 0$

$\Rightarrow$ $2x(x+5)-3(x+5) = 0$

$\Rightarrow$ $(x+5)(2x-3) = 0$

$\therefore$ $x = -5, x = \frac{3}{2}$

$\therefore$ Zeroes of given polynomial are

$\sqrt{2}, -\sqrt{2}, -5$ and $\frac{3}{2}$.

7. Sum of zeroes $S = 1+(-3) = -2$ ...(*i*)

Product of zeroes $P = 1 \times (-3) = -3$ ...(*ii*)

Quadratic polynomial

$$p(x) = x^2 - Sx + P = x^2-(-2)x-3 = x^2+2x-3$$

Here, $a = 1, b = 2, c = -3$

$$\frac{-b}{a} = \frac{-2}{1} = -2$$

$\therefore$ Sum of zeroes $= \frac{-b}{a}$ [Using (*i*)]

Also $\frac{c}{a} = \frac{-3}{1} = -3$

$\Rightarrow$ Product of zeroes $= \frac{c}{a}$ [Using (*ii*)]

8. $p(x) = \sqrt{3}x^2-8x+4\sqrt{3}$

$= \sqrt{3}x^2-6x-2x+4\sqrt{3} = \sqrt{3}x(x-2\sqrt{3})-2(x-2\sqrt{3})$

$\Rightarrow p(x) = (x-2\sqrt{3})(\sqrt{3}x-2)$

If $p(x) = 0$

$\Rightarrow (x-2\sqrt{3}).(\sqrt{3}x-2) = 0$

$\Rightarrow x-2\sqrt{3} = 0$ or $\sqrt{3}x-2 = 0$

$\Rightarrow x = 2\sqrt{3}$ or $x = \frac{2}{\sqrt{3}}$

9.
$$\begin{array}{r|l} & 3x^2+5x-2 \\ \hline 2x+1 & 6x^3+13x^2+x-2 \\ & \underline{\overset{-}{6x^3} \overset{-}{+} 3x^2} \\ & 10x^2+x-2 \\ & \underline{\overset{-}{10x^2} \overset{-}{+} 5x} \\ & -4x-2 \\ & \underline{\overset{+}{-4x} \overset{+}{-} 2} \\ & 0 \end{array}$$

$\therefore$ Quotient $= 3x^2+5x-2$ and remainder $= 0$

10.
$$\begin{array}{r|l} & 2x^2+5 \\ \hline 3x^2+4x+1 & 6x^4+8x^3+17x^2+21x+7 \\ & \underline{\overset{-}{6x^4}\overset{-}{+}8x^3\overset{-}{+}2x^2} \\ & 15x^2+21x+7 \\ & \underline{\overset{-}{15x^2}\overset{-}{+}20x\overset{-}{+}5} \\ & x+2 \end{array}$$

but given remainder is $ax+b$.

then $ax+b = x+2 \Rightarrow a = 1, b = 2$.

11. $\because$ $-\sqrt{3}$ and $\sqrt{3}$ are zeroes of

$p(x) = 2x^3+x^2-6x-3$

$\therefore$ $\{(x-(-\sqrt{3})\}(x-\sqrt{3}) = x^2-3$ is a factor of $p(x)$

For other factor,

$$\begin{array}{r|l} & 2x+1 \\ \hline x^2-3 & 2x^3+x^2-6x-3 \\ & \underline{\overset{-}{2x^3} \qquad \overset{+}{-6x}} \\ & x^2-3 \\ & \underline{\overset{-}{x^2}\overset{+}{-}3} \\ & \times \end{array}$$

For other zeroes, $2x+1 = 0 \Rightarrow x = \frac{-1}{2}$

$\therefore$ All zeroes are $-\sqrt{3}, \sqrt{3}$ and $\frac{-1}{2}$

12. Since 3 is a zero of the polynomial $2x^2+x+k$

$\therefore$ $p(3) = 0$

$\Rightarrow$ $p(x) = 2x^2+x+k$

$\Rightarrow$ $p(3) = 2(3)^2+3+k$

$\Rightarrow$ $0 = 18+3+k$

$\Rightarrow$ $k = -21$

13. $\because$ -1 and 2 are zeroes of $p(x) = 2x^3-x^2-5x-2$

$\therefore$ $\{x-(-1)\}(x-2) = x^2-x-2$ is a factor of $p(x)$

For other zeroes,

$$\begin{array}{r|l} & 2x+1 \\ \hline x^2-x-2 & 2x^3-x^2-5x-2 \\ & \underline{\overset{-}{2x^3}\overset{+}{-}2x^2\overset{+}{-}4x} \\ & x^2-x-2 \\ & \underline{\overset{-}{x^2}\overset{+}{-}x\overset{+}{-}2} \\ & \times \end{array}$$

For other zeroes $2x+1 = 0 \Rightarrow x = -\frac{1}{2}$

$\therefore$ Other zero $= -\frac{1}{2}$

14.
$$\begin{array}{r|l} & x^2+x-3 \\ \hline x^2-x+1 & x^4-3x^2+4x+5 \\ & \underline{\overset{-}{x^4}\overset{-}{+}x^2 \quad \overset{+}{-}x^3} \\ & x^3-4x^2+4x+5 \\ & \underline{\overset{-}{x^3}\overset{+}{-}x^2\overset{-}{+}x} \\ & -3x^2+3x+5 \\ & \underline{\overset{+}{-3x^2}\overset{-}{+}3x\overset{+}{-}3} \\ & +8 \end{array}$$

Quotient $= x^2+x-3$

Remainder $= 8$

15. $p(x) = x^2+(a+1)x+b$

$\because$ 2 is a zero of $p(x)$

$\Rightarrow p(2) = 0 \Rightarrow 2^2 + (a+1)\,2 + b = 0$

$\Rightarrow \quad 2a + b = -6 \quad ...(i)$

Also, $\quad p(-3) = 0$

$\Rightarrow \quad (-3)^2 + (a+1)(-3) + b = 0$

$\Rightarrow \quad -3a + b = -6 \quad ...(ii)$

Solving (*i*) and (*ii*), we get

$a = 0, b = -6$

16. $p(x) = 7x - x^3 - 6$

$\because$ 1 is a zero of $p(x) \Rightarrow x - 1$ is a factor of $p(x)$

For other zeroes,

$$\begin{array}{r|l} & -x^2 - x + 6 \\ \hline x - 1 & -x^3 + 7x - 6 \\ & \underline{\mp x^3 \qquad \pm x^2} \\ & -x^2 + 7x - 6 \\ & \underline{\mp x^2 \pm x} \\ & 6x - 6 \\ & \underline{6x - 6} \\ & 0 \end{array}$$

$\therefore$ Other factor $= -x^2 - x + 6$

For other zeroes,

$-x^2 - x + 6 = 0$

$\Rightarrow \quad x^2 + x - 6 = 0$

$\Rightarrow \quad (x+3)(x-2) = 0$

$\Rightarrow \quad x = -3, x = 2$

$\therefore$ Other zeroes are -3 and 2.

17. Let α, β are zeroes of $2x^2 - 5x - 3$.

$\therefore \alpha + \beta = \frac{5}{2}, \alpha\beta = \frac{-3}{2}$

A.T.Q.,

2α and 2β are zeroes of $x^2 + px + q$

$\therefore \quad 2\alpha + 2\beta = -p$

$\Rightarrow \quad 2 \times (\alpha + \beta) = -p$

$2 \times \frac{5}{2} = -p \Rightarrow p = -5$

and $\quad 2\alpha \times 2\beta = q \Rightarrow 4\,\alpha\beta = q$

$\Rightarrow \quad 4 \times \left(\frac{-3}{2}\right) = q \Rightarrow q = -6$

18. $p(x) = x^2 - 7x + 12.$

For zeroes $\quad p(x) = 0$

$\Rightarrow \quad x^2 - 7x + 12 = 0$

$\Rightarrow \quad x^2 - 3x - 4x + 12 = 0$

$\Rightarrow \quad x(x-3) - 4(x-3) = 0$

$\Rightarrow \quad (x-3)(x-4) = 0$

$\Rightarrow \quad x - 3 = 0$ or $x - 4 = 0$

$\Rightarrow \quad x = 3$ or $x = 4$

$\therefore$ Zeroes are 3 and 4.

Now, $a = 1, b = -7, c = 12$

$\frac{-b}{a} = \frac{-(-7)}{1} = 7$

Sum of zeroes $= 3 + 4 = 7$

$\Rightarrow \quad$ Sum of zeroes $= \frac{-b}{a}$

Also $\frac{c}{a} = \frac{12}{1} = 12$

Product of zeroes $= 3 \times 4 = 12$

$\Rightarrow \quad$ Product of zeroes $= \frac{c}{a}$

19. $p(x) = 4x^2 + 24x + 36$

For zeroes, $\quad p(x) = 0$

$\Rightarrow \quad 4x^2 + 24x + 36 = 0$

$\Rightarrow \quad 4(x^2 + 6x + 9) = 0$

$\Rightarrow \quad 4(x^2 + 3x + 3x + 9) = 0$

$\Rightarrow \quad (x+3)(x+3) = 0$

$\Rightarrow \quad x + 3 = 0$ or $x + 3 = 0$

$\Rightarrow \quad x = -3, x = -3$

$\therefore$ Zeroes are $-3, -3$.

Now $\quad a = 4, b = 24, c = 36$

$\frac{-b}{a} = \frac{-24}{4} = -6$

Sum of zeroes $= -3 + (-3) = -6$

$\Rightarrow \quad$ Sum of zeroes $= \frac{-b}{a}$

Also, $\quad \frac{c}{a} = \frac{36}{4} = 9$

and $\quad$ Product of zeroes $= (-3) \times (-3) = 9$

$\Rightarrow \quad$ Product of zeroes $= \frac{c}{a}$

20. $p(x) = 100x^2 - 81$

For zeroes, $\quad p(x) = 0$

$\Rightarrow \quad 100x^2 - 81 = 0$

$\Rightarrow \quad (10x)^2 - (9)^2 = 0$

$\Rightarrow \quad (10x - 9)(10x + 9) = 0$

$\Rightarrow \quad 10x - 9 = 0$ or $10x + 9 = 0$

$\Rightarrow \quad x = \frac{9}{10}$ or $x = \frac{-9}{10}$

Now $\quad a = 100, b = 0, c = -81$

$\therefore \quad \frac{-b}{a} = \frac{0}{100} = 0$

Sum of zeroes $= \frac{9}{10} + \left(\frac{-9}{10}\right) = 0$

Also $\quad \frac{c}{a} = \frac{-81}{100}$

Product of zeroes $= \frac{9}{10} \times \left(\frac{-9}{10}\right) = \frac{-81}{100}$

$\Rightarrow \quad$ Product of zeroes $= \frac{c}{a}$

21. $p(x) = 25x^2 + 5x$

For zeroes $\quad p(x) = 0$

$25x^2 + 5x = 0$

$\Rightarrow \quad 5x(5x + 1) = 0$

$\Rightarrow \quad 5x = 0$ or $5x + 1 = 0$

$\Rightarrow \quad x = 0$ or $x = \frac{-1}{5}$

$\therefore$ Zeroes are $0, -\frac{1}{5}$.

Now, $\quad a = 25, b = 5, c = 0$

$\frac{-b}{a} = \frac{-5}{25} = \frac{-1}{5}$

Sum of zeroes $= 0 + \left(\frac{-1}{5}\right) = \frac{-1}{5}$

$\Rightarrow \quad$ Sum of zeroes $= \frac{-b}{a}$

Also, $\quad \frac{c}{a} = \frac{0}{25} = 0$

and $\quad$ product of zeroes $= 0 \times \left(\frac{-1}{5}\right) = 0$

$\Rightarrow \quad$ Product of zeroes $= \frac{c}{a}$

22. $p(y) = 4\sqrt{3}y^2 + 5y - 2\sqrt{3}$

For zeroes of $p(y)$,

$$p(y) = 0$$

$$\Rightarrow \quad 4\sqrt{3}y^2 + 5y - 2\sqrt{3} = 0$$

$$\Rightarrow \quad 4\sqrt{3}y^2 + 8y - 3y - 2\sqrt{3} = 0$$

$$\Rightarrow \quad 4y(\sqrt{3}y+2) - \sqrt{3}(\sqrt{3}y+2) = 0$$

$$\Rightarrow \quad (\sqrt{3}y+2)(4y-\sqrt{3}) = 0$$

$$\Rightarrow \quad y = \frac{-2}{\sqrt{3}}, y = \frac{\sqrt{3}}{4}$$

$$\text{Sum of zeroes} = \frac{-2}{\sqrt{3}} + \frac{\sqrt{3}}{4} = \frac{-5}{4\sqrt{3}} = -\frac{b}{a}$$

$$\text{Product of zeroes} = \frac{(-2)(\sqrt{3})}{(\sqrt{3})(4)} = \frac{-1}{2} = \frac{-2\sqrt{3}}{4\sqrt{3}} = \frac{c}{a}$$

23. Zeroes are 3 and – 5.

S = Sum of zeroes = 3 + (–5) = –2

P = Product of zeroes = 3 × (–5) = –15

Quadratic polynomial

$$p(x) = x^2 - (S)x + P$$
$$= x^2 + 2x - 15$$

24. $\text{Sum of zeroes} = -9 + \left(-\frac{1}{9}\right) = \frac{-81-1}{9} = \frac{-82}{9}$

$\text{Product of zeroes} = (-9) \times \left(-\frac{1}{9}\right) = 1$

Quadratic polynomial

$= x^2 - (\text{sum of zeroes})\, x + \text{product of zeroes}$

$= x^2 - \left(\frac{-82}{9}\right)x + 1 = 9x^2 + 82x + 9$

25. Quadratic Polynomial

$= x^2 - (\text{sum of zeroes})\, x + \text{product of zeroes}$

$= x^2 - 5x + 17$

26. Here sum of zeroes, S = 0

Product of zeroes, $P = \sqrt{15}$

Quadratic polynomial

$$p(x) = x^2 - (S)x + P$$
$$= x^2 - 0x + \sqrt{15} = x^2 + \sqrt{15}$$

27. One zero = 5, P = Product of zeroes = 30

$\therefore$ Other zeroes $= \frac{30}{5} = 6$

$\therefore$ Sum of zeroes, S = 5 + 6 = 11

Quadratic polynomial

$$p(x) = x^2 - (S)x + P$$
$$p(x) = x^2 - 11x + 30$$

28. One zero = 7, S = Sum of zeroes = – 18

$\therefore$ Other zeroes = – 18 – 7 = – 25

$\therefore$ Product of zeroes P = 7 × (– 25) = –175

Quadratic polynomial

$$p(x) = x^2 - (S)x + P$$
$$= x^2 - (-18)x - 175$$
$$= x^2 + 18x - 175$$

29. Sum of zeroes, $S = (3+\sqrt{5}) + (3-\sqrt{5}) = 6$

Product of zeroes, $P = (3+\sqrt{5}).(3-\sqrt{5}) = 9 - 5 = 4$

Polynomial $p(x) = x^2 - (S)x + P$

$= x^2 - 6x + 4$

30. One zero = – 8

Sum of zeroes = 0

$\therefore$ Other zero = 0 – (–8) = 8

Product of zeroes = 8 × (–8) = – 64

Polynomial $p(x) = x^2 - (S)x + P = x^2 - 64$

31. One zero = – 5

Product of zeroes = 0

$\therefore$ Other zero $= \frac{0}{-5} = 0$

Sum of zeroes = – 5 + 0 = – 5

Polynomial $p(x) = x^2 - (S)x + P$

$= x^2 + 5x$

32. $p(x) = 2x^3 - 17x^2 + 38x - 15$

$$p\left(\frac{1}{2}\right) = 2\times\left(\frac{1}{2}\right)^3 - 17\times\left(\frac{1}{2}\right)^2 + 38\times\frac{1}{2} - 15$$

$$= \frac{1}{4} - \frac{17}{4} + 19 - 15 = -4 + 19 - 15 = 0$$

$\therefore \frac{1}{2}$ is a zero of $p(x)$.

$$p(3) = 2(3)^3 - 17(3)^2 + 38(3) - 15$$
$$= 54 - 153 + 114 - 15 = 0$$

$\therefore$ 3 is a zero of $p(x)$.

33. $p(x) = x^3 + x^2 + x - 2$

$q(x) = x^2 + 2x + 1$

Remainder $= 2x - 1$

Now $p(x) = q(x) \times g(x) + \text{remainder}$

$$\Rightarrow \quad x^3 + x^2 + x - 2 = (x^2 + 2x + 1).\ g(x) + 2x - 1$$

$$\Rightarrow \quad g(x) = \frac{x^3 + x^2 + x - 2 - (2x - 1)}{x^2 + 2x + 1}$$

$$= \frac{x^3 + x^2 - x - 1}{x^2 + 2x + 1}$$

$$\begin{array}{r|l} & x - 1 \\ x^2 + 2x + 1 & x^3 + x^2 - x - 1 \\ & x^3 + 2x^2 + x \\ & - \quad - \quad - \\ \hline & -x^2 - 2x - 1 \\ & -x^2 - 2x - 1 \\ & + \quad + \quad + \\ \hline & \times \end{array}$$

So, $g(x) = x - 1$.

34.

$$\begin{array}{r|l} & 2y^3 + 5y^2 + 2y - 7 \\ 3y^2 + 5 & 6y^5 + 15y^4 + 16y^3 + 4y^2 + 10y - 35 \\ & 6y^5 \quad + 10y^3 \\ & - \quad\quad - \\ \hline & 15y^4 + 6y^3 + 4y^2 + 10y - 35 \\ & 15y^4 \quad + 25y^2 \\ & - \quad\quad - \\ \hline & 6y^3 - 21y^2 + 10y - 35 \\ & 6y^3 \quad + 10y \\ & - \quad\quad - \\ \hline & -21y^2 - 35 \\ & -21y^2 - 35 \\ & + \quad + \\ \hline & \times \end{array}$$

$\because$ Remainder = 0

Hence, $3y^2 + 5$ is a factor of $6y^5 + 15y^4 + 16y^3 + 4y^2 + 10y - 35$

35.

$$\begin{array}{r|l} & 12x^2 + 10xy - 2 \\ 2xy & 24x^3y + 20x^2y^2 - 4xy \\ & 24x^3y \\ & - \\ \hline & 20x^2y^2 - 4xy \\ & 20x^2y^2 \\ & - \\ \hline & -4xy \\ & \mp 4xy \\ \hline & 0 \end{array}$$

Quotient $= 12x^2 + 10xy - 2$, remainder = 0

36.
$$\begin{array}{r|l} & 5x^2 - 9x + 3 \\ \hline 2x^2 + 7x - 1 & 10x^4 + 17x^3 - 62x^2 + 30x - 3 \\ & 10x^4 + 35x^3 - 5x^2 \\ & \underline{- \quad - \quad +} \\ & -18x^3 - 57x^2 + 30x - 3 \\ & -18x^3 - 63x^2 + 9x \\ & \underline{+ \quad + \quad -} \\ & 6x^2 + 21x - 3 \\ & 6x^2 + 21x - 3 \\ & \underline{- \quad - \quad +} \\ & \times \end{array}$$

37.
$$\begin{array}{r|l} & ax + b \\ \hline x + c & ax^2 + bx + acx + bc \\ & ax^2 + acx \\ & \underline{- \quad -} \\ & bx + bc \\ & bx + bc \\ & \underline{- \quad -} \\ & 0 \end{array}$$

$\therefore$ Quotient $= ax + b$, remainder $= 0$

38.
$$\begin{array}{r|l} & 8y^2 + 20y + 50 \\ \hline 8y - 20 & 64y^3 - 1000 \\ & 64y^3 - 160y^2 \\ & \underline{- \quad +} \\ & 160y^2 - 1000 \\ & 160y^2 - 400y \\ & \underline{- \quad +} \\ & 400y - 1000 \\ & 400y - 1000 \\ & \underline{- \quad +} \\ & \times \end{array}$$

Quotient $= 8y^2 + 20y + 50$, remainder $= 0$

39.
$$\begin{array}{r|l} & 5y^3 - 2y^2 + \frac{5}{3}y \\ \hline 3y - 2 & 15y^4 - 16y^3 + 9y^2 - \frac{10}{3}y \\ & 15y^4 - 10y^3 \\ & \underline{- \quad +} \\ & -6y^3 + 9y^2 - \frac{10}{3}y \\ & -6y^3 + 4y^2 \\ & \underline{+ \quad -} \\ & 5y^2 - \frac{10}{3}y \\ & 5y^2 - \frac{10}{3}y \\ & \underline{- \quad +} \\ & \times \end{array}$$

$\therefore$ Quotient $= 5y^3 - 2y^2 + \frac{5}{3}y$, remainder $= 0$

40.
$$\begin{array}{r|l} & 2y^3 - 5y + \frac{5}{2} \\ \hline 2y - 5 & 4y^4 - 10y^3 - 10y^2 + 30y - 15 \\ & 4y^4 - 10y^3 \\ & \underline{- \quad +} \\ & -10y^2 + 30y - 15 \\ & -10y^2 + 25y \\ & \underline{+ \quad -} \\ & 5y - 15 \\ & 5y - \frac{25}{2} \\ & \underline{- \quad +} \\ & -\frac{5}{2} \end{array}$$

$\because$ Remainder $\neq 0$

$\therefore$ $2y - 5$ is not a factor of $4y^4 - 10y^3 - 10y^2 + 30y - 15$.

41. One zero $= 2+\sqrt{5}$

Sum of zeroes $= 4$

$\therefore$ Other zeroes $= 4 - (2+\sqrt{5}) = 2-\sqrt{5}$

Product of zeroes $= (2+\sqrt{5})(2-\sqrt{5}) = -1$

Polynomial is $x^2 - 4x - 1$.

42. One zeroes $= \sqrt{5}$, Product of zeroes, P $= -2\sqrt{5}$

$$\therefore \quad \text{Other zero} = \frac{-2\sqrt{5}}{\sqrt{5}} = -2$$

$\therefore$ Sum of zeroes, S $= -2 + \sqrt{5}$

Quadratic polynomial

$$\begin{aligned} p(x) &= x^2 - (\text{S})x + \text{P} \\ &= x^2 - (-2 + \sqrt{5})x + (-2\sqrt{5}) \\ &= x^2 + (2 - \sqrt{5})x - 2\sqrt{5} \end{aligned}$$

43. $p(x) = \sqrt{x^2 - 4x + 3} + \sqrt{x^2 - 9} - \sqrt{4x^2 - 14x + 6}$

$$p(3) = \sqrt{3^2 - 4\times3 + 3} + \sqrt{3^2 - 9} - \sqrt{4\times3^2 - 14\times3 + 6}$$
$$= 0 + 0 + 0 = 0$$

$\therefore$ 3 is a zero of $p(x)$.

44. $\alpha + \beta = -\frac{b}{a}$

$$\Rightarrow \quad \alpha + \beta = \frac{k+6}{1}$$

$$\Rightarrow \quad \alpha\beta = \frac{c}{a} \Rightarrow \alpha\beta = \frac{2(2k-1)}{1}$$

Now, $\quad \alpha + \beta = \frac{1}{2}\alpha\beta$

$$\Rightarrow \quad k + 6 = \frac{1}{2} \times 2(2k-1) \Rightarrow k + 6 = 2k - 1$$

$$\Rightarrow \quad 6 + 1 = 2k - k \Rightarrow 7 = k$$

45. $p(x) = ax^2 - 5x + c$

$$m + n = -\frac{(-5)}{a} = \frac{5}{a}, \; m.n = \frac{c}{a}$$

Now $\quad m + n = 10$

$$\Rightarrow \quad \frac{5}{a} = 10 \Rightarrow a = \frac{1}{2}$$

and $\quad m.n = 10$

$$\Rightarrow \quad \frac{c}{a} = 10$$

$$\Rightarrow \quad c = 10\,a$$

$$\Rightarrow \quad c = 10 \times \frac{1}{2} = 5$$

$$\therefore \quad a = \frac{1}{2}, c = 5$$

46. Sum of zeroes, S $= \frac{3-\sqrt{3}}{5} + \frac{3+\sqrt{3}}{5} = \frac{3-\sqrt{3}+3+\sqrt{3}}{5} = \frac{6}{5}$

Product of zeroes

$$\text{P} = \left(\frac{3-\sqrt{3}}{5}\right) \times \left(\frac{3+\sqrt{3}}{5}\right) = \frac{3^2 - (\sqrt{3})^2}{5\times5} = \frac{6}{25}$$

Polynomial $\quad p(x) = x^2 - (\text{S})x + \text{P}$

$$= x^2 - \frac{6}{5}x + \frac{6}{25} = k\,(25x^2 - 30x + 6)$$

47. $p(x) = 4x^2 - 2x + k - 4$, Here $a = 4$, $b = -2$, $c = k - 4$.

Product of zeroes $= \dfrac{c}{a}$

$\Rightarrow \quad \alpha \times \dfrac{1}{\alpha} = \dfrac{k-4}{4}$

$\Rightarrow \quad 1 = \dfrac{k-4}{4} \Rightarrow k - 4 = 4 \therefore k = 8$

48. $\alpha + \beta = -\dfrac{b}{a}$

$\Rightarrow \quad \alpha + \beta = -5$

and $\quad \alpha\beta = \dfrac{c}{a} \Rightarrow \alpha\beta = 5$

Now $\quad \alpha^{-1} + \beta^{-1} = \dfrac{1}{\alpha} + \dfrac{1}{\beta} = \dfrac{\alpha+\beta}{\alpha\beta} = \dfrac{-5}{5} = -1$

49. $p(x) = x^2 + 7x + 7$

Here, $\quad a = 1, b = 7, c = 7$

$\therefore$ α, β are both zeroes of $p(x)$

$\therefore \quad \alpha + \beta = \dfrac{-b}{a} = -7 \qquad \ldots(i)$

$\alpha\beta = \dfrac{c}{a} = 7 \qquad \ldots(ii)$

Now $\quad \dfrac{1}{\alpha} + \dfrac{1}{\beta} - 2\alpha\beta = \dfrac{\beta+\alpha}{\alpha\beta} - 2\alpha\beta$

$= \dfrac{-7}{7} - 2 \times 7 = -1 - 14 = -15$

50. $\sqrt{3}\,x^2 + 10x + 7\sqrt{3} = \sqrt{3}\,x^2 + 3x + 7x + 7\sqrt{3}$

$= \sqrt{3}\,x(x + \sqrt{3}) + 7(x + \sqrt{3})$

$= (\sqrt{3}\,x + 7)(x + \sqrt{3})$

For zeroes of the polynomial,

$(\sqrt{3}x + 7)(x + \sqrt{3}) = 0$

$\Rightarrow \quad \sqrt{3}\,x + 7 = 0 \text{ or } x + \sqrt{3} = 0$

$\Rightarrow \quad \sqrt{3}\,x = -7 \text{ or } x = -\sqrt{3}$

$\Rightarrow \quad x = -\dfrac{7}{\sqrt{3}}, -\sqrt{3}$

51. One zero $= 3 - \sqrt{5}$

P = Product of zeroes = 4

$\therefore \quad$ Other zero $= \dfrac{4}{3-\sqrt{5}} = \dfrac{4(3+\sqrt{5})}{(3-\sqrt{5})(3+\sqrt{5})}$

$= \dfrac{4(3+\sqrt{5})}{9-5} = \dfrac{4(3+\sqrt{5})}{4} = 3 + \sqrt{5}$

$\therefore$ Sum of zeroes, S $= 3 - \sqrt{5} + 3 + \sqrt{5} = 6$

Quadratic polynomial

$p(x) = x^2 - (\text{S})x + \text{P}$

$= x^2 - 6x + 4$

52.

$$\begin{array}{r|l} & 2x^2 + 4 \\ x^3 - 4x + 1 & 2x^5 - 4x^3 + 2x^2 + 5x + 1 \\ & \underset{-}{2x^5} \underset{+}{-} 8x^3 \underset{-}{+} 2x^2 \\ \hline & 4x^3 + 5x + 1 \\ & \underset{-}{4x^3} \underset{+}{-} 16x \underset{-}{+} 4 \\ \hline & 21x - 3 \end{array}$$

Remainder $= 21x - 3 \neq 0$

$\therefore$ $g(x)$ is not a factor of $p(x)$.

53. $p(x) = x^2 - 2x - 8$

For zeroes, $p(x) = 0$

$\Rightarrow \quad x^2 - 2x - 8 = 0 \Rightarrow x^2 - 4x + 2x - 8 = 0$

$\Rightarrow \quad x(x-4) + 2(x-4) = 0 \Rightarrow (x-4)(x+2) = 0$

$\Rightarrow \quad x - 4 = 0 \text{ or } x + 2 = 0$

$\Rightarrow \quad x = 4 \text{ or } x = -2$

$\therefore$ Zeroes are 4, – 2

Now, here $a = 1$, $b = -2$, $c = -8$

$\dfrac{-b}{a} = \dfrac{-(-2)}{1} = 2$

and $\quad$ sum of zeroes $= 4 + (-2) = 2$

$\therefore \quad$ Sum of zeroes $= \dfrac{-b}{a}$

Also $\quad \dfrac{c}{a} = \dfrac{-8}{1} = -8$

and $\quad$ product of zeroes $= 4 \times (-2) = -8$

$\therefore \quad$ Product of zeroes $= \dfrac{c}{a}$

54. $p(s) = 4s^2 - 4s + 1$

For zeroes of $p(s)$,

$p(s) = 0 \Rightarrow 4s^2 - 4s + 1 = 0$

$\Rightarrow \quad (2s - 1)^2 = 0$

$\Rightarrow \quad 2s - 1 = 0 \text{ or } 2s - 1 = 0 \Rightarrow s = \dfrac{1}{2}, s = \dfrac{1}{2}$

$\therefore$ Zeroes are $\dfrac{1}{2}, \dfrac{1}{2}$.

Now, here $a = 4$, $b = -4$, $c = 1$

$\dfrac{-b}{a} = \dfrac{-(-4)}{4} = 1$

and $\quad$ sum of zeroes $= \dfrac{1}{2} + \dfrac{1}{2} = 1$

$\Rightarrow \quad$ Sum of zeroes $= \dfrac{-b}{a}$

Also, $\quad \dfrac{c}{a} = \dfrac{1}{4}$

and $\quad$ product of zeroes $= \dfrac{1}{2} \times \dfrac{1}{2} = \dfrac{1}{4}$

$\therefore \quad$ Product of zeroes $= \dfrac{c}{a}$

55. $p(t) = t^2 - 15$

For zeroes, $p(t) = 0$

$\Rightarrow \quad t^2 - 15 = 0 \Rightarrow t^2 = 15 \Rightarrow t = \pm\sqrt{15}$

$\therefore$ Zeroes are $\sqrt{15}, -\sqrt{15}$.

Now, here $a = 1$, $b = 0$, $c = -15$

$\dfrac{-b}{a} = \dfrac{0}{1} = 0$

and $\quad$ sum of zeroes $= \sqrt{15} + (-\sqrt{15}) = 0$

$\Rightarrow \quad$ Sum of zeroes $= \dfrac{-b}{a}$

Also $\dfrac{c}{a} = \dfrac{-15}{1} = -15$ and product of zeroes $= \sqrt{15} \times (-\sqrt{15}) = -15$

$\Rightarrow \quad$ Product of zeroes $= \dfrac{c}{a}$

56. $p(x) = x^3 - 4x^2 + 5x - 2$

$p(2) = 2^3 - 4\times2^2 + 5\times2 - 2 = 0$

$p(1) = 1^3 - 4\times1^2 + 5\times1 - 2 = 0$

$\therefore$ 2, 1, 1 are zeroes of $p(x)$.

Here, $a = 1, b = -4, c = 5, d = -2$

(*i*) $\alpha+\beta+\gamma = 2+1+1 = 4$

$\Rightarrow -\dfrac{b}{a} = \dfrac{-(-4)}{1} = 4 \Rightarrow \alpha+\beta+\gamma = -\dfrac{b}{a}$

(*ii*) $\alpha\beta+\beta\gamma+\alpha\gamma = 2\times1+1\times1+2\times1 = 5$

$\dfrac{c}{a} = 5 \Rightarrow \alpha\beta+\beta\gamma+\gamma\alpha = \dfrac{c}{a}$

(*iii*) $\alpha\beta\gamma = 2\times1\times1 = 2$

and $-\dfrac{d}{a} = \dfrac{-(-2)}{1} = 2$

$\alpha\beta\gamma = -\dfrac{d}{a}$

57. S = Sum of zeroes = 2

T = Sum of the product of zeroes taken two at a time = –7

P = product of zeroes = –14

A cubic polynomial

$p(x) = x^3 - (S)x^2 + (T)x - P$

$= x^3 - (2)x^2 + (-7)x - (-14) = x^3 - 2x^2 - 7x + 14$

58. (*i*) Divisor $x^2 - 2$, dividend $x^3 - 3x^2 + 5x - 3$, quotient $x - 3$

$$\begin{array}{r} x^3 - 3x^2 + 5x - 3 \\ x^3 \qquad\ \ - 2x \\ \hline -3x^2 + 7x - 3 \\ -3x^2 \qquad\ \ + 6 \\ \hline 7x - 9 \end{array}$$

$\therefore$ Quotient $= x - 3$, remainder $= 7x - 9$

(*ii*) Divisor $x^2 - x + 1$, dividend $x^4 - 3x^2 + 4x + 5$, quotient $x^2 + x - 3$

$$\begin{array}{r} x^4 - 3x^2 + 4x + 5 \\ x^4 + x^2 - x^3 \\ \hline x^3 - 4x^2 + 4x + 5 \\ x^3 - x^2 + x \\ \hline -3x^2 + 3x + 5 \\ -3x^2 + 3x - 3 \\ \hline 8 \end{array}$$

$\therefore$ Quotient $= x^2 + x - 3$, remainder $= 8$

(*iii*) Divisor $-x^2 + 2$, dividend $x^4 - 5x + 6$, quotient $-x^2 - 2$

$$\begin{array}{r} x^4 - 5x + 6 \\ x^4 \qquad - 2x^2 \\ \hline 2x^2 - 5x + 6 \\ 2x^2 \qquad - 4 \\ \hline -5x + 10 \end{array}$$

Quotient $= -x^2 - 2$, remainder $= -5x + 10$

59. (*i*) Divisor $x^3 - 3x + 1$, dividend $x^5 - 4x^3 + x^2 + 3x + 1$, quotient $x^2 - 1$

$$\begin{array}{r} x^5 - 4x^3 + x^2 + 3x + 1 \\ x^5 - 3x^3 + x^2 \\ \hline -x^3 + 3x + 1 \\ -x^3 + 3x - 1 \\ \hline 2 \end{array}$$

Quotient $= x^2 - 1$, remainder $= 2 \neq 0$

$\therefore$ $x^3 - 3x + 1$ is not a factor of given polynomial.

(*ii*) Divisor $t^2 + 3t + 1$, dividend $2t^4 + 3t^3 - 2t^2 - 9t - 12$, quotient $2t^2 - 3t + 5$

$$\begin{array}{r} 2t^4 + 3t^3 - 2t^2 - 9t - 12 \\ 2t^4 + 6t^3 + 2t^2 \\ \hline -3t^3 - 4t^2 - 9t - 12 \\ -3t^3 - 9t^2 - 3t \\ \hline 5t^2 - 6t - 12 \\ 5t^2 + 15t + 5 \\ \hline -21t - 17 \end{array}$$

$\therefore$ Remainder $= -21t - 17 \neq 0$

$\Rightarrow$ $t^2 + 3t + 1$ is not a factor of given polynomial.

60. $p(x) = x^3 - 3x^2 + x + 2$

$q(x) = x - 2$

$r(x) = -2x + 4$

According to division algorithm,

$p(x) = g(x)\,q(x) + r(x)$

$\Rightarrow x^3 - 3x^2 + x + 2 = g(x)\,.\,(x - 2) + (-2x + 4)$

$\Rightarrow x^3 - 3x^2 + x + 2 + 2x - 4 = g(x)\,(x - 2)$

$\Rightarrow \dfrac{x^3 - 3x^2 + 3x - 2}{x - 2} = g(x)$

$\Rightarrow g(x) = x^2 - x + 1$

SHORT ANSWER TYPE QUESTIONS-II

1. $2+\sqrt{3}$ and $2-\sqrt{3}$ are zeroes of $p(x)$.

$\Rightarrow (x - 2 - \sqrt{3})\,(x - 2 + \sqrt{3})$ is a factor of $p(x)$.

$\Rightarrow x^2 - 4x + 1$ is a factor of $p(x)$.

Divisor $x^2 - 4x + 1$, dividend $2x^4 - 9x^3 + 5x^2 + 3x - 1$, quotient $2x^2 - x - 1$

$$\begin{array}{r} 2x^4 - 9x^3 + 5x^2 + 3x - 1 \\ 2x^4 - 8x^3 + 2x^2 \\ \hline -x^3 + 3x^2 + 3x - 1 \\ -x^3 + 4x^2 - x \\ \hline -x^2 + 4x - 1 \\ -x^2 + 4x - 1 \\ \hline 0 \end{array}$$

$\Rightarrow$ Other factor is $2x^2 - x - 1$.

For other zeroes, $2x^2 - x - 1 = 0$

$\Rightarrow 2x^2 - 2x + x - 1 = 0$

$\Rightarrow 2x(x - 1) + 1(x - 1) = 0$

$\Rightarrow (x - 1)\,(2x + 1) = 0$

$\Rightarrow x = 1, x = -\dfrac{1}{2}$

2. $p(x) = 6x^4 + 8x^3 - 5x^2 + ax + b$

$g(x) = 2x^2 - 5$

Dividing by $2x^2 - 5$

$$\begin{array}{r|l} & 3x^2+4x+5 \\ 2x^2-5 & 6x^4+8x^3-5x^2+ax+b \\ & \underline{\overset{-}{6x^4} \quad \overset{+}{-15x^2}} \\ & 8x^3+10x^2+ax+b \\ & \underline{\overset{-}{8x^3} \quad \overset{+}{-20x}} \\ & 10x^2+20x+ax+b \\ & \underline{\overset{-}{10x^2} \quad \overset{+}{-25}} \\ & 20x+ax+b+25 \end{array}$$

$\because$ $p(x)$ is divisible by $g(x)$

$\therefore \quad (20+a)x+(b+25) = 0$

Then $\quad 20+a = 0$

i.e., $\quad a = -20, b = -25$

3.

$$\begin{array}{r|l} & 2x^2+5 \\ 3x^2+4x+1 & 6x^4+8x^3+17x^2+21x+7 \\ & \underline{\overset{-}{6x^4}+\overset{-}{8x^3}+\overset{-}{2x^2}} \\ & 15x^2+21x+7 \\ & \underline{\overset{-}{15x^2}+\overset{-}{20x}+\overset{-}{3}} \\ & x+2 \end{array}$$

Quotient $= 2x^2+5$, remainder $= x+2$

4. $p(x) = 4x^4-5x^3-39x^2-46x-2$

$q(x) = x^2-3x-5$, $r(x) = -5x+8$

According to divison algorithm,

$p(x) = g(x) \cdot q(x) + r(x)$

$\Rightarrow \quad p(x)-r(x) = g(x) \cdot q(x)$

$\Rightarrow \quad \dfrac{p(x)-r(x)}{q(x)} = g(x)$

$\Rightarrow g(x) = \dfrac{(4x^4-5x^3-39x^2-46x-2)-(-5x+8)}{x^2-3x-5}$

$= \dfrac{4x^4-5x^3-39x^2-41x-10}{x^2-3x-5} = 4x^2+7x+2$

5. Here $\quad p(x) = 4x^2+4x-3$

$\therefore \quad p\left(\dfrac{1}{2}\right) = 4\left(\dfrac{1}{2}\right)^2+4\left(\dfrac{1}{2}\right)-3 = 4\times\dfrac{1}{4}+4\times\dfrac{1}{2}-3$

$= 1+2-3=0$

Similarly, $p\left(-\dfrac{3}{2}\right) = 4\left(-\dfrac{3}{2}\right)^2+4\left(\dfrac{-3}{2}\right)-3$

$= 4\times\dfrac{9}{4}+4\times\left(\dfrac{-3}{2}\right)-3 = 9-6-3=0$

$\therefore \quad \dfrac{1}{2}$ and $\dfrac{-3}{2}$ are zeroes of polynomial.

Now sum of zeroes $= \dfrac{1}{2}+\dfrac{-3}{2} = \dfrac{-2}{2} = -1$ ...(*i*)

Product of zeroes $= \dfrac{1}{2}\times\left(\dfrac{-3}{2}\right) = \dfrac{-3}{4}$ ...(*ii*)

In equation $4x^2+4x-3$,

$a=4, b=4, c=-3$

$\therefore \quad \dfrac{-b}{a} = \dfrac{-4}{4} = -1 =$ Sum of roots [From (*i*)]

$\dfrac{c}{a} = \dfrac{-3}{4} =$ Product of roots [From(*ii*)]

6. $p(x) = 2x^3+9x^2-x-b$

$2x+3$ is a factor of $p(x)$.

$\therefore \; p\left(\dfrac{-3}{2}\right) = 0$

$\Rightarrow \quad 2\left(\dfrac{-3}{2}\right)^3+9\left(\dfrac{-3}{2}\right)^2-\left(\dfrac{-3}{2}\right)-b=0$

$\Rightarrow \quad 2\times\left(\dfrac{-27}{8}\right)+\dfrac{81}{4}+\dfrac{3}{2}-b=0$

$\Rightarrow \quad b = \dfrac{-27}{4}+\dfrac{81}{4}+\dfrac{3}{2} = 15$

7. $P(x) = 2x^2+px+4$

One zero = 2

Let other zero $= a$

Now, sum of zeroes $= -\dfrac{p}{2}$ ...(*i*)

and product of zeroes $= \dfrac{4}{2} = 2$

$\Rightarrow \quad 2\times a = 2$

$\therefore \quad$ Other zero = 1

$\therefore \quad$ Sum of zeroes $= 2+1 = 3$ ...(*ii*)

From (*i*) and (*ii*),

$-\dfrac{p}{2} = 3 \Rightarrow p = -6$

8. $\quad \alpha+\beta = 6, \alpha\beta = k$

Now $\quad (\alpha+\beta)^2-2\alpha\beta = 40$

$\Rightarrow \quad 6^2-2k = 40$

$\Rightarrow \quad 36-2k = 40$

$\Rightarrow \quad -2k = 40-36$

$\Rightarrow \quad -2k = 4$

$\Rightarrow \quad k = \dfrac{4}{-2} = -2$

9. $\alpha+\beta = \dfrac{-5}{2}, \alpha\beta = \dfrac{k}{2}$.

Putting these value in $(\alpha+\beta)^2-\alpha\beta = 24$, we get

$\left(\dfrac{-5}{2}\right)^2-\dfrac{k}{2} = 24$

$\Rightarrow \quad \dfrac{25}{4}-\dfrac{k}{2} = 24$

$\Rightarrow \quad \dfrac{-k}{2} = 24-\dfrac{25}{4}$

$\Rightarrow \quad \dfrac{-k}{2} = \dfrac{96-25}{4}$

$\Rightarrow \quad k = \dfrac{-71}{4}\times 2 = \dfrac{-71}{2}$

10. $\because$ 3 is a root (zero) of $p(x)$

$\Rightarrow \quad 2(3)^2-3\times 3+p=0$

$\Rightarrow \quad 18-9+p=0 \Rightarrow p=-9$

Now $p(x) = 2x^2-3x-9 = 2x^2-6x+3x-9$

$= 2x(x-3)+3(x-3)$

$= (x-3)(2x+3)$

For roots of polynomial,

$$p(x) = 0 \Rightarrow (x-3)(2x+3) = 0$$

$\Rightarrow \quad x = 3$ or $x = -\frac{3}{2}$, Other root $= -\frac{3}{2}$

11. $P(x) = 3x^2 - 2kx + 2m$,

Since 2 is zeroes of $p(x)$, then

$$P(2) = 3(2)^2 - 4k + 2m$$

$$= 12 - 4k + 2m$$

$\therefore \quad P(2) = 0$

$\Rightarrow 12 - 4k + 2m = 0$

$\Rightarrow \quad 4k - 2m = 12$ ...(*i*)

Since 3 is zeroes of $p(x)$, then

$$P(3) = 3(3)^2 - 2 \times 3k + 2m = 0$$

$\Rightarrow 27 - 6k + 2m = 0$

$\Rightarrow \quad 6k - 2m = 27$ ...(*ii*)

From (*i*) and (*ii*), $k = \frac{15}{2}$, $m = 9$.

12. $\alpha + \beta = 6$, $\alpha\beta = a$

Now $\quad 3\alpha + 2\beta = 20$

$\Rightarrow \quad \alpha + 2\alpha + 2\beta = 20$

$\Rightarrow \quad \alpha + 2(\alpha + \beta) = 20$

$\Rightarrow \quad \alpha + 2 \times 6 = 20$

$\Rightarrow \quad \alpha = 20 - 12 = 8$

$\therefore \quad \beta = -2$

Now $\quad \alpha\beta = a$

$\therefore \quad 8 \times (-2) = a$

$\Rightarrow \quad a = -16$

13. Let $\alpha = 7\beta$ and $\frac{-b}{a} = \frac{8}{3}$, $\frac{c}{a} = \frac{-2k-1}{3}$

$\therefore \quad \alpha + \beta = 7\beta + \beta = \frac{8}{3}$

i.e., $\quad 8\beta = \frac{8}{3} \Rightarrow \beta = \frac{8}{3} \times \frac{1}{8}$

$\Rightarrow \quad \beta = \frac{1}{3}$

Since α is 7β,

$\therefore \quad \alpha = 7 \times \frac{1}{3} = \frac{7}{3}$

Now $\alpha \times \beta = 7\beta \times \beta = \frac{-2k-1}{3}$

i.e., $\quad 7\beta^2 = \frac{-2k-1}{3}$

$\Rightarrow \quad 7\left(\frac{1}{3}\right)^2 = \frac{-2k-1}{3} \Rightarrow \frac{7}{9} = \frac{-2k-1}{3}$

$\Rightarrow \quad \frac{7}{9} \times 3 = -2k - 1$

$\Rightarrow \quad \frac{7}{3} = -2k - 1$

$\therefore \quad -2k = \frac{7}{3} + 1 \Rightarrow -2k = \frac{7+3}{3}$

$\Rightarrow \quad k = \frac{10}{3} \times \left(\frac{-1}{2}\right) = \frac{-5}{3}$

14. (*i*)

$$\begin{array}{r|l} & 3x^3 - 5x^2 + 2x + 4 \\ \hline x^2 + x + 1 & 3x^5 - 2x^4 + x^2 - 2 \\ & 3x^5 + 3x^4 \quad + 3x^3 \\ & - \quad - \quad\quad - \\ \hline & -5x^4 - 3x^3 + x^2 - 2 \\ & -5x^4 - 5x^3 - 5x^2 \\ & + \quad + \quad + \\ \hline & +2x^3 + 6x^2 - 2 \\ & +2x^3 + 2x^2 \quad + 2x \\ & - \quad - \quad - \\ \hline & +4x^2 - 2x - 2 \\ & +4x^2 + 4x + 4 \\ & - \quad - \quad - \\ \hline & -6x - 6 \end{array}$$

Checking

$$p(x) = q(x) \times g(x) + r(x)$$

$$= (x^2 + x + 1)(3x^3 - 5x^2 + 2x + 4) + (-6x - 6)$$

$$= x^2(3x^3 - 5x^2 + 2x + 4) + x(3x^3 - 5x^2 + 2x + 4) + 1(3x^3 - 5x^2 + 2x + 4) - 6x - 6$$

$$= 3x^5 - 5x^4 + 2x^3 + 4x^2 + 3x^4 - 5x^3 + 2x^2 + 4x + 3x^3 - 5x^2 + 2x + 4 - 6x - 6$$

$= 3x^5 - 2x^4 + x^2 - 2$ Hence verified.

(*ii*)

$$\begin{array}{r|l} & 3x^3 + 2x^2 - 9x - \frac{1}{2} \\ \hline 2x^2 - 3 & 6x^5 + 4x^4 - 27x^3 - 7x^2 - 27x - 6 \\ & 6x^5 \quad - 9x^3 \\ & - \quad\quad + \\ \hline & 4x^4 - 18x^3 - 7x^2 - 27x - 6 \\ & 4x^4 \quad - 6x^2 \\ & - \quad\quad + \\ \hline & -18x^3 - x^2 - 27x - 6 \\ & -18x^3 \quad + 27x \\ & + \quad\quad - \\ \hline & -x^2 - 54x - 6 \\ & -x^2 \quad + \frac{3}{2} \\ & + \quad\quad - \\ \hline & -54x - \frac{15}{2} \end{array}$$

$$p(x) = 6x^5 + 4x^4 - 27x^3 - 7x^2 - 27x - 6$$

$$q(x) = 2x^2 - 3, \; g(x) = 3x^3 + 2x^2 - 9x - \frac{1}{2}$$

$$r(x) = -54x - \frac{15}{2}$$

Now

$$p(x) = q(x).g(x) + r(x)$$

$$= (2x^2 - 3)(3x^3 + 2x^2 - 9x - \frac{1}{2}) + (-54x - \frac{15}{2})$$

$$= 6x^5 + 4x^4 - 18x^3 - x^2 - 9x^3 - 6x^2 + 27x + \frac{3}{2} - 54x - \frac{15}{2}$$

$= 6x^5 + 4x^4 - 27x^3 - 7x^2 - 27x - 6$ Hence verified.

15. Let $r(x) = ax + b$, when $p(x)$ is divided by $(x-2)(x-3)$.

According to division algorithm,

$\Rightarrow \quad p(x) = (x-2)(x-3) \,.\, q(x) + r(x)$

$\Rightarrow \quad p(x) = (x-2)(x-3)\, q(x) + ax + b$

Now, $\quad p(2) = 1$

$\Rightarrow \quad (2-2)(2-3)\, q(2) + a \times 2 + b = 1$

$\Rightarrow \quad 2a + b = 1$ ... (*i*)

Also, $\quad p(3) = 3$

$\Rightarrow \quad (2-3)(3-3)\,q(3) + a \times 3 + b = 3$

$\Rightarrow \quad 3a + b = 3$ ... (*ii*)

Solving (*i*) and (*ii*), we get

$a = 2,\ b = -3$

$\therefore \quad r(x) = 2x - 3$

16. Since $\sqrt{\frac{5}{3}}$ and $-\sqrt{\frac{5}{3}}$ are two zeroes of $p(x)$.

$\therefore \quad \left(x - \sqrt{\frac{5}{3}}\right)\left(x + \sqrt{\frac{5}{3}}\right) = \left(x^2 - \frac{5}{3}\right) = \frac{1}{3}(3x^2 - 5)$

$\therefore \quad (3x^2 - 5)$ is a factor of $3x^4 + 6x^3 - 2x^2 - 10x - 5$.

Now by division algorithm,

$$\begin{array}{r|l} & x^2 + 2x + 1 \\ 3x^2 - 5 & 3x^4 + 6x^3 - 2x^2 - 10x - 5 \\ & 3x^4 - 5x^2 \\ & - + \\ \hline & 6x^3 + 3x^2 - 10x - 5 \\ & 6x^3 - 10x \\ & - + \\ \hline & 3x^2 - 5 \\ & 3x^2 - 5 \\ & - + \\ \hline & \times \end{array}$$

For other zeroes, $x^2 + 2x + 1 = 0$

$\Rightarrow \quad (x+1)(x+1) = 0$

$\Rightarrow \quad x = -1,\ x = -1$

$\therefore$ Zeroes are $\sqrt{\frac{5}{3}}, -\sqrt{\frac{5}{3}}, -1$ and -1

17. Since zeroes of polynomial are 3 and 1

$\therefore \quad (x-3)(x-1) = x^2 - x - 3x + 3$

$= x^2 - 4x + 3$

$x^2 - 4x + 3$ is the factor of $x^4 - 7x^3 + 17x^2 - 17x + 6$.

By division algorithm, we get

$$\begin{array}{r|l} & x^2 - 3x + 2 \\ x^2 - 4x + 3 & x^4 - 7x^3 + 17x^2 - 17x + 6 \\ & x^4 - 4x^3 + 3x^2 \\ & - + - \\ \hline & -3x^3 + 14x^2 - 17x + 6 \\ & -3x^3 + 12x^2 - 9x \\ & + - + \\ \hline & 2x^2 - 8x + 6 \\ & 2x^2 - 8x + 6 \\ & - + - \\ \hline & \times \end{array}$$

So, another factor of polynomial is $x^2 - 3x + 2$

By splitting, we get $x(x-2) - 1(x-2)$ *i.e.*, $(x-1)(x-2)$

So, other zeroes are 1 and 2.

LONG ANSWER TYPE QUESTIONS

1. $p(x) = x^4 - 7x^2 + 12$

$\because \sqrt{3}$ and $-\sqrt{3}$ are the zeroes of $p(x)$,

$\therefore \left(x - \sqrt{3}\right)\left(x + \sqrt{3}\right)$ is a factor of $p(x) \Rightarrow x^2 - 3$ is a factor of $p(x)$.

For other factor of $p(x)$,

$$\begin{array}{r|l} & x^2 - 4 \\ x^2 - 3 & x^4 - 7x^2 + 12 \\ & x^4 - 3x^2 \\ & - + \\ \hline & -4x^2 + 12 \\ & -4x^2 + 12 \\ & + - \\ \hline & \times \end{array}$$

$\therefore$ Other factor is $x^2 - 4$.

For other zeroes, $x^2 - 4 = 0$

$\Rightarrow (x-2)(x+2) = 0$

$\Rightarrow x = 2, -2$

$\therefore$ Other zeroes are 2 and -2.

2. $p(x) = x^4 - 3x^3 - x^2 + 9x - 6$

$\because \sqrt{3}$ and $-\sqrt{3}$ are the zeroes of $p(x)$

$\therefore \left(x - \sqrt{3}\right)\left(x + \sqrt{3}\right)$ is a factor of $p(x)$

$\Rightarrow x^2 - 3$ is factor of $p(x)$

For other factor

$$\begin{array}{r|l} & x^2 - 3x + 2 \\ x^2 - 3 & x^4 - 3x^3 - x^2 + 9x - 6 \\ & x^4 - 3x^2 \\ & - + \\ \hline & -3x^3 + 2x^2 + 9x - 6 \\ & -3x^3 + 9x \\ & + - \\ \hline & 2x^2 - 6 \\ & 2x^2 - 6 \\ & - + \\ \hline & \times \end{array}$$

$\therefore$ Other factor of $p(x)$ is $x^2 - 3x + 2$.

for other zeroes, $x^2 - 3x + 2 = 0$

$\Rightarrow \quad (x-2)(x-1) = 0$

$\Rightarrow \quad x = 2,\ x = 1$

$\therefore$ Other zeroes are 1 and 2.

3.

$$\begin{array}{r|l} & 2x^2 - x - 1 \\ x^2 - 4x + 1 & 2x^4 - 9x^3 + 5x^2 + 3x - 8 \\ & 2x^4 - 8x^3 + 2x^2 \\ & - + - \\ \hline & -x^3 + 3x^2 + 3x - 8 \\ & -x^3 + 4x^2 - x \\ & + - + \\ \hline & -x^2 + 4x - 8 \\ & -x^2 + 4x - 1 \\ & + - + \\ \hline & -7 \end{array}$$

Checking

$p(x) = 2x^4 - 9x^3 + 5x^2 + 3x - 8$

$q(x) = 2x^2 - x - 1,\ r(x) = -7$

$g(x) = x^2 - 4x + 1$

Now, $g(x) \times q(x) + r(x)$

$= (x^2 - 4x + 1)(2x^2 - x - 1) + (-7)$
$= 2x^4 - x^3 - x^2 - 8x^3 + 4x^2 + 4x + 2x^2 - x - 1 - 7$
$= 2x^4 - 9x^3 + 5x^2 + 3x - 8 = p(x)$
$\therefore p(x) = g(x) \cdot q(x) + r(x)$

4. $p(x) = x^3 - 3\sqrt{5}x^2 - 5x + 15\sqrt{5}$

Since $x - \sqrt{5}$ is a factor of $p(x)$. For other factor

$$\begin{array}{r|l} & x^2 - 2\sqrt{5}x - 15 \\ x - \sqrt{5} & x^3 - 3\sqrt{5}x^2 - 5x + 15\sqrt{5} \\ & x^3 - \sqrt{5}x^2 \\ & - \quad + \\ \hline & -2\sqrt{5}x^2 - 5x + 15\sqrt{5} \\ & -2\sqrt{5}x^2 + 10x \\ & + \quad - \\ \hline & -15x + 15\sqrt{5} \\ & -15x + 15\sqrt{5} \\ & + \quad - \\ \hline & \times \end{array}$$

$\therefore$ Other factor $= x^2 - 2\sqrt{5}x - 15$

For other zeroes, $x^2 - 2\sqrt{5}x - 15 = 0$

$\Rightarrow x^2 - 3\sqrt{5}x + \sqrt{5}x - 15 = 0$

$\Rightarrow x(x - 3\sqrt{5}) + \sqrt{5}(x - 3\sqrt{5}) = 0$

$\Rightarrow (x - 3\sqrt{5})(x + \sqrt{5}) = 0$

$\Rightarrow x = 3\sqrt{5}, x = -\sqrt{5}$

$\therefore$ All zeroes of $p(x)$ are $\sqrt{5}$, $3\sqrt{5}$ and $-\sqrt{5}$.

5. $P(x) = x^4 - 6x^3 + 16x^2 - 25x + 10$, $g(x) = x^2 - 2x + k$

$r(x) = x + a$. let $q(x) = px^2 + qx + r$

$\because \quad P(x) = g(x)\,q(x) + r(x)$

$\therefore \quad x^4 - 6x^3 + 16x^2 - 25x + 10$

$= (x^2 - 2x + k)(px^2 + qx + r) + x + a$

$\Rightarrow \quad P(x) = px^4 + qx^3 + rx^2 - 2px^3 - 2qx^2 - 2rx + kpx^2 + kqx + kr + x + a.$

$\Rightarrow \quad P(x) = px^4 + (q - 2p)x^3 + (r - 2q + kp)x^2 + (kq - 2r + 1)x + kr + a$

$\Rightarrow x^4 - 6x^3 + 16x^2 - 25x + 10 = px^4 + (q - 2p)x^3 + (r - 2q + kp)x^2 + (kq - 2r + 1)x + kr + a]$

Equating the coefficients, we get

$p = 1, q - 2p = -6 \Rightarrow q - 2 \times 1 = -6 \Rightarrow q = -4$

Also $kq - 2r + 1 = -25$

$\Rightarrow \quad k \times (-4) - 2r = -26 \Rightarrow -4k - 2r = -26$

$\Rightarrow \quad 2k + r = 13$ ...(*i*)

Using

$r - 2q + kp = 16$

$\Rightarrow \quad r - 2 \times (-4) + k \times 1 = 16 \Rightarrow r + k = 8$ ...(*ii*)

From (*i*) and (*ii*) we get $r = 3, k = 5$

$\because \quad kr + a = 10 \Rightarrow 5 \times 3 + a = 10 \Rightarrow a = -5$

6.

$$\begin{array}{r|l} & 2x^2 + 2x - 1 \\ 4x^2 + 3x - 2 & 8x^4 + 14x^3 - 2x^2 + 8x - 12 \\ & 8x^4 + 6x^3 - 4x^2 \\ & - \quad - \quad + \\ \hline & 8x^3 + 2x^2 + 8x - 12 \\ & 8x^3 + 6x^2 - 4x \\ & - \quad - \quad + \\ \hline & -4x^2 + 12x - 12 \\ & -4x^2 - 3x + 2 \\ & + \quad + \quad - \\ \hline & 15x - 14 \end{array}$$

Remainder $= 15x - 14$

$\therefore$ If we subtract $15x - 14$ or add $-15x + 14$ then remainder will be 0.

Then $4x^2 + 3x - 2$ will be a factor of given polynomial.

NCERT EXEMPLAR PROBLEMS

1. (*i*) No,

Let, $p(x)$ = dividend, $g(x)$ = divisor,

$q(x)$ = quotient and $r(x)$ = remainder.

By division algorithm, $p(x) = g(x) \cdot q(x) + r(x)$

Here $p(x)$ has degree 6, $g(x)$ has degree 5 and $q(x)$ has degree 2.

Degree of the product of $g(x)$ and $q(x)$ is 7 and greater than $p(x)$.

So, its not possible.

(*ii*) $g(x)$ has degree greater than $p(x)$, degree $g(x) >$ degree $p(x)$.

(*iii*) degree of $g(x) \leq$ degree of $p(x)$.

2. (*i*) False, if zeroes are +ve, then sum of zeroes is +ve.

$\Rightarrow -\frac{b}{a}$ is + ve $\Rightarrow \frac{b}{a}$ is –ve

$\Rightarrow$ a and b are of opposite sign.

Also, $\frac{c}{a}$ is +ve $\therefore$ c and a are of same sign.

$\Rightarrow$ a and c are of same sign opposite to that of b.

(*ii*) False, if quadratic polynomial is a perfect square, then it will have exactly one zeroes and it will intersect (touch) x-axis only at one point.

(*iii*) True, any cubic polynomial with two zeroes as zero is $(x - 0)(x - 0)(x - a) = x^3 - ax^2$

3. $4x^2 + 5\sqrt{2}x - 3 = 4x^2 + 6\sqrt{2}x - \sqrt{2}x - 3$

$= 2\sqrt{2}x(\sqrt{2}x + 3) - 1(\sqrt{2}x + 3)$

$= (2\sqrt{2}x - 1)(\sqrt{2}x + 3)$

$\Rightarrow x = \frac{1}{2\sqrt{2}}, x = -\frac{3}{\sqrt{2}}$ are zeroes of the polynomial.

4. $2s^2 - (1 + 2\sqrt{2})s + \sqrt{2} = 2s^2 - s - 2\sqrt{2}s + \sqrt{2}$

$= s(2s - 1) - \sqrt{2}(2s - 1)$

$= (2s - 1)(s - \sqrt{2})$

$\Rightarrow s = \frac{1}{2}, \sqrt{2}$ are zeroes of the polynomial.

5. $v^2 + 4\sqrt{3}v - 15 = v^2 + 5\sqrt{3}v - \sqrt{3}v - 15$

$= v(v + 5\sqrt{3}) - \sqrt{3}(v + 5\sqrt{3})$

$= (v - \sqrt{3})(v + 5\sqrt{3})$

$\Rightarrow v = \sqrt{3}, -5\sqrt{3}$ are zeroes of the polynomial.

6. $y^2 + \frac{3}{2}\sqrt{5}y - 5 = \frac{1}{2}(2y^2 + 3\sqrt{5}y - 10)$

$= \frac{1}{2}(2y^2 + 4\sqrt{5}y - \sqrt{5}y - 10)$

$= \frac{1}{2}[2y(y + 2\sqrt{5}) - \sqrt{5}(y + 2\sqrt{5})]$

$$= \frac{1}{2}(y+2\sqrt{5})\,(2y-\sqrt{5})$$

$\Rightarrow \qquad y = -2\sqrt{5}, \frac{\sqrt{5}}{2}$ are zeroes of the polynomial.

7. $7y^2 - \frac{11}{3}y - \frac{2}{3} = \frac{1}{3}(21y^2 - 11y - 2)$

$$= \frac{1}{3}(21y^2 - 14y + 3y - 2)$$

$$= \frac{1}{3}[7y(3y-2) + 1(3y-2)]$$

$$= \frac{1}{3}(3y-2)\,(7y+1)$$

$\Rightarrow y = \frac{2}{3}, \frac{-1}{7}$ are zeroes of the polynomial.

8. $x^3 - 6x^2 + 3x + 10$

Here $a = 1, b = -6, c = 3, d = 10$

Here $\alpha = a, \beta = a + b$ and $\gamma = a + 2b$

According to relationships,

$$\alpha + \beta + \gamma = -\frac{b}{a}$$

$$a + a + b + a + 2b = \frac{6}{1}$$

$\Rightarrow \qquad 3a + 3b = 6$

$\Rightarrow \qquad a + b = 2 \qquad \ldots (i)$

and $\quad \alpha\beta + \beta\gamma + \gamma\alpha = \frac{c}{a}$

$\Rightarrow a(a + b) + (a + b)\,(a + 2b) + (a + 2b)\,a = 3$

Putting the value of $a + b = 2$ from (*i*), we get

$$2a + 2(a + 2b) + (a + 2b)a = 3$$

Here $a = 2 - b$

$\Rightarrow 2(2 - b) + 2(2 - b + 2b) + (2 - b + 2b)\,(2 - b) = 3$

$\Rightarrow \qquad 4 - 2b + 2(2 + b) + (2 + b)\,(2 - b) = 3$

$\Rightarrow \qquad 4 - 2b + (2 + b)\,[2 + 2 - b] = 3$

$\Rightarrow \qquad 4 - 2b + (2 + b)\,(4 - b) = 3$

$\Rightarrow \qquad 4 - 2b + 8 - 2b + 4b - b^2 = 3$

$\Rightarrow \qquad 12 - b^2 = 3 \Rightarrow 12 - 3 = b^2$

$\Rightarrow \qquad 9 = b^2 \Rightarrow b = \pm 3$

Either $b = +3$ or $b = -3$

$\Rightarrow \qquad a = 2 - 3$ or $a = 2 - (-3)$

$\Rightarrow \qquad a = -1$ or $a = 5$

So $\quad \alpha = -1, \beta = 2, \gamma = 5.$ or $\quad \alpha = 5, \beta = 2, \gamma = -1.$

$\therefore \quad \alpha\beta\gamma = -10 \qquad \therefore \ \alpha\beta\gamma = -10$

So, $\ \alpha\beta\gamma = -10$

9. Let $g(x) = x^2 + 2x + k$ is a factor of $p(x) = 2x^4 + x^3 - 14x^2 + 5x + 6$, then remainder is zero when we divide $p(x)$ by $g(x)$.

$$\begin{array}{l}
\qquad\qquad\qquad 2x^2 - 3x - (8 + 2k) \\
x^2 + 2x + k \,\big)\, 2x^4 + x^3 - 14x^2 + 5x + 6 \\
\qquad\qquad\quad \underline{\underset{-}{}2x^4 \underset{-}{+} 4x^3 \underset{-}{+} 2kx^2} \\
\qquad\qquad\qquad -3x^3 - (14 + 2k)x^2 + 5x + 6 \\
\qquad\qquad\qquad \underline{\underset{+}{-}3x^3 \underset{+}{-} 6x^2 \qquad \underset{+}{-} 3kx} \\
\qquad\qquad\qquad -(8 + 2k)x^2 + (5 + 3k)x + 6 \\
\qquad\qquad\qquad \underline{\underset{+}{-}(8 + 2k)x^2 \underset{+}{-} (16 + 4k)x \underset{+}{-} (8k + 2k^2)} \\
\qquad\qquad\qquad\qquad \underline{(21 + 7k)x + (2k^2 + 8k + 6)}
\end{array}$$

Because $g(x)$ is a factor of $p(x)$, so remainder is zero.

$\Rightarrow \qquad 21 + 7k = 0 \Rightarrow 7k = -21 \Rightarrow k = -\frac{21}{7} = -3$

and $\quad 2k^2 + 8k + 6 = 0 \Rightarrow k^2 + 4k + 3 = 0$

$\Rightarrow \quad k^2 + 3k + k + 3 = 0 \Rightarrow k(k + 3) + 1\,(k + 3) = 0$

$\Rightarrow \quad (k + 3)\,(k + 1) = 0 \Rightarrow k = -3, -1$

So; if $k = -3$

Then $g(x) = x^2 + 2x - 3 = x^2 + 3x - x - 3$

$\qquad = (x + 3)\,(x - 1) \Rightarrow x = -3, 1$ are zeroes.

and $\ p(x) = 2x^4 + x^3 - 14x^2 + 5x + 6 = g(x)\,.\,p(x)$

$\qquad = (x^2 + 2x - 3)\,\{2x^2 - 3x - (8 + 2k)\}$

$\qquad = (x^2 + 2x - 3)\,(2x^2 - 3x - 2)\ (\because k = -3)$

Either $x^2 + 2x - 3 = 0$ or $2x^2 - 3x - 2 = 0$

$\Rightarrow \qquad x = -3, 1 \quad$ or $x = 2, -\frac{1}{2}$

So; $1, 2, -\frac{1}{2}, -3$ are all the zeroes of the given polynomial.

10. If all the zeroes of $q(x)$ are also the zeroes of $p(x)$.

Then $q(x)$ is a factor of $p(x)$ and if we divide $p(x)$ with $q(x)$, then the remainder must be zero. So

$$\begin{array}{l}
\qquad\qquad\qquad x^2 - 3x + 2 \\
x^3 + 2x^2 + a \,\big)\, x^5 - x^4 - 4x^3 + 3x^2 + 3x + b \\
\qquad\qquad\quad \underline{\underset{-}{}x^5 \underset{-}{+} 2x^4 \qquad \underset{-}{+} ax^2} \\
\qquad\qquad\qquad -3x^4 - 4x^3 + (3 - a)x^2 + 3x + b \\
\qquad\qquad\qquad \underline{\underset{+}{-}3x^4 \underset{+}{-} 6x^3 \qquad\qquad \underset{+}{-} 3ax} \\
\qquad\qquad\qquad 2x^3 + (3 - a)x^2 + (3 + 3a)x + b \\
\qquad\qquad\qquad \underline{\underset{-}{}2x^3 \underset{-}{+} 4x^2 \qquad\qquad \underset{-}{+} 2a} \\
\qquad\qquad\qquad \underline{(-1 - a)x^2 + (3 + 3a)x + b - 2a}
\end{array}$$

So; $\ -1 - a = 0$ or $3 + 3a = 0$ or $b - 2a = 0$

$\Rightarrow a = -1$ or $a = -1$ or $b = 2a \Rightarrow b = -2.$

So for $a = -1$, $b = -2$ $q(x)$ has all the zeroes of the polynomial $p(x)$.

By division algorithm,

$$x^5 - x^4 - 4x^3 + 3x^2 + 3x - 2 = (x^3 + 2x^2 - 1)\,(x^2 - 3x + 2)$$

$$x^2 - 3x + 2 = 0 \Rightarrow (x - 2)\,(x - 1) = 0 \Rightarrow x = 1, 2$$

All zeroes of $p(x)$ but not $g(x)$.

HOTS QUESTIONS

1. $5z^2 + 13z - p$

Let two zeroes are α and $\frac{1}{\alpha}$ then

product of zeroes $\alpha \cdot \frac{1}{\alpha} = \frac{-p}{5}$

$\Rightarrow \qquad -p = 5 \quad$ *i.e.,* $\quad p = -5$

2. Product of two zeroes = 3. Let α, β, γ are zeroes of polynomial $2x^3 + 3x^2 - 5x - 6$.

then $\ \alpha \cdot \beta \cdot \gamma = -\left(\frac{d}{a}\right) = -\left(\frac{-6}{2}\right) = 3$

but $\qquad \alpha\beta = 3$ then, $\ 3\gamma = 3 \ \Rightarrow \gamma = 1$

3. Dividing $x^4 + 2x^3 - 4x^2 + 6x - 3$ by $x^2 - x + 1$:

Quotient: $x^2 + 3x - 2$

$$x^4 + 2x^3 - 4x^2 + 6x - 3$$
$$\underline{\; x^4 - x^3 + x^2 \;}$$
$$3x^3 - 5x^2 + 6x - 3$$
$$\underline{\; 3x^3 - 3x^2 + 3x \;}$$
$$-2x^2 + 3x - 3$$
$$\underline{\; -2x^2 + 2x - 2 \;}$$
$$x - 1$$

Remainder $= x - 1$

$\therefore$ $x - 1$ should be subtracted

So that $p(x)$ is completely divisible by $g(x)$.

4. $y^4 + 4y^2 + 5$

$\Rightarrow y^4 + 5y^2 - y^2 + 5$

$\Rightarrow y^2(y^2 + 5) - 1(y^2 - 5)$ since factors are not possible

$\therefore$ polynomial have no zeroes.

5. $S = 2\sqrt{3}$, $P = 5$

Quadratic polynomial $= x^2 - 2\sqrt{3}x + 5$

6. $x^2 + 2x + 1 = (x + 1)^2$ For zeroes $(x + 1)^2 = 0$, *i.e.*, $x = -1, -1$

7. Let α, β, γ zeroes of polynomial $x^3 - 12x^2 + 39x + a$ are in AP then

$$2\beta = \alpha + \gamma \quad \text{but } \alpha + \beta + \gamma = 12$$
$$\alpha\beta\gamma = -a \quad \Rightarrow 3\beta = 12 \Rightarrow \beta = 4.$$
$$4\alpha\gamma = -a \quad \Rightarrow \alpha\gamma = \frac{a}{4}$$

Now $\alpha\beta + \beta\gamma + \gamma\alpha = 39$

$$\beta(\alpha + \gamma) + \gamma\alpha = 39$$
$$\Rightarrow \beta \cdot 2\beta + \left(\frac{-a}{4}\right) = 39$$
$$\Rightarrow 2 \cdot (4)^2 + \left(\frac{-a}{4}\right) = 39$$
$$\Rightarrow \frac{-a}{4} = 39 - 32 = 7 \Rightarrow a = -28$$

8. Dividing $2x^3 + 5x^2 - 14x + 10$ by $2x - 3$:

Quotient: $x^2 - 4x - 1$

$$2x^3 + 5x^2 - 14x + 10$$
$$\underline{\; 2x^3 - 3x^2 \;}$$
$$8x^2 - 14x + 10$$
$$\underline{\; 8x^2 - 12x \;}$$
$$-2x + 10$$
$$\underline{\; -2x + 3 \;}$$
$$7$$

$p(x)$ is divisible by $g(x)$ if $p(x) = 0$.

$\therefore$ -7 should be added.

9. $x^2 + 5x + 6 = x^2 + 3x + 2x + 6$

$= x(x + 3) + 2(x + 3) = (x + 2)(x + 3) \Rightarrow x = -2, -3$

Now $a = 1, b = 5, c = 6$.

$$\text{Sum of roots } = (-2) + (-3) = -5 = \frac{-5}{1} = \frac{b}{a}$$

$$\text{Product of roots } = (-2)(-3) = 6 = \frac{6}{1} = \frac{c}{a}$$

10. Since zeroes of polynomial $x^3 - ax^2 + bx - c$ are in AP then $2\beta = \alpha + \gamma$

$$\alpha + \beta + \gamma = a \quad \Rightarrow 3\beta = a \Rightarrow \beta = \frac{a}{3}$$

$$\alpha\beta\gamma = c \Rightarrow \frac{a}{3}\alpha\gamma = c$$
$$\Rightarrow \quad \alpha\gamma = \frac{3c}{a}$$
$$\Rightarrow \quad \alpha\beta + \beta\gamma + \gamma\alpha = b$$
$$\beta(\alpha + \gamma) + \gamma\alpha = b$$
$$\Rightarrow \quad 2\beta^2 + \frac{3c}{a} = b$$
$$\Rightarrow \quad 2 \cdot \frac{a^2}{9} + \frac{3c}{a} = b$$
$$\Rightarrow \quad \frac{2a^3 + 27c}{9a} = b$$
$$\Rightarrow \quad 2a^3 + 27c = 9ab$$
$$\Rightarrow \quad 2a^3 - 9ab + 27c = 0$$

Proved.

11. $p(1) = 0$

$\Rightarrow$ $L(1)^4 + M(1)^3 + N(1)^2 + R(1) + P = 0$

$\Rightarrow$ $L + M + N + R + P = 0$ ... (*i*)

Also $p(-1) = 0$

$\Rightarrow$ $L - M + N - R + P = 0$

$\Rightarrow$ $L + N + P = M + R$ ... (*ii*)

From (*i*) and (*ii*),

$2(M + R) = 0$

$\Rightarrow$ $M + R = 0$

From (*ii*), $L + N + P = 0$

Hence, $L + N + P = M + R = 0$

12. $f(x) = -2x^2 + 4x$

x	1	2	0	3	–1
$f(x)$	2	0	0	–6	–6

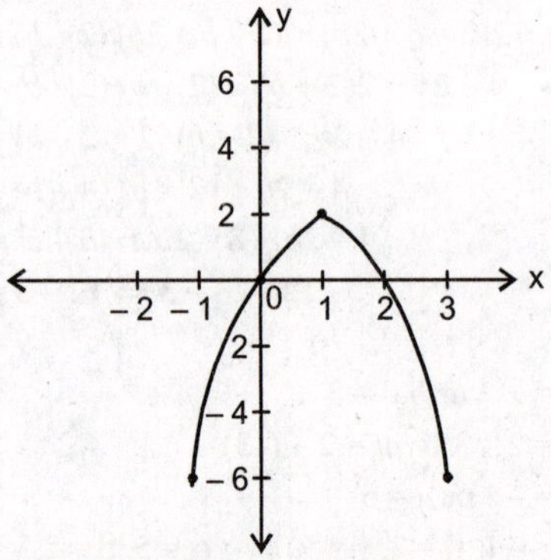

13. Since $x + a$ is a factor of $x^2 + px + q$

then $(-a)^2 - pa + q = 0 \Rightarrow a^2 = Pa - q$ ... (*i*)

also $(x + a)$ is a factor of $x^2 + mx + n$ then we get

$(-a)^2 - am + n = 0$

$\Rightarrow$ $a^2 - am + n = 0$

$a^2 = am - n$ ... (*ii*)

From eq. (*i*) and (*ii*), we get

$$am - n = ap - q$$
$$\Rightarrow \quad am - ap = n - q$$

Hence, $a = \left[\dfrac{n - q}{m - p}\right]$

14. Cubic polynomial

$P(x) = x^3 -$ (sum of zeroes) $x^2 +$ (sum of product of zeroes taken two at a time) $x -$ product of zeroes.

$\therefore\ p(x) = x^3 - (3)x^2 + \left(\frac{-1}{2}\right)x - \frac{5}{4}$

$= (4x^3 - 12x^2 - 2x - 5) = k(4x^2 - 12x^2 - 2x - 5)$

15. S = Sum of zeroes

$= \frac{2+\sqrt{5}}{2} + \frac{2-\sqrt{5}}{2} + 4 = \frac{2+\sqrt{5}+2-\sqrt{5}}{2} + 4 = 6$

T = Sum of product of zeroes taken two at a time

$= \left(\frac{2+\sqrt{5}}{2}\right)\left(\frac{2-\sqrt{5}}{2}\right) + \left(\frac{2-\sqrt{5}}{2}\right) \times 4 + \left(\frac{2+\sqrt{5}}{2}\right) \times 4$

$= \frac{-1}{4} + 4 - 2\sqrt{5} + 4 + 2\sqrt{5} = \frac{31}{4}$

P = Product of zeroes = $\left(\frac{2+\sqrt{5}}{2}\right) \times \left(\frac{2-\sqrt{5}}{2}\right) \times 4 = -1$

Cubic polynomial $p(x) = x^3 - (S)x^2 + (T)x - P$

$= x^3 - 6x^2 + \frac{31}{4}x + 1$

$= \frac{1}{4}(4x^3 - 24x^2 + 31x + 4)$

$= k(4x^3 - 24x^2 + 31x + 4)$

16. Here, $a = k$, $b = 0$, $c = -5$, $d = 9$

$\alpha + \beta + \gamma = -\frac{b}{a} = 0$

$\alpha\beta\gamma = -\frac{d}{a} = -\frac{9}{k}$

$\because\ \alpha + \beta + \gamma = 0$

$\therefore\ \alpha^3 + \beta^3 + \gamma^3 = 3\alpha\beta\gamma$

$\Rightarrow\ 27 = 3 \times \left(\frac{-9}{k}\right)$

$\Rightarrow\ k = -1$

17. $p(x) = x^3 - 12x^2 + 44x + c$

$\alpha + \beta + \gamma = \frac{-(-12)}{1} = 12$

Also $\alpha + \gamma = 2\beta$ ($\because$ α, β, γ are in AP)

$\Rightarrow\ 2\beta + \beta = 12 \Rightarrow \beta = 4$

Now $\alpha . \beta . \gamma = -c$

$\Rightarrow\ \alpha . \gamma . 4 = -c$

$\Rightarrow\ \alpha . \gamma = -\frac{c}{4}$

Also $\alpha\beta + \beta\gamma + \alpha\gamma = 44$

$\Rightarrow\ \beta(\alpha + \gamma) + \left(-\frac{c}{4}\right) = 44$

$\Rightarrow\ \beta \times 2\beta - \frac{c}{4} = 44$

$\Rightarrow\ 4 \times 2 \times 4 - \frac{c}{4} = 44 \Rightarrow -\frac{c}{4} = 44 - 32$

$\Rightarrow\ -\frac{c}{4} = 12 \Rightarrow -c = 48 \Rightarrow c = -48$

18. $p(x) = ax^3 + 3x^2 - bx - 6$

$\because\ -1$ is a zeroes $\therefore p(-1) = 0$

$\Rightarrow\ a(-1)^3 + 3(-1)^2 - b(-1) - 6 = 0$

$\Rightarrow\ -a + 3 + b - 6 = 0 \Rightarrow -a + b = 3$ ...(*i*)

Also -2 is another zeroes.

$\therefore\ p(-2) = 0$

$\Rightarrow\ a(-2)^3 + 3(-2)^2 - b(-2) - 6 = 0$

$\Rightarrow\ -8a + 12 + 2b - 6 = 0$

$\Rightarrow\ -8a + 2b = -6 \Rightarrow -2(4a - b) = -6$

$\Rightarrow\ 4a - b = 3$ ...(*ii*)

Solving equation (*i*) and (*ii*), we get

$a = 2$ and $\Rightarrow b = 5$

$\therefore\ p(x) = 2x^3 + 3x^2 - 5x - 6$

Let 3rd zero = k

Sum of zeroes = $\frac{-3}{2}$

$\Rightarrow\ -1 + (-2) + k = \frac{-3}{2} \Rightarrow k = \frac{-3}{2} + 3 = \frac{3}{2}$

$\therefore$ 3rd zero = $\frac{3}{2}$

19. $p(x) = x^3 - 2x^2 + qx - r$

Here, $a = 1$, $b = -2$, $c = q$, $d = -r$

Sum of zeroes = $\frac{-b}{a}$

$\Rightarrow\ \alpha + \beta + \gamma = \frac{-(-2)}{1} = 2$

$\Rightarrow\ 0 + \gamma = 2$ ($\because\ \alpha + \beta = 0$)

$\Rightarrow\ \gamma = 2$ ...(*i*)

Also, $\alpha\beta + \beta\gamma + \alpha\gamma = \frac{c}{a}$

$\Rightarrow\ \alpha\beta + \gamma(\alpha + \beta) = \frac{q}{1}$

$\Rightarrow\ \alpha\beta + \gamma \times 0 = q \Rightarrow \alpha\beta = q$ ...(*ii*)

and $\alpha . \beta . \gamma = \frac{-d}{a}$

$\Rightarrow\ q . 2 = -(-r)$ (Using (*i*) and (*ii*))

$\Rightarrow\ 2q = r.$

20. $P(x) = x^3 + px^2 + qx + 2$

Here, $a = 1$, $b = p$, $c = q$, $d = 2$

Now $\alpha + \beta + \gamma = \frac{-b}{a} = -p$ ...(*i*)

$\alpha\beta + \beta\gamma + \alpha\gamma = \frac{c}{a} = q$

$\Rightarrow\ \alpha\beta + \gamma(\beta + \alpha) = q$ ...(*ii*)

and $\alpha . \beta . \gamma = \frac{-d}{a} = -2$

$\Rightarrow\ \alpha . \beta . \gamma = -2$ ...(*iii*)

Also, $\alpha\beta + 1 = 0 \Rightarrow \alpha\beta = -1$

$\therefore$ (*iii*) becomes $-1 \times \gamma = -2 \Rightarrow \gamma = 2$

Substituting in (*i*) we get

$\alpha + \beta + 2 = -p$

$\Rightarrow\ \alpha + \beta = -p - 2$

Substituting these value in (*ii*), we get

$-1 + 2(-p - 2) = q$

$\Rightarrow\ -1 - 2p - 4 = q$

$\Rightarrow\ 2p + q + 5 = 0$

VALUE BASED QUESTIONS

1. Number of honest people = 12

Number of dishonest people = 8

$\therefore$ Sum of roots = 12 + 8 = 20

Product of roots = 12 × 8 = 96

Quadratic polynomial = $k(x^2 - 20x + 96)$

Value preferred : Honesty.

2. (*a*) For zeroes of $x^2 - 300x + 22500$

$$x^2 - 300x + 22500 = 0$$
$$\Rightarrow (x - 150)(x - 150) = 0$$
$$\therefore \quad x = 150, x = 150$$

(*b*) Honesty.

3. (*a*) Yes, here $a = 2, b = -4, c = 5$

Sum of zeroes $= \frac{-b}{a} = \frac{4}{2} = 2$

$\therefore$ Navdeep is right.

(*b*) Sincerity, Brilliance.

MULTIPLE CHOICE QUESTIONS

1. (*d*) Sum of zeroes, S = 3
Product of zeroes, P = – 2
$\therefore$ Quadratic polynomial
$p(x) = x^2 - (S)x + P = x^2 - 3x - 2$

2. (*c*) $(x + 1)$ is factor of $2x^3 + ax^2 + 2bx + 1$
$\therefore \quad 2(-1)^3 + a(-1)^2 - 2b + 1 = 0$
$\Rightarrow \quad a - 2b = 1$ ... (*i*), also $2a - 3b = 4$ ... (*ii*)
Solving (*i*) and (*ii*),
$a = 5, b = 2$

3. (*c*) $f(x) = (x - 2)^2 + 4 = (x - 2)^2 + (2)^2$
Sum of two perfect square is zero if each of them is zero
$\therefore$ No value of x.

4. (*a*) $f(x) = 4x^2 - 12x + 9$
For zeroes, $4x^2 - 12x + 9 = 0$
$\Rightarrow (2x)^2 - 2 \times 2x \times 3 + (3)^2 = 0$
$\Rightarrow (2x - 3)^2 = 0$
$\Rightarrow 2x - 3 = 0$ or $2x - 3 = 0$
$\Rightarrow x = \frac{3}{2}$ or $x = \frac{3}{2}$
$\therefore$ Zeroes are $\frac{3}{2}, \frac{3}{2}$

5. (*b*) Zero of $p(x)$
Let $p(x) = ax + b$
Put $x = k$
$P(k) = ak + b = 0$
$\therefore$ k is zero of $p(x)$.

6. (*d*) $\frac{-b}{a}$,
$p(x) = ax + b$
For zeroes of $p(x)$, $p(x) = 0$
$ax + b = 0$
$\Rightarrow x = \frac{-b}{a}$

7. (*c*) Parabola, because it is a second degree equation.

8. (*b*) Intersects x-axis.

9. (*a*) Zero.

10. (*c*) Maximum number of zeroes of a polynomial = degree of the polynomial.

11. (*d*) Product of zeroes $= \frac{c}{a}$

12. (*d*) Sum of zeroes $= -\frac{b}{a}$

13. (*b*) If $a + b + c = 0$ then zeroes are 1 and $\frac{c}{a}$.

14. (*c*) If $a + c = b$ then zeroes are –1 and $-\frac{c}{a}$.

15. (*b*) Let other zero = k
$\Rightarrow 6 + k = 0 \Rightarrow k = -6$
Product of zeroes $= 6 \times (-6) = -36$
$\therefore$ Polynomial $= x^2 - 36$

16. (*b*) Let other zero = k
$\therefore k \times 5 = 0 \Rightarrow k = 0$
$\Rightarrow$ Polynomial $= x^2 - 5x$

17. (*c*) Division algorithm for polynomials.

18. (*c*) Sum of degrees of $p(x)$ and $q(x)$.

19. (*c*) Equal to degree of $p(x)$.

20. (*b*) $\because$ Degree of $p(x)$ is equal to sum of degree of $g(x)$ and $q(x)$.
$\therefore$ Degree of $q(x) = 3$

21. (*a*) Here $p(3) = 0$
$\Rightarrow 4(3)^2 - 6 \times 3 - m = 0$
$\Rightarrow 36 - 18 - m = 0$
$\Rightarrow m = 18$
$\therefore$ Value of m is exactly divisible by 9.

22. (*a*) Product of zeroes = 7
$\Rightarrow 2k^2 - 1 = 7$
$\Rightarrow 2k^2 = 8$
$\Rightarrow k^2 = 4 \Rightarrow k = \pm 2$

23. (*d*)

$$\begin{array}{r|l} & x^2 - 2x + (p-1) \\ x^2 + 2x + 5 & x^4 + px^2 + q \\ & x^4 + 5x^2 \quad + 2x^3 \\ & - \quad - \qquad - \\ \hline & -2x^3 + (p-5)x^2 + q \\ & -2x^3 - 4x^2 \quad - 10x \\ & + \quad + \qquad + \\ \hline & (p-1)x^2 + 10x + q \\ & (p-1)x^2 + 2(p-1)x + 5p - 5 \\ & - \quad - \quad - \quad + \\ \hline & (12-2p)x + q - 5p + 5 \end{array}$$

For remainder to be zero
$12 - 2p = 0 \Rightarrow p = 6$
$q - 5p + 5 = 0 \Rightarrow q = +25$

24. (*b*) $p(x) = x^6 + 1$
when divided by $x - 1$
remainder $= p(1)$
$\therefore \quad p(1) = 1^6 + 1 = 2$

25. (*a*) $x - 2$ is a factor of $x^3 - 3x^2 + 4x - 4$
$\because$ remainder is zero
Similarly $x + 1$ is a factor of $2x^3 + 4x + 6$
but $x - 1$ is not a factor of $x^5 + x^4 - x^3 + x^2 - x + 1$

∵ remainder is not zero

∴ Statements 1 and 2 are correct.

26. **(a)**

$$\begin{array}{r|l} & x^2+x-2 \\ x+2 & x^3+3x^2-4 \\ & \underline{-x^3 \pm 2x^2} \\ & x^2-4 \\ & \underline{-x^2 \quad \pm 2x} \\ & -2x-4 \\ & -2x-4 \\ & \underline{+ \quad +} \\ & 0 \end{array}$$

Since quotient $= x^2 - mx - 2$

∴ $x^2 - mx - 2 = x^2 + x - 2$

On comparison of coefficient of x, we get $m = -1$.

27. **(b)** Remainder $= f(\sqrt{2}) = 5 \times \sqrt{2} - 10$

$=$ an irrational number

28. **(b)** $p(x) = x^3 + 7kx^2 - 4kx + 12$

$p(-3) = (-3)^3 + 7k(-3)^2 - 4k(-3) + 12$

∵ $x + 3$ is a factor

∴ $p(-3) = 0$

$\Rightarrow -27 + 63k + 12k + 12 = 0$

$\Rightarrow 75k - 15 = 0 \Rightarrow k = \frac{1}{5}$

29. **(b)** ∵ Zeroes are reciprocal of each other

∴ Product of zeroes $= 1 \Rightarrow \frac{k}{2} = 1 \Rightarrow k = 2$

30. **(b)** ∵ For zeroes $z^2 - 27 = 0$

$\Rightarrow z^2 = 27 \Rightarrow z = \pm\sqrt{27}$

$\Rightarrow z = \pm 3\sqrt{3}$

NCERT EXEMPLAR PROBLEMS

1. **(a)** $(k-1)x^2 + kx + 1$

One zero is -3, so it must satisfy the equation and make it zero.

∴ $(k-1)(-3)^2 + k(-3) + 1 = 0$

$\Rightarrow 9k - 9 - 3k + 1 = 0$

$\Rightarrow 6k - 8 = 0$

$\Rightarrow k = \frac{8}{6} = \frac{4}{3}$

2. **(d)** $x^2 + (a+1)x + b$

∵ $x = 2$ is a zero and $x = -3$ is another zero

∴ $(2)^2 + (a+1)2 + b = 0$ and $(-3)^2 + (a+1)(-3) + b = 0$

$\Rightarrow 4 + 2a + 2 + b = 0$ and $9 - 3a - 3 + b = 0$

$\Rightarrow 2a + b = -6$ ...(*i*) and $-3a + b = -6$ ...(*ii*)

Solving (*i*) and (*ii*), we get $5a = 0 \Rightarrow a = 0$ and $b = -6$.

3. **(d)** ∵ $x^2 - 3x - 10$, $2x^2 - 6x - 20$, $\frac{1}{2}x^2 - \frac{3}{2}x - 5$, $3x^2 - 9x - 30$ etc., have zeroes -2 and 5.

4. **(b)** ∵ $\alpha\beta + \beta\gamma + \gamma\alpha = \frac{c}{a}$

Let, $\alpha = 0$

So, $0 + \beta\gamma + 0 = \frac{c}{a}$. $\Rightarrow \beta\gamma = \frac{c}{a}$

5. **(b)** If k = positive, here $b = c$.

6. **(c)** c and a have the same sign.

7. **(a)** Has no linear term and the constant term is negative.

8. **(d)** ∵ It crosses the x-axis in three points.

Formative Assessment

Activity/Project 1

- **Topic Covered :** Polynomials
- **Objective :** Factorisation of a polynomial.
- **Skills Developed :** Understanding, Expression, Creativity
- **Time Required :** 15 Min.
- **Method :**

Step I. Consider the following polynomials :

1. $3x + 5$ 2. $x^2 + 2x + 1$ 3. $3x^2 + 4x + 1$ 4. $x^2 + x + 1$

Step II. Write the factors of every polynomial.

Result :

Polynomial 1	No factors
Polynomial 2	$(x + 1)(x + 1)$
Polynomial 3	$(3x + 1)(x + 1)$
Polynomial 4	No factors

➢ **Observation :**

1. Factor of every polynomials are of degree less than the degree of given polynomial. Degree of polynomial 1 is one therefore its factors are of degree zero. Polynomial of degree zero is a constant, therefore a linear expression can not be expressed as a product of constant terms.
2. A quadratic expression can be expressed as product of two linear factors if its discriminant ($D = b^2 - 4ac$) ≥ 0. Discriminant of polynomial $x^2 + x + 1$ is $(1)^2 - 4 \times 1 \times 1 = -3$ which is less than zero. Therefore, it can not be factorised.

Class Work

Which of the following polynomial can be factorised ?

(*i*) $7x^2 - 12x + 5$ (*ii*) $6x^2 - 11x + 5$ (*iii*) $5x^2 + 4x + 6$ (*iv*) $x^2 - x + 1$

Homework

$(x + 1)$ is one factor of $x^3 + 2x^2 + 2x + 1$. Find the other factors. Explain your answer.

Quiz Questions

1. $x = 1$ is a zero of $(x + 1)(x + 2)$. State True/False.
2. $x = -2$ is a zero of $(x + 2) + x^2 - 1$. State True/False.

Oral Questions

1. Write a linear polynomial.
2. Why linear polynomial can not be factorised ?

Activity/Project 2

- **Topic Covered :** Polynomials
- **Objective :** Degree of remainder when a polynomial is divided by another polynomial.
- **Skills Developed :** Analysis, Problem solving, Creativity
- **Time Required :** 15 Min.
- **Method :**

If $p(x)$ is divided by $g(x)$ then $q(x)$ is quotient and $r(x)$ is remainder. Use division algorithm to complete the given table :

$p(x)$	$g(x)$	$q(x)$	$r(x)$
1. $5x^2 + 4x - 1$	$x - 1$		
2. $4x^4 - 10x^3 - 10x^2 + 30x - 15$	$2x - 5$		
3. $x^3 - 3x^2 + 5x - 3$	$x^2 - 2$		
4. $x^4 - 5x + 6$	$2 - x^2$		

- **Observation :**
  1. When $r(x) = 0$, $g(x)$ is a factor of $p(x)$.
  2. When $r(x) \neq 0$, degree of $r(x)$ is less than $g(x)$.
- **Conclusion :**

  Degree of remainder is always less than the degree of divisor.

Class Work

1. When $x^3 - 3x^2 + x + 1$ is divided by $x + 1$, then degree of remainder is one. State True / False. Explain your answer.
2. If $p(x)$ is divided by a polynomial of degree two, then quotient is of degree two and remainder of degree zero. Write the degree of $p(x)$.

Homework

Write two polynomials which when divided by $x^2 - 1$ leave remainder $x + 2$.

Quiz Questions

1. If $p(x) = g(x) \cdot q(x)$, then $g(x)$ is a factor of $p(x)$. State True/False.
2. Degree of $r(x)$ is equal to $g(x)$. State True/False.

Oral Questions

1. Find remainder when $x^2 + 2x + 1$ is divided by $x + 1$.
2. Find $q(x)$ when $g(x) = x - 1$ divides $2x + 3$.

Suggested Activity/Assignment

Write division algorithm for polynomials and verify it by taking a suitable example.

Activity/Project 3

- **Topic Covered :** Polynomials
- **Objective :** Understanding division algorithm using long division.
- **Skills Developed :** Analysis, Problem solving, Creativity
- **Time Required :** 15 Min.
- **Method :**

  Divide $4x^4 + 5x^3 + 7x^2 + 3x + 1$ by $x + 2$ using long division method and fill the empty boxes.

$$
\begin{array}{r|l}
 & \square x^3 + \square x^2 + \square x + \square \\
x + 2 & 4x^4 + 5x^3 + 7x^2 + 3x + 1 \\
 & \square x^4 + \square x^3 \\
\hline
 & \quad \square x^3 + 7x^2 + 3x + 1 \\
 & \quad \square x^3 + \square x^2 \\
\hline
 & \qquad \square x^2 + 3x + 1 \\
 & \qquad \square x^2 + \square x \\
\hline
 & \qquad\quad \square x + 1 \\
 & \qquad\quad \square x + \square \\
\hline
 & \qquad\qquad \square \\
\hline
\end{array}
$$

■ **Observation :**

$p(x) = g(x) \cdot q(x) + r(x)$

Class Work

Give example of polynomial $p(x)$, $g(x)$, $r(x)$ and $q(x)$ which satisfy the division algorithm and

1. Degree of $p(x)$ = degree of $r(x)$.
2. Degree of $p(x)$ = degree of $q(x)$.

Homework

Divide $x^4 + 3x^3 + 2$ by $x - 2$ and verify division algorithm.

Quiz Questions

1. If $g(x)$ is a factor of $p(x)$ then degree of $g(x)$ is always more than degree of $p(x)$. State True/False.
2. If degree of $g(x)$ is two then degree of $r(x)$ may be one. State True/False.

Oral Questions

1. Write a quadratic polynomial with zeroes 2 and 1.
2. Find the zeroes of $2x - 1$.

Suggested Activity/Assignment

Divide a polynomial of degree four by a polynomial of degree two and verify the division algorithm.

Pair of Linear Equations in Two Variables

CHAPTER 3

Basic Concepts with Examples

- **LINEAR EQUATION.** An equation of the form $ax + by + c = 0$ or $ax + by = d$, where a, b, c, d are real numbers, $a^2 + b^2 \neq 0$ and x, y are variables, is called linear equation in two variables.
- **METHODS OF SOLVING SIMULTANEOUS PAIR OF LINEAR EQUATIONS IN TWO VARIABLES:**

(*i*) Graphical Method (*ii*) Algebraic Method

Let the system of linear equation in two variables be $a_1x + b_1y = c_1$ and $a_2x + b_2y = c_2$.

(*i*) Graphical Method:

(*a*) If $\frac{a_1}{a_2} \neq \frac{b_1}{b_2}$, then graph will represent two intersecting lines. Point of intersection is the solution of the system.

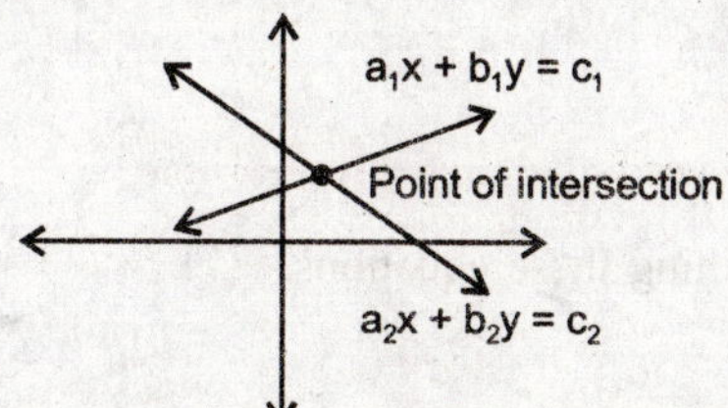

Intersecting lines represent unique solution.

(*b*) If $\frac{a_1}{a_2} = \frac{b_1}{b_2} \neq \frac{c_1}{c_2}$, then graph will represent two parallel lines. System has no common solution.

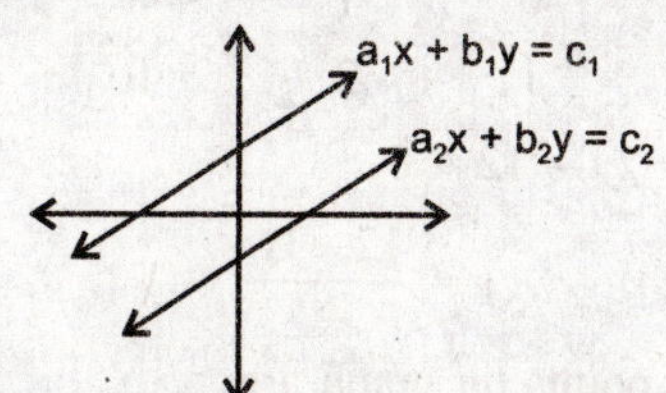

Parallel lines represent no solution.

(*c*) If $\frac{a_1}{a_2} = \frac{b_1}{b_2} = \frac{c_1}{c_2}$, then graph will represent two coinciding lines. System has infinitely many solutions.

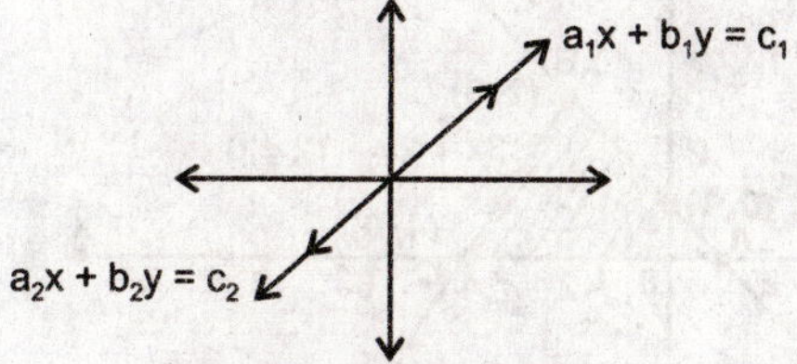

Coincidents lines represent infinitely many solutions.

(*ii*) Algebraic Method:

Let the system of linear equation in two variables be $a_1x + b_1y = c_1$ and $a_2x + b_2y = c_2$.

(*a*) If $\frac{a_1}{a_2} \neq \frac{b_1}{b_2}$, then system has a unique solution and is known as consistent.

(*b*) If $\frac{a_1}{a_2} = \frac{b_1}{b_2} \neq \frac{c_1}{c_2}$, then system has no solution and is known as inconsistent.

(*c*) If $\frac{a_1}{a_2} = \frac{b_1}{b_2} = \frac{c_1}{c_2}$, then system has infinitely many solutions and is known as dependent consistent.

1. For what values of a and b, the following system of linear equations have an infinite number of solutions? $2x + 3y = 7$; $(a - b)x + (a + b)y = 3a + b - 2$

Sol. Given equations are

$2x + 3y = 7$ and $(a - b)x + (a + b)y = 3a + b - 2$

Here $a_1 = 2$, $b_1 = 3$, $c_1 = -7$ and $a_2 = (a - b)$, $b_2 = (a + b)$, $c_2 = -(3a + b - 2)$

For system to have infinite solutions,

$$\frac{a_1}{a_2} = \frac{b_1}{b_2} = \frac{c_1}{c_2}$$

$$\Rightarrow \quad \frac{2}{a-b} = \frac{3}{a+b} = \frac{-7}{-(3a+b-2)}$$

$$\Rightarrow \quad \frac{2}{a-b} = \frac{3}{a+b}$$

$$\Rightarrow \quad 2a + 2b = 3a - 3b$$

$$\Rightarrow \quad a = 5b \qquad \dots (i)$$

and

$$\frac{3}{a+b} = \frac{-7}{-(3a+b-2)}$$

$$\Rightarrow \quad 9a + 3b - 6 = 7a + 7b$$

$$\Rightarrow \quad 2a - 4b = 6 \qquad \dots (ii)$$

From (*i*) and (*ii*),

$\Rightarrow \quad 10b - 4b = 6 \quad \Rightarrow \quad 6b = 6$

$\Rightarrow \quad b = 1$

$\Rightarrow$ As $\quad a = 5b \quad \Rightarrow \quad a = 5$

Hence for system to have infinite solutions $a = 5$, $b = 1$.

2. Find the value of k for which the given system has unique solution.

$$2x + 3y - 5 = 0; \quad kx - 6y - 8 = 0$$

Sol. Given equations are

$2x + 3y - 5 = 0$ and $kx - 6y - 8 = 0$

Here $a_1 = 2, b_1 = 3, c_1 = -5$

$a_2 = k, b_2 = -6, c_2 = -8$

For the system to have unique solution,

$$\frac{a_1}{a_2} \neq \frac{b_1}{b_2}$$

$$\frac{2}{k} \neq \frac{3}{-6} \Rightarrow k \neq -4$$

3. Find the value of k for which the following system of equation has no solution.

$$kx - 5y = 2; \quad 6x + 2y = 7$$

Sol. Given equations are

$kx - 5y = 2$ and $6x + 2y = 7$

Here $a_1 = k, b_1 = -5, c_1 = -2$

$a_2 = 6, b_2 = 2, c_2 = -7$

For the system to have no solution,

$$\frac{a_1}{a_2} = \frac{b_1}{b_2} \neq \frac{c_1}{c_2}$$

$$\frac{k}{6} = \frac{-5}{2} \neq \frac{-2}{-7} \Rightarrow \frac{k}{6} = \frac{-5}{2} \Rightarrow k = -15$$

GRAPHICAL METHOD:

- **Method of solving simultaneous linear equations in two variables:**

  Step 1. Get three solutions of each of given linear equations.

  Step 2. Plot these points on graph in order to draw the lines representing these equations.

  Step 3. Get the solution of equations.

4. The path of a train A is given by $2x + y - 4 = 0$ and the path of another train B is given by the equation $4x + 2y - 16 = 0$. Represent this situation graphically.

Sol. Solution table for $2x + y - 4 = 0$

x	0	1	2
y	4	2	0

Solution table for $4x + 2y - 16 = 0$

x	4	3	2
y	0	2	4

Plotting these points on graph, we get

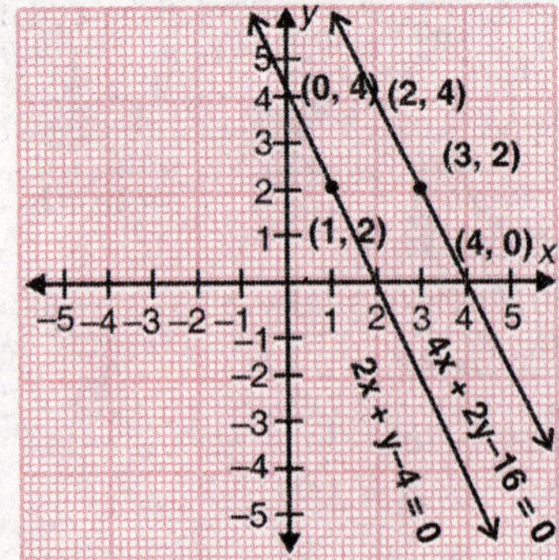

5. Draw the graph of $x - y + 1 = 0$ and $3x + 2y - 12 = 0$ and show that there is a unique solution for these equations. Calculate the area bounded by these lines and x-axis. **[Delhi 2002]**

Sol. Consider equations:

$$x - y + 1 = 0$$

$\Rightarrow \quad y = x + 1$

Some points on graph are :

x	0	1	–1
y	1	2	0

$3x + 2y - 12 = 0$

$\Rightarrow \quad y = \dfrac{12 - 3x}{2}$

Some points on graph are:

x	0	2	4
y	6	3	0

Plotting these points on graph, we get

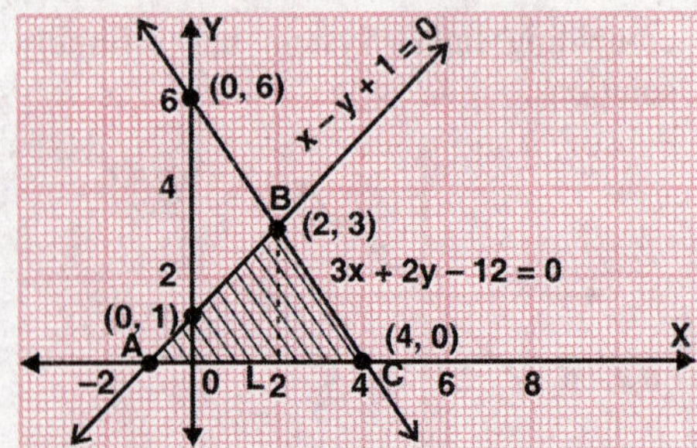

We find that the graph of equations $x - y + 1 = 0$ and $3x + 2y - 12 = 0$ intersects at point B(2, 3). So the system of equation has unique solution, *i.e.*, (2, 3).

Shaded portion is the area bounded by the lines and the x-axis.

$$\text{Area} = \frac{1}{2}\text{AC} \times \text{BL}$$

$$= \frac{1}{2} \times 5 \times 3 = 7.5 \text{ sq. units}$$

6. Show graphically that the system of equations $x + 2y = 5$; $3x + 6y = 15$ have infinitely many solutions.

Sol. Consider equations:

$$x + 2y = 5$$

$$\Rightarrow \quad x = 5 - 2y$$

Some points on graph are :

x	5	3	1
y	0	1	2

$$3x + 6y = 15$$

$$\Rightarrow \quad x = \frac{15-6y}{3}$$

Some points on graph are :

x	5	3	1
y	0	1	2

Plotting these points on graph, we get

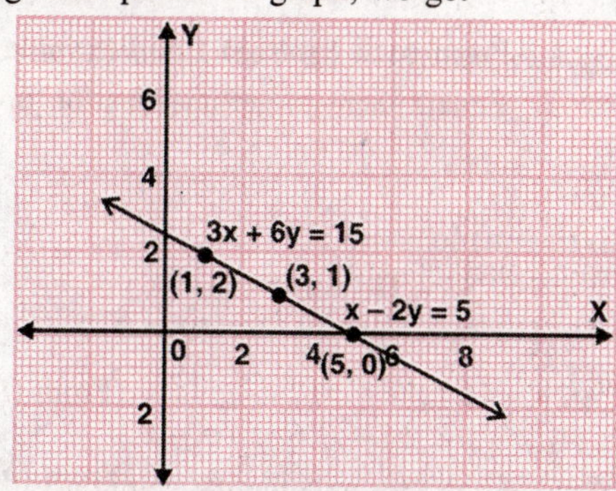

We find that the graph of equations $x + 2y = 5$ and $3x + 6y = 15$ are coinciding to each other. So, the system of equations have infinite solutions.

ALGEBRAIC METHOD:

- **Method of solving simultaneous linear equations in two variables using Substitution.**

Step 1. From one equation find the value of one variable, (say y) in terms of other variable, *i.e.*, x.

Step 2. Substitute the value of variable obtained in step 1, in the other equation to get an equation in one variable.

Step 3. Solve the equation obtained in step 2 to get the value of one variable.

Step 4. Substitute the value of variable so obtained in any given equation to find the value of other variable.

7. Solve the system of equation using substitution:

$x + y = 6$; $x - y = 2$

Sol. Consider equation $\quad x + y = 6 \quad$ ... (*i*)

$\Rightarrow \quad x = 6 - y \quad$ ... (*ii*)

Substitute the value of x in $x - y = 2$

$6 - y - y = 2 \quad \Rightarrow \quad 6 - 2y = 2$

$\Rightarrow \quad 2y = 6 - 2 \quad \Rightarrow \quad y = 4/2 = 2$

Substitute the value of y in (*ii*)

$x = 6 - 2 = 4$

The solution is $x = 4$, $y = 2$

- **Method of solving simultaneous linear equations in two variables using Elimination method by Equating the Coefficients:**

Step 1. Obtain the two equations.

Step 2. Multiply the equations so as to make the coefficients of one of the variables equal, which is to be eliminated.

Step 3. (*i*) If the coefficients of the variable to be eliminated are having same sign, then subtract the equation obtained in step 2.

(*ii*) If the coefficients of the variable to be eliminated are having opposite sign, then add the equation obtained in step 2.

Step 4. Solve the equation obtained in step 3 to obtain the value of one variable.

Step 5. Substitute the value of variable so obtained in any of the given equation to find the value of other variable.

8. Solve the system of equations:

$3(a + 3b) = 11ab$; $3(2a + b) = 7ab$.

Sol. Given equations are:

$3(a + 3b) = 11ab \quad \Rightarrow 3a + 9b = 11ab \quad$ ... (*i*)

and $3(2a + b) = 7ab \quad \Rightarrow 6a + 3b = 7ab \quad$ ... (*ii*)

Multiply (*i*) by 2 and subtract (*ii*) from it.

$$\begin{array}{rcl} 6a + 18b & = & 22ab \\ 6a + 3b & = & 7ab \\ - \quad - & & - \\ \hline 15b & = & 15ab \end{array} \quad \Rightarrow 15b - 15ab = 0$$

$\Rightarrow 15b(1 - a) = 0$

$\Rightarrow$ Either $15b = 0$ or $1 - a = 0 \Rightarrow b = 0$ or $a = 1$.

When $b = 0$, Put in (*i*) $3a = 0 \quad \Rightarrow a = 0$

$\therefore$ Solution is $a = 0, b = 0$.

When $a = 1$, Put in (i)

$3 + 9b = 11b \Rightarrow 2b = 3 \Rightarrow b = 3/2$

$\therefore$ Solution is $a = 1, b = 3/2$

■ Solving system of linear equations by Method of Comparison :

Step 1. From each equation find the value of one variable in terms of other.

Step 2. Equate them to get an equation in one variable and solve.

9. Solve : $x - 2y + 11 = 0$; $2x - 3y + 18 = 0$

Sol. Given equations are :

$$x - 2y + 11 = 0 \quad \ldots (i)$$

$$2x - 3y + 18 = 0 \quad \ldots (ii)$$

$$x = 2y - 11 \quad \text{[From } (i)] \quad \ldots (iii)$$

$$x = \frac{3y-18}{2} \quad \text{[From } (ii)] \quad \ldots (iv)$$

Equating (iii) and (iv)

$$2y - 11 = \frac{3y-18}{2} \Rightarrow 4y - 22 = 3y - 18$$

$$\Rightarrow 4y - 3y = 22 - 18 \Rightarrow y = 4$$

Substituting $y = 4$ in (iii), we get

$$x = 2 \times 4 - 11 \Rightarrow -3$$

Solution is $x = -3, y = 4$.

■ Method of Cross Multiplication for solving system of linear equations :

For system of linear equations $a_1x + b_1y + c_1 = 0$ and $a_2x + b_2y + c_2 = 0$ when , $\frac{a_1}{a_2} \neq \frac{b_1}{b_2}$

the unique solution is given by $x = \frac{b_1c_2 - b_2c_1}{a_1b_2 - a_2b_1}$ and $y = \frac{a_2c_1 - c_2a_1}{a_1b_2 - a_2b_1}$

10. Solve the following system of equations by cross multiplication method :

$$\frac{x}{a} + \frac{y}{b} = a + b; \quad \frac{x}{a^2} + \frac{y}{b^2} = 2$$

Sol. The given equations are :

$$\frac{x}{a} + \frac{y}{b} = a + b \Rightarrow bx + ay = ab\,(a + b)$$

$$\Rightarrow \quad bx + ay - ab(a + b) = 0 \quad \ldots (i)$$

and $\frac{x}{a^2} + \frac{y}{b^2} = 2 \Rightarrow b^2x + a^2y = 2a^2b^2$

$$\Rightarrow \quad b^2x + a^2y - 2a^2b^2 = 0 \quad \ldots (ii)$$

From (i) and (ii),

$$\frac{x}{\begin{matrix} a & -ab(a+b) \\ a^2 & -2a^2b^2 \end{matrix}} = \frac{-y}{\begin{matrix} b & -ab(a+b) \\ b^2 & -2a^2b^2 \end{matrix}} = \frac{1}{\begin{matrix} b & a \\ b^2 & a^2 \end{matrix}}$$

$$\Rightarrow \frac{x}{-2a^3b^2 + a^3b(a+b)} = \frac{-y}{-2a^2b^3 + ab^3(a+b)} = \frac{1}{a^2b - ab^2}$$

$$\Rightarrow \frac{x}{-a^3b(2b-a-b)} = \frac{-y}{-ab^3(2a-a-b)} = \frac{1}{ab(a-b)}$$

$$\Rightarrow \frac{x}{-a^3b(b-a)} = \frac{y}{ab^3(a-b)} = \frac{1}{ab(a-b)}$$

$$\Rightarrow x = \frac{a^3b(a-b)}{ab(a-b)} = a^2 \ ; \ y = \frac{ab^3(a-b)}{ab(a-b)} = b^2$$

The solution is $x = a^2, y = b^2$.

■ To solve the Word Problems, the following points should be kept in mind :

(i) Reduce the given conditions in terms of x and y as mathematical statements.

(ii) Solve these as simultaneous equations.

11. Five years ago, I was thrice as old as my son and ten years later, I shall be twice as old as my son. How old are we now ?

Sol. Suppose my age is x years and my son's age is y years. Then, according to problem

$$x - 5 = 3(y - 5) \quad \ldots (i)$$

and $\quad x + 10 = 2(y + 10) \quad \ldots (ii)$

$$x - 3y = -10 \quad \text{[From } (i)] \quad \ldots (iii)$$

$$x - 2y = 10 \quad \text{[From } (ii)] \quad \ldots (iv)$$

Solving (iii) and (iv), we get $x = 50, y = 20$

Hence, my age is 50 years and my son's age is 20 years.

12. Seven times a two digit number is equal to four times the number obtained by reversing the order of digits and the sum of the digits is 3. Find the number.

Sol. Suppose x is the digit at the ten's place and y is at the unit's place of the given number.

Then the given number is $10x + y$ and if we reverse the order of the digits, the new number becomes $10y + x$.

Then, according to the problem

$7(10x + y) = 4(10y + x)$ ... (*i*) and $x + y = 3$ ... (*ii*)

$70x + 7y = 40y + 4x \Rightarrow 2x - y = 0$ [From (*i*)] ... (*iii*)

By solving (*ii*) and (*iii*), we get $x = 1$ and $y = 2$

Hence, the number is 12.

13. A sailor goes 8 km downstream in 40 minutes and comes back in one hour. Determine the speed of the sailor in still water and speed of the current.

Sol. Let speed of sailor = x km/hr

and speed of current = y km/hr

$\therefore$ speed upstream = $(x - y)$ km/hr

speed downstream = $(x + y)$ km/hr

According to the given condition,

$$\frac{8}{x+y} = \frac{40}{60} \Rightarrow \frac{8}{x+y} = \frac{2}{3} \Rightarrow x + y = 12 \quad \text{... }(i)$$

$$\text{and } \frac{8}{x-y} = 1 \Rightarrow x - y = 8 \quad \text{... }(ii)$$

Adding (*i*) and (*ii*), we get

$2x = 20 \Rightarrow x = 10$

Putting $x = 10$ in (*i*), we get $y = 2$

$\therefore$ Speed of sailor = 10 km/hr

Speed of current = 2 km/hr

14. Points A and B are 70 km apart on a highway. A car starts from A and another car starts from B at the same time. If they travel in the same direction, they meet in 7 hours, but if they travel towards each other they meet in one hour. What are their respective speeds?

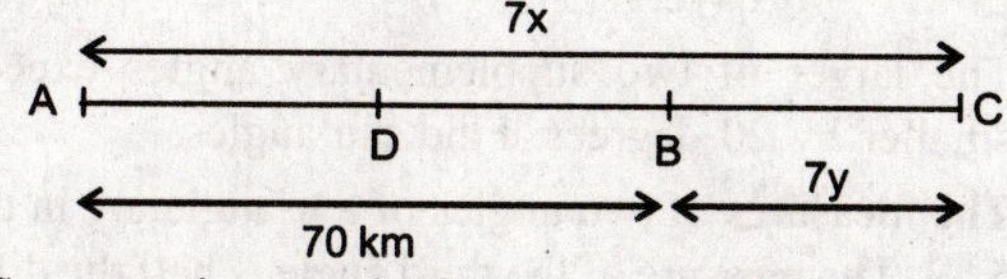

Sol. Suppose the speed of car A = x km/hr

and the speed of car B = y km/hr

Then according to problem,

$7x - 70 = 7y$ ... (*i*)

and $x + y = 70$ ... (*ii*)

From (*i*), $7x - 7y = 70 \Rightarrow x - y = 10$ ... (*iii*)

From (*ii*) and (*iii*), we get $x = 40, y = 30$

$\therefore$ Speed of car A = 40 km/hr

Speed of car B = 30 km/hr

15. A man has ₹ 50 in 50 p and 25 p coins. All the 50 p coins are worth as much as all the 25 coins. How many coins of each he has ?

Sol. Let number of 50 p coins = x

and number of 25 p coins = y

According to the given condition

$50x + 25y = 5000 \Rightarrow 2x + y = 200$ ... (*i*)

and $50x = 25y \Rightarrow 2x = y$ ... (*ii*)

From (*i*) and (*ii*), we get

$2x + 2x = 200$

$\Rightarrow 4x = 200 \Rightarrow x = 50$

Put $x = 50$ in (*ii*), we get $y = 100$

$\therefore$ 50 p coins = 50

25 p coins = 100

16. 10 students of class X took part in a Mathematics quiz. If the number of girls is 4 more than the number of boys, find the number of boys and girls who took part in the quiz. Form a pair of linear equations of the above problem and find its solution graphically.

Sol. Let number of boys is x and number of girls is y.

$\therefore x + y = 10$

Also $y = x + 4$

$\Rightarrow x - y = -4$

Hence pair of linear equations of the problem is

$x + y = 10$ ... (*i*)

$x - y = -4$ ... (*ii*)

Solution table for $x + y = 10$:

x	0	10	5
y	10	0	5

Solution table for $x - y = -4$:

x	0	2	6
y	4	6	10

Plotting the points on graph, we get

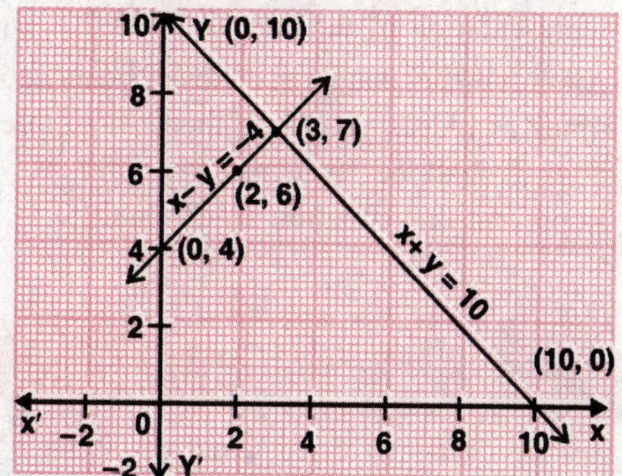

From the graph, we observe that two lines intersect at $x = 3, y = 7$.

$\therefore$ Number of boys = 3, Number of girls = 7

Summative Assessment

Subjective Questions

VERY SHORT ANSWER TYPE QUESTIONS

This category is given for better understanding of the topic and for EXTRA PRACTICE

PREVIOUS YEARS' QUESTIONS

1. Find the value of k so that the following system of equations has no solution :
$3x - y - 5 = 0,\ 6x - 2y + k = 0$ [Delhi 2008]

2. Find the value of k so that the following system of equation has infinite solutions :
$3x - y - 5 = 0,\ 6x - 2y + k = 0$ [Delhi 2008]

3. Find the number of solutions of the following pair of linear equations :
$$x + 2y - 8 = 0$$
$$2x + 4y = 16$$ [AI 2009]

4. Write whether the following pair of linear equations is consistent or not.
$x + y = 14,\ x - y = 4$ [Foreign 2009]

PRACTICE QUESTIONS

5. The coach of cricket team buys 3 bats and 6 balls for ₹ 3900. Later she buys another bat and 2 more balls of the same kind for ₹ 1300. Represent this situation algebraically.

6. The cost of 2 kg of apples and 1 kg of grapes on a day was found to be ₹ 160. Also, the cost of 4 kg of apples and 2 kg of grapes is ₹ 300. Represent this situation algebraically.

7. Find whether the following pair of equations has no solution, unique solution or infinitely many solutions.
$$5x - 8y + 1 = 0;$$
$$3x - \frac{24}{5}y + \frac{3}{5} = 0$$

8. Given the linear equation $3x + 4y - 8 = 0$, write another linear equation in two variables such that the geometrical representation of the pair so formed is parallel lines.

9. Aftab tells his daughter, "Seven years ago, I was seven times as old as you were then. Also, three years from now, I shall be three times as old as you will be." Represent this situation algebraically.

10. Find c, if the system $cx + 3y + (3 - c) = 0$ and $12x + cy - c = 0$ has infinitely many solutions.

11. Check whether the pair of equations $x + 3y = 6$, $2x - 3y = 12$ is consistent.

12. For which values of p, does the pair of equations given below has unique solution ?
$$4x + py + 8 = 0$$
$$2x + 2y + 2 = 0$$

13. Sum of two numbers is 35 and their difference is 13. Find the numbers.

14. If the system of equations $6x - 2y = 3$ and $kx - y = 2$ has a unique solution, find k.

15. Half the perimeter of a rectangular garden, whose length is 12 m more than its width is 60 m. Find the dimensions of the garden.

16. Which type of solution will equations $x + 2y = 4$ and $2x + y = 5$ have ?

17. The larger of two supplementary angles exceeds the smaller by 20 degrees. Find the angles.

18. The measures of two angles of a triangle are in the ratio 5 : 3. The measure of the third angle is half the difference of the measure of the above two angles. Find the measure of each angle.

SHORT ANSWER TYPE QUESTIONS-I

PREVIOUS YEARS' QUESTIONS

1. Solve for x and y :
$47x + 31y = 63,\ 31x + 47y = 15$

Or

Solve for x and y :
$\frac{ax}{b} - \frac{by}{a} = a + b\ ;\ ax - by = 2ab$ [Delhi 2006, AI 2009]

2. Solve the system of equations :

$$\frac{bx}{a} - \frac{ay}{b} + a + b = 0$$

and $$bx - ay + 2ab = 0$$

Or

The sum of the digits of a two-digit number is 12. The number obtained by interchanging the two digits exceeds the given number by 18. Find the number. [AI 2006]

3. Solve the system of equations for x :

$$\frac{b^2x}{a} - \frac{a^2y}{b} = ab\ (a + b)$$

and $$b^2x - a^2y = 2a^2b^2$$

Or

A man sold a table and a chair together for ₹ 850 at a loss of 10% on the table and a gain of 10% on the chair. By selling them together for ₹ 950, he would have made a gain of 10% on the table and loss of 10% on the chair. Find the cost price of each.

[Foreign 2006]

4. Solve the following equations for x and y :

$$mx - ny = m^2 + n^2$$
$$x + y = 2m$$

Or

Abdul travels 300 km by train and 200 km by taxi, it takes him 5 hours 30 minutes. But if, he travels 260 km by train and 240 km by taxi, he takes 6 minutes longer. Find the speed of the train and that of the taxi.

[Delhi 2006C]

5. Solve the following equations for x and y :

$$\frac{a^2}{x} - \frac{b^2}{y} = 0; \quad \frac{a^2b}{x} + \frac{b^2a}{y} = a + b \qquad x, y \neq 0.$$

Or

The sum of the numerator and the denominator of a fraction is 12. If the denominator is increased by 3, the fraction becomes $\frac{1}{2}$. Find the fraction. [AI 2006C]

6. Solve for x and y :

$$x + \frac{6}{y} = 6, \quad 3x - \frac{8}{y} = 5$$

Or

Solve for x and y :

$$\frac{x+1}{2} + \frac{y-1}{3} = 8 \; ; \; \frac{x-1}{3} + \frac{y+1}{2} = 9$$ [Delhi 2007]

7. Solve for x and y :

$$8x - 9y = 6xy \; ; \; 10x + 6y = 19xy$$

Or

Solve for x and y :

$$4x + \frac{y}{3} = \frac{8}{3}; \quad \frac{x}{2} + \frac{3y}{4} = -\frac{5}{2}$$ [AI 2007]

8. Find the value of k for which the pair of linear equations $kx + 3y = k - 2$ and $12x + ky = k$ has no solution. [Delhi 2009]

9. Without drawing the graph, find out wheather the lines representing the following pair of linear equations intersect at a point, are parallel or coincident.

$$18x - 7y = 24 \; ; \; \frac{9}{5}x - \frac{7}{10}y = \frac{9}{10}$$ [Foreign 2009]

10. Find the value of m for which the pair of linear equations $2x + 3y - 7 = 0$ and $(m - 1)x + (m + 1)y = (3m - 1)$ has infinitely many solutions. [AI 2010]

11. For what value of k will the following pair of linear equations have no solution ?

$2x + 3y = 9; 6x + (k - 2)y = (3k - 2)$ [Foreign 2010]

12. For what value of p will the following pair of linear equations have infinitely many solutions ?

$(p - 3)x + 3y = p; px + py = 12$ [Foreign 2010]

13. Find the values of a and b for which the following pair of linear equations has infinitely many solutions:

$2x + 3y = 7; (a + b)x + (2a - b)y = 21$

[Foreign 2010]

14. Find the value of 'k' for which the system of equations $kx + 3y = 1; 12x + ky = 2$ has no solution.

[CBSE (CCE) 2011]

15. Solve : $99x + 101y = 499$
$101x + 99y = 501$ [CBSE (CCE) 2011]

16. For what value of p will the following system of equations has no solution; $(2p - 1)x + (p - 1)y = 2p + 1$; $y + 3x - 1 = 0$ [CBSE (CCE) 2011]

Practice Questions

17. Find the value of k for which the system of equations has no solution.
$3x - 4y + 7 = 0, kx + 3y - 5 = 0$

18. Find the value of k for which the system of equations has a unique solution.
$x - ky = 2, 3x + 2y = -5$

19. Determine k for which the system of equations has infinite solutions : $4x + y = 3, 8x + 2y = 5k$

20. Determine, by drawing graphs, whether the following system of linear equations has a unique solution or not:
$2x - 3y = 6, x + y = 1$

21. For what value of k will the system of linear equations has infinite number of solutions ?
$kx + 4y = k - 4, 16x + ky = k$

22. Determine the value of k for which the following system of equations has no solution :
$kx + 2y - 1 = 0, 5x - 3y + 2 = 0$

23. Determine the value of k for which the following system of linear equations has infinite number of solutions : $(k-3)x + 3y = k$, $kx + ky = 12$

24. Solve for x and y : $x - y = 0$, $2x - y = 2$

25. Given below are three equations. Two of them have infinite solution and two have unique solution. State the two pairs:
(i) $3x - 2y = 4$ (ii) $6x + 2y = 4$
(iii) $9x - 6y = 12$

26. Solve the following systems of linear equations in x and y :
(i) $x + y - (a + b) = 0$; $ax - by - (a^2 - b^2) = 0$;
(ii) $\frac{x}{a} + \frac{y}{b} - 2 = 0$; $ax - by + b^2 - a^2 = 0$

27. Find the values of a and b for which the following system of linear equations has infinite number of solutions :
$2x - 3y = 7$, $(a + b)x - (a + b - 3)y = 4a + b$

28. The sum of the digits of a two digit number is 15. The number obtains by interchanging the digits exceeds the given number by 9. Find the number.

29. Two years ago, a father was five times as old as his son. Two years later, his age will be 8 years more than three times the age of the son. Find the present ages of father and son.

30. If 1 is added to each of numerator and denominator of a fraction, it becomes 2/3. However, if 1 is subtracted from each of numerator and denominator it becomes 3/5. Find the fraction.

31. If $(x + 3)$ is a factor of $x^3 + ax^2 - bx + 6$ and $a + b = 7$, find the values of a and b.

32. Show that the system of equations is consistent and dependent : $x - 5y = 6$, $2x - 10y = 12$.

33. Solve for x and y : $\frac{x}{2} + y = 0.8$ and $\frac{7}{x + \frac{y}{2}} = 10$.

34. Solve for x and y :
$7(y + 3) - 2(x + 2) = 14$, $4(y - 2) + 3(x - 3) = 2$

35. For what value of k will the equations $x + 2y + 7 = 0$, $2x + ky + 14 = 0$ represent coincident lines ?

36. Find the value of k for which the system $3x + ky = 7$, $2x - 5y = 1$ will have infinitely many solutions.

37. For what value of k, the system of equations :
$2x - ky + 3 = 0$, $4x + 6y - 5 = 0$ is consistent ?

38. For what value of k, the system of equations $kx - 3y + 6 = 0$, $4x - 6y + 15 = 0$ represents parallel lines ?

39. Solve for x and y :
$2x + 3y = 7$, $6x + 9y = 11$

40. Solve for x and y :
$0.4x + 3y = 1.2$, $7x - 2y = \frac{17}{6}$

41. Solve for x and y : $5x + \frac{4}{y} = 9$, $7x - \frac{2}{y} = 5$, $y \neq 0$

42. Solve for x and y :
$\frac{3}{x} + 4y = 7$, $-\frac{2}{x} + 7y = 5$, $x \neq 0$

43. Solve for x, y and z :
$x + y = 4$, $y + z = 1$, $z + x = 1$

44. Solve for x, y and z :
$x + y + z = 9$, $2y - z = 2$, $z - x = 2$

45. Solve for x and y : $\frac{m}{x} - \frac{n}{y} = a$, $px - qy = 0$

46. Solve for x and y : $\frac{4}{x} - \frac{9}{y} = -1$, $\frac{15x}{2} - 5y = 0$

47. Solve for x and y :
$149x - 330y = -511$, $-330x + 149y = -32$

48. Solve for x and y : $\frac{4}{x} + 5y = 7$, $\frac{3}{x} + 4y = 5$, $x \neq 0$

49. Solve for x and y : $\sqrt{7}x + \sqrt{11}y = 0$, $\sqrt{3}x - \sqrt{5}y = 0$

50. Solve for x and y : $\sqrt{5}x - \sqrt{7}y = 0$, $\sqrt{3}y + \sqrt{13}x = 0$

51. Solve for x and y : $3x - 4y - 1 = 0 = 2x - \frac{8}{3}y + 5$

52. Determine the value of c for which the following system of linear equations has no solution :
$cx + 3y = 3$, $12x + cy = 6$.

53. For what value of k, the following system of equations $2x + ky = 1$, $3x - 5y = 7$ has
(i) a unique solution
(ii) no solution

54. If $(x - 4)$ is a factor of $x^3 + ax^2 + 2bx - 24$ and $a - b = 8$, find the values of a and b.

55. For what value of k the following pair of linear equation has no solution ?
$2x + ky = 1$
$6x + 4y = 3$

56. For what value of a the following pair of linear equation has infinitely many solution ?
$ax - 3y = 1$
$-12x + ay = 2$

57. For what value of a the following pair of linear equation has infinitely many solutions ?
$2x + ay = 8$
$ax + 8y = a$

58. For what value of k the following pair of linear equations has unique solution ?
$7x + 8y = k$
$9x - 4y = 12$

59. For what value of k the following pair of linear equation has unique solution ?
$kx + 3y = 3$
$12x + ky = 6$

NCERT QUESTION

60. Solve the following systems of linear equations in x and y :
(i) $px + qy - p + q = 0$, $qx - py - p - q = 0$
(ii) $\frac{x}{a} - \frac{y}{b} = 0$, $ax + by = a^2 + b^2$

SHORT ANSWER TYPE QUESTIONS-II

PREVIOUS YEARS' QUESTIONS

1. Draw the graphs of the equations
$4x - y - 8 = 0$ and $2x - 3y + 6 = 0$
Also, determine the vertices of the triangle formed by the lines and x-axis. [Delhi 2006]

2. Draw the graphs of the equations :
$4x - 3y - 6 = 0; \quad x + 3y - 9 = 0$
Determine the co-ordinates of the vertices of the triangle formed by the lines and the y-axis. [Foreign 2006]

3. Solve the following system of linear equations graphically:
$3x - 2y - 1 = 0; \quad 2x - 3y + 6 = 0$
Shade the region bounded by the lines and x-axis. [Delhi 2006C]

4. Solve the following system of equations graphically for x and y :
$3x + 2y = 12; \; 5x - 2y = 4$
Find the co-ordinates of the points where the lines meet the y-axis. [AI 2006C]

5. Solve for x and y :
$(a - b)x + (a + b)y = a^2 - 2ab - b^2$
$(a + b)(x + y) = a^2 + b^2$ [AI 2008]
Or
Solve for x and y :
$37x + 43y = 123, \; 43x + 37y = 117$

6. Represent the following pair of equations graphically and write the coordinates of points where the lines intersect y-axis :
$x + 3y = 6, \; 2x - 3y = 12$ [Foreign 2008]

7. Solve the following system of equations for x and y :
$$\frac{5}{x-1} + \frac{1}{y-2} = 2, \quad \frac{6}{x-1} - \frac{3}{y-2} = 1$$
[Delhi 2008C, Delhi 2009]
Or
For what values of a and b does the following pair of linear equations have an infinite numbers of solutions ?
$2x + 3y = 7, \; a(x + y) - b(x - y) = 3a + b - 2$ [Delhi 2008 C]

8. Solve the following pairs of equations for x and y :
$$\frac{15}{x-y} + \frac{22}{x+y} = 5, \quad \frac{40}{x-y} + \frac{55}{x+y} = 13$$
$x \neq y, x \neq -y$ [AI 2008 C]

9. For what values of k will the following pair of linear equations have infinitely many solutions ?
$2x - 3y = 7, \; (k + 1)x + (1 - 2k)y = 5k - 4$ [AI 2008C]

10. Solve the following pair of equations :
$$\frac{10}{x+y} + \frac{2}{x-y} = 4; \quad \frac{15}{x+y} - \frac{5}{x-y} = -2$$ [Delhi 2009]

11. Solve for x and y,
$$\frac{ax}{b} - \frac{by}{a} = a + b; \; ax - by = 2ab$$ [AI 2009]

12. The sum of numerator and denominator of a fraction is 3 less than twice the denominator. If each of the numerator and denominator is decreased by 1, the fraction becomes $\frac{1}{2}$. Find the fraction.
Or
Solve the following pair of equations :
$$\frac{4}{x} + 3y = 8; \; \frac{6}{x} - 4y = -5.$$ [Delhi 2010]

13. Solve the following pair of linear equations for x and y :
$$\frac{b}{a}x + \frac{a}{b}y = a^2 + b^2; \; x + y = 2ab$$ [AI 2010]

14. Solve the following pair of linear equations for x and y :
$2(ax - by) + (a + 4b) = 0$;
$2(bx + ay) + (b - 4a) = 0$
Or
A number consists of two digits. Where the number is divided by the sum of its digits, the quotient is 7. If 27 is subtracted from the number, the digits interchange their places, find the number. [Foreign 2010]

15. The sum of the digits of a two digit number is 12. The number obtained by interchanging the two digits exceeds the given number by 18. Find the number. [CBSE (CCE) 2011]

16. In the figure below, ABCDE is a pentagon with BE || CD and BC || DE. BC is perpendicular to CD. If the perimeter of ABCDE is 21 cm, find the value of x and y.

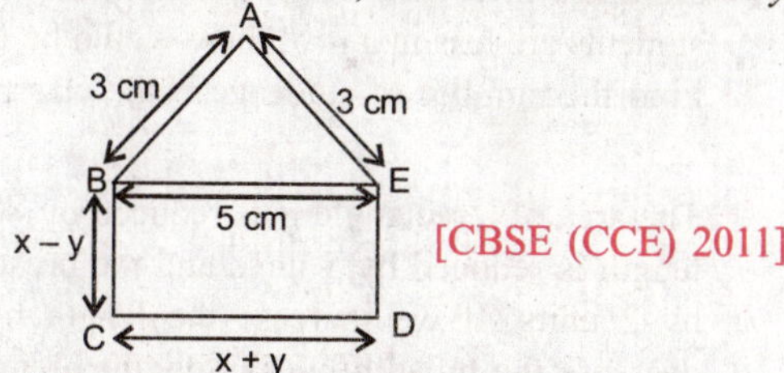

[CBSE (CCE) 2011]

17. The sum of the numerator and denominator of a fraction is 12. If 1 is added to both the numerator and the denominator the fraction becomes $\frac{3}{4}$. Find the fraction. [CBSE (CCE) 2011]

18. 4 men and 6 boys can finish a piece of work in 5 days while 3 men and 4 boys can finish it in 7 days. Find the time taken by 1 man alone or that by 1 boy alone. [CBSE (CCE) 2011]

19. A man travels 600 km partly by train and partly by car. It takes 8 hours and 40 minutes if he travels 320 km by train and the rest by car. It would take 30 minutes more if he travels 200 km by train and the rest by car. Find the speed of the train and the car separately. [CBSE (CCE) 2011]

20. Solve for the following pair of equations for x and y :
$$\frac{a^2}{x} - \frac{b^2}{y} = 0; \; \frac{a^2 b}{x} + \frac{b^2 a}{y} = a + b, \; x \neq 0; \; y \neq 0$$

21. 2 tables and 3 chairs together cost ₹ 2000 whereas 3 tables and 2 chairs together cost ₹ 2500. Find the total cost of 1 table and 5 chairs.

22. 5 books and 7 pens together cost ₹ 79 whereas 7 books and 5 pens together cost ₹ 77. Find the total cost of 1 book and 2 pens.

23. Use a single graph paper and draw the graph of the following equations :

$$2y - x = 8 \; ; \; 5y - x = 14; \; y - 2x = 1$$

Obtain the vertices of the triangle so obtained.

24. The sum of the digits of a two digit number is 15. The number is decreased by 27 if the digits are reversed. Find the number.

25. In a two digit number, the tens digit is three times the unit digit. When the number is decreased by 54, the digits are reversed. Find the number.

26. A part of monthly expenses of a family is constant and the remaining varies with the price of wheat. When the rate of wheat is ₹ 250 a quintal, the total monthly expenses of the family are ₹ 1000 and when it is ₹ 240 a quintal, the total monthly expenses are ₹ 980. Find the total monthly expenses of the family when the cost of wheat is ₹ 350 a quintal.

27. The sum of the digits of a two digit number is 8 and the difference between the number and that formed by reversing the digits is 18. Find the number.

28. Students of a class are made to stand in rows. If 4 students are extra in a row, there would be two rows less. If 4 students are less in a row, there would be four more rows. Find the number of students in the class.

Or

The area of a rectangle gets reduced by 80 sq. units if its length is reduced by 5 units and the breadth is increased by 2 units. If we increase the length by 10 units and decrease the breadth by 5 units, the area is increased by 50 sq. units. Find the length and breadth of the rectangle.

29. Solve graphically :

$$x + 4y = 10, \quad y - 2 = 0$$

30. Solve graphically :

$$2x - 3y = 6, \quad x - 6 = 0$$

31. Determine graphically the vertices of a triangle, the equations of whose sides are given below :

$$2y - x = 8, \; 5y - x = 14, \; - 2x + y = 1$$

32. Solve for x and y :

$$\frac{6}{x+y} = \frac{7}{x-y} + 3, \quad \frac{1}{2(x+y)} = \frac{1}{3(x-y)},$$

where $x + y \neq 0$ and $x - y \neq 0$.

33. A takes 3 hours more than B to walk 30 km. But if A doubles his pace, he is ahead of B by 3/2 hours. Find their speed of walking.

34. A train covered a certain distance at a uniform speed. If the train would have been 6 km/hr. faster, it would have taken 4 hours less than the scheduled time. And if the train were slower by 6 km/hr., it would have taken 6 hours more than the scheduled time. Find the length of the journey.

35. 8 men and 12 boys can finish a piece of work in 10 days while 6 men and 8 boys can finish it in 14 days. Find the time taken by one man alone and that by one boy alone to finish the work.

36. On selling a tea set at 5% loss and a lemon set at 15% gain, a crockery seller gains ₹ 7. If he sells the tea set at 5% gain and lemon set at 10% gain, he gains ₹ 13. Find the actual price of the tea set.

37. In an examination paper, one mark is awarded for every correct answer while 1/4 mark is deducted for every wrong answer. A student answered 120 questions and got 90 marks. How many questions did he answer correctly ?

38. Ratio between the girls and boys in a class of 40 students is 2 : 3. Five new students joined the class. How many of them must be boys so that the ratio between girls and boys becomes 4 : 5 ?

39. If you travel by an autorickshaw, the fare for the first kilometre is different from the rate per km for the remaining distance. The total fare for a distance of 20 km is ₹ 37.70 and that for a distance of 26 km is ₹ 48.50. Find the auto fare for the first kilometre and for each successive kilometre.

40. One kilogram of tea and 4 kg of sugar together cost ₹ 220. If the price of sugar increases by 50% and the price of tea increases by 10%, the cost would be ₹ 266. Find the original cost per kilogram of each.

41. If three times the larger of the two numbers is divided by the smaller one, we get 4 as quotient and 3 as remainder. Also, if seven times the smaller number is divided by the larger one, we get 5 as quotient and 1 as remainder. Find the numbers.

42. In a parallelogram, one angle is 4/5th of the adjacent angle. Determine the angles of the parallelogram.

43. A man bought 4 horses and 9 cows for ₹ 1340. He sells the horses at a profit of 10% and the cows at a profit of 20% and his whole gain is ₹ 188. What price did he pay for the horse ?

44. The age of a father is equal to the sum of the ages of his 5 children. After 15 years, sum of the ages of the children will be twice the age of the father. Find the age of father.

45. The largest angle of a triangle is equal to the sum of the other angles. The smallest angle is 1/4 of the largest angle. Find the angles of the triangle.

46. A car goes uphill at the rate of 30 km an hour and downhill at the rate of 50 km an hour. After 15 hours, it has covered 650 km. How long does it go downhill and uphill respectively ?

47. In a den (a small enclosure), there are rabbits and pheasants (game birds). They have between them 35 heads and 98 feet. How many rabbits and pheasants are there ?

48. A man when asked how many hens and buffaloes he has, told that his animals have 120 eyes and 180 legs. How many hens and buffaloes has he ?

49. Ten years ago, the sum of the ages of two sons was one third of their father's age. One son is two years older than the other and sum of their present ages is 14 years less than the father's present age. Find the present ages of all.

50. If twice the son's age in years is added to the father's age, the sum is 70. But if twice the father's age is added to the son's age, the sum is 95. Find the ages of father and son.

51. A fraction reduces by 4/5 when 2 is subtracted from the numerator and 3 is added to the denominator. But it reduces to 12/17 if 6 is added to the numerator and the denominator is multiplied by 2. Find the fraction.

52. In a ΔABC, $\angle A = x°$, $\angle B = 3x°$ and $\angle C = y°$. If $3y° - 5x° = 30°$ prove that the triangle is right angled.

53. A man invested ₹ 30,000 in two types of bonds. On one, he earns 5% and on the other, he gets 7%. If his total earnings are ₹ 2000, find his investment in each type of bond.

54. The sum of digits of two digit number is 9. Also, nine times this number is twice the number obtained by reversing the order of digits of the number. Find the number.

55. The sum of the digits of a two digit number is 9. The number obtained by reversing the order of digits of the given number exceeds the given number by 27. Find the given number.

56. The sum of a two digit number and the number formed by reversing the order of digits is 154. If the two digits differ by 4, find the number.

57. Find the four angles of a cyclic quadrilateral ABCD in which $\angle A = (2x - 1)°$, $\angle B = (y + 5)°$, $\angle C = (2y + 15)°$ and $\angle D = (4x - 7)°$.

58. A and B are two points 150 km apart on a highway. Two cars start from A and B at the same time. If they move in the same direction, they meet in 15 hours. But if they move in the opposite directions, they meet in one hour. Find their speeds.

59. If in a rectangle, the length is increased and breadth is reduced each by 2 units, the area gets reduced by 28 square units. If however, the length is reduced by 1 and the breadth is increased by 2 units, the area increases by 33 square units. Find the area of the rectangle.

60. A man starts his job with a certain monthly salary and earns a fixed increment every year. If his salary was ₹ 1500 after 4 years of service and ₹ 1800 after 10 years of service, what was his starting salary and what is the annual increment ?

61. Solve for x and y : $a^2x + b^2y = c^2$; $b^2x + a^2y = d^2$

62. A villager went to a hotel in a town with his big family. They consumed 23 idlies, 18 pooris, 7 dosas and 19 vadas. The bill came to them was ₹ 108. On next day, they consumed 34 idlies, 8 vadas, 22 pooris and 7 dosas. The bill came to them was ₹ 114. If one idli costs the same as a vada, what is the cost of one poori ?

63. The incomes of X and Y are in the ratio of 8 : 7 and their expenditures are in the ratio of 19 : 16. If each saves ₹ 1250, find their income.

64. A and B each has a certain number of mangoes. A says to B, "if you give 30 of your mangoes, I will have twice as many as left with you." B replies "if you give me 10, I will have thrice as many as left with you". How many mangoes does each have ?

65. There are two examination rooms A and B. If 10 candidates are sent from A to B, the number of students in each room is same. If 20 candidates are sent from B to A, the number of students in A is double the number of students in B. Find the number of students in each room.

66. A bag contains 94 coins of 50 paise and 25 paise denominations. If the total worth of these coins be ₹ 29.75, find the number of coins of each kind.

67. Solve the following system of equations graphically : Also find the points where the lines represented by the given equations intersect the x-axis.

(*i*) $3x - 5y + 1 = 0,\ 2x - y + 3 = 0$

(*ii*) $2x + 3y = 6,\ 3x - \frac{5}{2}y = 2$

(*iii*) $3x + 2y = 14,\ x - 4y = -7$

(*iv*) $x = 5,\ 3x - 6y = 15$

(*v*) $3x - 4y - 1 = 0,\ 2x - \frac{8}{3}y - 4 = 0$

(*vi*) $x - 5y = 6,\ 2x - 10y = 10.$

68. Solve for x and y :

$$\frac{xy}{x+y} = \frac{1}{5},\ \frac{xy}{x-y} = \frac{1}{7},$$

$x + y \neq 0$ and $x \neq y$.

69. Solve for x and y :

$$\frac{57}{x+y} + \frac{6}{x-y} = 5,$$

$$\frac{38}{x+y} + \frac{21}{x-y} = 9,\ x + y \neq 0,\ x - y \neq 0.$$

70. Solve for x and y :

$$\frac{xy}{x+y} = \frac{1}{2},\ \frac{xy}{x-y} = \frac{1}{6},$$

$x + y \neq 0,\ x \neq y$.

71. Seven times a given two digit number is equal to four times, the number obtained by interchanging the digits and the difference of the digits is 3. Find the number.

72. Solve the following system of linear equations graphically: $2x + y = 10,\ 4x - y = 8$

Does the point (1, –4) lie on any of the lines ? Write its equation.

73. Draw the graph $x - y + 1 = 0$ and $2x + y - 10 = 0$. Shade the region bounded by these lines and x-axis. Find the area of the shaded region.

74. Solve the following system of linear equations graphically: $x - y = 1$, $2x + y = 8$. Shade the area bounded by these two lines and the y-axis.

75. Solve graphically for x and y : $2x + 3y = 12$, $x - y = 1$. Shade the region between the two lines and x-axis.

76. A man travels 600 km partly by train and partly by car. If he covers 400 km by train and rest by car, it takes him 6 hours and 30 minutes. But if he travels 200 km by train and rest by car. He takes half an hour longer. Find the speed of the train and that of car.

77. The taxi charge in a city comprise of a fixed charge together with the charge for the distance covered. For a journey of 10 km the charge paid is ₹ 75 and for a journey of 15 km charge paid is ₹ 110. What will a person have to pay for travelling a distance of 25 km ?

78. Father's age is three times the sum of ages of his two children. After 5 years, his age will be twice the sum of age of two children. Find the age of father.

79. 5 years hence, the age of father shall be three times the age of his son while 5 years earlier, the age of the father was 7 times the age of his son. Find their present ages.

80. Solve for x and y : $\frac{1}{7x} + \frac{1}{6y} = 3$, $\frac{1}{2x} - \frac{1}{3y} = 5$.

81. Solve for x and y : $\frac{5}{x+1} - \frac{2}{y-1} = \frac{1}{2}$,

$\frac{10}{x+1} + \frac{2}{y-1} = \frac{5}{2}$, $x \neq -1$ and $y \neq 1$.

82. Solve for x and y : $\frac{x+y}{xy} = 2$, $\frac{x-y}{xy} = 6$

83. Solve for x and y : $\frac{2}{y} + \frac{3}{x} = \frac{7}{xy}$, $\frac{6}{y} + \frac{9}{x} = \frac{11}{xy}$,

$x \neq 0$, $y \neq 0$.

NCERT Questions

84. A part of monthly hostel charges is fixed and the remaining depends on the number of days one has taken food in the mess. When a student A takes food for 20 days, he has to pay ₹ 1,000 as hostel charges whereas a student B, who takes food for 26 days, pays ₹ 1,180 as hostel charges. Find the fixed charge and the cost of food per day.

85. Solve the following pairs of linear equations :

(*i*) $(a - b)x + (a + b)y = a^2 - b^2 - 2ab$

$(a + b)(x + y) = a^2 + b^2$

(*ii*) $px + qy = p - q$

$qx - py = p + q$

(*iii*) $ax + by = c$, $bx + ay = 1 + c$

(*iv*) $\frac{x}{a} - \frac{y}{b} = 0$

$ax + by = a^2 + b^2$

86. Ritu can row downstream 20 km in 2 hrs and upstream 4 km in 2 hrs. Find her speed of rowing in still water and the speed of the current.

87. In a ΔABC, $\angle C = 3\angle B = 2(\angle A + \angle B)$. Find the three angles.

88. The area of rectangle gets reduced by 9 square units, if its length is reduced by 5 units and the breadth is increased by 3 units. If we increase the length by 3 units and breadth by 2 units, then the area is increased by 67 square units. Find the length and breadth of the rectangle.

89. The students of a class are made to stand in rows. If 3 students are extra in a row, there would be 1 row less. If 3 students are less in a row, there would be 2 rows more. Find the number of students in the class.

LONG ANSWER TYPE QUESTIONS

Previous Years' Questions

1. Solve the equations graphically :

$2x + y = 2$; $2y - x = 4$

What is the area of the triangle formed by the two lines and the line $y = 0$? [CBSE (CCE) 2011]

2. Draw the graphs of the following equations :

$x + y = 5$, $x - y = 5$

(*i*) Find the solution of the equations from the graph.

(*ii*) Shade the triangular region formed by the lines and the y-axis. [CBSE (CCE) 2011]

3. For what value of k will the pair of equations have no solution ?

$3x + y = 1$

$(2k - 1)x + (k - 1)y = 2k + 1$ [CBSE (CCE) 2012]

4. Solve for x and y :

$\frac{5}{x-1} + \frac{1}{y-2} = 2$; $\frac{6}{x-1} - \frac{3}{y-2} = 1$ [CBSE (CCE) 2012]

5. Solve the following pair of linear equations graphically.

$x + 3y = 6$; $2x - 3y = 12$

Also find the area of the triangle formed by the lines representing the given equations with y-axis.

[CBSE (CCE) 2012]

NCERT EXEMPLAR PROBLEMS

SHORT ANSWER QUESTIONS WITH REASONING

1. Do the following pair of linear equations have no solution? Justify your answer.

(*i*) $x = 2y$; $y = 2x$

(*ii*) $3x + y - 3 = 0$; $2x + \frac{2}{3}y = 2$

2. Do the following equations represent a pair of coincident lines ? Justify your answer.

(*i*) $3x + \frac{1}{7}y = 3$; $7x + 3y = 7$

(*ii*) $\frac{x}{2} + y + \frac{2}{5} = 0$; $4x + 8y + \frac{5}{16} = 0$

3. Are the following pair of linear equations consistent ? Justify your answer.

(*i*) $\frac{3}{5}x - y = \frac{1}{2}$; $\frac{1}{5}x - 3y = \frac{1}{6}$

(*ii*) $2ax + by = a$, $4ax + 2by - 2a = 0$; $a, b \neq 0$

4. For the pair of equations

$\lambda x + 3y = -7$; $2x + 6y = 14$

to have infinitely many solutions, the value of λ should be 1. Is the statement true ? Give reasons.

5. The line represented by $x = 7$ is parallel to the x-axis. Justify whether the statement is true or not.

SHORT ANSWER QUESTIONS

6. For which value(s) of λ, do the pair of linear equations $\lambda x + y = \lambda^2$ and $x + \lambda y = 1$ have

(*i*) no solution ?

(*ii*) infinitely many solutions ?

(*iii*) a unique solution ?

7. If $2x + y = 23$ and $4x - y = 19$, find the values of $5y - 2x$ and $\frac{y}{x} - 2$.

8. Find the values of x and y in the following rectangle :

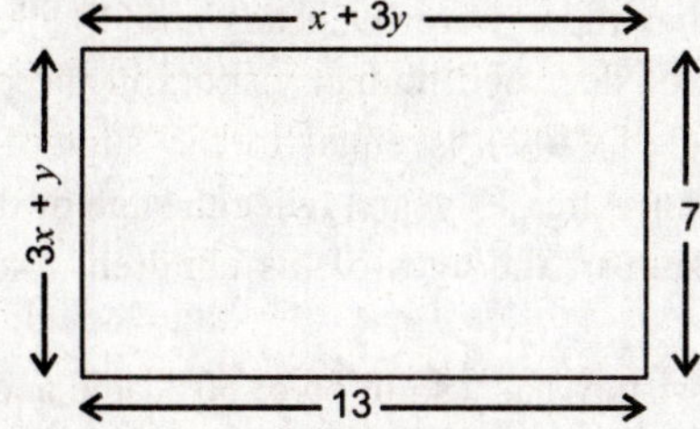

9. Solve the following pairs of equations :

(*i*) $x + y = 3.3$; $\frac{0.6}{3x - 2y} = -1$, $3x - 2y \neq 0$

(*ii*) $4x + \frac{6}{y} = 15$; $6x - \frac{8}{y} = 14$, $y \neq 0$

10. Find the solution of the pair of equations $\frac{x}{10} + \frac{y}{5} - 1 = 0$ and $\frac{x}{8} + \frac{y}{6} = 15$. Hence, find λ, if $y = \lambda x + 5$.

11. A shopkeeper gives books on rent for reading. She takes a fixed charge for the first two days, and an additional charge for each day thereafter. Latika paid ₹ 22 for a book kept for six days, while Anand paid ₹ 16 for the book kept for four days. Find the fixed charges and the charge for each extra day.

LONG ANSWER QUESTIONS

12. Draw the graphs of the equations $x = 3$, $x = 5$ and $2x - y - 4 = 0$. Also find the area of the quadrilateral formed by the lines and the x-axis.

13. A two-digit number is obtained by either multiplying the sum of the digits by 8 and then subtracting 5 or by multiplying the difference of the digits by 16 and then adding 3. Find the number.

14. Vijay had some bananas, and he divided them into two lots A and B. He sold the first lot at the rate of ₹ 2 for 3 bananas and the second lot at the rate of ₹ 1 per banana, and got a total of ₹ 400. If he had sold the first lot at the rate of ₹ 1 per banana, and the second lot at the rate of ₹ 4 for 5 bananas, his total collection would have been ₹ 460. Find the total number of bananas he had.

HIGHER ORDER THINKING SKILLS (HOTS) QUESTIONS

1. Solve the following systems of linear equation :

(*i*) $a(x + y) + b(x - y) = a^2 - ab + b^2$

$a(x + y) - b(x - y) = a^2 + ab + b^2$

(*ii*) $\frac{x}{a} + \frac{y}{b} = a + b$, $\frac{x}{a^2} + \frac{y}{b^2} = 2$

2. Solve for u and v :

$2(3u - v) = 5uv$

$2(u + 3v) = 5uv$

3. Solve for x and y :

$$3x - \frac{y+7}{11} + 2 = 10,\ 2y + \frac{x+11}{7} = 10$$

4. Solve for x and y : $x - y = 0.9, \frac{11}{x+y} = 2$

5. Solve for x and y :

$$148x + 231y = 527$$
$$231x + 148y = 610$$

6. Solve for x and y :

$$631x + 279y = 910$$
$$279x + 631y = 910$$

7. Solve for x and y :

$$254x + 309y = -55$$
$$309x + 254y = 55$$

8. Solve for x and y :

$$\frac{148}{x} + \frac{231}{y} = \frac{527}{xy}$$
$$\frac{231}{x} + \frac{148}{y} = \frac{610}{xy},\ x \neq y,\ y = 0$$

9. Solve for x and y :

$$3x + 2y = 2x + y + 3 = 4x + 3y - 3$$

10. Solve for x and y :

$$6x + 3y = 8x + 9y - 5 = 10x + 12y - 8$$

11. A railway half ticket costs half the full fare and the reservation charge is the same on half ticket as on full ticket. One reserved first class ticket from Mumbai to Ahemdabad costs ₹ 216 and one full and one half reserved first class tickets costs ₹ 327. What is the basic first class full fare and what is the reservation charge ?

12. Solve for x and y :

(*i*) $bx + ay = a + b$,

$$ax\left(\frac{1}{a-b} - \frac{1}{a+b}\right) + by\left(\frac{1}{b-a} - \frac{1}{b+a}\right) = \frac{2a}{a+b}$$

(*ii*) $x\left(a - b + \frac{ab}{a-b}\right) = y\left(a + b - \frac{ab}{a+b}\right),\ x + y = 2a^2$

13. A man walks a certain distance with certain speed. If he walks 1/2 km/hr. faster, he takes 1 hr. less. But if he walks 1 km/hr. slower, he takes 3 more hours. Find the distance covered by the man and his original rate of walking.

14. A boat goes 24 km upstream and 28 km downstream in 6 hrs. It goes 30 km upstream and 21 km downstream in $6\frac{1}{2}$ hrs. Find the speed of the boat in still water and also speed of stream.

15. The area of a rectangle remains the same if the length is increased by 7 m and the breadth is decreased by 3 m. The area remains unaffected also if the length is decreased by 7 m and breadth is increased by 5 m. Find the dimensions of the rectangle.

16. A man wished to give ₹ 12 to each person and found that he fell short of ₹ 6 when he wanted to give to all the persons present. He, therefore, distributed ₹ 9 to each person and found that ₹ 9 were left over. How much money did he have and how many persons were there ?

17. The expenses of a lunch are partly constant and partly proportional to the number of guests. The expenses amount to ₹ 65 for 7 guests and ₹ 97 for 11 guests. How much the expenses for 18 guests will amount to ?

18. The sum total of the ages of father and son is 55 years. If the father was to live till his son's age equals his present age, the total of their ages would be 93 years. Find their present ages.

19. A number consists of three digits whose sum is 17. The middle one exceeds the sum of the other two by 1. If the digits be reversed, the number is diminished by 396. Find the number.

20. Two candles of equal height but different thickness are lighted. The first burns off in 6 hours and the second in 8 hours. How long, after lighting both, will the first candle be half the height of the second ?

21. A person invested some amount at the rate of 10% simple interest and some other amount at the rate of 12% simple interest. He received yearly interest of ₹ 130. But if he had interchanged the amounts invested, he would have received ₹ 4 more as interest. How much amount did he invest at different rates ?

22. In a deer park, the number of heads and the number of legs of deer and human visitors were counted and it was found that there were 39 heads and 132 legs. Find the number of deer and human visitors in the park.

23. The age of father is equal to the sum of ages of his 6 children. After 15 years, twice the age of the father will be the sum of the ages of his children. Find the age of father.

24. A lab assistant has a solution of 50% acid and other which has 25% acid. How much of each should be mixed to make 10 litres of a 40% acid solution ?

25. Solve for x and y :

$$\frac{2}{3x+2y} + \frac{3}{3x-2y} = \frac{17}{5}$$
$$\frac{5}{3x+2y} + \frac{1}{3x-2y} = 2$$

26. Draw the graph of the following equations $2y - x = 8$, $5y - x = 14$, $y - 2x = 1$ and obtain the vertices of the triangle.

27. The result of dividing a number of two digits by the number with the digits reversed is 7/4. If the sum of the digits is 12, find the number.

28. If from twice the greater of two positive numbers 16 is subtracted, the result is half the other number. If from half the greater number 1 is subtracted, the result is still half the other number. What are the numbers ?

29. After covering a distance of 30 km with uniform speed, there occurs some defect in a train engine and thereafter, its speed is reduced to 4/5 of its original speed. Consequently, the train reaches its destination late by 45 minutes. Had it happened after covering 18 km more, the train would have reached 36 minutes late. Find the speed of the train and distance of the journey.

30. The ages of two friends Sunaina and Tanishtha differ by 2 years. Sunaina's father is twice as old as Sunaina and Tanishtha is twice as old as his brother Shiva. The ages of Sunaina's father and Shiva differ by 40 years. Find the ages of Sunaina and Tanishtha.

VALUE BASED QUESTIONS

1. Mohit and Sahil are driving on two roads represented by the equation $2x + 3y = 7$ and $4x + 6y = 12$.
They drive within the speed limit.
(*a*) Will they meet at some point ? Justify your answer.
(*b*) What value is depicted by Mohit and Sahil ?

2. 5 books and 7 pens together cost ₹ 285. 4 books and 4 pens together cost ₹ 220. Monu purchased 3 books and 5 pens and calculated total cost to be ₹ 195. He paid ₹ 195 to the shopkeeper. Shopkeeper rechecked and returned some money to Monu.
(*a*) How much money did shopkeeper returned to Monu?
(*b*) What value is depicted by the shopkeeper ?

3. A test consists of 'True' or 'False' questions. One mark is awarded for every correct answer while $\frac{1}{4}$ mark is deducted for every wrong answer. A student knew answers to some of the questions. Rest of the questions he attempted by cheating. He answered 120 questions and got 90 marks. If answer to all questions he attempted by cheating were wrong, then how many questions did he answer correctly ? How the habit of cheating will affect his character building ?

4. A man wished to donate some money to a group of poor people. he decided to give ₹ 120 to each person and found that he fell short of ₹ 60, when he wanted to give to all the people present. He, therefore, distributed ₹ 90 to each person and found that ₹ 90 were left over. How much money did he have and how many people were there ?
What value is depicted by the man ?

5. A trust has ₹ 30,000. It is invested in two different types of bonds. The first bond pays 5% interest per annum which will be given to orphanage and second bond pays 7% interest per annum which will be given to an N.G.O. for cancer aid. Trust obtains an annual interest of ₹ 1800.
(*a*) How much amount is invested in each type of bond ?
(*b*) Which value is depicted by trust ?

Multiple Choice Questions

PREVIOUS YEARS' QUESTIONS

1. One equation of a pair of dependent linear equations is $-5x + 7y = 2$, the second equation can be : [CBSE (CCE) 2011]
(*a*) $10x + 14y + 4 = 0$ (*b*) $-10x - 14y + 4 = 0$
(*c*) $-10x + 14y + 4 = 0$ (*d*) $10x - 14y = -4$

2. The pair of linear equations $2x - 3y = 1$ and $3x - 2y = 4$ has : [CBSE (CCE) 2011]
(*a*) One solution (*b*) Two solutions
(*c*) No solution (*d*) Many solutions

3. If $x = a$, $y = b$ is the solution of the pair of equation $x - y = 2$ and $x + y = 4$, then the respective values of a and b are [CBSE (CCE) 2012]
(*a*) 3, 5 (*b*) 5, 3
(*c*) 3, 1 (*d*) −1, −3

4. Two lines are given to be parallel. The equation of one of the lines is $4x + 3y = 14$. The equation of the second line can be [CBSE (CCE) 2012]
(*a*) $3x + 4y = 14$ (*b*) $8x + 6y = 28$
(*c*) $12x + 9y = 42$ (*d*) $-12x = 9y$

5. The pair of equations $x = 4$ and $y = 3$ graphically represents lines which are [CBSE (CCE) 2012]
(*a*) parallel (*b*) intersecting at (3, 4)
(*c*) coincident (*d*) intersecting at (4, 3)

PRACTICE QUESTIONS

6. $ax + by + c = 0$ where a, b, c are real numbers is called a linear equation in two variables x and y if
(*a*) $a \neq b$ (*b*) $a^2 = b^2$
(*c*) $a^2 + b^2 = 0$ (*d*) $a^2 + b^2 \neq 0$

7. Every solution of $ax + by + c = 0$ is a pair of values
(*a*) one for a and other for b
(*b*) one for a and other for c
(*c*) one for b and other for c
(*d*) one for x and other for y

8. Graphically $ax + by + c = 0$ represents a line. Every solution of the equation is a point

(a) on the line representing it
(b) not on the line representing it
(c) on the x-axis
(d) on the y-axis

9. Every point on the line representing the linear equation in two variables

(a) may not be a solution of the equation
(b) is a solution of the equation
(c) is a solution if it is also a point on x-axis
(d) is a solution of the equation if it is also a point on y-axis

10. $a_1x + b_1y + c_1 = 0$ and $a_2x + b_2y + c_2 = 0$, where $a_1, b_1, c_1, a_2, b_2, c_2$ are all real numbers and $a_1^2 + b_1^2 \neq 0, a_2^2 + b_2^2 \neq 0$, is called a

(a) family of two different straight lines
(b) family of two coincident lines
(c) pair of linear equations in two variables
(d) none of these

11. For given two lines in a plane, which of the following is not possible ?

(a) the two lines will intersect at a point
(b) the two lines will be parallel
(c) the two lines will be coincident
(d) the two lines will neither intersect nor parallel

12. Match the Column :

(1) $2x + 3y = 40$ $6x + 5y = 10$	**(A) Coincident lines**
(2) $2x + 3y = 40$ $6x + 9y = 50$	**(B) Intersecting lines**
(3) $2x + 3y = 10$ $4x + 6y = 20$	**(C) Parallel lines**

(a) 1 – A, 2 – B, 3 – C (b) 1 – B, 2 – A, 3 – C
(c) 1 – B, 2 – C, 3 – A (d) 1 – C, 2 – A, 3 – B

13. Aseem went to a stationery shop and purchased 3 pens and 5 pencils for ₹ 40. His cousin Manik bought 4 pencils and 5 pens for ₹ 58. If cost of 1 pen is ₹ x and 1 pencil is ₹ y, then which of the following represent the situation algebraically ?

(a) $3x + 5y = 40, 4x + 5y = 58$
(b) $3x + 4y = 40, 5x + 5y = 58$
(c) $3x + 5y = 40, 5x + 4y = 58$
(d) $3x + 5y = 40, 4x + 3y = 58$

14. Match the Column :

(1) $2x + 5y = 10$ $3x + 4y = 7$	**(A) Unique solution**
(2) $2x + 5y = 10$ $6x + 15y = 20$	**(B) Infinitely many solutions**
(3) $5x + 2y = 10$ $10x + 4y = 20$	**(C) No common solution**

(a) 1 – A, 2 – B, 3 – C (b) 1 – B, 2 – C, 3 – A
(c) 1 – C, 2 – B, 3 – A (d) 1 – A, 2 – C, 3 – B

15. Match the Column :

(1) $2x + 5y = 7$ $3x + 4y = 7$	**(A) Inconsistent pair of equations**
(2) $2x + 5y = 7$ $4x + 10y = 7$	**(B) Consistent pair of equations**
(3) $2x + 5y = 7$ $4x + 10y = 14$	**(C) Dependent consistent pair of equations**

(a) 1 – A, 2 – C, 3 – B (b) 1 – B, 2 – C, 3 – A
(c) 1 – B, 2 – A, 3 – C (d) 1 – C, 2 – A, 3 – B

16. The pair of linear equations $x = y$ and $x + y = 0$ has

(a) no common solution (b) infinitely many solutions
(c) unique solution (d) none of these

17. The pair of linear equations $x = 2$ and $x = 5$ has

(a) no common solution (b) infinitely many solutions
(c) unique solution (d) none of these

18. The pair of linear equations $3x + 4y = 12$ and $y = 4$ has

(a) no common solution (b) infinitely many solutions
(c) unique solution (d) none of these

19. The pair of linear equations $2x + 3y = 5$ and $4x + 6y = 10$ is

(a) inconsistent (b) consistent
(c) dependent consistent (d) none of these

20. There are two positive numbers such that sum of twice the first and thrice the second is 39, while the sum of thrice the first and twice the second is 36. The larger of the two is

(a) 6 (b) 8
(c) 9 (d) 10

21. The difference between a two digit number and the number obtained by interchanging the digits is 27. What is the difference between the two digits of the number?

(a) 9 (b) 6
(c) 12 (d) 3

22. Which of the following can not be the difference between a two digit number and the number obtained by interchanging the digits ?

(a) 72 (b) 36
(c) 54 (d) 48

23. The area of a trapezium is 1400 cm². Its altitude is 50 cm. Find the two bases, if the number of cms in each base is an integer divisible by 9. The number of solution to this problem is

(a) one (b) two
(c) three (d) none of these

24. For what value of k, the pair of equations $2x + 3y + 5 = 0$ and $kx + 4y = 10$, has a unique solution?

(a) $k = \frac{8}{3}$ (b) $k \neq \frac{8}{3}$
(c) $k = 3$ (d) $k \neq 3$

25. **If the pair of equations $x + y = \sqrt{2}$ and $x \sin\theta + y \cos\theta = 1$ has infinitely many solutions, then θ =**

(a) 30° (b) 45°
(c) 60° (d) 90°

26. **The value of k for which the lines $3x + 4y = 5$, $5x + 4y = 4$ and $kx + 4y = 6$ meet at a point is**

(a) 1 (b) 2
(c) 3 (d) 4

27. **The equations $ax + by + c = 0$ and $dx + ey + c = 0$ represent the same straight line if**

(a) $ad = be$ (b) $ac = bd$
(c) $bc = ad$ (d) $ab = de$

28. **The pair of equations $ax + 2y = 7$ and $3x + by = 16$ represent parallel lines if**

(a) $a = b$ (b) $3a = 2b$
(c) $2a = 3b$ (d) $ab = 6$

29. **The value of a for which the lines $x = 1$, $y = 2$ and $ax + 2y - 6 = 0$ are concurrent is**

(a) 1 (b) 2
(c) 3 (d) 4

30. **Equation of line $L_1 : a_1 x + b_1 y + c_1 = 0$,**
Equation of line $L_2 : a_2 x + b_2 y + c_2 = 0$,
Equation of line
$L_3 : (a_1x + b_1y + c_1) + (a_2x + b_2y + c_2) = 0$
if $\frac{a_1}{a_2} = \frac{b_1}{b_2} = \frac{c_1}{c_2}$, then line L_3 is

(a) parallel to line L_1 (b) parallel to line L_2
(c) is coincident with L_2 or L_1 (d) none of these

31. **A two digit number is k times the sum of its digits. The number formed by interchanging the digits is the sum of digits multiplied by**

(a) $9 - k$ (b) $11 - k$
(c) $k - 1$ (d) $k + 1$

32. **Given the pair of equations $ax + (a - 1) y = 1$ and $(a + 1)x - ay = 1$, for which one of the following values of a is there no common solution of the given pair of equations ?**

(a) $\pm\sqrt{2}$ (b) $\pm\frac{1}{\sqrt{2}}$
(c) ± 2 (d) $1 \pm \frac{1}{2}$

33. **$y = a + \frac{b}{x}$ where a, b are real numbers, if $y = 1$ when $x = -1$ and $y = 5$ when $x = -5$, then $a + b$ equals**

(a) – 1 (b) 0
(c) 11 (d) 10

NCERT Exemplar Problems

1. The pair of equations $y = 0$ and $y = -7$ has
   (a) one solution
   (b) two solutions
   (c) infinitely many solutions
   (d) no solution
2. The pair of equations $x = a$ and $y = b$ graphically represents lines which are
   (a) parallel
   (b) intersecting at (b, a)
   (c) coincident
   (d) intersecting at (a, b)
3. A pair of linear equations which has a unique solution $x = 2$, $y = -3$ is
   (a) $x + y = -1$; $2x - 3y = -5$
   (b) $2x + 5y = -11$; $4x + 10y = -22$
   (c) $2x - y = 1$; $3x + 2y = 0$
   (d) $x - 4y - 14 = 0$; $5x - y - 13 = 0$
4. The father's age is six times his son's age. Four years hence, the age of the father will be four times his son's age. The present ages, in years, of the son and the father are, respectively
   (a) 4 and 24 (b) 5 and 30
   (c) 6 and 36 (d) 3 and 24

ANSWERS

Very Short Answer Type Questions

1. $k \neq -10$ **2.** $k = -10$ **3.** Infinitely many solutions **4.** Consistent
5. $x + 2y = 1300$, $3x + 6y = 3900$ **6.** $2x + y = 160$, $4x + 2y = 300$ **7.** Infinitely many solutions
8. $3x + 4y + k = 0$ where k is any real number and $k \neq -8$ **9.** $x - 7y + 42 = 0$; $x - 3y - 6 = 0$ **10.** $c = 6$
11. Yes, consistent **12.** $p \neq 4$ **13.** 24 and 11 **14.** $k \neq 3$ **15.** 36 m, 24 m
16. Consistent and have a unique solution **17.** 100°, 80° **18.** 20°, 60°, 100°

Short Answer Type Questions-I

1. $x = 2, y = -1$ **Or** $x = b, y = -a$ **2.** $x = -a, y = b$ **Or** 57
3. $x = a^2, y = -b^2$ **Or** Cost of table = ₹ 700, Cost of chair = ₹ 200
4. $x = m + n, y = m - n$ **Or** Speed of train = 100 km/h. Speed of taxi = 80 km/h. **5.** $x = a^2, y = b^2$ **Or** $\frac{5}{7}$

6. $x = 3, y = 2$ **Or** $x = 7, y = 13$ **7.** $x = 3/2, y = 2/3$ **Or** $x = 1, y = -4$ **8.** $k = \pm 6$
9. coincident lines **10.** $m = 5$ **11.** $k = 11$ **12.** $p = 6$ **13.** $a = 5, b = 1$
14. $k = -6$ **15.** $x = 3, y = 2$ **16.** $p = 2/5$ **17.** $k = -2.25$ **18.** $k \neq -2/3$
19. $k = 6/5$ **20.** Unique Solution **21.** $k = 8$ **22.** $k = -10/3$ **23.** $k = 6$
24. $x = 2, y = 2$ **25.** Infinite solution equations : (*i*) and (*iii*), Unique solution equations : (*i*) and (*ii*)
26. (*i*) $x = a, y = b$ (*ii*) $x = a, y = b$ **27.** $a = -5, b = -1$ **28.** 78 **29.** 42 years, 10 years
30. 7/11 **31.** $a = 0, b = 7$ **33.** $x = 0.4, y = 0.6$ **34.** $x = 5, y = 1$ **35.** $k = 4$
36. There is no value of k **37.** $k \neq -3$ **38.** $k = 2$ **39.** No solution
40. $x = 1/2, y = 1/3$ **41.** $x = 1, y = 1$ **42.** $x = 1, y = 1$ **43.** $x = 2, y = 2, z = -1$
44. $x = 2, y = 3, z = 4$ **45.** $x = (mp - nq)/ap, y = (mp - nq)/aq$ **46.** $x = 2, y = 3$ **47.** $x = 1, y = 2$
48. $x = 1/3, y = -1$ **49.** $x = 0, y = 0$ **50.** $x = 0, y = 0$ **51.** No solution **52.** $c = -6$
53. (*i*) $k \neq -10/3$ (*ii*) $k = -10/3$ **54.** $a = 1, b = -7$ **55.** No value of k **56.** $a = -6$
57. No value of a **58.** k is any real no. **59.** $k \neq \pm 6$ **60.** (*i*) $x = 1, y = -1$ (*ii*) $x = a, y = b$

Short Answer Type Questions-II

1. (–3, 0), (2, 0), (3, 4) **2.** (0, 3), (3, 2), (0, –2) **3.** $x = 3, y = 4$ **4.** $x = 2, y = 3$
5. $x = a + b, y = \frac{-2ab}{a+b}$ **Or** $x = 1, y = 2$ **6.** (0, 2), (0 – 4) **7.** $x = 4, y = 5$ **Or** $a = 5, b = 1$
8. $x = 8, y = 3$ **9.** $k = 5$ **10.** $x = 3, y = 2$ **11.** $x = b, y = -a$
12. $\frac{4}{7}$ **Or** $x = 2, y = 2$ **13.** $x = ab, y = ab$ **14.** $x = -\frac{1}{2}, y = 2$ **Or** 63 **15.** 57 **16.** $x = 5, y = 0$
17. 5/7 **18.** man = 35 days, boys = 70 days **19.** train = 80 km/h, car = 60 km/h
20. $x = a^2, y = b^2$ **21.** ₹ 1700 **22.** ₹ 20 **23.** (1, 3), (–4, 2) and (2, 5)
24. 96 **25.** 93 **26.** ₹ 1200 **27.** 53
28. 96 **Or** Length = 40, Breadth = 30 **29.** $x = 2, y = 2$ **30.** $x = 6, y = 2$
31. (–4, 2), (1, 3), (2, 5) **32.** $x = -5/4, y = -1/4$ **33.** A = 10/3 km/hr., B = 5 km/hr. **34.** 720 km.
35. One man = 140 days, One boy = 280 days **36.** Tea set = ₹ 100 **37.** 96
38. 1 **39.** ₹ 3.50, ₹ 1.80 **40.** ₹ 160, ₹ 15 **41.** 25, 18
42. 100°, 80° **43.** ₹ 200 **44.** 45 years **45.** $22\frac{1}{2}°, 67\frac{1}{2}°$ and 90°
46. Downhill – 10 hours; Uphill – 5 hours **47.** 14 rabbits **48.** Buffaloes = 30 ; Hens = 30
49. 17 years and 15 years, 46 years **50.** Father's age = 40 yrs, Son's age = 15 yrs
51. 18/17 **52.** ∠A = 30°, ∠B = 90°, ∠C = 60° **53.** 5% : ₹ 5,000, 7% : ₹ 25,000
54. 18 **55.** 36 **56.** 95 or 59
57. ∠A = 65°, ∠B = 55°, ∠C =115°, ∠D =125° **58.** 80 km/hr, 70 km/hr. **59.** 253 square units
60. ₹ 1300 and ₹ 50 **61.** $x = \frac{a^2c^2 - b^2d^2}{a^4 - b^4}, y = \frac{a^2d^2 - b^2c^2}{a^4 - b^4}$ **62.** ₹ 1.50
63. X's income = ₹ 6000, Y's income = ₹ 5250 **64.** A = 34, B = 62 **65.** 100, 80
66. 50 paise coin = 25, 25 paise coin = 69
67. (*i*) $x = -2, y = -1$, (–1/3, 0) (–3/2, 0), (*ii*) $x = 3/2, y = 1$, (3, 0), (2/3, 0), (*iii*) $x = 3, y = 5/2$, (14/3, 0) (–7, 0), (*iv*) $x = 5, y = 0$, (5, 0) (–5/2, 0) (*v*) no solution, (1/3, 0) (2.7, 0) (*vi*) no solution, (6, 0) (5, 0)
68. $x = -1, y = 1/6$ **69.** $x = 11, y = 8$ **70.** $x = -1/2, y = 1/4$ **71.** 36
72. (3, 4); yes; $3x - 4y = 1$ **73.** 12 sq. unit **74.** $x = 3, y = 2$ **75.** $x = 3, y = 2$
76. Speed of train 100 km/h, Speed of car 80 km/h **77.** ₹ 180 **78.** 45 years
79. 40 years, 10 years **80.** $x = 1/14, y = 1/6$ **81.** $x = 4, y = 5$ **82.** $x = 1/2, y = 1/4$
83. No solution **84.** Fixed Charges = ₹ 400, Cost of food = ₹ 30 per day
85. (*i*) $x = a + b, y = \frac{-2ab}{a+b}$ (*ii*) $x = 1, y = -1$ (*iii*) $x = \frac{c}{a+b} - \frac{b}{a^2 - b^2}, y = \frac{c}{a+b} + \frac{a}{a^2 - b^2}$ (*iv*) $x = \frac{a}{a^2 + b^2}, y = \frac{b}{a^2 + b^2}$
86. 6 km/hr., 4 km/hr. **87.** 20°, 40°, 120° **88.** 17 and 9 **89.** 36

Long Answer Type Questions

1. 5 sq. units **2.** $x = 5, y = 0$ **3.** $k = 2$ **4.** $x = 4, y = 5$
5. area = 18 sq. units.

NCERT Exemplar Problems

1. (*i*) No (*ii*) No **2.** (*i*) No (*ii*) No **3.** (*i*) Yes (*ii*) Yes **4.** No **5.** Not True

6. (*i*) ± 1 (*ii*) $1, \omega, \omega^2$ (*iii*) $\neq \pm 1$ **7.** 31, –5/7 **8.** 1, 4 **9.** (*i*) (1.2, 2.1), (*ii*) (3, 2)

10. $-\frac{1}{2}$ **11.** 10, 3 **12.** 8 sq. units **13.** 83 **14.** 500

Higher Order Thinking Skills (Hots) Questions

1. (*i*) $x = b^2/2a$, $y = (2a^2 + b^2)/2a$ (*ii*) $x = a^2, y = b^2$ **2.** $u = 2, v = 1$ **3.** $x = 3, y = 4$

4. $x = 3.2, y = 2.3$ **5.** $x = 2, y = 1$ **6.** $x = 1, y = 1$ **7.** $x = 1, y = -1$ **8.** $x = 1, y = 2$

9. Infinite solutions **10.** $x = \frac{1}{2}, y = \frac{2}{3}$ **11.** Full basic fare = ₹ 210, Reservation charge = ₹ 6

12. (*i*) $x = \frac{b^2 + a^2}{ab + b^2}, y = \frac{2b}{a+b}$ (*ii*) $\frac{a^3 - b^3}{a}, \frac{a^3 + b^3}{a}$ **13.** Distance = 36 km; Speed = 4 km/hr.

14. Speed of stream = 4 km/hr., Speed of boat = 10 km/hr. **15.** length = 28 m, breadth = 15 m

16. ₹ 54 and number of persons = 5 **17.** ₹ 153 **18.** 37 yrs, 18 yrs

19. 692 **20.** 4 hours 48 minutes **21.** ₹ 500 at 12%, ₹ 700 at 10% **22.** (27, 12)

23. 60 years **24.** 6 ltrs of 50%, 4 ltrs of 25% **25.** $x = 1, y = 1$ **26.** (–4, 2), (1, 3), (2, 5)

27. 84 **28.** 10, 8 **29.** 30 km/ hr and 120 km **30.** 26 yrs and 24 yrs.

Value Based Questions

1. (*a*) No (*b*) obey traffic rules **2.** (*a*) ₹ 20 (*b*) Honesty **3.** 96, violate the habit of honesty.

4. ₹ 540, 5 people; Concern for others **5.** (*a*) ₹ 15000 (*b*) helping poor people.

Multiple Choice Questions

1. (*d*) **2.** (*a*) **3.** (*c*) **4.** (*d*) **5.** (*d*) **6.** (*d*)

7. (*d*) **8.** (*a*) **9.** (*b*) **10.** (*c*) **11.** (*d*) **12.** (*c*)

13. (*c*) **14.** (*d*) **15.** (*c*) **16.** (*c*) **17.** (*a*) **18.** (*c*)

19. (*c*) **20.** (*c*) **21.** (*d*) **22.** (*d*) **23.** (*d*) **24.** (*b*)

25. (*b*) **26.** (*a*) **27.** (*d*) **28.** (*d*) **29.** (*b*) **30.** (*c*)

31. (*b*) **32.** (*b*) **33.** (*c*)

NCERT Exemplar Problems

1. (*d*) **2.** (*d*) **3.** (*b*) **4.** (*c*)

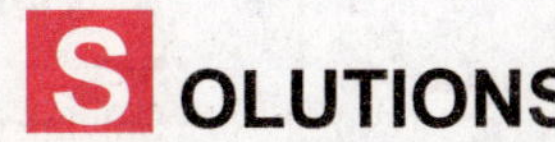

VERY SHORT ANSWER TYPE QUESTIONS

1. Here $a_1 = 3, b_1 = -1, c_1 = -5,$

and $a_2 = 6, b_2 = -2, c_2 = k.$

For no solution,

$$\frac{a_1}{a_2} = \frac{b_1}{b_2} \neq \frac{c_1}{c_2} \Rightarrow \frac{3}{6} = \frac{-1}{-2} \neq \frac{-5}{k}$$

$$\Rightarrow \frac{1}{2} \neq \frac{-5}{k} \Rightarrow k \neq -10$$

2. Here, $a_1 = 3, \ b_1 = -1, \ c_1 = -5$

$a_2 = 6, \ b_2 = -2, \ c_2 = k$

For infinite solution,

$$\frac{a_1}{a_2} = \frac{b_1}{b_2} = \frac{c_1}{c_2}$$

$$\frac{3}{6} = \frac{-1}{-2} = \frac{-5}{k}$$

$$\frac{1}{2} = \frac{-5}{k}$$

$$k = -10$$

$\therefore$ For $k = -10$, equations has infinite solutions.

3.
$$x + 2y - 8 = 0 \quad \text{... (i)}$$
$$2x + 4y - 16 = 0 \quad \text{... (ii)}$$

Here, $a_1 = 1, b_1 = 2, c_1 = -8$

and $a_2 = 2, b_2 = 4, c_2 = -16$

Now, $\frac{a_1}{a_2} = \frac{1}{2}, \frac{b_1}{b_2} = \frac{2}{4} = \frac{1}{2}, \frac{c_1}{c_2} = \frac{-8}{-16} = \frac{1}{2}$

$$\Rightarrow \frac{a_1}{a_2} = \frac{b_1}{b_2} = \frac{c_1}{c_2}$$

$\therefore$ Given pair of linear equations has infinite many solutions.

4. $x + y = 14$
$x - y = 4$

Here $\frac{a_1}{a_2} = 1, \frac{b_1}{b_2} = -1$

since $\frac{a_1}{a_2} \neq \frac{b_1}{b_2}$

$\Rightarrow$ The equations have unique solution.

$\therefore$ Pair of linear equations is consistent.

5. Let cost of 1 bat be ₹ x
and cost of 1 ball be ₹ y
A.T.Q., $3x + 6y = 3900$
$x + 2y = 1300$

6. Let cost of 1 kg of apples be ₹ x.
Let cost of 1 kg of grapes be ₹ y.
$\therefore$ A.T.Q.,
$$2x + 1y = 160$$
$$4x + 2y = 300$$

7. Here $a_1 = 5, b_1 = -8, c_1 = 1$

and $a_2 = 3, b_2 = \frac{-24}{5}, c_2 = \frac{3}{5}$

$$\frac{a_1}{a_2} = \frac{5}{3} \quad ...(i)$$

$$\frac{b_1}{b_2} = \frac{-8}{-24/5} = \frac{5}{3} \quad ...(ii)$$

and $$\frac{c_1}{c_2} = \frac{1}{3/5} = \frac{5}{3} \quad ...(iii)$$

From (*i*), (*ii*) and (*iii*)

$$\frac{a_1}{a_2} = \frac{b_1}{b_2} = \frac{c_1}{c_2}$$

$\therefore$ The pair of equations has infinitely many solutions.

8. Lines are parallel when $\frac{a_1}{a_2} = \frac{b_1}{b_2} \neq \frac{c_1}{c_2}$

$\therefore$ One of the linear equation in two variables can be
$6x + 8y + k = 0$ where k is constant not equal to -16.

9. Let present age of Aftab be x years.
and age of his daughter be y years.
A.T.Q., $x - 7 = 7(y - 7)$
$\Rightarrow$ $x - 7y = -42$ ...(*i*)
and $x + 3 = 3(y + 3)$
$\Rightarrow$ $x - 3y = 6$ ...(*ii*)

10. Here $a_1 = c, b_1 = 3, c_1 = 3 - c$
and $a_2 = 12, b_2 = c, c_2 = -c$
For infinite many solutions

$$\frac{a_1}{a_2} = \frac{b_1}{b_2} = \frac{c_1}{c_2}$$

$\Rightarrow$ $\frac{c}{12} = \frac{3}{c} = \frac{3-c}{-c}$

$\Rightarrow$ $\frac{c}{12} = \frac{3}{c}$ and $\frac{3}{c} = \frac{3-c}{-c}$

$\Rightarrow$ $c^2 = 36$ and $-3c = 3c - c^2$

$\Rightarrow$ $c = \pm 6$ and $c^2 - 6c = 0$

$\Rightarrow$ $c = \pm 6$ and $c = 0$ or $c = 6$

$\therefore$ $c = 6$

11. Since $\frac{a_1}{a_2} = \frac{1}{2}, \frac{b_1}{b_2} = \frac{3}{-3} = -1$

$\therefore$ $\frac{a_1}{a_2} \neq \frac{b_1}{b_2}$

$\therefore$ system is consistent.

12. For unique solution,

$$\frac{a_1}{a_2} \neq \frac{b_1}{b_2} \Rightarrow \frac{4}{2} \neq \frac{p}{2}$$

$\Rightarrow$ $p \neq 4$

Therefore, for all values of p other than 4, the given pair of equations has unique solution.

13. Let the two numbers be x and y.
A.T.Q.,
$x + y = 35$ ... (*i*)
$x - y = 13$ ... (*ii*)
Adding equation (*i*) and (*ii*), we get
$2x = 48$
$x = 24$
Putting value of x in (*i*), we get
$24 + y = 35$
$y = 35 - 24$
$y = 11$
So, numbers are 24 and 11.

14. Since system of equation has unique solution.

$\therefore$ $\frac{a_1}{a_2} \neq \frac{b_1}{b_2}$

i.e., $\frac{6}{k} \neq \frac{-2}{-1}$

$-6 \neq -2k$

$k \neq 3$

15. Let length be x m and width be y m
$\Rightarrow$ $x = y + 12$ ...(*i*)
Also, half the perimeter $= 60$
$\Rightarrow$ $x + y = 60$ ...(*ii*)
Solving (*i*) and (*ii*), we get $x = 36, y = 24$

16. Since, $a_1 = 1, b_1 = 2, c_1 = -4$
$a_2 = 2, b_2 = 1, c_2 = -5$

$\therefore$ $\frac{a_1}{a_2} = \frac{1}{2}, \frac{b_1}{b_2} = \frac{2}{1}$

$\Rightarrow$ $\frac{a_1}{a_2} \neq \frac{b_1}{b_2}$

$\therefore$ System of equations has unique solution.

17. Let larger angle be x and smaller be y
$\therefore$ $x + y = 180°$ ...(*i*)
Also, $x = y + 20°$ ...(*ii*)
Solving (*i*) and (*ii*), we get
$x = 100°$ and $y = 80°$

18. Let two angles are $5x$ and $3x$.

A.T.Q. 3rd angle $= \frac{1}{2}(5x - 3x) = x$

Now, $5x + 3x + x = 180°$

$\Rightarrow$ $9x = 180° \Rightarrow x = 20°$

Angles are 20°, 3 × 20°, 5 × 20° *i.e.,* 20°, 60°, 100°.

SHORT ANSWER TYPE QUESTIONS-I

1.
$$47x + 31y = 63 \quad ...(i)$$
$$31x + 47y = 15 \quad ...(ii)$$

Adding $78x + 78y = 78$

$\Rightarrow \quad x + y = 1 \quad ...(iii)$

Subtracting (*ii*) from (*i*), we get

$$16x - 16y = 48$$

$\Rightarrow \quad x - y = 3 \quad ...(iv)$

Solving (*iii*) and (*iv*), we get

$x = 2, y = -1$

Or

$$\frac{ax}{b} - \frac{by}{a} = a + b$$

$\Rightarrow \quad \dfrac{a^2x - b^2y}{ab} = a + b$

$\Rightarrow \quad a^2x - b^2y = ab(a + b)$

$\Rightarrow \quad a^2x - b^2y = a^2b + ab^2 \quad ...(i)$

and $\quad ax - by = 2ab \quad ...(ii)$

Multiply equation (*ii*) by *a* and subtracting (*ii*) from (*i*), we get

$$a^2x - b^2y = a^2b + ab^2$$
$$\underset{-}{a^2x} \underset{+}{-aby} = \underset{-}{2a^2b}$$
$$-b^2y + aby = a^2b + ab^2 - 2a^2b$$
$$by(a - b) = ab^2 - a^2b$$
$$yb(a - b) = ab(b - a)$$
$$y = \frac{ab(b-a)}{b(a-b)}$$
$$y = -a$$

Putting value of *y* in (*ii*), we get

$$ax - b(-a) = 2ab$$
$$ax = 2ab - ab$$
$$ax = ab$$
$$x = \frac{ab}{a} = b$$

The solution is $x = b, y = -a$

2.
$$\frac{bx}{a} - \frac{ay}{b} = -a - b$$

$\Rightarrow \quad \dfrac{b^2x - a^2y}{ab} = -a - b$

$\Rightarrow \quad b^2x - a^2y = -a^2b - ab^2 \quad ...(i)$

and $\quad bx - ay + 2ab = 0$

$\Rightarrow \quad bx - ay = -2ab \quad ...(ii)$

Multiply equation (*ii*) by *b* and subtracting (*ii*) from (*i*), we get

$$b^2x - a^2y = -a^2b - ab^2$$
$$\underset{-}{b^2x} \underset{+}{-aby} = \underset{+}{-2ab^2}$$
$$-a^2y + aby = -a^2b - ab^2 + 2ab^2$$
$$ay(b - a) = -a^2b + ab^2$$
$$ay(b - a) = ab(b - a)$$
$$y = \frac{ab(b-a)}{b(b-a)}$$
$$y = b$$

Putting value of *y* in (*ii*), we get

$$bx - a(b) = -2ab$$
$$bx = -2ab + ab$$
$$bx = -ab$$
$$x = -\frac{ab}{b} = -a$$

$\therefore$ The solution is $x = -a, y = b$.

Or

Let unit's digit be *x* and ten's digit be *y*.

A.T.Q.,

$$x + y = 12 \quad ...(i)$$

Number $= 10y + x$

Again A.T.Q.,

$$(10x + y) - (10y + x) = 18$$
$$9x - 9y = 18$$
$$x - y = 2 \quad ...(ii)$$

From (*i*) and (*ii*),

$$x - y = 2$$
$$x + y = 12$$
$$\therefore \quad x = 7, y = 5$$

Number = 57

3.
$$\frac{b^2x}{a} - \frac{a^2y}{b} = ab(a + b)$$

$\Rightarrow \quad \dfrac{b^3x - a^3y}{ab} = ab(a + b)$

$\Rightarrow \quad b^3x - a^3y = a^2b^2(a + b)$

$\Rightarrow \quad b^3x - a^3y = a^3b^2 + a^2b^3 \quad ...(i)$

and $\quad b^2x - a^2y = 2a^2b^2 \quad ...(ii)$

Multiplying equation (*ii*) by *b* and then subtracting from (*i*), we get

$$b^3x - a^3y = a^3b^2 + a^2b^3$$
$$\underset{-}{b^3x} \underset{+}{-a^2by} = \underset{-}{2a^2b^3}$$
$$a^2by - a^3y = a^3b^2 + a^2b^3 - 2a^2b^3$$
$$a^2y(b - a) = a^3b^2 - a^2b^3$$
$$ay(b - a) = a^2b^2(a - b)$$
$$y = \frac{a^2b^2(a-b)}{a^2(b-a)}$$
$$y = -b^2$$

Putting value of *y* in (*ii*), we get

$$b^2x - a^2(-b^2) = 2a^2b^2$$
$$b^2x = 2a^2b^2 - a^2b^2$$
$$x = \frac{a^2b^2}{b^2}$$
$$x = a^2$$

Or

Let the cost price of one table be ₹ *x* and one chair be ₹ *y*.

Loss on table = 10%

$\therefore$ S.P. of one table $= x - \dfrac{10}{100}x = \dfrac{9}{10}x$

Gain on chair = 10%

$\therefore$ S.P. of one chair $= y + \dfrac{10}{100}y = \dfrac{11}{10}y$

A.T.Q.,

$$\frac{9x}{10}+\frac{11}{10}y = 850 \quad ...(i)$$

Again,

Gain on table = 10% and loss on chair = 10% then total selling price = 950

Again A.T.Q., $\frac{11}{10}x+\frac{9}{10}y = 950 \quad ...(ii)$

Adding eq. (*i*) and (*ii*), we get

$$\frac{20}{10}x + \frac{20}{10}y = 1800$$

$$2x + 2y = 1800$$

$$2(x + y) = 1800$$

$$x + y = 900 \quad ...(iii)$$

Substracting eq. (*ii*) from (*i*),

$$-\frac{2x}{10} + \frac{2y}{10} = -100$$

$$-\frac{2}{10}(x - y) = -100$$

$$x - y = 500 \quad ...(iv)$$

Solving (*iii*) and (*iv*), we get $x = 700$, $y = 200$

4.
$$mx - ny = m^2 + n^2 \quad ...(i)$$

$$x + y = 2m \quad ...(ii)$$

Multiplying equation (*ii*) by *m* and subtracting from (*i*), we get

$$mx - ny = m^2 + n^2$$

$$\underset{-}{mx} + \underset{-}{my} = \underset{-}{2m^2}$$

$$-(m + n)y = m^2 + n^2 - 2m^2$$

$$y = \frac{n^2 - m^2}{-(m+n)}$$

$$y = -\frac{(n-m)(n+m)}{m+n}$$

$$y = m - n$$

Putting value of *y* in (*i*), we get

$$mx - n(m - n) = m^2 + n^2$$

$$mx - nm + n^2 = m^2 + n^2$$

$$mx = m^2 + nm$$

$$x = \frac{m(m+n)}{m}$$

$$x = m + n$$

The solution is $x = m + n$, $y = m - n$

Or

Let the speed of the train be *x* km/hr.

Speed of the taxi be *y* km/hr.

Case-I Time taken to cover distance 300 km by train and 200 km by taxi.

A.T.Q.,

$$\frac{300}{x}+\frac{200}{y} = 5\frac{1}{2} \text{ hours} \Rightarrow \frac{300}{x}+\frac{200}{y} = \frac{11}{2} \text{ hours} \quad ...(i)$$

Case-II Time taken to cover distance 260 km by train and 240 km by taxi.

A.T.Q.,

$$\frac{260}{x}+\frac{240}{y} = 5+\frac{36}{60} \Rightarrow \frac{260}{x}+\frac{240}{y} = \frac{28}{5} \quad ...(ii)$$

Let $\frac{1}{x} = A$, $\frac{1}{y} = B$

$\therefore$ Equation (*i*) becomes

$$300A + 200B = \frac{11}{2} \quad ...(iii)$$

and equation (*ii*) become

$$260A + 240B = \frac{28}{5}$$

$$\Rightarrow 240B = \frac{28}{5} - 260A$$

$$B = \frac{\left(\frac{28}{5} - 260A\right)}{240} \quad ...(iv)$$

$\therefore$ Equation (*iii*) becomes

$$300A + \frac{200\left(\frac{28}{5} - 260A\right)}{240} = \frac{11}{2}$$

$$\Rightarrow 300A + \frac{5\left(\frac{28}{5} - 260A\right)}{6} = \frac{11}{2}$$

$$\Rightarrow \frac{1800A + 28 - 1300A}{6} = \frac{11}{2}$$

$$500A + 28 = 33$$

$$500A = 5$$

$$\Rightarrow A = \frac{1}{100}$$

When $A = \frac{1}{100}$, equation (*iv*) becomes

$$B = \frac{\frac{28}{5} - 260 \times \frac{1}{100}}{240}$$

$$B = \frac{\frac{15}{5}}{240} \Rightarrow B = \frac{1}{80}$$

When $A = \frac{1}{100} \Rightarrow \frac{1}{x} = \frac{1}{100} \Rightarrow x = 100$

and when $B = \frac{1}{80} \Rightarrow \frac{1}{y} = \frac{1}{80} \Rightarrow y = 80$

$\therefore$ Speed of train = 100 km/h., speed of car = 80 km/h.

5.
$$\frac{a^2}{x} - \frac{b^2}{y} = 0$$

$$\frac{a^2b}{x} + \frac{b^2a}{y} = a + b$$

Let $\frac{1}{x} = u$, $\frac{1}{y} = v$

$$a^2u - b^2v = 0 \quad ...(i)$$

and
$$a^2bu + b^2av = a + b \quad ...(ii)$$

Multiply (*i*) by *b* and then subtract equation (*ii*) from it.

$$a^2bu - b^3v = 0$$

$$\underset{-}{a^2bu} + \underset{-}{b^2av} = \underset{-}{a} + \underset{-}{b}$$

$$-b^2av - b^3v = -a - b$$

$$-b^2v(a + b) = -(a + b)$$

$$v = \frac{(a+b)}{b^2(a+b)}$$

$$v = \frac{1}{b^2}$$

Putting value of v in equation (*i*), we get

$$a^2u - b^2\left(\frac{1}{b^2}\right) = 0$$

$$a^2u - 1 = 0$$

$$u = \frac{1}{a^2}$$

Since, $\frac{1}{x} = u$ and $\frac{1}{y} = v$

$\therefore \frac{1}{x} = \frac{1}{a^2}$ and $\frac{1}{y} = \frac{1}{b^2}$

$$x = a^2 \text{ and } y = b^2$$

The solution is $x = a^2$ and $y = b^2$

Or

Let numerator and denominator of fraction be x and y.

A.T.Q., $\quad x + y = 12 \quad \text{...}(i)$

and $\quad \frac{x}{y+3} = \frac{1}{2}$

$$2x = y + 3$$

or $\quad 2x - y = 3 \quad \text{...}(ii)$

Adding (*i*) and (*ii*), we get

$$x + y = 12$$
$$2x - y = 3$$
$$\overline{3x = 15}$$

$$x = \frac{15}{3}$$

$$x = 5$$

Putting value of x in equation (*i*), we get

$$y = 12 - x$$
$$y = 12 - 5$$
$$y = 7$$

The fraction is $\frac{5}{7}$.

6. $\quad x + \frac{6}{y} = 6 \quad \text{...}(i)$

$\quad 3x - \frac{8}{y} = 5 \quad \text{...}(ii)$

Multiplying equation (*i*) by 3, we get

$\quad 3x + \frac{18}{y} = 18 \quad \text{...}(iii)$

Subtracting (*ii*) from (*iii*), we get

$$3x + \frac{18}{y} = 18$$
$$3x - \frac{8}{y} = 5$$
$$\underline{- \quad + \quad -}$$

$$\frac{26}{y} = 13 \Rightarrow y = 2, x = 3$$

Or

$\quad \frac{x+1}{2} + \frac{y-1}{3} = 8 \quad \text{...}(i)$

$\quad \frac{x-1}{3} + \frac{y+1}{2} = 9 \quad \text{...}(ii)$

From (*i*)

$$\frac{x+1}{2} + \frac{y-1}{3} = 8 \Rightarrow \frac{3(x+1)+2(y-1)}{6} = 8$$

$\Rightarrow \quad 3x + 3 + 2y - 2 = 48 \Rightarrow 3x + 2y = 47 \quad \text{...}(iii)$

From (*ii*)

$$\frac{x-1}{3} + \frac{y+1}{2} = 9 \Rightarrow \frac{2(x-1)+3(y+1)}{6} = 9$$

$\Rightarrow \quad 2x - 2 + 3y + 3 = 54 \Rightarrow 2x + 3y = 53 \quad \text{...}(iv)$

Solving (*iii*) and (*iv*), we can get the required result as $x = 7$, $y = 13$.

7. $\quad 8x - 9y = 6xy, 10x + 6y = 19xy$

Dividing by xy, we get

$$\frac{8}{y} - \frac{9}{x} = 6 \text{ and } \frac{10}{y} + \frac{6}{x} = 19$$

Let $\frac{1}{y} = v$ and $\frac{1}{x} = u$

$\therefore$ Equation becomes

$\quad 8v - 9u = 6 \quad \text{...}(i)$

$\quad 10v + 6u = 19 \quad \text{...}(ii)$

Multiplying equation (*i*) by 2 and (*ii*) by 3 and then adding both, we get

$$16v - 18u = 12$$
$$30v + 18u = 57$$
$$\overline{46v = 69}$$

$$v = \frac{69}{46} = \frac{3}{2}$$

Putting value of v in equation (*i*), we get

$$8\left(\frac{3}{2}\right) - 9u = 6$$

$$12 - 9u = 6$$
$$-9u = 6 - 12$$
$$u = \frac{-6}{-9}$$
$$u = \frac{2}{3}$$

Since $\frac{1}{y} = v$ and $\frac{1}{x} = u$

$\frac{1}{y} = \frac{3}{2}$ and $\frac{1}{x} = \frac{2}{3}$

$$y = \frac{2}{3} \text{ and } x = \frac{3}{2}$$

The solution is $u = \frac{3}{2}$ and $y = \frac{2}{3}$

Or

$$4x + \frac{y}{3} = \frac{8}{3}, \frac{x}{2} + \frac{3y}{4} = -\frac{5}{2}$$

$$3\left(4x + \frac{y}{3} = \frac{8}{3}\right), 4\left(\frac{x}{2} + \frac{3y}{4} = -\frac{5}{2}\right)$$

$\quad 12x + y = 8 \quad \text{...}(i)$

$\quad 2x + 3y = -10 \quad \text{...}(ii)$

Multiply equation (*i*) by 3 and then subtract equation (*ii*) from (*i*).

$$36x + 3y = 24$$
$$2x + 3y = -10$$
$$\underline{- \quad - \quad +}$$
$$34x = 34$$
$$x = 1$$

Putting value of x in equation (*ii*), we get

$$2 \times 1 + 3y = -10$$
$$3y = -10 - 2$$

$$y = -\frac{12}{3}$$

$$y = -4$$

The solution is $x = 1, y = -4$

8. Since pair of linear equations has no solution

then $\frac{k}{12} = \frac{3}{k} \neq \frac{k-2}{k}$

i.e., $k^2 = 36 \Rightarrow k = \pm 6$

9. $18x - 7y = 24$

$$\frac{9}{5}x - \frac{7}{10}y = \frac{9}{10} \text{ or } \frac{18x-7y}{10} = \frac{9}{10}$$

Here $\frac{a_1}{a_2} = \frac{18}{18} = 1$

$$\frac{b_1}{b_2} = \frac{-7}{-7} = 1 \text{ and } \frac{c_1}{c_2} = \frac{24}{9} = \frac{8}{3}$$

Since $\frac{a_1}{a_2} = \frac{b_1}{b_2} \neq \frac{c_1}{c_2}$

$\Rightarrow$ The lines are parallel.

10. $2x + 3y - 7 = 0$

$$(m-1)x + (m+1)y = (3m-1)$$

For infinitely many solutions, the condition is

$$\frac{a_1}{a_2} = \frac{b_1}{b_2} = \frac{c_1}{c_2}$$

$$\frac{2}{m-1} = \frac{3}{m+1} = \frac{7}{3m-1}$$

$$\Rightarrow \frac{2}{m-1} = \frac{3}{m+1} \quad \text{or} \quad \frac{3}{m+1} = \frac{7}{3m-1}$$

$2(m+1) = 3(m-1)$, $\quad 3(3m-1) = 7(m+1)$

$2m + 2 = 3m - 3$, $\quad 9m - 3 = 7m + 7$

$2 + 3 = 3m - 2m$, $\quad 9m - 7m = 7 + 3$

$m = 5$, $\quad 2m = 10$

$m = 5$

Hence, for $m = 5$

$$\frac{a_1}{a_2} = \frac{b_1}{b_2} = \frac{c_1}{c_2}$$

11. Given equations are $2x + 3y = 9$

and $6x + (k-2)y = (3k-2)$

For no solution,

$$\frac{2}{6} = \frac{3}{k-2} \neq \frac{9}{3k-2} \quad ...(i)$$

Consider, $\frac{2}{6} = \frac{3}{k-2}$

$\Rightarrow 2k - 4 = 18 \Rightarrow 2k = 22 \Rightarrow k = 11$

From (*i*), we have $\frac{2}{6} = \frac{3}{9} \neq \frac{9}{31}$

Hence, for $k = 11$, pair of equations has no solution.

12. Consider equations $(p-3)x + 3y = p$

and $px + py = 12$

For infinitely many solutions,

$$\frac{p-3}{p} = \frac{3}{p} = \frac{p}{12} \quad ...(i)$$

Consider, $\frac{3}{p} = \frac{p}{12} \Rightarrow p^2 = 36 \Rightarrow p = \pm 6$

For $p = 6$, from (*i*)

$$\frac{3}{6} = \frac{3}{6} = \frac{6}{12}, \text{ true}$$

For $p = -6$, from (*i*)

$$\frac{-9}{-6} = \frac{3}{-6} = \frac{-6}{12}, \text{ false.}$$

Hence, for $p = 6$, pair of linear equations has infinitely many solutions.

13. Consider equations $2x + 3y = 7$

and $(a+b)x + (2a-b)y = 21$

For infinitely many solutions,

$$\frac{2}{a+b} = \frac{3}{2a-b} = \frac{7}{21} \quad ...(i)$$

$$\Rightarrow \frac{2}{a+b} = \frac{1}{3} \text{ and } \frac{3}{2a-b} = \frac{1}{3}$$

$$\Rightarrow a + b = 6 \text{ and } 2a - b = 9$$

Solving for a and b, we get

$$a = 5, b = 1$$

Substituting in (*i*), we get

$$\frac{2}{5+1} = \frac{3}{10-1} = \frac{7}{21}$$

$$\Rightarrow \frac{1}{3} = \frac{1}{3} = \frac{1}{3}, \text{ true}$$

Hence, for $a = 5$ and $b = 1$, pair of equations has infinite solutions.

14. $kx + 3y = 1$, $12x + ky = 2$, $a_1 = k, b_1 = 3, c_1 = -1$

$a_2 = 12, b_2 = k, c_2 = -2$

For no solution, $\frac{a_1}{a_2} = \frac{b_1}{b_2} \neq \frac{c_1}{c_2}$

$$\frac{k}{12} = \frac{3}{k} \neq \frac{1}{2} \Rightarrow k^2 = 36 \text{ and } \frac{3}{k} \neq \frac{1}{2}$$

$\Rightarrow k = \pm 6$ and $k \neq 6$, $\therefore k = -6$

15.

$$99x + 101y = 499 \quad ...(i)$$

$$101x + 99y = 501 \quad ...(ii)$$

Adding $200x + 200y = 1000$

$$x + y = 5 \quad ...(iii)$$

Also, $99x + 101y = 499$

$$101x + 99y = 501$$

$$- \quad - \quad -$$

$$-2x + 2y = -2$$

$$\Rightarrow x - y = 1 \quad ...(iv)$$

Solving (*iii*) and (*iv*) we get $x = 3, y = 2$

16. $(2p-1)x + (p-1)y = 2p + 1 \quad ...(i)$

$y + 3x = 1 \quad ...(ii)$

Now,

$$a_1 = 2p-1,\ b_1 = p-1,\ c_1 = -(2p+1)$$

$$a_2 = 1,\ b_2 = 3,\ c_2 = -1$$

For no solution,

$$\frac{a_1}{a_2} = \frac{b_1}{b_2} \neq \frac{c_1}{c_2}$$

$$\Rightarrow \quad \frac{2p-1}{1} = \frac{p-1}{3} \neq -\frac{2p+1}{1}$$

$$\frac{2p-1}{1} = \frac{p-1}{3} \text{ and } \frac{p-1}{3} \neq 2p+1$$

$$\Rightarrow \quad 6p-3 = p-1 \text{ and } p-1 \neq 6p+3$$

$$p = \frac{2}{5},\ p \neq -\frac{4}{5} \quad \therefore p = 2/5$$

17. $3x - 4y + 7 = 0$

$kx + 3y - 5 = 0$

For no solution, $\frac{a_1}{a_2} = \frac{b_1}{b_2} \neq \frac{c_1}{c_2}$

$$\frac{3}{k} = \frac{-4}{3} \neq \frac{7}{-5}$$

$$\Rightarrow \quad \frac{3}{k} = \frac{-4}{3}$$

$$\Rightarrow \quad -4k = 9$$

$$\Rightarrow \quad k = -\frac{9}{4} = -2.25$$

18. For unique solution,

$$\frac{a_1}{a_2} \neq \frac{b_1}{b_2}$$

i.e. $\frac{1}{3} \neq \frac{-k}{2}$

$$k \neq \frac{-2}{3}$$

19. For infinite many solutions

$$\frac{a_1}{a_2} = \frac{b_1}{b_2} = \frac{c_1}{c_2}$$

$$\therefore \quad \frac{4}{8} = \frac{1}{2} = \frac{-3}{-5k}$$

$$\frac{1}{2} = \frac{3}{5k}$$

$$5k = 6$$

$$k = \frac{6}{5}$$

20. $2x - 3y = 6$

Solution table for $2x - 3y = 6$ is

x	3	0	–3
y	0	–2	–4

Solution table for $x + y = 1$ is

x	1	0	2
y	0	1	–1

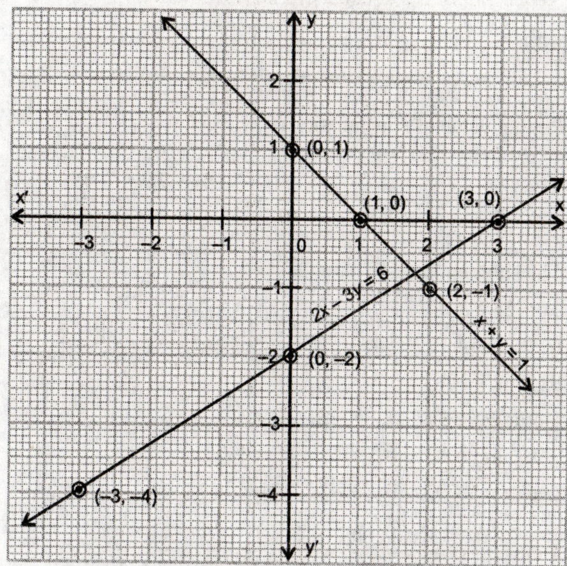

Clearly from the graph, these equations have unique solution.

21. For infinite number of solutions, $\frac{a_1}{a_2} = \frac{b_1}{b_2} = \frac{c_1}{c_2}$,

$$\frac{k}{16} = \frac{4}{k} = \frac{k-4}{k}$$

$$\Rightarrow \quad \frac{k}{16} = \frac{4}{k} \Rightarrow k^2 = 64$$

$$\Rightarrow \quad k = \sqrt{64} \Rightarrow k = \pm 8$$

Taking $\frac{4}{k} = \frac{k-4}{k}$

$$\Rightarrow \quad 4k = k(k-4)$$

$$\Rightarrow \quad 4k = k^2 - 4k$$

$$\Rightarrow \quad k^2 = 8k$$

$$\Rightarrow \quad k(k-8) = 0$$

$$\Rightarrow \quad k = 0 \text{ and } k = 8$$

$k = 8$ satisfies both the equality.

$\therefore$ It is solution.

22. For no solution,

$$\frac{a_1}{a_2} = \frac{b_1}{b_2} \neq \frac{c_1}{c_2} \Rightarrow \frac{k}{5} = \frac{2}{-3} \neq \frac{-1}{2} \Rightarrow k = \frac{-10}{3}$$

23. $(k-3)x + 3y = k$; $kx + ky = 12$

For infinite number of solutions, $\frac{a_1}{a_2} = \frac{b_1}{b_2} = \frac{c_1}{c_2}$

$$\Rightarrow \quad \frac{a_1}{a_2} = \frac{(k-3)}{k};\ \frac{b_1}{b_2} = \frac{3}{k};\ \frac{c_1}{c_2} = \frac{k}{12}$$

Taking $\frac{a_1}{a_2} = \frac{b_1}{b_2} \Rightarrow \frac{k-3}{k} = \frac{3}{k} \Rightarrow k = 6$ ...(*i*)

Taking $\frac{b_1}{b_2} = \frac{c_1}{c_2} \Rightarrow \frac{3}{k} = \frac{k}{12}$

$\Rightarrow \quad k^2 = 36 \Rightarrow k = \pm 6$ ...(*ii*)

From (*i*) and (*ii*), we get $k = 6$.

24. $x - y = 0$ ...(*i*)

$2x - y = 2$ ...(*ii*)

Subtracting equation (*ii*) from (*i*), we get

$$\begin{array}{r} x - y = 0 \\ 2x - y = 2 \\ - \quad + \quad - \\ \hline -x \quad = -2 \Rightarrow x = 2 \end{array}$$

Putting value of x in equation (*i*), we get

$$2 - y = 0$$

$$-y = -2$$

$$y = 2$$

The solution is $x = 2$, $y = 2$.

25. $3x - 2y = 4$ ...(*i*)

$6x + 2y = 4$ ...(*ii*)

$9x - 6y = 12$ ...(*iii*)

Here $a_1 = 3$, $b_1 = -2$, $c_1 = 4$,

$a_2 = 6$, $b_2 = 2$, $c_2 = 4$,

and $a_3 = 9$, $b_3 = -6$, $c_3 = 12$

Here $\frac{a_1}{a_3} = \frac{3}{9} = \frac{1}{3}$, $\frac{b_1}{b_3} = \frac{-2}{-6} = \frac{1}{3}$, $\frac{c_1}{c_3} = \frac{4}{12} = \frac{1}{3}$

$\Rightarrow \frac{a_1}{a_3} = \frac{b_1}{b_3} = \frac{c_1}{c_3}$

$\therefore$ Equations (*i*) and (*iii*) have infinite solutions.

Also $\frac{a_1}{a_2} = \frac{3}{6} = \frac{1}{2}$, $\frac{b_1}{b_2} = \frac{-2}{2} = -1$,

$\Rightarrow \frac{a_1}{a_2} \neq \frac{b_1}{b_2}$

$\therefore$ Equations (*i*) and (*ii*) have a unique solution.

26. (*i*) $x + y - (a + b) = 0$ ...(*i*)

$ax - by - a^2 + b^2 = 0$ ...(*ii*)

Multiplying equation (*i*) by a, we get

$ax + ay - a^2 - ab = 0$ ...(*iii*)

On subtracting eq. (*ii*) from eq. (*iii*), we get

$$\begin{array}{r} ax + ay - a^2 - ab = 0 \\ ax - by - a^2 + b^2 = 0 \\ - \quad + \quad + \quad - \\ \hline ay + by - b^2 - ab = 0 \end{array}$$

$\Rightarrow (a + b)y = ab + b^2$

$$y = \frac{b(a+b)}{(a+b)} = b$$

Now substituting the value of y in eq. (*i*), we get

$$x = a$$

(*ii*) $\frac{x}{a} + \frac{y}{b} - 2 = 0$

$\Rightarrow bx + ay - 2ab = 0$ ... (*i*)

$ax - by + b^2 - a^2 = 0$ ... (*ii*)

Equation (*i*) × a, we get

$abx + a^2y - 2a^2b = 0$ ... (*iii*)

Equation (*ii*) × b, we get $abx - b^2y + b^3 - a^2b = 0$

$$\begin{array}{r} - \quad + \quad - \quad + \\ \hline a^2y + b^2y - b^3 - a^2b = 0 \end{array}$$

Subtracting

$(a^2 + b^2)\,y = b^3 + a^2 b$

$(a^2 + b^2)\,y = b(b^2 + a^2)$

$\Rightarrow y = b$

When $y = b$, equation (*i*) becomes.

$bx + ab - 2ab = 0$

$bx = ab \Rightarrow x = a$

$\therefore x = a, y = b.$

27. Here $a_1 = 2$, $b_1 = -3$, $c_1 = -7$

and $a_2 = (a + b)$, $b_2 = -(a + b - 3)$, $c_2 = -(4a + b)$

For infinite solutions

$$\frac{a_1}{a_2} = \frac{b_1}{b_2} = \frac{c_1}{c_2}$$

$$\Rightarrow \frac{2}{a+b} = \frac{-3}{-(a+b-3)} = \frac{-7}{-(4a+b)}$$

Taking $\frac{2}{a+b} = \frac{3}{a+b-3}$

$\Rightarrow 2a + 2b - 6 = 3a + 3b$

$\Rightarrow a + b + 6 = 0$ ...(*i*)

and $\frac{3}{a+b-3} = \frac{7}{4a+b}$

$\Rightarrow 12a + 3b = 7a + 7b - 21$

$\Rightarrow 5a - 4b + 21 = 0$ ...(*ii*)

Solving (*i*) and (*ii*), we get

$a = -5$, $b = -1$

28. Let digit at unit's place be x

and digit at ten's place be y

$\therefore$ Number $= 10y + x$

Number obtained by interchanging the digits $= 10x + y$

According to the question

$x + y = 15$ ...(*i*)

and $10x + y = 10y + x + 9$

$\Rightarrow 9x - 9y = 9$

$\Rightarrow x - y = 1$ ...(*ii*)

Solving equation (*i*) and (*ii*), we get

$x = 8$ and $y = 7$

$\therefore$ Number is 78.

29. Let present age of father be x years

and age of son be y years

According to the question,

Two years ago, $x - 2 = 5(y - 2)$

$\Rightarrow x - 2 = 5y - 10$

$\Rightarrow x - 5y = -8$ ...(*i*)

and two years later, $x + 2 = 3(y + 2) + 8$

$\Rightarrow \quad x + 2 = 3y + 6 + 8$

$\Rightarrow \quad x - 3y = 12$ ...(*ii*)

Solving eq. (*i*) and (*ii*), we get

$x = 42$ yrs. and $y = 10$ yrs.

$\therefore$ Present age of father = 42 years

and present age of son = 10 years

30. Let numerator be x and denominator be y

$\therefore$ Fraction $= \dfrac{x}{y}$

According to the question,

$\dfrac{x+1}{y+1} = \dfrac{2}{3}$

$\Rightarrow 3x + 3 = 2y + 2$

$\Rightarrow 3x - 2y = -1$ ...(*i*)

and $\dfrac{x-1}{y-1} = \dfrac{3}{5}$

$\Rightarrow 5x - 5 = 3y - 3$

$\Rightarrow 5x - 3y = 2$ ...(*ii*)

Solving eq. (*i*) and (*ii*), we get

$x = 7, \; y = 11$

$\therefore$ Fraction is $\dfrac{7}{11}$.

31. Since $x + 3$ is factor of $p(x) = x^3 + ax^2 - bx + 6$

$\therefore \quad p(-3) = 0$

$(-3)^3 + a(-3)^2 - b(-3) + 6 = 0$

$-27 + 9a + 3b + 6 = 0$

$9a + 3b - 21 = 0$

$9a + 3b = 21$ ... (*i*)

and $a + b = 7$ ... (*ii*)

Multiply equation (*ii*) by 3 and then subtract equation (*ii*) from (*i*).

$$\begin{array}{r} 9a + 3b = 21 \\ 3a + 3b = 21 \\ - \quad - \quad\quad - \\ \hline 6a = 0 \end{array}$$

$a = 0$

Putting value of a in equation (*ii*), we get

$0 + b = 7$

$b = 7$

32. Here, $a_1 = 1, \quad b_1 = -5, \quad c_1 = 6$

$a_2 = 2, \quad b_2 = -10, \quad c_2 = 12$

$\dfrac{a_1}{a_2} = \dfrac{1}{2}, \dfrac{b_1}{b_2} = \dfrac{-5}{-10} = \dfrac{1}{2}, \dfrac{c_1}{c_2} = \dfrac{6}{12} = \dfrac{1}{2}$

$\Rightarrow \quad \dfrac{a_1}{a_2} = \dfrac{b_1}{b_2} = \dfrac{c_1}{c_2}$

Hence, the system of equations is consistent and dependent.

33. $\dfrac{x}{2} + y = 0.8 \quad \Rightarrow x + 2y = 1.6$

$\Rightarrow \quad x = 1.6 - 2y$ ...(*i*)

and $\dfrac{7}{x + \dfrac{y}{2}} = 10 \quad \Rightarrow \dfrac{14}{2x + y} = 10$

$\Rightarrow \quad 20x + 10y = 14 \Rightarrow 10x + 5y = 7$ ...(*ii*)

Using equation (*i*), we get

$10(1.6 - 2y) + 5y = 7 \Rightarrow 16 - 20y + 5y = 7$

$\Rightarrow \quad -15y = -9$

$y = \dfrac{9}{15} = 0.6$

When $y = 0.6$, equation (*i*) becomes $x = 1.6 - 2 \times 0.6 = 0.4$

34. $7(y + 3) - 2(x + 2) = 14$

$\Rightarrow \quad 7y + 21 - 2x - 4 = 14$

$\Rightarrow \quad 7y - 2x = -3$

$\Rightarrow \quad x = \dfrac{7y + 3}{2}$ ...(*i*)

and $4(y - 2) + 3(x - 3) = 2$

$\Rightarrow \quad 4y - 8 + 3x - 9 = 2$

$\Rightarrow \quad 4y + 3x = 19$ ...(*ii*)

Using equation (*i*), we get

$4y + 3\left(\dfrac{7y + 3}{2}\right) = 19$

$\Rightarrow \quad \dfrac{8y + 21y + 9}{2} = 19$

$\Rightarrow \quad 29y + 9 = 38$

$29y = 29 \Rightarrow y = 1$

When $y = 1$, eq. (*i*) becomes

$x = \dfrac{7 \times 1 + 3}{2} = 5$

$\therefore \quad x = 5, y = 1$

35. For coincident lines, we have infinite many solutions.

$\therefore \quad \dfrac{a_1}{a_2} = \dfrac{b_1}{b_2} = \dfrac{c_1}{c_2}$

Here, $\dfrac{a_1}{a_2} = \dfrac{1}{2}, \dfrac{b_1}{b_2} = \dfrac{2}{k}, \dfrac{c_1}{c_2} = \dfrac{7}{14}$

$\dfrac{1}{2} = \dfrac{2}{k}$ | $\dfrac{2}{k} = \dfrac{7}{14}$

$k = 4$ | $7k = 28 \Rightarrow k = \dfrac{28}{7} = 4$

$\therefore$ For $k = 4$, lines will concide.

36. For infinite many solutions, $\dfrac{a_1}{a_2} = \dfrac{b_1}{b_2} = \dfrac{c_1}{c_2}$

i.e. $\dfrac{3}{2} = \dfrac{k}{-5} = \dfrac{7}{1}$

$\Rightarrow \quad \dfrac{3}{2} = \dfrac{k}{-5}$ and $\dfrac{k}{-5} = \dfrac{7}{1}$

$\Rightarrow \quad -\dfrac{15}{2} = k$ and $k = -35$

$\therefore$ There is no common value of k which satisfies the equality.

37. For unique (consistent solution), $\dfrac{a_1}{a_2} \neq \dfrac{b_1}{b_2}$

i.e. $\frac{2}{4} \neq \frac{-k}{6}$

$\Rightarrow \frac{2\times 3}{2} \neq -k \Rightarrow k \neq -3$

38. For no solution, $\frac{a_1}{a_2}=\frac{b_1}{b_2}\neq\frac{c_1}{c_2}$

i.e. $\frac{k}{4}=\frac{-3}{-6}$

$\Rightarrow \frac{k}{4}=\frac{1}{2} \Rightarrow k=\frac{4}{2}=2$

39. $2x+3y=7, \quad 6x+9y=11$

Since, $\frac{a_1}{a_2}=\frac{2}{6}=\frac{1}{3}$

and $\frac{b_1}{b_2}=\frac{3}{9}=\frac{1}{3}, \ \frac{c_1}{c_2}=\frac{7}{11}$

$\frac{a_1}{a_1}=\frac{b_1}{b_2}\neq\frac{c_1}{c_2}$

$\therefore$ System of equations has no solution.

40.
$$\frac{4x}{10}+3y = \frac{12}{10}$$
$$\Rightarrow \frac{4x+30y}{10} = \frac{12}{10}$$
$$\Rightarrow 4x+30y = 12$$
$$\Rightarrow 2x+15y = 6 \quad \text{... }(i)$$

and $7x-2y = \frac{17}{6}$

Multiply this equation with 6 on both sides.

$$\Rightarrow 42x-12y = 17 \quad \text{... }(ii)$$

Multiply equation (*i*) by 21 and substract (*ii*) from it.

$$42x+315y = 126$$
$$42x-12y = 17$$
$$\underline{-\quad + \quad -}$$
$$327y = 109$$
$$y = \frac{109}{327}=\frac{1}{3}$$

Putting value of y in equation (*i*), we get

$$2x+15\left(\frac{1}{3}\right) = 6$$
$$\Rightarrow 2x+5 = 6$$
$$\Rightarrow 2x = 6-5$$
$$2x = 1$$
$$x = \frac{1}{2}$$

41. Let $\frac{1}{y}=z \Rightarrow 5x+4z=9$ ...(*i*)

and $7x-2z = 5$ ...(*ii*)

Multiply equation (*ii*) by 2 and add in equation (*i*), find value of x and z. Also use $y=\frac{1}{z}$. $\left[\because \frac{1}{y}=z\right]$

42.
$$\frac{3}{x}+4y = 7$$
$$-\frac{2}{x}+7y = 5$$

Let $\frac{1}{x} = u$

$$3u+4y = 7 \quad \text{... }(i)$$
$$-2u+7y = 5 \quad \text{... }(ii)$$

Multiply equation (*i*) by 2 and (*ii*) by 3 and then add both the equations.

$$6u+8y = 14$$
$$-6u+21y = 15$$
$$29y = 29$$
$$y = 1$$

Putting value of y in (*i*), we get

$$3u+4(1) = 7$$
$$3u = 7-4$$
$$u = \frac{3}{3}$$
$$u = 1$$

Since $\frac{1}{x} = u$

$\therefore \frac{1}{x} = 1$

$\Rightarrow x = 1$

The solution is $x=1, y=1$.

43.
$$x+y=4 \quad \text{...}(i)$$
$$y+z=1 \quad \text{...}(ii)$$
$$z+x=1 \quad \text{...}(iii)$$

Subtract equation (*ii*) from (*i*), we get $x-z=3$ ...(*iv*)

Add (*iii*) and (*iv*), we get $2x=4 \Rightarrow x=2$

Putting $x=2$, in eq. (*i*) and (*iii*) respectively, we get $y=2$ and $z=-1$.

44.
$$x+y+z=9 \quad \text{... }(i)$$
$$2y-z=2 \quad \text{... }(ii)$$
$$z-x=2 \quad \text{... }(iii)$$

Adding equation (*i*) and (*ii*), we get

$$x+y+z=9$$
$$2y-z=2$$
$$x+3y=11 \quad \text{... }(iv)$$

Adding equation (*ii*) and (*iii*), we get

$$2y-z=2$$
$$z-x=2$$
$$2y-x=4 \quad \text{... }(v)$$

Solving (*iv*) and (*v*), we get

$$x+3y=11$$
$$-x+2y=4$$
$$+5y=15$$

$$y = 3$$

Putting value of y in equation (*iv*),

$$x + 3(3) = 11$$
$$x = 11 - 9$$
$$x = 2$$

Putting value of x in equation (*iii*)

$$z - x = 2$$
$$z - 2 = 2$$
$$z = 4$$

The solution is $x = 2, y = 3, z = 4$.

45. $px - qy = 0$ ...(*i*) and $\frac{m}{x} - \frac{n}{y} = a$ ...(*ii*)

Expressing x in terms of y in equation (*i*),we get

$$x = \frac{qy}{p}$$

Putting the value of x in equation (*ii*), we get

$$\frac{mp}{qy} - \frac{n}{y} = a \Rightarrow \frac{mp - nq}{qy} = a$$

$$\Rightarrow \frac{mp - nq}{aq} = y. \text{ Hence, } x = \frac{mp - nq}{ap}$$

{Substituting in equation (*i*)}

46. $\frac{4}{x} - \frac{9}{y} = -1 \Rightarrow 4y - 9x = -xy$

$\Rightarrow 9x - 4y = xy$ ... (*i*)

$\frac{15x}{2} - 5y = 0 \Rightarrow 15x - 10y = 0 \Rightarrow 3x - 2y = 0$ ... (*ii*)

Solve equations (*i*) and (*ii*) and get the result as $x = 2, y = 3$.

47. $149x - 330y = -511$ ...(*i*)

$-330x + 149y = -32$ ...(*ii*)

Adding (*i*) and (*ii*), we get

$-181x - 181y = -543$

$\Rightarrow \quad x + y = 3$ ...(*iii*)

Subtracting (*ii*) from (*i*), we get

$479x - 479y = -479 \Rightarrow x - y = -1$ ...(*iv*)

Adding (*iii*) and (*iv*), we get

$2x = 2 \Rightarrow x = 1$ and $y = 2$

48.
$$\frac{4}{x} + 5y = 7$$
$$\frac{3}{x} + 4y = 5$$

Let $\frac{1}{x} = u$

Equation becomes

$4u + 5y = 7$ ... (*i*)

$3u + 4y = 5$ ... (*ii*)

Multiply equation (*i*) by 3 and equation (*ii*) by 4 and then subtract (*ii*) from (*i*).

$$12u + 15y = 21$$
$$12u + 16y = 20$$
$$- \quad - \quad -$$
$$-y = 1$$
$$y = -1$$

Putting value of y in equation (*i*), we get

$$4u + 5(-1) = 7$$
$$\Rightarrow \quad 4u = 7 + 5$$
$$\Rightarrow \quad 4u = 12$$
$$\Rightarrow \quad u = \frac{12}{4}$$
$$\Rightarrow \quad u = 3$$

Since, $\frac{1}{x} = u$

$\therefore \quad \frac{1}{x} = 3$

$\Rightarrow \quad x = \frac{1}{3}$

The solution is $x = \frac{1}{3}, y = -1$.

49. $\sqrt{7}x + \sqrt{11}y = 0$ ...(*i*)

$\sqrt{3}x - \sqrt{5}y = 0$ ...(*ii*)

Multiplying equation (*i*) by $\sqrt{3}$ and equation (*ii*) by $\sqrt{7}$, we get

$\sqrt{21}x + \sqrt{33}y = 0$ ...(*iii*)

and $\sqrt{21}x - \sqrt{35}y = 0$ ...(*iv*)

Subtracting $\quad - \quad +$

$$\sqrt{33}y + \sqrt{35}y = 0$$
$$\Rightarrow \quad (\sqrt{33} + \sqrt{35})y = 0$$
$$\Rightarrow \quad y = 0$$

Putting the value of y in eq. (*i*), we get

$$x = 0$$

50. $\sqrt{5}x - \sqrt{7}y = 0$ ... (*i*)

$\sqrt{13}x + \sqrt{3}y = 0$ ... (*ii*)

Multiply equation (*i*) by $\sqrt{13}$, equation (*ii*) by $\sqrt{5}$ and then subtract equation (*ii*) from (*i*).

$$\sqrt{65}x - \sqrt{91}y = 0$$
$$\sqrt{65}x + \sqrt{15}y = 0$$
$$- \quad - \quad -$$
$$-\sqrt{91}y - \sqrt{15}y = 0$$
$$(-\sqrt{91} - \sqrt{15})y = 0$$
$$\Rightarrow \quad y = 0$$

Putting value of y in equation (*i*), we get

$$\sqrt{5}x - \sqrt{7}(0) = 0$$
$$\sqrt{5}x = 0$$
$$\Rightarrow \quad x = 0$$

The solution is $x = 0, y = 0$.

51. Here $a_1 = 3, \quad b_1 = -4, \quad c_1 = -1$

and $a_2 = 2, \quad b_2 = \frac{-8}{3}, \quad c_2 = 5$

$$\frac{a_1}{a_2} = \frac{3}{2}, \frac{b_1}{b_2} = \frac{-4}{-8/3} = \frac{3 \times 4}{8} = \frac{3}{2}, \frac{c_1}{c_2} = -\frac{1}{5}$$

$$\Rightarrow \frac{a_1}{a_2} = \frac{b_1}{b_2} \neq \frac{c_1}{c_2}$$

$\Rightarrow$ Hence, the given pair of equations has no solution.

52. $cx + 3y = 3$

$12x + cy = 6$

For no solution, $\frac{a_1}{a_2} = \frac{b_1}{b_2} \neq \frac{c_1}{c_2}$

$\Rightarrow \quad \frac{c}{12} = \frac{3}{c} \neq \frac{3}{6} \quad \Rightarrow \frac{c}{12} = \frac{3}{c}$

$\Rightarrow \quad c^2 = 36 \quad \Rightarrow c = \pm 6$ ...(i)

or $\quad \frac{3}{c} \neq \frac{3}{6}$

$c \neq \frac{18}{3} \neq 6$

$\Rightarrow \quad c \neq 6$ ...(ii)

$\therefore$ From (i) and (ii), $c = -6$

53. (i) For unique solution, $\frac{a_1}{a_2} \neq \frac{b_1}{b_2} \Rightarrow \frac{2}{3} \neq \frac{k}{-5} \Rightarrow \frac{-10}{3} \neq k$

(ii) For no solution, $\frac{a_1}{a_2} = \frac{b_1}{b_2} \neq \frac{c_1}{c_2}$

$\Rightarrow \frac{2}{3} = \frac{k}{-5} \neq \frac{1}{7} \Rightarrow k = -\frac{10}{3}$

54. As $(x - 4)$ is a factor of $p(x) = x^3 + ax^2 + 2bx - 24$, therefore, $p(4) = 0$

$\Rightarrow (4)^3 + a(4)^2 + 2b(4) - 24 = 0$

$\Rightarrow 64 + 16a + 8b - 24 = 0$

$\Rightarrow 16a + 8b + 40 = 0$

$\Rightarrow 2a + b + 5 = 0$ ...(i)

Also given, $a - b = 8$ ...(ii)

Adding (i) and (ii), we get

$3a = 3 \Rightarrow a = 1$

Substituting $a = 1$ in (ii), we get

$1 - b = 8 \Rightarrow b = -7$

Hence, $a = 1$ and $b = -7$

55. For no solution,

$\Rightarrow \quad \frac{a_1}{a_2} = \frac{b_1}{b_2} \neq \frac{c_1}{c_2}$

$\Rightarrow \quad \frac{2}{6} = \frac{k}{4} \neq \frac{1}{3}$

$\Rightarrow \quad \frac{2}{6} = \frac{k}{4}$ and $\frac{k}{4} \neq \frac{1}{3}$

$\Rightarrow \quad \frac{1}{3} = \frac{k}{4}$ and $\frac{k}{4} \neq \frac{1}{3}$

$\therefore$ No value of k.

56. For infinitely many solutions,

$\frac{a_1}{a_2} = \frac{b_1}{b_2} = \frac{c_1}{c_2}$

$\Rightarrow \quad \frac{a}{-12} = \frac{-3}{a} = \frac{1}{2}$

$\Rightarrow \quad \frac{a}{-12} = \frac{-3}{a}$ and $\frac{-3}{a} = \frac{1}{2}$

$\Rightarrow \quad a^2 = 36$ and $a = -6$

$\Rightarrow \quad a = \pm 6$ and $a = -6$

For $a = -6$, pair of given linear equations has infinitely many solutions.

57. For infinite numbers of solution

$\frac{2}{a} = \frac{a}{8} = \frac{8}{a}$

$\Rightarrow \quad \frac{2}{a} = \frac{a}{8}$ and $\frac{a}{8} = \frac{8}{a}$

$a^2 = 16$ and $a^2 = 64$

$\therefore$ The system has not infinite solutions for any values of a.

58. $7x + 8y = k \Rightarrow 7x + 8y - k = 0$ ... (i)

Also $\quad 9x - 4y = 12 \Rightarrow 9x - 4y - 12 = 0$ ... (ii)

Here, $a_1 = 7, b_1 = 8, c_1 = -k,$

$a_2 = 9, b_2 = -4, c_2 = -12$

Now, $\frac{a_1}{a_2} = \frac{7}{9}, \quad \frac{b_1}{b_2} = \frac{8}{-4} = -2$

$\because \quad \frac{a_1}{a_2} \neq \frac{b_1}{b_2}$

Therefore, the solution is unique for any values of k.

59. For unique solutions $\frac{k}{12} \neq \frac{3}{k}$

$\Rightarrow k^2 \neq 36 \Rightarrow k \neq \pm 6.$

60. (i) $px + qy - p + q = 0 \Rightarrow px + qy - (p - q) = 0$ ... (i)

$qx - py - p - q = 0 \Rightarrow qx - py - (p + q) = 0$ ... (ii)

Here, $a_1 = p, b_1 = q, c_1 = -(p - q)$

$a_2 = q, b_2 = -p, c_2 = -(p + q)$

$$\begin{array}{cccccccc} & x & & y & & 1 & \\ q & & -(p-q) & & p & & q \\ -p & & -(p+q) & & q & & -p \end{array}$$

$$\frac{x}{-(p+q).q - [(-p).\{-(p-q)\}]} = \frac{y}{-(p-q).q - [-(p+q).p]}$$

$= \frac{1}{-p^2 - q^2}$

$\Rightarrow \frac{x}{-pq - q^2 - [p^2 - pq]} = \frac{y}{-pq + q^2 - (-p^2 - pq)}$

$= \frac{1}{-p^2 - q^2}$

$\Rightarrow \frac{x}{-pq - q^2 - p^2 + pq} = \frac{y}{-pq + q^2 + p^2 + pq} = \frac{1}{-p^2 - q^2}$

$\Rightarrow \frac{x}{-q^2 - p^2} = \frac{y}{p^2 + q^2} = \frac{1}{-p^2 - q^2}$

$\Rightarrow x = \frac{-q^2 - p^2}{-p^2 - q^2}$ and $y = \frac{p^2 + q^2}{-p^2 - q^2} = \frac{p^2 + q^2}{-(p^2 + q^2)} = -1$

$= 1$

$\therefore x = 1, y = -1$

(ii) $\frac{x}{a} - \frac{y}{b} = 0 \Rightarrow \frac{bx - ay}{ab} = 0 \Rightarrow bx - ay = 0$ ... (i)

Also, $ax + by = a^2 + b^2$ ... (ii)

Now, solve (i) and (ii), we get $x = a$, $y = b$.

SHORT ANSWER TYPE QUESTIONS-II

1.

$$4x - y - 8 = 0,$$
$$2x - 3y + 6 = 0$$
$$\Rightarrow \quad -y = -4x + 8,$$
$$\Rightarrow \quad -3y = -2x - 6$$
$$\Rightarrow \quad y = 4x - 8,$$
$$\Rightarrow \quad 3y = 2x + 6$$

Solution table for $4x - y - 8 = 0$ is

x	0	1	2
y	–8	–4	0

Solution table for $2x - 3y + 6 = 0$ is

x	0	3	–3
y	2	4	0

Vertices of the triangle formed by lines and x-axis are (2, 0), (3, 4) and (–3, 0).

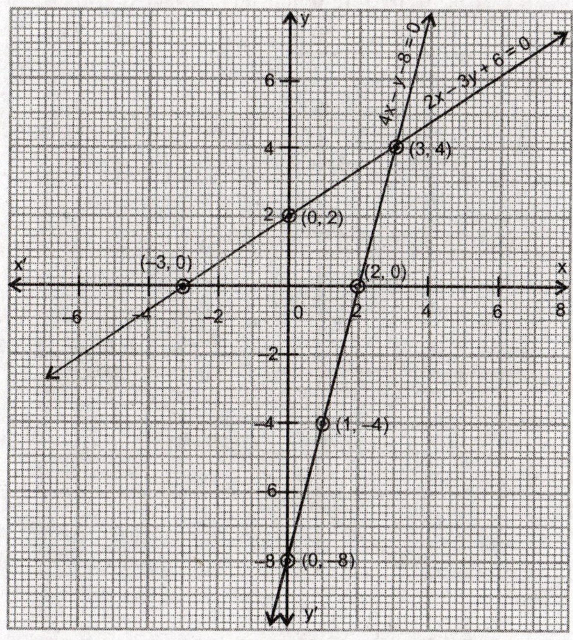

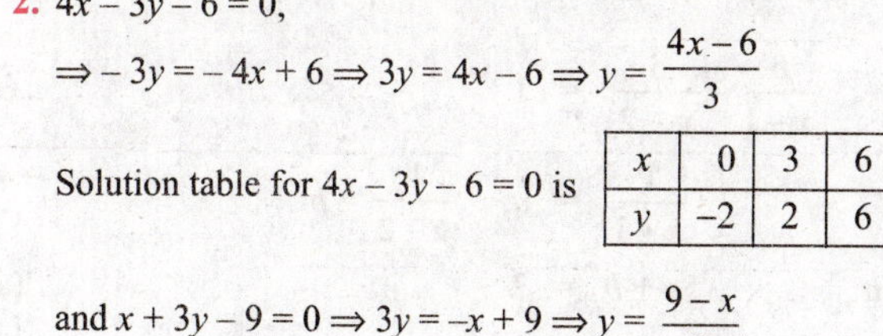

2. $4x - 3y - 6 = 0,$

$\Rightarrow -3y = -4x + 6 \Rightarrow 3y = 4x - 6 \Rightarrow y = \frac{4x - 6}{3}$

Solution table for $4x - 3y - 6 = 0$ is

x	0	3	6
y	–2	2	6

and $x + 3y - 9 = 0 \Rightarrow 3y = -x + 9 \Rightarrow y = \frac{9 - x}{3}$

Solution table for $x + 3y - 9 = 0$ is

x	0	3	6
y	3	2	1

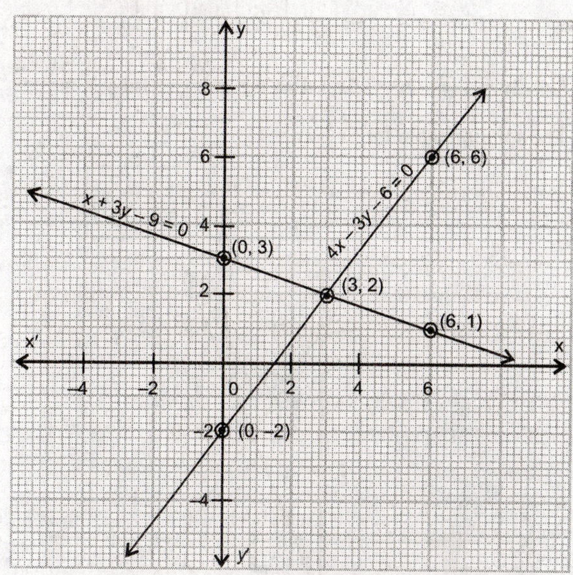

The vertices of triangle formed is (0, 3), (3, 2) and (0, –2).

3. $3x - 2y - 1 = 0$

$\Rightarrow -2y = -3x + 1 \Rightarrow 2y = 3x - 1 \Rightarrow y = \frac{3x - 1}{2}$

The solution table for $3x - 2y - 1 = 0$

x	0	1	–1
y	–1/2	1	–2

and $2x - 3y + 6 = 0 \Rightarrow -3y = -2x - 6 \Rightarrow 3y = 2x + 6$

$\Rightarrow y = \frac{2x+6}{3}$

The solution table for $2x - 3y + 6 = 0$

x	0	3	–3
y	2	4	0

Plotting the points on graph, we get

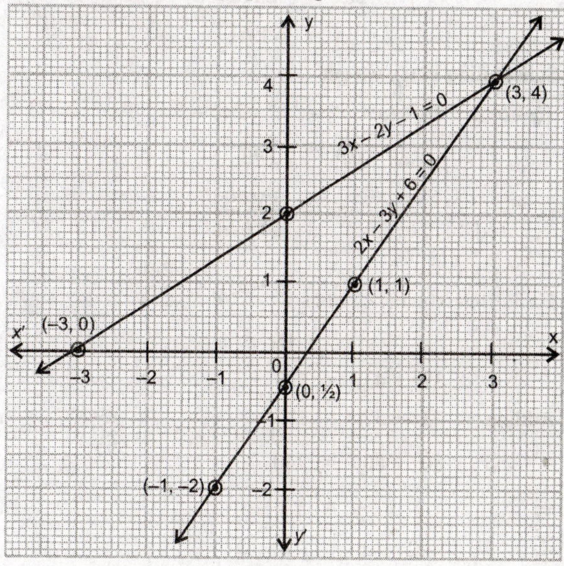

4. $3x + 2y = 12$

$\Rightarrow 2y = 12 - 3x \Rightarrow y = \dfrac{12-3x}{2}$

The solution table for $3x + 2y = 12$ is

x	0	1	2
y	6	4.5	3

$5x - 2y = 4 \Rightarrow -2y = -5x + 4 \Rightarrow y = \dfrac{-5x+4}{-2} \Rightarrow y = \dfrac{5x-4}{2}$

The solution table for $5x - 2y = 4$

x	0	1	2
y	-2	1/2	3

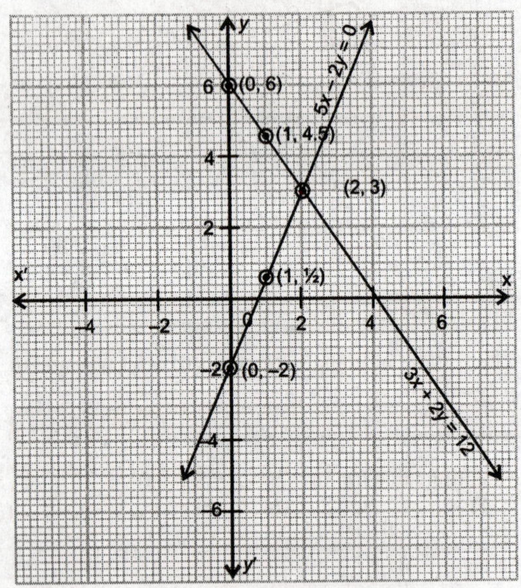

Points where lines meet the y-axis are (0, 6), (0, –2).

5. $(a-b)x + (a+b)y = a^2 - 2ab - b^2$ ... (i)

$(a+b)x + (a+b)y = a^2 + b^2$ ... (ii)

$\underline{-\qquad\quad -\qquad\qquad - \quad -}$

$(a-b)x - (a+b)x = -2ab - 2b^2$

$-2bx = -2b(a+b)$

$x = a + b$

Substituting in (i), we get

$(a-b)(a+b) + (a+b)y = a^2 - 2ab - b^2$

$(a+b)y = -2ab$

$y = \dfrac{-2ab}{(a+b)}$

$\therefore \quad x = a + b, y = \dfrac{-2ab}{a+b}$

Or

$37x + 43y = 123$ ... (i)

Adding $43x + 37y = 117$ ... (ii)

$80x + 80y = 240$

$\Rightarrow \quad x + y = 3$ ... (iii)

Now $37x + 43y = 123$

$43x + 37y = 117$

Subtracting $\underline{-\qquad -\qquad -}$

$-6x + 6y = 6$

$\Rightarrow \quad x - y = -1$ ... (iv)

Solving (iii) and (iv),

$x = 1, y = 2$

6. $x + 3y = 6$

$\Rightarrow \quad 3y = 6 - x$

$y = \dfrac{6-x}{3}$

The solution table for $x + 3y = 6$ is

x	0	3	6
y	2	1	0

and $2x - 3y = 12$

$\Rightarrow \quad -3y = 12 - 2x$

$3y = 2x - 12$

$y = \dfrac{2x-12}{3}$

The solution table for $2x - 3y = 12$ is

x	0	3	6
y	-4	-2	0

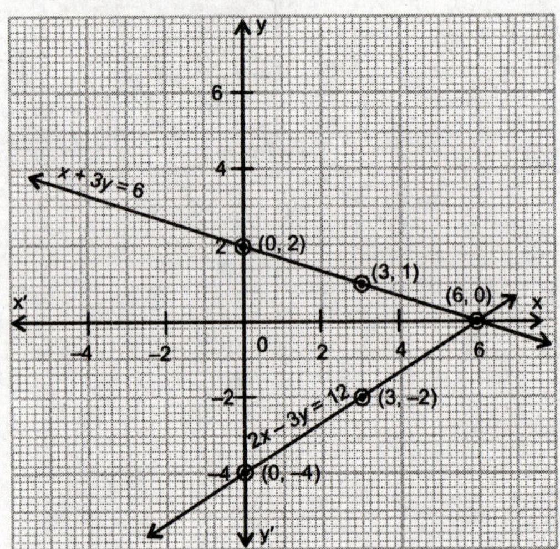

7. $\dfrac{5}{x-1} + \dfrac{1}{y-2} = 2$ ...(i)

$\dfrac{6}{x-1} - \dfrac{3}{y-2} = 1$ ...(ii)

Let $\dfrac{1}{x-1} = a, \quad \dfrac{1}{y-2} = b$

then $5a + b = 2$ ...(iii)

$6a - 3b = 1$ ...(iv)

Multiplying (iii) by 3, then adding in (iv)

$15a + 3b = 6$

$6a - 3b = 1$

$$21a = 7 \Rightarrow a = \frac{1}{3}$$

i.e., $\frac{1}{x-1} = \frac{1}{3}$

or $x - 1 = 3 \Rightarrow x = 4$

Again $b = -5a + 2 = 2 - \frac{5}{3} = \frac{1}{3}$

then $\frac{1}{y-2} = \frac{1}{3}$

or $y - 2 = 3 \Rightarrow y = 5$

$$x = 4, \quad y = 5$$

Or

$$2x + 3y = 7 \quad \text{... } (i)$$

$$a(x + y) - b(x - y) = 3a + b - 2$$

$\Rightarrow$ $ax + ay - bx + by = 3a + b - 2$

$\Rightarrow$ $(a - b)x + (a + b)y = 3a + b - 2$ ... (*ii*)

The system has infinite number of solutions.

$\therefore \quad \frac{a_1}{a_2} = \frac{b_1}{b_2} = \frac{c_1}{c_2} \Rightarrow \frac{2}{a-b} = \frac{3}{a+b} = \frac{7}{-(3a+b-2)}$

$\Rightarrow$ $2(a + b) = 3(a - b)$ and $3(3a + b - 2) = 7(a + b)$

$\Rightarrow$ $2a + 2b = 3a - 3b$ $\Rightarrow 9a + 3b - 6 = 7a + 7b$

$\Rightarrow$ $a - 5b = 0$ ... (*i*) $\Rightarrow 2a - 4b = 6 \Rightarrow a - 2b = 3$ ... (*ii*)

Solving equation (*i*) and (*ii*), we get result as $a = 5$, $b = 1$.

8. $\frac{15}{x-y} + \frac{22}{x+y} = 5, \quad \frac{40}{x-y} + \frac{55}{x+y} = 13$

Let $\frac{1}{x-y} = u, \quad \frac{1}{x+y} = v$

$$15u + 22v = 5 \quad \text{...}(i)$$

$$40u - 55v = 13 \quad \text{...}(ii)$$

Multiply equation (*i*) by 5 and (*ii*) by 2 and then subtract equation (*ii*) from (*i*).

$$75u + 110v = 25$$

$$80u + 110v = 26$$

$$- \quad - \quad -$$

$$-5u = -1$$

$$u = \frac{-1}{-5} \Rightarrow u = \frac{1}{5}$$

Putting value of u in equation (*i*), we get

$$15\left(\frac{1}{5}\right) + 22v = 5$$

$\Rightarrow$ $3 + 22v = 5$

$\Rightarrow$ $22v = 5 - 3$

$\Rightarrow$ $22v = 2$

$\Rightarrow$ $v = \frac{2}{22} \Rightarrow v = \frac{1}{11}$

Since $\frac{1}{x-y} = u$ and $\frac{1}{x+y} = v$

$\therefore \quad \frac{1}{x-y} = \frac{1}{5}$ and $\frac{1}{x+y} = \frac{1}{11}$

$\therefore \quad x - y = 5$ ...(*iii*)

$x + y = 11$ ...(*iv*)

Adding equation (*iii*) and (*iv*), we get

$$2x = 16$$

$$x = 8$$

Putting value of x in equation (*iii*), we get

$$8 - y = 5$$

$$-y = 5 - 8$$

$$-y = -3$$

$$y = 3$$

The solution is $x = 8$, $y = 3$.

9. For infinitely many solutions $\frac{a_1}{a_2} = \frac{b_1}{b_2} = \frac{c_1}{c_2}$

$$\frac{2}{k+1} = \frac{-3}{1-2k} = \frac{7}{5k-4}$$

$\Rightarrow$ $\frac{2}{k+1} = \frac{-3}{1-2k}$ and $\frac{-3}{1-2k} = \frac{7}{5k-4}$

$\Rightarrow$ $2(1 - 2k) = -3(k + 1)$ and $-3(5k - 4) = 7(1 - 2k)$

$2 - 4k = -3k - 3$ and $-15k + 12 = 7 - 14k$

$-4k + 3k = -3 - 2$ and $-15k + 14k = 7 - 12$

$-k = -5$ and $-k = -5$

$k = 5$ and $k = 5$

For $k = 5$, pair of linear equation has infinitely many solutions.

10. $\frac{10}{x+y} + \frac{2}{x-y} = 4$, let $\frac{1}{x+y} = a$

$\frac{15}{x+y} - \frac{5}{x-y} = -2$, let $\frac{1}{x-y} = b$

then $10a + 2b = 4$ ...(*i*) × 3

$15a - 5b = -2$ ...(*ii*) × 2

After multiplication, we get

$$30a + 6b = 12$$

$$30a - 10b = -4$$

$$- \quad + \quad + \quad \text{(On subtracting)}$$

$$16b = 16 \Rightarrow b = 1$$

then $10a = 4 - 2 = 2 \Rightarrow a = \frac{1}{5}$

now $\frac{1}{x-y} = b = 1$ or $x - y = 1$ ...(*iii*)

or $\frac{1}{x+y} = \frac{1}{5}$ or $x + y = 5$ ...(*iv*)

On adding (*iii*) and (*iv*),

$$2x = 6$$

$$\Rightarrow \quad x = 3 \text{ and } y = 2$$

11. $$\frac{ax}{b} - \frac{by}{a} = a + b \quad \text{... } (i)$$

$$ax - by = 2ab \quad \text{... } (ii)$$

Multiplying equation (*ii*) with $\frac{1}{b}$, we get

$$\frac{ax}{b} - y = 2a \quad \text{... } (iii)$$

Now subtracting equation (*iii*) from (*i*)

$$\frac{ax}{b} - \frac{by}{a} = a + b$$

$$\frac{ax}{b} - y = 2a$$

$$- \quad + \quad -$$

$$y - \frac{by}{a} = -a + b$$

$$\Rightarrow \quad \frac{ay - by}{a} = -a + b$$

$$\Rightarrow \quad y\left(\frac{a-b}{a}\right) = -(a-b)$$

$$\Rightarrow \quad y = -(a-b) \times \frac{a}{a-b}$$

$$\Rightarrow \quad y = -a$$

Substituting the value of y in equation (*iii*), we get

$$\frac{ax}{b} - (-a) = 2a$$

$$\Rightarrow \quad \frac{ax}{b} = a \Rightarrow x = a \times \frac{b}{a} = b$$

$$\therefore \quad x = b, y = -a$$

12. Let numerator be x and denominator be y

$$\therefore \quad \text{Fraction} = \frac{x}{y}$$

A.T.Q. $\quad x + y = 2y - 3$

and $\quad x - y = -3 \quad \text{...}(i)$

Again A.T.Q.

$$\frac{x-1}{y-1} = \frac{1}{2}$$

$$2x - 2 = y - 1$$

$$2x - y = 1 \quad \text{...}(ii)$$

Multiply (*i*) by 2 and solving (*ii*) and (*iii*)

$$2x - 2y = -6 \quad \text{...}(iii)$$

$$2x - 2y = -6$$

$$2x - y = 1$$

$$- \quad + \quad -$$

$$-y = -7$$

$$y = 7$$

From (*i*),

$$x - y = -3$$

$$x - 7 = -3$$

$$x = 4$$

$$\therefore \quad \text{Fraction} = \frac{4}{7}$$

Or

$$\frac{4}{x} + 3y = 8, \ \frac{6}{x} - 4y = -5$$

$$4 + 3xy = 8x \quad \text{...}(i)$$

$$6 - 4xy = -5x \quad \text{...}(ii)$$

Multiply (*i*) by 4 and (*ii*) by 3 and then solving

$$16 + 12xy = 32x$$

$$18 - 12xy = -15x \quad \text{(On adding)}$$

$$34 = 17x$$

$$x = \frac{34}{17} = 2$$

From (*i*),

$$4 + 3 \times 2y = 8 \times 2$$

$$4 + 6y = 16$$

$$6y = 12$$

$$y = 2$$

$$x = 2, y = 2$$

13. $$\frac{b}{a}x + \frac{a}{b}y = a^2 + b^2$$

$$\Rightarrow \quad b^2x + a^2y = (a^2 + b^2)ab \quad \text{...}(i)$$

$$x + y = 2ab \quad \text{...}(ii)$$

Multiply (*ii*) by b^2 and subtracting from (*i*), we get

$$b^2x + a^2y = (a^2 + b^2)ab$$

$$b^2x + b^2y = 2ab^3$$

$$- \quad - \quad -$$

$$(a^2 - b^2)y = (a^2 + b^2 - 2b^2)ab$$

$$(a^2 - b^2)y = (a^2 - b^2)ab$$

$$y = \frac{(a^2 - b^2)ab}{a^2 - b^2}$$

$$y = ab$$

$$x + y = 2ab$$

$$x + ab = 2ab$$

$$x = ab$$

$$x = ab, y = ab$$

14. Consider equations :

$2(ax - by) + (a + 4b) = 0$; and $2(bx + ay) + (b - 4a) = 0$

$$\Rightarrow \quad 2ax - 2by = -a - 4b \quad \text{...}(i)$$

and $$2bx + 2ay = 4a - b \quad \text{...}(ii)$$

Multiply (*i*) by a and (*ii*) by b and adding, we get

$$2(a^2 + b^2)x = (-a - 4b)a + b(4a - b)$$

$$= -a^2 - 4ab + 4ab - b^2$$

$$= -(a^2 + b^2)$$

$$\Rightarrow \quad x = -\frac{1}{2}$$

Substituting in (*i*), we get

$$-a-2by = -a-4b$$

$\Rightarrow \quad -2by = -4b \Rightarrow y = 2$

$\therefore$ solution is $x = -\frac{1}{2}$ and $y = 2$.

Or

Let digit at unit place be x, and at tenth place be y.

$\therefore \quad$ Number $= 10y + x$

According to the question,

$$\frac{10y+x}{y+x} = 7 \Rightarrow 6x - 3y = 0$$

$\Rightarrow \quad 2x - y = 0 \quad$ ...(*i*)

Again according to the question,

$(10y + x) - 27 = 10x + y$

$9x - 9y = -27 \Rightarrow x - y = -3 \quad$ ...(*ii*)

Solving for x & y, we get

$x = 3$ and $y = 6$

$\therefore$ Number is 63.

15. Let digit at tenth place be x

and digit at unit place be y

$\therefore$ Number is $10x + y$.

A.T.Q., $\quad x + y = 12 \quad$ ...(*i*)

and $\quad 10y + x = 10x + y + 18$

$\Rightarrow \quad 9y - 9x = 18$

$y - x = 2 \quad$ ...(*ii*)

Solving equation (*i*) and (*ii*), we get

$x = 5, y = 7$

Number $= 57$

16. $\because$ CD = BE

$\Rightarrow x + y = 5 \quad$ ...(*i*)

Also, AB + BC + CD + DE + AE = 21

$3 + x - y + x + y + x - y + 3 = 21$

$3x - y = 15 \quad$ ...(*ii*)

Solving equation (*i*) and (*ii*) $x = 5, y = 0$

17. Let numerator $= x$ and denominator $= y$

A.T.Q., $x + y = 12 \quad$ ...(*i*)

Also, $\quad \frac{x+1}{y+1} = \frac{3}{4} \Rightarrow 4x + 4 = 3y + 3$

$\Rightarrow \quad 4x - 3y = -1 \quad$ ...(*ii*)

Solving (*i*) and (*ii*), $x = 5, y = 7$

$\therefore$ Fraction is $\frac{5}{7}$.

18. Let a man takes x days to finish the work and a boy takes y days to finish the work.

$\therefore$ One man's one day's work $= \frac{1}{x}$

and one boy's one day's work $= \frac{1}{y}$

A.T.Q.,

$$4 \times \frac{1}{x} + 6 \times \frac{1}{y} = \frac{1}{5} \Rightarrow \frac{4}{x} + \frac{6}{y} = \frac{1}{5} \quad ...(i)$$

and $\quad \frac{3}{x} + \frac{4}{y} = \frac{1}{7} \quad$ ...(*ii*)

Solving (*i*) and (*ii*), we get

$x = 35, y = 70$

19. Let speed of train be x km/hr

and speed of car be y km/hr

ATQ,

$$\frac{320}{x} + \frac{280}{y} = \frac{26}{3} \quad ...(i)$$

$$\frac{200}{x} + \frac{400}{y} = \frac{55}{6} \quad ...(ii)$$

Solving (*i*) & (*ii*) for x and y,

$x = 80$ and $y = 60$

20.

$$\frac{a^2}{x} - \frac{b^2}{y} = 0 \quad ...(i)$$

$$\frac{a^2b}{x} + \frac{b^2a}{y} = a + b \quad ...(ii)$$

$(b) \times (i)$ becomes

$$\frac{a^2b}{x} - \frac{b^3}{y} = 0 \quad ...(iii)$$

Now $\quad \frac{a^2b}{x} + \frac{b^2a}{y} = a + b \quad$ ...(*ii*)

$$\frac{a^2b}{x} - \frac{b^3}{y} = 0 \quad ...(iii)$$

Subtracting $\quad - \quad +$

$$\frac{b^2a}{y} + \frac{b^3}{y} = a + b$$

$$\frac{b^2(a+b)}{y} = a + b \Rightarrow y = b^2$$

Substituting in (*i*), we get $x = a^2$

$\therefore \quad x = a^2, y = b^2$

21. Let the cost of each table = ₹ x

and that of each chair = ₹ y

Then, $2x + 3y = 2000$ and $3x + 2y = 2500$

The above equations can be written as

$3(2x + 3y = 2000) \Rightarrow \quad 6x + 9y = 6000 \quad$ ...(*i*)

and $2(3x + 2y = 2500) \Rightarrow \quad 6x + 4y = 5000 \quad$ ...(*ii*)

Subtracting (*ii*) from (*i*), we get

$5y = 1000 \quad \Rightarrow \quad y = 200$

Substituting $y = 200$ in (*i*), we get

$2x + 3 \times 200 = 2000$

$\Rightarrow \quad 2x = 2000 - 600 = 1400 \quad \Rightarrow \quad x = 700$

Hence the cost of each table = ₹ 700

And cost of each chair = ₹ 200

$\therefore$ Total cost of 1 table and 5 chairs = 700 + 5 × 200 = ₹ 1700

22. Let the cost of 1 book be ₹ x

and the cost of 1 pen be ₹ y

A.T.Q.,

$$5x + 7y = 79 \quad \text{... }(i)$$

$$7x + 5y = 77 \quad \text{... }(ii)$$

Multiply equation (*i*) by 5 and equation (*ii*) by 7 and then subtract equation (*ii*) from (*i*).

$$25x + 35y = 395$$

$$49x + 35y = 539$$

$$- \quad - \quad -$$

$$-24x = -144$$

$$x = \frac{-144}{-24} = 6$$

Putting value of x in equation (*i*), we get

$$5(6) + 7y = 79$$

$$7y = 79 - 30$$

$$y = \frac{49}{7} = 7$$

Cost of a book is ₹ 6 and cost of a pen is ₹ 7.

Total cost of 1 book and 2 pen = 6 + 2 × 7 = ₹ 20

23. $2y - x = 8$

x	0	– 8	2	4	– 4
y	4	0	5	6	2

$5y - x = 14$

x	1	–4	6
y	3	2	4

$y - 2x = 1$

x	2	0	1
y	5	1	3

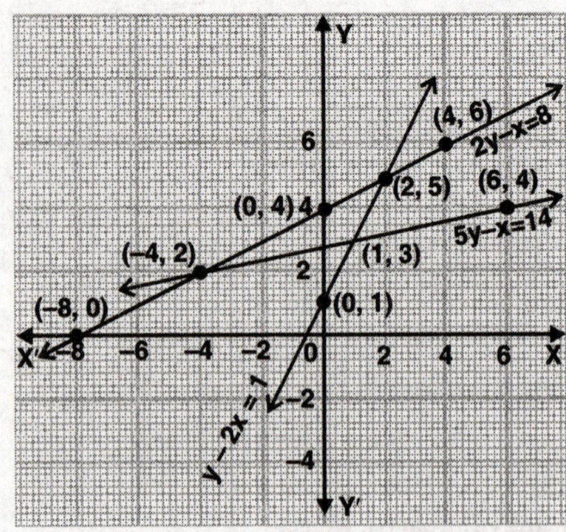

The vertices of the triangle formed are (1, 3), (–4, 2) and (2, 5).

24. Let the digits at unit place be y and tenth place be x.

Number $= 10x + y$

A.T.Q.

$$x + y = 15 \quad \text{... }(i)$$

and $\quad 10x + y - 27 = 10y + x$

$\Rightarrow \quad 10x - x + y - 10y = 27$

$$9x - 9y = 27$$

$$x - y = 3 \quad \text{... }(ii)$$

Solving (*i*) and (*ii*), we get

$$2x = 18$$

$$x = 9$$

Putting value of x in (*i*), we get

$$9 + y = 15$$

$$y = 15 - 9 = 6$$

So, the number is 10(9) + 6 = 96.

25. Let the digits at unit place be x and tens place be y.

Number = $10y + x$

The digits are reversed, then the number = $10x + y$

According to the question,

$$y = 3x \quad \text{...}(i)$$

A.T.Q.,

$\Rightarrow \quad (10y + x) - 54 = 10x + y$

$\Rightarrow \quad 10y + x - 10x - y = 54$

$\Rightarrow \quad 9y - 9x = 54$

$\Rightarrow \quad y - x = 6 \quad \text{...}(ii)$

From (*i*) and (*ii*),

$$3x - x = 6$$

$$2x = 6 \quad \text{or} \quad x = 3 \text{ and } y = 9$$

$\therefore$ Number = 93.

26. Let the constant expenditure be ₹ x

And consumption of wheat be y quintals

Then total expenditure = $x + y$ × Rate per quintal

A.T.Q.,

$$1000 = x + 250y \quad \text{...}(i)$$

$$980 = x + 240y \quad \text{...}(ii)$$

Subtracting (*ii*) from (*i*), we get $y = 2$.

Putting $y = 2$ in (*i*), we get

$$1000 = x + 2 \times 250 \Rightarrow 1000 = x + 500$$

$\Rightarrow \quad x = 1000 - 500 = 500$

Total expenses when the price of wheat is ₹ 350 per quintal = 500 + 350 × 2 = 500 + 700 = ₹ 1,200

27. Let the digit at unit place be y and tenth place be x.

Number $= 10x + y$

A.T.Q.

$$x + y = 8 \quad \text{... }(i)$$

and $\quad 10x + y - 10y - x = 18$

$\Rightarrow \quad 9x - 9y = 18$

$\Rightarrow \quad x - y = 2 \quad \text{... }(ii)$

Adding (*i*) and (*ii*), we get

$$2x = 10$$

$$x = 5$$

Putting $x = 5$ in equation (*i*)

$$5 + y = 8 \Rightarrow y = 3$$

So, the number is $10 \times 5 + 3 = 53$.

28. Let number of students in a row be x and number of rows be y.

$\Rightarrow$ Total number of students $= x \, . \, y$

From condition 1 : $(x + 4)(y - 2) = xy$

$\Rightarrow$ $xy - 2x + 4y - 8 = xy$

$\Rightarrow$ $-2x + 4y = 8$ ...(*i*)

From condition 2 : $(x - 4)(y + 4) = xy$

$\Rightarrow$ $xy + 4x - 4y - 16 = xy$

$\Rightarrow$ $4x - 4y = 16$ ...(*ii*)

Adding (*i*) and (*ii*), we get

$2x = 24 \Rightarrow x = 12$

Substituting in (*i*), we get $y = 8$.

$\therefore$ Total number of students $= xy = 12 \times 8 = 96$

Or

Let length of rectangle be l units, breadth be b units.

$\therefore$ $(l - 5)(b + 2) = lb - 80$ ...(*i*)

$(l + 10)(b - 5) = lb + 50$ ...(*ii*)

On solving, we get $l = 40$ units and $b = 30$ units.

29. $x + 4y = 10$

$\Rightarrow$ $4y = 10 - x$

$\Rightarrow$ $y = \dfrac{10 - x}{4}$

The solution table of equation $x + 4y = 10$ is

x	2	−2	6
y	2	3	1

and $y - 2 = 0$

$\Rightarrow$ $y = 2$

The unique solution is $x = 2$, $y = 2$.

30. $2x - 3y = 6$

$\Rightarrow$ $-3y = 6 - 2x$

$\Rightarrow$ $3y = 2x - 6$

$\Rightarrow$ $y = \dfrac{2x - 6}{3}$

The solution table of the equation $2x - 3y = 6$ is

x	0	3	−3
y	−2	0	−4

and $x - 6 = 0$

$\Rightarrow$ $x = 6$

Plotting the points on graph, we get

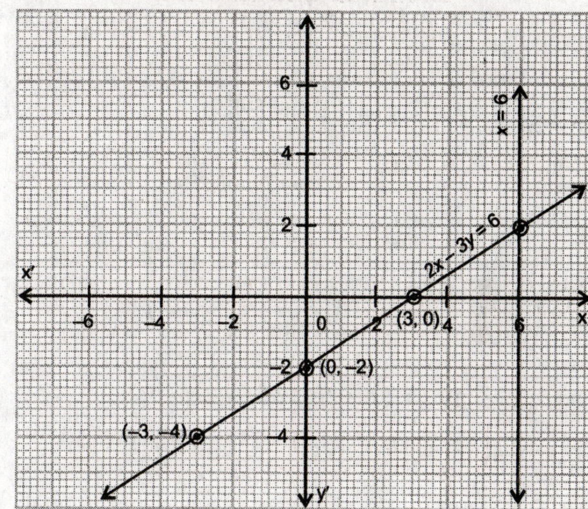

The unique solution is $x = 6$, $y = 2$.

31. $2y - x = 8$

$\Rightarrow$ $2y = 8 + x$

$\Rightarrow$ $y = \dfrac{8 + x}{2}$

The solution table for $2y - x = 8$ is

x	0	2	−2
y	4	5	3

$5y - x = 14$

$\Rightarrow$ $5y = 14 + x$

$\Rightarrow$ $y = \dfrac{14 + x}{5}$

The solution table for $5y - x = 14$

x	1	6	−4
y	3	4	2

$-2x + y = 1$

$\Rightarrow$ $y = 1 + 2x$

x	0	1	−1
y	1	3	−1

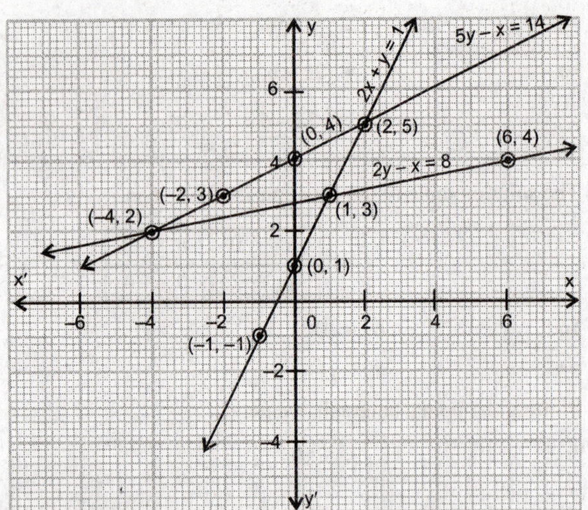

Vertices of Δ are (2, 5), (1, 3) and (–4, 2).

32. $\frac{6}{x+y} = \frac{7}{x-y} + 3$ ...(*i*) $\quad \frac{1}{2(x+y)} = \frac{1}{3(x-y)}$ ...(*ii*)

Put $\frac{1}{x+y} = A$ and $\frac{1}{x-y} = B$

Equation (*i*) and (*ii*) becomes,

$6A = 7B + 3 \Rightarrow 6A - 7B = 3$ ...(*iii*)

and $\frac{1}{2}A = \frac{1}{3}B \Rightarrow 3A = 2B \Rightarrow 3A - 2B = 0$ .. .(*iv*)

On solving (*iii*) and (*iv*), we get

$A = -\frac{2}{3}$ and $B = -1$

$\Rightarrow \frac{1}{x+y} = -\frac{2}{3}$ and $\frac{1}{x-y} = -1$

$\Rightarrow x + y = -\frac{3}{2}$ and $x - y = -1$

$\Rightarrow x = -\frac{5}{4}$ and $y = -\frac{1}{4}$

33. Let speed of A = x km/hr and speed of B = y km/hr.

According to the question,

$\frac{30}{x} - \frac{30}{y} = 3$ ...(*i*) and $\frac{30}{2x} - \frac{30}{y} = -\frac{3}{2}$ ...(*ii*)

On solving, we get $x = \frac{10}{3}$ km/hr. and $y = 5$ km/hr.

34. Let usual speed be x km/hr and usual time be y hr.

Then distance = $x \cdot y$ km

According to the question,

$(x + 6)(y - 4) = xy \Rightarrow xy - 4y + 6x - 24 = xy$

$\Rightarrow 4x - 6y = -24 \Rightarrow 2x - 3y = -12$ ...(*i*)

and $(x - 6)(y + 6) = xy \Rightarrow xy + 6x - 6y - 36 = xy$ ...(*ii*)

$\Rightarrow x - y = 6$

On solving, we get $y = 24$ hr. and $x = 30$ km/hr

∴ Distance = 720 km

35. Let 1 man finishes work in x days.

∴ 1 man's 1 day's work = $\frac{1}{x}$

Let 1 boy finishes work in y days.

∴ 1 boy's 1 day's work = $\frac{1}{y}$

According to the question,

$\frac{8}{x} + \frac{12}{y} = \frac{1}{10}$ ...(*i*) and $\frac{6}{x} + \frac{8}{y} = \frac{1}{10}$ ...(*ii*)

On solving, we get $x = 140$ and $y = 280$

36. Let C.P. of tea set be ₹ x

$\text{Loss} = \frac{5}{100} \times x = ₹\ \frac{x}{20}$

C.P. of lemon set be ₹ y

$\text{Profit} = \frac{15}{100} \times y = ₹\ \frac{3y}{20}$

According to the question,

$\frac{3y}{20} - \frac{x}{20} = 7 \Rightarrow 3y - x = 140$ ...(*i*)

Also, if gain on tea set = $\frac{5}{100} \times x = \frac{x}{20}$

and gain on lemon set = $\frac{10}{100} \times y = \frac{y}{10}$

According to the question,

$\frac{x}{20} + \frac{y}{10} = 13 \Rightarrow x + 2y = 260$ ...(*ii*)

On solving equations (*i*) and (*ii*), we get

$x = 100$ and $y = 80$

∴ Cost of tea set = ₹ 100

37. Let number of correct answers = x and number of wrong answers = y

According to the question,

$x + y = 120$ ...(*i*)

and $x - \frac{1}{4}y = 90$ ...(*ii*)

On solving, we get $x = 96$

38. Let number of girls = $2x$ and number of boys = $3x$

∴ $2x + 3x = 40 \Rightarrow x = 8$

∴ Number of girls = 2 × 8 = 16

and number of boys = 3 × 8 = 24

Let out of 5 students, y are boys.

∴ Number of girls = $5 - y$

A.T.Q.,

$$\frac{16 + 5 - y}{24 + y} = \frac{4}{5}$$

$$5(21 - y) = 4(24 + y)$$

$$\Rightarrow 105 - 5y = 96 + 4y$$

$$9y = 105 - 96$$

$$9y = 9 \Rightarrow y = 1$$

∴ There must be one boy among five new students.

39. Let the fare for first kilometre = ₹ x

and fare for remaining distance = ₹ y

According to the question,

$$x + 19y = 37.70 \quad ...(i)$$

$$x + 25y = 48.50 \quad ...(ii)$$

On solving, we get $x = 3.50$ and $y = 1.80$

40. Let price of sugar = ₹ x / kg and price of tea = ₹ y / kg

$\therefore \quad 4x + y = 220 \quad ...(i)$

New price of sugar $= x + \frac{50}{100}x =$ ₹ $\frac{3}{2}x$ / kg

New price of tea $= y + \frac{10}{100}y =$ ₹ $\frac{11}{10}y$ / kg

According to the question,

$$4 \times \frac{3}{2}x + \frac{11}{10}y = 266 \Rightarrow 6x + \frac{11}{10}y = 266 \quad ...(ii)$$

Solving equations (*i*) and (*ii*), we get

$x = 15$ and $y = 160$

41. Let larger number = x

and smaller number = y

A.T.Q. $\quad 3x = 4y + 3$

$$\Rightarrow \quad x = \frac{4y+3}{3} \quad ...(i)$$

and $\quad 7y = 5x + 1 \quad ...(ii)$

Using (*i*) in (*ii*), we get

$$7y = 5\left(\frac{4y+3}{3}\right) + 1$$

$$7y = \frac{20y+15+3}{3}$$

$$\Rightarrow \quad 21y = 20y + 18$$

$$y = 18$$

When $y = 18$, eq. (*i*) becomes $x = \frac{4 \times 18 + 3}{3}$

$$\Rightarrow \quad x = 25$$

$\therefore$ Numbers are 25 and 18.

42. Let one angle be x and its adjacent angle be $y°$

Now $x + y = 180° \Rightarrow y = 180° - x^2 \quad ...(i)$

According to the question,

$$x = \frac{4}{5}y$$

$\Rightarrow 5x = 4y \Rightarrow 5x - 4y = 0 \quad ...(ii)$

Using eq. (*i*), we get

$$5x - 4(180° - x) = 0$$

$$\Rightarrow \quad 5x - 720° + 4x = 0$$

$$9x = 720°$$

$$x = \frac{720°}{9} = 80°$$

$$\therefore \quad y = 100°$$

43. Let cost price of 1 horse = ₹ x

and cost price of 1 cow = ₹ y

$\therefore \quad 4x + 9y = 1340 \quad ...(i)$

S.P. of one horse $= x + \frac{10}{100}x = \frac{11}{10}x$

S.P. one cow $= y + \frac{20}{100}y = \frac{6y}{5}$

$\therefore$ Total selling price $= 4 \times \frac{11}{10}x + 9 \times \frac{6y}{5}$

A.T.Q.

$$\frac{44}{10}x + \frac{54y}{5} = 1340 + 188$$

$$\Rightarrow \quad 44x + 108y = 15280 \quad ...(ii)$$

Eq. (*i*) × 12, we get

$$48x + 108y = 16080 \quad ...(iii)$$

$$44x + 108y = 15280 \quad \text{[From eq.}(ii)\text{]}$$

$$- \quad - \quad -$$

Also subtracting

$$4x = 800$$

$$\Rightarrow \quad x = 200$$

$\therefore$ Cost price of horse = ₹ 200

44. Let age of father = x years and

sum of the ages of 5 children = y years

$\Rightarrow x = y \quad ...(i)$

After 15 years, father's age = $x + 15$ and sum of ages of 5 children $= y + 75$

According to the question,

$$y + 75 = 2(x + 15) \Rightarrow 2x - y = 45 \quad ...(ii)$$

Using eq. (*i*), we get

$$2x - x = 45 \Rightarrow x = 45$$

$\therefore$ Age of father = 45

45. Let largest angle = x

$\therefore$ smallest angle $= \frac{1}{4}x$

Let third angle = y

A.T.Q.,

$$x = y + \frac{1}{4}x \Rightarrow x - \frac{1}{4}x = y$$

$$\Rightarrow \quad \frac{3}{4}x = y \quad ...(i)$$

Also, sum of three angles of a triangle = 180°

$$\Rightarrow \quad x + y + \frac{1}{4}x = 180°$$

$$\Rightarrow \quad x + \frac{3}{4}x + \frac{1}{4}x = 180° \quad \text{[Using eq. }(i)\text{]}$$

$$\Rightarrow \quad 2x = 180°$$

$$\Rightarrow \quad x = 90° \quad ...(i)$$

$\therefore$ Other angles are $\frac{3}{4}\times 90°$, $\frac{1}{4}\times 90°$ = 67.5°, 22.5°.

$\therefore$ Angles are 90°, 67.5°, 22.5°

46. Let time taken for uphill journey = x hrs

and time taken for downhill journey = y hrs

$\therefore \quad x + y = 15$ ...(i)

and $30x + 50y = 650$ ...(ii)

Solving (i) and (ii), we have

Downhill journey = 10 hours and uphill journey = 5 hours

47. Let the number of rabbits = x

and number of pheasants = y

According to the question,

$x + y = 35$

$x = 35 - y$ ... (i) and $4x + 2y = 98$ ...(ii)

Using eq. (i), we get

$$4(35 - y) + 2y = 98$$

$$\Rightarrow 140 - 4y + 2y = 98 \Rightarrow 2y = 42$$

$$y = 21$$

When $y = 21$, eq. (i) becomes

$$x = 35 - 21 = 14$$

$\therefore$ Number rabbits = 14, Number of pheasants = 21

48. Number of lions = x

Number of buffaloes = y

$\therefore \quad 2x + 2y = 120$

$2x + 4y = 180$

$\therefore \quad y = 30, x = 30$

49. Let father's present age = x years

Sum of ages of two sons = y years

$\therefore \quad y = x - 14$ ...(i)

Ten years ago father's age = $(x - 10)$ years

Sum of ages of two sons = $(y - 20)$ years.

$\therefore \quad y - 20 = \frac{1}{3}(x - 10) \Rightarrow 3y - 60 = x - 10$

$\Rightarrow x - 3y = -50$ ...(ii)

Using eq. (i), we get

$$x - 3(x - 14) = -50$$

$$\Rightarrow x - 3x + 42 = -50$$

$$-2x = -92 \Rightarrow x = 46$$

$\therefore$ Age of father = 46 years

Putting the value of x in (i), we get

$y = 46 - 14 = 32$, i.e. sum of sons age = 32 years

But they differ by 2 years, therefore, elder's age = 17 years and younger's age = 15 years.

50. Let age of father = x years and son's age = y years.

According to the question,

$2y + x = 70$

$\Rightarrow x = 70 - 2y$ ... (i)

and $y + 2x = 95$ ...(ii)

Using eq (i), we get

$$y + 2(70 - 2y) = 95$$

$$\Rightarrow y + 140 - 4y = 95 \Rightarrow 3y = 45 \Rightarrow y = 15$$

When $y = 15$, eq. (i) becomes

$$x = 70 - 2 \times 15 = 40$$

$\therefore$ Age of father = 40 years. Age of son = 15 years

51. Let numerator = x and denominator = y

$\therefore$ Fraction = $\frac{x}{y}$

$\frac{x-2}{y+3} = \frac{4}{5}$ ...(i) $\quad \frac{x+6}{2y} = \frac{12}{17}$ ...(ii)

From equation (i)

$5x - 10 = 4y + 12 \Rightarrow 5x - 4y = 22$ ...(iii)

From equation (ii)

$17x + 102 = 24y \Rightarrow 17x - 24y = -102$ ...(iv)

Solving equation (iii) and (iv), we get

$x = 18, y = 17 \quad \therefore$ Fraction is $\frac{18}{17}$.

52. Equations are

$3y° - 5x° = 30°$ ...(i)

$x° + 3x° + y° = 180°$ ...(ii)

Solve for $x°$ and $y°$.

53. Let amount invested at 5% = ₹ x

and amount invested at 7% = ₹ y

$\Rightarrow \quad x + y = 30000$ ...(i)

Also, $\frac{5}{100}x + \frac{7}{100}y = 2000$

$\Rightarrow \quad 5x + 7y = 200000$ ...(ii)

Solving (i) and (ii), we get

$x = 5000, y = 25000$

54. Let digit at unit place = x

and digit at tenth place = y

$\Rightarrow \quad x + y = 9$

Number = $10y + x$

A.T.Q., $\quad 9(10y + x) = 2(10x + y)$

$\Rightarrow \quad 90y + 9x = 20x + 2y$

$\Rightarrow \quad 11x - 88y = 0$

$\Rightarrow \quad x - 8y = 0$

Solving (i) and (ii), we get

$y = 1, x = 8$

$\therefore$ number = 18

55. Let the tenth place digit be x.

and unit place digit be y.

Number = $10x + y$

A.T.Q.,

$x + y = 9$ ... (i)

and $\quad 10x + y - 27 = 10y + x$

$\Rightarrow \quad 9x - 9y = 27$

$\Rightarrow \quad x - y = 3$ ... (ii)

Adding (*i*) and (*ii*),

$$2x = 12$$
$$x = 6$$

Putting value of x in equation (*ii*), we get

$$6 - y = 3$$
$$\Rightarrow \quad -y = 3 - 6$$
$$\Rightarrow \quad -y = -3$$
$$\Rightarrow \quad y = 3$$

So, the given number is 63.

56. Let digit at unit place be x

and digit at ten's place be y

$$\therefore \quad \text{Number} = 10y + x$$

A.T.Q.,

$$(10y + x) + (10x + y) = 154$$
$$\Rightarrow \quad 11x + 11y = 154$$
$$\Rightarrow \quad x + y = 14 \quad ...(i)$$
$$\text{and} \quad x - y = \pm 4$$
$$\Rightarrow \quad x - y = 4 \quad ...(ii)$$
$$\text{or} \quad x - y = -4 \quad ...(iii)$$

Solving (*i*) and (*ii*), we get

$$x = 9, y = 5$$
$$\therefore \quad \text{Number} = 59$$

and solving (*i*) and (*iii*), we get

$$x = 5, y = 9$$
$$\therefore \quad \text{Number} = 95$$

57. Given : $\angle A = (2x - 1)°$, $\angle B = (y + 5)°$,

$\angle C = (2y + 15)°$, $\angle D = (4x - 7)°$

In a cyclic quadrilateral ABCD,

$\angle A + \angle C = 180°$ ...(*i*) and $\angle B + \angle D = 180°$ ...(*ii*)

Putting the value of $\angle A$ and $\angle C$ in (*i*), we get

$$(2x - 1) + (2y + 15) = 180°$$
$$\Rightarrow 2x + 2y + 14 = 180° \Rightarrow 2x + 2y = 166° \quad ...(iii)$$

Dividing the equation by 2, we get

$$x + y = 83°$$

Putting the value of $\angle B$ and $\angle D$ in (*ii*), we get

$$(y + 5) + (4x - 7) = 180° \Rightarrow y + 4x - 2 = 180° \quad ...(iv)$$
$$y + 4x = 182$$

Solve (*iii*) and (*iv*) and proceed.

58. Let speed of A be x km/hr and speed of B be y km/hr.

Distance covered by A in 15 hrs = $15x$ km

Distance covered by B in 15 hrs = $15y$ km

$\therefore \quad 15x - 15y = 150 \quad \Rightarrow \quad x - y = 10$ ...(*i*)

Also distance covered by A in 1 hr = x km

Distance covered by B in 1 hr = y km

$\therefore \quad x + y = 150$ ...(*ii*)

Solve (*i*) and (*ii*) and proceed.

59. Let x and y be the length and width of a rectangle

$\therefore \quad \text{Area} = xy$ ...(*i*)

Case I : When length is increased and breadth is reduced each by 2 units.

$$\Rightarrow \quad (x + 2)(y - 2) = xy - 28$$
$$xy - 2x + 2y - 4 = xy - 28$$
$$-2x + 2y = -24$$
$$x - y = 12 \quad ...(ii)$$

Case II : When length is reduced by 1 and the breadth is increased by 2 units.

$$\Rightarrow \quad (x - 1)(y + 2) = xy + 33$$
$$xy + 2x - y - 2 = xy + 33$$
$$2x - y = 35 \quad ...(iii)$$

Solving (*ii*) and (*iii*), we get

$$x = 23$$

Putting $x = 23$ in (*ii*), we get $y = 11$

$\therefore$ Required Area = $xy = 23 \times 11 = 253$ sq. units

60. Let the man starts his job by monthly salary be ₹ x.

Let fixed increment every year be ₹ y.

According to the question,

Salary after 4 years

$$x + 4y = 1500 \quad ...(i)$$

Salary after 10 years

$$x + 10y = 1800 \quad ...(ii)$$

Solve (*i*) and (*ii*) and get value of x and y.

61.
$$a^2x + b^2y = c^2 \quad ...(i)$$
$$b^2x + a^2y = d^2 \quad ...(ii)$$

Multiply equation (*i*) by b^2 and (*ii*) by a^2 and then subtracting (*i*) from (*ii*),

$$a^2b^2x + a^4y = a^2d^2$$
$$a^2b^2x + b^4y = b^2c^2$$
$$- \quad - \quad -$$
$$(a^4 - b^4)y = a^2d^2 - b^2c^2$$
$$y = \frac{a^2d^2 - b^2c^2}{a^4 - b^4}$$

Putting value of y in equation (*i*), we get

$$a^2x + b^2\left(\frac{a^2d^2 - b^2c^2}{a^4 - b^4}\right) = c^2$$
$$a^2x = c^2 - b^2\left(\frac{a^2d^2 - b^2c^2}{a^4 - b^4}\right)$$
$$a^2x = \frac{c^2(a^4 - b^4) - b^2(a^2d^2 - b^2c^2)}{a^4 - b^4}$$
$$a^2x = \frac{c^2a^4 - c^2b^4 - b^2a^2d^2 + b^4c^2}{a^4 - b^4}$$
$$x = \frac{c^2a^4 - b^2a^2d^2}{a^2(a^4 - b^4)}$$

$$x = \frac{a^2(c^2a^2 - b^2d^2)}{a^2(a^4 - b^4)}$$

$$x = \frac{c^2a^2 - b^2d^2}{a^4 - b^4}$$

62. Let cost of idly be ₹ x, cost of poori be ₹ y, cost of dosa be ₹ z and cost of vada = ₹ x

[∵ cost of vada = cost of idly]

According to the question,

$23x + 18y + 7z + 19x = 108$

$\Rightarrow 42x + 18y + 7z = 108$ ...(*i*)

and $34x + 8x + 22y + 7z = 114$

$\Rightarrow 42x + 22y + 7z = 114$ ...(*ii*)

Subtracting (*i*) from (*ii*), we get $4y = 6 \Rightarrow y = \frac{3}{2}$

∴ Cost of poori = ₹ 1.50.

63. Let income of X be ₹ $8x$ and income of Y be ₹ $7x$.

Expenditure of X = $19y$ and Expenditure of Y = $16y$

According to the question,

$8x - 19y = 1250$ ....(*i*) and $7x - 16y = 1250$ ...(*ii*)

Now proceed.

64. Let number of mangoes with A be x and number of mangoes with B be y

According to the question,

$x + 30 = 2(y - 30) \Rightarrow x + 30 = 2y - 60$

$\Rightarrow x - 2y = -90$ ...(*i*)

and $y + 10 = 3(x - 10) \Rightarrow y + 10 = 3x - 30$

$\Rightarrow 3x - y = 40$ ...(*ii*)

Now solve equations (*i*) and (*ii*).

65. Let number of students in room A = x and number of students in room B = y

Case I. When 10 students are sent from room A to room B.

Number of students in Room A = $x - 10$

Number of students in Room B = $y + 10$

A.T.Q.

$x - 10 = y + 10$

$\Rightarrow x - y = 20$ ...(*i*)

Case II. When 20 students are sent from room B to room A.

Number of students in room A = $x + 20$

Number of students in room B = $y - 20$

A.T.Q.

$x + 20 = 2(y - 20)$

$\Rightarrow x + 20 = 2y - 40$

$\Rightarrow x - 2y = -60$ (*ii*)

Subtracting (*ii*) from (*i*), we get

$$\begin{array}{r} x - y = 20 \\ x - 2y = -60 \\ - \quad + \quad\quad + \\ \hline y = 80 \end{array}$$

When $y = 80$, eq. (*i*) becomes

$x - 80 = 20 \Rightarrow x = 100$

∴ Number of students in room A = 100

∴ Number of students in room B = 80

66. Let number of coins of 50 paise be x. and number of coins of 25 paise be y.

∴ $x + y = 94$ ...(*i*)

and $\frac{50}{100}x + \frac{25}{100}y = 29.75$ ...(*ii*)

Now solve equations (*i*) and (*ii*).

67. (*i*) $3x - 5y + 1 = 0$

$\Rightarrow -5y = -3x - 1$

$\Rightarrow y = \frac{3x + 1}{5}$

The solution table for $3x - 5y + 1$ is

x	3	−7	8
y	2	−4	5

$2x - y + 3 = 0$

$\Rightarrow -y = -2x - 3$

$\Rightarrow y = 2x + 3$

The solution table for $2x - y + 3 = 0$ is

x	1	0	−1
y	5	3	1

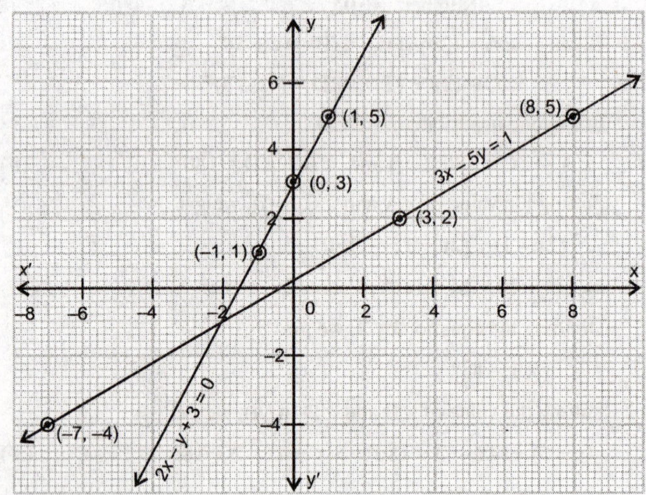

(−2, −1) is the point of intersection, lines intersect x-axis at (−1/3, 0) and (−3/2, 0).

(*ii*) $2x + 3y = 6$

The solution table for $2x + 3y = 6$ is

x	0	3	6
y	2	0	−2

$3x - \frac{5}{2}y = 2$

The solution table for $3x - \frac{5}{2}y = 2$ is

x	0	2 / 3	4
y	–4 / 5	0	4

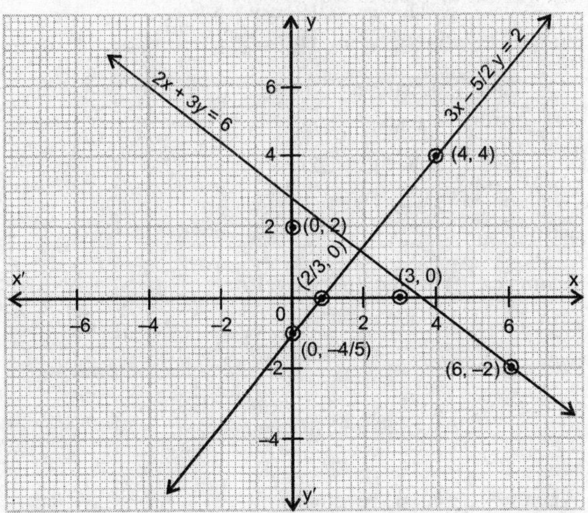

$\therefore$ The points where lines intersect x-axis are (3, 0) and (2/3, 0).

(*iii*) $3x + 2y = 14$, $x - 4y = -7$

Table for eq. $3x + 2y = 14$

x	0	4	2
y	7	1	4

Table for $x - 4y = -7$

x	–7	–3	1
y	0	1	2

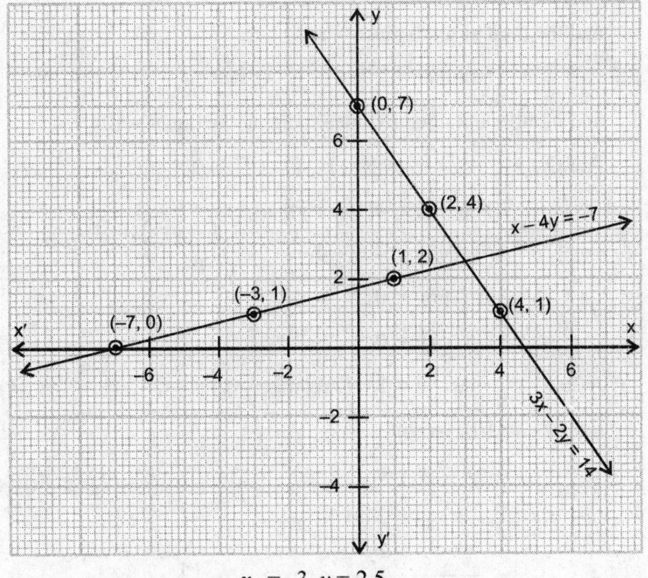

$x = 3, y = 2.5$

Points of intersection with x-axis are (–7, 0) and $\left(\frac{14}{3}, 0\right)$.

(*iv*) Table for $x = 5$

x	5	5	5
y	0	1	2

Table for $3x - 6y = 15$

x	5	3	1
y	0	–1	–2

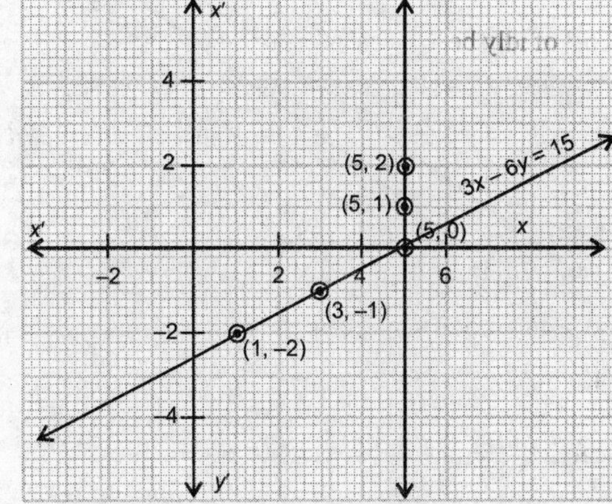

Solution $x = 5$, $y = 0$. Point of intersection with x-axis is (5, 0)

(*v*) Table for $3x - 4y - 1 = 0$

x	–1	3	–5
y	–1	2	–4

Table for $2x - \frac{8}{3}y - 4 = 0$

x	2	6	–2
y	0	3	–3

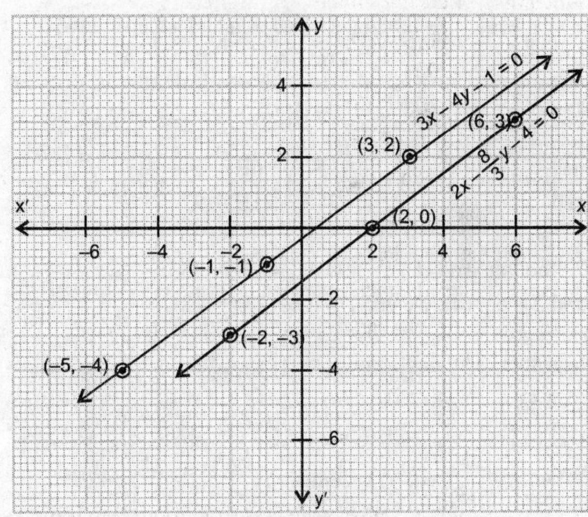

Lines intersect x-axis at (1/3, 0) and (2, 0).

(*vi*) $x - 5y = 6$

$2x - 10y = 10$

Table for $x = 5y + 6$

x	1	6	–4
y	–1	0	–2

Table for $2x = 10y + 10$

x	0	5	–5
y	–1	0	–2

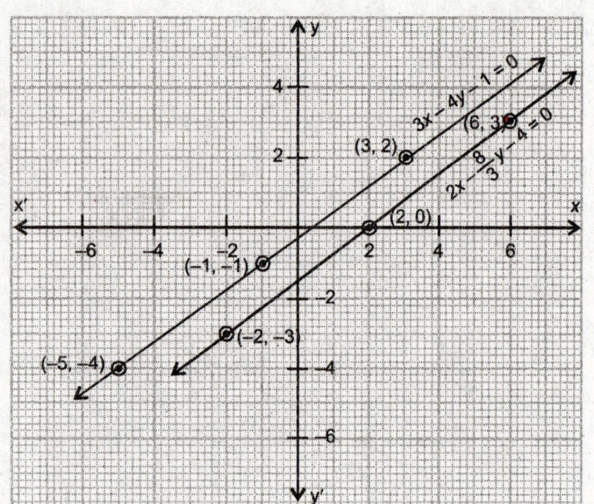

68. $\frac{xy}{x+y} = \frac{1}{5} \Rightarrow \frac{x+y}{xy} = \frac{5}{1}$

$\Rightarrow \quad \frac{1}{y} + \frac{1}{x} = 5 \qquad ...(i)$

and $\frac{xy}{x-y} = \frac{1}{7} \Rightarrow \frac{x-y}{xy} = 7$

$\Rightarrow \quad \frac{1}{y} - \frac{1}{x} = 7 \qquad ...(ii)$

Now solve equation (*i*) and (*ii*) by assuming

$\frac{1}{y} = u$, $\frac{1}{x} = v$, and get value of u and v then x and y.

69. Let $\frac{1}{x+y} = \text{A}$ and $\frac{1}{x-y} = \text{B}$

$\Rightarrow \quad 57\text{A} + 6\text{B} = 5 \qquad ...(i)$

and $\quad 38\text{A} + 21\text{B} = 9 \qquad ...(ii)$

$7 \times$ eq. (*i*), we get $\quad 399\text{ A} + 42\text{ B} = 35 \qquad ...(iii)$

and $2 \times$ eq. (*ii*), we get $\quad 76\text{ A} + 42\text{ B} = 18 \qquad ...(iv)$

$\quad - \quad - \quad -$

$$323\text{ A} = 17$$

$$\text{A} = \frac{17}{323} = \frac{1}{19}$$

When $\text{A} = \frac{1}{19}$, eq. (*i*) becomes

$$57 \times \frac{1}{19} + 6\text{B} = 5 \Rightarrow 6\text{B} = 2$$

$$\text{B} = \frac{1}{3}$$

When $\quad \text{A} = \frac{1}{19}$

$\Rightarrow \quad \frac{1}{x+y} = \frac{1}{19} \Rightarrow x + y = 19 \qquad ...(v)$

and when $\text{B} = \frac{1}{3}$

$\Rightarrow \quad \frac{1}{x-y} = \frac{1}{3} \Rightarrow x - y = 3 \qquad ...(vi)$

Adding eq. (*v*) and eq. (*vi*), we get

$$x + y = 19$$
$$x - y = 3$$
$$2x = 22 \Rightarrow x = 11$$

When $x = 11$, eq. (*v*) becomes

$$11 + y = 19 \Rightarrow y = 8$$

$\therefore \quad x = 11, y = 8$

70. $\frac{xy}{x+y} = \frac{1}{2}$

$x + y = 2xy \qquad ...(i)$

and $\quad \frac{xy}{x-y} = \frac{1}{6} \Rightarrow x - y = 6xy \qquad ...(ii)$

Adding eq. (*i*) and eq. (*ii*), we get

$$x + y = 2xy$$
$$x - y = 6xy$$
$$2x = 8xy \Rightarrow y = \frac{1}{4}$$

When $y = \frac{1}{4}$, eq. (*i*) becomes

$$x + \frac{1}{4} = 2x \times \frac{1}{4}$$

$\Rightarrow \quad x - \frac{1}{2}x = -\frac{1}{4}$

$\Rightarrow \quad \frac{1}{2}x = -\frac{1}{4} \Rightarrow x = -\frac{1}{2}$

$\therefore \quad x = -\frac{1}{2}, y = \frac{1}{4}$

71. Let the digit at unit place $= x$

and the digit at tenth place $= y$

$\therefore \quad$ Number $= 10y + x$

and number obtained by interchanging the digits $= 10x + y$

A.T.Q.,

$$7(10y + x) = 4(10x + y)$$

$\Rightarrow \quad 70y + 7x = 40x + 4y$

$$33x - 66y = 0$$

$x - 2y = 0 \qquad ...(i)$

Also, $\quad x - y = 3$ (keeping $x > y$) $\qquad ...(ii)$

Eq. (*i*) – eq.(*ii*), we get

$$x - 2y = 0$$
$$x - y = 3$$
$$- \quad + \quad -$$
$$-y = -3 \Rightarrow y = 3$$

When $\quad y = 3$, eq. (*ii*) becomes

$$x - 3 = 3 \Rightarrow x = 6$$

$\therefore \quad$ Number $= 10\,y + x = 10 \times 3 + 6 = 36$

72. Table for $2x + y = 10$

x	5	4	3
y	0	2	4

Table for $4x - y = 8$

x	2	3	1
y	0	4	–4

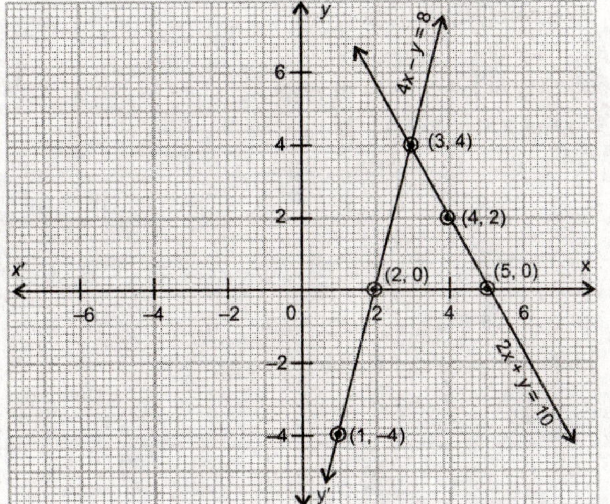

Common solution $x = 3, y = 4$. $(1, -4)$ lies on the line $4x - y = 8$

73. $x - y + 1 = 0$ and $2x + y - 10 = 0$

$\Rightarrow$ $y = x + 1$ and $\Rightarrow$ $y = 10 - 2x$

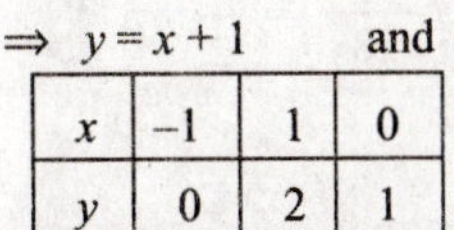

x	–1	1	0
y	0	2	1

x	2	4	5
y	6	2	0

Plotting these points on the graph, we get

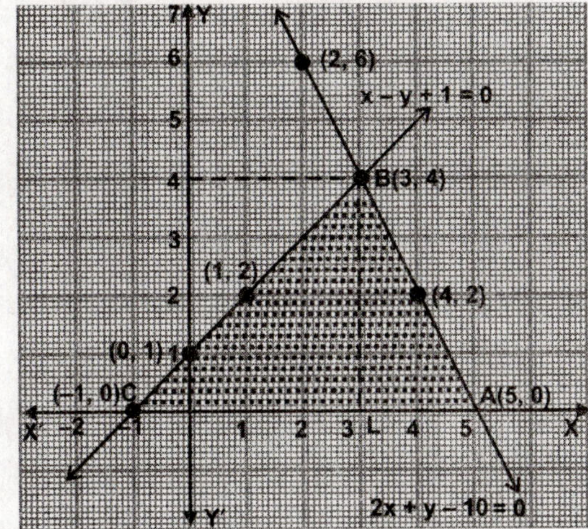

Shaded portion is the area bounded by the lines and x-axis.

$$\text{Area} = \frac{1}{2}\,AC \times BL = \frac{1}{2} \times 6 \times 4 = 12 \text{ sq. units}$$

74. $x - y = 1$ $\qquad$ $2x + y = 8$

$x = 1 + y$ $\qquad$ $x = \dfrac{8 - y}{2}$

x	2	3	4
y	1	2	3

x	3	2	1
y	2	4	6

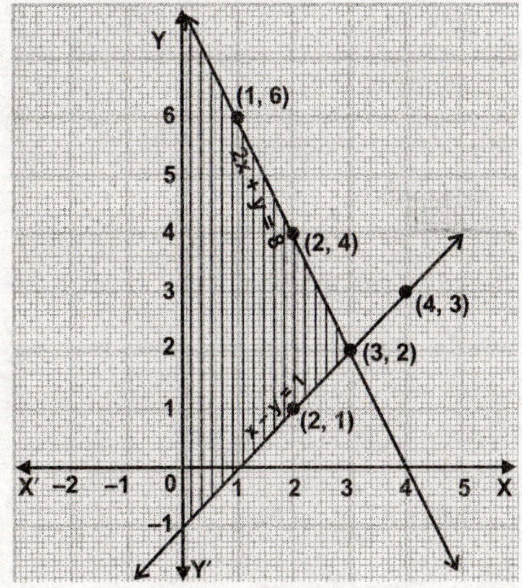

75. $2x + 3y = 12$ and $x - y = 1$

$x = \dfrac{12 - 3y}{2}$ and $x = 1 + y$

x	3	0	6
y	2	4	0

x	2	3	4
y	1	2	3

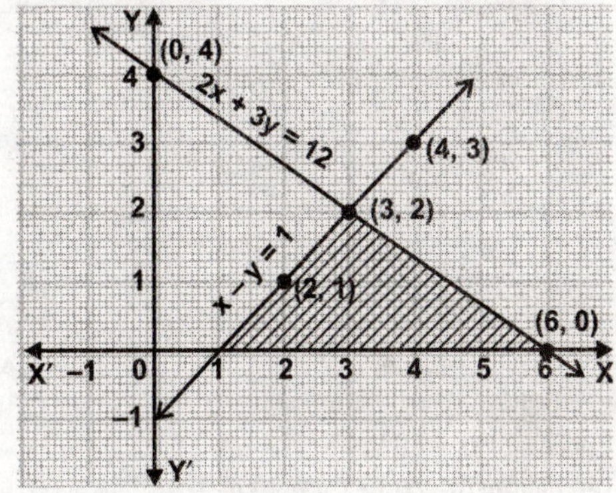

From the graph we get $x = 3, y = 2$

76. Total distance = 600 km

Let speed of train be x km/h and speed of car be y km/h.

Case I : When 400 km covers by train and the rest by car

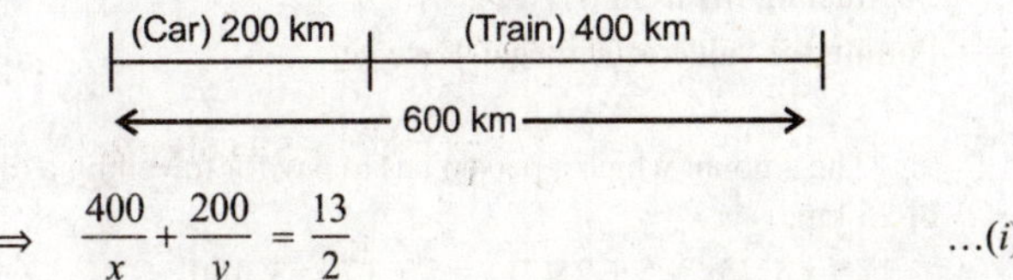

$$\Rightarrow \quad \frac{400}{x} + \frac{200}{y} = \frac{13}{2} \qquad \ldots(i)$$

Case II : When 200 km covers by train and the rest by car.

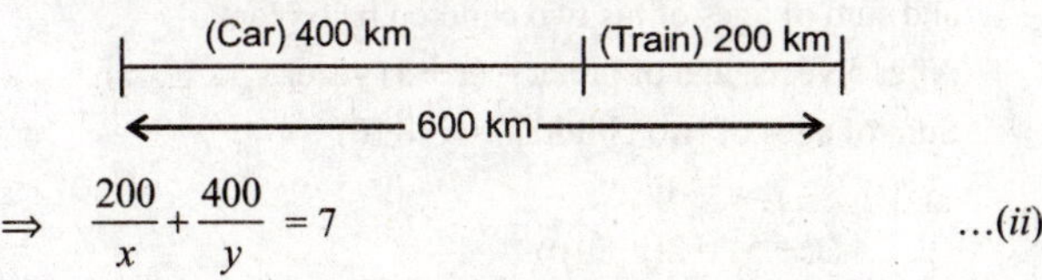

$$\Rightarrow \quad \frac{200}{x} + \frac{400}{y} = 7 \qquad \ldots(ii)$$

Putting $\dfrac{1}{x} = p$ and $\dfrac{1}{y} = q$ in (*i*) and (*ii*), we get

$$400p + 200q = \frac{13}{2} \quad ...(iii)$$

$$200p + 400q = 7 \quad ...(iv)$$

$$- \quad - \quad -$$

$$200p - 200q = -\frac{1}{2}$$

$$p - q = -\frac{1}{400}$$

$$\therefore \quad p = q - \frac{1}{400} \quad ...(v)$$

Putting $p = q - \frac{1}{400}$ in (*iii*), we get

$$400\left(q - \frac{1}{400}\right) + 200q = \frac{13}{2}$$

$$\therefore \quad 400q - 1 + 200q = \frac{13}{2}$$

or $$600q = \frac{13}{2} + 1 = \frac{15}{2}$$

$$\therefore \quad q = \frac{15}{2} \times \frac{1}{600} = \frac{1}{80}$$

Putting $q = \frac{1}{80}$ in (*v*), we get

$$p = \frac{1}{80} - \frac{1}{400} = \frac{5-1}{400} = \frac{4}{400}$$

$$p = \frac{1}{100}$$

$$\therefore \quad x = \frac{1}{p} = 100 \text{ km/h and}$$

$$y = \frac{1}{q} = 80 \text{ km/h}$$

$\therefore$ Speed of train = 100 km/h and

Speed of car = 80 km/h

77. Let fixed charge of taxi be ₹ x per km and running charge of taxi be ₹ y per km.

According to the question,

$$x + 10y = 75 \quad ...(i)$$

$$x + 15y = 110 \quad ...(ii)$$

Subtracting (*ii*) from (*i*), we get $y = 7$

Putting the value of y in eq. (*i*), we get

$$x + 70 = 75 \quad \Rightarrow \quad x = 5$$

$\therefore$ The amount which a person has to pay for travelling a distance of 25 km.

$$= x + 25y = 5 + 25(7) = 5 + 175 = ₹\ 180.$$

78. Let the age of father be x years.

and sum of ages of his two children be y years.

After 5 years, age of father = $(x + 5)$ years.

Sum of ages of two children = $(y + 10)$

$$x = 3y \quad ...(i)$$

$$(x + 5) = 2(y + 10) \quad ...(ii)$$

Solve for x and y.

79. Let present age of son = x years

and present age of father = y years

According to the question, 5 years hence

Age of son = $(x + 5)$ years

Age of father = $(y + 5)$ years

So, $3(x + 5) = y + 5 \Rightarrow 3x - y = -10 \quad ...(i)$

5 years earlier,

Age of son = $(x - 5)$ years

Age of father = $(y - 5)$ years

So, $7(x - 5) = y - 5 \Rightarrow 7x - y = 30 \quad ...(ii)$

Solve (*i*) and (*ii*) and get the ages.

80. $$\frac{1}{7x} + \frac{1}{6y} = 3 \quad ...(i)$$

$$\frac{1}{2x} - \frac{1}{3y} = 5 \quad ...(ii)$$

Multiplying equation (*ii*) by $\frac{1}{2}$, we get

$$\frac{1}{4x} - \frac{1}{6y} = \frac{5}{2} \quad ...(iii)$$

$$\frac{1}{7x} + \frac{1}{6y} = 3 \quad ...(i)$$

Adding eq. (*i*) and (*iii*), we get

$$\frac{1}{4x} + \frac{1}{7x} = \frac{5}{2} + 3 \Rightarrow \frac{7+4}{28x} = \frac{11}{2}$$

$$\Rightarrow \quad \frac{11}{28x} = \frac{11}{2}$$

$$\Rightarrow \quad 28x = 2 \Rightarrow x = \frac{1}{14}$$

Putting the value of x in eq. (*i*), we get

$$y = \frac{1}{6}$$

81. Put $\frac{1}{x+1} = \text{A}$ and $\frac{1}{y-1} = \text{B}$, we get

$$5\text{A} - 2\text{B} = \frac{1}{2} \quad ...(i) \quad \text{and } 10\text{A} + 2\text{B} = \frac{5}{2} \quad ...(ii)$$

On solving these equations, we get

$$\text{A} = \frac{1}{5} \text{ and B} = \frac{1}{4}$$

$$\therefore \quad \frac{1}{x+1} = \frac{1}{5} \text{ and } \frac{1}{y-1} = \frac{1}{4}$$

$$\Rightarrow \quad x = 4 \text{ and } y = 5$$

82. $$\frac{x+y}{xy} = 2 \quad \Rightarrow x + y = 2xy \quad ...(i)$$

$$\frac{x-y}{xy} = 6 \quad \Rightarrow x - y = 6xy \quad ...(ii)$$

Adding eqs. (*i*) and (*ii*), we get

$$x + y = 2xy$$

$$x - y = 6xy$$

$$2x = 8xy \Rightarrow y = \frac{1}{4}$$

When $y = \frac{1}{4}$, eq. (*i*) becomes

$$x + \frac{1}{4} = 2x \times \frac{1}{4} \Rightarrow x - \frac{1}{2}x = \frac{-1}{4}$$

$$\Rightarrow \quad \frac{1}{2}x = \frac{-1}{4}$$

$$x = \frac{-1}{2}$$

$$\therefore \quad x = \frac{-1}{2} \text{ and } y = \frac{1}{4}$$

83. $\frac{2}{y} + \frac{3}{x} = \frac{7}{xy}$

Multiplying by xy, we get

$2x + 3y = 7$ ...(*i*)

Similarly, $\frac{6}{y} + \frac{9}{x} = \frac{11}{xy}$ becomes $6x + 9y = 11$ ...(*ii*)

Now, $\frac{a_1}{a_2} = \frac{b_1}{b_2} \neq \frac{c_1}{c_2}$

84. Let fixed charges = ₹ x

Food charges per day = ₹ y

A.T.Q. $\quad x + 20y = 1000$ ...(*i*)

and $\quad x + 26y = 1180$ ...(*ii*)

Solving (*i*) and (*ii*), we get

$y = 30$ and $x = 400$

85. (*i*) $\quad (a - b)x + (a + b)y = a^2 - b^2 - 2ab$ ...(*i*)

$(a + b)(x + y) = a^2 + b^2$

$\Rightarrow \quad (a + b)x + (a + b)y = a^2 + b^2$ ...(*ii*)

Eq. (*i*) – (*ii*), we get

$$\begin{aligned} (a - b)x + (a + b)y &= a^2 - b^2 - 2ab \\ (a + b)x + (a + b)y &= a^2 + b^2 \\ - \quad\quad - \quad\quad & \quad - \; - \end{aligned}$$

$(a - b)x - (a + b)x = - 2b^2 - 2ab$

$\Rightarrow \quad -2bx = -2b(b + a)$

$\Rightarrow \quad x = (a + b)$

When $x = (a + b)$, eq. (*i*) becomes

$(a - b)(a + b) + (a + b)y = a^2 - b^2 - 2ab$

$\Rightarrow \quad a^2 - b^2 + (a + b)y = a^2 - b^2 - 2ab$

$(a + b)y = -2ab$

$$y = \frac{-2ab}{a + b}$$

$$\therefore \quad x = a + b, y = \frac{-2ab}{a + b}$$

(*ii*) $\quad px + qy = p - q$ ...(*i*)

$qx - py = p + q$ ...(*ii*)

$q \times$ eq. (*i*), we get, $\quad pqx + q^2y = qp - q^2$ ...(*iii*)

$p \times$ eq. (*ii*), we get, $\quad pqx - p^2y = p^2 + pq$ ...(*iv*)

Subtracting $\quad - \quad + \quad\quad - \quad -$

$q^2y + p^2y = - q^2 - p^2$

$(q^2 + p^2)y = - (q^2 + p^2)$

$y = - 1$

When $y = -1$, (*i*) becomes

$px + q \times -1 = p - q$

$\Rightarrow \quad px = p \Rightarrow x = 1$

$\therefore \quad x = 1, y = -1$

(*iii*) $\quad ax + by = c$ ...(*i*)

$bx + ay = 1 + c$ ...(*ii*)

$b \times$ eq. (*i*), we get $abx + b^2y = bc$ ...(*iii*)

$a \times$ eq. (*ii*), we get $abx + a^2y = a + ac$ ...(*iv*)

Subtracting $\quad - \quad - \quad\quad - \; -$

$b^2y - a^2y = bc - a - ac$

$\Rightarrow \quad (b^2 - a^2)y = bc - a - ac$

$$y = \frac{bc - a - ac}{b^2 - a^2}$$

Now,

$a \times$ eq. (*i*), we get $a^2x + aby = ac$ ...(*v*)

$b \times$ eq. (*ii*), we get $b^2x + aby = b + bc$

Subtracting $\quad - \quad - \quad\quad - \; -$

$a^2x - b^2x = ac - b - bc$

$\Rightarrow \quad (a^2 - b^2)x = ac - b - bc$

$$x = \frac{ac - b - bc}{a^2 - b^2}$$

$$\therefore \quad x = \frac{ac - b - bc}{a^2 - b^2}, y = \frac{bc - a - ac}{a^2 - b^2}$$

(*iv*) $\quad \frac{x}{a} - \frac{y}{b} = 0 \Rightarrow \frac{bx - ay}{ab} = 0$

$\Rightarrow \quad bx - ay = 0$ ...(*i*)

$ax + by = a^2 + b^2$ ...(*ii*)

$a \times$ eq. (*i*), we get, $\quad abx - a^2y = 0$ ...(*iii*)

$b \times$ eq. (*ii*), we get, $\quad abx + b^2y = a^2b + b^3$

Subtracting $\quad - \quad - \quad\quad - \; -$

$- a^2y - b^2y = - a^2b - b^3$

$(- a^2 - b^2)y = b(- a^2 - b^2)$

$\Rightarrow \quad y = b$

When $y = b$, eq. (*i*) becomes

$bx - ab = 0$

$\Rightarrow \quad bx = ab \Rightarrow x = a$

$\therefore \quad x = a, y = b$

86. Let speed of boat in still water = x km/h

speed of current = y km/h

$\therefore$ Downstream speed $= (x + y)$ km/h

Upstream speed $= (x - y)$ km/h

Now, distance travelled in downstream = 20 km

$$\therefore \quad \text{time} = \frac{20}{x + y}$$

$$\Rightarrow \quad \frac{20}{x + y} = 2 \Rightarrow x + y = \frac{20}{2}$$

$\Rightarrow \quad x + y = 10$

Distance travelled in upstream = 4 km

$$\therefore \quad \text{time} = \frac{4}{x - y} \Rightarrow \frac{4}{x - y} = 2$$

$\Rightarrow \quad x - y = 2$ ...(*ii*)

Solving (*i*) and (*ii*), we get

$x = 6$ and $y = 4$

$\therefore$ speed of boat = 6 km/hr

speed of current = 4 km/hr

87.

$$\angle C = 3\angle B \quad ...(i)$$

and $$\angle C = 2(\angle A + \angle B) \quad ...(ii)$$

Also, $$\angle A + \angle B + \angle C = 180° \quad ...(iii)$$

[Sum of three angles of Δ]

$\Rightarrow \quad \angle A + \angle B + 3\angle B = 180°$ [Using eq. (i)]

$\Rightarrow \quad \angle A + 4\angle B = 180° \quad ...(iv)$

Also $\angle A + \angle B + 2(\angle A + \angle B) = 180°$ [Using (ii) in eq. (iii)]

$\Rightarrow \quad 3\angle A + 3\angle B = 180° \quad ...(iv)$

$\Rightarrow \quad \angle A + \angle B = 60° \quad ...(v)$

Eq. (iv) – eq. (v), we get

$$\angle A + 4\angle B = 180°$$
$$\angle A + \angle B = 60°$$
$$- \quad - \quad -$$
$$3\angle B = 120°$$

$\Rightarrow \quad \angle B = 40°$

When $\angle B = 40°$, eq (iv) becomes

$$\angle A + 4 \times 40° = 180°$$

$\Rightarrow \quad \angle A = 180° - 160° = 20°$

From eq (i) $\quad \angle C = 3 \times 40° \Rightarrow \angle C = 120°$

88. Let length of the rectangle = x units

and breadth of the rectangle = y units

$\therefore \quad$ Area = $x.y$ sq. units

(i) When length = $(x - 5)$ units

and breadth = $(y + 3)$ units

Area = $xy - 9$

$\Rightarrow \quad (x-5)(y+3) = xy - 9$

$\Rightarrow \quad xy + 3x - 5y - 15 = xy - 9$

$\Rightarrow \quad 3x - 5y = 6 \quad ...(i)$

(ii) When length = $(x + 3)$ units

breadth = $(y + 2)$ units

then area = $xy + 67$

$\Rightarrow \quad (x+3)(y+2) = xy + 67$

$\Rightarrow \quad xy + 2x + 3y + 6 = xy + 67$

$\Rightarrow \quad 2x + 3y = 61 \quad ...(ii)$

2 × eq. (i) we get $\quad 6x - 10y = 12 \quad ...(iii)$

3 × eq. (ii) we get $\quad 6x + 9y = 183 \quad ...(iv)$

Subtracting $\quad - \quad - \quad -$

$$-19y = -171$$
$$y = \frac{-171}{-19} = 9$$

When $y = 9$, eq (i) becomes

$$3x - 5 \times 9 = 6$$

$\Rightarrow \quad 3x = 51 \Rightarrow x = 17$

$\therefore$ Length = 17 units, breadth = 9 units

89. Let number of students in a row = x

and number of rows = y

$\therefore \quad$ Total number of students = xy

A.T.Q.,

If number of students in a row = $x + 3$

then number of rows = $y - 1$

$\Rightarrow \quad (x+3)(y-1) = xy$

$\Rightarrow \quad xy - x + 3y - 3 = xy$

$\Rightarrow \quad -x + 3y - 3 = 0$

$\quad x - 3y + 3 = 0 \quad ...(i)$

Also, if number of students in a row = $x - 3$

then number of rows = $y + 2$

$\Rightarrow \quad (x-3)(y+2) = xy$

$\Rightarrow \quad xy + 2x - 3y - 6 = xy$

$\Rightarrow \quad 2x - 3y - 6 = 0 \quad ...(ii)$

Solving (i) and (ii), we get

$$x = 9, y = 4$$

$\therefore \quad$ Number of students = 36

LONG ANSWER TYPE QUESTIONS

1. $2x + y = 2$ ... (i), $2y - x = 4$ ... (ii)

From (i), $\quad 2x + y = 2$

x	1	0	2
y	0	2	–2

From (ii), $\quad 2y - x = 4$

x	0	–4	2
y	2	0	3

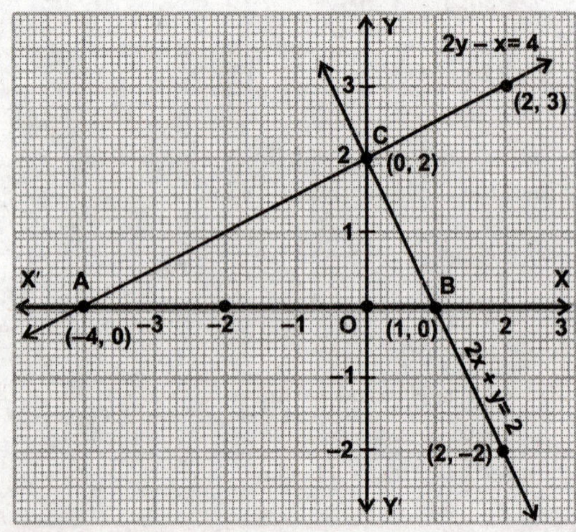

Area $\Delta = \frac{1}{2} AB \times CO = \frac{1}{2} \times 5 \times 2 = 5$ square units

2. Proceed as in Q.1 to obtain the solution. Graph is shown below.

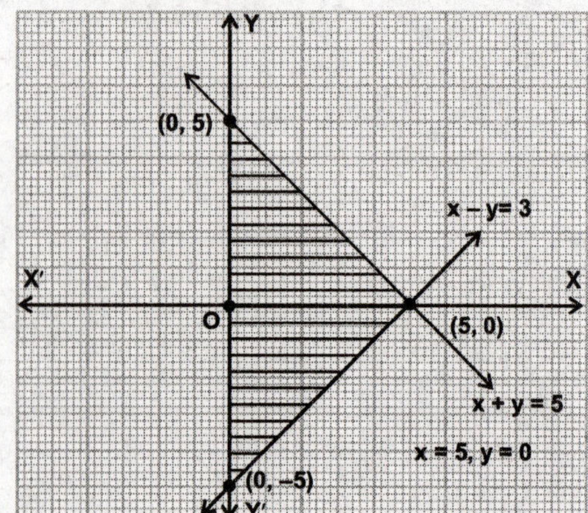

3. $a_1 = 3, b_1 = 1, c_1 = 1$

$a_2 = 2k - 1, b_2 = k - 1, c_2 = 2k + 1$

For no solution $\frac{a_1}{a_2} = \frac{b_1}{b_2} \neq \frac{c_1}{c_2}$

$$\Rightarrow \quad \frac{3}{2k-1} = \frac{1}{k-1} \neq \frac{1}{2k+1}$$

$$\Rightarrow \quad \frac{3}{2k-1} = \frac{1}{k-1} \quad \text{and} \quad \frac{1}{k-1} \neq \frac{1}{2k+1}$$

$$3k - 3 = 2k - 1 \quad \text{and} \quad 2k + 1 \neq k - 1$$

$$k = 2 \qquad k \neq -2$$

$$\therefore \quad k = 2$$

4. Let $\frac{1}{x-1} = \text{A}$ and $\frac{1}{y-2} = \text{B}$

Then given equations become

$$5\text{A} + \text{B} = 2 \ldots (i) \quad \text{and} \quad 6\text{A} - 3\text{B} = 1 \quad \ldots (ii)$$

Solving (*i*) and (*ii*), we get

$$\text{A} = \frac{1}{3} \quad \text{and} \quad \text{B} = \frac{1}{3}$$

$$\Rightarrow \quad \frac{1}{x-1} = \frac{1}{3} \quad \text{and} \quad \frac{1}{y-2} = \frac{1}{3}$$

$$\Rightarrow \quad x - 1 = 3 \quad \text{and} \quad y - 2 = 3$$

$$x = 4 \qquad y = 5$$

5. $x + 3y = 6$

$x = 6 - 3y$

x	0	3	6
y	2	1	0

$2x = 12 + 3y$

x	0	3	6
y	–4	–2	0

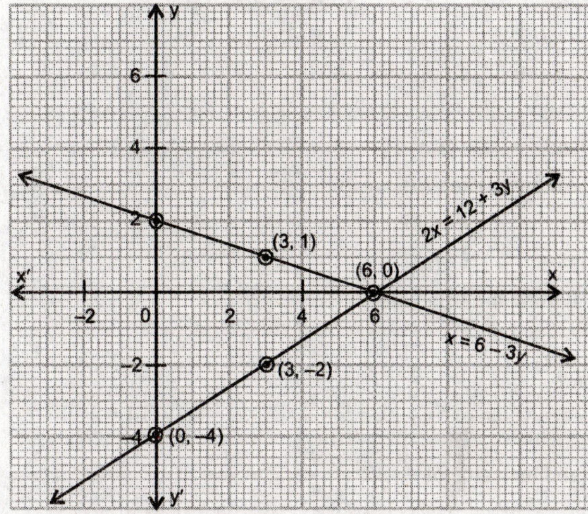

Proceed and find the area as given in Q. No. 1.

NCERT EXEMPLAR PROBLEMS

1. (*i*) No, $x - 2y = 0$

$2x - y = 0$

$\Rightarrow$ Unique solution $\left(\because \frac{a_1}{a_2} \neq \frac{b_1}{b_2}\right)$

(*ii*) No, $3x + y = 3$

$6x + 2y = 6$

$\Rightarrow$ Infinitely many solutions $\left(\because \frac{a_1}{a_2} = \frac{b_1}{b_2} = \frac{c_1}{c_2}\right)$

2. (*i*) No $21x + y = 21$

$7x + 3y = 7$ $\left(\because \frac{a_1}{a_2} \neq \frac{b_1}{b_2}\right)$

(*ii*) No, $5x + 10y = -4$

$4x + 8y = -\frac{5}{16}$ $\left(\because \frac{a_1}{a_2} = \frac{b_1}{b_2} \neq \frac{c_1}{c_2}\right)$

3. (*i*) Yes, $6x - 10y = 5$

$6x - 90y = 5$ consistent $\left(\because \frac{a_1}{a_2} \neq \frac{b_1}{b_2}\right)$

(*ii*) Yes, $2ax + by = a$

$4ax + 2by = 2a$

consistent but dependent $\left(\because \frac{a_1}{a_2} = \frac{b_1}{b_2} = \frac{c_1}{c_2}\right)$

4. No, for infinite many solutions

$$\frac{\lambda}{2} = \frac{3}{6} = -\frac{7}{14}$$

$$\Rightarrow \quad \frac{\lambda}{2} = \frac{1}{2} = -\frac{1}{2}$$

but $\frac{1}{2} \neq -\frac{1}{2}$

$\therefore$ No value of λ will satisfy the condition for infinite many solutions.

5. Not true, $x = 7$ will be parallel to y-axis.

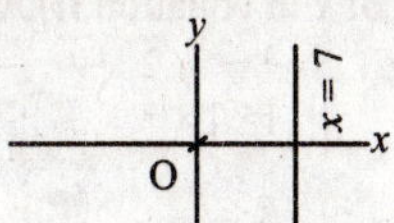

6. $\because$ $\lambda x + y = \lambda^2$

$x + \lambda y = 1$

(*i*) For no solution, $\frac{a_1}{a_2} = \frac{b_1}{b_2} \neq \frac{c_1}{c_2}$

$$\Rightarrow \quad \frac{\lambda}{1} = \frac{1}{\lambda} \neq \frac{\lambda^2}{1}$$

$$\Rightarrow \quad \frac{\lambda}{1} = \frac{1}{\lambda}$$

$$\Rightarrow \quad \lambda^2 = 1 \text{ or } 1 \neq \lambda^3$$

$$\Rightarrow \quad \lambda = \pm 1$$

(*ii*) For infinitely many solutions, $\frac{a_1}{a_2} = \frac{b_1}{b_2} = \frac{c_1}{c_2}$

$$\Rightarrow \quad \frac{\lambda}{1} = \frac{1}{\lambda} = \frac{\lambda^2}{1}$$

$$\Rightarrow \quad \lambda^2 = 1 \quad \text{or} \quad \lambda^3 = 1$$
$$\Rightarrow \quad \lambda = \pm 1 \text{ or } \quad \lambda = 1, \omega, \omega^2$$

where ω and ω^2 are cube roots of unity.

(*iii*) For a unique solution, $\frac{a_1}{a_2} \neq \frac{b_1}{b_2}$

$$\Rightarrow \quad \frac{\lambda}{1} \neq \frac{1}{\lambda}$$
$$\Rightarrow \quad \lambda^2 \neq 1$$
$$\Rightarrow \quad \lambda \neq \pm 1$$

7. $\because \quad 2x + y = 23$ ...(*i*)

and $4x - y = 19$ ...(*ii*)

Adding (*i*) and (*ii*), we get

$$6x = 42$$
$$\Rightarrow \quad x = 7$$

Putting the value of x in (*i*), we get

$$2 \times 7 + y = 23$$
$$\Rightarrow \quad y = 9$$

then value of $5y - 2x = 5 \times 9 - 2 \times 7 = 45 - 14 = 31$

and the value of $\frac{y}{x} - 2 = \frac{9}{7} - 2 = -\frac{5}{7}$

8. Solving $x + 3y = 13$ and $3x + y = 7$ we get $x = 1, y = 4$.

9. (*i*) $x + y = 3.3$ ...(*i*)

$3x - 2y = -0.6$ ...(*ii*)

By Elimination Method,

$$2x + 2y = 6.6$$
$$3x - 2y = -0.6$$
$$5x = 6 \quad \Rightarrow x = \frac{6}{5} = 1.2$$
$$\frac{6}{5} + y = 3.3$$
$$\Rightarrow \quad y = 3.3 - 1.2 = 2.1$$

Putting value of x in equation (*i*), we get $y = 2.1$

(*ii*) $4x + \frac{6}{y} = 15\ ] \times 3$ ... (*iii*)

$6x - \frac{8}{y} = 14\ ] \times 2$ ... (*iv*)

$$12x + \frac{18}{y} = 45$$
$$12x - \frac{16}{y} = 28$$
$$- \quad + \quad -$$
$$\frac{34}{y} = 17$$
$$2 = y$$
$$4x + \frac{6}{2} = 15$$
$$4x = 15 - 3 = 12$$
$$x = \frac{12}{4} = 3$$

10. $\frac{x}{10} + \frac{y}{5} - 1 = 0$, and $\frac{x}{8} + \frac{y}{6} = 15$

$\Rightarrow \quad x + 2y = 10$ ...(*i*)

and $\frac{3x + 4y}{24} = 15$

$\Rightarrow \quad 3x + 4y = 360$ ... (*ii*)

By Elimination method,

$$2x + 4y = 20$$
$$3x + 4y = 360$$
$$- \quad - \quad -$$
$$\Rightarrow \quad -x = -340$$
$$x = 340$$

Putting the value of x in (*i*)

$$340 + 2y = 10$$
$$\Rightarrow \quad 2y = -330 \Rightarrow y = -165$$

Given $y = \lambda x + 5$

$$-165 = 340\lambda + 5$$
$$-170 = 340\lambda$$
$$\frac{-1}{2} = \lambda$$

11. Let fixed charge be ₹ x.

and charge of extra days be ₹ y per day.

A.T.Q.,

$x + 4y = 22$ ...(*i*)

$x + 2y = 16$ ...(*ii*)

By Elimination Method

$$x + 4y = 22$$
$$x + 2y = 16$$
$$- \quad - \quad -$$
$$2y = 6 \Rightarrow y = 3$$

Putting value of y in equation (*i*), we get

$$x + 4y = 22 \Rightarrow x + 4 \times 3 = 22$$
$$x = 22 - 12$$
$$x = 10$$

12. $2x - y = 4$

x	2	3	5
y	0	2	6

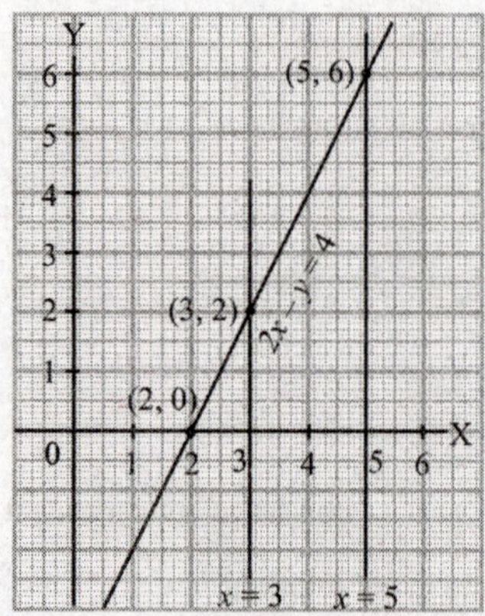

Quadrilateral is like trapezium whose parallel sides are 2 units and 6 units. Distance between parallel sides is 2 units.

So, area of trapezium $= \frac{1}{2} \times$ (Sum of parallel sides) $\times$ (Distance between parallel sides)

$= \frac{1}{2}(2+6) \times 2 = 8$ sq. units

13. Let digit at one's place $= x$

and digit at ten's place $= y$

so number $= 10y + x$

A.T.Q.

$$10y + x = 8(x+y) - 5 \quad \text{... (i)}$$

$$10y + x = 16(y - x) + 3 \quad \text{... (ii)}$$

Equation (*i*) becomes

$$10y + x = 8x + 8y - 5$$

$$7x - 2y = 5] \times 3 \quad \text{...(iii)}$$

Equation (*ii*) becomes

$$10y + x = 16y - 16x + 3$$

$$17x - 6y = 3 \quad \text{...(iv)}$$

By Elimination method

$$21x - 6y = 15$$

$$17x - 6y = 3$$

$$- \quad + \quad -$$

$$4x = 12 \Rightarrow x = 3$$

By equation (*iii*), we have

$$7(3) - 2y = 5 \Rightarrow 21 - 5 = 2y$$

$$\Rightarrow 16 = 2y \Rightarrow 8 = y$$

$\therefore$ Number $= 83$.

14. Let bananas in lot A $= x$

and bananas in lot B $= y$

A.T.Q.,

$$\frac{x}{3} \times 2 + y = 400 \quad \text{... (i)}$$

$$x + \frac{4}{5}y = 460 \quad \text{... (ii)}$$

Equations (*i*) and (*ii*) become

$$2x + 3y = 1200] \times 5 \quad \text{... (iii)}$$

$$5x + 4y = 2300] \times 2 \quad \text{... (iv)}$$

By Elimination method

$$10x + 15y = 6000$$

$$10x + 8y = 4600$$

$$- \quad - \quad -$$

$$7y = 1400 \Rightarrow y = 200$$

Putting in equation (*iii*),

$$2x + 3y = 1200$$

$$\Rightarrow 2x + 600 = 1200$$

$$\Rightarrow 2x = 600$$

$$\Rightarrow x = 300$$

$\therefore$ Total number of bananas $= 500$

HOTS QUESTIONS

1. (*i*) $a(x+y) + b(x-y) = a^2 - ab + b^2$ ...(*i*)

$a(x+y) - b(x-y) = a^2 + b^2 + ab$ ...(*ii*)

Rearrange (*i*) $\times (a-b)$ and (*ii*) $\times (a+b)$

$$(a-b)\,[(a+b)x + (a-b)y = a^2 - ab + b^2]$$

$$(a+b)\,[(a-b)x + (a+b)y = a^2 + ab + b^2]$$

$$(a^2 - b^2)x + (a-b)^2 y = (a-b)(a^2 - ab + b^2) \quad \text{...(iii)}$$

$$(a^2 - b^2)x + (a+b)^2 y = (a+b)(a^2 + ab + b^2) \quad \text{...(iv)}$$

On subtracting (*iv*) from (*iii*)

$$[(a-b)^2 - (a+b)^2]\,y = a^3 - a^2b + ab^2 - ba^2 + ab^2 - b^3 - b^3 - a^3 - a^2b - ab^2 - ba^2 - ab^2$$

$$-4aby = -4a^2b - 2b^3$$

$$-4aby = -2b[2a^2 + b^2]$$

$$y = \left[\frac{2a^2 + b^2}{2a}\right]$$

Put in equation (*i*), we have

$$(a+b)x = a^2 - ab + b^2 - (a-b)\left(\frac{2a^2 + b^2}{2a}\right)$$

$$2a(a+b)x = 2a^3 - 2a^2b + 2ab^2 - 2a^3 - ab^2 + 2a^2b + b^3$$

$$2a(a+b)x = ab^2 + b^3 = b^2(a+b)$$

$$x = \frac{b^2}{2a}$$

(*ii*) $\frac{x}{a} + \frac{y}{b} = a + b \qquad \frac{x}{a^2} + \frac{y}{b^2} = 2$

$$bx + ay = ab(a+b) \quad \text{...(i)}$$

$$b^2x + a^2y = 2a^2b^2 \quad \text{...(ii)}$$

Multiplying (*i*) by (*a*),

$$abx + a^2y = a^2b(a+b) \quad \text{...(iii)}$$

Subtracting (*iii*) from (*ii*)

$$(b^2 - ab)x = 2a^2b^2 - a^3b - a^2b^2$$

$$b(b-a)x = a^2b^2 - a^3b$$

$$b(b-a)x = a^2b(b-a)$$

$$x = a^2$$

On substituting in (*ii*),

$$a^2y = 2a^2b^2 - b^2a^2$$

$$\Rightarrow a^2y = a^2b^2 \Rightarrow y = b^2$$

2. $2(3u - v) = 5uv$ ...(*i*)

$2(u + 3v) = 5uv$ ...(*ii*)

Dividing both the equations by uv on both sides,

$$\frac{6u}{uv} - \frac{2v}{uv} = 5$$

$$\frac{6}{v} - \frac{2}{u} = 5 \quad \text{...(iii)}$$

$$\frac{2u}{uv} + \frac{6v}{uv} = 5$$

$$\frac{2}{v} + \frac{6}{u} = 5 \quad \text{...(iv)}$$

Multiplying (*ii*) by 3

$$\frac{18}{v} - \frac{6}{u} = 15$$

Adding to (*iv*), we get

$$\frac{20}{v} = 20 \Rightarrow v = 1$$

Put v in (*iv*), we get

$$\frac{6}{u} = 3 \Rightarrow u = 2$$

3. $$3x - \frac{y+7}{11} + 2 = 10$$

$$\Rightarrow \frac{33x-(y+7)+22}{11} = 10$$

$$33x - y - 7 + 22 = 110$$

$$\Rightarrow 33x - y = 95 \quad ...(i)$$

$$2y + \frac{x+11}{7} = 10$$

$$\Rightarrow \frac{14y+x+11}{7} = 10$$

$$\Rightarrow 14y + x + 11 = 70$$

$$14y + x = 59$$

$$14y = 59 - x$$

$$\Rightarrow y = \frac{59-x}{14}$$

$\therefore$ Eq (*i*) becomes

$$33x - \left(\frac{59-x}{14}\right) = 95$$

$$\Rightarrow \frac{462x - 59 + x}{14} = 95$$

$$\Rightarrow 463x - 59 = 1330$$

$$463x = 1389 \Rightarrow x = 3$$

When $x = 3$

Eq (*i*) becomes

$$33 \times 3 - y = 95$$

$$\Rightarrow y = 4$$

4. $$x - y = 0.9$$

$$\frac{11}{x+y} = 2 \Rightarrow x + y = 5.5$$

On adding both equations

$$2x = 6.4 \Rightarrow x = 3.2 \Rightarrow y = 2.3$$

5. $148x + 231y = 527$

$231x + 148y = 610$

On adding the given equations, we get

$$379x + 379y = 1137$$

$$\Rightarrow x + y = 3 \quad ...(i)$$

On subtracting the given equations, we get

$$-83x + 83y = -83$$

$$\Rightarrow -x + y = -1 \quad ...(ii)$$

On adding equations (*i*) and (*ii*)

$$2y = 2 \Rightarrow y = 1 \Rightarrow x = 2$$

6. $631x + 279y = 910$

$279x + 631y = 910$

On adding the given equations, we get

$$910x + 910y = 1820 \Rightarrow x + y = 2$$

On subtracting the given equations, we get

$$352x - 352y = 0 \Rightarrow x - y = 0$$

On solving $x = 1, y = 1$

7. $$254x + 309y = -55$$

$$309x + 254y = 55$$

On adding the given equations, we get

$$563x + 563y = 0 \Rightarrow x + y = 0$$

On subtracting the given equations, we get

$$55x - 55y = 110 \Rightarrow x - y = 2$$

On solving both equations, we have

$$2x = 2 \Rightarrow x = 1, y = -1$$

8. $$\frac{148}{x} + \frac{231}{y} = \frac{527}{xy}$$

$$148y + 231x = 527 \quad ...(i)$$

Similarly, $231y + 148x = 610$ ...(*ii*)

Now solution is same as question no. 5, above.

9. $$3x + 2y = 2x + y + 3 = 4x + 3y - 3$$

On solving 1st and 2nd

$$x + y = 3 \quad ...(i)$$

On solving 2nd and 3rd

$$2x + 2y = 6$$

$$x + y = 3 \quad ...(ii)$$

Infinitely many solutions.

10. $$6x + 3y = 8x + 9y - 5$$

$$2x + 6y = 5 \quad ...(i)$$

$$8x + 9y - 5 = 10x + 12y - 8$$

$$2x + 3y = 3 \quad ...(ii)$$

On solving (*i*) and (*ii*),

$$3y = 2 \Rightarrow y = \frac{2}{3}$$

$$2x = 5 - 6 \times \frac{2}{3} = 5 - 4 = 1$$

$$x = \frac{1}{2}$$

11. Let full fare = ₹ x and reservation charges = ₹ y.

A.T.Q., $$x + y = 216$$

$$\Rightarrow y = 216 - x \quad ...(i)$$

and $$(x+y) + \left(\frac{1}{2}x + y\right) = 327$$

$$\Rightarrow \frac{3}{2}x + 2y = 327$$

$$\Rightarrow 3x + 4y = 654 \quad ...(ii)$$

Using (*i*), we get

$$3x + 4(216 - x) = 654$$

$$3x + 864 - 4x = 654$$

$$x = 210$$

When $x = 210$, eq (*i*) becomes

$$y = 216 - 210 = 6$$

$\therefore$ Full fare = ₹ 210 and reservation charges = ₹ 6

12. (*i*) $$ax\left(\frac{1}{a-b} - \frac{1}{a+b}\right) + by\left(\frac{1}{b-a} - \frac{1}{b+a}\right) = \frac{2a}{a+b}$$

$$\Rightarrow ax\left(\frac{a+b-a+b}{(a-b)(a+b)}\right) + by\left(\frac{b+a-b+a}{(b-a)(b+a)}\right) = \frac{2a}{a+b}$$

$$\Rightarrow \quad \frac{2abx}{a^2-b^2} - \frac{2aby}{a^2-b^2} = \frac{2a}{a+b}$$

$$\frac{2ab}{a^2-b^2}(x-y) = \frac{2a}{a+b}$$

$$x - y = \frac{2a}{a+b} \times \frac{a^2-b^2}{2ab}$$

$$\Rightarrow \quad x - y = \frac{a-b}{b}$$

$$\Rightarrow \quad x = \frac{a-b}{b} + y \quad ...(i)$$

and $\quad bx + ay = a + b \quad ...(ii)$

Using eq. (*i*), we get

$$b\left(\frac{a-b}{b} + y\right) + ay = a + b$$

$$\Rightarrow \quad a - b + by + ay = a + b$$

$$(a + b) = 2b$$

$$y = \frac{2b}{a+b}$$

$\therefore$ When $y = \dfrac{2b}{a+b}$, eq. (*i*) becomes

$$x = \frac{a-b}{b} + \frac{2b}{a+b}$$

$$\Rightarrow \quad x = \frac{a^2 - b^2 + 2b^2}{b(a+b)} = \frac{a^2+b^2}{b(a+b)}$$

(*ii*) $$x\left(a - b + \frac{ab}{a-b}\right) = y\left(a + b - \frac{ab}{a+b}\right)$$

$$\Rightarrow \quad x\left(\frac{(a-b)^2 + ab}{a-b}\right) = y\left(\frac{(a+b)^2 - ab}{a+b}\right)$$

$$\Rightarrow \quad x\left(\frac{a^2+b^2-ab}{a-b}\right) = y\left(\frac{a^2+b^2+ab}{a+b}\right)$$

$$x(a^2 + b^2 - ab)(a + b) = y(a^2 + b^2 + ab)(a - b)$$

$$\Rightarrow \quad x(a^3 + b^3) = y(a^3 - b^3) \quad ...(i)$$

Also, $\quad x + y = 2a^2$

$$y = 2a^2 - x \quad ...(ii)$$

Using (*ii*), eq (*i*) becomes

$$x(a^3 + b^3) = (2a^2 - x)(a^3 - b^3)$$

$$x(a^3 + b^3) = 2a^2(a^3 - b^3) - x(a^3 - b^3)$$

$$\Rightarrow \quad x(a^3 + b^3) + x(a^3 - b^3) = 2a^2(a^3 - b^3)$$

$$\Rightarrow \quad 2a^3x = 2a^2(a^3 - b^3)$$

$$x = \frac{2a^2(a^3-b^3)}{2a^3} = \frac{a^3-b^3}{a}$$

When $x = \dfrac{a^3-b^3}{a}$, eq. (*ii*) becomes

$$y = 2a^2 - \frac{a^3-b^3}{a}$$

$$\Rightarrow \quad y = \frac{2a^3 - a^3 + b^3}{a}$$

$$\Rightarrow \quad y = \frac{a^3+b^3}{a}$$

13. Let speed of the man = x km/hr
and time taken = y hr
$\therefore$ distance = xy km

(*i*) When speed = $\left(x + \frac{1}{2}\right)$ km/hrs.

Then time taken = $(y - 1)$ hrs.

$$\Rightarrow \quad \text{Distance} = \left(x + \frac{1}{2}\right)(y-1)$$

$$\Rightarrow \quad \left(x + \frac{1}{2}\right)(y-1) = xy$$

$$\Rightarrow \quad xy - x + \frac{1}{2}y - \frac{1}{2} = xy$$

$$\Rightarrow \quad -x + \frac{1}{2}y = \frac{1}{2}$$

$$\Rightarrow \quad -2x + y = 1$$

$$\Rightarrow \quad y = 1 + 2x \quad ...(i)$$

(*ii*) When speed = $(x - 1)$ km/hr
Then time taken = $(y + 3)$ hrs.

$$\therefore \quad \text{Distance} = (x - 1)(y + 3)$$

$$\Rightarrow \quad (x - 1)(y + 3) = xy$$

$$\Rightarrow \quad xy + 3x - y - 3 = xy$$

$$3x - y = 3 \quad ...(ii)$$

Using eq. (*i*), we get

$$3x - (1 + 2x) = 3$$

$$x - 1 = 3 \Rightarrow x = 4$$

When $x = 4$ eq. (*i*) becomes

$$y = 1 + 2 \times 4 = 9$$

$$\therefore \quad \text{speed} = 4 \text{ km/hr}$$

$$\text{distance} = xy = 4 \times 9 = 36 \text{ km}$$

14. Let speed of boat in still water = x km/hr
and speed of stream = y km/hr.
$\therefore$ Downstream speed of boat = $(x + y)$ km/hr
and upstream speed of boat = $(x - y)$ km/hr

According to the question,

$$\frac{24}{x-y} + \frac{28}{x+y} = 6 \quad ...(i)$$

and $$\frac{30}{x-y} + \frac{21}{x+y} = 6\frac{1}{2} \quad ...(ii)$$

Now proceed further.

15. Let length = x m, breadth = y m

$$\therefore \quad \text{Area} = x.y \text{ m}^2$$

(*i*) When length = $(x + 7)$ m, breadth = $(y - 3)$ m then area = xym²

$$\Rightarrow \quad (x + 7)(y - 3) = xy$$

$$\Rightarrow \quad xy - 3x + 7y - 21 = xy$$

$$\Rightarrow \quad -3x + 7y = 21 \quad ...(i)$$

(*ii*) When length $(x - 7)$ m, breadth = $(y + 5)$ m
then area = xy m²

$$\Rightarrow \quad (x - 7)(y + 5) = xy$$

$$\Rightarrow \quad xy + 5x - 7y - 35 = xy$$

$$5x - 7y = 35 \quad ...(ii)$$

Adding eq. (*i*) and eq. (*ii*), we get

$$\begin{aligned} -3x + 7y &= 21 \\ 5x - 7y &= 35 \\ \hline 2x &= 56 \\ \Rightarrow \quad x &= 28 \end{aligned}$$

When $x = 28$

eq (*i*) becomes

$$\begin{aligned} -3 \times 28 + 7y &= 21 \\ \Rightarrow \quad 7y &= 21 + 84 \\ y &= \frac{105}{7} = 15 \end{aligned}$$

$\therefore$ Length = 28 m, breadth = 15 m.

16. Let number of persons = x, Share per person = ₹ y

$\therefore$ Total money = ₹ xy

According to the question,

$12 \times x = xy + 6 \quad \Rightarrow \quad 12x - 6 = xy$ ...(*i*)

and $9x = xy - 9 \quad \Rightarrow \quad 9x + 9 = xy$ ...(*ii*)

Equating (*i*) and (*ii*), we get

$12x - 6 = 9x + 9 \quad \Rightarrow \quad 3x = 15$

$\Rightarrow \quad x = 5$ and $xy = 54$

17. Let fixed expenses = ₹ x

and proportional charges = ₹ y

A.T.Q., $x + 7y = 65$ ...(*i*)

and $x + 11y = 97$ ...(*ii*)

Subtracting

$$\begin{aligned} -4y &= -32 \\ y &= 8 \end{aligned}$$

When $y = 8$, eq (*i*) becomes

$$\begin{aligned} x + 7 \times 8 &= 65 \\ x &= 9 \end{aligned}$$

Now, expenses for 18 guests

$$\begin{aligned} &= x + 18y \\ &= 9 + 18 \times 8 \\ &= 9 + 144 = ₹\ 153 \end{aligned}$$

18. Let age of father = x years and age of son = y years.

$\therefore \quad x + y = 55$ ...(*i*)

When son's age = x years then father's age

$= x + (x - y) = 2x - y$

According to the question,

$x + 2x - y = 93 \quad \Rightarrow \quad 3x - y = 93$

Now solve equation (*i*) and (*ii*).

19. Let digit at unit's place = x and

digit at hundred's place = y

$\therefore$ Middle digit = $x + y + 1$

Number = $100y + 10(x + y + 1) + x$

Number obtained by reversing the digits

$= 100x + 10(x + y + 1) + y$

According to the question,

$x + y + (x + y + 1) = 17 \Rightarrow 2x + 2y = 16$ ...(*i*)

and $100x + 10(x + y + 1) + y$

$= 100y + 10(x + y + 1) + x - 396$ ...(*ii*)

$\Rightarrow \quad 99x - 99y = -396 \quad \Rightarrow \quad x - y = -4$

Solve equation (*i*) and (*ii*), we get

$x = 2, \ y = 6$ and

Number = 692

20. Let height of each candle = x unit.

Height of 1st candle burnt in 1 hr = $\frac{x}{6}$ unit

and height of 2nd candle burnt in 1 hr = $\frac{x}{8}$ unit

Let the required time = y hrs.

Length of 1st candle burnt after y hrs = $\frac{y \times x}{6}$ unit

Height of 1st candle left = $\left(x - \frac{xy}{6}\right)$

Length of 2nd candle burnt after y hrs = $\left(\frac{y \times x}{8}\right)$ unit

Height of 2nd candle left = $\left(x - \frac{xy}{8}\right)$

A.T.Q.,

Height of 1st candle = $\frac{1}{2}$ Height of 2nd candle

$$\begin{aligned} \Rightarrow \quad x - \frac{xy}{6} &= \frac{1}{2}\left(x - \frac{xy}{8}\right) \\ \Rightarrow \quad x\left(1 - \frac{y}{6}\right) &= \frac{1}{2}x\left(1 - \frac{y}{8}\right) \\ 1 - \frac{y}{6} &= \frac{1}{2}\left(1 - \frac{y}{8}\right) \\ \Rightarrow \quad 2 - \frac{y}{3} &= 1 - \frac{y}{8} \\ 2 - 1 &= \frac{y}{3} - \frac{y}{8} \\ 1 &= \frac{8y - 3y}{24} \\ \Rightarrow \quad 24 &= 5y \\ y &= \frac{24}{5} \\ y &= 4.8 \text{ hrs.} = 4\text{hrs. } 48 \text{ minutes} \end{aligned}$$

21. Let ₹ x be invested at 12% simple interest

and ₹ y be invested at 10% simple interest.

The yearly interest = $\frac{x \times 12}{100} + \frac{y \times 10}{100} = 130$

$$\begin{aligned} \Rightarrow \quad \frac{12x}{100} + \frac{10y}{100} &= 130 \\ \Rightarrow \quad 12x + 10y &= 13{,}000 \\ \Rightarrow \quad 6x + 5y &= 6{,}500 \quad \text{...}(i) \end{aligned}$$

If the amount invested are interchanged,

Then, yearly interest = $\frac{10x}{100} + \frac{12y}{100} = 134$

$\Rightarrow \quad 5x + 6y = 6{,}700$ ...(*ii*)

Multiply (*i*) by 5 and (*ii*) by 6, we get

$\Rightarrow \quad 30x + 25y = 32{,}500$ ...(*iii*)

$\Rightarrow \quad 30x + 36y = 40{,}200$ ...(*iv*)

On solving (*iii*) and (*iv*), we get value of x and y.

22. Let number of human be x and deer be y.

then $x + y = 39$ ...(*i*)

and $2x + 4y = 132$

$x + 2y = 66$ ...(*ii*)

On solving (*i*) and (*ii*)

$-y = -27 \Rightarrow y = 27, x = 12$

23. Let age of father be x and sum of ages of 6 children be y.

Then $x = y$

After 15 years age of father will be $(x + 15)$ and the sum of ages of six children wull be $(y + 90)$

According to question,

$2(x + 15) = y + 90$

$2x + 30 = y + 90$

but $x = y$ then

$2x - x = 90 - 30 = 60$

$x = 60$ years

24. Let x L of 50% acid be mixed with y L. of 25% acid.

then $x + y = 10$ ...(*i*)

as $\frac{50x}{100} + \frac{25y}{100} = \frac{40 \times 10}{100}$

therefore $2x + y = 16$ ...(*ii*)

On solving, $x = 6$ L, $y = 4$ L

25. $\frac{2}{3x+2y} + \frac{3}{3x-2y} = \frac{17}{5}$

$\frac{5}{3x+2y} + \frac{1}{3x-2y} = 2$

Let $\frac{1}{3x+2y} = a$, $\frac{1}{3x-2y} = b$

$2a + 3b = \frac{17}{5}$ ...(*i*)

$5a + b = 2$ ...(*ii*)

On multiplying (*ii*) by 3

$15a + 3b = 6$ ...(*iii*)

Subtracting (*iii*) from (*i*)

$-13a = \frac{17}{5} - 6$

$-13a = \frac{-13}{5} \Rightarrow a = \frac{1}{5}$

i.e., $3x + 2y = 5$ ...(*iii*)

again $b = 2 - 5 \cdot \frac{1}{5} = 1$, $\therefore$ $3x - 2y = 1$ ...(*iv*)

Adding (*iii*) and (*iv*),

$6x = 6 \Rightarrow x = 1$

Hence, $2y = 5 - 3 = 2 \Rightarrow y = 1$

26. $2y - x = 8$, $y = \frac{x+8}{2}$

x	0	–2	–4
y	4	3	2

$5y - x = 14$, $x = 5y - 14$

x	1	6	–4
y	3	4	2

$y - 2x = 1$, $y = 2x + 1$

x	0	–1	1
y	1	–1	3

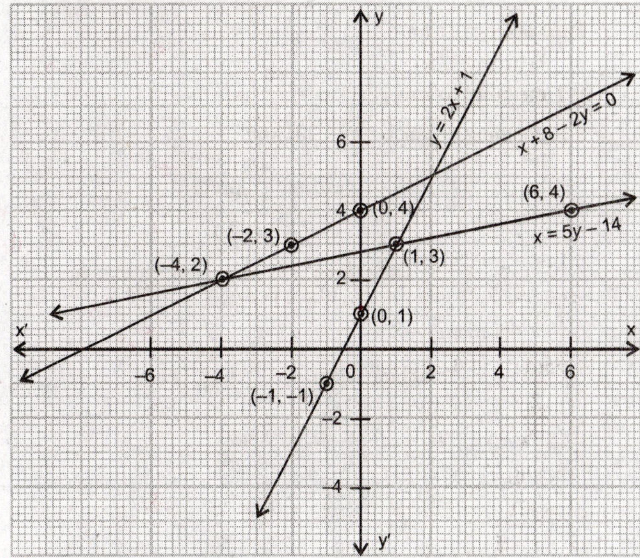

27. Let digit at unit place $= x$

and digit at tens place $= y$

$\therefore$ Number $= 10y + x$

A.T.Q.,

$\frac{10y + x}{10x + y} = \frac{7}{4}$

$\Rightarrow$ $40y + 4x = 70x + 7y$

$66x - 33y = 0$

$2x - y = 0$ ...(*i*)

and $x + y = 12$ ...(*ii*)

Eq. (*i*) + eq. (*ii*), we get $2x - y = 0$

$x + y = 12$

$3x = 12$

$\Rightarrow$ $x = 4$

When $x = 4$, eq. (*i*) becomes

$2 \times 4 - y = 0 \Rightarrow y = 8$

$\therefore$ Number $= 10 \times 8 + 4 = 84$

28. Let greater number $= x$

and smaller number $= y$

A.T.Q.,

$2x - 16 = \frac{1}{2}y$

$\Rightarrow$ $4x - 32 = y$

$\Rightarrow$ $4x - y = 32$ ...(*i*)

and $\frac{1}{2}x - 1 = \frac{1}{2}y$

$\Rightarrow$ $x - 2 = y$

$\Rightarrow$ $x - y = 2$ ...(*ii*)

(*i*) – (*ii*) we get $4x - y = 32$

$x - y = 2$

Subtracting $-\quad+\quad-$

$3x = 30 \Rightarrow x = 10$

When $x = 10$, eq. (*ii*) becomes

$$10 - y = 2 \Rightarrow y = 8$$

$\therefore$ Numbers are 10 and 8.

29. Let speed of the train = x km/h
and distance = y km

$\therefore$ Time taken = $\frac{y}{x}$ hr.

(I) Time taken to cover 30 km = $\frac{30}{x}$ hr

Remaining distance = $(y - 30)$ km.

$$\text{speed} = \frac{4}{5} \text{ km/hr}$$

$$\therefore \quad \text{Time taken} = \frac{y-30}{\frac{4}{5}x} = \frac{5y-150}{4x} \text{ hr}$$

A.T.Q.,

$$\text{Total time taken} = \left(\frac{y}{x} + \frac{45}{60}\right) \text{hr}$$

$$\Rightarrow \frac{30}{x} + \frac{5y-150}{4x} = \frac{y}{x} + \frac{3}{4}$$

$$\Rightarrow \frac{30}{x} + \frac{5y-150}{4x} - \frac{y}{x} = \frac{3}{4}$$

$$\frac{120 + 5y - 150 - 4y}{4x} = \frac{3}{4}$$

$$y - 30 = \frac{3}{4} \times 4x$$

$$y - 30 = 3x \Rightarrow 3x - y + 30 = 0$$

$$\Rightarrow \quad y = 3x + 30 \quad \text{...}(i)$$

(II) Distance covered with uniform speed = 30 + 18 = 48 km

$$\therefore \quad \text{Time taken} = \frac{48}{x} \text{ hr}$$

Remaining distance = $y - 48$

$$\text{Speed} = \frac{4}{5}x$$

$$\text{Time taken } \frac{y-48}{\frac{4}{5}x} = \frac{5y-240}{4x} \text{ hr}$$

A.T.Q.,

$$\text{Total time taken} = \frac{y}{x} + \frac{36}{60}$$

$$\Rightarrow \frac{48}{x} + \frac{5y-240}{4x} = \frac{y}{x} + \frac{36}{60}$$

$$\Rightarrow \frac{48}{x} + \frac{5y-240}{4x} - \frac{y}{x} = \frac{36}{60}$$

$$\frac{192 + 5y - 240 - 4y}{4x} = \frac{36}{60}$$

$$\Rightarrow \frac{y-48}{4x} = \frac{12}{20}$$

$$\frac{y-48}{4x} = \frac{3}{5}$$

$$5y - 240 = 12x$$

$$\Rightarrow \quad 12x - 5y + 240 = 0 \quad \text{...}(ii)$$

$$\Rightarrow \quad 12x - 5(3x + 30) + 240 = 0 \quad \text{[Using eq. (i)]}$$

$$12x - 15x - 150 + 240 = 0$$

$$-3x + 90 = 0 \Rightarrow x = 30$$

When $x = 30$, eq (i) becomes

$$y = 3 \times 30 + 30 = 120$$

$$\therefore \quad \text{Speed} = 30 \text{ km/hr, distance} = 120 \text{ km}$$

30. Let age of Sunaina = x years
and age of Tanishtha = y years

A.T.Q.,

$$x - y = 2 \text{ or } y - x = 2 \quad \text{...}(ii)$$

Age of Sunaina's father = $2x$

Age of Tanishta = 2 × age of Shiva

$$\Rightarrow \quad y = 2 \times \text{age of Shiva}$$

$$\Rightarrow \quad \text{Shiva's age} = \frac{y}{2}$$

A.T.Q.,
$$2x - \frac{y}{2} = 40$$

$$\Rightarrow \quad 4x - y = 80 \quad \text{...}(iii)$$

Taking eq. (i) and eq. (iii)

$$4x - y = 80 \quad \text{...}(iii)$$
$$x - y = 2 \quad \text{...}(i)$$
$$- \quad + \quad -$$

$$3x = 78 \Rightarrow x = 26$$

When $x = 26$, eq. (i) becomes

$$26 - y = 2 \Rightarrow y = 24$$

$\therefore$ Age of Sunaina = 26 years

Age of Tanishtha = 24 years.

VALUE BASED QUESTIONS

1. **(a)** $a_1 = 2, b_1 = 3, c_1 = 7, a_2 = 4, b_2 = 6, c_2 = 12$

Now, $\frac{a_1}{a_2} = \frac{2}{4} = \frac{1}{2}, \frac{b_1}{b_2} = \frac{3}{6} = \frac{1}{2}, \frac{c_1}{c_2} = \frac{7}{12}$

$$\Rightarrow \frac{a_1}{a_2} = \frac{b_1}{b_2} \neq \frac{c_1}{c_2}$$

$\therefore$ Roads are parallel. They will never meet at one point.

(b) Obey traffic rules.

2. **(a)** Let cost of one book be ₹x
and cost of one pen be ₹y

A.T.Q. $\quad 5x + 7y = 285 \quad \text{...}(i)$

$\quad 4x + 4y = 220 \quad \text{...}(ii)$

Solving (i) and (ii) we get

$$x = 50, y = 5$$

Cost of 3 books and 5 pens = 3 × 50 + 5 × 5
= ₹175

Money returned = ₹195 – ₹175 = ₹20

(b) Honesty.

3. Let number of questions whose answer is known to the student be x
and questions attempted by cheating be y

A.T.Q.

$$\therefore \quad x + y = 120 \quad \text{...}(i)$$

Marks got = 90

$$\Rightarrow \quad x - \frac{1}{4}y = 90 \Rightarrow 4x - y = 360 \quad ...(ii)$$

Solving (*i*) and (*ii*) we get

$x = 96$ and $y = 24$

∴ He answer correctly = 96 questions.

His cheating habit will be violating the value of honesty in his character. He should not guess the answer as well as that may fetch him negative marking for wrong guess. Cheating may get him marks in this exam but this habit may not let him develop an integrity of character in the long run.

4. Let amount of money be ₹ x

and number of person be y.

A.T.Q.

$$120 \times y = x + 60 \quad ...(i)$$

and

$$90 \times y = x - 90 \quad ...(ii)$$

solving (*i*) and (*ii*)

We get $x = 540$ and $y = 5$.

∴ Money he have = ₹ 540

and number of people = 5

Value : concern for others

5. (*a*) Let amount invested be ₹ x and ₹ y.

A.T.Q.

$$x + y = 30000 \quad ...(i)$$

$$\frac{5}{100}x + \frac{7}{100}y = 1800$$

$$5x + 7y = 180000 \quad ...(ii)$$

From equation (*i*) we get $y = 30000 - x$

substituting in equation (*ii*) we get

$$5x + 7(30000 - x) = 180000$$

$$-2x = -30000$$

$$x = 15000$$

$$\therefore \quad y = 15000$$

amount invested is ₹ 15000 in each type of bond.

(*b*) Helping poor people.

MULTIPLE CHOICE QUESTIONS

1. (*d*) $10x - 14y = -4$

2. (*a*) $a_1 = 2, b_1 = -3, c_1 = -1$

$a_2 = 3, b_2 = -2, c_2 = -4$

$$\frac{a_1}{a_2} = \frac{2}{3}, \frac{b_1}{b_2} = \frac{3}{2} \Rightarrow \frac{a_1}{a_2} \neq \frac{b_1}{b_2}$$

∴ Unique solution or one solution

3. (*c*)

$$x - y = 2 \quad ...(i)$$

$$x + y = 4 \quad ...(ii)$$

Adding

$$2x = 6 \Rightarrow x = 3 \quad ...(i)$$

$$\therefore \quad y = 1$$

$$\Rightarrow \quad a = 3, b = 1$$

4. (*d*)

$$4x + 3y = 14$$

Considering

$$12x + 9y = 0$$

$$a_1 = 12, b_1 = 9, c_1 = 0$$

$$a_2 = 4, b_2 = 3, c_2 = 14$$

$$\frac{a_1}{a_2} = \frac{12}{4} = 3, \frac{b_1}{b_2} = \frac{9}{3} = 3$$

$$\frac{c_1}{c_2} = \frac{0}{14} = 0$$

$$\Rightarrow \quad \frac{a_1}{a_2} = \frac{b_1}{b_2} \neq \frac{c_1}{c_2}$$

∴ Lines are parallel.

5. (*d*) When lines $x = 4, y = 3$ will intersect, then x coordinate = 4

y-coordinate = 3

6. (*d*) ∴ $a^2 + b^2 \neq 0$ if $a = 0$ and $b = 0$.

For $ax + by + c = 0$ to be a linear equation in two variables, a and b are simultaneously not equal to zero.

7. (*d*) Solution means set of values of variables satisfying the given equation.

8. (*a*) Solution set of $ax + by + c = 0$ is collinear with other points on the line.

9. (*b*) Solution sets of a linear equation in two variables lies on the line representing it.

10. (*c*) Definition of pair of linear equations in two variables.

11. (*d*) Lines in a plane will either intersect or parallel.

12. (*c*) (1) $\frac{2}{6} \neq \frac{3}{5}$ (2) $\frac{2}{6} = \frac{3}{9} \neq \frac{40}{50}$ (3) $\frac{2}{4} = \frac{3}{6} = \frac{10}{20}$

13. (*c*) $3x + 5y = 40, 5x + 4y = 58$

14. (*d*) (1) $\frac{2}{3} \neq \frac{5}{4}$ ∴ unique solution.

(2) $\frac{2}{6} = \frac{5}{15} \neq \frac{10}{20}$ ∴ no solution.

(3) $\frac{5}{10} = \frac{2}{4} = \frac{10}{20}$ ∴ infinite many solutions.

15. (*c*) (1) $\frac{2}{3} \neq \frac{5}{4}$ ∴ consistent pair.

(2) $\frac{2}{4} = \frac{5}{10} \neq \frac{7}{7}$ ∴ inconsistent pair.

(3) $\frac{2}{4} = \frac{5}{10} = \frac{7}{14}$ ∴ dependent and consistent.

16. (*c*) Equations are $x - y = 0, x + y = 0$.

$\frac{1}{1} \neq \frac{-1}{1}$

∴ unique solution.

17. (*a*) Equations represent parallel lines. Hence, no common solution.

18. (*c*) $y = 4$ is a line parallel to x-axis and $3x + 4y = 12$ intersects the x-axis.

∴ It also intersects $y = 4$. Hence, unique solution.

19. (*c*) Compare $\frac{a_1}{a_2}, \frac{b_1}{b_2}$ and $\frac{c_1}{c_2}$.

20. (*c*) Let first integer = x and second integer = y

$\therefore\ 2x + 3y = 39$ and $3x + 2y = 36$

Solving both the equations, we get $x = 6$ and $y = 9$

21. (*d*) Let number is $10x + y$ (x, y are the digits).

Number obtained by interchanging the digits = $10y + x$

$\therefore\ 10x + y - (10y + x) = 27$

$9(x - y) = 27 \quad \therefore x - y = 3$

22. (*d*) Difference between a two digit number and number obtained by interchanging the order of digits is always divisible by 9. 48 is not divisible by 9.

23. (*d*) Let bases are $9x$ and $9y$.

$\therefore\quad \text{Area} = \frac{1}{2} \times 50 \times (9x + 9y) = 1400$

$\Rightarrow\quad 9(x + y) = \frac{1400}{25} = 56$

Which is not a multiple of 9.

24. (*b*) Unique solution, $\frac{2}{k} \neq \frac{3}{4} \Rightarrow k \neq \frac{8}{3}$.

25. (*b*) $\frac{\sin\theta}{1} = \frac{\cos\theta}{1} = \frac{1}{\sqrt{2}}$

$\Rightarrow\quad \sin\theta = \frac{1}{\sqrt{2}},\ \cos\theta = \frac{1}{\sqrt{2}},$

$\therefore\quad \theta = 45°$

26. (*a*) $\because$ lines meet at a point

$\Rightarrow$ solution set of $3x + 4y = 5$ and $5x + 4y = 4$ also satisfies the equation $kx + 4y = 6$.

Solving, we get $k = 1$.

27. (*d*) $\because\quad \frac{a}{d} = \frac{b}{e} = \frac{c}{c}$

$\Rightarrow\quad \frac{a}{d} = 1,\ \frac{b}{e} = 1 \Rightarrow a = d, b = e$

$\Rightarrow\quad ab = de$

28. (*d*) For parallel lines, $\frac{a}{3} = \frac{2}{b} \neq \frac{7}{16} \Rightarrow ab = 6$

29. (*b*) Common solution of $x = 1$ and $y = 2$ is $(1, 2)$.

$\Rightarrow\quad a \times 1 + 2 \times 2 - 6 = 0 \Rightarrow a = 2$

30. (*c*) Lines L_1 and L_2 are coincident lines.

Line $L_3 : (a_1 + a_2)x + (b_1 + b_2)y + c_1 + c_2 = 0$

$\because\quad \frac{a_1}{a_2} = \frac{b_1}{b_2} = \frac{c_1}{c_2}$

$\Rightarrow\quad \frac{a_1}{a_2} + 1 = \frac{b_1}{b_2} + 1 = \frac{c_1}{c_2} + 1$

$\Rightarrow\quad \frac{a_1 + a_2}{a_2} = \frac{b_1 + b_2}{b_2} = \frac{c_1 + c_2}{c_2}$

$\therefore\ L_3$ and L_2 are coincident.

$\therefore$ Lines L_1, L_2 and L_3 are coincident.

31. (*b*) Let ten's digit be x and unit's digit be y.

$\therefore\quad \text{Number} = 10x + y$

$\Rightarrow\quad 10x + y = k(x + y) \quad \ldots(i)$

Let $\quad 10y + x = m(x + y) \quad \ldots(ii)$

$\Rightarrow$ Adding, we get

$11x + 11y = k(x + y) + m(x + y)$

$\Rightarrow (11 - k)x + (11 - k)y = m(x + y)$

$\Rightarrow\quad (11 - k)(x + y) = m(x + y)$

$\therefore\quad m = 11 - k$

32. (*b*) $\because\quad \frac{a}{a+1} = \frac{a-1}{-a} \neq 1$

$\Rightarrow\quad -a^2 = a^2 - 1 \Rightarrow 2a^2 = 1$

$\Rightarrow\quad a^2 = \frac{1}{2} \Rightarrow a = \pm\frac{1}{\sqrt{2}}.$

33. (*c*) A.T.Q., $\quad 1 = a + \frac{b}{-1} \Rightarrow a - b = 1 \quad \ldots(i)$

and $\quad 5 = a + \frac{b}{-5} \Rightarrow 25 = 5a - b \quad \ldots(ii)$

Solving (*i*) and (*ii*), we get the values of a and b as 6 and 5 respectively.

$\therefore\quad a + b = 11$

NCERT EXEMPLAR PROBLEMS

1. (*d*) No solution (parallel lines)
2. (*d*) Intersecting at (a, b)
3. (*b*) $2x + 5y = -11;\ 4x + 10y = -22$
4. (*c*) 6 and 36

Formative Assessment

Activity/Project 1

- **Topic Covered :** Solution of pair of linear equations in two variables.
- **Objective :** To develop algorithm for solving pair of linear equations in two variables by substitution method.
- **Skills Developed :** Analysis, comprehension expression, creativity.
- **Time Required :** 15 Min.
- **Method :** Consider pair of linear equations $a_1x + b_1y + c_1 = 0$ ...(*i*)

and $a_2x + b_2y + c_2 = 0$ ...(*ii*)

Step I : Select any of the equation and find the value of one of the variable in terms of other variable

e.g., $a_1x + b_1y + c_1 = 0$ From equation ... (*i*)

$\Rightarrow$ $a_1x = -c_1 - b_1y$

$\Rightarrow$ $x = \dfrac{-c_1 - b_1y}{a_1}$ ...(*iii*)

Step II : Substitute the value of this variable in other equation *i.e.*, substituting the value of x in equation (*ii*), we get,

$$a_2\left(\frac{-c_1 - b_1y}{a_2}\right) + b_2y + c_2 = 0$$

Step III : Solve the equation so obtained in one variable.

(*a*) If we get a true statement without any variable, we may conclude that pair of equations has infinite many solutions.

(*b*) If we get a false statement without any variable, then the pair of equations has no solution.

Step IV : Substitute the value of variable so obtained in the first equation to get the value of other variable.

- **Conclusion :**

We have substituted the value of one variable in terms of other variable to solve the pair of linear equations. That is why this method is known as substitution method.

Class Work :

Solve $5x + 3y + 5 = 0$, $3x + 5y + 7 = 0$ ...(*i*) and $2x + 3y = 9$, $4x + 6y = 18$ ...(*ii*) using substitution method.

Homework :

Solve $x + 2 = 0$ and $5x - 3y + 14 = 0$ using substitution method.

Quiz Questions :

A pair of linear equations in two variables have exactly one solution. State T/F.

Oral Questions :

Find one integral solution of $2x + 3y = 5$.

Suggested Activity/Assignment :

Solve the equations $4x + 3y = 12$ and $5x + 6y = 11$ graphically.

Activity/Project 2

- **Topic Covered :** Solution of pair of linear equations in two variables.
- **Objective :** To develop an algorithm for solving pair of linear equations using elimination method.
- **Skills Developed :** Analysis, comprehension, expression, creativity.
- **Time Required :** 15 Min.
- **Method :** Consider pair of linear equations in two variables

$a_1x + b_1y + c_1 = 0$ ...(*i*)

and $a_2x + b_2y + c_2 = 0$ ...(*ii*)

Step I : Multiply both the equations by some suitable non-zero constants to make the coefficients of one variable (x or y) numerically equal.

Step II : If equated coefficients of any variable (x or y) are of same sign, subtract the two equations else add the two equations.

Step III : (*a*) If we obtain a true statement involving no variable, then the given pair of linear equations has infinitely many solutions.

(*b*) If we obtain a false statement involving no variable, then the given pair of equations has no solution.

(*c*) Otherwise, solve the equation in one variable so obtained.

Step IV : Substitute the value of this variable in either of the given equation to get the value of other variable.

- **Conclusion :**

In this method, we eliminate one variable to get an equation in one variable. This method is known as elimination method.

Class Work :

Solve $2x + 3y = 7$ and $4x + 6y = 8$ using elimination method.

Homework :

Solve $2x + 5y = 18$ and $3x + \frac{15}{2}y = 12$ using elimination method.

Quiz Questions :

A pair of linear equations with unequal coefficients of x and y has infinite many solutions. State T/F.

Oral Questions :

Find the solution of $y = 1$ and $7x + 2y = 9$.

Suggested Activity/Assignment :

Write an algorithm to solve pair of linear equations in two variables using cross multiplication method.

Activity/Project 3

- **Topic Covered :** Linear equation in two variables.
- **Objective :** To find an integral solution of the equation in two variables $83x + 19y = 11$ using Euclid's algorithm.
- **Skills Developed :** Analysis, problem solving, creativity and application.
- **Time Required :** 15 Min.
- **Method :**

1. Take the coefficient of x and y *i.e.*, 83 and 19.
2. Now

$$83 = 19 \times 4 + 7 \quad \text{...(i)}$$
$$19 = 7 \times 2 + 5 \quad \text{...(ii)}$$
$$7 = 5 \times 1 + 2 \quad \text{...(iii)}$$
$$5 = 2 \times 2 + 1 \quad \text{...(iv)}$$
$$2 = 1 \times 2 + 0 \quad \text{...(v)}$$

3. From (*iv*) we get

$$1 = 5 - (2 \times 2)$$
$$\Rightarrow \quad 1 = 5 - 2\,(7 - 5) \quad \text{(From (iii))}$$
$$\Rightarrow \quad 1 = 5 - 2 \times 7 + 2 \times 5$$
$$\Rightarrow \quad 1 = 3 \times 5 - 2 \times 7$$
$$\Rightarrow \quad 1 = 3(19 - 7 \times 2) - 2 \times 7 \quad \text{(From (ii))}$$
$$\Rightarrow \quad 1 = 3 \times 19 - 6 \times 7 - 2 \times 7$$
$$\Rightarrow \quad 1 = 3 \times 19 - 8 \times 7$$
$$\Rightarrow \quad 1 = 3 \times 19 - 8(83 - 4 \times 19) \quad \text{(From (i))}$$

$\Rightarrow \quad 1 = 3 \times 19 - 8 \times 83 + 32 \times 19$

$\Rightarrow \quad 1 = -8 \times 83 + 35 \times 19$

Now comparing with $83x + 19y = 1$,

we get $x = -8$ and $y = 35$

Class Work :

Find integral solution of $17x + 15y = 1$.

Homework :

Find integral solution of $9x + 11y = 2$.

Quiz Questions :

Equation $4x + 6y = 5$ have no integral solution. State True/False.

Oral Questions :

How many non integral solutions are there of the equation $3x + 4y = 2$?

Suggested Activity/Assignment :

Solve $3x + 5y = 1$ and $x + 2y = 1$ using Euclid's algorithm.

4 CHAPTER

Similar Triangles

Basic Concepts with Examples

- **SIMILAR TRIANGLES.** Two triangles are said to be similar if (*i*) their corresponding angles are equal and (*ii*) their corresponding sides are proportional.
- All congruent triangles are similar but the similar triangles need not be congruent.
- Two polygons of the same numbers of sides are similar, if (*i*) their corresponding angles are equal and (*ii*) their corresponding sides are in the same ratio.
- **BASIC PROPORTIONALITY THEOREM.** In a triangle, a line drawn parallel to one side, to intersect the other sides in distinct points, divides the two sides in the same ratio.
- **CONVERSE OF BASIC PROPORTIONALITY THEOREM.** If a line divides any two sides of a triangle in the same ratio, the line must be parallel to the third side.

1. Using similar triangles, prove that the line drawn from the mid-point of one side of a triangle is parallel to another side bisects the third side.

Sol. Given. In ΔABC, $DE \parallel BC$ and $AD = DB$.

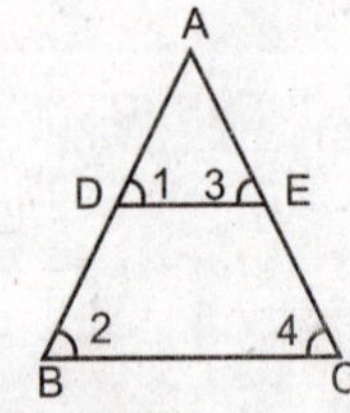

To Prove. $AE = EC$.

Proof. In ΔADE and ΔABC

$\angle 1 = \angle 2$ [Corresponding $\angle$s]

$\angle 3 = \angle 4$ [Corresponding $\angle$s]

[$\because DE \parallel BC$]

$\therefore \quad \Delta ABC \sim \Delta ADE$

$\therefore \quad \dfrac{AD}{DB} = \dfrac{AE}{EC}$ [Basic Proportionality Theorem]

But $\quad AD = DB$ [Given]

$\therefore \quad AE = EC$

2. P is the mid-point of side BC of ΔABC. Q is the mid-point of AP, BQ when produced meets AC at L. Prove that $AL = \frac{1}{3}AC$. [Delhi 2006C]

Sol. Given. In ΔABC, P is the mid-point of BC and Q the mid-point of AP . BQ when produced meets AC at L.

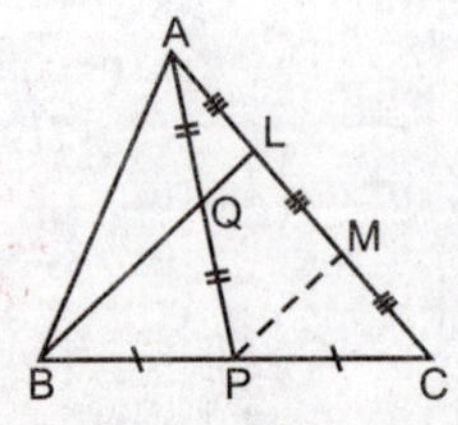

To Prove. $AL = \frac{1}{3}AC$.

Construction. Draw $PM \parallel BL$ meeting AC at M.

Proof. In ΔBLC, $PM \parallel BL$

$\Rightarrow \quad \dfrac{BP}{PC} = \dfrac{LM}{MC}$ [Basic Proportionality Theorem]

But $\quad BP = PC$

$\Rightarrow \quad \dfrac{LM}{MC} = 1 \Rightarrow LM = MC$ ...(*i*)

Consider ΔAPM, Q is the mid-point of AP, and $QL \parallel PM$

$\therefore \quad \dfrac{AQ}{QP} = \dfrac{AL}{LM}$ [Basic Proportionality Theorem]

But $\quad AQ = QP$

$\Rightarrow \quad \dfrac{AL}{LM} = 1 \Rightarrow AL = LM$ ...(*ii*)

From (*i*) and (*ii*),

$AL = LM = MC \Rightarrow AL = \frac{1}{3}AC$.

Hence proved.

SIMILARITY OF TRIANGLES

- **AAA-SIMILARITY.** If in two triangles, corresponding angles are equal, *i.e.*, the two corresponding angles are equal, then the triangles are similar.
- **SSS-SIMILARITY.** If the corresponding sides of two triangles are proportional, then they are similar.
- **SAS-SIMILARITY.** If in two triangles one pair of corresponding sides are proportional and the included angles are equal then the two triangles are similar.
- The ratio of the areas of similar triangles is equal to the ratio of the squares of their corresponding sides.

3. Two poles of height 'a' and 'b' ($b > a$) are 'c' metres apart. Prove that the height 'h' in metres of the point of intersection of the lines joining the top of each pole to the foot of the opposite pole is $\dfrac{ab}{a+b}$.

Sol. Given. Two poles AB = a and CD = b, ($b > a$).

BD = c, AD and BC, intersect at E (Say).

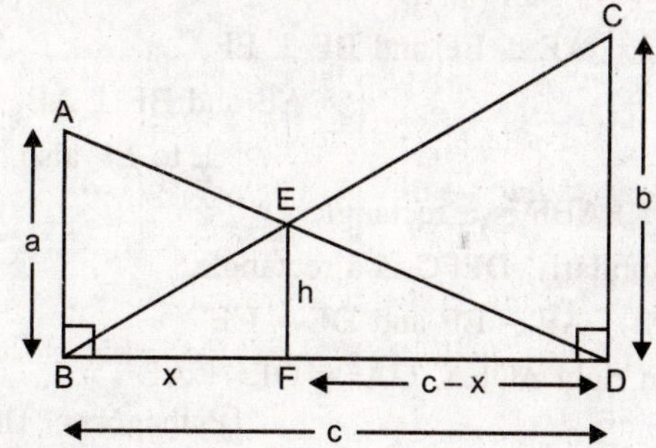

To Find. EF.

Consider Δs ABD and EFD, ∠ABD = ∠EFD

and ∠ADB = ∠EDF [Common]

∴ ΔABD ~ ΔEFD [AA-Similarity]

$$\therefore \quad \frac{EF}{AB} = \frac{FD}{BD} \Rightarrow \frac{h}{a} = \frac{c-x}{c} \quad ...(i)$$

Similarly ΔCDB ~ ΔEFB [AA-Similarity]

$$\therefore \quad \frac{EF}{CD} = \frac{BF}{BD} \Rightarrow \frac{h}{b} = \frac{x}{c} \Rightarrow x = \frac{hc}{b}.$$

Substituting in (*i*), we get $\dfrac{h}{a} = \dfrac{c - \frac{hc}{b}}{c} = \dfrac{b-h}{b}$

$\Rightarrow bh = ab - ah \Rightarrow (a + b)h = ab$

$\Rightarrow h = \dfrac{ab}{a+b}$.

4. A girl of height 90 cm is walking away from the base of a lamp post at a speed of 1.2 m/second. If the lamp is 3.6 m above the ground, find the length of her shadow after 4 seconds. [NCERT]

Sol. Let AB is the lamp post and CD be the girl after walking for 4 seconds away from the lamp post. DE is the shadow of the girl. Let DE = x m.

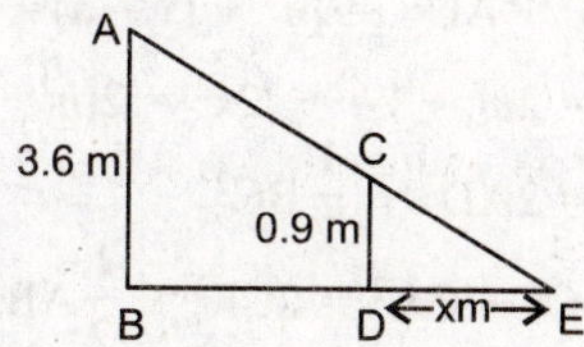

BD = Distance covered by girl in 4 seconds

= 1.2 × 4 = 4.8 m.

In ΔABE and ΔCDE,

∠B = ∠D [Each 90°]

∠E = ∠E [Common]

∴ ΔABE ~ ΔCDE [AA-Similarity]

$$\Rightarrow \frac{BE}{DE} = \frac{AB}{CD} \Rightarrow \frac{BD + DE}{DE} = \frac{3.6}{0.9}$$

$$\Rightarrow \frac{4.8 + x}{x} = \frac{3.6}{0.9}$$

$\Rightarrow 4.8 + x = 4x \Rightarrow 3x = 4.8 \Rightarrow x = 1.6$

∴ Shadow of girl after walking for 4 seconds is 1.6 m long.

5. In figure, CM and RN are respectively the medians of ΔABC and ΔPQR. If ΔABC ~ ΔPQR, prove that (*i*) ΔAMC ~ Δ PNR (*ii*) $\dfrac{CM}{RN} = \dfrac{AB}{PQ}$ (*iii*) ΔCMB ~ Δ RNQ

Sol. (*i*) ΔABC ~ ΔPQR

So, $\dfrac{AB}{PQ} = \dfrac{BC}{QR} = \dfrac{CA}{RP}$ ... (*i*)

and ∠A = ∠P, ∠B = ∠Q and ∠C = ∠R ... (*ii*)

But AB = 2 AM and PQ = 2 PN

(As CM and RN are medians)

So, from (*i*), $\dfrac{2AM}{2PN} = \dfrac{CA}{RP}$

i.e., $\dfrac{AM}{PN} = \dfrac{CA}{RP}$ ... (*iii*)

Also, ∠MAC = ∠NPR [From (*ii*)] ... (*iv*)

So, from (*iii*) and (*iv*),

ΔAMC ~ ΔPNR [SAS similarity] ... (*v*)

(*ii*) From (*v*), $\dfrac{CM}{RN} = \dfrac{CA}{RP}$ ... (*vi*)

But $\dfrac{CA}{RP} = \dfrac{AB}{PQ}$ [From (*i*)]... (*vii*)

∴ $\dfrac{CM}{RN} = \dfrac{AB}{PQ}$ [From (*vi*) and (*vii*)] ... (*viii*)

(*iii*) Again $\dfrac{AB}{PQ} = \dfrac{BC}{QR}$ [From (*i*)]

∴ $\dfrac{CM}{RN} = \dfrac{BC}{QR}$ [From (*viii*)] ... (*ix*)

Also, $\dfrac{CM}{RN} = \dfrac{AB}{PQ} = \dfrac{2BM}{2QN}$ *i.e.*, $\dfrac{CM}{RN} = \dfrac{BM}{QN}$ ... (*x*)

i.e., $\dfrac{CM}{RN} = \dfrac{BC}{QR} = \dfrac{BM}{QN}$ [From (*ix*) and (*x*)]

Therefore, ΔCMB ~ ΔRNQ [SAS similarity]

[**Note :** You can also prove part (*iii*) by following the same method as used for proving (*i*).]

- **PYTHAGORAS THEOREM.** In a right triangle, the square of the hypotenuse is equal to the sum of the squares of the other two sides.
- **CONVERSE OF PYTHAGORAS THEOREM.** In a triangle, if the square of one side is equal to the sum of the squares of the other two sides then the angle opposite to the first side is a right angle.

6. ABC is a right triangle, right angled at C. If p is length of the perpendicular from C to AB and AB = c, BC = a and CA = b, then prove that :

(i) $pc = ab$ (ii) $\frac{1}{p^2} = \frac{1}{a^2} + \frac{1}{b^2}$.

Sol. Given. ΔABC in which $\angle C = 90°$, $CD \perp AB$, $CD = p$, $BC = a$, $AB = c$ and $AC = b$.

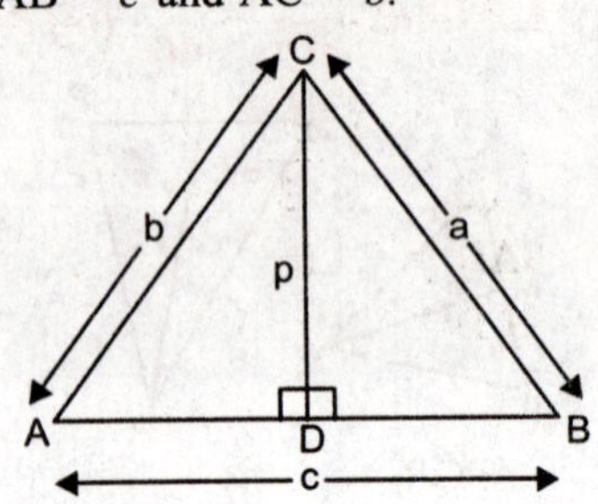

To Prove. (i) $pc = ab$ (ii) $\frac{1}{p^2} = \frac{1}{a^2} + \frac{1}{b^2}$

Proof. Area of right $\Delta ABC = \frac{1}{2} AB \times CD = \frac{1}{2} c \times p$ ...(i)

Again area of right $\Delta ABC = \frac{1}{2} AC \times BC = \frac{1}{2} b \times a$ ...(ii)

From eq. (i) and (ii), we get

$\Rightarrow \quad \frac{1}{2} cp = \frac{1}{2} ba \Rightarrow pc = ab$

or $\quad c = \frac{ab}{p}$ ...(iii)

Using Pythagoras theorem in right ΔACB

$c^2 = a^2 + b^2$ ...(iv)

Putting the value of c in (iv)

$$\frac{a^2b^2}{p^2} = a^2 + b^2 \Rightarrow \frac{1}{p^2} = \frac{a^2 + b^2}{a^2b^2} = \frac{a^2}{a^2b^2} + \frac{b^2}{a^2b^2} = \frac{1}{b^2} + \frac{1}{a^2}$$

$$\Rightarrow \quad \frac{1}{p^2} = \frac{1}{a^2} + \frac{1}{b^2}.$$

7. A point O in the interior of a rectangle ABCD is joined with each of the vertices A, B, C, D. Prove that $OA^2 + OC^2 = OB^2 + OD^2$.

Sol. Given. O is a point inside the rectangle ABCD and OA, OB, OC, OD are joined together.

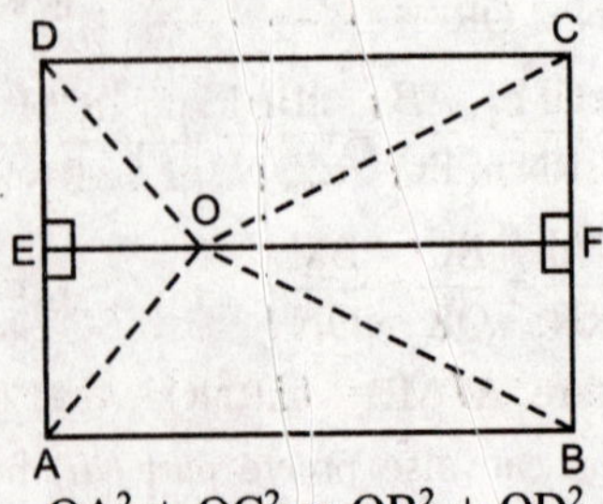

To Prove. $OA^2 + OC^2 = OB^2 + OD^2$.

Construction. Draw EF through O, parallel to AB meeting AD at E and BC at F.

Proof. $EF \parallel AB$ [Construction]

$\therefore \quad AE \perp EF$ and $BF \perp EF$

[$\because$ AE and BF $\perp$ AB, these will be $\perp$ to EF also, as EF $\parallel$ AB]

$\therefore$ EABF is a rectangle

Similarly, DEFC is a rectangle.

$\therefore \quad AE = BF$ and $DE = CF$ ...(i)

In right ΔOEA, $OA^2 = OE^2 + AE^2$

[Pythagoras Theorem] ...(ii)

In right ΔOFC, $OC^2 = CF^2 + OF^2$ ...(iii)

$\therefore \quad OA^2 + OC^2 = OE^2 + OF^2 + AE^2 + CF^2$ ...(iv)

[From (i) and (iii)]

Again In right ΔOFB, $OB^2 = OF^2 + BF^2$ ...(v)

In right ΔOED, $OD^2 = DE^2 + OE^2$ ...(vi)

$\therefore \quad OB^2 + OD^2 = OE^2 + OF^2 + DE^2 + BF^2$ ...(vii)

[From (v) and (vi)]

From (i), (iv) and (vii)

$OE^2 + OF^2 + AE^2 + CF^2 = OE^2 + OF^2 + DE^2 + BF^2$

$\Rightarrow \quad OA^2 + OC^2 = OB^2 + OD^2$.

8. Prove that three times the sum of the squares of the sides of a triangle is equal to four times the sum of the squares of the medians of the triangles.

Sol. Given. In ΔABC, D, E, F are the mid-points of sides BC, CA and AB respectively.

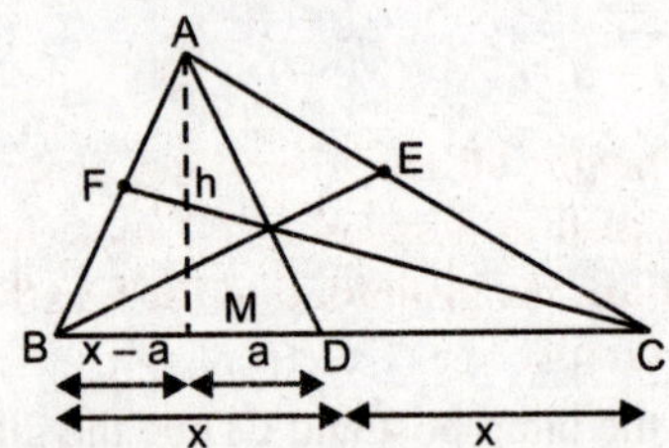

To Prove. $3(AB^2 + BC^2 + CA^2) = 4(AD^2 + BE^2 + CF^2)$.

Const. $AM \perp BC$

Proof. $AB^2 + AC^2 = [h^2 + (x - a)^2] + [h^2 + (x + a)^2]$

$= 2h^2 + 2a^2 + 2x^2 = 2(h^2 + a^2) + \frac{(2x)^2}{2}$

$= 2AD^2 + \frac{1}{2}BC^2$ ...(i)

Similarly, $BC^2 + CA^2 = 2CF^2 + \frac{1}{2}AB^2$ ...(ii)

and $AB^2 + BC^2 = 2BE^2 + \frac{1}{2}AC^2$ ...(iii)

Adding (i), (ii) and (iii), we get

$2(AB^2 + BC^2 + CA^2)$

$= 2(AD^2 + BE^2 + CF^2) + \frac{1}{2}(BC^2 + AB^2 + AC^2)$

or $\frac{3}{2}(AB^2 + BC^2 + CA^2) = 2(AD^2 + BE^2 + CF^2)$

$\Rightarrow 3(AB^2 + BC^2 + CA^2) = 4(AD^2 + BE^2 + CF^2)$.

9. In an equilateral triangle ABC, D is a point on the side BC such that $BD = \frac{1}{3}BC$. Prove that $9AD^2 = 7AB^2$.

Sol. Given. In an equilateral triangle ABC, D is a point on BC such that BC is trisected at D.

To Prove. $9AD^2 = 7AB^2$.

Construction. Draw $AL \perp BC$, meeting BC at L.

Proof. $AL \perp BC \Rightarrow BL = LC = \frac{1}{2}BC$ ...(i)

[In an equilateral triangle, the perpendicular from the vertex bisects the base]

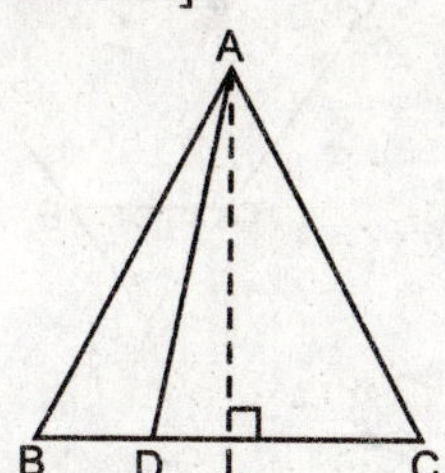

Also, $BD = \frac{1}{3}BC$ and $DC = \frac{2}{3}BC$ ... (ii)

Consider right ΔALD,

$AD^2 = AL^2 + DL^2$ ...(iii)

and right ΔALB, $AB^2 = AL^2 + BL^2$ ...(iv)

Subtracting (iii) from (iv), we get

$AB^2 - AD^2 = BL^2 - DL^2 = (BL + DL)(BL - DL)$

$AB^2 - AD^2 = \left(\frac{1}{2}BC + BL - BD\right)(BD)$

$AB^2 - AD^2 = \left(\frac{1}{2}BC + \frac{BC}{2} - \frac{BC}{3}\right)\left(\frac{BC}{3}\right)$

$= \left(\frac{2}{3}BC\right) \cdot \frac{BC}{3} = \frac{2}{9}BC^2$

$\Rightarrow 9AB^2 - 9AD^2 = 2AB^2$ [$\because$ BC = AB]

$\Rightarrow 7AB^2 = 9AD^2$.

Hence proved.

Summative Assessment

Subjective Questions

VERY SHORT ANSWER TYPE QUESTIONS

This category is given for better understanding of the topic and for EXTRA PRACTICE.

PREVIOUS YEARS' QUESTIONS

1. In the fig., P and Q are points on the sides AB and AC respectively of ΔABC such that AP = 3.5 cm, PB = 7 cm, AQ = 3 cm and QC = 6 cm. If PQ = 4.5 cm, find BC.

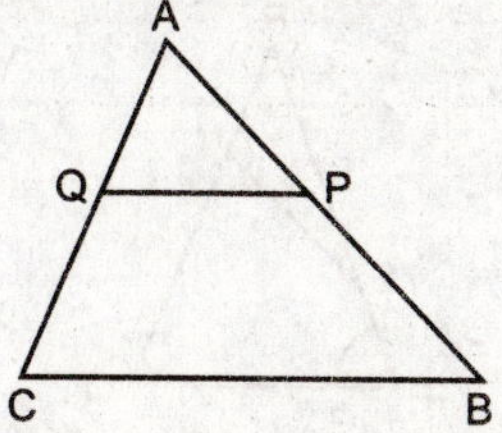

[Delhi 2008]

2. In the fig., PQ = 24 cm, QR = 26 cm, $\angle PAR = 90°$, PA = 6 cm and AR = 8 cm. Find $\angle QPR$.

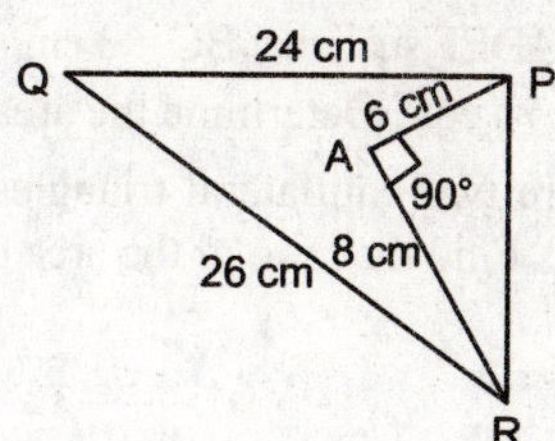

[Delhi 2008]

3. The lengths of the diagonals of a rhombus are 30 cm and 40 cm. Find the side of the rhombus. [Foreign 2008]

4. In the fig., PQ || BC and AP : PB = 1 : 2. Find $\frac{ar(\Delta APQ)}{ar(\Delta ABC)}$.

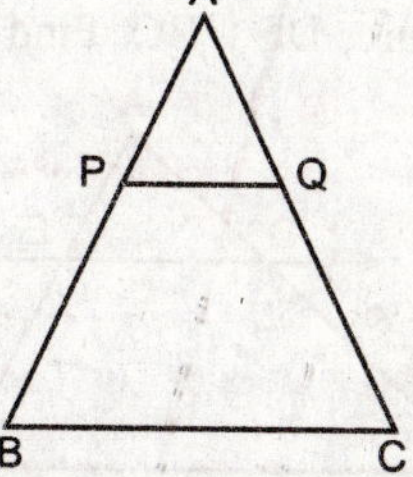

[Foreign 2008]

5. The perimeter of two similar triangles ABC and LMN are 60 cm and 48 cm respectively. If LM = 8 cm, then what is the length of AB ? [Delhi 2008C]

6. In ΔABC shown in figure, DE || BC. If BC = 8 cm, DE = 6 cm and area of ΔADE = 45 cm^2, what is the area of ΔABC ?

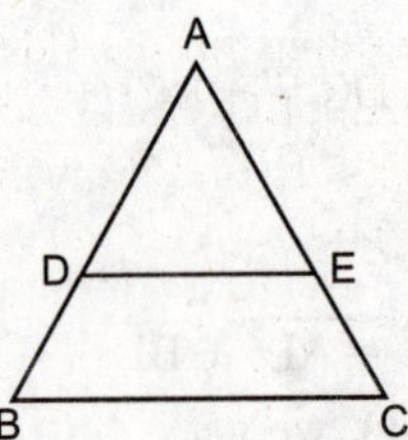

[AI 2008C]

7. In fig. $\angle M = \angle N = 46°$, express x in terms of a, b and c, where a, b and c are lengths of LM, MN and NK respectively.

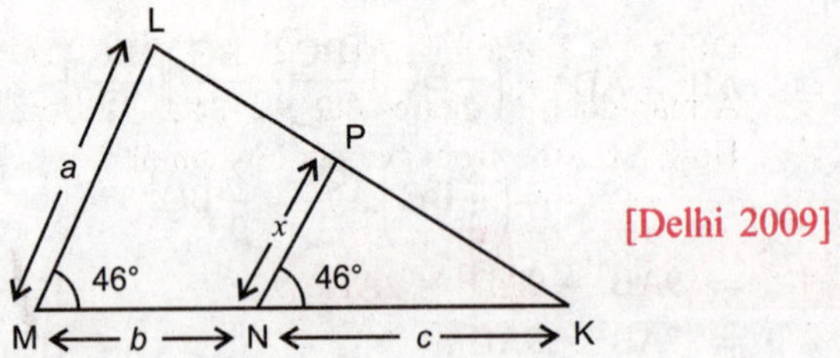

[Delhi 2009]

8. If the areas of two similar triangles are in ratio 25 : 64, write the ratio of their corresponding sides. [AI 2009]

9. In figure, S and T are points on the sides PQ and PR, respectively of ΔPQR, such that PT = 2 cm, TR = 4 cm and ST is parallel to QR. Find the ratio of the areas of ΔPST and ΔPQR.

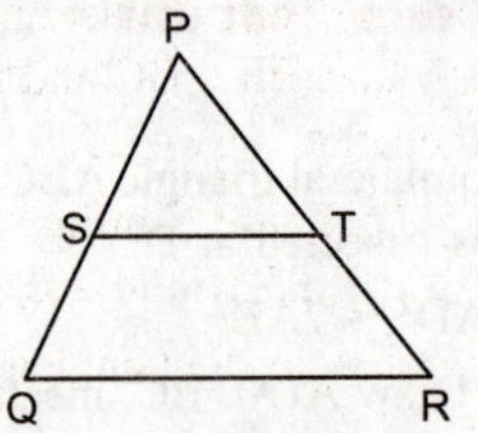

[Delhi 2010]

10. In figure, DE || BC in ΔABC such that BC = 8 cm, AB = 6 cm and DA = 1.5 cm. Find DE.

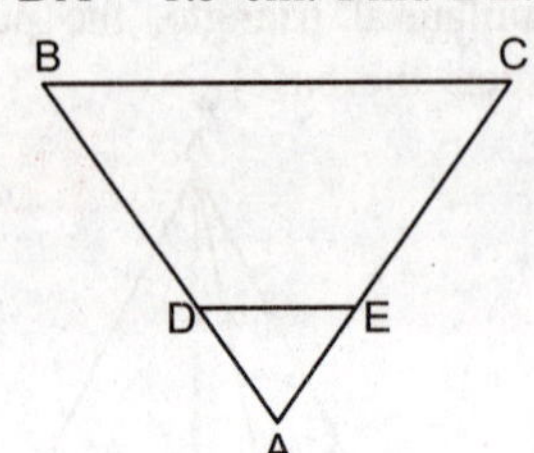

[AI 2010]

Practice Questions

11. State whether the following quadrilaterals are similar or not :

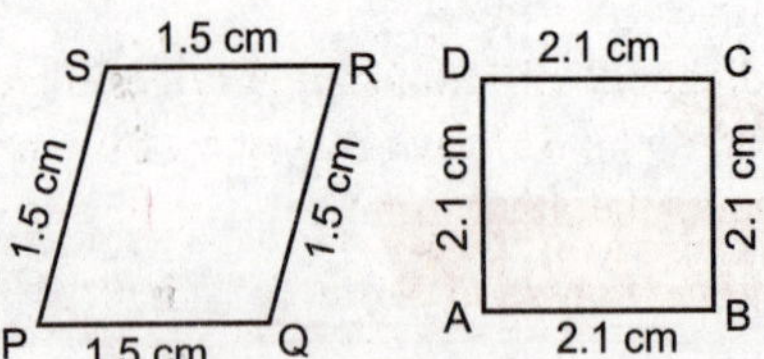

12. In the given figure, DE || BC. Find EC.

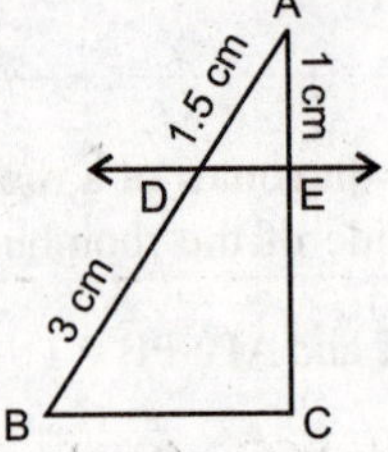

13. In the given figure, DE || BC. Find AD.

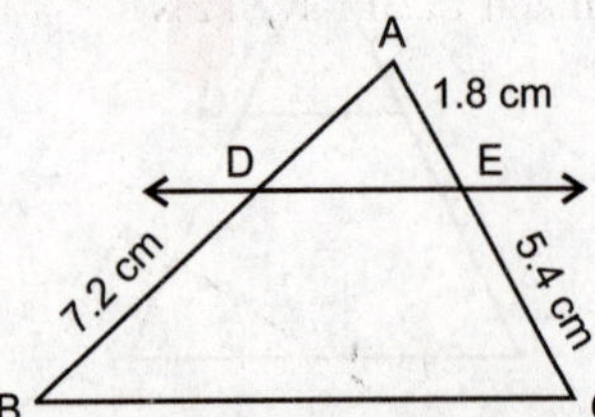

14. E and F are points on the sides PQ and PR respectively of a ΔPQR. If, PE = 4 cm, QE = 4.5 cm, PF = 8 cm, RF = 9 cm, state whether EF || QR.

15. E and F are points on the side PQ and PR respectively of a ΔPQR. If PE = 3.9 cm, EQ = 3 cm, PF = 3.6 cm and FR = 2.4 cm, state whether EF || QR.

16. In the given figure, $\Delta ODC \sim \Delta OBA$. $\angle BOC = 125°$ and $\angle CDO = 70°$, find $\angle DOC$.

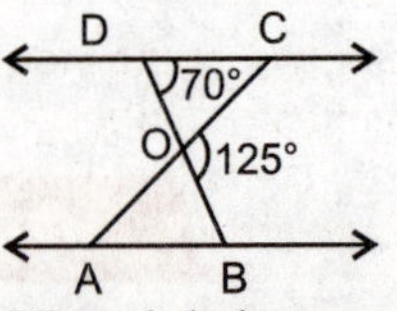

17. Let $\Delta ABC \sim \Delta DEF$ and their areas be respectively, 64 cm^2 and 121 cm^2. If EF = 15.4 cm, find BC.

18. A vertical stick 12 m long casts a shadow 8 m long on the ground. At the same time a tower casts the shadow 40 m long on the ground. Determine the height of the tower.

19. If $\angle P = \angle RTS$,
Then show that $\angle PQR = \angle RST$.

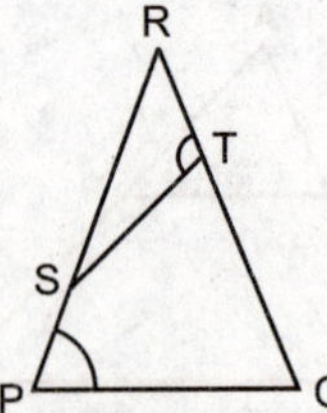

20. The perimeters of two similar triangles are 25 cm and 15 cm respectively. If one side of first triangle is 9 cm., what is the corresponding side of the other triangle ?

21. If ΔABC is similar to ΔDEF such that BC = 3 cm, EF = 4 cm and area of ΔABC = 54 cm^2. Determine the area of ΔDEF.

22. ΔABC and ΔBDE are two equilateral triangles such that D is mid-point of BC. Find the ratio of the area of triangles ABC and BDE.

23. If ΔABC and ΔDEF are two triangles such that $\frac{AB}{DE} = \frac{BC}{EF} = \frac{AC}{DF} = \frac{4}{7}$. Find $\frac{\text{area } \Delta ABC}{\text{area } \Delta DEF}$.

24. In ΔABC, D and E are points on sides AB and AC respectively such that DE || BC and AD : DB = 3 : 1. If EA = 6.6 cm then find AC.

25. If ΔABC and ΔDEF are similar triangles such that $\angle A = 57°$ and $\angle E = 83°$. Find $\angle C$.

26. ΔABC is a right triangle right angled at A and $AD \perp BC$. Show that $\frac{BD}{AD} = \frac{AB}{AC}$.

27. In ΔABC, D and E are points on the sides AB and AC respectively, such that DE || BC. If AD = x, DB = $x - 2$, AE = $x + 2$ and EC = $x - 1$, find the value of x.

28. In fig $\angle 1 = \angle 3$, $\angle 2 = \angle 4$

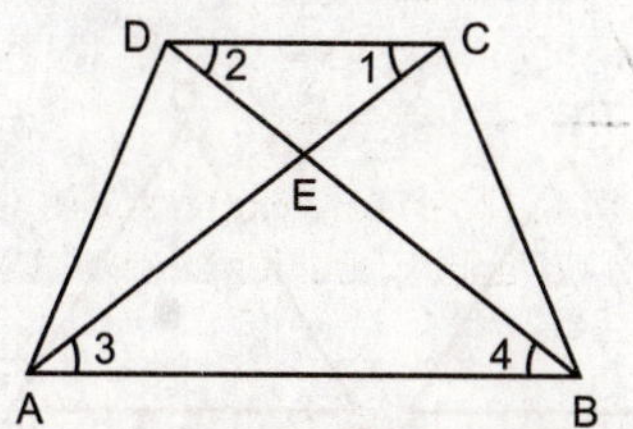

DE = 4, CE = $x + 1$, AE = $2x + 4$; BE = $4x - 2$. Find x.

29. A man goes 24 metres due west and 7 metres due north. How far is he from the starting point ?

SHORT ANSWER TYPE QUESTIONS-I

Previous Years' Questions

1. In ΔABC, $AD \perp BC$, prove that

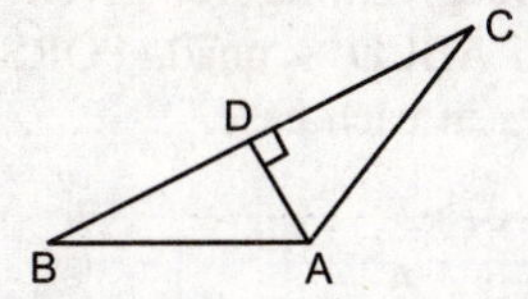

$AB^2 + CD^2 = AC^2 + DB^2$.

[AI 2006] [NCERT]

2. In figure, S and T trisect the side QR of a right triangle PQR. Prove that $8PT^2 = 3PR^2 + 5PS^2$. [Foreign 2009]

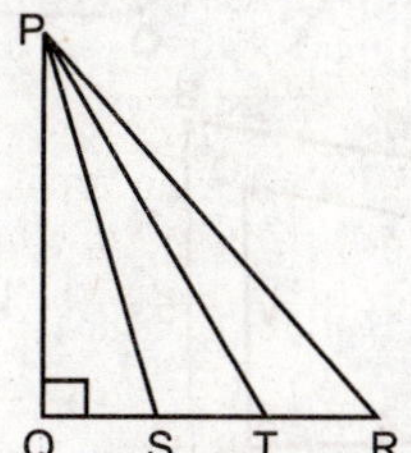

3. In fig., AB || DE and BD || EF, Prove that $DC^2 = CF \times AC$.

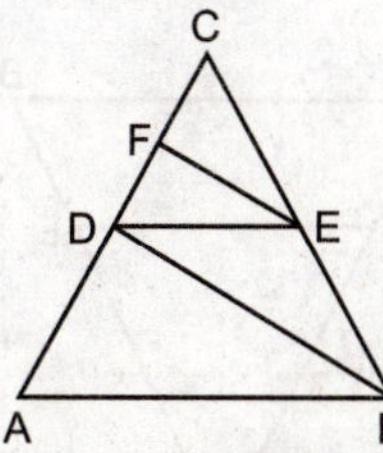

[Delhi 2007]

4. E is a point on the side AD produced of a ||gm ABCD and BE intersects CD at F. Show that $\Delta ABE \sim \Delta CFB$. [Foreign 2008]

5. Triangle ABC is right angled at B, and D is mid-point of BC. Prove that $AC^2 = 4AD^2 - 3AB^2$. [AI 2008C; Delhi 2010]

6. In figure, ΔABD is a right triangle, right angled at A and $AC \perp BD$. Prove that $AB^2 = BC.BD$.

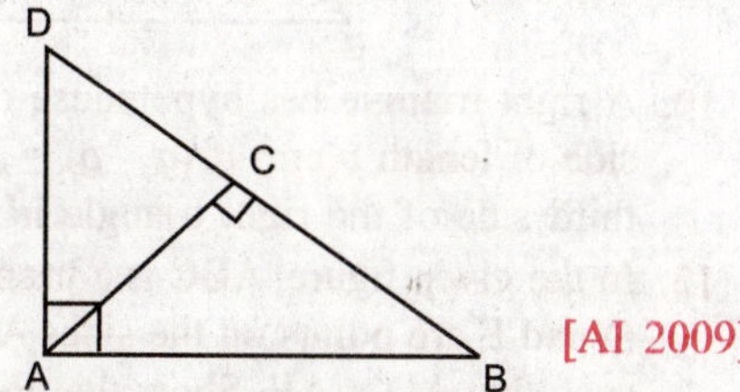

[AI 2009]

7. If one diagonal of a trapezium divides the other diagonal in the ratio 1 : 3. Prove that one of the parallel sides is three times the other. [CBSE (CCE) 2011]

8. In the given figure, if AB || DC, find the value of x.

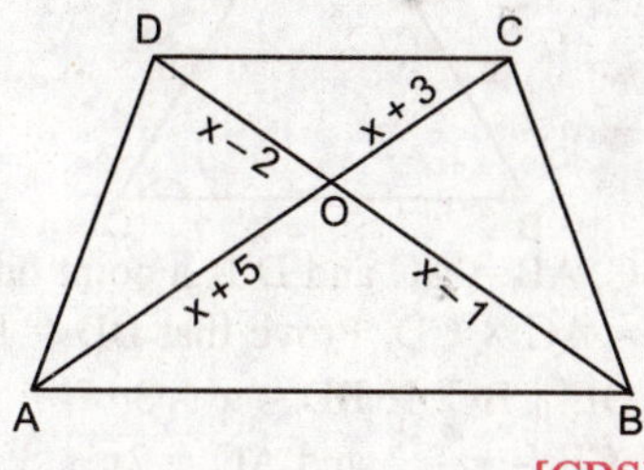

[CBSE (CCE) 2012]

9. In the given triangle PQR, $\angle QPR = 90°$, PQ = 24 cm and QR = 26 cm and in ΔPKR, $\angle PKR = 90°$ and KR = 8 cm find PK.

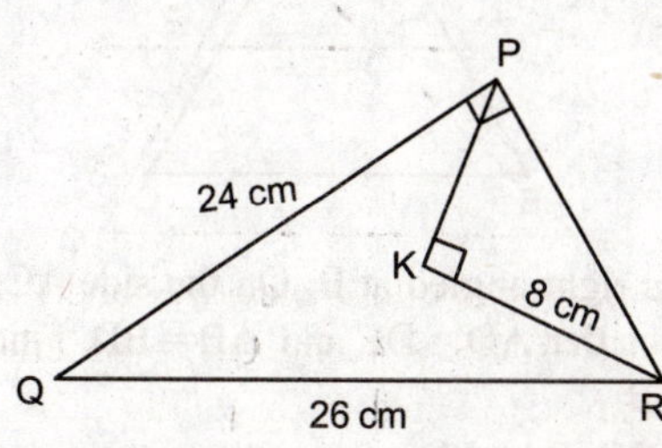

[CBSE (CCE) 2012]

PRACTICE QUESTIONS

10. In the given figure, ΔABC and ΔDEF are similar, BC = 3 cm, EF = 4 cm and area of $\Delta ABC = 54\ cm^2$. Determine the area of ΔDEF.

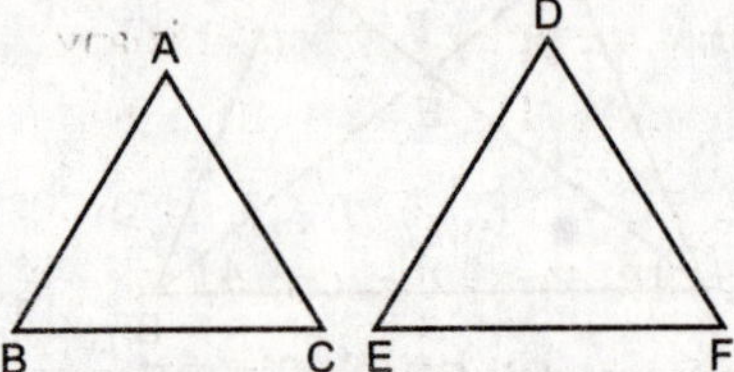

11. In the given figure, considering triangles BEP and CPD, prove that BP × PD = EP × PC.

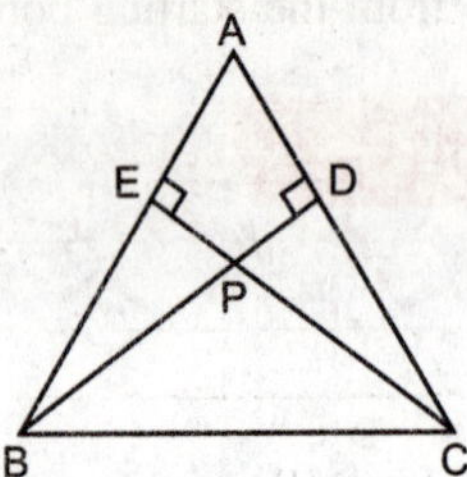

12. A right triangle has hypotenuse of length q cm and one side of length p cm. If $(q - p) = 2$, express the length of third side of the right triangle in terms of q.

13. In the given figure, ABC is a triangle in which AB = AC, D and E are points on the sides AB and AC respectively, such that AD = AE. Show that the points B, C, E and D are concyclic.

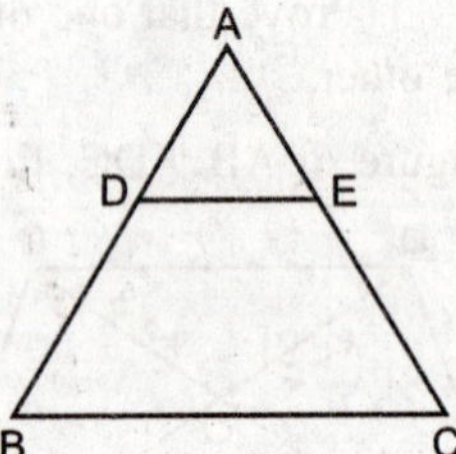

14. In a ΔABC, AB = AC and D is a point on side AC, such that $BC^2 = AC \times CD$. Prove that BD = BC.

15. In figure DE ∥ BC. If BD = $x - 3$ AB = $2x$. CE = $x - 2$ and AC = $2x + 3$ Find x.

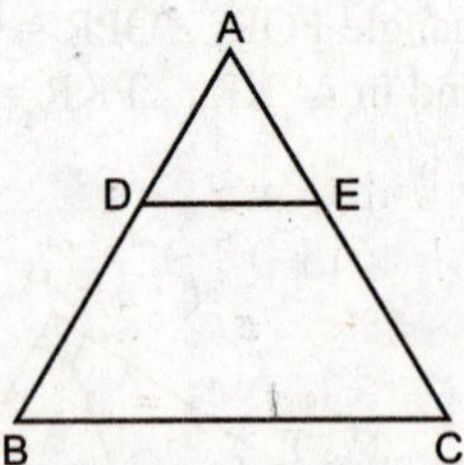

16. ΔABC is right angled at B. On the side AC, a point D is taken such that AD = DC and AB = BD. Find the measure of $\angle CAB$.

17. In a ΔABC, P and Q are points on the sides AB and AC respectively such that PQ is parallel to BC. Prove that median AD, drawn from A to BC, bisects PQ.

18. In the given figure, $\Delta ABC \sim \Delta DEF$. If AB = 2DE and area of ΔABC is 56 sq. cm, find the area of ΔDEF.

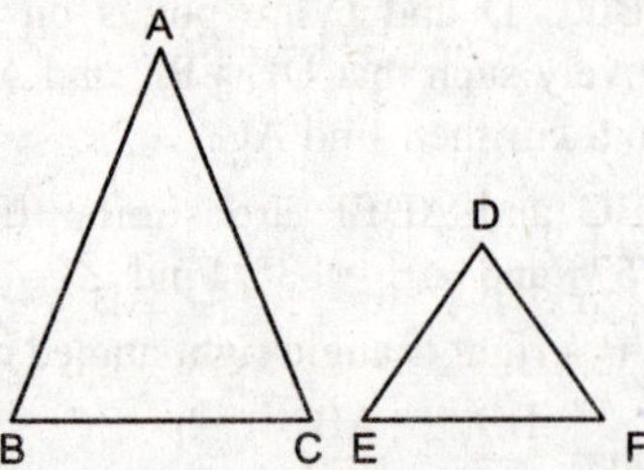

19. Two poles of height 7 m and 12 m stand on a plane ground. If the distance between their feet is 12 m, find the distance between their tops.

20. If three or more parallel lines are intersected by two transversals, prove that the intercepts made by them on the transversal are proportional.

21. In the given figure, given that $\Delta ABC \sim \Delta PQR$ and quad. ABCD ~ quad. PQRS. Determine the values of x, y, z in each case.

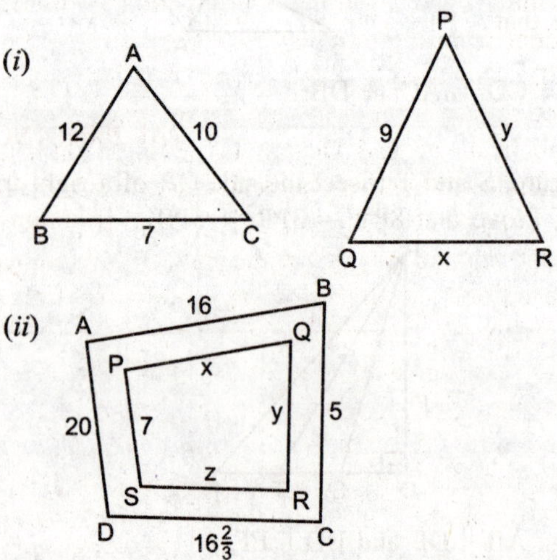

22. What value(s) of x will make DE ∥ AB in the given figure ?

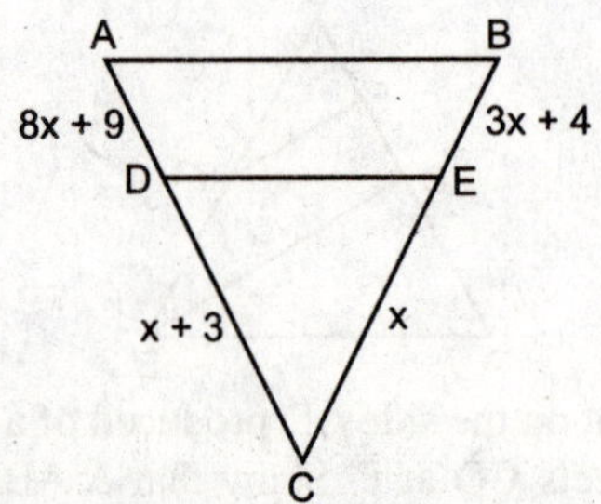

23. P and Q are points on the sides AB and AC respectively of a triangle ABC. If AP = 2 cm, PB = 4 cm, AQ = 3cm, QC = 6 cm, prove that BC = 3PQ.

24. A vertical stick 12 cm long casts a shadow 8 cm long on the ground. At the same time, a tower casts a shadow 40 m long on the ground. Determine the height of the tower.

25. In the figure, AP = 3 cm, AR = 4.5 cm, AQ = 6 cm, AB = 5 cm and AC = 10 cm. Find the length of AD.

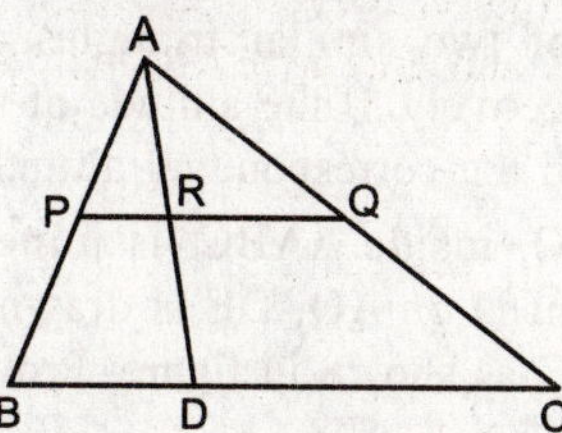

26. In the figure, if $\angle A = \angle CED$, AB = 9 cm, AD = 7 cm, CD = 8 cm and CE = 10 cm. Find DE.

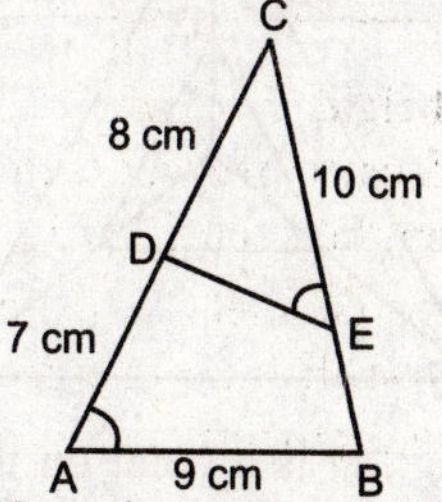

27. A man goes 150 m due east and then 200 m due north. How far is he from the starting point ?

28. A ladder 25 m long reaches a window of a building 20 m above the ground. Determine the distance of the foot of the ladder from the building.

29. A ladder is placed in such a way that its foot is at a distance of 5 m from a wall and its tip reaches a window 12 m above the ground. Determine the length of the ladder.

30. In a triangle AD is drawn perpendicular to BC. Prove that $(AB)^2 - (BD)^2 = (AC)^2 - (CD)^2$.

31. A ladder reaches a window which is 12 m above the ground on one side of the street, keeping its foot at the same point, the ladder is turned to the other side of the street to reach a window 9 m high. Find the width of the street if the length of the ladder is 15 m.

32. If D and E are respectively the points on the side AB and AC of a triangle ABC such that AD = 6 cm, BD = 9 cm, AE = 8 cm and EC = 12 cm, then show that DE || BC.

33. In the given figure, $\Delta ACB \sim \Delta AQP$. If BC = 8 cm, PQ = 4 cm, BA = 6.5 cm. AQ = 2.8 cm, find CA and PA.

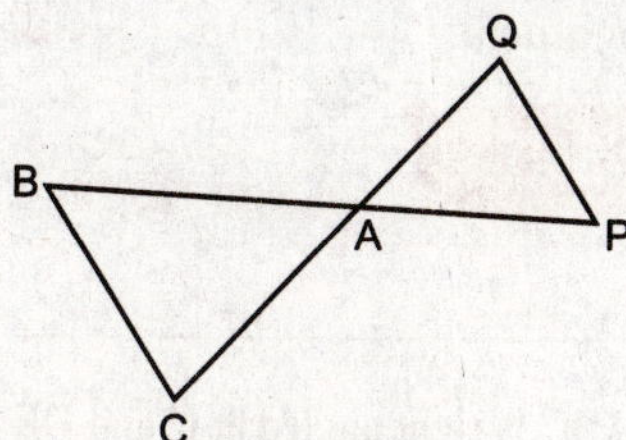

34. In the given figure, $\frac{AO}{OC} = \frac{BO}{OD} = \frac{1}{2}$ and AB = 4 cm, find the value of DC.

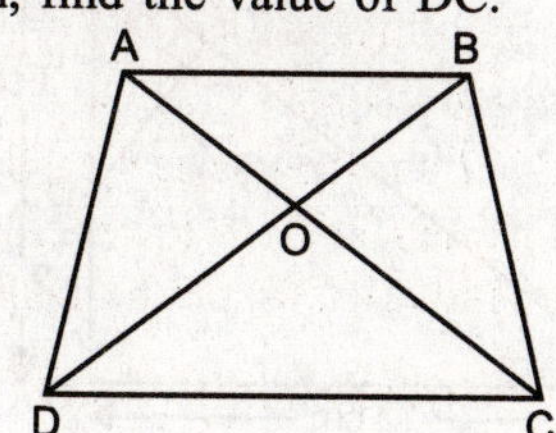

35. In an isosceles ΔABC, the base AB is produced both ways to P and Q such that $AP \times BQ = (AC)^2$. Prove that $\Delta ACP \sim \Delta BCQ$.

36. The diagonal BD of a parallelogram ABCD intersects the segment AE at the point F, where E is any point on the side BC, prove that DF × FE = BF × FA.

37. In a ΔPQR, PQ = PR, X is a point on PR such that $QR^2 = PR \times XR$. Prove that QX = QR.

38. In the given figure, DE || BC. If DE = 4 cm, BC = 8 cm and area of ΔADE = 25 sq. cm. Find the area of ΔABC.

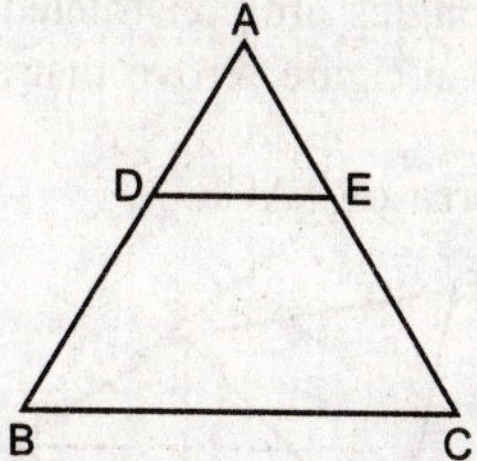

39. The altitudes AD and BE of a ΔABC intersect each other at O. Prove that $\frac{CD}{AC} = \frac{CE}{BC}$.

40. In a trapezium ABCD, AB || CD and AB = 2CD. O is the point of intersection of AC and DB. Find the ratio of areas of ΔAOB and ΔCOD.

41. $\angle A$ of an isosceles ΔABC is acute in which AB = AC, and $BD \perp AC$, prove that $(BC)^2 = 2AC \cdot CD$.

42. Prove in a right angled triangle if a perpendicular is drawn from the right angle to the hypotenuse, the area of square on the perpendicular is equal to the area of rectangle contained by the two segments of the hypotenuse.

43. In the given figure, $\Delta AMB \sim \Delta CMD$. Determine MD in terms of x, y and z.

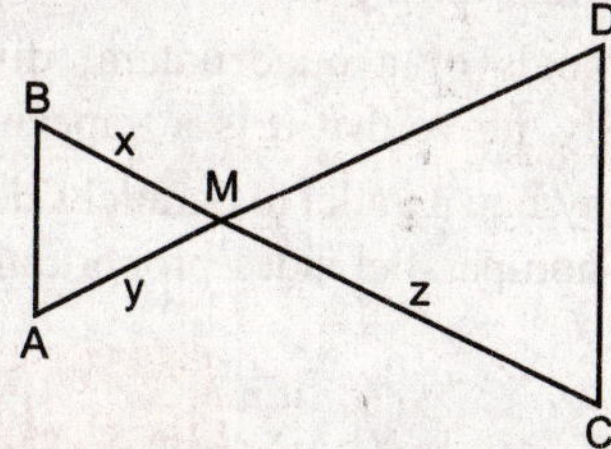

44. In a ΔABC, P and Q are points on sides AB and AC respectively such that PQ || BC. If AP = 4 cm, PB = 6 cm and PQ = 3 cm, determine BC.

45. In figure, $\angle 1 = \angle 2$ and $\angle 3 = \angle 4$. Show that PT × QR = PR × ST.

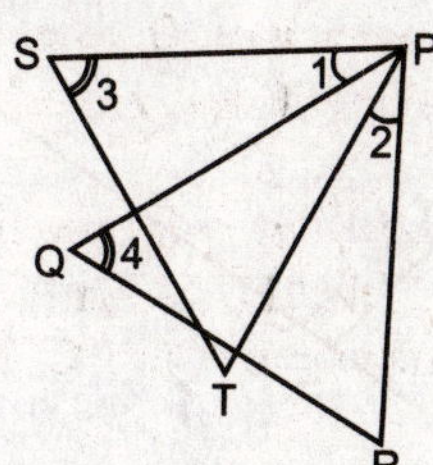

46. In figure, LM ∥ NQ and LN ∥ PQ. If $MP = \frac{1}{3}MN$, find the ratio of the areas of ΔLMN and ΔQNP.

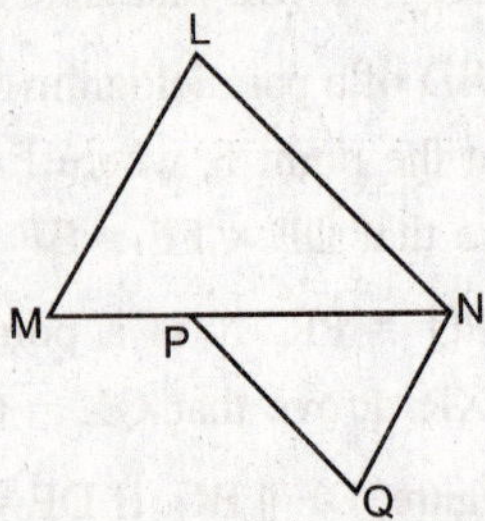

47. ABC is an isosceles triangle right angled at B. Two equilateral triangles are constructed with sides BC and AC as shown in figure. Prove that area of

$\Delta BCD = \frac{1}{2}$ area of ΔACE.

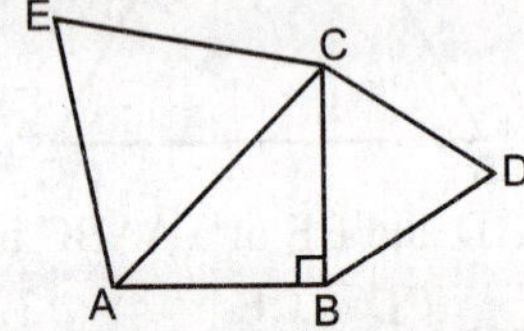

48. L and M are the mid-points of AB and BC respectively of ΔABC, right-angled at B. Prove that $4LC^2 = AB^2 + 4BC^2$.

49. The areas of two similar triangles are 121 cm^2 and 64 cm^2 respectively. If the median of the first triangle is 12.1 cm, find the corresponding median of the other.

50. The areas of two similar triangles are 100 cm^2 and 49 cm^2 respectively. If the altitude of the bigger triangle is 5 cm, find the corresponding altitude of the other.

51. Any point O, inside ΔABC, is joined to its vertices. From a point D on AO, DE is drawn so that DE ∥ AB and EF ∥ BC as shown in figure. Prove that DF ∥ AC.

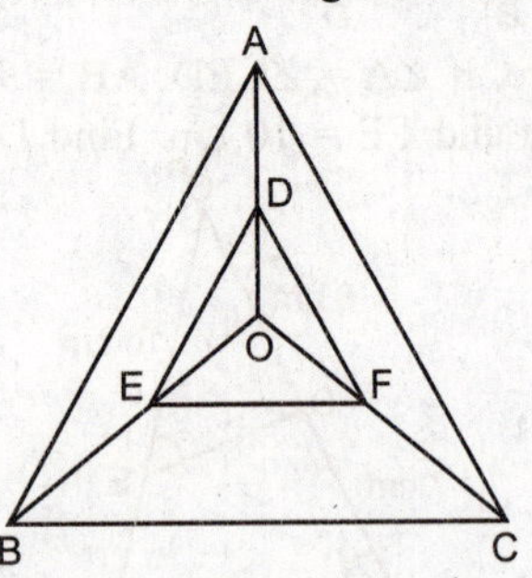

52. In the given fig., D is a point on the side BC of ΔABC such that ∠ADC = ∠BAC. Prove that $\frac{CA}{CD} = \frac{CB}{CA}$

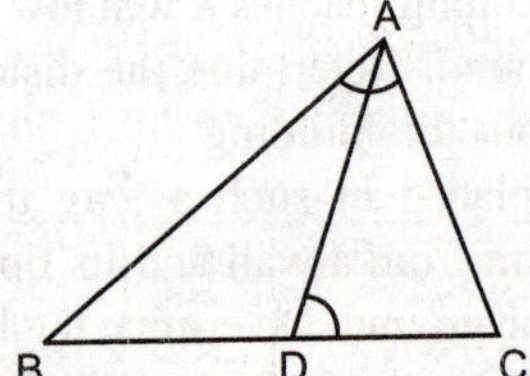

NCERT Questions

53. Using Basic Proportionality Theorem, prove that the line drawn from the mid-point of one side of a triangle parallel to another side bisects the third side.

54. Prove, using converse of Basic Proportionality Theorem, that the line joining the mid-points of two sides of a triangle is parallel to the third side.

55. If the diagonals of a quadrilateral divide each other proportionally, prove that it is a trapezium.

56. Prove that any line parallel to parallel sides of a trapezium divides the non-parallel sides proportionally.

57. M and N are points on the sides PQ and PR respectively of a ΔPQR. For each of the following cases state whether MN ∥ QR :

(*i*) PM = 4, QM = 4.5, PN = 4, NR = 4.5

(*ii*) PQ = 1.28, PR = 2.56, PM = 0.16, PN = 0.32.

58. Two poles of heights 6 m and 11 m stand vertically on a plane ground. If the distance between their feet is 12 m, determine the distance between their tops.

59. If the areas of two similar triangles are equal, prove that they are congruent. [CBSE (CCE) 2011]

SHORT ANSWER TYPE QUESTIONS-II

Previous Years' Questions

1. In fig., DB ⊥ BC, DE ⊥ AB and AC ⊥ BC. Prove that $\frac{BE}{DE} = \frac{AC}{BC}$.

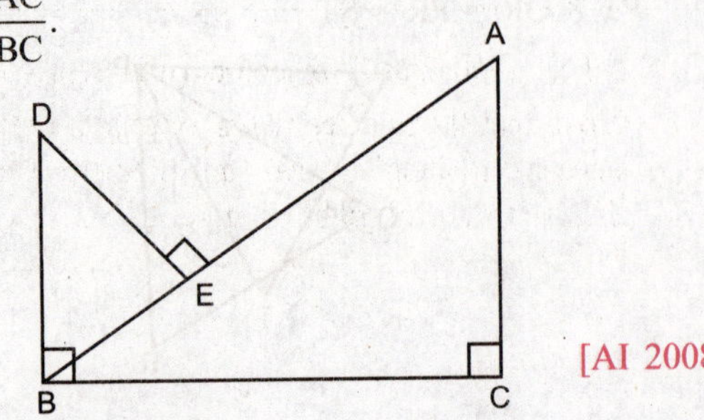

[AI 2008]

2. In figure, ΔABC is right angled at C and DE ⊥ AB. Prove that ΔABC ~ ΔADE and hence find the lengths of AE and DE.

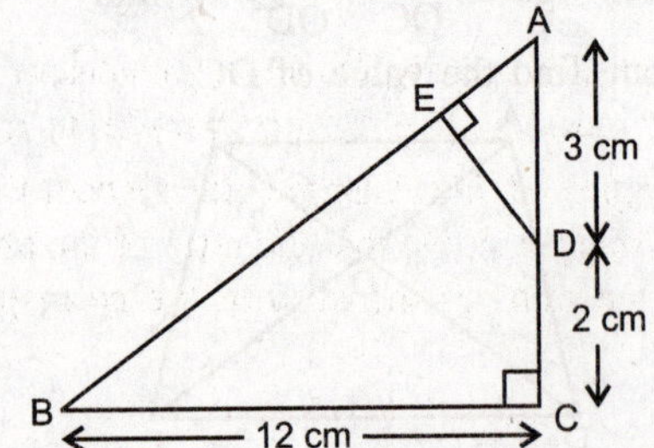

[Delhi 2009]

3. In figure, DEFG is a square and $\angle BAC = 90°$. Show that $DE^2 = BD \times EC$.

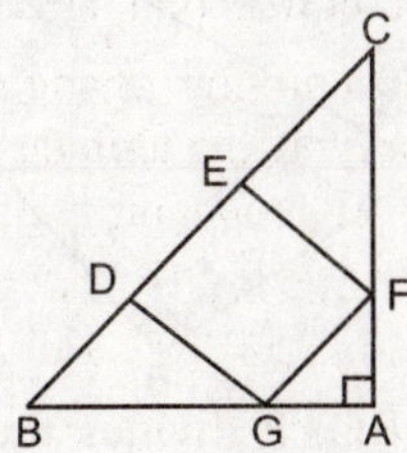

[Delhi 2009]

4. In figure, $AD \perp BC$ and $BD = \frac{1}{3}CD$. Prove that $2CA^2 = 2AB^2 + BC^2$. [AI 2009]

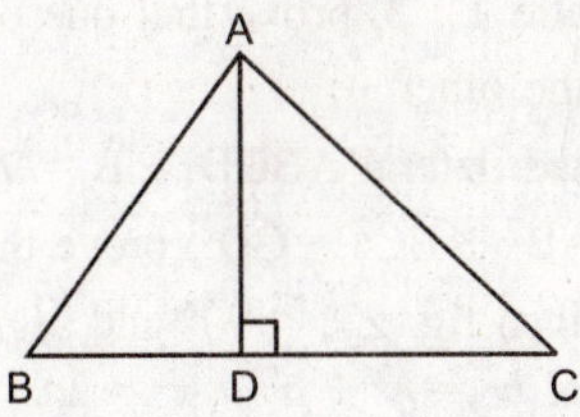

5. In figure, M is mid-point of side CD of a parallelogram ABCD. The line BM is drawn intersecting AC at L and AD produced at E. Prove that EL = 2BL.

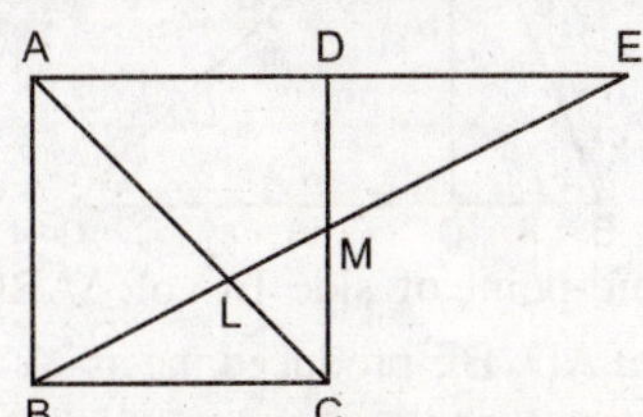

[AI 2009]

6. In figure, two triangles ABC and DBC lie on the same side of base BC. P is a point on BC such that PQ || BA and PR || BD. Prove that QR || AD.

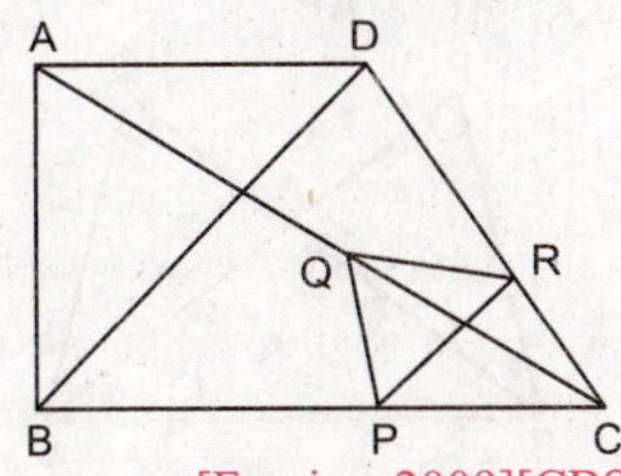

[Foreign 2009][CBSE (CCE) 2012]

7. In figure, ABC is an isosceles triangle in which AB = AC. E is a point on the side CB produced, such that $FE \perp AC$. If $AD \perp CB$, prove that : $AB \times EF = AD \times EC$.

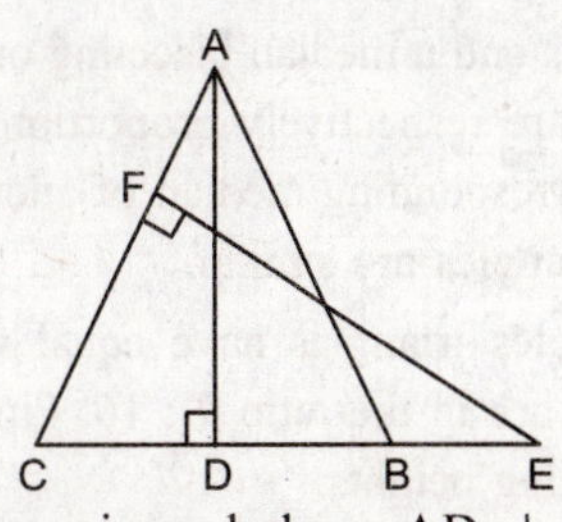

[AI 2010]

8. In the figure given below, $AD \perp BC$. Prove that $AB^2 + CD^2 = BD^2 + AC^2$: [CBSE (CCE) 2011]

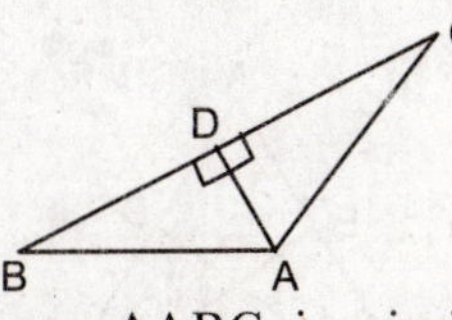

9. In given figure ΔABC is similar to ΔXYZ and AD and XE are angle bisectors of $\angle A$ and $\angle X$ respectively such that AD and XE in centimetres are 4 and 3 respectively, find the ratio of area of ΔABD and area of ΔXYE. [CBSE (CCE) 2011]

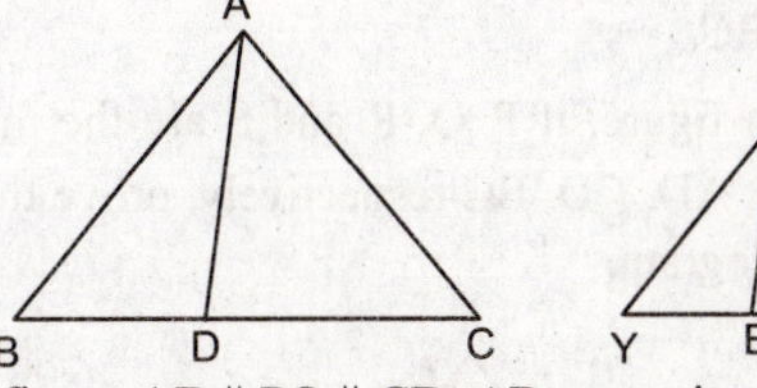

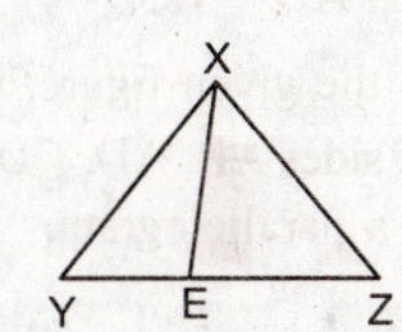

10. In figure AB || PQ || CD, AB = x units, CD = y units and PQ = z units, prove that $\frac{1}{x} + \frac{1}{y} = \frac{1}{z}$ [CBSE (CCE) 2011]

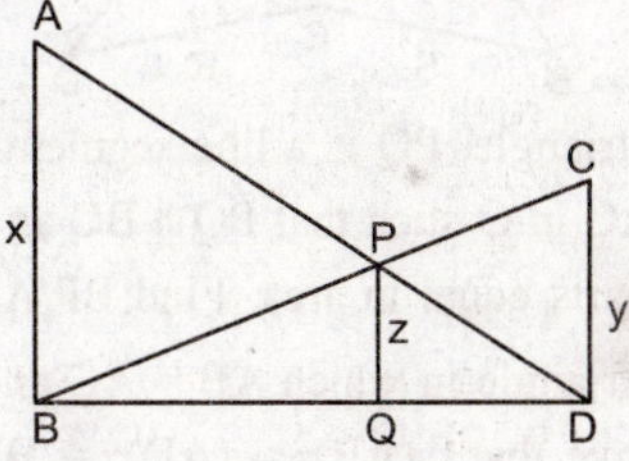

11. In the given figure PQ || BA; PR || CA. If PD = 12 cm. Find $BD \times CD$.

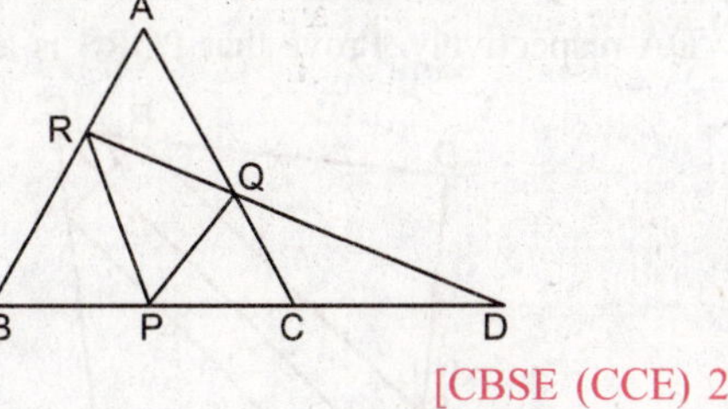

[CBSE (CCE) 2012]

Practice Questions

12. Prove that, if a line divides any two sides of a triangle in the same ratio, the line must be parallel to the third side.

13. If the angles of one triangle are respectively equal to another triangle, prove that the ratio of the corresponding sides is the same as the ratio of the corresponding :

(*i*) medians (*ii*) bisectors of angles (*iii*) altitudes.

14. Prove that the line segments joining the mid-points of the sides of a triangle, form four triangles each of which is similar to the original triangle.

15. If two sides and a median bisecting one of these sides of a triangle are respectively proportional to the two sides and the corresponding median of another triangle, prove that the triangles are similar.

16. Two isosceles triangles have equal vertical angles and their areas are in the ratio 9 : 16. Find the ratio of their corresponding heights.

17. In ΔABC, BD ⊥ AC, CE ⊥ AB. Prove that BP × PD = EP × PC.

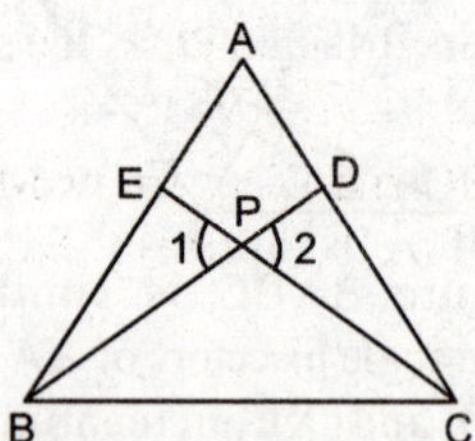

18. ABCD is a trapezium with AB || DC in which diagonals AC and BD intersect at E and ΔAED ~ ΔBEC. Prove that AD = BC.

19. In the given figure, if P, Q, R and S are the mid-points of sides AB, AD, CD, BC respectively, prove that PQRS is a parallelogram.

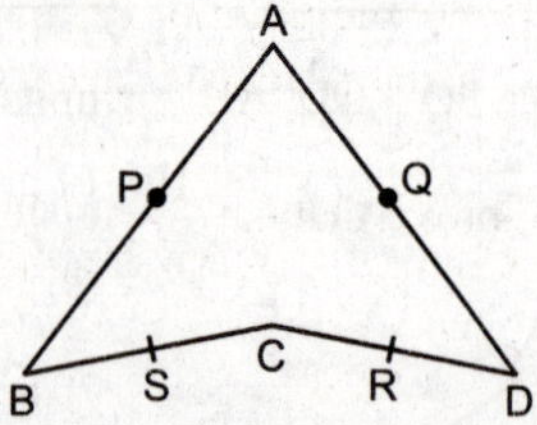

20. ABC is a triangle. PQ is a line segment intersecting AB in P and AC in Q such that PQ || BC and divides ΔABC into two parts equal in area. Find BP/AB.

21. ABC is a triangle in which AB = AC and D is any point in BC. Prove that : $(AB)^2 - (AD)^2 = BD \cdot CD$.

22. In the given fig. ABCD is a quadrilateral P, Q, R and S are the points of trisection of the sides AB, BC, CD and DA respectively. Prove that PQRS is a parallelogram.

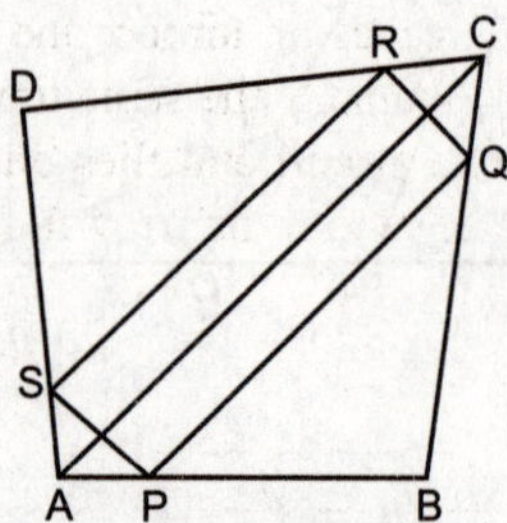

23. AD is the median of ΔABC, O is any point on AD. BO and CO produced meet AC and AB in E and F respectively. AD is produced to X such that OD = DX. Prove that AO : AX = AF : AB.

24. In a quadrilateral ABCD, ∠B = 90°, $AD^2 = AB^2 + CD^2 + BC^2$ Prove that ∠ACD = 90°

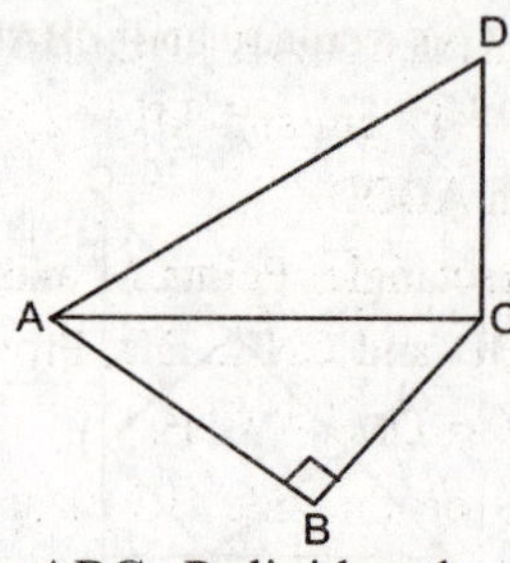

25. In a triangle ABC, P divides the side AB such that AP : PB = 1 : 2, Q is a point on AC such that PQ || BC. Find the ratio of the areas of ΔAPQ and trapezium BPQC.

26. If one diagonal of a trapezium divides the other diagonal in the ratio 1 : 2, prove that one of the parallel sides is double the other.

27. In a quadrilateral ABCD, ∠B = 90° and $AD^2 = AB^2 + BC^2 + CD^2$, prove that ∠ACD = 90°.

28. In the given fig. ∠B < 90° and segment AD ⊥ side BC, show that $b^2 = h^2 + a^2 + x^2 - 2ax$.

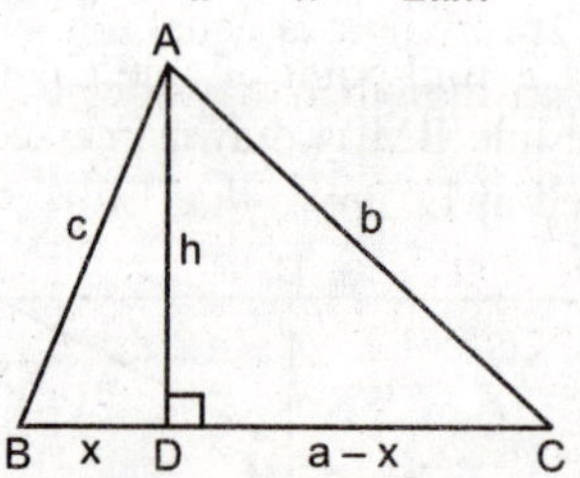

29. D is the mid-point of side BC of ΔABC and E is the mid-point of AD. BE produced meets AC at the point M. Prove that BE = 3EM.

30. In ΔABC, ∠A is obtuse, PB ⊥ AC and QC ⊥ AB. Prove that :

AB × AQ = AC × AP

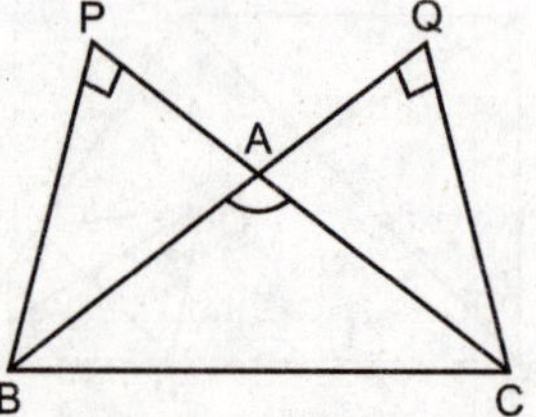

31. In a ΔABC, ∠ABC = 45°, prove that $(AC)^2 = (AB)^2 + (BC)^2 - 4$ ar (ΔABC).

32. Prove that sum of the squares of the diagonals of a parallelogram is equal to sum of the squares of its sides.

33. In a ΔABC, AD is a median. Prove that $(AB)^2 + (AC)^2 = 2(AD)^2 + 2(DC)^2$.

34. In a ΔABC, ∠C is an obtuse angle. AD ⊥ BC and $(AB)^2 = (AC)^2 + 3(BC)^2$. Prove that BC = CD.

35. In ΔABC, ∠A = 60°, prove that $(BC)^2 = (AB)^2 + (AC)^2 - AB \cdot AC$.

36. In a quadrilateral ABCD, given that ∠A + ∠D = 90°, prove that $(AC)^2 + (BD)^2 = (AD)^2 + (BC)^2$.

37. In a trapezium ABCD, AB || DC, BD ⊥ AD and AC ⊥ BC. If AD = BC = 15 cm and AB = 25 cm, find the area of the trapezium ABCD.

38. ABCD is a rectangle. Points M and N are on BD, such that AM ⊥ BD and CN ⊥ BD. Prove that

$BM^2 + BN^2 = DM^2 + DN^2$.

39. A point D is on the side BC of an equilateral triangle ABC such that $DC = \frac{1}{4}BC$. Prove that $(AD)^2 = 13(CD)^2$.

40. In the given figure, altitude BD is drawn to the hypotenuse AC of a right angled triangle ABC. The lengths of different line segments are marked. Find x and y.

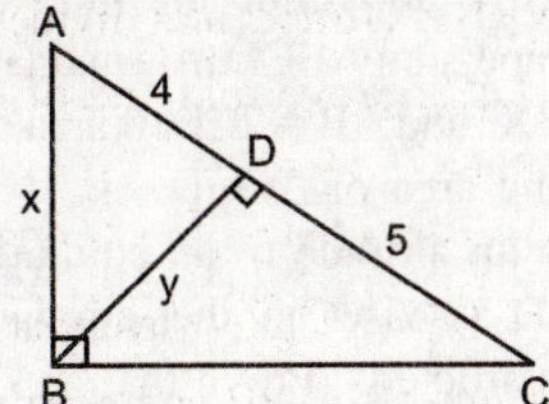

41. Prove that the area of a semicircle on the hypotenuse of a right-angled triangle is equal to the sum of the areas of semicircle on the other two sides.

42. In the figure, ∠AEF = ∠AFE and E is the mid-point of CA. Prove that $\frac{BD}{CD} = \frac{BF}{CE}$.

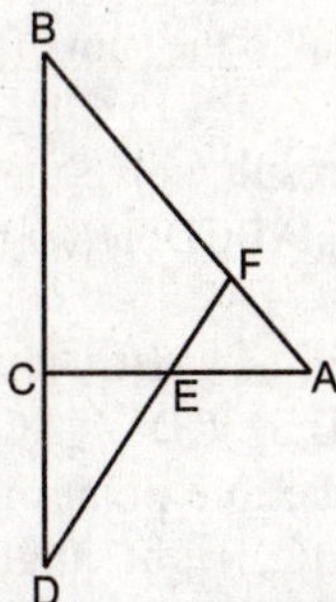

43. ABCD is a rectangle in which BC = 2AB. A point E lies on ray CD such that CE = 2BC. Prove that BE ⊥ AC.

44. In a ΔABC, AD is a median. X is a point on AD such that AX : XD = 2 : 3. Ray BX intersects AC in Y. Prove that BX = 4XY.

45. In an isosceles triangles ABC, AB = AC and BC is produced to D. Prove that $AD^2 - AC^2 = BD \cdot CD$.

46. In the given figure, ABC is a right triangle, right-angled at B. AD and CE are the two medians drawn from A and C respectively. If AC = 5 cm and $AD = \frac{3\sqrt{5}}{2}$ cm, find the length of CE.

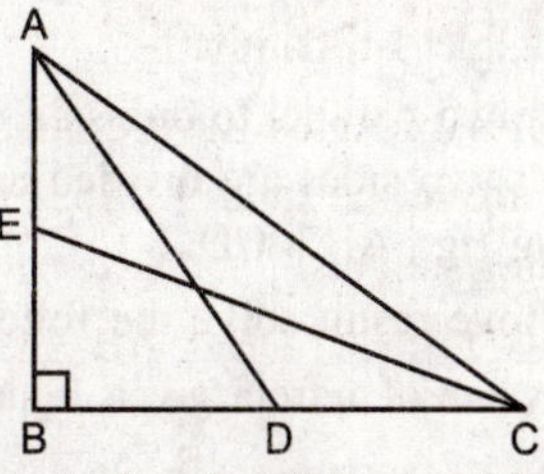

47. In an equilateral triangle PQR, the side QR is trisected at S. Prove that $9PS^2 = 7PQ^2$.

48. In figure, ABC and DBC are two triangles on the same base BC. If AD intersect BC at O. Prove that $\frac{ar\ \Delta ABC}{ar\ \Delta DBC} = \frac{AO}{DO}$.

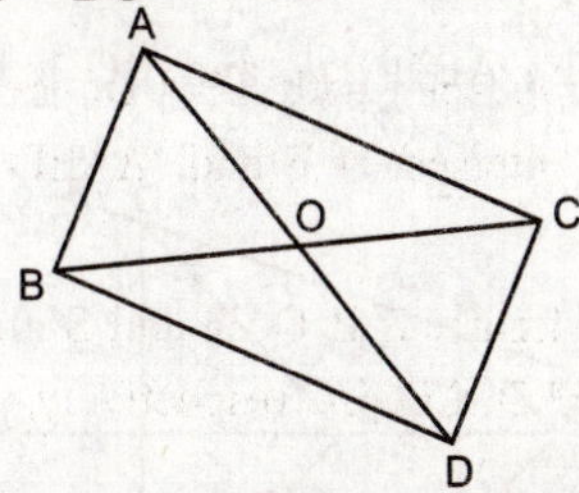

49. In figure P is the mid point of BC and Q is the mid point of AP. If BQ when produced meets AC at R, prove that

$AR = \frac{1}{3}AC$.

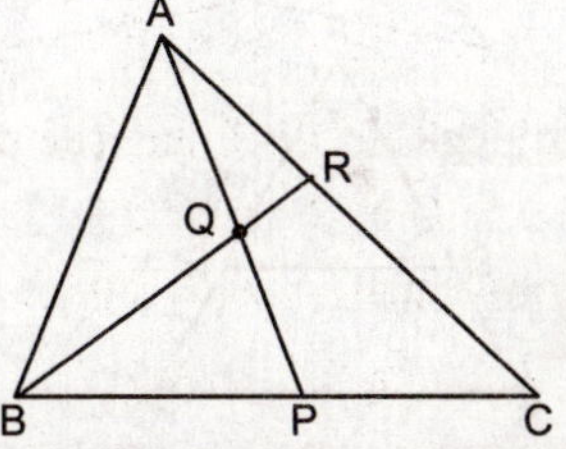

50. ABC is a triangle and PQ is a straight line meeting AB in P and AC in Q. If AP = 1 cm, PB = 3 cm, AQ = 1.5 cm, QC = 4.5 cm, prove that area of ΔAPQ is one-sixteenth of the area of ΔABC.

51. An aeroplane leaves an airport and flies due north at a speed of 1000 km/h. At the same time, another aeroplane leaves the same airport and flies due west at a speed of 1200 km/h. How far apart will be the two planes $1\frac{1}{2}$ hours ?

NCERT Questions

52. Prove that the ratio of the areas of two similar triangles is equal to the ratio of the squares of any two corresponding medians.

53. Prove that the area of the equilateral triangle described on the side of a square is half the area of the equilateral triangle described on its diagonal.

54. D, E, F are the mid-points of the sides BC, CA and AB respectively of a triangle ABC. Determine the ratio of the areas of ΔABC and ΔDEF.

55. In an equilateral triangle, prove that three times the square on one side is equal to four times the square on its altitude.

LONG ANSWER TYPE QUESTIONS

PREVIOUS YEARS' QUESTIONS

1. If a line is drawn parallel to one side of a triangle, prove that the other two sides are divided in the same ratio. In the following fig., AB || CD.

Using the above result solve the following :

(*i*) Find the value of x from given figure :

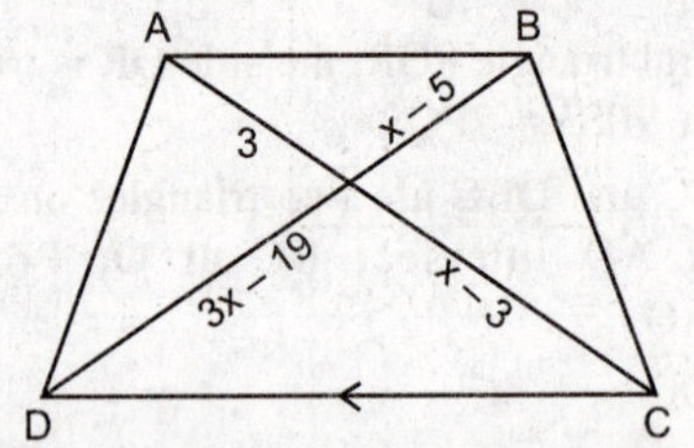

[Foreign 2006]

(*ii*) In the fig., AB || DE and BC || EF. Prove that AC || DF.

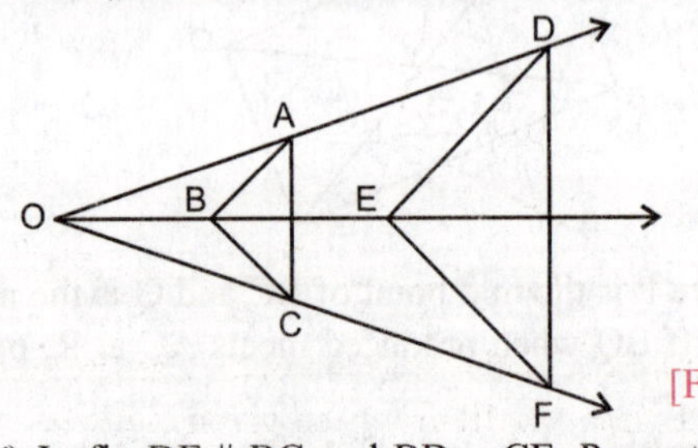

[Foreign 2008]

(*iii*) In fig DE || BC and BD = CE. Prove that ΔABC is an isosceles triangle.

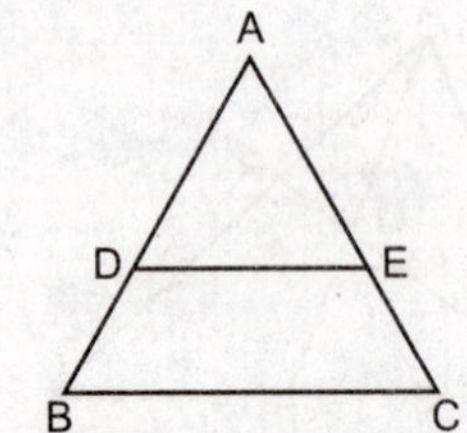

[Delhi 2009]

(*iv*) In figure, BA || QR, and CA || SR, prove that $\frac{QB}{BP} = \frac{SC}{CP}$.

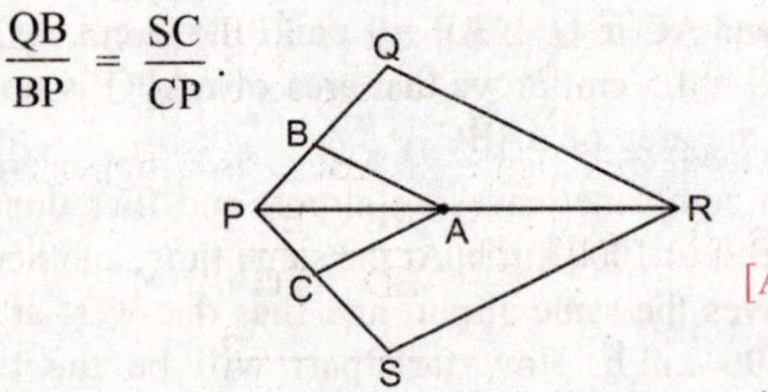

[AI 2010]

(*v*) In figure, PQ || AB and AQ || CB. Prove that $AR^2 = PR \cdot CR$.

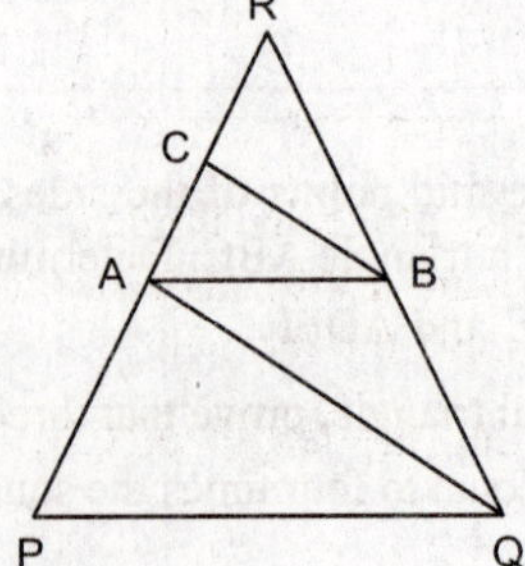

[Foreign 2010]

2. Prove that the ratio of the areas of two similar triangles is equal to the ratio of squares of their corresponding sides.

(*i*) Using the above, do the following : If D is a point on the AB of ΔABC such that AD : DB = 3 : 2 and E is a point on BC such that DE || AC, find the ratio of the areas of ΔABC and ΔBDE. [Delhi 2008 C]

(*ii*) Using the above, do the following : In a trapezium ABCD, AC and BD intersecting at O, AB || DC and AB = 2CD, if area of ΔAOB = 84 cm², find the area of ΔCOD. [Delhi 2009]

(*iii*) AD is an altitude of an equilateral triangle ABC. On AD as base, another equilateral triangle ADE is constructed. Prove that, area (ΔADE) : area (ΔABC) = 3 : 4 [AI 2010]

(*iv*) The area of two similar triangles are 49 cm² and 64 cm² respectively. If the difference of the corresponding altitudes is 10 cm, then find the lengths of altitudes (in centimetres). [CBSE (CCE) 2011]

3. Prove that in a right angle triangle the square on the hypotenuse is equal to the sum of the squares on the other two sides.

Using the above result, solve the following :

(*i*) In a rhombus ABCD, prove that $4AB^2 = AC^2 + BD^2$. [AI 2008 C]

(*ii*) Prove that in a ΔABC, if AD is perpendicular to BC, then $AB^2 + CD^2 = AC^2 + BD^2$. [Delhi 2009]

(*iii*) Point D is the mid-point of the side BC of a right triangle ABC, right angled at C. Prove that, $4AD^2 = 4AC^2 + BC^2$. [Delhi 2010]

(*iv*) In an equilateral triangle ABC, D is a point on side BC such that 4BD = BC. Prove that $16AD^2 = 13BC^2$. [CBSE (CCE) 2011]

4. In a triangle, if the square on one side is equal to the sum of the squares on the other two sides, prove that the angle opposite to first side is a right-angle.

(*i*) Using the above, do the following : ΔABC is an isosceles triangle in which AC = BC. If $AB^2 = 2AC^2$ then, prove that ΔABC is right triangle.

(*ii*) In an isosceles triangle PQR, PQ = QR and $PR^2 = 2PQ^2$. Prove that ∠Q is a right angle. [AI 2010]

5. In the given figure, BL and CM are medians of a triangle ABC right angled at A Prove that $4(BL^2 + CM^2) = 5BC^2$

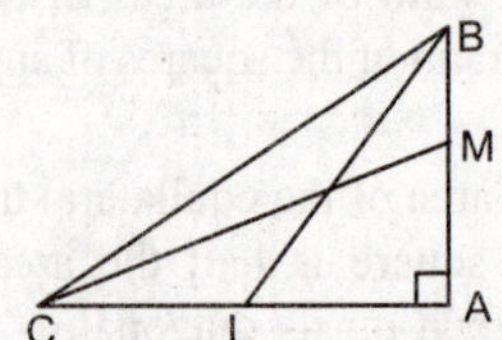

PRACTICE QUESTIONS

6. In a right triangle ABC, right angled at C, P and Q are points on the sides CA and CB respectively which divide these sides in the ratio 1 : 2. Prove that
   (*i*) $9AQ^2 = 9AC^2 + 4BC^2$
   (*ii*) $9BP^2 = 9BC^2 + 4AC^2$
   (*iii*) $9(AQ^2 + BP^2) = 13AB^2$.

7. In the given figure, perpendiculars OD, OE and OF are drawn to sides BC, CA and AB respectively from a point O in the interior of a ΔABC. Prove that :

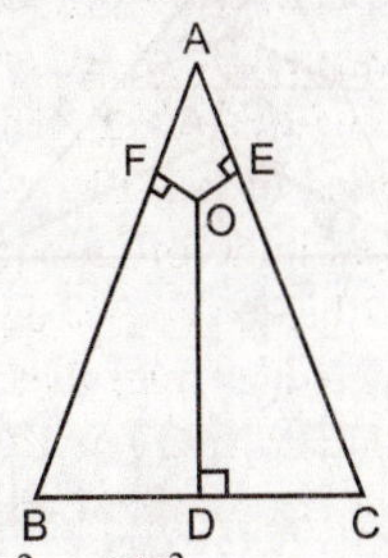

   (*i*) $AF^2 + BD^2 + CE^2 = OA^2 + OB^2 + OC^2 - OD^2 - OE^2 - OF^2$.
   (*ii*) $AF^2 + BD^2 + CE^2 = AE^2 + CD^2 + BF^2$.

8. ABCD is a parallelogram. AB is divided at P and CD at Q so that AP : PB = 3 : 2 and CQ : QD = 4 : 1. If PQ meets AC at R, then prove that $AR = \frac{3}{7}AC$.

9. If A is the area of a right angled triangle and *b* is one of the sides containing right angle. Prove that the length of altitude on the hypotenuse is $\frac{2Ab}{\sqrt{b^4 + 4A^2}}$.

10. ABCD is a trapezium in which AB is parallel to DC and the diagonals AC, BD cut at X. A line is drawn through C parallel to DA to cut DB, produced if necessary at Y. Prove that :
    (*i*) ΔAXD, ΔBXC are equal in area
    (*ii*) ΔAXD ~ ΔCXY
    (*iii*) $\frac{XB}{XY} = \frac{XA^2}{XC^2}$.

11. In the given figure, ΔPQR in which XY || QR, PX = 1 cm, XQ = 3 cm, YR = 4.5 cm, QR = 9 cm, find PY and XY. Further if the area of ΔPXY is 'A' cm², find in terms of A, the area of ΔPQR and area of trapezium XYRQ.

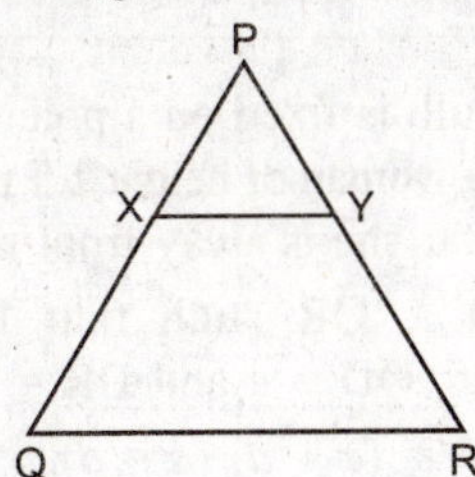

12. In the given figure, ABCD and AEFG are squares. Prove that :
    (*i*) $\frac{AF}{AG} = \frac{AC}{AD}$ (*ii*) ΔACF ~ ΔADG.

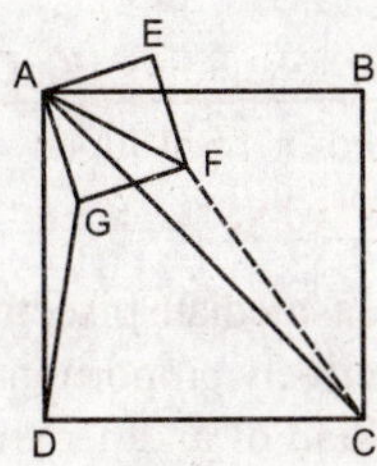

13. In figure, ΔFEC ≅ ΔGBD and ∠1 = ∠2. Prove that ΔADE ~ ΔABC.

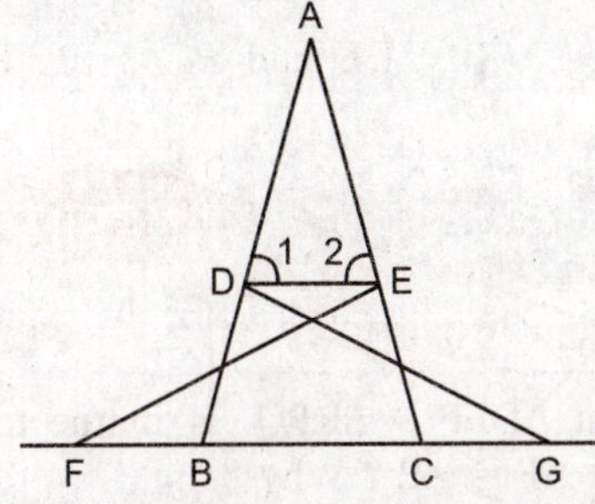

14. State and prove pythagoras theorem. Use the theorem and calculate area ΔPMR from the given figure.

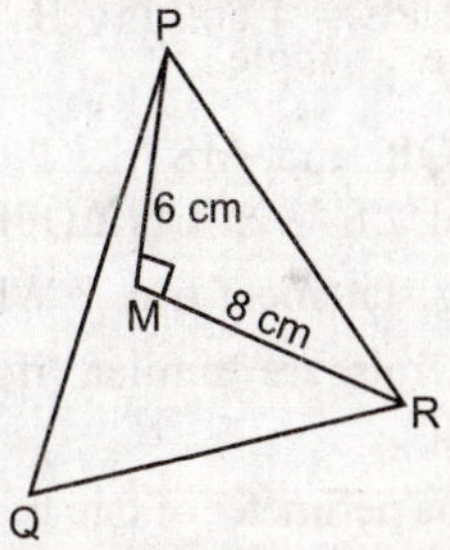

15. If a line is drawn parallel to one side of a triangle, the other two sides are divided in the same ratio, prove it. Use this result to prove the following :

    In the given figure, if ABCD is a trapezium in which AB || DC || EF, then $\frac{AE}{ED} = \frac{BF}{FC}$.

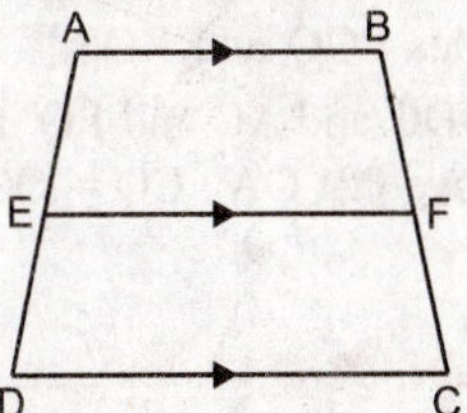

16. In a right triangle, prove that the square on the hypotenuse is equal to the sum of the squares on the other two sides. Using above, solve the following :

    From figure, find the length of CA, if CD ⊥ DB and AB ⊥ DB.

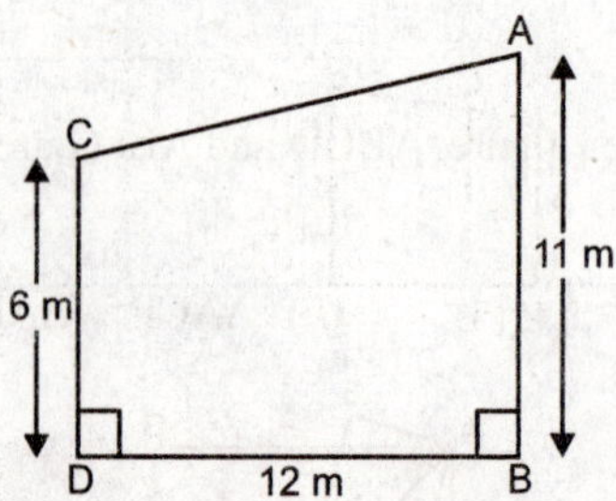

17. In a right-angled triangle, the square on the hypotenuse is equal to the sum of the squares on the other two sides. Using the above result, find the length of the second diagonal of a rhombus whose side is 5 cm and one of the diagonals is 6 cm.

NCERT Questions

18. If two sides and a median bisecting the third side of a triangle are respectively proportional to the corresponding sides and the median of another triangle, then prove that the two triangles are similar.
19. In figure, DE || AQ and DF || AR. Prove that EF || QR.

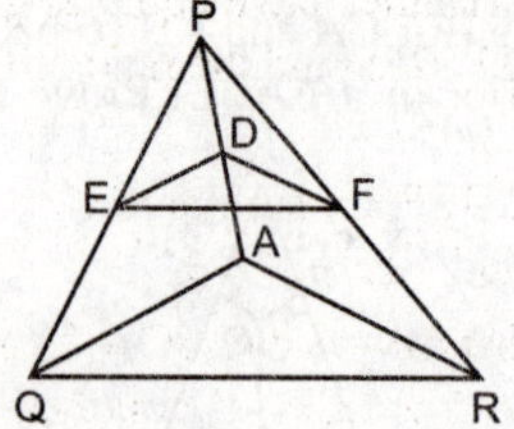

NCERT EXEMPLAR PROBLEMS

Short Answer Questions with Reasoning

1. It is given that $\Delta DEF \sim \Delta RPQ$. Is it true to say that $\angle D = \angle R$ and $\angle F = \angle P$? Why ?
2. A and B are respectively the points on the sides PQ and PR of a triangle PQR such that PQ = 12.5 cm, PA = 5 cm, BR = 6 cm and PB = 4 cm. Is AB || QR ? Give reasons for your answer.
3. In triangles PQR and MST, $\angle P = 55°$, $\angle Q = 25°$, $\angle M = 100°$ and $\angle S = 25°$. Is $\Delta QPR \sim \Delta TSM$? Why ?
4. Is the following statement true ? Why ?
"Two quadrilaterals are similar, if their corresponding angles are equal".
5. Two sides and the perimeter of one triangle are respectively three times the corresponding sides and the perimeter of the other triangle. Are the two triangles similar ? Why ?
6. The ratio of the corresponding altitudes of two similar triangles is $\frac{3}{5}$. Is it correct to say that ratio of their areas is $\frac{6}{5}$? Why ?
7. Is it true to say that if in two triangles, an angle of one triangle is equal to an angle of another triangle and two sides of one triangle are proportional to the two sides of the other triangle, then the triangles are similar ? Give reasons for your answer.

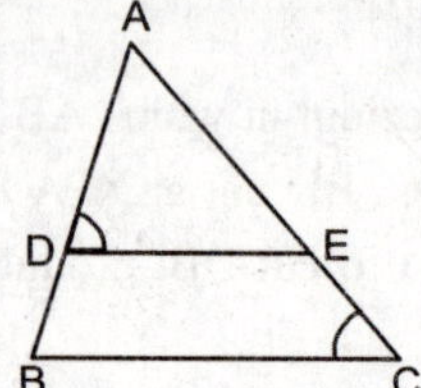

Short Answer Questions

8. Diagonals of a trapezium PQRS intersect each other at the point O, PQ || RS and PQ = 3 RS. Find the ratio of the areas of triangles POQ and ROS.
9. In fig. if AB || DC and AC and PQ intersect each other at the point O, prove that OA . CQ = OC . AP.

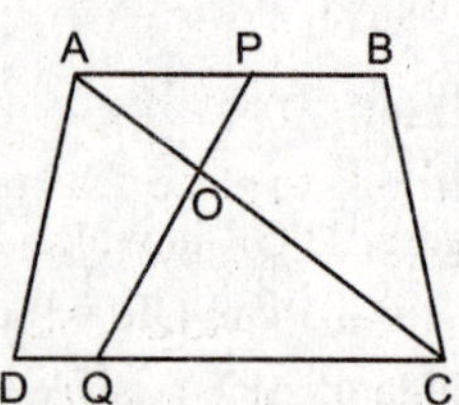

Long Answer Questions

10. For going to a city B from city A, there is a route via city C such that AC ⊥ CB, AC = $2x$ km and CB = $2(x + 7)$ km. It is proposed to construct a 26 km highway which directly connects the two cities A and B. Find how much distance will be saved in reaching city B from city A after the construction of the highway.
11. A street light bulb is fixed on a pole 6 m above the level of the street. If a woman of height 1.5 m casts a shadow of 3 m, find how far she is away from the base of the pole.
12. In ΔPQR, PD ⊥ QR such that D lies on QR. If PQ = a, PR = b, QD = c and DR = d, prove that $(a + b)(a - b) = (c + d)(c - d)$.

13. In a quadrilateral ABCD, $\angle A + \angle D = 90°$. Prove that $AC^2 + BD^2 = AD^2 + BC^2$.
[**Hint :** Produce AB and DC to meet at E.]

14. In fig., PA, QB, RC and SD are all perpendiculars to a line l, AB = 6 cm, BC = 9 cm, CD = 12 cm and SP = 36 cm. Find PQ, QR and RS.

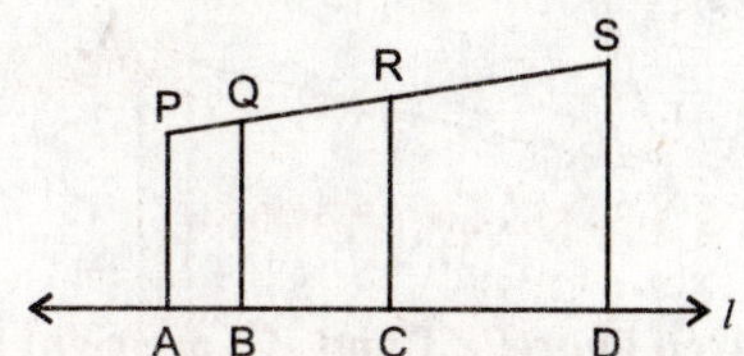

HIGHER ORDER THINKING SKILLS (HOTS) QUESTIONS

1. Two isosceles triangles have equal angles and their areas are in the ratio 81 : 25. Find the ratio of their corresponding heights.
2. If ΔABC and ΔDEF are similar such that 2AB = DE and BC = 8 cm. Find EF.
3. ΔABC is such that AB = 3 cm, BC = 2 cm and CA = 2.5 cm. If ΔDEF ~ ΔABC and EF = 4 cm, then find the perimeter of ΔDEF.
4. In ΔABC, a line XY parallel to BC cuts AB at X and AC at Y. If BY bisects ∠XYC then show that BC = CY.
5. In triangle ABC and DEF, $\angle A = \angle E = 50°$, $\frac{AB}{ED} = \frac{AC}{EF}$ and $\angle F = 65°$. Show that AB = AC.
6. In ΔABC, ∠A = 90°, AB = 5 cm and AC = 12 cm. If AD ⊥ BC, then find AD.
7. In ΔPQR, perpendicular PM from P on QR meets QR at M. If QM = 8 cm, MR = 2 cm and PM = 4 cm, then show that ∠P = 90°.
8. In ΔPQR, point M is on side PQ and point S is on the side PR such that QRSM is a trapezium. If MS : QR = 3 : 5, then find area (ΔPMS) : area (QRSM)
9. ABCD is a trapezium such that BC ∥ AD and AB = 4 cm. If the diagonals AC and BD intersect at E such that $\frac{AE}{EC} = \frac{DE}{BE} = \frac{1}{2}$, then find BC

VALUE BASED QUESTIONS

1. Kitchen garden of Ms. Sanjana is in the form of a triangle as shown. She wants to divide it in two parts; one triangle and one trapezium.

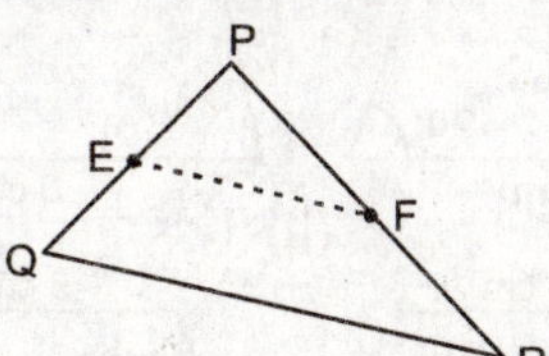

She takes PE = 4 m, QE = 4.5 m, PF = 8 m and RF = 9 m.

(*a*) Is EF ∥ QR ? Justify your answer.

(*b*) What values are depicted by Ms. Sanjana ?

2. An old man was trying to place a ladder 13 m long in such a way that it reaches a window of a building 12 m above the ground. Anoop decided to help the old man. He placed the foot of the ladder at such a distance from the wall that the top of ladder reached the window.

(*a*) Find the distance of the foot of ladder from the wall.

(*b*) What value is depicted by Anoop ?

3. Students of a school decided to participate in 'Save girl child' campaign. They decided to decorate a triangular path as shown

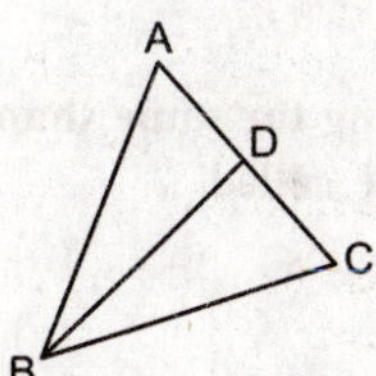

(*a*) If AB = AC, $BC^2 = AC \times CD$ prove that BD = BC.

(*b*) What value is depicted by the students ?

4. On occasion of independence day, Mira made a Rangoli in a design as shown

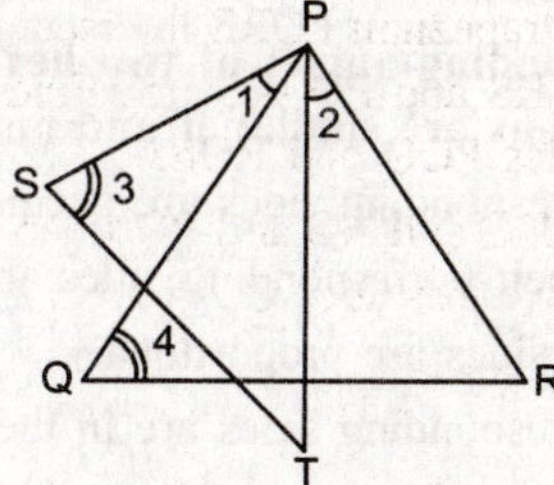

(*a*) If ∠1 = ∠2, ∠3 = ∠4, show that PT . QR = PR . ST.

(*b*) What value is depicted by Mira ?

Multiple Choice Questions

Previous Years' Questions

1. In the given figure, ∠T and ∠B are right angles. If the lengths of AT, BC and AS (in centimetres) are 15, 16 and 17 respectively, then the length of TC (in centimetres) is : [CBSE (CCE) 2011]

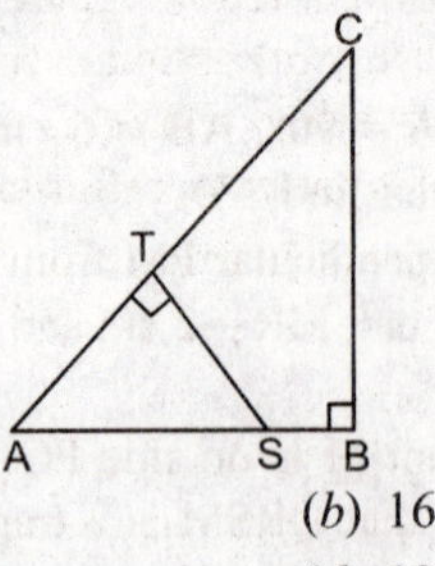

(*a*) 18 (*b*) 16
(*c*) 19 (*d*) 12

2. In the given figure, value of x (in cm) is : [CBSE (CCE) 2011]

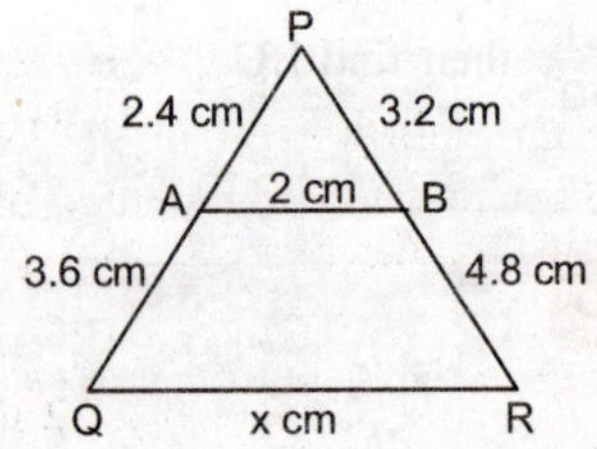

(*a*) 4 (*b*) 5
(*c*) 6 (*d*) 8

3. In the given figure, if ΔABC ~ ΔPQR [CBSE (CCE) 2012]

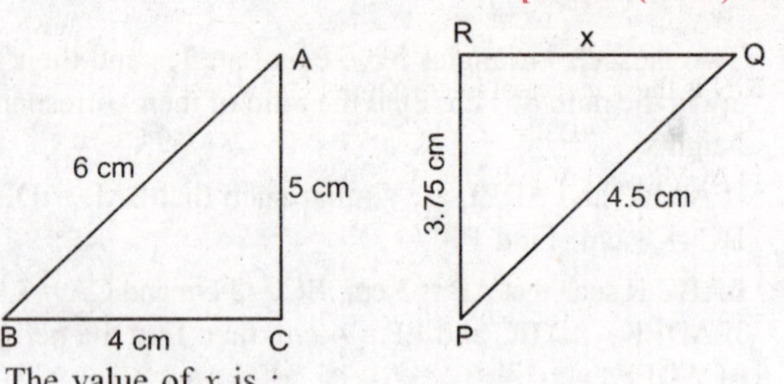

The value of x is :

(*a*) 2.5 cm (*b*) 3.5 cm
(*c*) 2.75 cm (*d*) 3 cm

Practice Questions

4. Two figures having the same shape and not necessarily the same size are called

(*a*) congruent figures (*b*) equal figures
(*c*) similar figures (*d*) symmetric figures

5. The height of mountains is found out using the idea of indirect measurements which is based on the

(*a*) principle of congruent figures
(*b*) principle of similarity of figures
(*c*) principle of equality of figures
(*d*) none of these.

6. The corresponding angles of two hexagons are equal. These hexagons are similar if and only if

(*a*) their corresponding sides are unequal
(*b*) two of their corresponding sides are proportional
(*c*) any four sides are proportional
(*d*) their corresponding sides are in the same ratio.

7. Which pair of the given quadrilaterals is similar ?

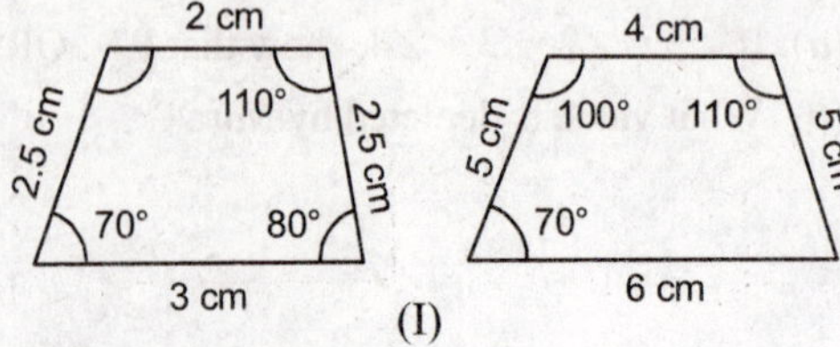

(I)

(II)

(III)

(IV)

(*a*) I (*b*) II
(*c*) III (*d*) IV

8. "If a line is drawn parallel to one side of a triangle to intersect the other two sides in distinct points, the other two sides are divided in the same ratio." This theorem is known as

(*a*) Pythagoras Theorem (*b*) Laplace Theorem
(*c*) Thales Theorem (*d*) Area Theorem

9. **"If a line divides any two sides of a triangle in the same ratio, then the line is parallel to the third side." This theorem is known as converse of**
   (*a*) Area Theorem
   (*b*) Basic Proportionality Theorem
   (*c*) Pythagoras Theorem
   (*d*) Laplace Theorem

10. **In the given figure, $\frac{AD}{DB} = \frac{AE}{EC}$ and $\angle ADE = 70°$, $\angle BAC = 50°$, then angle $\angle BCA =$**

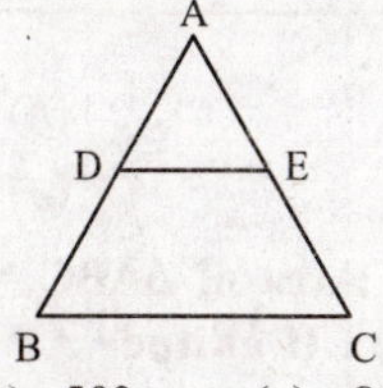

   (*a*) 70° (*b*) 50° (*c*) 80° (*d*) 60°

11. **In the given figure, MN || QR and PM = 3 cm, MQ = 4 cm, PN = 6 cm, PR = x cm, then x =**

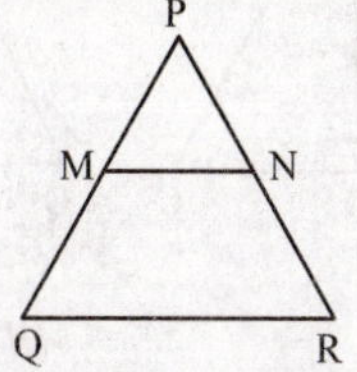

   (*a*) 8 cm (*b*) 6 cm (*c*) 10 cm (*d*) 14 cm

12. **In the given figure, AD = 1.28 cm, DB = 2.56 cm, AE = 0.64 cm. DE will be parallel to BC, if EC =**
   (*a*) 1.28 cm
   (*b*) 2.56 cm
   (*c*) 0.64 cm
   (*d*) 0.32 cm

13. **In ΔABC and ΔDEF, if ∠A = ∠D, ∠B = ∠E and ∠C = ∠F, then ΔABC and ΔDEF are similar. Symbolically we write**
   (*a*) ΔABC = ΔDEF (*b*) ΔABC ≅ ΔDEF
   (*c*) (*a*) or (*b*) (*d*) ΔABC ~ ΔDEF

14. **In ΔABC and ΔDEF, $\frac{AB}{EF} = \frac{AC}{DE}$, and ∠A = ∠E, then symbolically we write**
   (*a*) ΔABC ~ ΔDEF (*b*) ΔABC ~ ΔEDF
   (*c*) ΔABC ~ ΔEFD (*d*) ΔABC ≅ ΔEFD

15. **In ΔPQR and ΔMNS, $\frac{PQ}{NS} = \frac{QR}{MS} = \frac{PR}{MN}$, then symbolically we write**
   (*a*) ΔPQR ~ ΔMNS (*b*) ΔPQR ~ ΔSMP
   (*c*) ΔQRP ~ ΔNSM (*d*) ΔQRP ~ ΔSMN

16. **ABC and BDE are two equilateral triangles such that D is mid-point of BC. Ratio of the areas of triangles ABC and BDE is**
   (*a*) 2 : 1 (*b*) 1 : 4
   (*c*) 1 : 2 (*d*) 4 : 1

17. **Sides of two similar triangles are in the ratio 3 : 7. Areas of these triangles are in the ratio**
   (*a*) 9 : 35 (*b*) 9 : 49
   (*c*) 49 : 9 (*d*) 9 : 42

18. **Sides of triangles are (*i*) 3 cm, 4 cm, 6 cm. (*ii*) 4 cm, 5 cm, 6 cm. (*iii*) 7 cm, 24 cm, 25 cm (*iv*) 5 cm, 12 cm, 14 cm. Which of these is right triangle ?**
   (*a*) (*i*) (*b*) (*ii*) (*c*) (*iii*) (*d*) (*iv*)

19. **If in two right triangles, hypotenuse and one side of one triangle are proportional to the hypotenuse and one side of the other triangle, then the two triangles are similar. This may be reffered to as the**
   (*a*) AAA similarity criterion
   (*b*) ASS similarity criterion
   (*c*) SAS similarity criterion
   (*d*) RHS similarity criterion

20. **The medians of a right triangle which are drawn from the vertices of the acute angles are 5 cm and $2\sqrt{10}$ cm. The hypotenuse of the triangle is**
   (*a*) 10 cm (*b*) $2\sqrt{40}$ cm
   (*c*) $\sqrt{13}$ cm (*d*) $2\sqrt{13}$ cm

21. **If two poles 2 m and 8 m high are 100 m apart, then the height of the point of intersection of the line joining the top of each pole to the foot of the opposite pole is**
   (*a*) 1.4 m (*b*) 1.6 m
   (*c*) 1.7 m (*d*) 1.2 m

22. **In the given figure, it is given that ∠C = 90°, AD = DB, DE ⊥ AB, AB = 20 cm and AC = 12 cm. The area of quadrilateral ADEC is**
   (*a*) 75 cm²
   (*b*) $58\frac{1}{2}$ cm²
   (*c*) $37\frac{1}{2}$ cm²
   (*d*) 48 cm²

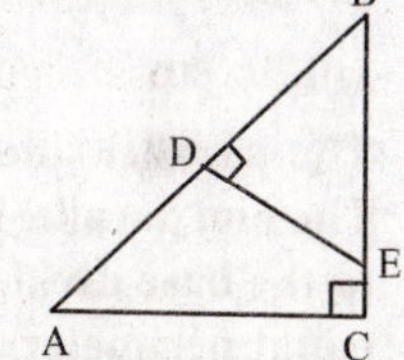

23. **Instead of walking along two adjacent sides of a rectangular field, a boy took a short cut along the diagonal of the field and saved a distance equal to $\frac{1}{2}$ of the longer side. The ratio of the shorter side of the rectangle to the longer side is**
   (*a*) $\frac{1}{2}$ (*b*) $\frac{3}{4}$
   (*c*) $\frac{4}{3}$ (*d*) $\frac{5}{7}$

24. **In a right triangle shown below, AB + AD = BC + CD, if AB = x, BC = h and CD = d, then x equal**
   (*a*) $\frac{hd}{2h+d}$
   (*b*) $d - h$
   (*c*) $h + d$
   (*d*) $\frac{1}{2}h$

25. In right triangle ABC, BC = a, AC = b, AB = c, CD ⊥ AB. If area of ΔABC = A, then CD =

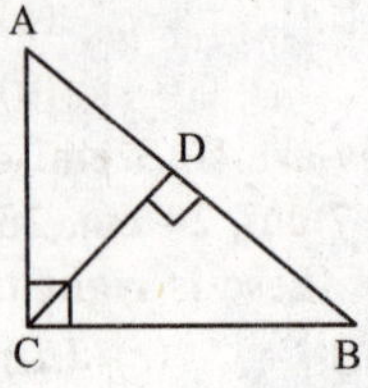

(a) $\frac{2A}{c}$ (b) $\frac{A}{c}$

(c) $\frac{A}{2c}$ (d) $\frac{c}{A}$

26. In the given figure, ABC is an equilateral triangle. DE is parallel to BC such that area of quadrilateral DBCE is equal to one half the area of ΔABC. If BC = 2 cm, then DE =

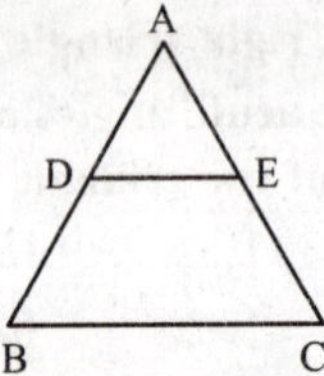

(a) 1 cm (b) $1\frac{1}{2}$ cm

(c) $\sqrt{2}$ cm (d) none of these

27. Side AC of right triangle ABC is divided into 8 equal parts. Seven line segments parallel to BC are drawn from the points of divisions. If BC = 10 cm, then the sum of lengths of the seven line segments is

(a) 33 cm (b) 34 cm

(c) 36 cm (d) 35 cm

28. The parallel sides of a trapezoid are 3 cm and 9 cm. The non parallel sides are 4 cm and 6 cm. A line parallel to the base divides the trapezoid into two trapezoids of equal perimeters. The ratio in which each of the non-parallel sides is divided is

(a) 4 : 5 (b) 3 : 2

(c) 4 : 1 (d) 3 : 1

29. The sides of a triangle are 30, 70 and 80 units. If an altitude is droped upon the side of length 80 units, the larger segment cut off on this side is

(a) 62 units (b) 63 units

(c) 64 units (d) 65 units

30. In the given figure, ABC is a triangle and GHED is a rectangle. BC = 12 cm, HE = 6 cm, FC = BF and altitude AF is 24 cm. The area of rectangle is

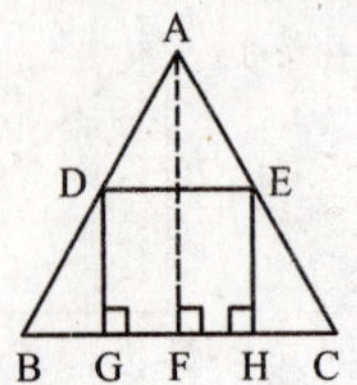

(a) 56 cm² (b) 54 cm²

(c) 60 cm² (d) 72 cm²

31. In given figure, MN || QR. If PM = x cm, MQ = 10 cm, PN = (x – 2) cm, NR = 6 cm, then x =

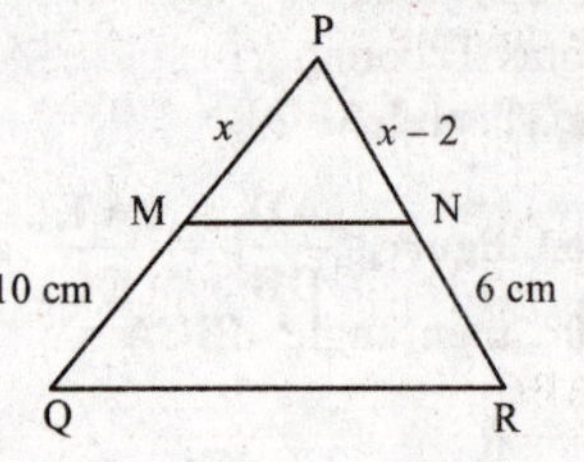

(a) 4 cm (b) 5 cm

(c) 6 cm (d) 8 cm

32. ΔABC ~ ΔPQR. Area of ΔABC = 81 cm² and area of ΔPQR = 121 cm². If altitude AD = 9 cm, then PM =

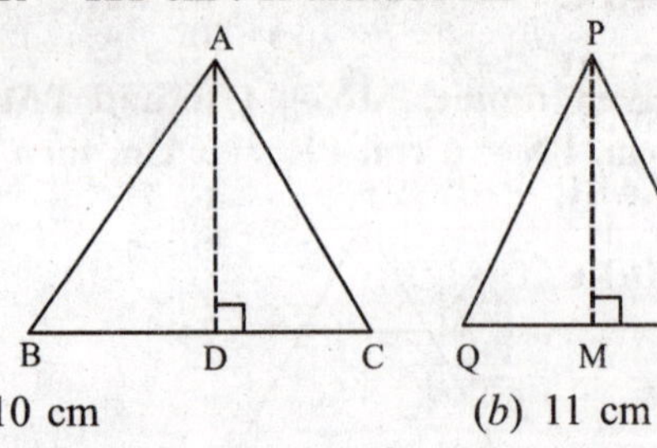

(a) 10 cm (b) 11 cm

(c) 12 cm (d) 15 cm

33. In given figure, AD = 3 cm, AE = 5 cm, BD = 4 cm, CE = 4 cm, CF = 2 cm, BF = 2.5 cm, then

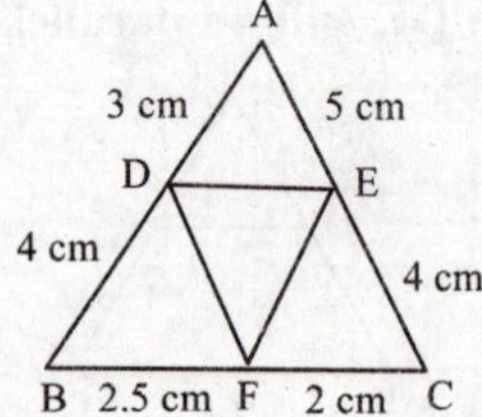

(a) DE || BC (b) DF || AC

(c) EF || AB (d) none of these

34. Match the column :

1. The ratio of any two corresponding sides in equiangular triangles is always the same	**(A) Baudhayana**
2. The diagonal of a rectangle produces by itself the same area as produced by its both sides (*i.e.,* length and breadth)	**(B) Thales**
3. In a right triangle, the square of the hypotenuse is equal to the sum of the squares of the other two sides.	**(C) Pythagoras**
	(D) Aryabhatta

(*a*) 1 → A, 2 → B, 3 → C
(*b*) 1 → B, 2 → A, 3 → C
(*c*) 1 → B, 2 → C, 3 → A
(*d*) 1 → B, 2 → A, 3 → D

35. Match the column :

1. In ΔABC and ΔPQR $\frac{AB}{PQ} = \frac{AC}{PR}$, ∠A = ∠P ⇒ ΔABC ~ ΔPQR	**(A) AA similarity criterion**
2. In ΔABC and ΔPQR ∠A = ∠P, ∠B = ∠Q ⇒ ΔABC ~ ΔPQR	**(B) SAS similarity criterion**
3. In ΔABC and ΔPQR $\frac{AB}{PQ} = \frac{AC}{PR} = \frac{BC}{QR}$ ⇒ ΔABC ~ ΔPQR	**(C) SSS similarity criterion**
4. In ΔABC, DE ∥ BC ⇒ $\frac{AD}{BD} = \frac{AE}{CE}$	**(D) BPT**

(*a*) 1 → A, 2 → B, 3 → C, 4 → D
(*b*) 1 → D, 2 → A, 3 → C, 4 → B
(*c*) 1 → B, 2 → A, 3 → C, 4 → D
(*d*) 1 → C, 2 → B, 3 → D, 4 → A.

36. The value of x in the given figure

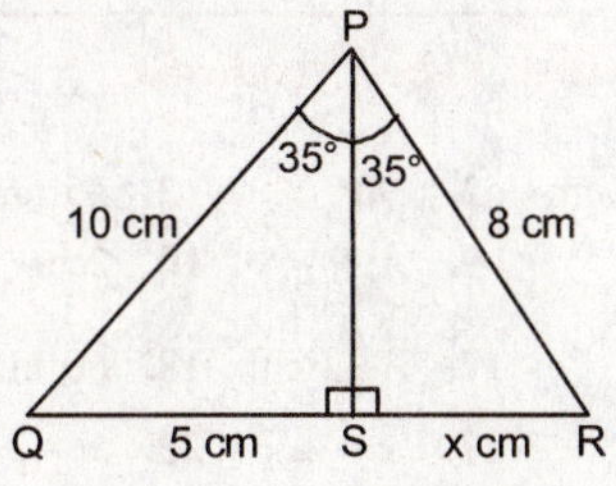

(*a*) 5 cm (*b*) 4 cm
(*c*) 2 cm (*d*) 3 cm

37. In ΔABC, AB = $6\sqrt{3}$ cm, AC = 12 cm and BC = 6 cm. The angle B is

(*a*) 120° (*b*) 60°
(*c*) 90° (*d*) 45°

38. If quadrilateral ABCD and PQRS are similar then x is equal to

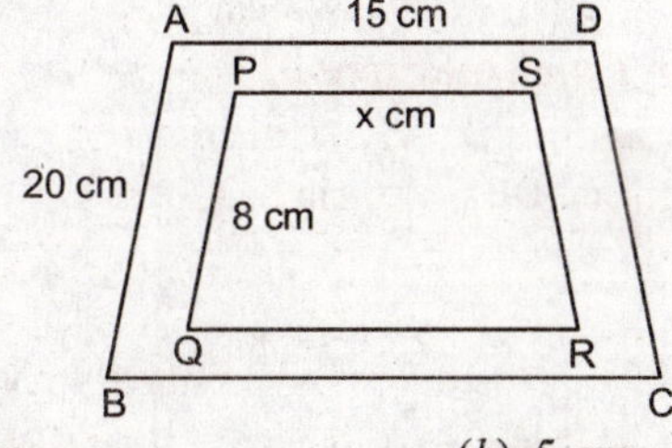

(*a*) 4 cm (*b*) 5 cm
(*c*) 6 cm (*d*) 7 cm

NCERT EXEMPLAR PROBLEMS

MULTIPLE CHOICE QUESTIONS

1. The lengths of the diagonals of a rhombus are 16 cm and 12 cm. Then, the length of the side of the rhombus is

(*a*) 9 cm (*b*) 10 cm
(*c*) 8 cm (*d*) 20 cm

2. If ΔABC ~ ΔEDF and ΔABC is not similar to ΔDEF, then which of the following is not true ?

(*a*) BC . EF = AC . FD (*b*) AB . EF = AC . DE
(*c*) BC . DE = AB . EF (*d*) BC . DE = AB . FD

3. If in two triangles ABC and PQR, $\frac{AB}{QR} = \frac{BC}{PR} = \frac{CA}{PQ}$, then

(*a*) ΔPQR ~ ΔCAB (*b*) ΔPQR ~ ΔABC
(*c*) ΔCBA ~ ΔPQR (*d*) ΔBCA ~ ΔPQR

4. If in two triangles DEF and PQR, ∠D = ∠Q and ∠R = ∠E, then which of the following is not true ?

(*a*) $\frac{EF}{PR} = \frac{DF}{PQ}$ (*b*) $\frac{DE}{PQ} = \frac{EF}{RP}$

(*c*) $\frac{DE}{QR} = \frac{DF}{PQ}$ (*d*) $\frac{EF}{RP} = \frac{DE}{QR}$

5. In triangles ABC and DEF, ∠B = ∠E, ∠F = ∠C and AB = 3DE. Then, the two triangles are

(*a*) congruent but not similar
(*b*) similar but not congruent
(*c*) neither congruent nor similar
(*d*) congruent as well as similar

6. If in triangles ABC and DEF, $\frac{AB}{DE} = \frac{BC}{FD}$, then they will be similar, when **[CBSE (CCE) 2011]**

(*a*) ∠B = ∠E (*b*) ∠A = ∠D
(*c*) ∠B = ∠D (*d*) ∠A = ∠F

7. If S is a point on side PQ of a ΔPQR such that PS = QS = RS, then

(*a*) PR . QR = RS^2 (*b*) $QS^2 + RS^2 = QR^2$
(*c*) $PR^2 + QR^2 = PQ^2$ (*d*) $PS^2 + RS^2 = PR^2$.

ANSWERS

Very Short Answer Type Questions

1. BC = 13.5 cm **2.** 90° **3.** 25 cm **4.** 1 : 9 **5.** 10 cm **6.** ar (ΔABC) = 80 cm^2 **7.** $x = \dfrac{ac}{b+c}$
8. 5 : 8 **9.** 1: 9 **10.** 2cm **11.** Not **12.** 2 cm **13.** 2.4 cm **14.** Yes **15.** No
16. 55° **17.** 11.2 cm **18.** 60 m **20.** 5.4 cm **21.** 96 cm^2 **22.** 4 : 1 **23.** $\dfrac{16}{49}$ **24.** 8.8 cm
25. 40° **27.** $x = 4$ **28.** $x = 3$ **29.** 25 m

Short Answer Type Questions-I

8. $x = 7$, **9.** 6 cm **10.** 96 sq. cm **12.** $2\sqrt{(q-1)}$ **15.** 9 cm **16.** ∠CAB = 60°
18. Area of ΔDEF = 14 sq. cm **19.** 13 m **21.** (*i*) $x = \dfrac{21}{4}$, $y = \dfrac{15}{2}$ (*ii*) $x = \dfrac{28}{5}$, $y = \dfrac{7}{4}$, $z = \dfrac{35}{6}$
22. 2 **24.** 60 m **25.** 7.5 cm **26.** 6 cm **27.** 250 m
28. 15 m **29.** 13 m **31.** 21 m **33.** PA = 3.25 cm, CA = 5.6 cm
34. DC = 8 cm **38.** 100 sq. cm **40.** 4 : 1 **43.** $\dfrac{xz}{y}$ **44.** 7.5 cm
46. 9 : 4 **49.** 8.8 cm **50.** 3.5 cm **57.** (*i*) Yes (*ii*) Yes **58.** 13 m

Short Answer Type Questions-II

2. AE = $\dfrac{15}{13}$ cm, DE = $\dfrac{36}{13}$ cm **9.** 16 : 9 **11.** 144 cm^2 **16.** 3 : 4
20. $\dfrac{\sqrt{2}-1}{\sqrt{2}}$ **25.** 1 : 8 **37.** 192 cm^2 **40.** $x = 6$, $y = \sqrt{20}$ **46.** CE = $2\sqrt{5}$ cm
51. 2343.07 km **54.** 4 : 1

Long Answer Type Questions

1. (*i*) $x = 8$ Or 9 **2.** (*i*) 25 : 4 (*ii*) 21 cm^2 (*iv*) 70 cm, 80 cm. **4.** (*i*) ∠PKR = 90°
11. PY = 1.5 cm, XY = $\dfrac{9}{4}$ cm, Area of ΔPQR = 16 A cm^2, Area of trapezium XYRQ = 15 A cm^2.
14. 24 cm^2 **16.** 13 m **17.** 8 cm.

NCERT Exemplar Problems

1. No **2.** Yes **3.** No **4.** No **5.** Yes **6.** No **7.** No
8. 9 : 1 **10.** 8 km **11.** 9 m **14.** PQ = 8 cm, QR = 12 cm, RS = 16 cm.

Higher Order Thinking Skills (HOTS) Questions

1. 9 : 5 **2.** 16 cm **3.** 15 cm **6.** $\dfrac{60}{13}$ cm **8.** 9 : 16 **9.** 8 cm

Value Based Questions

1. (*a*) Yes (*b*) Environmentally sensitive **2.** (*a*) 5 m, (*b*) Respect and care for old persons
3. (*b*) Respect for women **4.** (*b*) Patriotism

Multiple Choice Questions

1. (*c*) **2.** (*b*) **3.** (*d*) **4.** (*c*) **5.** (*b*) **6.** (*d*) **7.** (*c*) **8.** (*c*) **9.** (*b*) **10.** (*d*)
11. (*d*) **12.** (*a*) **13.** (*d*) **14.** (*c*) **15.** (*d*) **16.** (*d*) **17.** (*b*) **18.** (*c*) **19.** (*d*) **20.** (*d*)
21. (*b*) **22.** (*b*) **23.** (*b*) **24.** (*a*) **25.** (*a*) **26.** (*c*) **27.** (*d*) **28.** (*c*) **29.** (*d*) **30.** (*b*)
31. (*b*) **32.** (*b*) **33.** (*c*) **34.** (*b*) **35.** (*c*) **36.** (*b*) **37.** (*c*) **38.** (*c*)

NCERT Exemplar Problems

1. (*b*) **2.** (*c*) **3.** (*a*) **4.** (*b*) **5.** (*d*) **6.** (*c*) **7.** (*d*)

SOLUTIONS

VERY SHORT ANSWER TYPE QUESTIONS

1. $\frac{AP}{PB} = \frac{3.5}{7} = \frac{1}{2}$ ...(i)

$\frac{AQ}{QC} = \frac{3}{6} = \frac{1}{2}$ ...(ii)

From (i) and (ii), we have

$\frac{AP}{PB} = \frac{AQ}{QC}$

$\therefore$ PQ ∥ BC

$\therefore$ $\angle AQP = \angle ACB$

and $\angle APQ = \angle ABC$ (Corresponding angles)

$\therefore$ $\Delta AQP \sim \Delta ACB$ (AA similarity)

$\Rightarrow \frac{PQ}{BC} = \frac{AQ}{AC} = \frac{AQ}{AQ+QC}$

(By definition of SSS similarity)

$\Rightarrow \frac{4.5}{BC} = \frac{3}{9}$

$\Rightarrow$ BC = 13.5 cm.

2. In right angled ΔPAR, $PR^2 = AP^2 + AR^2$

$\Rightarrow PR^2 = 6^2 + 8^2$

$\Rightarrow PR = \pm 10$ cm

$\Rightarrow$ PR = 10 cm (Neglecting negative value)

Now in ΔQPR,

$QR^2 = 26^2 = 676$ ...(i)

$PQ^2 + PR^2 = 24^2 + 10^2 = 676$ ...(ii)

From (i) and (ii),

$QR^2 = PQ^2 + PR^2$

According to converse of Pythagoras theorem

$\angle QPR = 90°$

3. In a rhombus,

$(\text{Side})^2 = \left(\frac{1}{2}d_1\right)^2 + \left(\frac{1}{2}d_2\right)^2$

[∵ Diagonals of a rhombus bisect each other at right angle]

$= \left(\frac{1}{2}\times 30\right)^2 + \left(\frac{1}{2}\times 40\right)^2$

$\text{Side}^2 = 225 + 400 = 625$

$\Rightarrow$ Side = 25 cm.

4.

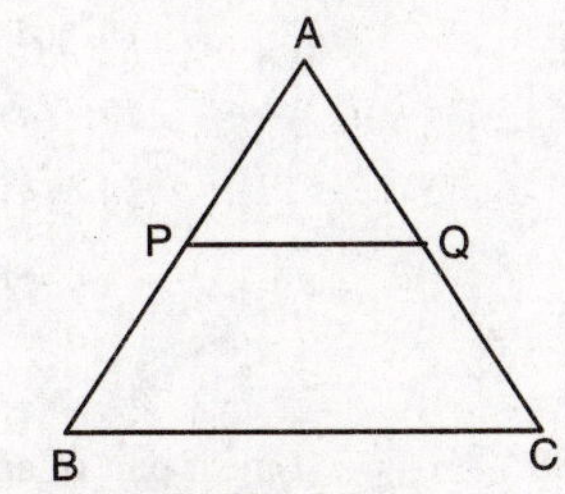

In Δ ABC, PQ ∥ BC

$\therefore \frac{AP}{AB} = \frac{AQ}{AC}$

Now in ΔAPQ and ΔABC,

$\frac{AP}{AB} = \frac{AQ}{AC}$ (As proved)

$\angle A = \angle A$ (Common angle)

$\therefore$ $\Delta APQ \sim \Delta ABC$ (SAS similarity)

$\Rightarrow \frac{ar(\Delta APQ)}{ar(\Delta ABC)} = \frac{AP^2}{AB^2} = \frac{AP^2}{(AP+PB)^2}$

$= \frac{1^2}{3^2} = \frac{1}{9}$

5. Given that $\Delta ABC \sim \Delta LMN$

$\therefore \frac{\text{Perimeter of } \Delta ABC}{\text{Perimeter of } \Delta LMN} = \frac{AB}{LM} = \frac{BC}{MN} = \frac{AC}{LN}$

Let AB = x cm

$\Rightarrow \frac{60}{48} = \frac{x}{8}$

$\Rightarrow x = \frac{60}{48}\times 8 = 10$ cm

$\therefore$ AB = 10 cm

6.

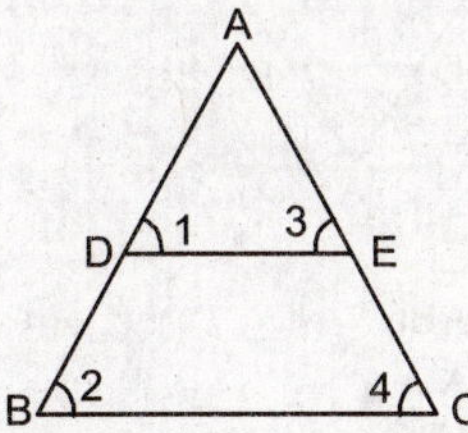

Given : In ΔABC

DE ∥ BC, BC = 8 cm,

DE = 6 cm

and area of $\Delta ADE = 45\text{cm}^2$,

To find : ar (ΔABC).

Proof : In ΔABC and ΔADE

DE ∥ BC (Given)

$\Rightarrow \angle 1 = \angle 2$ (Corresponding angles)

and $\angle 3 = \angle 4$ (Corresponding angles)

$\angle A = \angle A$ (Common)

$\Rightarrow \Delta ADE \sim \Delta ABC$ (AA similarity)

$\Rightarrow \frac{ar(\Delta ABC)}{ar(\Delta ADE)} = \frac{AB^2}{AD^2} = \frac{BC^2}{DE^2}$

$\Rightarrow \frac{ar(\Delta ABC)}{45} = \frac{8^2}{6^2}$

$\Rightarrow ar(\Delta ABC) = \frac{64}{36}\times 45 = 80 \text{ cm}^2$

7. In Δ LMK and ΔPNK

$\angle M = \angle N$ (each 46°)

$\angle K = \angle K$ (common)

$\therefore$ $\Delta LMK \sim \Delta PNK$

$\Rightarrow \frac{LM}{PN} = \frac{MK}{NK} = \frac{LK}{PK}$

$\Rightarrow \frac{a}{x} = \frac{b+c}{c} \Rightarrow x = \frac{ac}{b+c}$

8. Ratio of area of two similar Δ is equal to the ratio of the squares of their corresponding sides

∴ ratio of their corresponding sides = $\frac{\sqrt{25}}{\sqrt{64}} = \frac{5}{8}$

9. S and T are points on the sides PQ and PR of ΔPQR and PT = 2 cm, TR = 4 cm, ST ∥ QR

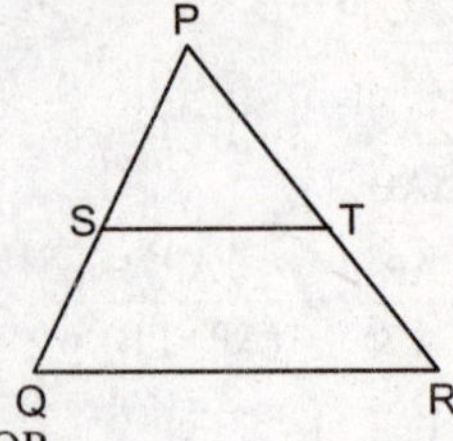

In ΔPST and ΔPQR

$\angle S = \angle Q$ (Corresponding angles)

$\angle P = \angle P$ (Common)

∴ ΔPST ~ ΔPQR (AA similarity)

∴ $\frac{ar(\Delta PST)}{ar(\Delta PQR)} = \frac{PT^2}{PR^2} = \frac{2^2}{6^2} = \frac{4}{36} = \frac{1}{9}$

10. DE ∥ BC.

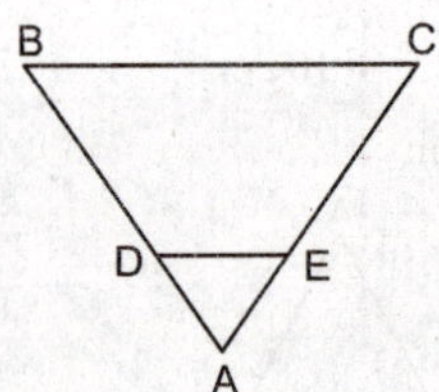

In ΔADE and ΔABC,

$\angle ADE = \angle ABC$ (Corresponding angles)

$\angle A = \angle A$ (Common)

∴ ΔADE ~ ΔABC (AA similarity)

⇒ $\frac{AD}{AB} = \frac{DE}{BC}$

Now $\frac{1.5}{6} = \frac{DE}{8}$

$DE = \frac{1.5 \times 8}{6} = 2$ cm

11. No, as their corresponding angles are not equal.

12.

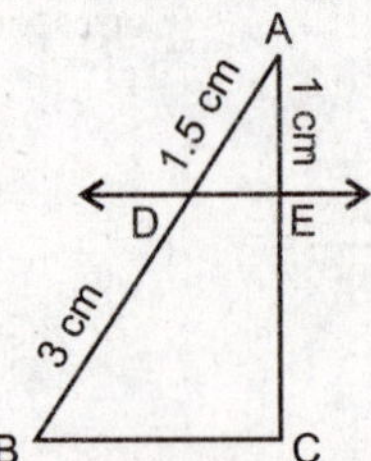

Given : DE ∥ BC,

To find: EC

Proof : In ΔABC

DE ∥ BC (Given)

⇒ $\frac{AD}{BD} = \frac{AE}{EC}$ (By BPT)

⇒ $\frac{1.5}{3} = \frac{1}{EC}$

⇒ $EC = \frac{3 \times 1}{1.5} = 2$ cm

13. ∵ DE ∥ BC

∴ $\frac{AD}{BD} = \frac{AE}{CE}$ (from B.P.T.)

⇒ $\frac{AD}{7.2} = \frac{1.8}{5.4}$ ⇒ AD = 2.4 cm.

14.

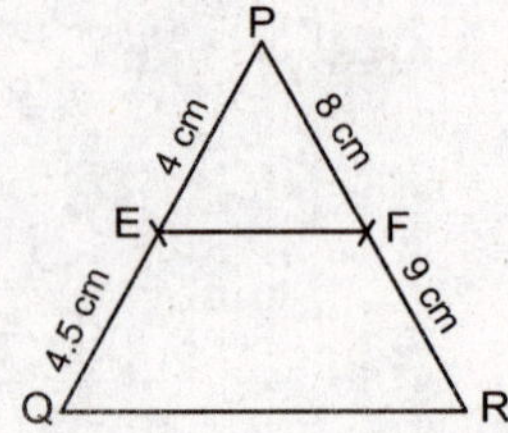

In ΔPQR,

$\frac{PE}{QE} = \frac{4}{4.5} = \frac{40}{45} = \frac{8}{9}$ ...(*i*)

$\frac{PF}{FR} = \frac{8}{9}$ ...(*ii*)

From (*i*) and (*ii*),

$\frac{PE}{QE} = \frac{PF}{FR}$

∵ $\frac{PE}{QE} = \frac{PF}{FR}$

∴ EF ∥ QR (By converse of BPT)

15.

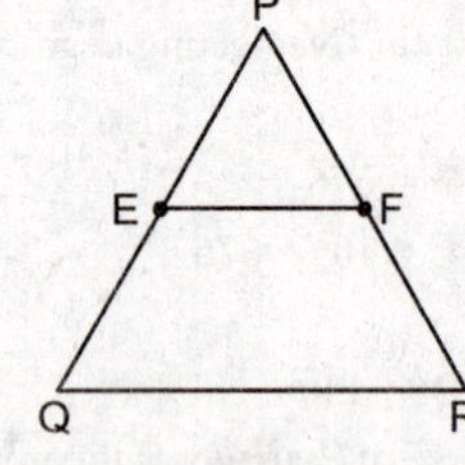

$\frac{PE}{EQ} = \frac{3.9}{3} = \frac{13}{10}$ ...(*i*)

$\frac{PF}{FR} = \frac{3.6}{2.4} = \frac{3}{2}$ ...(*ii*)

From (*i*) and (*ii*)

$\frac{PE}{EQ} \neq \frac{PF}{FR}$

∴ EF not parallel to QR

16.

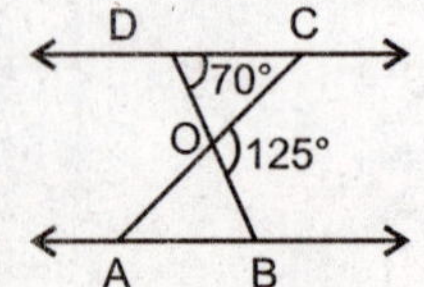

Given : ΔODC ~ ΔOBA

$\angle BOC = 125°$

and $\angle CDO = 70°$

To find : $\angle DOC$

Proof : $\angle DOC + \angle BOC = 180°$ (Linear pair of angles)

⇒ $\angle DOC = 180° - 125° = 55°$

17. Since $\Delta ABC \sim \Delta DEF$

$$\frac{ar(\Delta ABC)}{ar(\Delta DEF)} = \frac{64}{121} = \frac{BC^2}{EF^2}$$

$$\Rightarrow \quad \frac{BC^2}{(15.4)^2} = \frac{64}{121}$$

$$\Rightarrow \quad BC = \sqrt{\frac{64}{121}} \times 15.4 = \frac{8}{11} \times 15.4$$

$$= 8 \times 1.4 = 11.2 \text{ cm}$$

18.

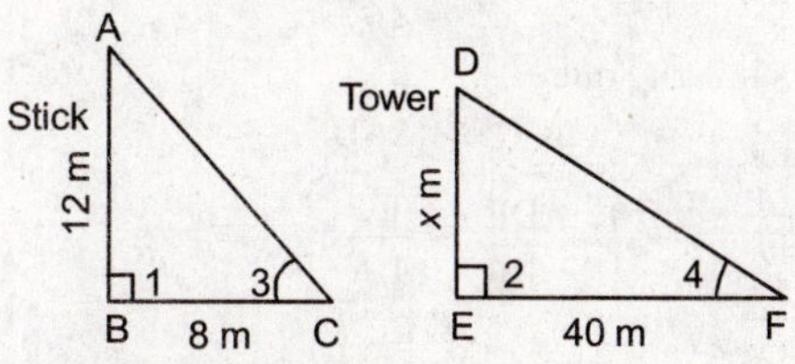

Let height of the tower be x m.

$\angle 1 = \angle 2 = (90°$ each)

$\angle 3 = \angle 4$ (Angle of inclination at the same time)

$\Rightarrow \Delta ABC \sim \Delta DEF$ (AA similiarity)

$$\Rightarrow \quad \frac{AB}{BC} = \frac{DE}{EF}$$

$$\Rightarrow \quad \frac{12}{8} = \frac{x}{40}$$

$$\Rightarrow \quad x = \frac{12}{8} \times 40 = 60 \text{ m}$$

Height of the tower DE = 60 m.

19.

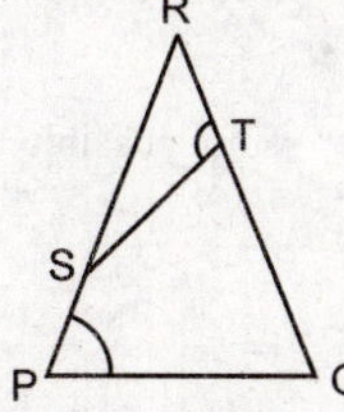

In ΔRPQ and ΔRTS

$\angle P = \angle RTS$ (Given)

$\angle R = \angle R$ (Common angle)

Hence, Δ's are similar

$\angle PQR = \angle RST$

20. Let corresponding sides of two similar Δ's are a, b, c and d, e, f respectively and $a = 9$cm.

$\because$ Δ's are similar

$$\frac{a}{d} = \frac{b}{e} = \frac{c}{f}$$

$$\Rightarrow \quad \frac{a+b+c}{d+e+f} = \frac{a}{d}$$ (Using property of proportion)

$$\frac{25}{15} = \frac{9}{d}$$

$$\Rightarrow \quad d = \frac{9 \times 15}{25} = \frac{27}{5} \text{ cm} = 5.4 \text{ cm}$$

21. $\Delta ABC \sim \Delta DEF$

$$\frac{ar(\Delta ABC)}{ar(\Delta DEF)} = \frac{BC^2}{EF^2}$$

$$\Rightarrow \quad \frac{54}{ar(\Delta DEF)} = \frac{3^2}{4^2} \quad \Rightarrow \frac{54 \times 16}{9} = ar(\Delta DEF)$$

$$\Rightarrow \quad 96 \text{ cm}^2 = ar(\Delta DEF)$$

22.

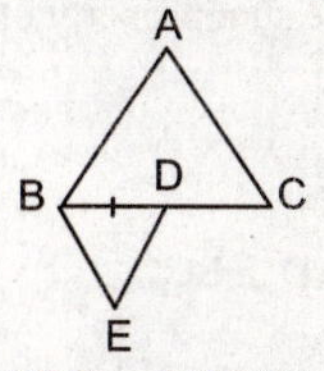

In ΔABC and ΔDBE,

$\Rightarrow \angle A = \angle D$ (Corresponding $\angle s$ of $\sim \Delta s$)

$\angle B = \angle B = 60°$ (Equilateral triangles)

$\angle C = \angle E = 60°$ (AA similarity)

$\Delta ABC \sim \Delta DBE$

$$\frac{ar(\Delta ABC)}{ar(\Delta DBE)} = \frac{AB^2}{BD^2} = \frac{AB^2}{\left(\frac{1}{2}BC\right)^2}$$

(D is the mid point of BC)

$$= \frac{AB^2 \times 4}{BC^2} = 4:1 \ (\because AB = BC)$$

23. $\because \quad \frac{AB}{DE} = \frac{BC}{EF} = \frac{AC}{DF}$

$\therefore \quad \Delta ABC \sim \Delta DEF$

$$\Rightarrow \quad \frac{\text{area } \Delta ABC}{\text{area } \Delta DEF} = \frac{AB^2}{DE^2}$$

$$= \left(\frac{AB}{DE}\right)^2 = \left(\frac{4}{7}\right)^2 = \frac{16}{49}$$

24.

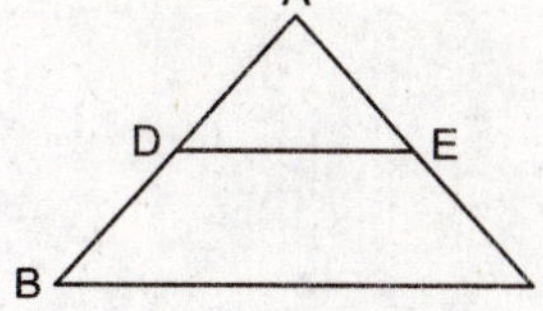

$AD : DB = 3 : 1$

In ΔABC, $DE \parallel BC$

$$\Rightarrow \quad \frac{AD}{DB} = \frac{AE}{EC}$$ (By BPT)

$$\Rightarrow \quad \frac{3}{1} = \frac{AE}{EC}$$

$$\Rightarrow \quad \frac{3}{1} = \frac{6.6}{EC}$$

$$\Rightarrow \quad EC = \frac{6.6}{3} = 2.2 \text{ cm}$$

$$AC = AE + EC = 6.6 + 2.2 = 8.8 \text{ cm}$$

25.

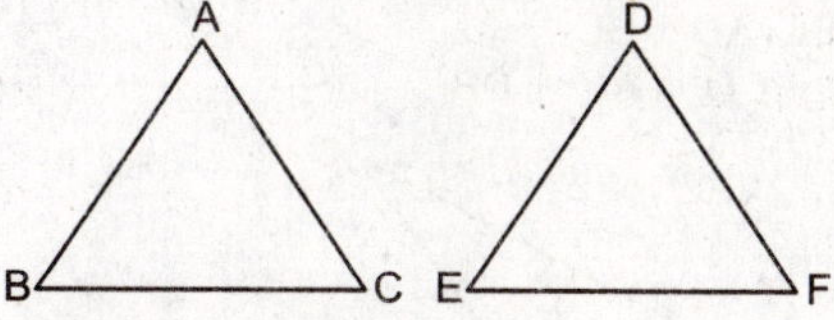

In ΔABC and ΔDEF,

$\Delta ABC \sim \Delta DEF$ (Given)

$\Rightarrow \angle A = \angle D$

$\angle B = \angle E$

and $\angle C = \angle F$

$\angle A = 57°, \angle B = 83°$

But $\angle A + \angle B + \angle C = 180°$ (Angle sum property of triangle)

$\Rightarrow \angle C = 180° - \angle A - \angle B$

$= 180° - 57° - 83°$

$\angle C = 180° - 140° = 40°$

26. **Given :** ΔABC right angled at A and AD ⊥ BC.

To prove : $\frac{BD}{AD} = \frac{AB}{AC}$

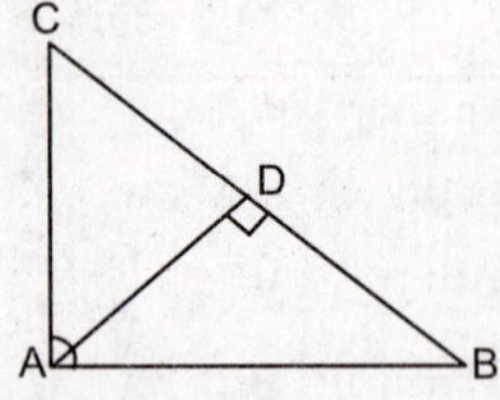

Proof : In ΔABC and ΔDCA.

$\angle A = \angle ADC$ (Each 90°)

$\angle C = \angle C$ (Common angles)

$\therefore$ ΔACB ~ ΔDCA (AA similarity)

$\therefore \angle B = \angle DAC$ (3rd angle of similar Δs)

In ΔABD and ΔCAD,

$\angle B = \angle DAC$ (As proved above)

$\angle BDA = \angle ADC$ (Each 90°)

$\therefore$ ΔABD ~ ΔCAD (AA similarity)

$\therefore \frac{BD}{AD} = \frac{AB}{AC}$ Hence proved.

27.

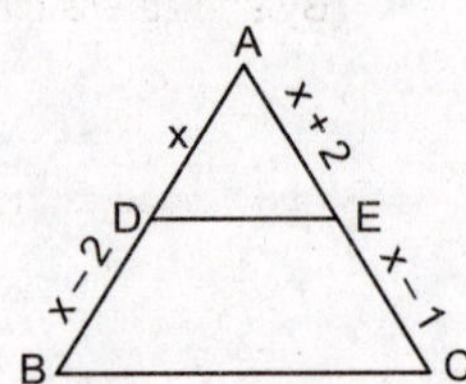

In ΔABC, DE || BC (Given)

$\frac{AD}{DB} = \frac{AE}{EC}$ (By BPT)

$\Rightarrow \frac{x}{x-2} = \frac{x+2}{x-1}$

$\Rightarrow x(x-1) = (x+2)(x-2)$

$\Rightarrow x^2 - x = x^2 - 2^2$

$\Rightarrow x^2 - x = x^2 - 4$

$\Rightarrow x = 4$

28.

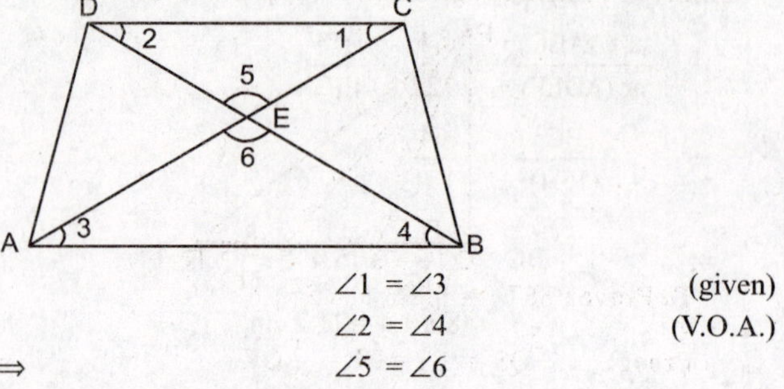

$\angle 1 = \angle 3$ (given)

$\angle 2 = \angle 4$ (V.O.A.)

$\Rightarrow \angle 5 = \angle 6$

Using AAA similarity rule

ΔCDE ~ ΔABE

$\frac{DE}{BE} = \frac{EC}{EA}$

$\frac{4}{4x-2} = \frac{x+1}{2x+4}$

$\Rightarrow 8x + 16 = 4x^2 - 2x + 4x - 2$

$\Rightarrow 8x + 16 = 4x^2 + 2x - 2$

$\Rightarrow 4x^2 + 2x - 8x - 2 - 16 = 0$

$\Rightarrow 4x^2 - 6x - 18 = 0$

$\Rightarrow 2x^2 - 3x - 9 = 0$

$\Rightarrow 2x^2 - 6x + 3x - 9 = 0$

$\Rightarrow 2x(x-3) + 3(x-3) = 0$

$\Rightarrow (x-3)(2x+3) = 0$

Either $x - 3 = 0$

$\Rightarrow x = 3$

or $2x + 3 = 0$

$x = \frac{-3}{2}$ is not possible

Hence, $x = 3$

29.

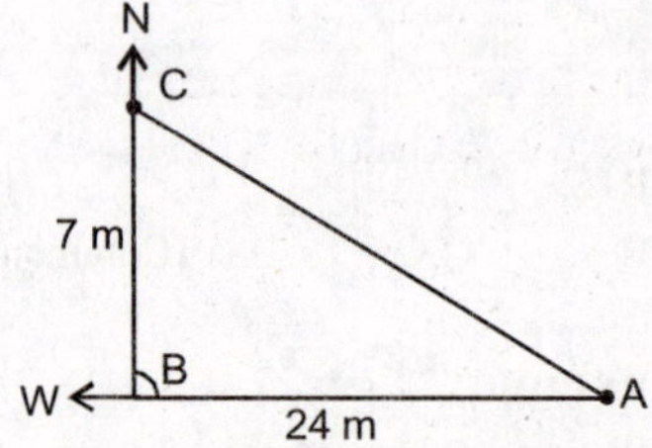

$AC^2 = AB^2 + BC^2$ (By Pythagoras theorem)

$= (24)^2 + 7^2 = 576 + 49 = 625$

$AC = 25$ m

He is 25 m away from his starting point.

SHORT ANSWER TYPE QUESTIONS-I

1. **Given :** In ΔABC, AD ⊥ BC

To prove : $AB^2 + CD^2 = AC^2 + DB^2$

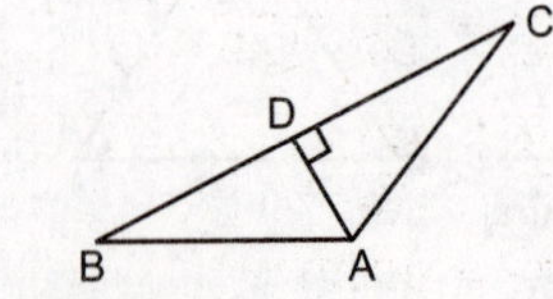

Proof : In ΔADB, $\angle D = 90°$

$\Rightarrow AB^2 = AD^2 + BD^2$ (By Pythagoras theorem) ...(*i*)

In ΔADC, $\angle D = 90°$

$\Rightarrow AC^2 = AD^2 + CD^2$ (By Pythagoras theorem)

$\Rightarrow CD^2 = AC^2 - AD^2$ ...(*ii*)

Adding (*i*) and (*ii*) we get

$AB^2 + CD^2 = AD^2 + BD^2 + AC^2 - AD^2$

$= BD^2 + AC^2$ Hence proved.

2. Given : Right triangle PQR. S and T trisect QR.

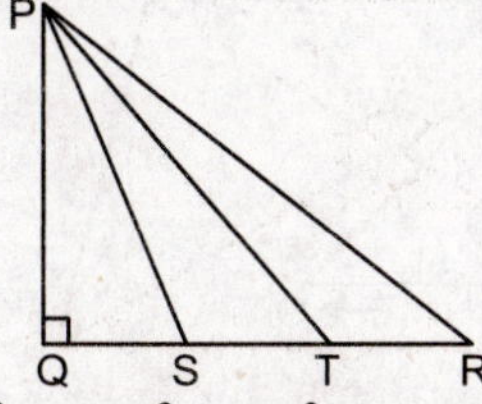

To Prove : $8PT^2 = 3PR^2 + 5PS^2$

Proof : $QS = ST = TR = \frac{1}{3}QR$ (Given) ...(*i*)

In right triangle PQS,

$PS^2 = PQ^2 + QS^2$ (By Pythagoras theorem) ...(*ii*)

In right triangle PQT,

$PT^2 = PQ^2 + QT^2$ (By Pythagoras theorem) ...(*iii*)

In right triangle PQR,

$PR^2 = PQ^2 + QR^2$ (By Pythagoras theorem) ...(*iv*)

Subtracting (*iii*) from (*ii*), we get

$PS^2 - PT^2 = QS^2 - QT^2$

$\Rightarrow \quad PS^2 - PT^2 = \left(\frac{1}{3}QR\right)^2 - \left(\frac{2}{3}QR\right)^2$

$= \frac{1}{9}QR^2 - \frac{4}{9}QR^2$ [From (*i*)]

$\Rightarrow \quad 3PS^2 - 3PT^2 = -QR^2$ ...(*v*)

Subtracting (*iv*) from (*iii*), we get

$PT^2 - PR^2 = QT^2 - QR^2$

$= \left(\frac{2}{3}QR\right)^2 - QR^2$ [From (*i*)]

$\Rightarrow \quad PT^2 - PR^2 = \frac{4}{9}QR^2 - QR^2$

$\Rightarrow \quad 9PT^2 - 9PR^2 = -5QR^2$ ...(*vi*)

Substituting for $(-QR^2)$ from (*v*) in (*vi*). we get

$\Rightarrow \quad 9PT^2 - 9PR^2 = 5(3PS^2 - 3PT^2)$

$\Rightarrow \quad 9PT^2 - 9PR^2 = 15PS^2 - 15PT^2$

$\Rightarrow \quad 24PT^2 = 15PS^2 + 9PR^2$

$\Rightarrow \quad 8PT^2 = 5PS^2 + 3PR^2$ Hence proved.

3. Given : ΔABC in which DE ∥ AB and BD ∥ EF.

To prove : $DC^2 = CF \times AC$

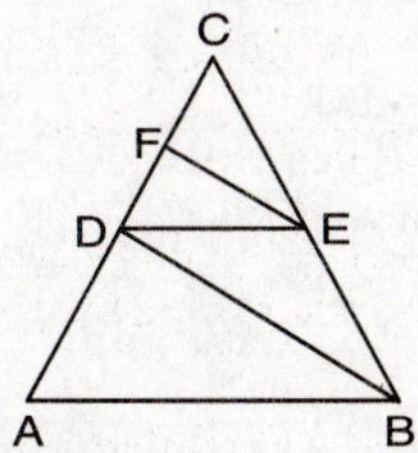

Proof. In Δ ABC, DE ∥ AB

$\Rightarrow \quad \frac{CD}{AC} = \frac{CE}{BC}$ ... (*i*)

(Cor. of Basic Proportionality Theorem)

Again in Δ CDB, EF ∥ BD

$\Rightarrow \quad \frac{CF}{CD} = \frac{CE}{CB}$ ... (*ii*)

(Cor. of Basic Proportionality Theorem)

From (*i*) and (*ii*), we get

$\frac{CD}{AC} = \frac{CF}{CD} \Rightarrow CD^2 = CF \times AC$ Hence proved.

4. Given : ABCD is a ∥ gm AD is produced and BE intersect CD at F.

To prove : ΔABE ~ ΔCFB

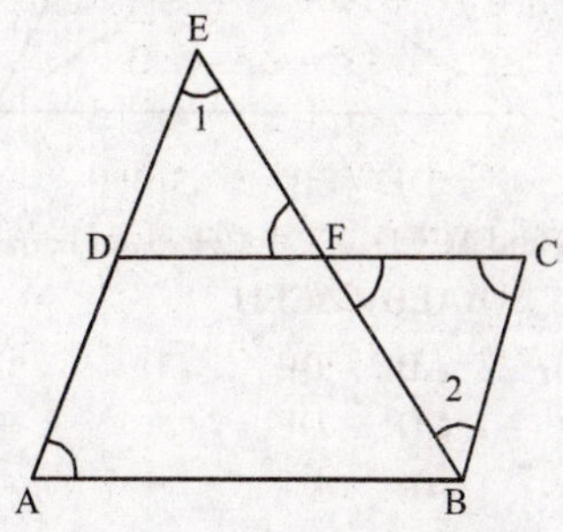

Proof : In ΔABE and ΔCFB, ∠1 = ∠2

(∵ AE ∥ BC (Alternate interior angles))

∠A = ∠C (Opposite angles of a ∥ gm.)

∴ ΔABE ~ ΔCFB (AA similarity)

Hence proved.

5.

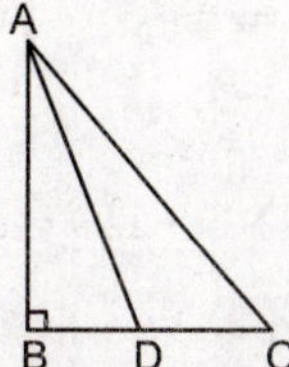

Given : ΔABC with ∠B = 90°

D is the mid-point of BC

To prove : $AC^2 = 4AD^2 - 3AB^2$

Proof : In ΔABC, ∠B = 90° (Given)

$AC^2 = AB^2 + BC^2$ (By Pythagoras theorem)

$= AB^2 + (2BD)^2$

$= AB^2 + 4BD^2$ ...(*i*)

In ΔABD, $AD^2 = AB^2 + BD^2$ (Using Pythagoras theorem)

$\Rightarrow \quad BD^2 = AD^2 - AB^2$ ...(*ii*)

From (*i*) and (*ii*) we get

$AC^2 = AB^2 + 4(AD^2 - AB^2)$

$= AB^2 + 4AD^2 - 4AB^2$

$AC^2 = 4AD^2 - 3AB^2$ Hence proved.

6.

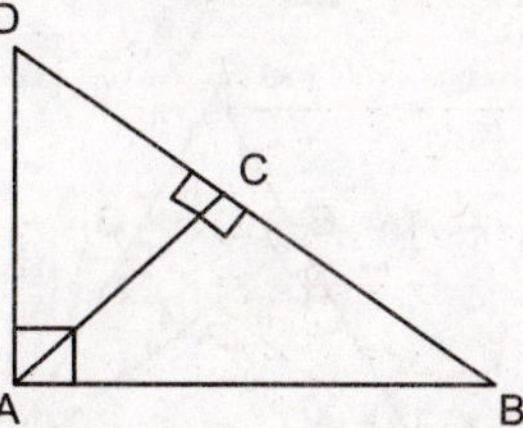

In Δ DAB, and ΔACB

∠DAB = ∠ACB = 90°

∠B = ∠B (common)

∴ ΔDAB ~ ΔACB

$\Rightarrow \quad \frac{AD}{AC} = \frac{AB}{BC} = \frac{BD}{AB}$

$\Rightarrow \quad \frac{AB}{BC} = \frac{BD}{AB}$

$\Rightarrow \quad AB^2 = BC.BD$ Hence proved

7. 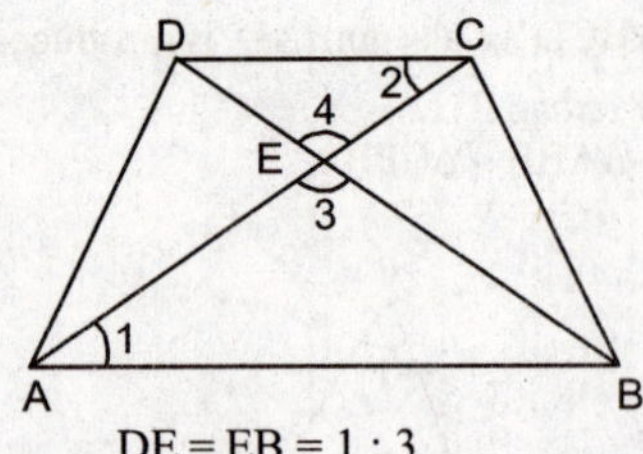

$$DE = EB = 1:3$$

In Δ AEB and ΔCED, $\angle 1 = \angle 2$ (all angles) $\angle 3 = \angle 4$ (V-O-A)

$\therefore \quad \Delta AEB \sim \Delta CED$

$\Rightarrow \quad \dfrac{AB}{CD} = \dfrac{BE}{DE}$

$\Rightarrow \quad \dfrac{AB}{CD} = \dfrac{3}{1} \qquad [\because DE : BE = 1:3]$

$\Rightarrow \quad AB = 3CD$

8. Given : AB $\parallel$ DC $\therefore$ $\Delta DOC \sim \Delta BOA$

$\therefore \quad \dfrac{OD}{OB} = \dfrac{OC}{OA} \Rightarrow \dfrac{x-2}{x-1} = \dfrac{x+3}{x+5}$

$(x-2)(x+5) = (x+3)(x-1) \Rightarrow x^2 + 3x - 10 = x^2 + 2x - 3$

$\Rightarrow \quad x = 7$

9. In right Δ QPR,

$PR^2 = QP^2 + PR^2$

$\Rightarrow \quad (26)^2 = (24)^2 + PR^2$

$\Rightarrow \quad PR^2 = 100$

In right ΔPKR,

$QR^2 = PK^2 + KR^2$

$\Rightarrow \quad 100 = PK^2 + 8^2$

$\Rightarrow \quad PK = 6$ cm

10. $\because \quad \Delta ABC \sim \Delta DEF$

$\therefore \quad \dfrac{\text{area } \Delta ABC}{\text{area } \Delta DEF} = \dfrac{BC^2}{EF^2}$

$\Rightarrow \quad \dfrac{54}{\text{area } \Delta DEF} = \dfrac{3^2}{4^2}$

$\Rightarrow \quad \text{Area } \Delta DEF = \dfrac{54 \times 4^2}{3^2} = 96 \text{ cm}^2$

11. Given : ΔBEP and ΔCPD

To prove : BP $\times$ PD = EP $\times$ PC

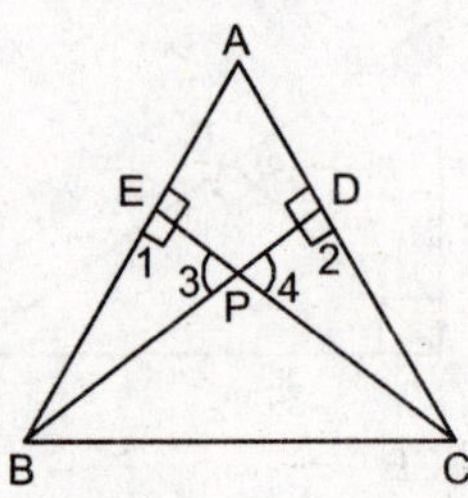

Proof : In ΔBEP and ΔCPD

$\angle 1 = \angle 2 = 90°$

$\angle 3 = \angle 4$ (Vertically opposite angles)

$\Rightarrow \quad \Delta BEP \sim \Delta CDP$ (By AA similarity)

or $\quad \dfrac{BP}{CP} = \dfrac{EP}{DP}$

$\Rightarrow \quad \dfrac{BP}{EP} = \dfrac{CP}{DP}$

$\Rightarrow \quad BP \times PD = EP \times CP$ Hence proved.

12. 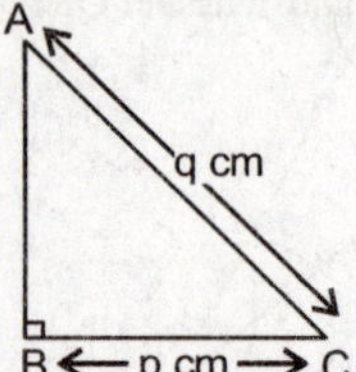

Given : In ΔABC, $\angle B = 90°$

$AC = q$ cm, $BC = p$ cm

$q - p = 2$

To find : AB in term of q.

Proof : In ΔABC, $\angle B = 90°$ (Given)

$\Rightarrow \quad AC^2 = AB^2 + BC^2$ (By Pythagoras theorem)

$\Rightarrow \quad q^2 = AB^2 + p^2$

$\Rightarrow \quad AB^2 = q^2 - p^2$

$\Rightarrow \quad = q^2 - (q-2)^2 (\because q - p = 2 \text{ given} \Rightarrow q - 2 = p)$

$= q^2 - (q^2 + 2^2 - 4q) = q^2 - q^2 - 4 + 4q = 4q - 4$

$\Rightarrow \quad AB = \sqrt{4q-4} = \sqrt{4(q-1)}$

$AB = 2\sqrt{(q-1)}$ cm

13. Given : In ΔABC, AB = AC, AD = AE

To prove : Points B, C, E and D are concyclic

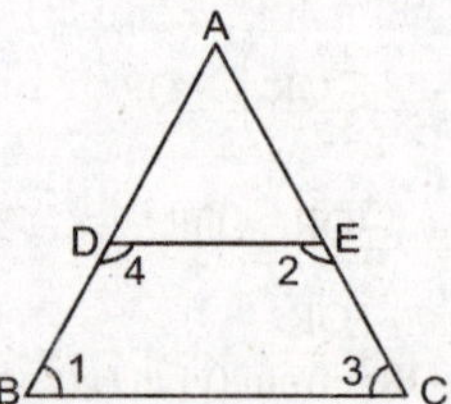

Proof : To prove that points B, C, E and D are concyclic only we need to prove $\angle 1 + \angle 2 = 180°$ and $\angle 3 + \angle 4 = 180°$.

In ΔABC, AB = AC

$\therefore \quad \angle 1 = \angle 3$ (Opp. angles of equal sides are equal) ...(*i*)

and $\quad AD = AE$ ...(*ii*)

Subtracting (*ii*) from (*i*) we get

$AB - AD = AC - AE$

$\Rightarrow \quad DB = EC$ ...(*iii*)

$[\because AB - AD = DB \text{ and } AC - AE = EC]$

Dividing (*ii*) by (*iii*) we get

$\dfrac{AD}{DB} = \dfrac{AE}{EC}$

$\Rightarrow \quad BE \parallel BC$ (By converse of BPT)

$\Rightarrow \quad \angle 1 + \angle 4 = 180°$ [Cointerior angles are supplementary]

$\Rightarrow \quad \angle 3 + \angle 4 = 180°$

$\because \quad \angle 1 = \angle 3$ (Proved above)

But $\angle 1 + \angle 2 + \angle 3 + \angle 4 = 360°$ ... (*iv*)

(Sum of the angles of a quadrilateral)

From (*iii*) and (*iv*),

$\angle 1 + \angle 2 + 180° = 360°$

$\Rightarrow \quad \angle 1 + \angle 2 = 180°$

$\therefore$ Points B, C, D and E are concylic. Hence proved.

14. Given : In a ΔABC, AB = AC

$BC^2 = AC \times CD$

To prove : BD = BC

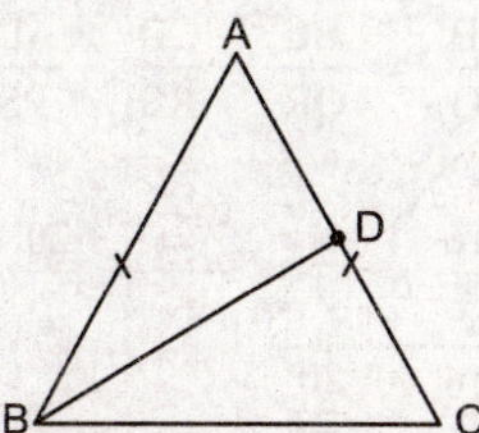

Proof : In ΔBCD and ΔACB

and $BC^2 = AC \times CD$

$\Rightarrow \frac{BC}{CD} = \frac{AC}{BC}$ ...(*i*)

$\angle C = \angle C$ (Common)

$\Rightarrow \Delta ACB \sim \Delta BCD$ (SAS similarity)

$\Rightarrow \frac{AB}{BD} = \frac{BC}{CD}$ ...(*ii*)

$\Rightarrow \frac{AB}{BD} = \frac{AC}{BC}$ [From (*i*)]

$\Rightarrow \frac{1}{BD} = \frac{1}{BC}$ (As AB = AC given)

$\Rightarrow BC = BD$ Hence proved.

15. In ΔABC, $DE \parallel BC$

$\therefore \frac{AD}{BD} = \frac{AE}{CE}$

$\Rightarrow \frac{AB - BD}{BD} = \frac{AC - CE}{CE}$

$\Rightarrow \frac{2x-(x-3)}{x-3} = \frac{2x+3-(x-2)}{x-2}$

$\Rightarrow \frac{x+3}{x-3} = \frac{x+5}{x-2} \Rightarrow (x-2)(x+3) = (x+5)(x-3)$

$\Rightarrow x^2 + x - 6 = x^2 + 2x - 15$

$\Rightarrow x = 9 \ \therefore \ x = 9$ cm

16. Given : In ΔABC $\angle B = 90°$

D is a point on AC such that

AD = DC and AB = BD

To find : $\angle CAB$

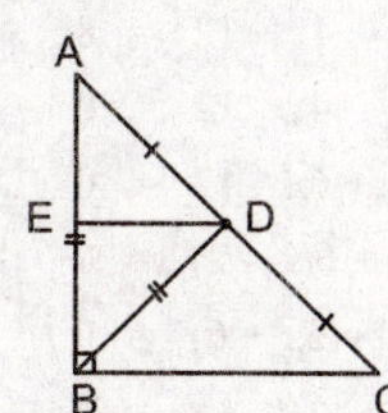

Proof : Given AD = DC ...(*i*)

⇒ D is the mid-point of AC [Circumcentre of a right triangle lies at the mid-point of hypotenuse]

⇒ D is the circumcentre of ΔABC

[∵ D being circumcenter of ΔABC, point is equal distant from A, B and C]

$\Rightarrow BD = AD = DC = \frac{1}{2}AC$

Given that $BD = AB \Rightarrow AB = \frac{1}{2}AC$

$\Rightarrow \angle ACB = 30°$ [∵ side opposite to 30° angle is half the hypotenuse]

$\Rightarrow \angle CAB = 180° - (30° + 90°)$ (ASP)

$= 180° - 120° = 60°$

17. Given : ΔABC in which $PQ \parallel BC$ and AD is median.

To prove : Median AD bisects PQ at E.

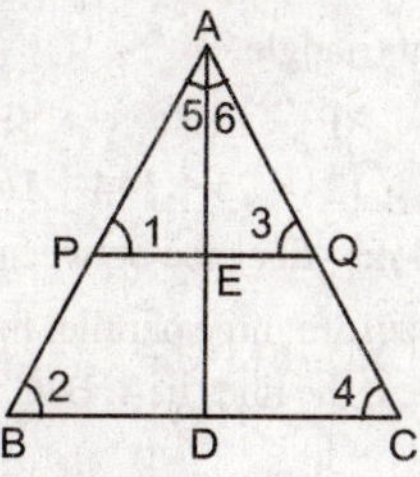

Proof : In ΔAPE and ΔABD

$\angle 5 = \angle 5$ (Common)

$\angle 1 = \angle 2$ (Corresponding angles)

$\therefore \Delta APE \sim \Delta ABD$ (By AA corollary)

Hence, $\frac{PE}{BD} = \frac{AE}{AD}$ ...(*i*)

(Corresponding sides of similar triangles are proportional.)

In ΔAEQ and ΔADC,

$\angle 6 = \angle 6$ (Common)

$\angle 3 = \angle 4$ (Corresponding angles)

$\Rightarrow \Delta AEQ \sim \Delta ADC$ (By AA corollary)

Hence $\frac{EQ}{DC} = \frac{AE}{AD}$ ...(*ii*)

(Corresponding sides of similar triangles)

From (*i*) and (*ii*) we have

$\frac{PE}{BD} = \frac{EQ}{DC}$

$\Rightarrow \frac{PE}{BD} = \frac{EQ}{BD}$ [AD is the median]

$\Rightarrow PE = EQ$

Median AD Bisects PQ at E. Hence proved.

18. $\because \Delta ABC \sim \Delta DEF$

$\therefore \frac{\text{Area } \Delta ABC}{\text{Area } \Delta DEF} = \frac{AB^2}{DE^2}$

$\Rightarrow \frac{56}{\text{area } \Delta DEF} = \frac{(2DE)^2}{DE^2}$ ($\because$ AB = 2DE)

$\Rightarrow \frac{56}{\text{area } \Delta DEF} = 4$

$\Rightarrow \text{area } \Delta \text{ DEF} = \frac{56}{4} = 14 \text{ cm}^2$

19.

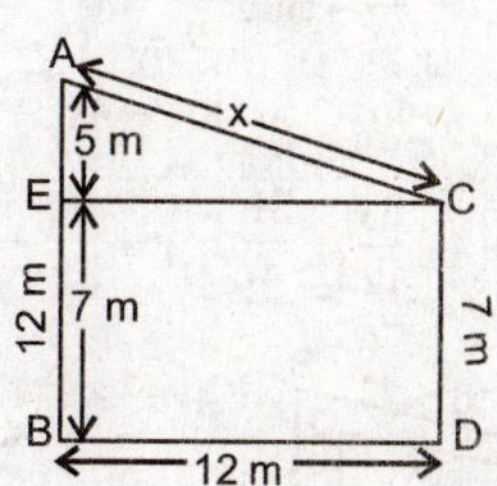

Given : Two poles AB = 12 m, CD = 7 m, BD = 12 m

To find : AC

Construction : Draw $CE \parallel BD$

Proof : BDCE is a rectangle

$\Rightarrow EB = CD = 7$ m

$\Rightarrow \qquad EC = BD = 12 \text{ m}$ (Opposite sides of a rectangle)

$AE = AB - BE = 12 \text{ m} - 7 \text{ m} = 5 \text{ m}$

and ΔAEC is a right triangle

$\Rightarrow \qquad AC^2 = AE^2 + EC^2$ (By Pythagoras theorem)

$\Rightarrow \qquad AC^2 = 5^2 + 12^2 = 25 + 144 = 169$

$\Rightarrow \qquad AC = 13$ m = distance between their tops.

20. Given : Let l, m and n are three parallel lines, t_1 and t_2 are any two transversals which cut the lines in A, B, C and P, Q, R respectively.

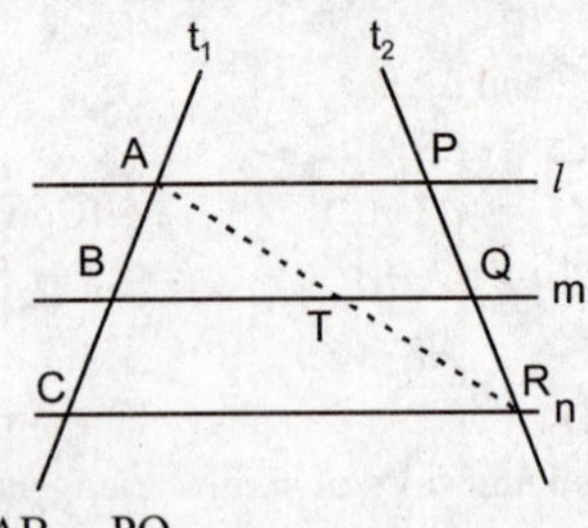

To Prove : $\dfrac{AB}{BC} = \dfrac{PQ}{QR}$

Construction : Joined AR which meets the line m in T.

Proof : Now, in ΔACR, $BT \parallel CR$

$\therefore \qquad \dfrac{AB}{BC} = \dfrac{AT}{TR}$ (By B.P. theorem) ... (*i*)

Also, in ΔAPR, $TQ \parallel AP$

$\therefore \qquad \dfrac{AT}{TR} = \dfrac{PQ}{QR}$ (By B.P. theorem) ... (*ii*)

Using (*i*) and (*ii*), we get

$\dfrac{AB}{BC} = \dfrac{PQ}{QR}$ Hence proved.

21. (*i*) $\Delta ABC \sim \Delta PQR$

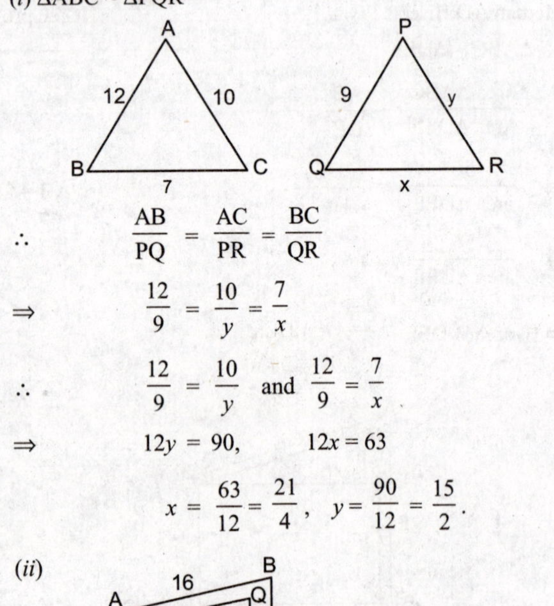

$\therefore \qquad \dfrac{AB}{PQ} = \dfrac{AC}{PR} = \dfrac{BC}{QR}$

$\Rightarrow \qquad \dfrac{12}{9} = \dfrac{10}{y} = \dfrac{7}{x}$

$\therefore \qquad \dfrac{12}{9} = \dfrac{10}{y}$ and $\dfrac{12}{9} = \dfrac{7}{x}$

$\Rightarrow \qquad 12y = 90, \qquad 12x = 63$

$x = \dfrac{63}{12} = \dfrac{21}{4}, \quad y = \dfrac{90}{12} = \dfrac{15}{2}.$

(*ii*)

Given : □ABCD ~ □PQRS

$\Rightarrow \qquad \dfrac{AB}{PQ} = \dfrac{BC}{QR} = \dfrac{CD}{RS} = \dfrac{AD}{PS}$

$\Rightarrow \qquad \dfrac{16}{x} = \dfrac{5}{y} = \dfrac{16\frac{2}{3}}{z} = \dfrac{20}{7}$

$\Rightarrow \qquad \dfrac{16}{x} = \dfrac{20}{7}$

or $\qquad x = \dfrac{16 \times 7}{20} = \dfrac{28}{5}$

$\dfrac{5}{y} = \dfrac{20}{7}$

$\Rightarrow \qquad y = \dfrac{5 \times 7}{20} = \dfrac{7}{4}$

$\dfrac{16\frac{2}{3}}{z} = \dfrac{20}{7} \Rightarrow \dfrac{50}{3z} = \dfrac{20}{7}$

$\Rightarrow \qquad \dfrac{7 \times 50}{20 \times 3} = z \Rightarrow z = \dfrac{35}{6}$

22. Given : ΔABC

Proof : DE will be parallel to AB

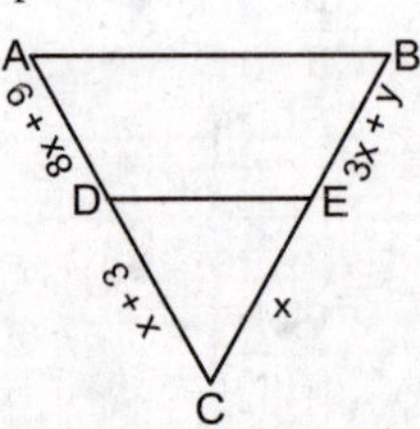

Only, if $\qquad \dfrac{CD}{AD} = \dfrac{CE}{BE}$ [Converse of BPT]

$\Rightarrow \qquad \dfrac{x+3}{8x+9} = \dfrac{x}{3x+4}$

$\Rightarrow \qquad (x + 3)(3x + 4) = x(8x + 9)$

$\Rightarrow \qquad 3x^2 + 9x + 4x + 12 = 8x^2 + 9x$

$\Rightarrow \qquad 5x^2 - 4x - 12 = 0$

$\Rightarrow \qquad 5x^2 - 10x + 6x - 12 = 0$

$\Rightarrow \qquad 5x(x - 2) + 6(x - 2) = 0$

$\Rightarrow \qquad (x - 2)(5x + 6) = 0$

$\Rightarrow$ either $\qquad x = 2$

or $\qquad 5x = -6$

$\Rightarrow \qquad x = \dfrac{-6}{5}$ (Impossible)

$\Rightarrow \qquad x = 2$

if $x = 2$ then $DE \parallel AB$.

23. Given : In ΔABC,

AP = 2 cm, PB = 4 cm, AQ = 3 cm, QC = 6 cm.

To Prove : BC = 3PQ.

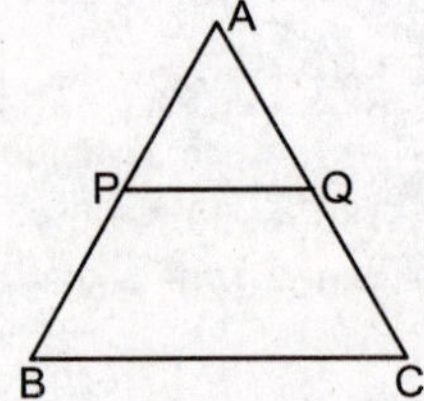

Proof : $\dfrac{AP}{PB} = \dfrac{2}{4} = \dfrac{1}{2}$ ...(i)

$\dfrac{AQ}{QC} = \dfrac{3}{6} = \dfrac{1}{2}$ ...(ii)

From (i) and (ii) $\dfrac{AP}{PB} = \dfrac{AQ}{QC}$

$\Rightarrow$ PQ || BC

In ΔABC and ΔAPQ.

$\dfrac{AB}{AP} = \dfrac{AC}{AQ}$ ($\because$ PQ || BC)

$\angle A = \angle A$ (Common)

$\therefore$ ΔABC ~ ΔAPQ, (SAS similarity)

$\Rightarrow \dfrac{AB}{AP} = \dfrac{BC}{PQ}$

$\Rightarrow \dfrac{AP+PB}{AP} = \dfrac{BC}{PQ}$

$\Rightarrow \dfrac{2+4}{2} = \dfrac{BC}{PQ}$

$\Rightarrow \dfrac{BC}{PQ} = 3$

$\Rightarrow$ BC = 3PQ. Hence proved.

24.

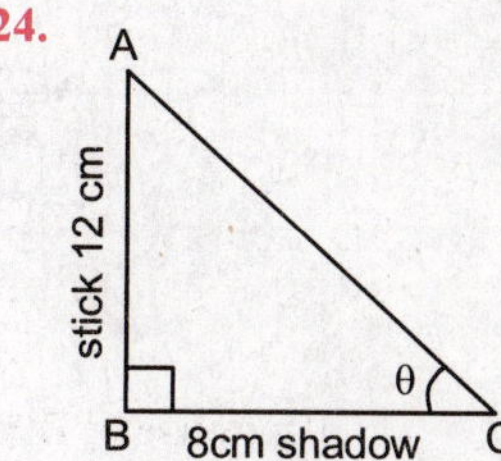

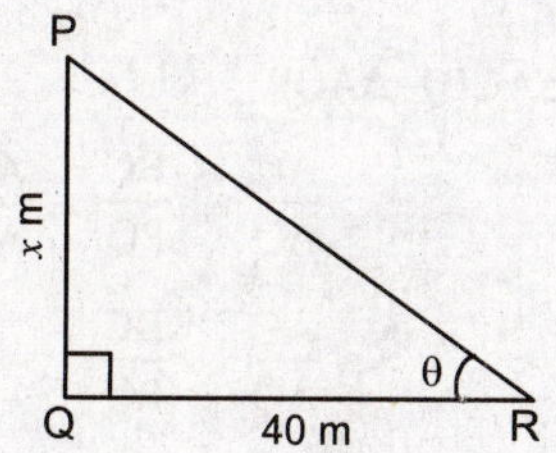

Let AB is vertical stick

$\therefore$ AB = 12 cm and BC is shadow $\therefore$ BC = 8cm

Let PQ = x = tower and QR is shadow of the tower = 40 m

In ΔABC and ΔPQR

$\angle ABC = \angle PQR = 90°$

$\angle C = \angle R$ (angle of elevation of sun)

$\therefore$ ΔABC ~ ΔPQR

$\Rightarrow \dfrac{AB}{PQ} = \dfrac{BC}{QR} \Rightarrow \dfrac{12}{8} = \dfrac{x}{40} \Rightarrow x = 60$ m

25.

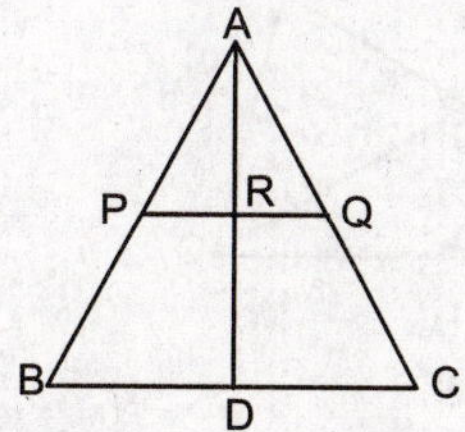

Given : AP = 3 cm, AR = 4.5 cm,

AQ = 6 cm, AB = 5 cm

AC = 10 cm

To find : AD.

Proof : In ΔABC, $\dfrac{AP}{AB} = \dfrac{3}{5}$ ...(i)

$\dfrac{AQ}{AC} = \dfrac{6}{10} = \dfrac{3}{5}$ ...(ii)

From (i) and (ii),

$\dfrac{AP}{AB} = \dfrac{AQ}{AC}$

$\Rightarrow$ PQ ∥ BC (By converse of BPT)

In ΔABD, PR ∥ BD

$\Rightarrow \dfrac{AP}{AB} = \dfrac{AR}{AD}$ (By BPT)

$\Rightarrow \dfrac{3}{5} = \dfrac{4.5}{AD}$

$\Rightarrow AD = \dfrac{4.5 \times 5}{3} = 7.5$ cm

26.

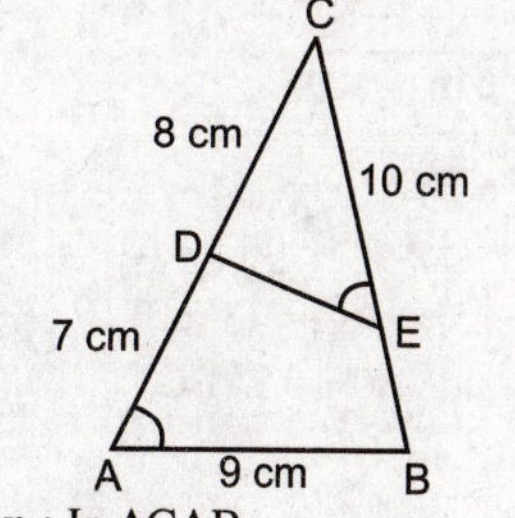

Given : In ΔCAB,

$\angle A = \angle CED$

AB = 9 cm

AD = 7 cm

CD = 8 cm

and CE = 10 cm

To find : DE

Proof : In ΔCED and ΔCAB

$\angle C = \angle C$ (Common)

$\angle CED = \angle CAB$ (Given)

Using AA similarity rule

ΔCAB ~ ΔCED

$\Rightarrow \dfrac{AB}{DE} = \dfrac{AC}{CE}$

$\Rightarrow \dfrac{9}{DE} = \dfrac{(AD+CD)}{10}$

$\Rightarrow \dfrac{9}{DE} = \dfrac{8+7}{10}$

$\Rightarrow \dfrac{9 \times 10}{15} = DE$

$\Rightarrow$ DE = 6 cm

27. In right ΔABC, $AC^2 = AB^2 + BC^2$

$AC^2 = (150)^2 + (200)^2$

$AC^2 = 22500 + 40000$

$AC^2 = 62500$

$\Rightarrow$ AC = 250 m

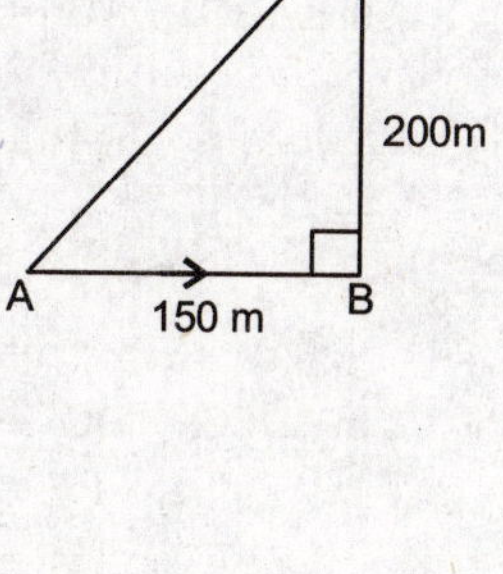

28.

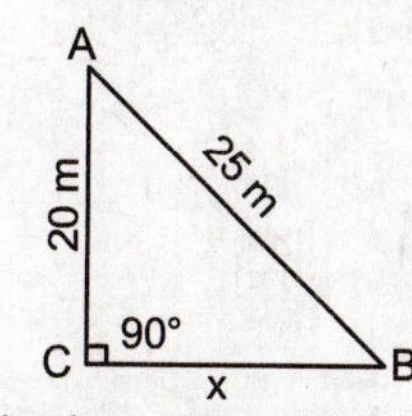

Let AC be the wall and AB be the ladder

$\Rightarrow \angle ACB = 90°$

Now in ΔACB,

$AB^2 = AC^2 + CB^2$

$\Rightarrow (25)^2 = 20^2 + x^2$

$\Rightarrow \quad 625 = 400 + x^2$

$\Rightarrow \quad x^2 = 625 - 400 = 225$

$\Rightarrow \quad x = 15$ m

$\therefore$ Distance of the foot of the ladder from the building is 15 m.

29.

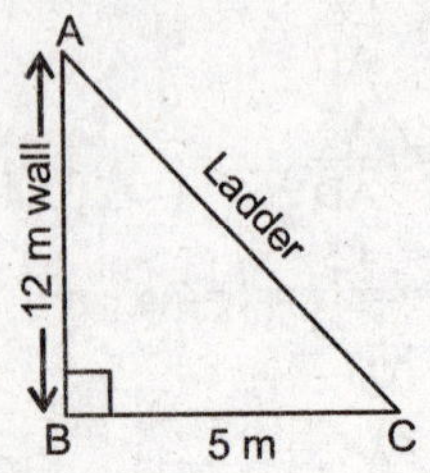

In right ΔABC, $AC^2 = AB^2 + BC^2$

$\Rightarrow \quad AC^2 = (12)^2 + (5)^2$

$\Rightarrow \quad AC^2 = 144 + 25 \Rightarrow AC = 13$ m

30. Given : AD be the ⊥ to the side BC of ΔABC.

To prove : $AB^2 - BD^2 = AC^2 - CD^2$

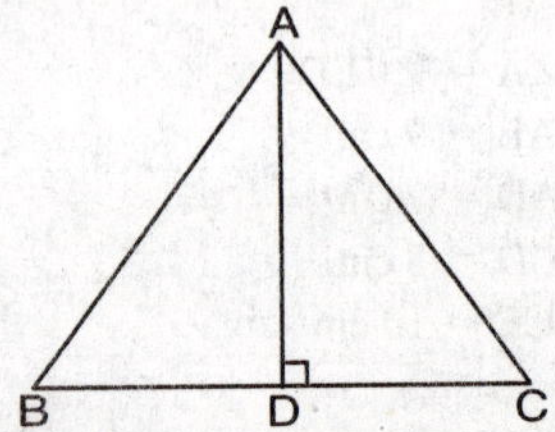

Proof : ∴ There are two right triangles formed.

In right triangle ABD,

$AD^2 = AB^2 - BD^2$ (By Pythagoras theorem)...(*i*)

In right triangle ACD,

$AD^2 = AC^2 - CD^2$ ... (*ii*)

Equating (*i*) and (*ii*), we get

$(AB)^2 - (BD)^2 = (AC)^2 - (CD)^2$ Proved.

31.

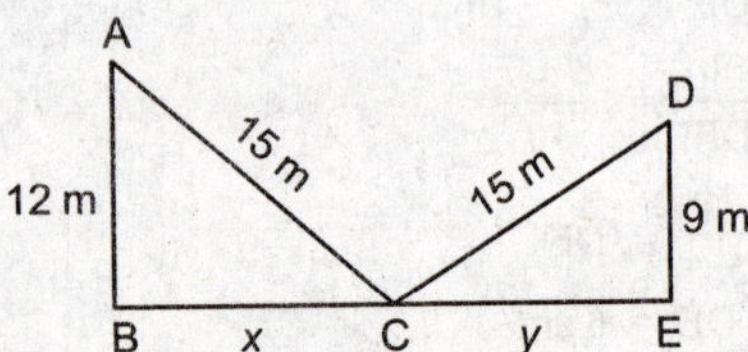

In ΔABC, $AC^2 = AB^2 + BC^2$ (By Pythagoras theorem)

$15^2 = 12^2 + x^2$ ...(*i*)

$\Rightarrow \quad x^2 = 225 - 144$

$x^2 = 81$

$x = 9$ m

In ΔDEC, $DC^2 = DE^2 + CE^2$

$15^2 = 9^2 + y^2$

$y^2 = 225 - 81 = 144$

$y = 12$ m ...(*ii*)

From (*i*) and (*ii*),

Width of the street= BE = BC + CE

$= x + y = 9$ m $+ 12$ m $= 21$ m.

32.

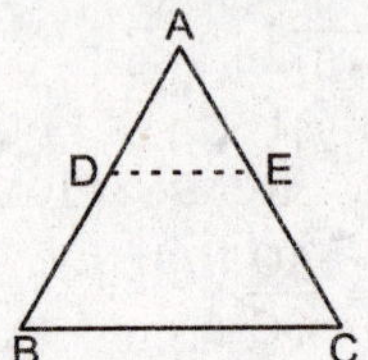

Now

$$\frac{AD}{BD} = \frac{6}{9} = \frac{2}{3} \quad ...(i)$$

$$\frac{AE}{EC} = \frac{8}{12} = \frac{2}{3} \quad ...(ii)$$

From (*i*) and (*ii*) $\frac{AD}{BD} = \frac{AE}{EC}$

∴ DE || BC (Using converse of B.P.T.)

33.

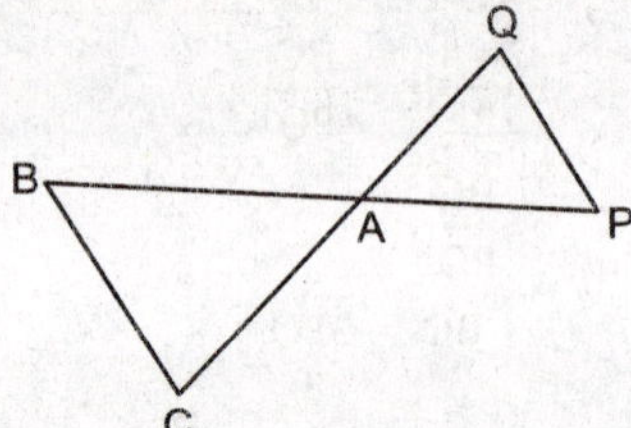

∵ ΔACB ~ ΔAQP

$\therefore \quad \frac{AC}{AQ} = \frac{BC}{PQ} = \frac{AB}{AP}$

$\Rightarrow \quad \frac{AC}{AQ} = \frac{BC}{PQ}$

$\Rightarrow \quad \frac{AC}{2.8} = \frac{8}{4}$

$\Rightarrow \quad AC = 5.6$ cm

Also $\quad \frac{BC}{PQ} = \frac{AB}{AP} \Rightarrow \frac{8}{4} = \frac{6.5}{AP}$

$\Rightarrow \quad AP = \frac{6.5}{2} = 3.25$ cm

34. Given :

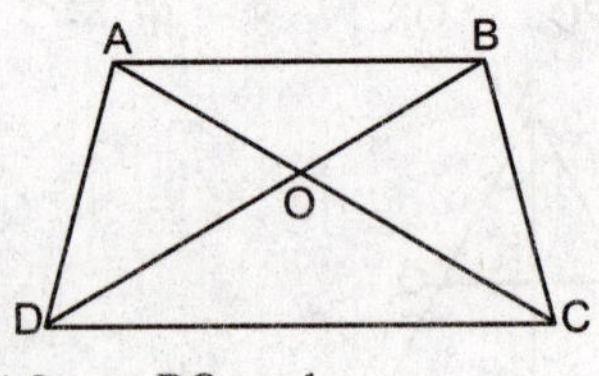

$$\frac{AO}{OC} = \frac{BO}{OD} = \frac{1}{2}$$

and $\quad AB = 4$ cm

To find : DC

Proof : In ΔAOB and ΔCOD,

$$\frac{AO}{OC} = \frac{BO}{OD}$$

and $\quad \angle AOB = \angle COD$

$\therefore \quad \Delta AOB \sim \Delta COD$ (SAS similarity)

$\Rightarrow \quad \frac{AO}{OC} = \frac{BO}{OD} = \frac{AB}{CD}$

$\Rightarrow \quad \frac{1}{2} = \frac{4}{CD}$

$\Rightarrow \quad CD = 8$ cm.

35. **Given :** ΔABC, in which $AC = BC$ and AB is produced to P and Q. Such that $AP \times BQ = (AC)^2$

To prove : $\Delta ACP \sim \Delta BCQ$

Proof : $\angle 1 + \angle 3 = \angle 2 + \angle 4$ (Each linear pair)

$AC = BC$ (Given)

As $\angle 1 = \angle 2$ (Opposite angles of isosceles Δ)

$\therefore \angle 3 = \angle 4$

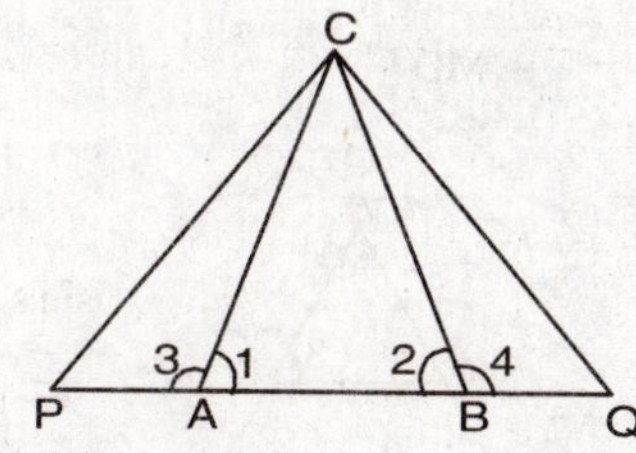

Also, $AP \times BQ = (AC)^2$

$\Rightarrow \quad AP \times BQ = AC \times BC$ [$\because AC = BC$]

$\Rightarrow \quad \frac{AP}{AC} = \frac{BC}{BQ}$

$\angle 3 = \angle 4$ (Proved above)

$\therefore \quad \Delta ACP \sim \Delta BCQ$ (SAS similarity)

36. **Given :** ABCD is a || gm and diagonal BD intersect AE at F.

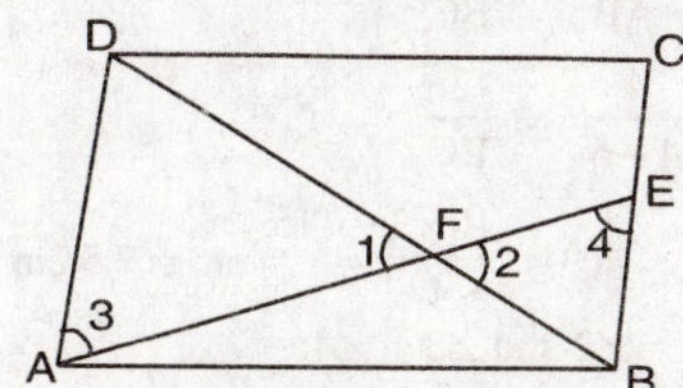

To prove : $DF \times FE = BF \times FA$

Proof : In ΔAFD and ΔEFB,

$\angle 1 = \angle 2$ (VOA's)

and $\angle 3 = \angle 4$ (Alternate $\angle s$)

$\therefore \quad \Delta AFD \sim \Delta EFB$ (AA similarity)

$\Rightarrow \quad \frac{DF}{BF} = \frac{AF}{FE} = \frac{AD}{BE}$

$\Rightarrow \quad DF \times FE = BF \times FA$ Hence proved.

37. **Given :** In ΔPQR, $PQ = PR$ and $QR^2 = PR \times XR$

To prove : $QX = QR$

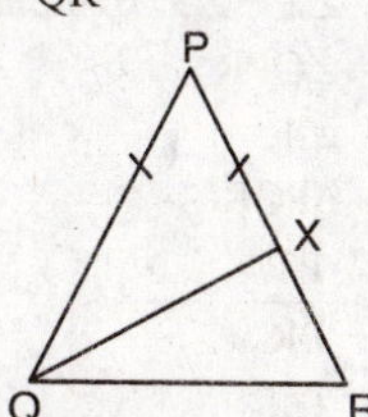

Proof : $QR^2 = PR \times XR$

$\Rightarrow \quad \frac{QR}{PR} = \frac{XR}{QR}$

and $\angle R = \angle R$

$\therefore \quad \Delta QRP \sim \Delta XRQ$ (SAS similarity)

$\Rightarrow \quad \frac{QR}{XR} = \frac{PR}{QR} = \frac{QP}{XQ} \Rightarrow \frac{PR}{QR} = \frac{QP}{XQ}$

Also, $PR = QP$ (Given)

$\therefore \quad QR = XQ.$ Hence proved.

38.

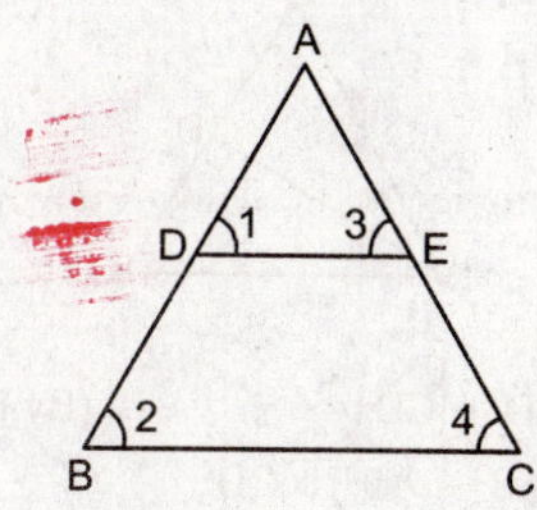

$\because \quad DE \parallel BC$

$\therefore \quad \angle 1 = \angle 2$ and $\angle 3 = \angle 4$ (Corresponding angles)

$\therefore \quad \Delta ADE \sim \Delta ABC$

$\Rightarrow \quad \frac{\text{area } \Delta ABC}{\text{area } \Delta ADE} = \frac{BC^2}{DE^2} \Rightarrow \frac{\text{area } \Delta ABC}{25} = \frac{64}{16}$

$\Rightarrow \quad \text{Area } \Delta ABC = 100 \text{ cm}^2$

39.

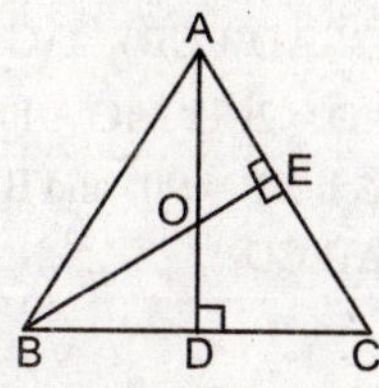

In ΔBEC, ΔADC

$\angle C = \angle C$ (common)

$\angle BEC = \angle ADC = 90°$

$\therefore \quad \Delta BEC \sim \Delta ADC$

$\Rightarrow \quad \frac{BE}{AD} = \frac{CE}{CD} = \frac{BC}{AC} \Rightarrow \frac{CE}{CD} = \frac{BC}{AC}$

$\Rightarrow \quad CE.AC = BC.CD$

$\Rightarrow \quad \frac{CE}{BC} = \frac{CD}{AC}$

40. **Given :** A trapezium ABCD, $AB \parallel CD$ and $AB = 2$ CD

O is the point of intersection of AC and DB.

To find: The ratio of areas of ΔAOB and ΔCOD.

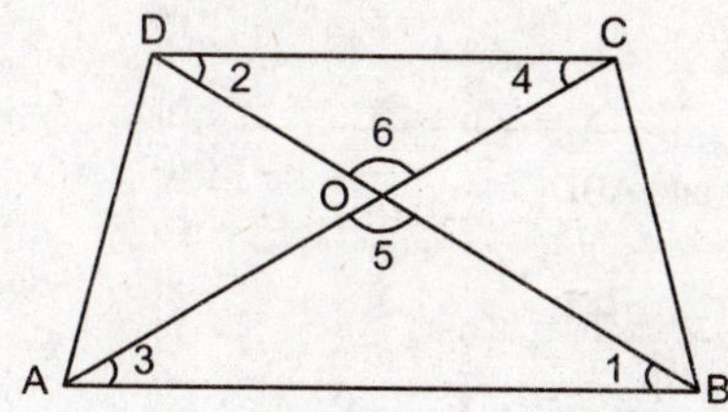

Proof : In ΔAOB and ΔCOD, $AB \parallel CD$ given

$\Rightarrow \quad \angle 3 = \angle 4$ (Alternate angles)

$\angle 1 = \angle 2$ (Alternate angles)

$\Rightarrow \quad \Delta AOB \sim \Delta COD$ (AA similarity)

$\Rightarrow \quad \frac{ar(\Delta AOB)}{ar(\Delta COD)} = \frac{AB^2}{CD^2} = \frac{(2CD)^2}{(CD)^2} = \frac{4CD^2}{CD^2} = 4:1$

41. **Given :** ΔABC in which $AB = AC$

$\angle A$ is an acute angles $BD \perp AC$

To prove : $(BC)^2 = 2AC.\,CD$

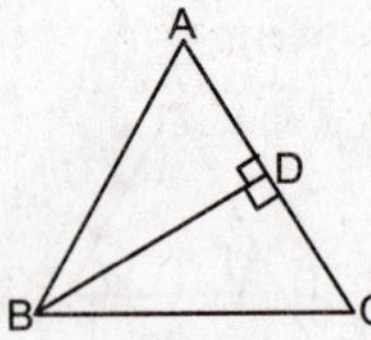

Proof : In ΔBDC,

$BC^2 = BD^2 + CD^2$ (By Pythagoras theorem)

$BC^2 = (AB^2 - AD^2) + CD^2$

In right ΔABD,

$AB^2 = AD^2 + BD^2$ (By Pythagoras theorem)

$BD^2 = AB^2 - AD^2$

$BC^2 = AB^2 + CD^2 - AD^2$

$= AB^2 + (CD - AD)(CD + AD)$

$= AC^2 + (CD - AD)(AC)$

$= AC\,(AC + CD - AD)$

$= AC\,(AC - AD + CD) = AC\,(CD + CD)$

$= AC\,.\,2CD = 2CD\,.\,AC.$ Proved.

42. **Given :** ΔABC in which $\angle B = 90°$ and $BD \perp AC$

To prove : $BD^2 = AD.CD$

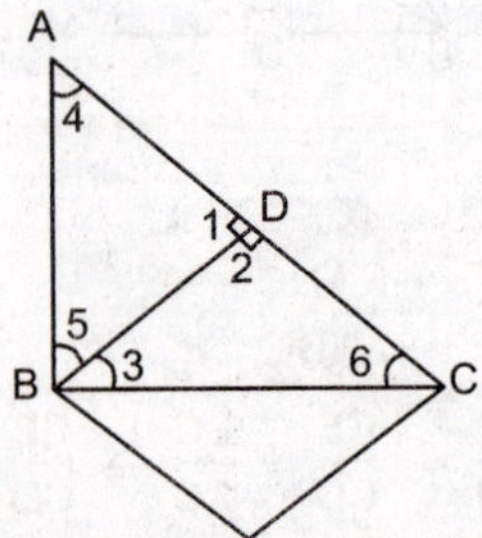

Proof : In ΔABC and ΔBDC

$\angle B = \angle D = 90°$

$\angle C = \angle C$ (Common)

$\Rightarrow \angle 3 = \angle A = \angle 4$ (ASP) ...(*i*)

In ΔABC and ΔADB

$\angle B = \angle D = 90°$

$\angle A = \angle A$ (Common)

$\Rightarrow \angle 5 = \angle 6 = \angle C$ ...(*ii*)

In ΔADB and ΔBDC

$\angle 1 = \angle 2 = 90°$

$\angle 5 = \angle 6$ [From (*ii*)]

$\angle 3 = \angle 4$ [From (*i*)]

Using AA similarity

$\Delta BDC \sim \Delta ADB$

$\Rightarrow \frac{BD}{CD} = \frac{AD}{BD}$

$\Rightarrow BD^2 = AD.CD$ Hence proved.

43.

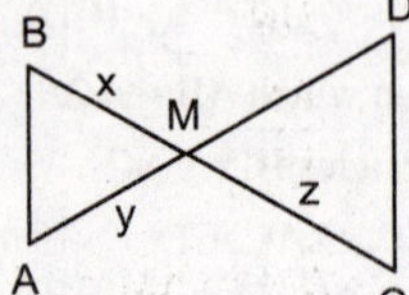

Given : $\Delta AMB \sim \Delta CMD$

To determine : MD in terms of x, y and z.

Proof : $\Delta AMB \sim \Delta CMD$ given

$\Rightarrow \frac{BM}{AM} = \frac{MD}{CM}$

$\frac{BM}{MD} = \frac{AM}{CM}$

(Corresponding sides of similar triangles are proportional)

$\Rightarrow \frac{x}{y} = \frac{MD}{z}$

$\Rightarrow \frac{xz}{y} = MD$

44.

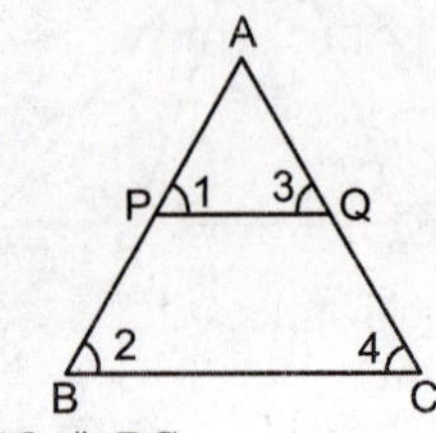

$\because PQ \parallel BC$

$\Rightarrow \angle 1 = \angle 2$

$\angle 3 = \angle 4$ (Corresponding angles)

$\therefore \Delta APQ \sim \Delta ABC$

$\Rightarrow \frac{AP}{AB} = \frac{PQ}{BC}$

$\Rightarrow \frac{4}{4+6} = \frac{3}{BC}$

$\Rightarrow BC = \frac{10\times 3}{4} = \frac{15}{2}$ cm = 7.5 cm

45. **Given :** $\angle 1 = \angle 2$ and $\angle 3 = \angle 4$.

To prove : PT.QR = PR.ST

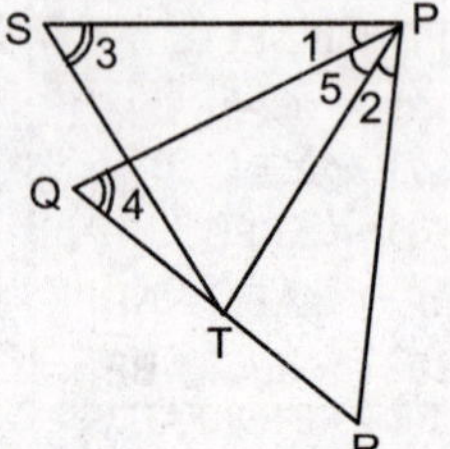

Proof : $\angle 1 = \angle 2$ (Given) ...(*i*)

$\Rightarrow \angle 1 + \angle 5 = \angle 2 + \angle 5$ [Adding $\angle 5$ to both sides of (*i*)]

$\Rightarrow \angle SPT = \angle QPR$

$\angle 3 = \angle 4$ (Given)

$\Rightarrow \Delta PST \sim \Delta PQR$ (AA similarity)

$\Rightarrow \frac{ST}{QR} = \frac{PT}{PR}$

$\Rightarrow \frac{ST}{PT} = \frac{QR}{PR}$

$\Rightarrow PR.ST = QR.PT$ Hence proved.

46. **Given :** In figure $LM \parallel NQ$ and $LN \parallel PQ$

$MP = \frac{1}{3} MN$

To find : $\frac{ar\,(\Delta LMN)}{ar\,(\Delta QNP)}$

Proof : $LM \parallel NQ$

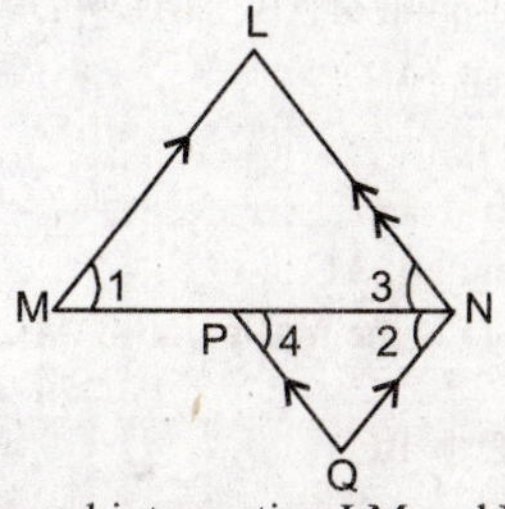

MN is the transversal intersecting LM and NQ

$\Rightarrow \quad \angle 1 = \angle 2$ (Alternate angles) ...(*i*)

LN ∥ PQ

MN is the transversal intersecting PQ and LN

$\Rightarrow \quad \angle 3 = \angle 4$ (Alternate angles) ...(*ii*)

In ΔLMN and ΔQNP

$\angle 1 = \angle 2$ [From (*i*)]

$\angle 3 = \angle 4$ [From (*ii*)]

$\Rightarrow \quad \Delta MNL \sim \Delta QNP$ (AA similarity)

$MP = \frac{1}{3} MN$ (Given)

$\Rightarrow \quad NP = MN - MP$

$\Rightarrow \quad NP = MN - \frac{1}{3} MN$

$\Rightarrow \quad \frac{3MN - MN}{3} = \frac{2\,MN}{3}$

$$\frac{ar\,(\Delta LMN)}{ar\,(\Delta QNP)} = \frac{MN^2}{NP^2} = \frac{(MN)^2}{\left(2\frac{MN}{3}\right)^2}$$

ar ΔLMN : ar ΔQNP = 9 : 4

47. Let AB = x, BC = x

Area of equilateral triangle of side $a = \frac{\sqrt{3}}{4} a^2$

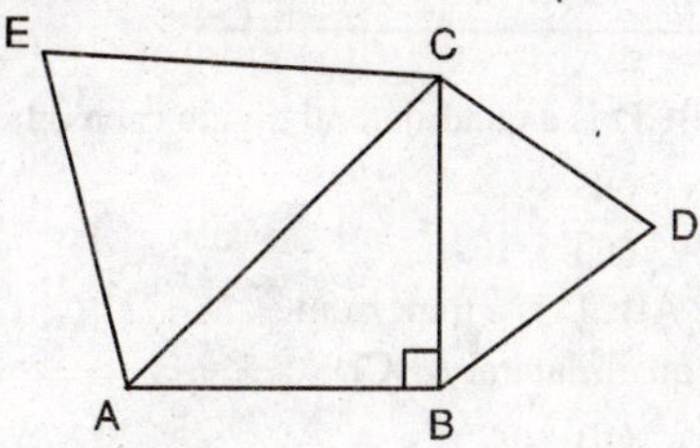

Area of $\Delta BCD = \frac{\sqrt{3}}{4} x^2$ ...(*i*)

Now side $AC = \sqrt{x^2 + x^2} = \sqrt{2x^2} = x\sqrt{2}$

(By Pythagoras theorem)

Area of $\Delta ACE = \frac{\sqrt{3}}{4}(\sqrt{2}x)^2 = 2.\frac{\sqrt{3}}{4}x^2$

$= 2$. Area of ΔBCD [From (*i*)]

$\therefore$ ar $\Delta BCD = \frac{1}{2}$ Area of ΔACE. Hence proved

48. Given : L and M are the mid-points of AB and BC and $\angle B = 90°$

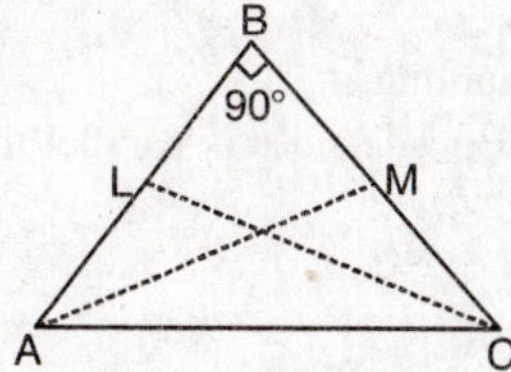

To prove : $4LC^2 = AB^2 + 4BC^2$

Construction : Join AM and LC.

Proof : In rt. ΔCLB,

$LC^2 = LB^2 + BC^2$ (By Pythagoras theorem)

$LC^2 = \left(\frac{1}{2}AB\right)^2 + BC^2$ (L is the mid-point of AB)

$LC^2 = \frac{1}{4}AB^2 + BC^2$

$\therefore \quad 4LC^2 = AB^2 + 4BC^2$ Hence proved.

49. Let the median of the other triangle be x

A.T.Q.

$$\frac{121}{64} = \frac{(12.1)^2}{x^2}$$

(Ratio of the areas of two similar triangles is equal to the ratio of the squares of their corresponding medians)

$\Rightarrow \quad x^2 = \frac{(12.1)^2 \times 64}{121}$

$\Rightarrow \quad x = \sqrt{\frac{(12.1)^2 \times 64}{121}}$

$= \frac{12.1 \times 8}{11} = 1.1 \times 8 = 8.8$ cm

$\therefore$ Median of the other triangle = 8.8 cm.

50.

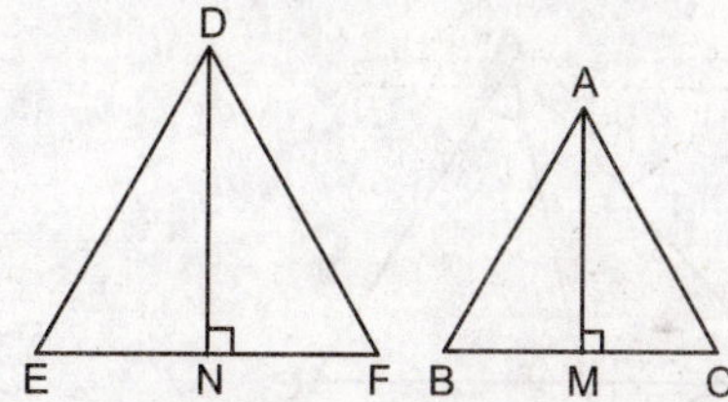

Let Area of ΔDEF = 100 cm^2

and area of ΔABC = 49 cm^2

and DN = 5 cm

$\because \quad \Delta ABC \sim \Delta DEF$

$\therefore \quad \frac{\text{Area } \Delta ABC}{\text{Area } \Delta DEF} = \frac{BC^2}{EF^2}$

$\Rightarrow \quad \frac{49}{100} = \frac{BC^2}{EF^2} \Rightarrow \frac{BC}{EF} = \frac{7}{10}$

Also $\frac{\text{Area } \Delta ABC}{\text{Area } \Delta DEF} = \frac{\frac{1}{2}BC \times AM}{\frac{1}{2}EF \times DN}$

$\frac{49}{100} = \frac{7}{10} \times \frac{AM}{DN}$

$\frac{7}{10} = \frac{AM}{DN}$

$\Rightarrow \quad \frac{7}{10} = \frac{AM}{5}$

$\Rightarrow \quad \frac{7 \times 5}{10} = AM \Rightarrow AM = 3.5$ cm

51. Given : A ΔABC, O is any point inside it. AO, OB, OC are joined. D is a point on OA. DE $\parallel$ AB meeting OB at E and EF $\parallel$ BC meeting OC at F.

To Prove : DF $\parallel$ AC.

Proof : In ΔOAB, DE $\parallel$ AB

$\Rightarrow \dfrac{OD}{AD} = \dfrac{OE}{EB}$ [Basic proportionality theorem] ...(*i*)

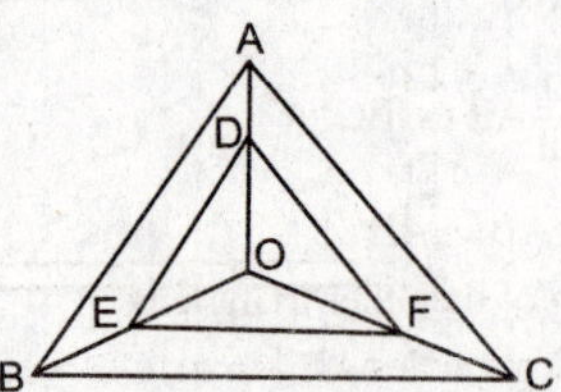

Again in ΔOBC, EF $\parallel$ BC

$\Rightarrow \dfrac{OE}{EB} = \dfrac{OF}{FC}$ [Basic proportionality theorem] ...(*ii*)

From (*i*) and (*ii*), we get

$$\frac{OD}{AD} = \frac{OF}{FC}$$

As in ΔOAC, $\dfrac{OD}{AD} = \dfrac{OF}{FC} \Rightarrow$ DF $\parallel$ AC. Hence proved

[$\because$ In a triangle if a line divides the two sides proportionally then it is parallel to the third side.]

52. Given : In ΔABC, D is a point on BC such that

$\angle BAC = \angle ADC$

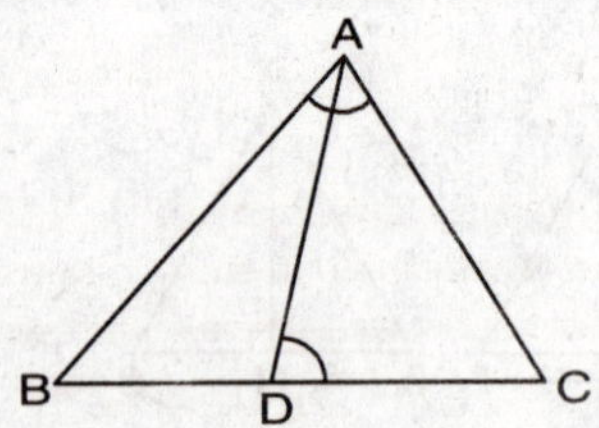

To prove : $\dfrac{CA}{CD} = \dfrac{CB}{CA}$

Proof : In ΔBAC and ΔADC,

$\angle BAC = \angle ADC$ (Given)

$\angle ACB = \angle ACD$ (Common)

$\therefore$ $\Delta BAC \sim \Delta ADC$ (AA similarity)

$\Rightarrow \dfrac{CA}{CD} = \dfrac{CB}{CA}$ Hence proved.

53. Given : Let D be the mid-point of the side AB and E be a point on the side AC in such a way that DE $\parallel$ BC.

To Prove : E be the mid-point of the side AC.

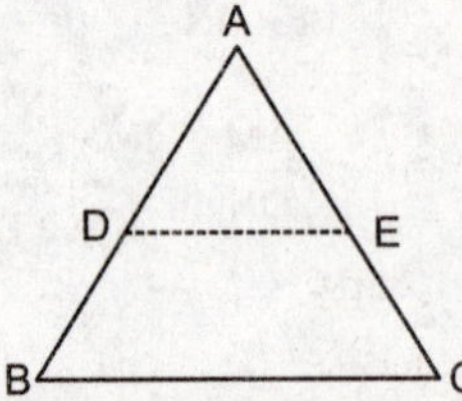

Proof : We know from the basic proportionality theorem that when a line drawn $\parallel$ to any one side of a triangle, it divides the other two sides in the same ratio.

$\therefore \dfrac{AD}{DB} = \dfrac{AE}{EC}$ ... (*i*)

But D be the mid-point of AB $\therefore$ AD = DB (Given)

Now, from (*i*), $\dfrac{AE}{EC} = 1$ $\left[\because \dfrac{AD}{DB} = 1\right]$

$\Rightarrow$ AE = EC

$\therefore$ E is the mid-point of AC. Proved

54. Given : ΔABC, D is the mid-point of AB, E is the mid-point of AC

To prove : DE $\parallel$ BC

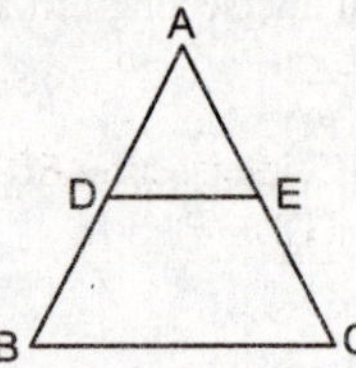

Proof : D is the mid-point of AB (Given)

$\Rightarrow \dfrac{AD}{DB} = 1$ ...(*i*)

E is the mid-point of AC (Given)

$\Rightarrow \dfrac{AE}{EC} = 1$ ...(*ii*)

From (*i*) and (*ii*)

$$\frac{AD}{DB} = \frac{AE}{EC}$$

$\Rightarrow$ DE $\parallel$ BC (By converse of BPT)

Hence proved.

55.

Given : ABCD is a quadrilateral divide each other proportionally

i.e., $\dfrac{AO}{OC} = \dfrac{OB}{OD}$

To prove : ABCD is a trapezium.

Proof : In quadrilateral ABCD,

$\dfrac{AO}{OC} = \dfrac{OB}{OD}$ (Given)

or $\dfrac{AO}{OB} = \dfrac{OC}{OD}$

In ΔAOB and ΔDOC,

$\angle 1 = \angle 2$ (Vertically opposite angles)

Using SAS similarity rule

$$\frac{AO}{OB} = \frac{OC}{OD}$$

$\Delta AOB \sim \Delta COD$

$\Rightarrow$ $\angle 3 = \angle 4$ (Corresponding angles of similar triangles)

But these angles are alt. int. $\angle$'s also

$\Rightarrow$ AB $\parallel$ DC

$\Rightarrow$ ABCD is a trapezium

[If one pair of opposite sides is parallel then quadrilateral is a trapezium]

Hence proved.

56.

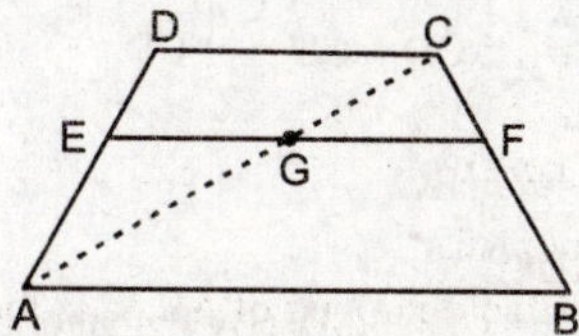

Given : A trapezium ABCD in which

AB || CD and EF || AB || CD

To prove : $\frac{AE}{DE} = \frac{BF}{CF}$

Construction : Join AC, intersecting EF at G

Proof : In ΔACD, EG || CD

$\therefore \quad \frac{AE}{DE} = \frac{AG}{GC}$ ...(*i*)

(Using B.P.T.)

In ΔABC, GF || AB

$\therefore \quad \frac{AG}{GC} = \frac{BF}{FC}$ ...(*ii*)

From (*i*) and (*ii*)

$\frac{AE}{DE} = \frac{BF}{FC}$

Hence proved

57. (*i*) **Given :** PM = 4, MQ = 4.5, PN = 4, NR = 4.5

Now, $\frac{PM}{MQ} = \frac{4}{4.5} = \frac{PN}{NR}$

$\therefore$ MN || QR (By BPT)

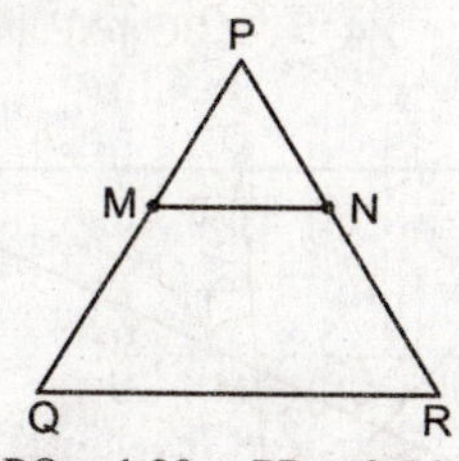

(*ii*) PQ = 1.28, PR = 2.56

PM = 0.16, PN = 0.32

Now, $\frac{PM}{PQ} = \frac{0.16}{1.28} = \frac{16}{128} = \frac{1}{8}$

$\frac{PN}{PR} = \frac{0.32}{2.56} = \frac{32}{256} = \frac{1}{8}$

$\therefore \quad \frac{PM}{PQ} = \frac{PN}{PR}$

$\therefore$ MN || QR

58.

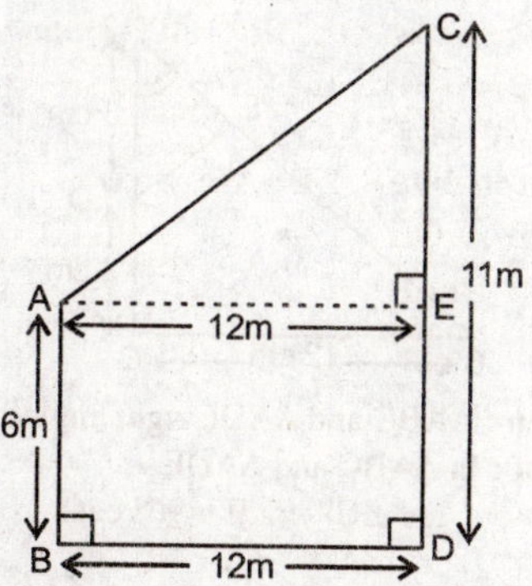

Let AB and CD are poles AB = 6 m, CD = 11 m

DEAB is parallelogram.

$\therefore$ DE = AB = 6 m

Now CE = CD – DE = 11 – 6 = 5

In right ΔAEC

$AC^2 = AE^2 + CE^2 \Rightarrow AC^2 = 12^2 + 5^2 \Rightarrow AC = 13\text{m}$

$\therefore$ Distance between their top = 13 m.

59. Given : Two similar triangles which are equal in area.

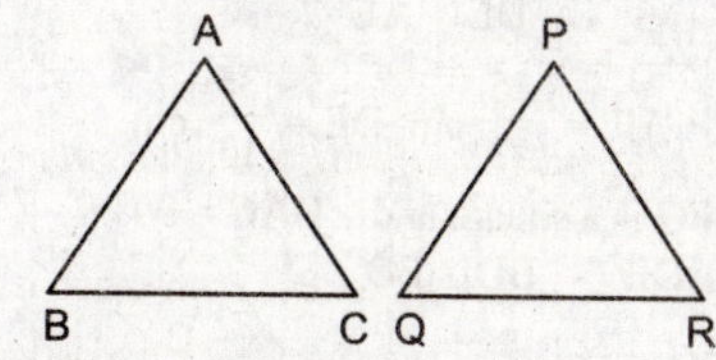

$\therefore$ ar (ΔABC) = ar (ΔPQR)

i.e., $\frac{ar(\Delta ABC)}{ar(\Delta PQR)} = 1$ ... (*i*)

To Prove : The triangles are congruent.

Proof : We know that the ratio of areas of two similar triangles is proportional to the ratio of the squares of their corresponding sides.

i.e., $\frac{ar(\Delta ABC)}{ar(\Delta PQR)} = \frac{AB^2}{PQ^2} = \frac{BC^2}{QR^2} = \frac{AC^2}{PR^2}$ ...(*ii*)

Put $\frac{ar(\Delta ABC)}{ar(\Delta PQR)} = 1$ in (*ii*), we get

$\frac{AB^2}{PQ^2} = \frac{BC^2}{QR^2} = \frac{AC^2}{PR^2} = 1 \Rightarrow \frac{AB^2}{PQ^2} = 1,$

$\therefore$ AB = PQ

Similarly, BC = QR and CA = PR

Since the corresponding sides are equal,

$\therefore$ ΔABC ≅ ΔPQR (By SSS congruency)

Hence Proved.

SHORT ANSWER TYPE QUESTIONS-II

1. Given : DB ⊥ BC, AC ⊥ BC and DE ⊥ AB.

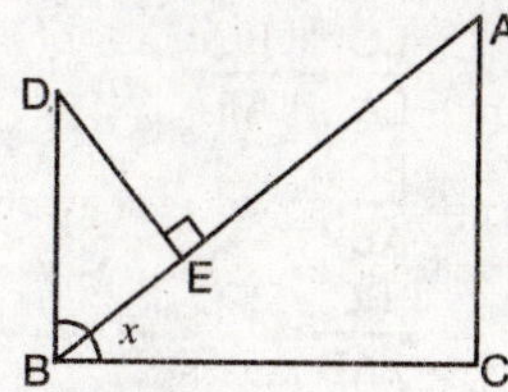

To Prove : $\frac{BE}{DE} = \frac{AC}{BC}$

Proof : ∠DEB = ∠ACB [Each 90°] ...(*i*)

Let ∠ABC = x

$\therefore$ ∠DBE = 90° – x

$\Rightarrow$ ∠BDE = x [∵ ∠DBE + ∠BDE = 90°]

$\therefore$ ∠ABC = ∠BDE ... (*ii*)

From (*i*) and (*ii*), we get

ΔABC ~ ΔBDE [By AA Similarity]

$\therefore \quad \frac{BE}{DE} = \frac{AC}{BC}$

2.

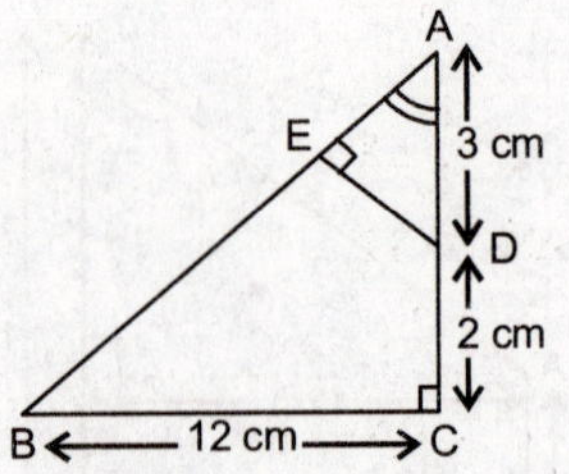

Given : ΔABC and ΔADE right angled at C and E.

Proof : In ΔABC and ΔADE

$\angle C = \angle E = 90°$ each

$\angle A = \angle A$ (Common angle)

$\Delta ABC \sim \Delta ADE$ (By AA similarity)

Since, $\Delta ABC \sim \Delta ADE$

In ΔABC, $AB^2 = AC^2 + BC^2$ (By Pythagoras theorem)

$AB^2 = 25 + 144 = 169$

$\Rightarrow AB = 13$

then, $\frac{AB}{AD} = \frac{BC}{DE} = \frac{AC}{AE}$

$\frac{13}{3} = \frac{12}{DE} = \frac{5}{AE}$

then, $AE = \frac{15}{13}$ cm, $DE = \frac{36}{13}$ cm

3. **Given :** DEFG is a square and $\angle BAC = 90°$.

To prove : $DE^2 = BD \times ED$

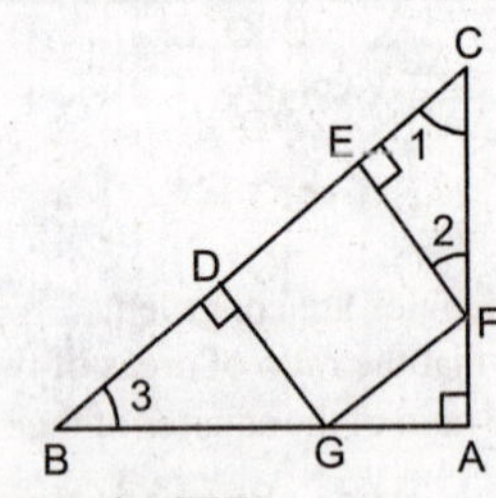

Proof : In ΔBDG and ΔCEF

$\angle 1 + \angle 3 = 90°$ (Because $\angle A = 90°$)

$\angle 1 + \angle 2 = 90°$ (Because $\angle CEF = 90°$)

then, $\angle 2 = \angle 3$ (On equating above equation)

and $\angle CEF = \angle BDG$ (90° each)

Hence, $\Delta BDG \sim \Delta FEC$ (AA Similarity)

$\frac{BD}{FE} = \frac{DG}{EC} = \frac{BG}{FC}$

Now DEFG is a square

$\therefore DE = EF = FG = DG$

then, $\frac{BD}{DE} = \frac{DE}{EC}$ (From first two ratios)

$DE^2 = BD . EC.$ Hence proved.

4.

Given : In ΔABC, $AD \perp BC$ and $BD = \frac{1}{3} CD$

To prove : $2CA^2 = 2AB^2 + BC^2$

Proof : $\because$ $BC = BD + CD$

and $BD = \frac{1}{3} CD$ (Given)

$\therefore BC = \frac{1}{3} CD + CD = \frac{4}{3} CD$

$\Rightarrow CD = \frac{3}{4} BC$ ...(*i*)

In right angled ΔADC,

$AC^2 = CD^2 + AD^2$ (By Pythagoras theorem) ...(*ii*)

In right ΔADB,

$AB^2 = AD^2 + BD^2$

$\Rightarrow AD^2 = AB^2 - BD^2$

Substituting in (*ii*), we get

$\Rightarrow AC^2 = CD^2 + AB^2 - BD^2$

$\Rightarrow AC^2 = CD^2 + AB^2 - \left(\frac{1}{3}CD\right)^2$ (Put $BD = \frac{1}{3}CD$)

$\Rightarrow AC^2 = CD^2 - \frac{1}{9} CD^2 + AB^2$

$\Rightarrow AC^2 = \frac{8}{9} CD^2 + AB^2$

$\Rightarrow AC^2 = \frac{8}{9}\left(\frac{3}{4}BC\right)^2 + AB^2$ [Using (*ii*)]

$\Rightarrow AC^2 = \frac{8}{9} \times \frac{9}{16} BC^2 + AB^2$

$\Rightarrow AC^2 = \frac{1}{2} BC^2 + AB^2$

$\Rightarrow 2AC^2 = BC^2 + 2AB^2$ Hence proved.

5. **Proof :** In ΔMDE and ΔMCB

$DM = CM$ (Given)

$\angle 1 = \angle 2$ (Vertically opposite)

$\angle 3 = \angle 4$ ($\because$ BC || AE and DC is a transversal) (Alt. int $\angle s$)

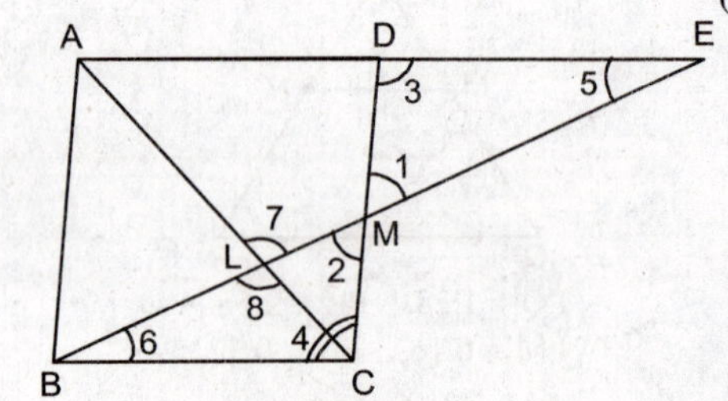

$\therefore \Delta MDE \cong \Delta MCB$ (ASA Congruency)

$DE = BC$ (CPCT) ... (*i*)

Also $BC = AD$ ... (*ii*)

(Opposite sides of the parallelogram)

$\therefore AD = DE$ [On equating (*i*) and (*ii*)]

Now, $AE = AD + DE$

$\Rightarrow AE = 2 AD$ (Put DE = AD)

In ΔBLC and ΔELA,

$\angle 5 = \angle 6$ (Alt. int. angles)

and $\angle 7 = \angle 8$ (Vertically opposite angles)

$\therefore \Delta BLC \sim \Delta ELA$ (AA similarity)

$\Rightarrow \frac{BL}{EL} = \frac{LC}{LA} = \frac{BC}{AE}$

$\Rightarrow \frac{BL}{EL} = \frac{BC}{AE}$

$\Rightarrow \frac{BL}{EL} = \frac{BC}{2AD}$

$\Rightarrow \frac{BL}{EL} = \frac{AD}{2AD}$ ($\because$ BC = AD)

$\Rightarrow \frac{BL}{EL} = \frac{1}{2} \Rightarrow EL = 2\,BL$

6. Given : In ΔABC, $PQ \parallel AB$ and $PR \parallel BD$

To prove : $QR \parallel AD$

Proof : By BPT

$$\frac{CP}{BP} = \frac{CQ}{AQ} \qquad \ldots(i)$$

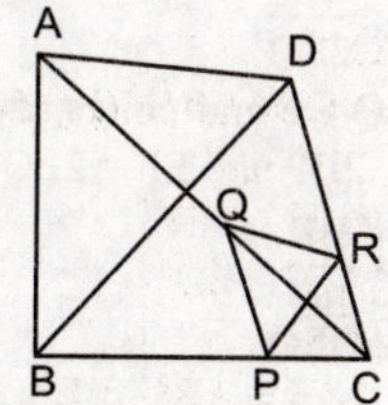

Now in ΔBCD, $PR \parallel BD$

$\Rightarrow$ By using BPT

$$\frac{CP}{BP} = \frac{CR}{RD} \qquad \ldots(ii)$$

From (*i*) and (*ii*),

$$\frac{CQ}{AQ} = \frac{CR}{RD}$$

$\Rightarrow$ By converse of BPT, $QR \parallel AD$

7. In ΔADB and ΔEFC,

$\angle D = \angle F$ (Each 90°)

and $\angle B = \angle C$ (Angles opp. to equal sides of a triangle are equal)

$\Rightarrow$ $\Delta ABD \sim \Delta ECF$ (AA similarity)

$\therefore$ $\frac{AB}{EC} = \frac{AD}{EF}$ (Corrs. sides of similar Δs)

$\therefore$ $AB \times EF = AD \times EC$ Hence proved.

8.

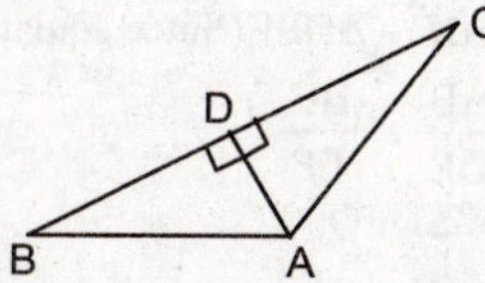

In right ΔADC,

$AC^2 = AD^2 + CD^2$

$\Rightarrow$ $AC^2 - CD^2 = AD^2$ ... (*i*)

In right ΔADB,

$\Rightarrow$ $AB^2 = AD^2 + BD^2$

$\Rightarrow$ $AB^2 - BD^2 = AD^2$ ... (*ii*)

From (*i*) and (*ii*)

$AC^2 - CD^2 = AB^2 - BD^2$

$\Rightarrow$ $AC^2 + BD^2 = AB^2 + CD^2$ Hence proved.

9.

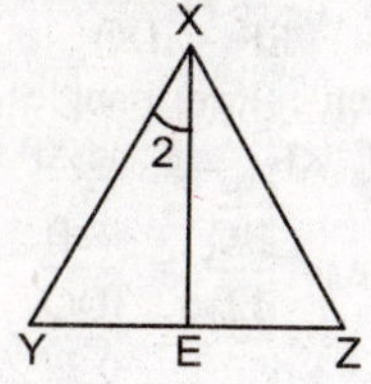

AD bisect $\angle A$ $\therefore$ $\angle 1 = \frac{1}{2}\angle A$

Similarly $\angle 2 = \frac{1}{2}\angle X$

$\because$ $\Delta ABD \sim \Delta XYZ$

$\therefore$ $\angle A = \angle X$

$\Rightarrow$ $\frac{1}{2}\angle A = \frac{1}{2}\angle X \Rightarrow \angle 1 = \angle 2$

Also $\angle B = \angle Y$

($\because$ $\Delta ABC \sim \Delta XYZ$, $\therefore$ $\Delta ABD \sim \Delta XYE$)

$$\frac{\text{Area } \Delta ABD}{\text{Area } \Delta XYE} = \frac{AD^2}{XE^2} = \frac{4^2}{3^2} = \frac{16}{9}$$

10.

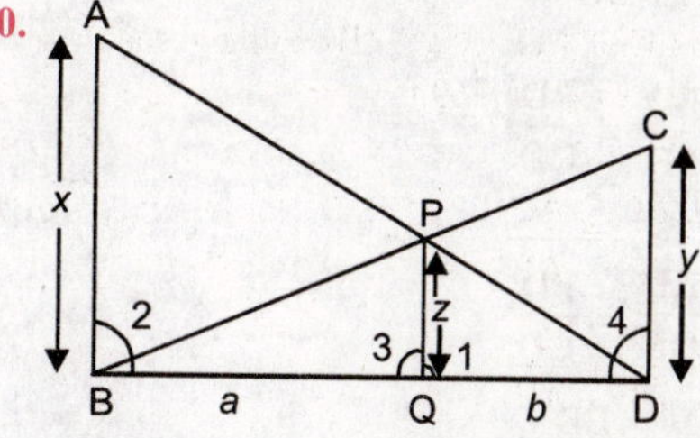

Let $BQ = a$ units, $DQ = b$ units

$\because$ $PQ \parallel AB$ $\therefore$ $\angle 1 = \angle 2$,

and $\angle ADB = \angle PDQ$

$\therefore$ $\Delta ADB \sim \Delta PDQ$

Similarly $\Delta BCD \sim \Delta BPQ$

$\because$ $\Delta ADB \sim \Delta PDQ$

$\therefore$ $\frac{AB}{PQ} = \frac{BD}{DQ}$

$$\frac{x}{z} = \frac{a+b}{b}$$

$$\frac{x}{z} = \frac{a}{b} + 1 \Rightarrow \frac{x}{z} - 1 = \frac{a}{b} \qquad \ldots(i)$$

Also $\Delta BCD \sim \Delta BPQ$

$\therefore$ $\frac{BD}{BQ} = \frac{CD}{PQ}$

$\Rightarrow$ $\frac{a+b}{a} = \frac{y}{z}$

$$1 + \frac{b}{a} = \frac{y}{z} \Rightarrow \frac{b}{a} = \frac{y}{z} - 1$$

$\Rightarrow$ $\frac{b}{a} = \frac{y-z}{z} \Rightarrow \frac{a}{b} = \frac{z}{y-z}$ ...(*ii*)

From (*i*) and (*ii*)

$$\frac{x}{z} - 1 = \frac{z}{y-z} \Rightarrow \frac{x}{z} = \frac{z}{y-z} + 1$$

$$\frac{x}{z} = \frac{z+y-z}{y-z}$$

$$\frac{x}{z} = \frac{y}{y-z}$$

$\Rightarrow$ $\frac{z}{x} = \frac{y-z}{y}$

$$\frac{z}{x} = 1 - \frac{z}{y}$$

$$z\left(\frac{1}{x}\right) = z\left(\frac{1}{z} - \frac{1}{y}\right)$$

$\Rightarrow$ $$\frac{1}{x} = \frac{1}{z} - \frac{1}{y}$$

$\Rightarrow$ $$\frac{1}{x} + \frac{1}{y} = \frac{1}{z}$$ (Hence proved)

11.

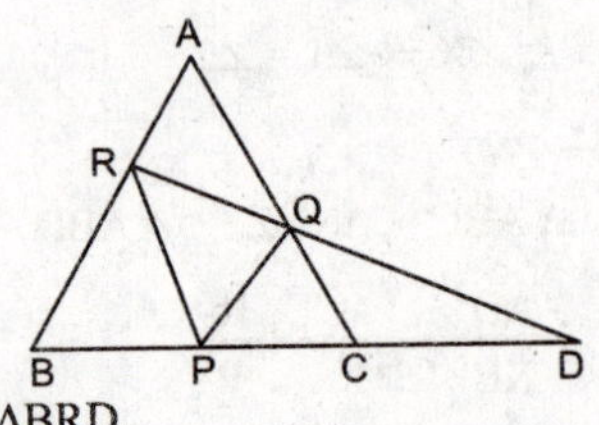

In ΔBRD,
BR || PQ (given)

$\therefore$ $$\frac{BD}{PD} = \frac{RD}{QD}$$ ...(*i*)

In ΔRDP, PR || QC (given)

$\therefore$ $$\frac{RD}{QD} = \frac{PD}{CD}$$ ...(*ii*)

From (*i*) and (*ii*) $\frac{PD}{CD} = \frac{BD}{PD}$

$\Rightarrow$ $BD \times CD = PD \times PD = 12 \times 12 = 144 \text{ cm}^2$

12.

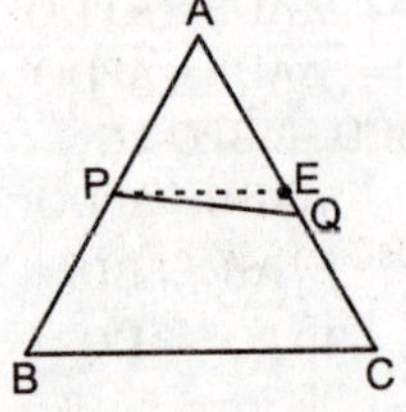

Given : In ΔABC

$$\frac{AP}{BP} = \frac{AQ}{QC}$$

To prove : PQ || BC
Construction : Let PE || BC
Proof : In ΔABC, PE || BC (Construction)

$\therefore$ $$\frac{AP}{PB} = \frac{AE}{EC}$$ ...(*i*)

Also $$\frac{AP}{BP} = \frac{AQ}{QC}$$ [(given (*ii*)]

From (*i*) and (*ii*) $\frac{AE}{EC} = \frac{AQ}{QC}$

$\Rightarrow$ E and Q coincides
$\therefore$ PQ || BC.

13. Given : ΔABC and ΔDEF

$\angle A = \angle D$
$\angle B = \angle E$
$\angle C = \angle F$

To prove : (*i*) $\frac{BC}{EF} = \frac{AP}{DQ}$ (AP and DQ are medians)

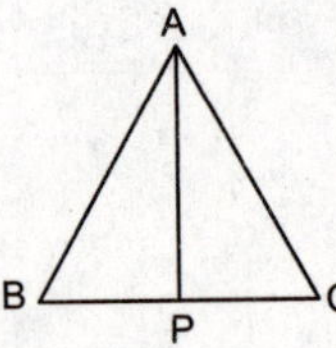

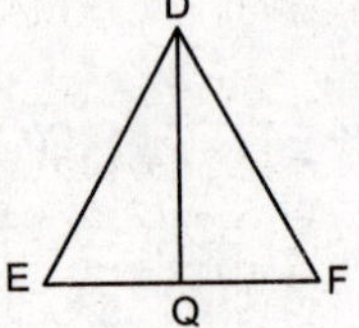

Proof : (*i*) Since, triangles which are equiangular are similar

$\therefore$ ΔABC ~ ΔDEF

$\Rightarrow$ $$\frac{AB}{DE} = \frac{BC}{EF}$$ ...(*i*)

$\Rightarrow$ $$\frac{AB}{DE} = \frac{2BP}{2EQ}$$

[$\because$ P and Q are mid-points of BC and EF respectively]

$\therefore$ BC = 2BP and EF = 2EQ

$\Rightarrow$ $$\frac{AB}{DE} = \frac{BP}{EQ}$$ ...(*ii*)

Now, in ΔABP and ΔDEQ,

$$\frac{AB}{DE} = \frac{BP}{EQ}$$ [From (*ii*)]

and $\angle B = \angle E$ (Given)

$\therefore$ ΔABP ~ ΔDEQ (By SAS similarity)

$\Rightarrow$ $$\frac{AB}{DE} = \frac{AP}{DQ}$$ ...(*iii*)

From (*i*) and (*iii*), we get

$$\frac{BC}{EF} = \frac{AP}{DQ}$$

Hence, the ratio of the corresponding sides is same as the ratio of corresponding medians.

(*ii*)

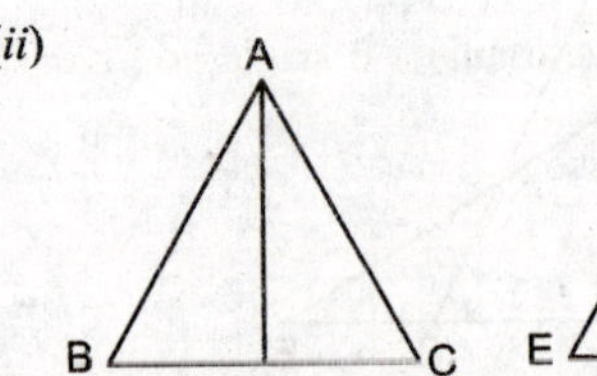

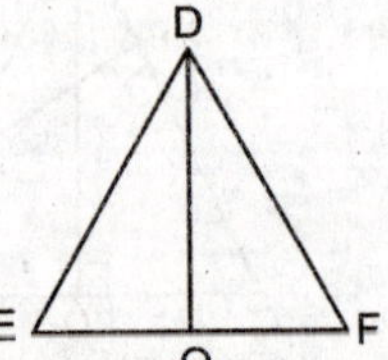

Given : Two triangles ΔABC and ΔDEF in which $\angle A = \angle D$, $\angle B = \angle E$, $\angle C = \angle F$ and AP, DQ are the bisectors of $\angle A$ and $\angle D$ respectively.

To prove : $\frac{BC}{EF} = \frac{AP}{DQ}$

Proof : ΔABC ~ ΔDEF (Since equiangular triangles are similar)

$\Rightarrow$ $$\frac{AB}{DE} = \frac{BC}{EF}$$ ...(*i*)

In ΔABP and ΔDEQ,

$\angle B = \angle E$ (Given)

and $\angle BAP = \angle EDQ$ ($\because \angle A = \angle D \Rightarrow \frac{1}{2}\angle A = \frac{1}{2}\angle D$)

$\Rightarrow$ $\angle BAP = \angle EDQ$

$\Rightarrow$ ΔABP ~ ΔEDQ (By AA corollary)

$\Rightarrow$ $$\frac{AB}{DE} = \frac{AP}{DQ}$$ ...(*ii*)

From (*i*) and (*ii*), we get

$$\frac{BC}{EF} = \frac{AP}{DQ}$$

(*iii*) **Given :** Two triangles ABC and DEF in which $\angle A = \angle D$, $\angle B = \angle E$, $\angle C = \angle F$ and AP ⊥ BC, DQ ⊥ EF

To prove : $\frac{BC}{EF} = \frac{AP}{DQ}$

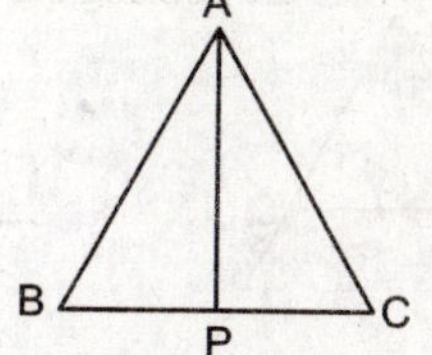

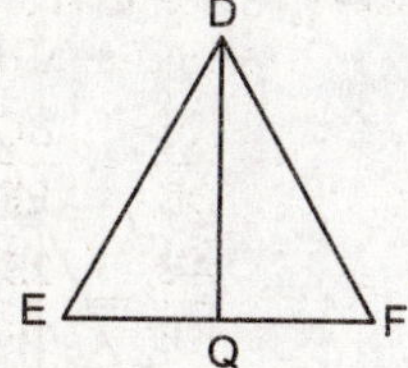

Proof : $\Delta ABC \sim \Delta DEF$

Since equiangular triangles are similar

$\Rightarrow \quad \frac{AB}{DE} = \frac{BC}{EF}$ ...(*i*)

In triangles APB and DQE,

$\angle APB = \angle DQE$ (Each equal to 90°)

$\angle B = \angle E$ (Given)

$\Delta APB \sim \Delta DQE$ (AA corollary)

$\Rightarrow \quad \frac{AB}{DE} = \frac{AP}{DQ}$ ...(*ii*)

From (*i*) and (*ii*),

We have $\frac{BC}{EF} = \frac{AP}{DQ}$

14. **Given :** In ΔABC D, E and F are the mid-points of the side of the ΔABC.

To prove : Each of the four triangles is similar to ΔABC.

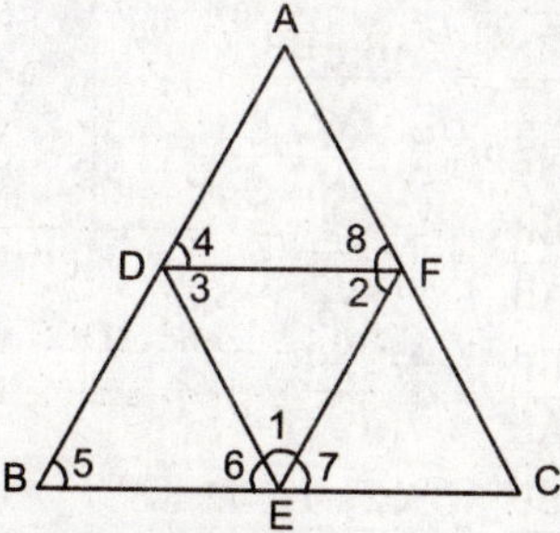

Proof : D is the mid-point of AB

E is the mid-point of BC

and F is the mid-point of AC

$\Rightarrow \quad DE \parallel AC$ [Mid-point theorem] ...(*i*)

$EF \parallel AB$ [Mid-point theorem] ...(*ii*)

$DF \parallel BC$ [Mid-point theorem]

From (*i*) and (*ii*),

□ADEF is a paralleogram.

BDFE and CFDE are also parallelogram

∵ ADEF is a parallelogram

$\Rightarrow \quad \angle A = \angle 1$

$\angle B = \angle 2$ and $\angle 3 = \angle C$ (Opp. angles of ||gm)

$\Rightarrow \quad \Delta ABC \sim \Delta EFD$ (AA similarity) ....(*i*)

Also, ∵ BDFE is a parallelogram

$\angle 5 = \angle 2$

$\angle 3 = \angle 6$

$\Rightarrow \quad \Delta BED \sim \Delta FDE$ ...(*ii*)

In parallelogram DECF,

$\angle 3 = \angle C$

$\angle 2 = \angle 7$

$\Rightarrow \quad \Delta FDE \sim \Delta ECF$ ...(*iii*)

and In parallelogram ADEF,

$\angle A = \angle 1$

$\angle 4 = \angle 2$

$\angle 3 = \angle 8$

$\Rightarrow \quad \Delta FDE \sim \Delta DFA$ ...(*iv*)

From (*i*), (*ii*), (*iii*) and (*iv*).

All triangles are similar to ΔABC.

15.

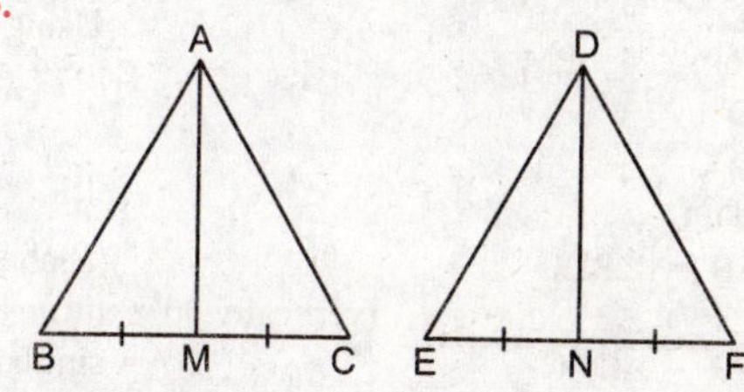

Given : In ΔABC and ΔDEF, M and N are mid points of BC and EF respectively

Also $\frac{AB}{DE} = \frac{BC}{EF} = \frac{AM}{DN}$

To prove : $\Delta ABC \sim \Delta DEF$

Proof : ∵ $\frac{AB}{DE} = \frac{BC}{EF} = \frac{AM}{DN}$

$\therefore \quad \frac{AB}{DE} = \frac{2BM}{2EN} = \frac{AM}{DN}$

$\Rightarrow \quad \frac{AB}{DE} = \frac{BM}{EN} = \frac{AM}{DN}$

$\Delta AMB \sim \Delta DNE$ (SSS similarity)

$\therefore \quad \angle B = \angle E$

Now, In ΔABC and ΔDEF

$\frac{AB}{DE} = \frac{BC}{CF}$ (given)

$\angle B = \angle E$ (Proved)

$\therefore \quad \Delta ABC \sim \Delta DEF$ (SAS similarity)

16. **Given :** In ΔABC and ΔPQR, $\angle A = \angle P$

$AB = AC$, $PQ = PR$ and $\frac{ar(\Delta ABC)}{ar(\Delta PQR)} = \frac{9}{16}$

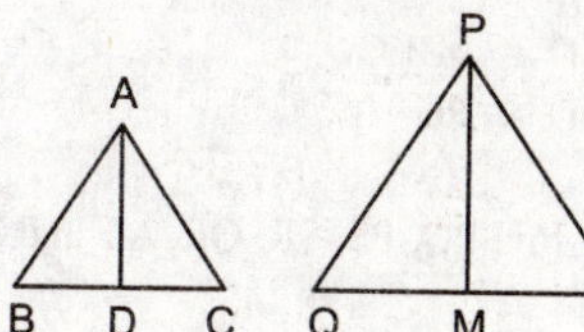

To Find : $\frac{AD}{PM}$

Proof : In ΔABC and PQR

$\frac{AB}{AC} = \frac{PQ}{PR}$

$\angle A = \angle P$

$\therefore \quad \Delta ABC \sim \Delta PQR$ (SAS similarity)

$\Rightarrow \quad \frac{ar(\Delta ABC)}{ar(\Delta PQR)} = \frac{BC^2}{QR^2}$

$\Rightarrow \quad \frac{BC^2}{QR^2} = \frac{9}{16}$

$\Rightarrow \quad \frac{BC}{QR} = \frac{3}{4}$ ...(*i*)

Also, $\frac{ar(\Delta ABC)}{ar(\Delta PQR)} = \frac{\frac{1}{2}BC \times AD}{\frac{1}{2}QR \times PM}$

$\Rightarrow \quad \frac{9}{16} = \frac{BC}{QR} \times \frac{AD}{PM}$

$\Rightarrow \quad \frac{9}{16} = \frac{3}{4} \times \frac{AD}{PM}$ [Using (*i*)]

$\Rightarrow \quad \frac{AD}{PM} = \frac{3}{4}$

17. In ΔEPB and ΔDPC

$\angle PEB = \angle PDC$ (each 90°)

$\angle 1 = \angle 2$ (vertically opposite angles)

$\therefore \quad \Delta EPB \sim \Delta DPC$ (AA similarity)

$\Rightarrow \quad \frac{EP}{DP} = \frac{PB}{PC} \Rightarrow EP.PC = PB.DP$

$\Rightarrow \quad EP.PC = BP.PD$ Hence proved

18.

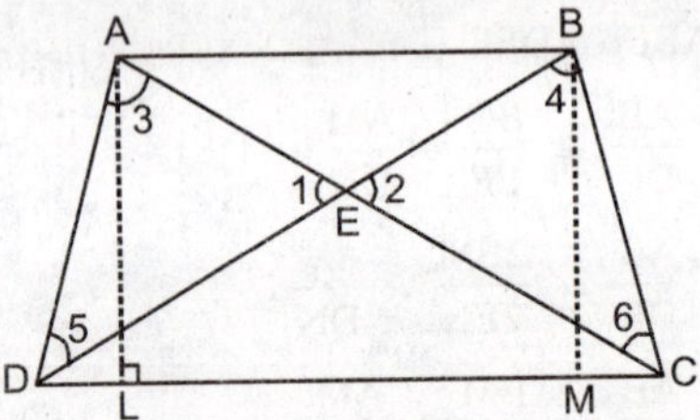

Given : A trapezium ABCD in which AB || DC

AC and BD intersect at E

$\Delta AED \sim \Delta BEC$

To prove : $AD = BC$

Proof : AB || DC (Given)

$\Rightarrow$ ar (ΔADC) = ar (ΔBCD)

(Triangles on the same base and between same parallel lines)

$\Rightarrow$ ar (ΔADC) – ar (ΔDEC) = ar (ΔBCD) – ar (ΔDEC)

$\Rightarrow$ ar (ΔAED) = ar (ΔBEC) ...(*i*)

Also, $(\Delta AED) \sim (\Delta BEC)$ ...(*ii*)

From (*i*) and (*ii*),

$\Delta AED \cong \Delta BEC$

$\Rightarrow \quad AD = BC$ (CPCT)

Hence proved.

19. Construction : Join PQ, PS, SR, QR, AC and BD.

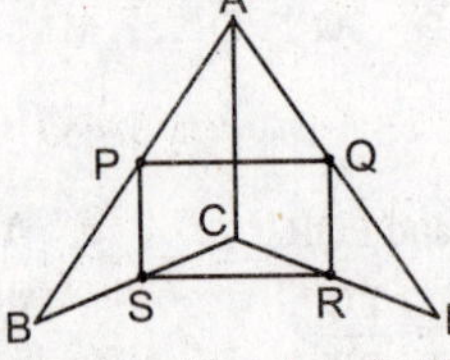

Proof : In ΔACB, P is mid-point of AB and S is mid-point of BC

$\therefore \quad PS \parallel AC$ ...(*i*)

Similarly, in ΔACD, QR || AC

From (*i*) and (*ii*), PS || QR

Similarly, PQ || SR

$\therefore$ PQRS is a parallelogram.

20. Given : ΔABC, in which PQ || BC

ar (ΔAPQ) = ar (quad. PQCB)

To find: BP : AB = ?

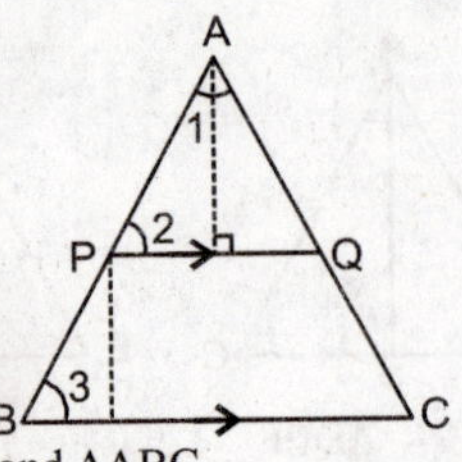

Proof : In ΔAPQ and ΔABC

$\angle 1 = \angle 1$ (Common angle)

$\angle 2 = \angle 3$ (Corresponding angles)

$\therefore \quad \Delta APQ \sim \Delta ABC$ (By AA corollary)

$\therefore \quad \frac{ar(\Delta APQ)}{ar(\Delta ABC)} = \frac{AP^2}{AB^2}$

(The areas of two similar triangles are in the ratio of the squares of their corresponding sides)

$\frac{ar(\Delta APQ)}{2 \times ar(\Delta APQ)} = \frac{AP^2}{AB^2}$

[ar (ΔABC) = 2 ar (ΔAPQ) $\because$ PQ divides ΔABC in two parts equal in area, given]

$\Rightarrow \quad \frac{1}{2} = \frac{AP^2}{AB^2} \Rightarrow \frac{1}{2} = \frac{AP^2}{AB^2}$

Taking square root, $\frac{1}{\sqrt{2}} = \frac{AP}{AB}$

$\Rightarrow \quad \frac{1}{\sqrt{2}} = \frac{AB - BP}{AB}$

$(\because AP = AB - BP)$

$\Rightarrow \quad \frac{BP}{AB} = 1 - \frac{1}{\sqrt{2}}$

$\Rightarrow \quad \frac{BP}{AB} = \frac{\sqrt{2}-1}{\sqrt{2}}$

$\Rightarrow \quad \frac{BP}{AB} = \frac{\sqrt{2}-1}{\sqrt{2}} \times \frac{\sqrt{2}}{\sqrt{2}}$

$\Rightarrow \quad \frac{BP}{AB} = \frac{2-\sqrt{2}}{2}$

$\therefore$ BP : AB = $2 - \sqrt{2} : 2$

21.

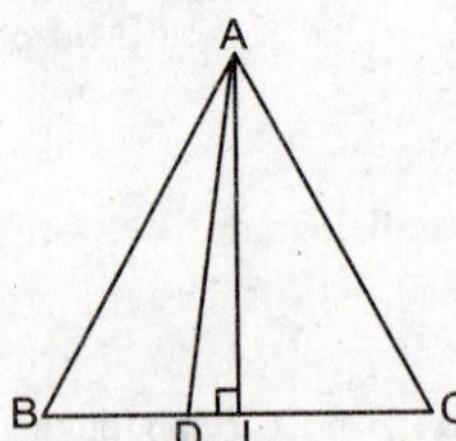

Given : A triangle ABC

In which AB = AC

D is any point on BC

To Prove : $(AB)^2 - (AD)^2 = BD.CD$

Construction : Draw $AL \perp BC$

Proof : In ΔABL,

$AB^2 = AL^2 + BL^2$ ...(*i*)

In ΔADL, $AD^2 = AL^2 + DL^2$ ...(*ii*)

Subtracting (*ii*) from (*i*),

$AB^2 - AD^2 = BL^2 - DL^2$

$$= (BL - DL)(BL + DL)$$
$$= BD (CL + DL)$$
$$= BD.\ CD$$

($\because$ AB = AC $\Rightarrow$ Altitude AL is the median also $\Rightarrow$ BL = CL)

22.

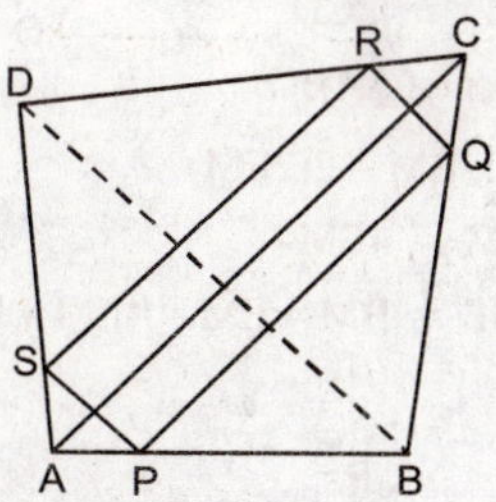

Join DB

$\because$ P is point of trisection of AB.

$\therefore \quad \dfrac{AP}{PB} = \dfrac{1}{2}$

Similarly $\quad \dfrac{CQ}{BQ} = \dfrac{1}{2}$

$\Rightarrow \quad \dfrac{AP}{PB} = \dfrac{CQ}{BQ}$

$\Rightarrow \quad PQ \parallel AC$ ...(*i*)

(Using converse of BPT in ΔABC)

Similarly SR $\parallel$ AC ...(*ii*)

From (*i*) and (*ii*)

$PQ \parallel SR$ ...(*iii*)

In ΔABD

$\dfrac{AP}{PB} = \dfrac{AS}{SD}$ ($\because$ P and S are points of trisection)

$\therefore \quad SP \parallel BD$ ...(*iv*)

Similarly RQ $\parallel$ BD ...(*v*)

From (*iv*) and (*v*)

$SP \parallel RQ$

Now in quadrilateral PQRS

PQ $\parallel$ RS, SP $\parallel$ RQ

$\therefore$ PQRS is a parallelogram.

23. Proof : In quadrilateral BXCO,

OD = DX (Given)

and BD = CD

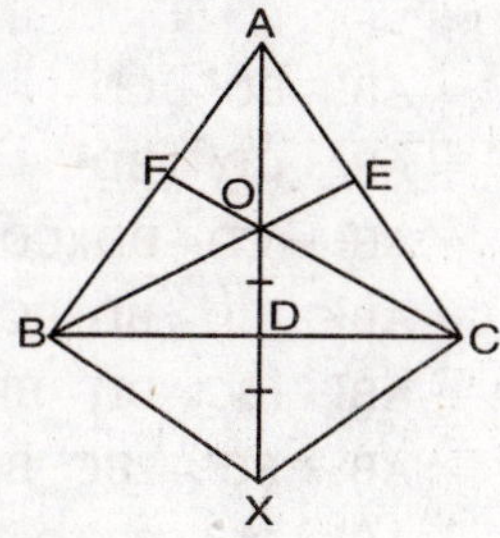

$\therefore$ BXCO is a $\parallel$gm

$\Rightarrow \quad BX \parallel CO$

$\Rightarrow \quad BX \parallel OF$

$\therefore \quad \dfrac{AO}{AX} = \dfrac{AF}{AB}$ Hence proved.

24.

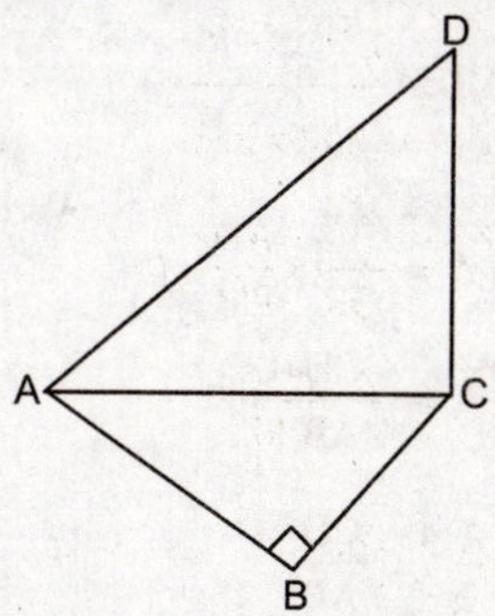

Proof : In right ΔABC

$AC^2 = AB^2 + BC^2$ ...(*i*)

Also

$AD^2 = AB^2 + BC^2 + CD^2$

$\Rightarrow \quad AD^2 = AC^2 + CD^2$ [Using (*i*)]

Now in ΔACD,

$AD^2 = AC^2 + CD^2$ (proved)

$\therefore \quad \angle ACD = 90°$ (Using converse of BPT)

25.

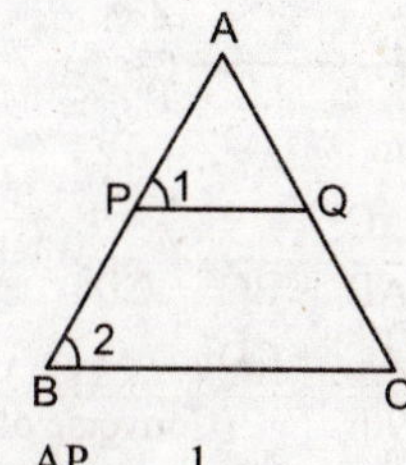

$\dfrac{AP}{PB} = \dfrac{1}{2}$

$\Rightarrow \quad \dfrac{PB}{AP} = \dfrac{2}{1}$

$\Rightarrow \quad \dfrac{PB}{AP} + 1 = \dfrac{2}{1} + 1$

$\dfrac{PB + AP}{AP} = \dfrac{2+1}{1} \Rightarrow \dfrac{AB}{AP} = \dfrac{3}{1}$

In ΔAPQ and ΔABC, $\angle A = \angle A$ (common)

$\angle 1 = \angle 2$ (corresponding angles)

$\therefore \quad \Delta APQ \sim \Delta ABC$

$\Rightarrow \quad \dfrac{\text{area}\,\Delta ABC}{\text{area}\,\Delta APQ} = \dfrac{AB^2}{AP^2} = \dfrac{9}{1}$

$\Rightarrow \quad \text{area}\,\Delta ABC = 9 \times \text{area}\,\Delta APQ$

Now $\dfrac{\text{Area}\,\Delta APQ}{\text{Area trapezium BCQP}} = \dfrac{\text{Area}\,\Delta APQ}{\text{Area}\,\Delta ABC - \text{Area}\,\Delta APQ}$

$\dfrac{\text{Area}\,\Delta APQ}{9 \times \text{Area}\,\Delta APQ - \text{Area}\,\Delta APQ} = \dfrac{\text{Area}\,\Delta APQ}{8 \cdot \text{Area}\,\Delta APQ} = \dfrac{1}{8}$

26.

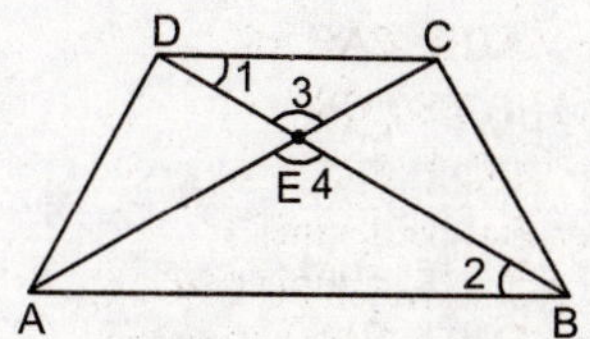

Given : In trapezium ABCD, AB $\parallel$ CD and DE : BE = 1 : 2

To prove : AB = 2 CD

In ΔDEC and ΔBEA

$\angle 1 = \angle 2$ (Corresponding angles)

$\angle 3 = \angle 4$ (Vertically opposite angles)

$\therefore$ $\Delta DEC \sim \Delta BEA$ (AA similarity)

$\Rightarrow$ $\frac{DE}{BE} = \frac{CD}{AB}$ $\left(\because \frac{DE}{BE} = \frac{1}{2}\right)$

$\Rightarrow$ $\frac{1}{2} = \frac{CD}{AB}$

$\Rightarrow$ AB = 2CD Hence proved

27. Given : ABCD is a quadrilateral, $\angle B = 90°$ and $AD^2 = AB^2 + BC^2 + CD^2$

To Prove : $\angle ACD = 90°$

Construction : Join A to C.

Proof : In ΔABC, $\angle B = 90°$

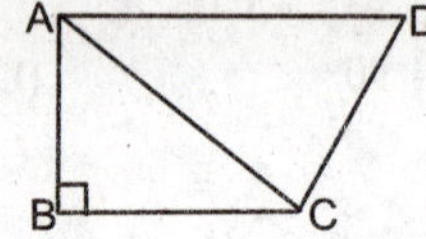

By Pythagoras theorem,

$AB^2 + BC^2 = AC^2$ ...(*i*)

$AD^2 = AB^2 + BC^2 + CD^2$ (Given)

$\therefore$ $AD^2 = AC^2 + CD^2$ [Using (*i*)]

$\therefore$ $\angle ACD = 90°$ (Converse of Pythagoras theorem)

Hence proved.

28. In ΔACD,

$\angle ADC = 90°$

$\therefore$ $AC^2 = AD^2 + CD^2$ (By Pythagoras theorem)

or $b^2 = h^2 + (a - x)^2$

$\Rightarrow$ $b^2 = h^2 + a^2 + x^2 - 2ax$ Hence proved.

29. Given : D is the mid-point of BC, E is the mid-point of AD. BE produced meets AC at M.

To Prove : BE = 3 EM.

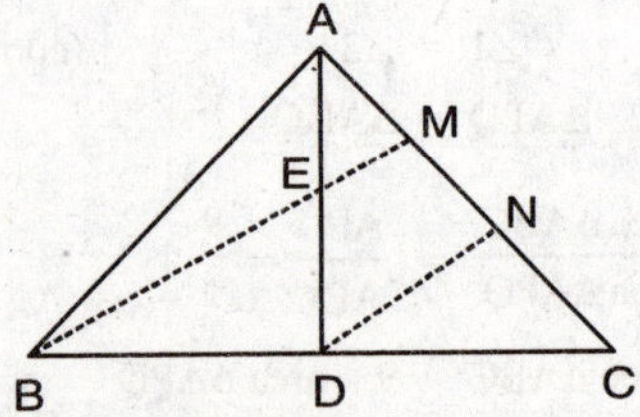

Construction : Draw DN || BM.

Proof : In ΔADN, EM || DN (Construction)

$\therefore$ $\Delta AEM \sim \Delta ADN$ (AA similarity)

$\therefore$ $\frac{EM}{DN} = \frac{AE}{AD}$ (By BPT)

But AD = 2AE

$\therefore$ $\frac{EM}{DN} = \frac{AE}{2AE}$

$\therefore$ DN = 2EM ...(*i*)

In ΔBCM, DN $\parallel$ BM (Construction)

$\therefore$ $\Delta CDN \sim \Delta CBM$ (AA similarity)

$\therefore$ $\frac{DN}{BM} = \frac{CD}{CB}$ (By BPT)

But, CB = 2CD

$\therefore$ $\frac{DN}{BM} = \frac{CD}{2CD}$

$\therefore$ BM = 2DN ...(*ii*)

From (*i*) and (*ii*), BM = 2(2EM)

$\therefore$ BM = 4EM

$\therefore$ BE = BM – EM = 4EM – EM = 3EM.

Hence proved.

30.

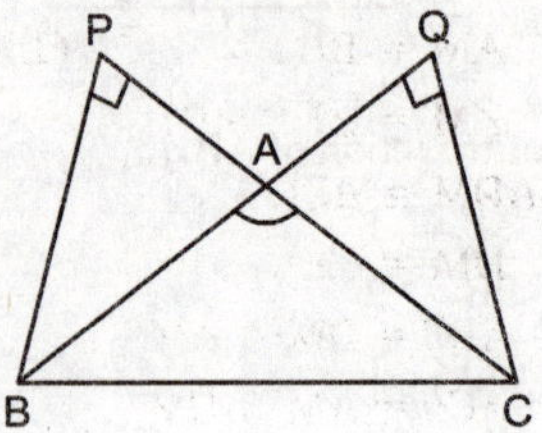

In ΔAPB and ΔAQC

$\angle BPA = \angle CQA$ (each 90°) ...(*i*)

$\angle BAP = \angle CAQ$ (Vertically opposite angles)

$\therefore$ $\Delta APB \sim \Delta AQC$ (AA similarity)

$\Rightarrow$ $\frac{AP}{AQ} = \frac{AB}{AC} \Rightarrow AC.AP = AB.AQ$

31. Construction : Draw AD $\perp$ BC

Proof : In right ΔADC,

$AC^2 = AD^2 + CD^2$ ...(*i*)

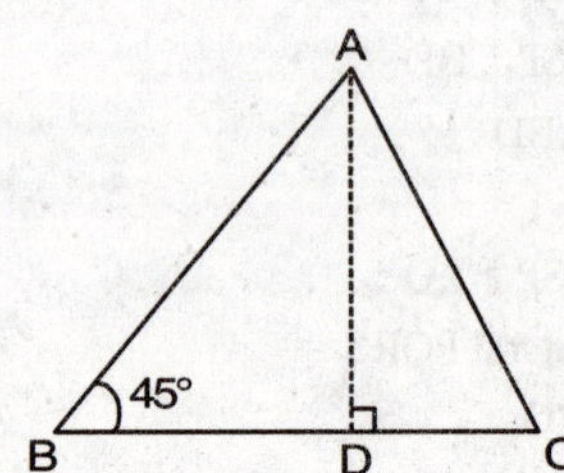

In right ΔADB,

$AB^2 = AD^2 + BD^2$

$\Rightarrow$ $AB^2 - BD^2 = AD^2$

Putting in (*i*), we get

$AC^2 = AB^2 - BD^2 + CD^2$

$\Rightarrow$ $AC^2 = AB^2 + CD^2 - BD^2$

$= AB^2 + (CD - BD)(CD + BD)$

$= AB^2 + (CD - BD)\ BC$

$= AB^2 + (BC - BD - BD)\ .\ BC$

$= AB^2 + BC^2 - 2BC\ .\ BD$

$= AB^2 + BC^2 - 2\ .\ AD\ .\ BC$

[$\because$ In ΔADB, AD = BD]

$= AB^2 + BC^2 - 4$ ar (ΔABC).

(ar $\Delta ABC = \frac{1}{2} BC \times AD$ $\therefore$ $BC \times AD = 2$ ar ΔABC)

32.

Given : A parallelogram ABCD

To prove : $AC^2 + BD^2 = AB^2 + BC^2 + CD^2 + AD^2$

Construction : Draw AM and BL ⊥ CD and DC produces.

Proof : In ΔADM and ΔBLC,

$AD = BC$ (Opp. sides of || gm)

$AM = BL$ (⊥ distance between || sides)

$\angle M = \angle L = 90°$

$\Rightarrow$ $\Delta ADM \cong \Delta BLC$ (RHS congruency)

$\Rightarrow$ $DM = CL$ (CPCT)

In ΔAMD, $AD^2 = DM^2 + AM^2$ ...(*i*)

In ΔAMC, $AC^2 = AM^2 + CM^2$ ...(*ii*)

$= (CD - DM)^2 + AM^2)$ ...(*iii*)

and In ΔBCL,

$BC^2 = BL^2 + CL^2$ ...(*iv*)

In ΔBLD, $BD^2 = BL^2 + DL^2$

$= BL^2 + (CD + CL)^2$ ...(*v*)

$AC^2 + BD^2 = AM^2 + CD^2 + DM^2 - 2CD \,.\, DM$

$+ BL^2 + CD^2 + CL^2 + 2CD \,.\, CL$

$= AD^2 + CB^2 + CD^2 + AB^2$

33.

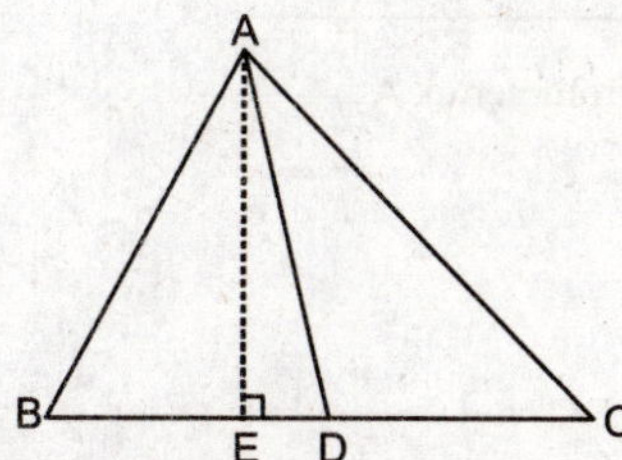

Given : In ΔABC, AD is a median.

To prove : $(AB)^2 + (AC)^2 = 2(AD)^2 + 2(DC)^2$

Construction: Draw AE ⊥ BC.

$\Rightarrow$ In ΔAEB, $AB^2 = AE^2 + BE^2$ ...(*i*)

In ΔAED, $AD^2 = AE^2 + DE^2$

$\Rightarrow$ $AE^2 = AD^2 - DE^2$ ...(*ii*)

From (*i*) and (*ii*)

$AB^2 = AD^2 - DE^2 + BE^2$ ...(*iii*)

In ΔAEC, $AC^2 = AE^2 + EC^2$

$= AD^2 - DE^2 + EC^2$ ...(*iv*)

From (*ii*) and (*iii*),

Adding (*iii*) and (*iv*),

$AB^2 + AC^2 = 2AD^2 - 2DE^2 + BE^2 + CE^2$

$= 2AD^2 - 2DE^2 + (BD - DE)^2 + (CD + DE)^2$

$= 2AD^2 - 2DE^2 + BD^2 + DE^2$

$- 2BD.DE + CD^2 + DE^2 + 2CD.DE$

$= 2AD^2 + CD^2 + CD^2 - 2DE(BD - CD)$

$= 2AD^2 + 2CD^2 - 2DE \times 0$ ($\because$ BD = CD)

$= 2AD^2 + 2CD^2$ Hence proved.

34.

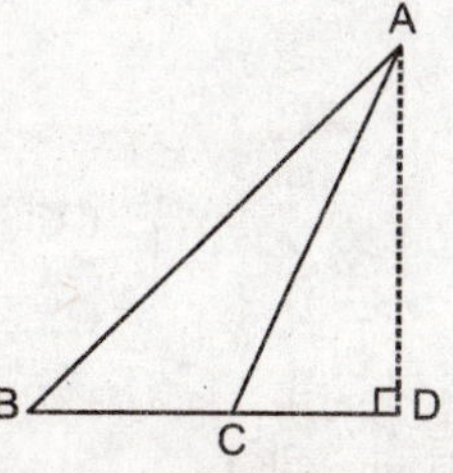

Given : A ΔABC, ∠C is an obtuse angle, AD ⊥ BC and $(AB)^2 = AC^2 + 3(BC)^2$

To prove : BC = CD

Proof : In ΔADB

$AB^2 = AD^2 + BD^2$ ...(*i*)

In ΔADC, $AC^2 = AD^2 + CD^2$

$\Rightarrow$ $AD^2 = AC^2 - CD^2$ ...(*ii*)

From (*i*) and (*ii*),

$AB^2 = AC^2 - CD^2 + BD^2$

$= AC^2 - CD^2 + (BC + CD)^2$

$= AC^2 - CD^2 + BC^2 + CD^2 + 2BC.CD$

$AB^2 = AC^2 + BC^2 + 2BC.CD$

$AC^2 + 3(BC)^2 = AC^2 + BC^2 + 2BC.CD$

[Given, $AB^2 = AC^2 + 3(BC)^2$]

$3(BC)^2 - (BC)^2 = 2BC.CD$

$\Rightarrow$ $2(BC)^2 = 2BC.CD$

$\Rightarrow$ $(BC)^2 = BC.CD$

$\Rightarrow (BC)^2 - BC.CD = 0$

$\Rightarrow BC.[BC - CD] = 0$

Either BC = 0

or BC − CD = 0

BC = CD. Hence proved.

35.

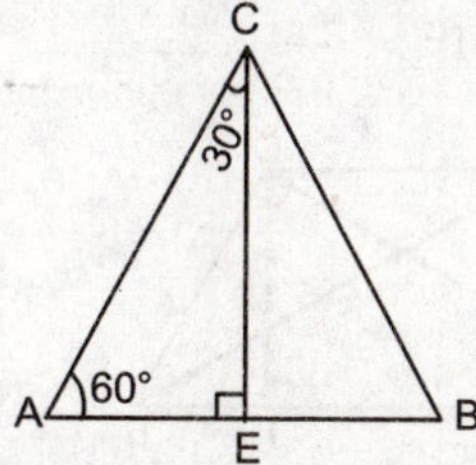

Given : In ΔABC, ∠A = 60°

To prove : $BC^2 = (AB)^2 + (AC)^2 - AB.AC$

Construction : Draw CE ⊥ AB

Proof : In ΔBEC

$BC^2 = CE^2 + BE^2$ ...(*i*)

In ΔACE, $AC^2 = CE^2 + AE^2$

$\Rightarrow$ $CE^2 = AC^2 - AE^2$ ...(*ii*)

From (*i*) and (*ii*),

$BC^2 = AC^2 - AE^2 + (AB - AE)^2$

($\because$ BE = AB − BE)

$BC^2 = AC^2 + AB^2 - 2AB.AE - 2AB.\frac{1}{2}AC$

Since side opposite to 30° angle is half the hypoteneuse.

$\therefore$ In ΔACE, $AE = \frac{1}{2}AC$

$\therefore$ $BC^2 = (AB)^2 + (AC)^2 - AB.AC$

36.

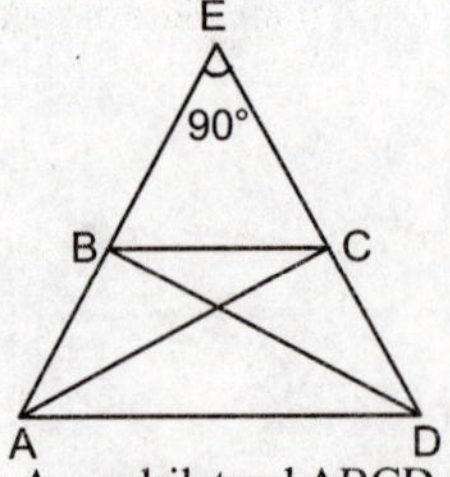

Given : A quadrilateral ABCD, given that

$\angle A + \angle D = 90°$

To Prove : $(AC)^2 + (BD)^2 = AD^2 + (BC)^2$

Construction : Produce AB and DC to intersect at E Join AC and BD.

Proof : In ΔADE

$\angle A + \angle D = 90°$ (Given) ...(i)

and $\angle A + \angle D + \angle E = 180°$ (Angle sum property) ...(ii)

From (i) and (ii),

$90° + \angle E = 180°$

$\Rightarrow \quad \angle E = 90°$ ...(iii)

$\Rightarrow \Delta$ AEC and ΔBDE are right triangles

In ΔAEC, $AC^2 = AE^2 + CE^2$ (By Pythagoras theorem) ...(iv)

In ΔBED, $BD^2 = BE^2 + DE^2$ ...(v)

Adding (iv) and (v),

$AC^2 + BD^2 = AE^2 + BE^2 + CE^2 + DE^2$

$= (AE^2 + DE^2) + (BE^2 + CE^2)$ ...(vi)

Since in ΔBEC, $\angle E = 90°$ from (iii)

So by Pythagoras theorem

$BC^2 = BE^2 + CE^2$ ...(vii)

and similarly ΔADE is also a right triangle as

$\angle AED = 90°$

$\Rightarrow \quad AD^2 = AE^2 + DE^2$ ...(viii)

From (vi), (vii) and (viii) we have

$AC^2 + BD^2 = BC^2 + AD^2$

37.

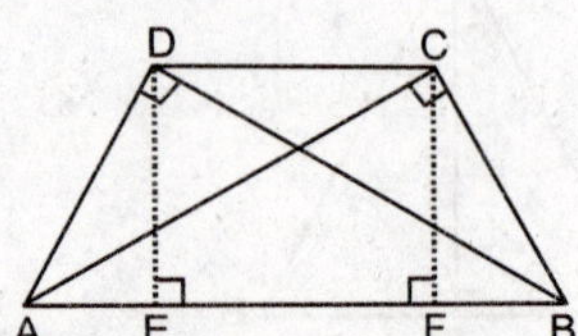

AB = 25 cm, AD = 15 cm and BC = 15cm

In right Δ ABD, $AB^2 = AD^2 + BD^2$

$\Rightarrow \quad (25)^2 = (15)^2 + (BD)^2$

$\Rightarrow \quad 625 = 225 + BD^2 \Rightarrow BD^2 = 400 \Rightarrow BD = 20$ cm

Also, Area $\Delta ABD = \frac{1}{2} AD \times BD = \frac{1}{2} \times 15 \times 20 = 150$ cm^2...(i)

Area $\Delta ABD = \frac{1}{2} AB.DE = \frac{1}{2} \times 25 \times DE$ ...(ii)

From (i) and (ii) $\frac{1}{2} \times 25 \times DE = 150 \Rightarrow DE = \frac{150 \times 2}{25} = 12$ cm.

In right ΔDEA, $AD^2 = AE^2 + DE^2 \Rightarrow (15)^2 = AE^2 + 12^2$

$\Rightarrow \quad 225 - 144 = AE^2 \Rightarrow AE^2 = 81 \Rightarrow AE = 9$ cm

$\because \Delta DEA \cong \Delta CFB \therefore AE = FB = 9$

$EF = AB - AE - BF$

$= 25 - 9 - 9 = 7$ cm

Now $\quad EF = CD = 7$cm

Area of trapezium $= \frac{1}{2} \times DE(AB + CD)$

$= \frac{1}{2} \times 12(25 + 7) = 192$ cm^2

38. In rt. ΔBMA, $BM^2 = AB^2 - AM^2$ ...(i)

In rt. ΔBNC, $BN^2 = BC^2 - CN^2$ ...(ii)

Adding (i) and (ii)

$BM^2 + BN^2 = AB^2 - AM^2 + BC^2 - CN^2$

ABCD is a rectangle,

i.e., $\quad AB = DC$ and $AD = BC$

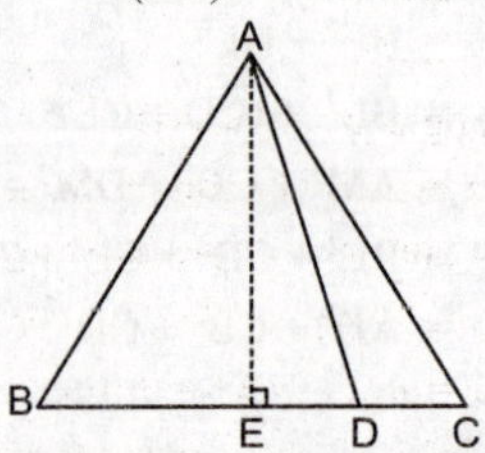

$BM^2 + BN^2 = DC^2 - AM^2 + AD^2 - CN^2$

$= DC^2 - CN^2 + AD^2 - AM^2$

$= DN^2 + DM^2$. Hence proved.

39. **Given :** Let the side of triangle = x

To prove : $AD^2 = 13(CD)^2$

If E is perpendicular from vertex A

$\therefore \quad EC = \frac{1}{2}x$

$DC = \frac{1}{4}x$ (Given)

$\therefore \quad ED = EC - DC = \frac{1}{2}x - \frac{1}{4}x = \frac{1}{4}x$

Also, $\quad AE = \frac{\sqrt{3}}{2}x$ (Height of equilateral triangle)

In ΔADE,

Now, $AD^2 = AE^2 + ED^2 = \left(\frac{\sqrt{3}}{2}x\right)^2 + \left(\frac{1}{4}x\right)^2$

$AD^2 = \frac{3x^2}{4} + \frac{x^2}{16} = \frac{13x^2}{16}$

$= 13.\left(\frac{x}{4}\right)^2 = 13\,(CD)^2$

40.

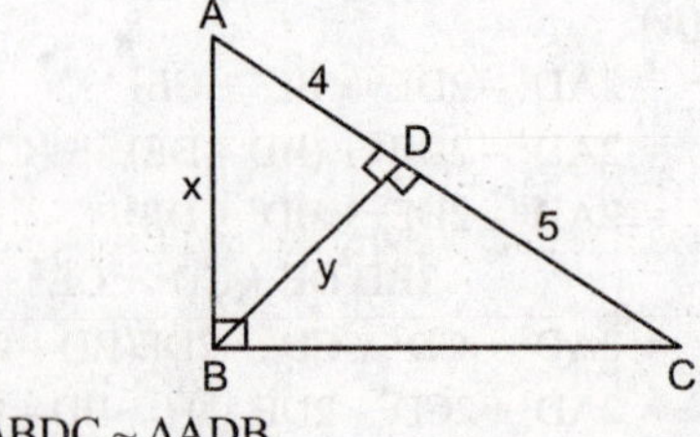

$\because \Delta BDC \sim \Delta ADB$

$$\therefore \quad \frac{BD}{AD} = \frac{CD}{BD} = \frac{BC}{AB}$$

$$\Rightarrow \quad \frac{y}{4} = \frac{5}{y} = \frac{BC}{x}$$

$$\Rightarrow \quad \frac{y}{4} = \frac{5}{y} \Rightarrow y^2 = 20 \Rightarrow y = 2\sqrt{5}$$

Also $\frac{5}{y} = \frac{BC}{x}$

$$BC = \frac{5x}{y} \Rightarrow BC = \frac{5x}{2\sqrt{5}}$$

In right Δ ABC, $AC^2 = BC^2 + AB^2$

$$\Rightarrow \quad 9^2 = \left(\frac{5x}{2\sqrt{5}}\right)^2 + x^2$$

$$81 = \frac{5}{4}x^2 + x^2 \Rightarrow 81 = \frac{9x^2}{4} \Rightarrow x^2 = 9 \times 4$$

$$\therefore \quad x = 6$$

41.

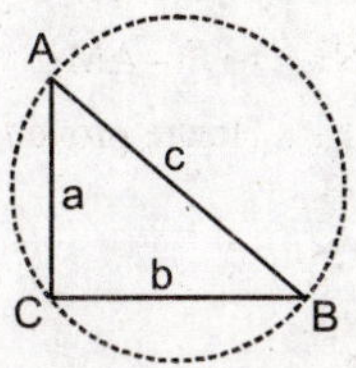

Given : A right angled triangle ABC.

Prove : That the area of a semicircle on the hypotenuse AC of triangle is equal to the sum of the areas of semicircles on the sides AC and BC.

Proof : Let AC = a and BC = b then AB = $\sqrt{a^2 + b^2}$ using Pythagoras, area of semicircle with AB as diameter

$$= \frac{1}{2}\pi\left(\frac{AB}{2}\right)^2$$

$$= \frac{\pi}{2}\left(\frac{\left(\sqrt{a^2+b^2}\right)}{2}\right)^2 = \frac{\pi}{8}(a^2+b^2)$$

Area of semicircle with AC as diameter + area of semicircle with BC as diameter

$$= \frac{\pi}{2}\left(\frac{a}{2}\right)^2 + \frac{\pi}{2}\left(\frac{b}{2}\right)^2 = \frac{\pi}{8}a^2 + \frac{\pi}{8}b^2$$

$$= \frac{\pi}{8}(a^2+b^2)$$

42.

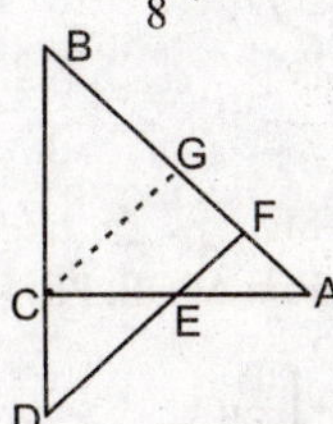

To prove : $\frac{BD}{CD} = \frac{BF}{CE}$

Construction : Draw CG || EF

Proof : In ΔACG, E is mid-point of AC and EF || GC.

$\therefore$ F is mid-point of AG

$\Rightarrow$ GF = AF ...(*i*)

Also, $\angle AFE = \angle AEF$ (Given)

$\therefore$ AF = AE (Opp. sides of equal angles) ...(*ii*)

From (*i*) and (*ii*),

AE = GF

and AE = EC (Given)

$\therefore$ GF = CE

In ΔBDF, CG || DF

$$\therefore \quad \frac{BD}{CD} = \frac{BF}{GF}$$

$$\Rightarrow \quad \frac{BD}{CD} = \frac{BF}{CE} \quad (\because GF = CE)$$

43. **To prove :** BE $\perp$ AC

Construction : Join AC and BE which intersect O.

Proof : Considering ΔABC and ΔBCE,

$$\frac{AB}{BC} = \frac{BC}{CE}$$

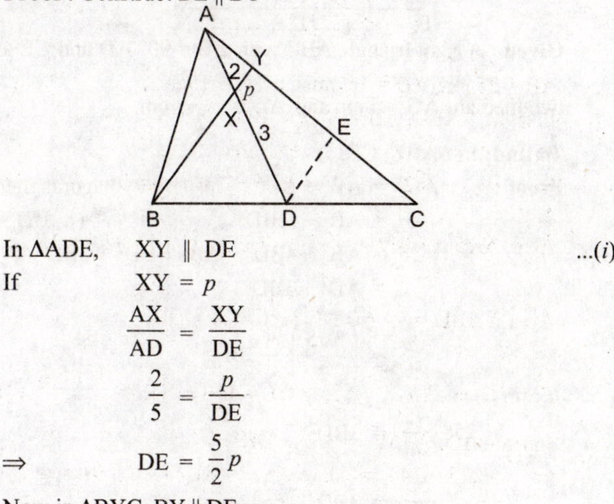

Also, $\angle ABC = \angle BCE = (90°$ each)

$\therefore$ $\Delta ABC \sim \Delta BCE$ (SAS similarity)

$\Rightarrow$ $\angle BAC = \angle CBE = \angle 1$

$\Rightarrow$ $\angle OBA = 90° - \angle 1$

Now in ΔAOB,

$$\angle AOB = 180° - \angle OAB - \angle OBA$$
$$= 180° - \angle 1 - (90° - \angle 1)$$
$$= 180° - \angle 1 - 90° + \angle 1 = 90°$$

Since $\angle AOB$ is 90°

$\Rightarrow$ BE $\perp$ AC.

44. **Given :** ΔABC, AD is median AX : XD = 2 : 3.

To prove : BX = 4XY

Proof : Construct DE || BY

A
Y
2
p
X
3
E
B
D
C

In ΔADE, XY || DE ...(*i*)

If XY = p

$$\frac{AX}{AD} = \frac{XY}{DE}$$

$$\frac{2}{5} = \frac{p}{DE}$$

$$\Rightarrow \quad DE = \frac{5}{2}p$$

Now in ΔBYC, BY || DE

$$\Rightarrow \quad \frac{DC}{BC} = \frac{DE}{BY}$$

$$\frac{1}{2} = \frac{\frac{5}{2}p}{BY}$$

$\Rightarrow$ $BY = 5p$

Now $BX = BY - XY = 5p - p = 4p$

Now $\frac{BX}{XY} = \frac{4p}{p} = 4$

$\Rightarrow$ $BX = 4XY$

45.

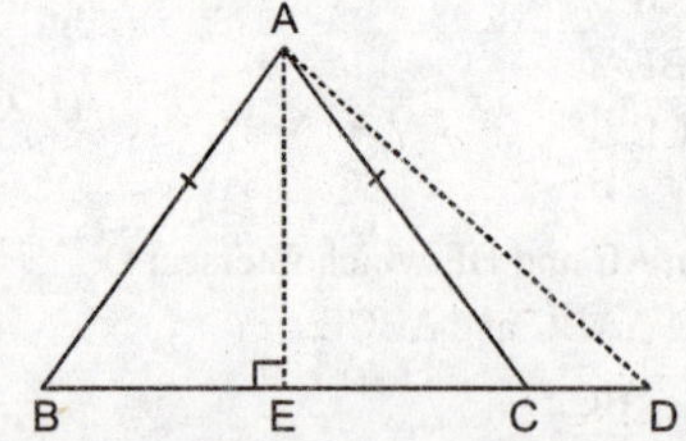

Given : In ΔABC, $AB = AC$, BC is produced to D.

To Prove : $AD^2 - AC^2 = BD.CD$

Construction : Join AD and draw $AE \perp BC$

Proof : $AE \perp BC$ (By construction)

In ΔABE and ΔACE,

$\because$ $AB = AC$ (Given)

$\angle AEB = \angle AEC$ (Each 90°)

$AE = AE$ (Common side) ...(*a*)

$\Delta ABE \cong \Delta ACE$ (RHS congruency)

$\therefore$ $BE = CE$ (CPCT)

In rt ΔAED, $AD^2 = AE^2 + DE^2$

In rt AEC, $AC^2 = AE^2 + CE^2$

$$AD^2 - AC^2 = DE^2 - CE^2$$
$$= (DE - CE)(DE + CE)$$
$$= CD\,(DE + BE) \text{ using } (a)$$
$$= CD.\,BD \quad (\because BE = CE)$$

46.

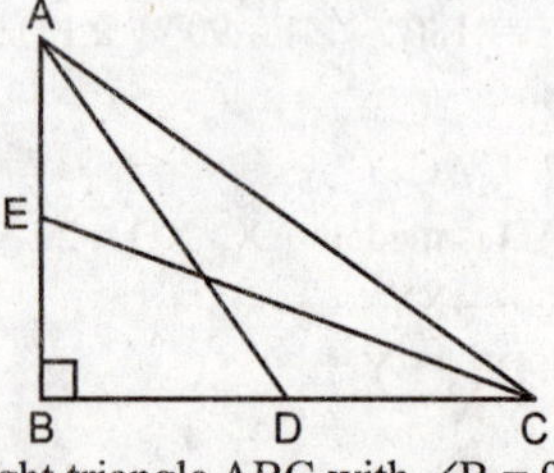

Given : A right triangle ABC with $\angle B = 90°$ AD and CE are the two medians $AC = 5$ cm and $AD = \frac{3\sqrt{5}}{2}$ cm.

To find: In ΔABC, CE

Proof : $AC^2 = AB^2 + BC^2$ (By Pythagoras theorem)

$5^2 = AB^2 + (2BD)^2$ $(\because BD = CD)$

$$25 = AB^2 + 4BD^2 = AB^2 + BD^2 + 3BD^2$$
$$= AD^2 + 3BD^2$$
$$= \left(\frac{3\sqrt{5}}{2}\right)^2 + 3BD^2$$
$$25 - \frac{45}{4} = 3BD^2$$
$$\frac{55}{4} = 3BD^2$$
$$BD^2 = \frac{55}{12}$$

$\Rightarrow$ $BD = \sqrt{\frac{55}{12}}$ cm

$BC = 2BD = 2\sqrt{\frac{55}{12}}$ cm

In ΔBCE, $CE^2 = BE^2 + BC^2$

$$= \left(\frac{1}{2}AB\right)^2 + BC^2$$
$$= \frac{1}{4}(AC^2 - BC^2) + BC^2$$
$$= \frac{1}{4} \times 5^2 - \frac{BC^2}{4} + BC^2$$
$$= \frac{25}{4} + \frac{3BC^2}{4}$$
$$= \frac{25}{4} + \frac{3}{4} \times 4 \times \frac{55}{12} = \frac{80}{4} = 20$$
$$CE = 2\sqrt{5} \text{ cm}$$

47.

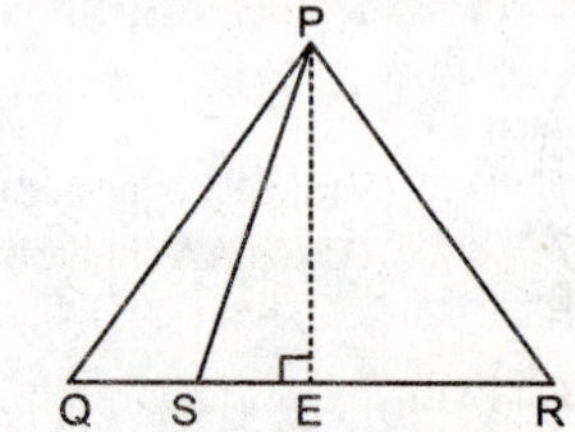

Given : A equilateral triangle PQR

$QS = \frac{1}{3}QR$

and $RS = \frac{2}{3}QR$

To prove : $9PS^2 = 7PQ^2$

Proof : $PQ = QR = PR = a$ (Say)

$QS = \frac{1}{3}QR = \frac{1}{3}a$

$\Rightarrow$ $SR = \frac{2}{3}QR = \frac{2}{3}a$

$PE \perp QR$

$\Rightarrow$ $QE = ER = \frac{1}{2}a$

$SE = QE - QS$

$SE = \frac{1}{2}a - \frac{1}{3}a = \frac{1}{6}a$

In ΔPSE,

$$PS^2 = PE^2 + SE^2$$
$$= PQ^2 - QE^2 + SE^2$$

(from ΔPQE, $PE^2 = PQ^2 - QE^2$)

$$= a^2 - \left(\frac{1}{2}a\right)^2 + \left(\frac{1}{6}a\right)^2$$
$$= a^2 - \frac{1}{4}a^2 + \frac{1}{36}a^2$$
$$= \frac{(36 - 9 + 1)}{36}a^2$$
$$= \frac{28}{36}a^2 = \frac{7}{9}PQ^2$$

$\Rightarrow$ $9PS^2 = 7PQ^2$ Hence proved.

48. Given : ΔABC and ΔDBC are two triangles on the same base BC. AD intersects BC at O.

To prove : $\frac{ar(\Delta ABC)}{ar(\Delta DBC)} = \frac{AO}{DO}$

Construction : Draw $AL \perp BC$ and $DM \perp BC$.

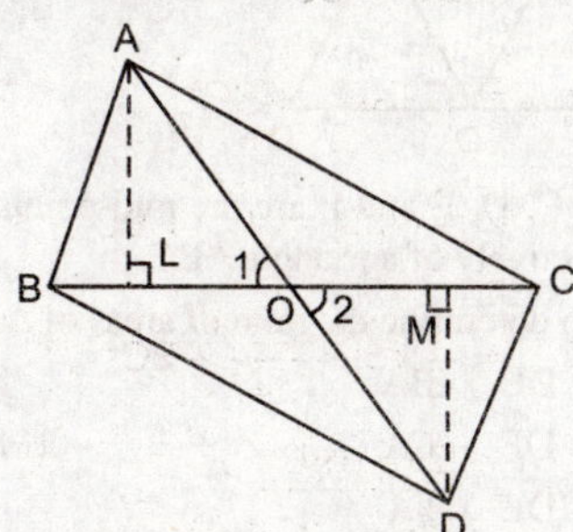

Proof : $ar\,\Delta ABC = \frac{1}{2} BC \times AL$ ...(*i*)

$ar\,\Delta BDC = \frac{1}{2} BC \times DM$ ...(*ii*)

In ΔALO and ΔDMO,

$\angle L = \angle M = \text{each } 90°$

$\angle 1 = \angle 2$ (Vertically opposite angles)

$\Delta ALO \sim \Delta DMO$ (Using AA similarity rule)

$\Rightarrow \quad \frac{AL}{DM} = \frac{AO}{DO}$ ...(*iii*)

(Corresponding sides of similar triangles)

From (*i*), (*ii*) and (*iii*), we get

$$\frac{ar(\Delta ABC)}{ar(\Delta BDC)} = \frac{\frac{1}{2} BC \times AL}{\frac{1}{2} BC \times DM} = \frac{AL}{DM} \quad ...(iv)$$

From (*iii*) and (*iv*),

$\frac{ar\,\Delta ABC}{ar\,\Delta BDC} = \frac{AO}{DO}$ Hence proved.

49.

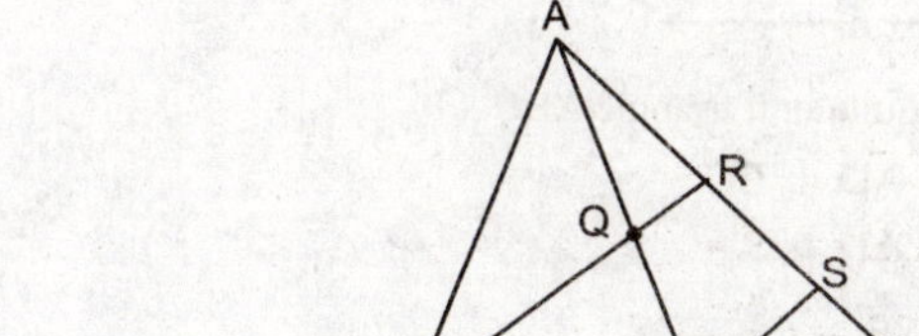

Given : In ΔABC, P is the mid point of BC. Q is mid point of AP.

To prove : $AR = \frac{1}{3} AC$

Construction : Draw $PS \parallel BR$

Proof : In ΔBCR, P is mid point of BC and $PS \parallel BR$

$\therefore$ S is mid piont of CR.

$\Rightarrow \quad CS = RS$ ...(*i*)

In ΔAPS, Q is mid point of AP and $PS \parallel QR$

$\Rightarrow \quad \frac{AQ}{PQ} = \frac{AR}{RS}$

$\Rightarrow \quad AQ = PQ \therefore \frac{AR}{RS} = 1 \Rightarrow AR = RS$ ...(*ii*)

From (*i*) and (*ii*) $CS = RS = AR$

Now $\quad AC = AR + RS + CS \Rightarrow AC = 3AR$

$\Rightarrow \quad AR = \frac{1}{3} AC$ Hence proved

50. Given : In ΔABC, line PQ meets AB at P and AC at Q. Also $AP = 1$ cm, $PB = 3$ cm, $AQ = 1.5$ cm, $QC = 4.5$ cm.

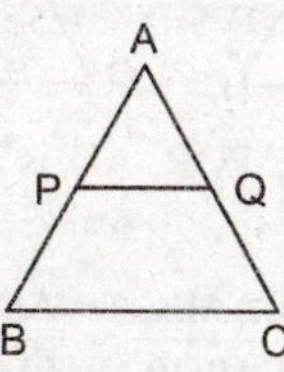

To Prove : $ar(\Delta APQ) = \frac{1}{16} ar(\Delta ABC)$

Proof : $\frac{AP}{PB} = \frac{1}{3}$

and $\frac{AQ}{QC} = \frac{1.5}{4.5} = \frac{1}{3}$

As $\frac{AP}{PB} = \frac{AQ}{QC}$

$\Rightarrow \quad PQ \parallel BC$

(Converse of Basic Proportionality theorem)

In Δs APQ and ABC

$\angle A$ is common.

$\angle APQ = \angle ABC$ (Corresponding angles)

$\therefore \quad \Delta APQ \sim \Delta ABC$ (AA similarity)

$\Rightarrow \quad \frac{ar(\Delta APQ)}{ar(\Delta ABC)} = \frac{AP^2}{AB^2}$ (Using above theorem)

$\Rightarrow \quad \frac{ar(\Delta APQ)}{ar(\Delta ABC)} = \frac{(1)^2}{(1+3)^2} = \frac{1}{16}$

$\Rightarrow \quad ar(\Delta APQ) = \frac{1}{16} ar(\Delta ABC)$

51.

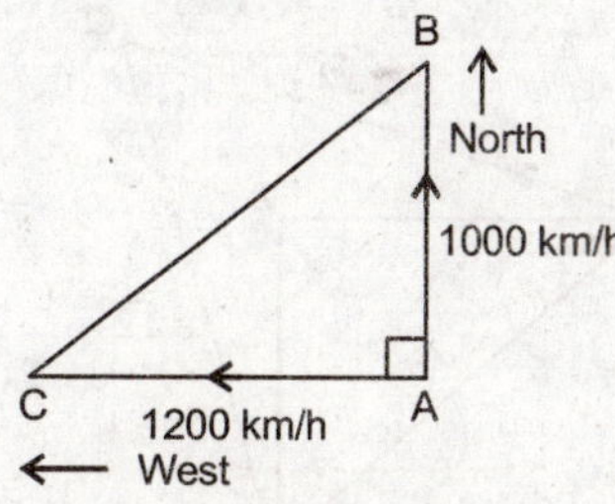

Let planes start from A. After $1\frac{1}{2}$ hr Ist plane reaches B and IInd plane reaches at C.

Distance between them = BC

AB = Distance covered by Ist plane in $1\frac{1}{2}$ hr.

$\therefore \quad AB = 1000 \times \frac{3}{2} = 1500$ km.

AC = distance covered by IInd plane in $1\frac{1}{2}$ hr

$\therefore \quad AC = 1200 \times \frac{3}{2} = 1800$ km.

Now $\quad BC^2 = AC^2 + AB^2$

$\Rightarrow \quad BC^2 = (1500)^2 + (1800)^2$

$\Rightarrow \quad BC^2 = 5490000$

$BC = \sqrt{5490000} = 2343.07$ km.

52.

Given : $\Delta ABC \sim \Delta PQR$, AD is median of ΔABC and PM is median of ΔPQR.

To prove : $\dfrac{\text{Area of } \Delta ABC}{\text{Area of } \Delta PQR} = \dfrac{AD^2}{PM^2}$

Proof : $\Delta ABC \sim \Delta PQR$

$$\Rightarrow \quad \frac{AB}{PQ} = \frac{BC}{QR} \quad ...(i)$$

[Corresponding sides of similar triangles are proportional and corresponding angles are equal]

and $\angle B = \angle Q$ ...(*ii*)

From (*i*) $\dfrac{AB}{PQ} = \dfrac{2 \times BD}{2 \times QM}$

$$\Rightarrow \quad \frac{AB}{PQ} = \frac{BD}{QM} \quad ...(iii)$$

From (*ii*) and (*iii*)

$\Delta ABD \sim \Delta PQM$ (SAS similarity)

$$\Rightarrow \quad \frac{AB}{PQ} = \frac{AD}{PM} \quad ...(iv)$$

Now $\dfrac{\text{area of } \Delta ABC}{\text{area of } \Delta PQR} = \dfrac{AB^2}{PQ^2} = \dfrac{AD^2}{PM^2}$

53.

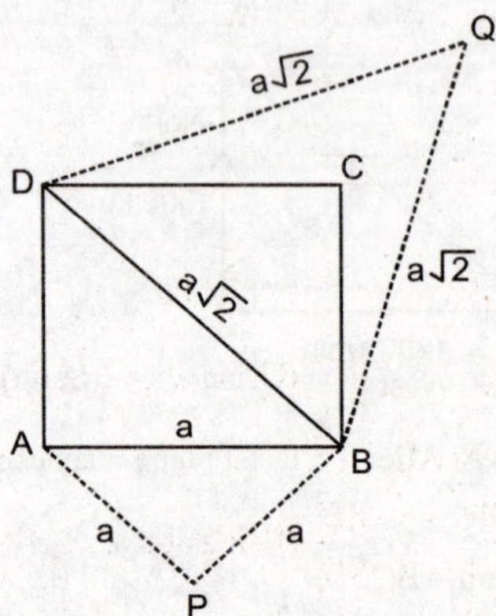

Given : A square ABCD, ΔBQD is an equilateral triangle described on the diagonal BD and ΔAPB is an equilateral triangle described on the side AB.

To prove : ar $(\Delta PAB) = \dfrac{1}{2}$ ar (ΔQBD)

Proof : ABCD is a square having side of length $= a$

Then the diagonal $BD = a\sqrt{2}$

$\Delta PAB \sim \Delta QBD$ (Equilateral triangles are similar)

$$\frac{\text{ar}(\Delta PAB)}{\text{ar}(\Delta QBD)} = \frac{AB^2}{BD^2} = \frac{a^2}{(a\sqrt{2})^2} = \frac{1}{2}$$

$$\Rightarrow \quad \text{ar}(\Delta PAB) = \frac{1}{2}\text{ar}(\Delta QBD)$$

54.

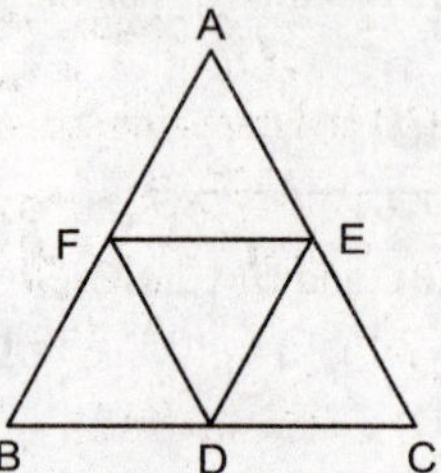

Given : ΔABC, D, E and F are the mid-points of sides BC, CA and AB respectively of a triangle ABC.

To prove : To determine the ratio of areas of ΔABC and ΔDEF.

Proof : $DE \parallel BA$,

$DF \parallel CA$ (Mid-point theorem)

$\Rightarrow$ $DE \parallel FA$

and $DF \parallel EA$

$\Rightarrow$ DEAF is a parallelogram

$\Rightarrow$ $\angle FDE = \angle A$ (Opp. $\angle$s of $\parallel$ gm)

and $\angle DEF = \angle B$

$\Rightarrow$ $\Delta DEF \sim \Delta ABC$

$$\Rightarrow \frac{\text{Area of } \Delta DEF}{\text{Area of } \Delta ABC} = \frac{DE^2}{AB^2} = \frac{\left(\frac{1}{2}AB\right)^2}{AB^2} = \frac{1}{4}$$

$$\text{or} \quad \frac{\text{ar}(\Delta ABC)}{\text{ar of } \Delta DEF} = \frac{AB^2}{DE^2} = \frac{AB^2}{\left(\frac{1}{2}AB\right)^2} = 4:1.$$

55.

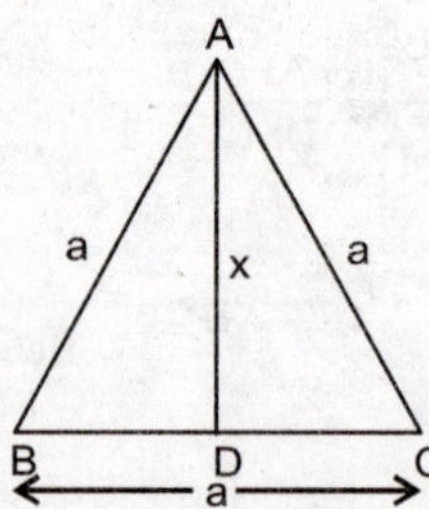

Given : An equilateral triangle ABC

$AD \perp BC$

Let $AD = x$

To prove : $3a^2 = 4x^2$

Proof : ΔABC is an equilateral triangle of side a (say)

$AD \perp BC$

$AD = x$ (say)

Now $BD = CD = \dfrac{1}{2}BC = \dfrac{1}{2}a$

In right angle ΔABD

$$AB^2 = AD^2 + BD^2$$

$$\Rightarrow \quad a^2 = x^2 + \left(\frac{1}{2}a\right)^2$$

$$\Rightarrow \quad a^2 = x^2 + \frac{1}{4}a^2$$

$$\Rightarrow \quad 4a^2 = 4x^2 + a^2$$

$$\Rightarrow \quad 3a^2 = 4x^2$$

LONG ANSWER TYPE QUESTIONS

1. (*i*) In the given figure, ABCD is a trapezium so, AB || CD.

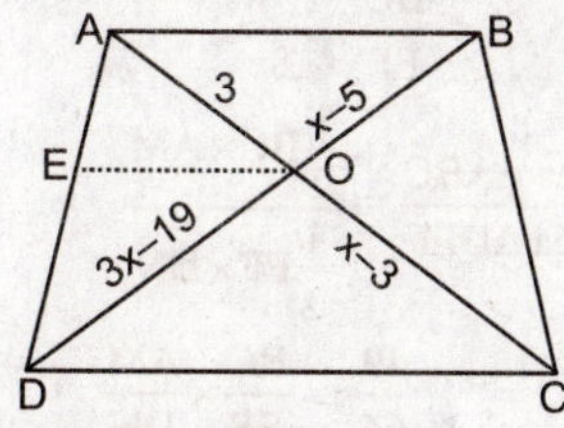

Draw OE || AB and as AB || CD

∴ OE || AB || CD ...(*i*)

In ΔDAB, OE || AB

$\Rightarrow \dfrac{DE}{EA} = \dfrac{DO}{OB}$ (Using above theorem) ...(*ii*)

In ΔDAC, OE || DC [From (*i*)]

$\Rightarrow \dfrac{DE}{EA} = \dfrac{OC}{AO}$ (Using above theorem) ...(*iii*)

From (*ii*) and (*iii*), we get

$\dfrac{DO}{OB} = \dfrac{OC}{AO}$

$\Rightarrow \dfrac{3x-19}{x-5} = \dfrac{x-3}{3}$

$\Rightarrow 9x - 57 = x^2 - 8x + 15$

$\Rightarrow x^2 - 17x + 72 = 0$

$\Rightarrow (x-9)(x-8) = 0$

$\Rightarrow x - 9 = 0$ or $x - 8 = 0$

$\Rightarrow x = 8$ or 9

(*ii*)

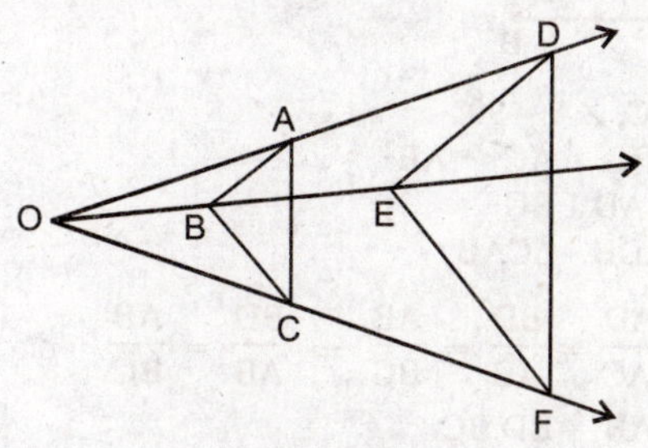

Given : AB || DE and BC || EF

To prove : AC || DF

Proof : In ΔDOE

AB ∥ DE (By BPT)

$\therefore \dfrac{OA}{AD} = \dfrac{OB}{BE}$ (Using above theorem) ...(*i*)

Also in ΔEOF,

BC ∥ EF

$\therefore \dfrac{OC}{CF} = \dfrac{OB}{BE}$ ...(*ii*)

Since RHS of (*i*) and (*ii*) are equal,

∴ Respective LHS are also equal.

i.e., $\dfrac{OA}{AD} = \dfrac{OC}{CF}$

∴ In ΔODF, AC || DF (Using converse of BPT)

Hence proved.

(*iii*)

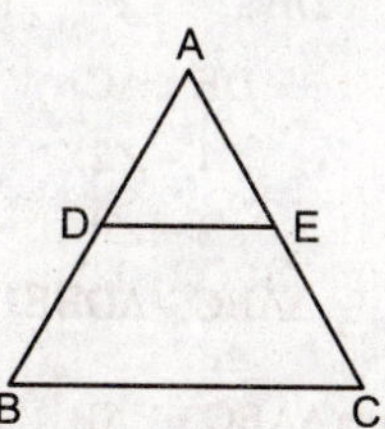

In Δ ABC

DE ∥ BC (given)

$\therefore \dfrac{AD}{BD} = \dfrac{AE}{CE}$ (Using converse of BPT)

∴ BD = CE (Given) ...(*i*)

∴ AD = AE ...(*ii*)

⇒ (*i*) + (*ii*) BD + AD = CE + AE

⇒ AB = AC

∴ ΔABC is isosceles Δ.

(*iv*) In Δ PRQ, AB || RQ

$\Rightarrow \dfrac{QB}{BP} = \dfrac{RA}{AP}$ ...(*i*)

In ΔPRS, CA || SR

$\therefore \dfrac{RA}{AP} = \dfrac{SC}{CP}$ ...(*ii*)

From (*i*) and (*ii*)

$\dfrac{QB}{BP} = \dfrac{SC}{CP}$

(*v*) In ΔPQR, AB || PQ

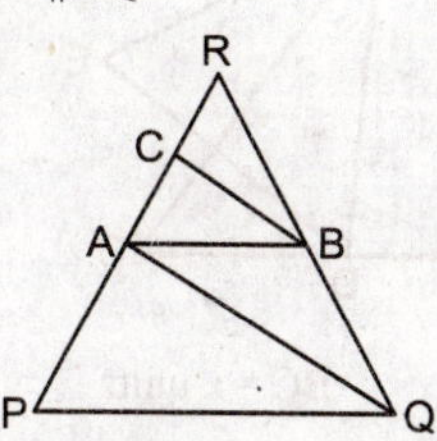

$\Rightarrow \dfrac{AR}{PR} = \dfrac{RB}{RQ}$ (Using above result) ...(*i*)

In ΔAQR, BC || AQ

$\Rightarrow \dfrac{CR}{AR} = \dfrac{RB}{RQ}$ (Using above result) ...(*ii*)

From (*i*) and (*ii*), we have

$\dfrac{AR}{PR} = \dfrac{CR}{AR} \Rightarrow AR^2 = PR \cdot CR$

2. (*i*)

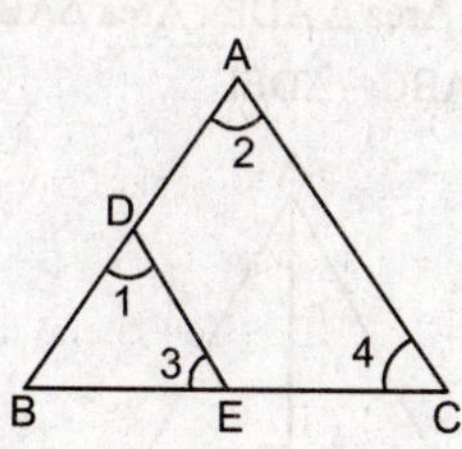

Now $\dfrac{AD}{DB} = \dfrac{3}{2} \Rightarrow \dfrac{AD}{DB} + 1 = \dfrac{3}{2} + 1$

$\Rightarrow \quad \dfrac{AD + DB}{DB} = \dfrac{5}{2} \Rightarrow \dfrac{AB}{DB} = \dfrac{5}{2}$

$\because \quad DE \parallel AC$

$\therefore \quad \angle 1 = \angle 2$ (corresponding angles)

$\angle 3 = \angle 4$

$\therefore \quad \Delta ABC \sim \Delta DBE$

$\Rightarrow \quad \dfrac{\text{area } \Delta ABC}{\text{area } \Delta DBE} = \dfrac{AB^2}{BD^2} = \left(\dfrac{5}{2}\right)^2 = \dfrac{25}{4}$

(*ii*)

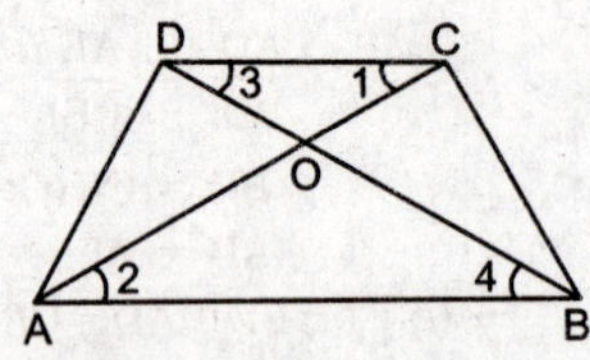

$AB \parallel CD$

$\therefore \quad \angle 1 = \angle 2$ (Alternate interior angles)

$\therefore \quad \angle 3 = \angle 4$

$\therefore \quad \Delta AOB \sim \Delta COD$

$\dfrac{\text{Area } \Delta AOB}{\text{Area } \Delta COD} = \dfrac{AB^2}{CD^2} = \dfrac{(2CD)^2}{CD^2}$

$\Rightarrow \quad \dfrac{84}{\text{Area}\Delta COD} = \dfrac{4}{1} \Rightarrow \text{Area } \Delta COD = \dfrac{84}{4} = 21 \text{ cm}^2$

(*iii*)

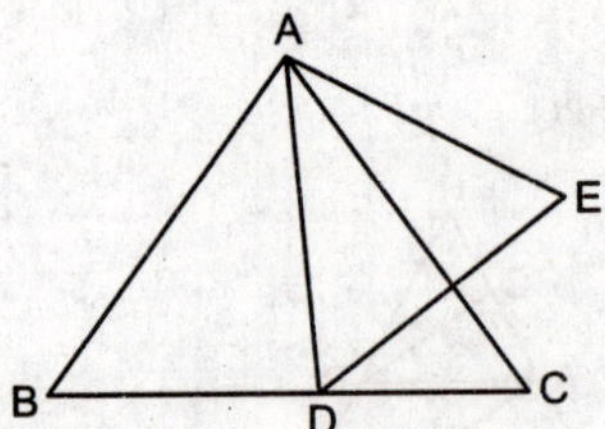

Let $\quad BC = x$ units

$\therefore \quad AD = \dfrac{\sqrt{3}}{2} \times x$ units

$\because$ ΔABC and ΔADE are equilateral triangles.

$\therefore \quad \Delta ADE \sim \Delta ABC$

$\therefore \quad \dfrac{\text{Area } \Delta ADE}{\text{Area } \Delta ABC} = \dfrac{AD^2}{BC^2} = \left(\dfrac{\frac{\sqrt{3}}{2}x}{x}\right)^2 = \dfrac{3}{4}$

$\therefore$ Area Δ ADE : Area ΔABC = 3 : 4

(*iv*) $\Delta ABC \sim \Delta DEF$

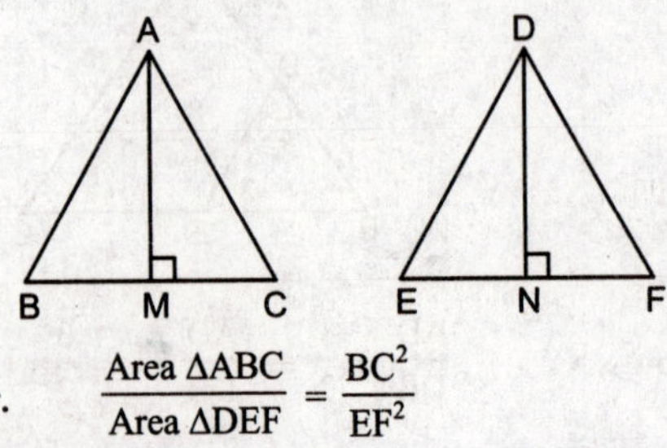

$\therefore \quad \dfrac{\text{Area } \Delta ABC}{\text{Area } \Delta DEF} = \dfrac{BC^2}{EF^2}$

$\Rightarrow \quad \dfrac{49}{64} = \dfrac{BC^2}{EF^2}$

$\Rightarrow \quad \dfrac{BC}{EF} = \dfrac{7}{8}$

Also $\quad \dfrac{\text{Area } \Delta ABC}{\text{Area } \Delta DEF} = \dfrac{\frac{1}{2} BC \times AM}{\frac{1}{2} EF \times DN}$

$\Rightarrow \quad \dfrac{49}{64} = \dfrac{BC}{EF} \times \dfrac{AM}{DN}$

$\Rightarrow \quad \dfrac{49}{64} = \dfrac{7}{8} \times \dfrac{AM}{DN}$

$\Rightarrow \quad \dfrac{7}{8} = \dfrac{AM}{DN}$

$\Rightarrow \quad DN = \dfrac{8}{7} AM$

Also $\quad AM - DN = 10$ (Given)

$\Rightarrow$ Not possible

$DN - AM = 10$

$\Rightarrow \quad \dfrac{8}{7} AM - AM = 10$

$\Rightarrow \quad \dfrac{1}{7} AM = 10$

$AM = 70$ cm

$\therefore \quad DN = 80$ cm

3.

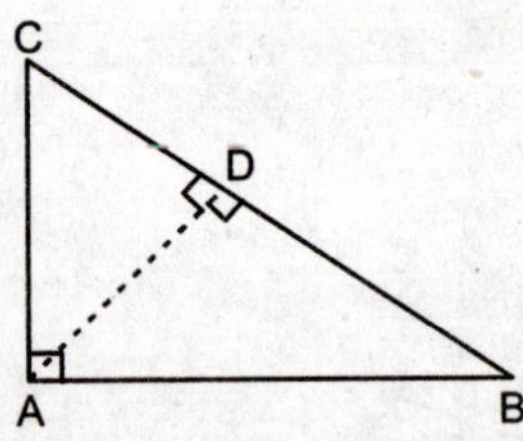

Given : In ΔABC, $\angle A = 90°$

To prove : $BC^2 = AC^2 + AB^2$

Construction : $AD \perp BC$

Proof : $\Delta ADB \sim \Delta CAB$

$\Rightarrow \quad \dfrac{AD}{AC} = \dfrac{BD}{AB} = \dfrac{AB}{BC} \Rightarrow \dfrac{BD}{AB} = \dfrac{AB}{BC}$

$\Rightarrow \quad AB^2 = BD.BC$ ...(*i*)

Also $\Delta ADC \sim \Delta BAC$

$\therefore \quad \dfrac{AD}{AB} = \dfrac{CD}{AC} = \dfrac{AC}{BC}$

$\Rightarrow \quad \dfrac{CD}{AC} = \dfrac{AC}{BC} \Rightarrow AC^2 = BC.CD$ ...(*ii*)

(*i*) + (*ii*)

$AB^2 + AC^2 = BC.CD + BC.BD$

$\Rightarrow \quad AB^2 + AC^2 = BC(CD + BD)$

$AB^2 + AC^2 = BC.BC = BC^2$

(*i*)

D C O A B

Given : ABCD is a rhombus

To prove : $4AB^2 = AC^2 + BD^2$

Proof : In rhombus ABCD, diagonal AC and BD intersect at O

$\therefore \quad \angle AOB = 90°$ and $AO = \frac{1}{2}AC$

$OB = \frac{1}{2}BD$.

In right ΔAOB, $AB^2 = AO^2 + BO^2$

$\Rightarrow \quad AB^2 = \left(\frac{1}{2}AC\right)^2 + \left(\frac{1}{2}BD\right)^2$

$\Rightarrow \quad AB^2 = \frac{1}{4}AC^2 + \frac{1}{4}BD^2$

$\Rightarrow \quad 4AB^2 = AC^2 + BD^2$ Hence proved

(*ii*)

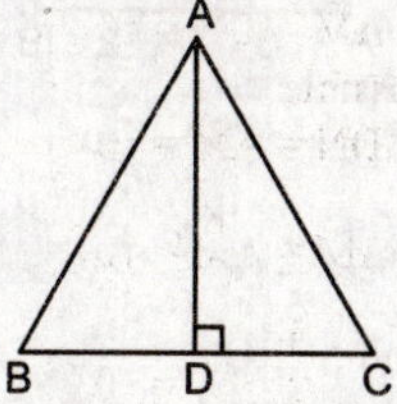

In right ΔADB

$AB^2 = AD^2 + BD^2$

$\Rightarrow AB^2 - BD^2 = AD^2$ ...(*i*)

Also in right Δ ADC, $AC^2 = AD^2 + CD^2$

$\Rightarrow AC^2 - CD^2 = AD^2$ ...(*ii*)

From (*i*) and (*ii*)

$AB^2 - BD^2 = AC^2 - CD^2$

$\Rightarrow AB^2 + CD^2 = AC^2 + BD^2$ (Hence proved)

(*iii*)

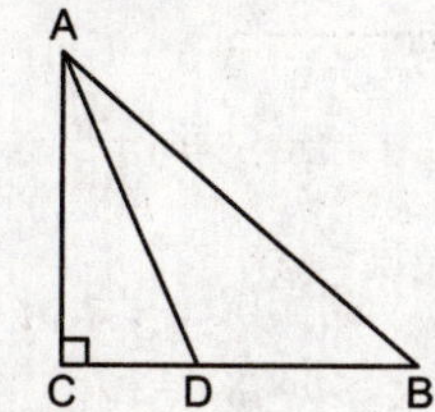

In right ΔACD

$AD^2 = AC^2 + CD^2$

$\Rightarrow \quad AD^2 = AC^2 + \left(\frac{1}{2}BC\right)^2$ ($\because$ D is mid point of BC)

$\Rightarrow \quad AD^2 = AC^2 + \frac{1}{4}BC^2$

$\Rightarrow \quad 4AD^2 = 4AC^2 + BC^2$.

(*iv*) In equilateral ΔABC . 4 BD = BC

Construction : Draw $AE \perp BC$. $\therefore$ $BE = \frac{1}{2}BC$.

In right Δ AED,

$AD^2 = DE^2 + AE^2$

$\Rightarrow \quad AE^2 = AD^2 - DE^2$ ... (*i*)

In right Δ AEB,

$AB^2 = AE^2 + BE^2$

$\Rightarrow \quad AB^2 = AD^2 - DE^2 + BE^2$ [Using (*i*)]

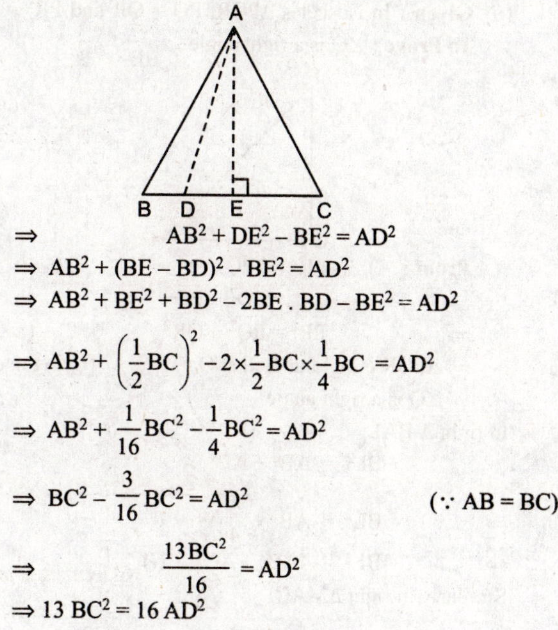

$\Rightarrow \quad AB^2 + DE^2 - BE^2 = AD^2$

$\Rightarrow AB^2 + (BE - BD)^2 - BE^2 = AD^2$

$\Rightarrow AB^2 + BE^2 + BD^2 - 2BE . BD - BE^2 = AD^2$

$\Rightarrow AB^2 + \left(\frac{1}{2}BC\right)^2 - 2\times\frac{1}{2}BC\times\frac{1}{4}BC = AD^2$

$\Rightarrow AB^2 + \frac{1}{16}BC^2 - \frac{1}{4}BC^2 = AD^2$

$\Rightarrow BC^2 - \frac{3}{16}BC^2 = AD^2$ ($\because$ AB = BC)

$\Rightarrow \quad \frac{13BC^2}{16} = AD^2$

$\Rightarrow 13\ BC^2 = 16\ AD^2$

4.

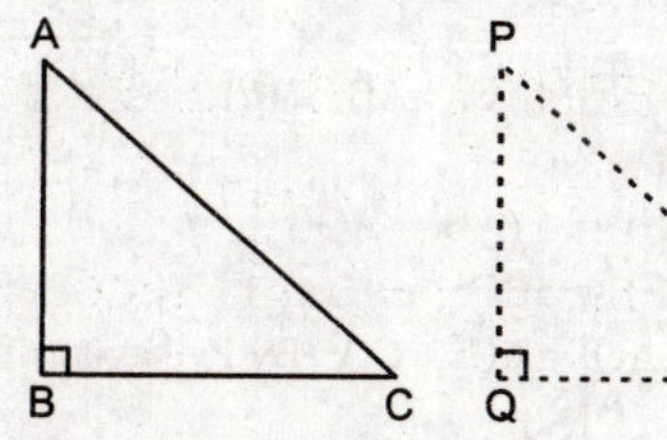

Given : In ΔABC, $AC^2 = AB^2 + BC^2$

To prove : $\angle B = 90°$

Construction : Construct a ΔPQR in which PQ = AB, QR = BC, and $\angle Q = 90°$

Proof : In right ΔPQR

$PR^2 = PQ^2 + QR^2$

$\Rightarrow \quad PR^2 = AB^2 + BC^2$ ($\because$ PQ = AB, QR = BC) ...(*i*)

Also $\quad AC^2 = AB^2 + BC^2$ (given) ...(*ii*)

From (*i*) and (*ii*)

$PR^2 = AC^2 \Rightarrow PR = AC$

Now in ΔABC and ΔPQR

AB = PQ (construction)

BC = QR

AC = PR (proved)

$\therefore \quad \Delta ABC \cong \Delta PQR$ (SSS)

$\therefore \quad \angle B = \angle Q$ (c.p.c.t.)

$\because \quad \angle Q = 90°$ (construction)

$\therefore \quad \angle B = 90°$ Hence proved

(*i*)

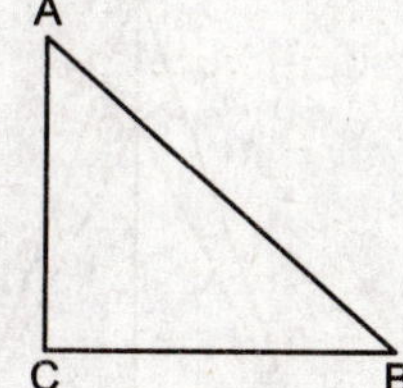

In ΔABC

$AB^2 = 2AC^2$ (given)

$\Rightarrow \quad AB^2 = AC^2 + AC^2$

$\Rightarrow \quad AB^2 = AC^2 + BC^2$ [$\because$ AC = BC (given)]

$\therefore \quad \angle C = 90°$ (Using above result)

(*ii*) **Given :** In isosceles ΔPQR, $PQ = QR$ and $PR^2 = 2PQ^2$.

To Prove : $\angle Q$ is a right angle.

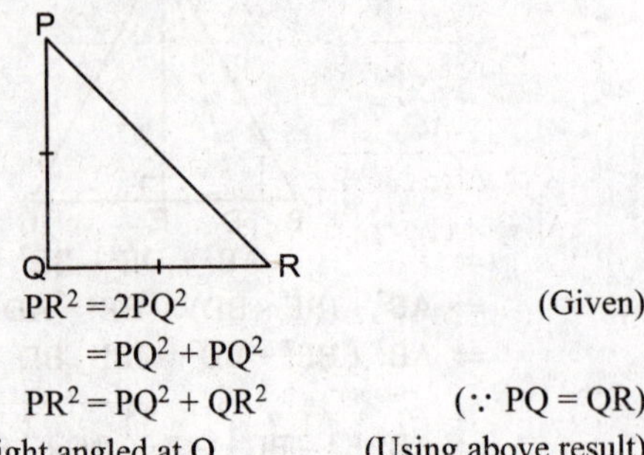

Proof : $PR^2 = 2PQ^2$ (Given)

$= PQ^2 + PQ^2$

$\Rightarrow$ $PR^2 = PQ^2 + QR^2$ ($\because PQ = QR$)

$\Rightarrow \Delta PQR$ is right angled at Q (Using above result)

$\therefore \angle Q$ is a right angle.

5. In right Δ BAL,

$$BL^2 = AB^2 + AL^2$$

$$\Rightarrow \quad BL^2 = AB^2 + \frac{1}{4}AC^2$$

$$\Rightarrow \quad 4BL^2 = 4AB^2 + AC^2 \quad ...(i)$$

Similarly in right ΔMAC,

$$CM^2 = AC^2 + \frac{1}{4}AB^2$$

$$\Rightarrow \quad 4CM^2 = 4AC^2 + AB^2 \quad ..(ii)$$

(*i*) + (*ii*), we get

$$4(BL^2 + CM^2) = 5(AB^2 + AC^2)$$

$$\Rightarrow 4(BL^2 + CM^2) = 5BC^2$$

6. (*i*) In ΔACQ, $AQ^2 = AC^2 + CQ^2$ (By Pythagoras theorem)

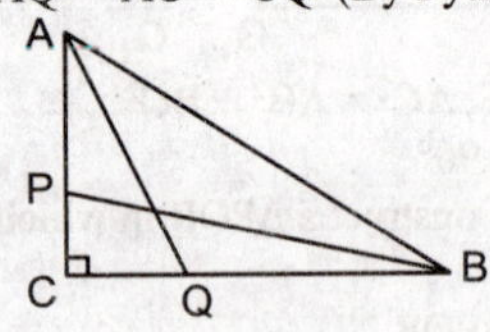

$$\Rightarrow \quad AQ^2 = AC^2 + \left(\frac{2}{3}BC\right)^2$$

$$\Rightarrow \quad AQ^2 = AC^2 + \frac{4}{9}BC^2$$

$$\Rightarrow \quad 9AQ^2 = 9AC^2 + 4BC^2 \quad ...(i)$$

(*ii*) Similarly, in ΔPCB,

$$9BP^2 = 9BC^2 + 4AC^2 \quad ...(ii)$$

(*iii*) Adding (*i*) and (*ii*), we get

$9(AQ^2 + BP^2) = 13AB^2$. ($\because AC^2 + BC^2 = AB^2$)

7. (*i*) **Given :** OD, OE and OF are perpendicular to the sides BC, AC and AB respectively.

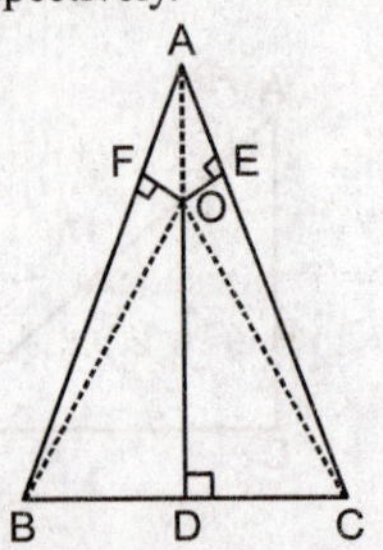

Construction : Join OA, OB and OC.

Proof. (*i*) In right angled ΔOAF

$$AF^2 = OA^2 - OF^2 \quad ...(i)$$

In right angled ΔOBD

$$BD^2 = OB^2 - OD^2 \quad ...(ii)$$

In right angled ΔOEC

$$CE^2 = OC^2 - OE^2 \quad ...(iii)$$

Adding (*i*), (*ii*) and (*iii*), we get

$$AF^2 + BD^2 + CE^2 = OA^2 + OB^2 + OC^2 - OF^2 - OD^2 - OE^2 \quad ...(iv)$$

(*ii*) From (*iv*), we have

$$AF^2 + BD^2 + CE^2 = (OA^2 - OE^2) + (OB^2 - OF^2) + (OC^2 - OD^2)$$

$= AE^2 + BF^2 + CD^2$ Proved.

8. Let $AB = DC = x$

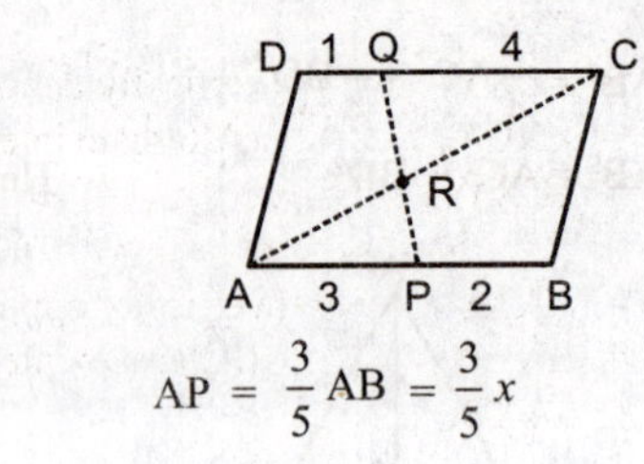

$$AP = \frac{3}{5}AB = \frac{3}{5}x$$

$$QC = \frac{4}{5}DC = \frac{4}{5}x$$

Now, in ΔARP and ΔCRQ

$\angle PAR = \angle RCQ$ (Alternate interior angles $\because AB \parallel CD$)

Also, $\angle ARP = \angle CRQ$ (Vertically opposite angles)

$\therefore$ $\Delta ARP \sim \Delta CRQ$ (AA similarity)

$$\Rightarrow \quad \frac{RC}{AR} = \frac{QC}{AP}$$

$$\Rightarrow \quad \frac{RC}{AR} = \frac{\frac{4}{5}x}{\frac{3}{5}x} = \frac{4}{3}$$

Using componendo,

$$\frac{RC + AR}{AR} = \frac{4+3}{3} \Rightarrow \frac{AC}{AR} = \frac{7}{3}$$

$$\Rightarrow \quad AR = \frac{3}{7}AC.$$

9. Area of $\Delta XYZ = \frac{1}{2} \times b \times XY = A$

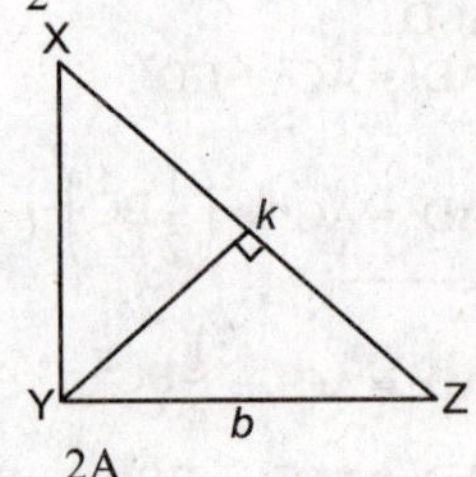

$$XY = \frac{2A}{b}$$

$$XZ = \sqrt{XY^2 + YZ^2} = \sqrt{\left(\frac{2A}{b}\right)^2 + b^2}$$

$$= \frac{\sqrt{4A^2 + b^4}}{b}$$

Area of $\Delta XYZ = \frac{1}{2} \times YK \times XZ = A$

$$\Rightarrow \frac{1}{2} \times YK \times \frac{\sqrt{4A^2 + b^4}}{b} = A$$

$$YK = \frac{2Ab}{\sqrt{4A^2 + b^4}}. \text{ Proved.}$$

10.

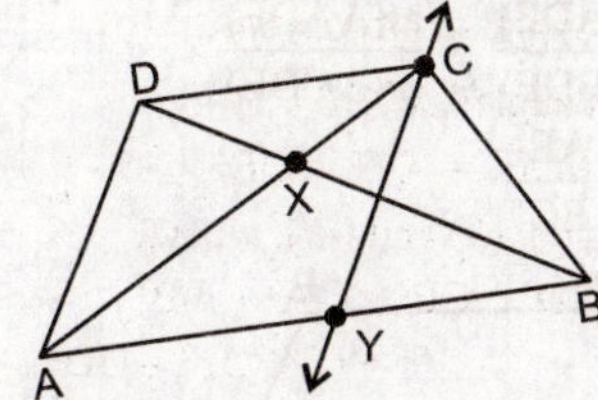

(*i*) ΔACD and ΔBCD are on same base CD and between parallel lines AB and CD

$\therefore$ Area ΔACD = Area ΔBCD

$\Rightarrow$ (Area ΔAXD + Area ΔXCD) = (Area ΔBXC + Area ΔBXC + Area ΔXCD)

$\Rightarrow$ Area ΔAXD = Area ΔBXC

(*ii*) In ΔAXD and ΔCXY

$\angle AXD = \angle CXY$ (vertically opposite angles)

$\angle DAX = \angle XCY$ (Alternate interior angles)

$\therefore$ $\Delta AXD \sim \Delta CXY$ ...(*i*)

(*iii*) In ΔCXD, AXB

$\angle CXD = \angle AXB$ (Vertically opposite angles)

$\angle DCX = \angle XAB$ (Alternate interior angles)

$\therefore$ $\Delta CXD \sim \Delta AXB$

$$\Rightarrow \frac{XC}{XA} = \frac{XD}{XB} = \frac{CD}{AB}$$

$$\Rightarrow \frac{XC}{XA} = \frac{XD}{XB} \Rightarrow XD = \frac{XC \cdot XB}{XA} \quad ...(ii)$$

Also ΔAXD ~ ΔCXY [From (*i*)]

$$\therefore \frac{XA}{XC} = \frac{XD}{XY} = \frac{AD}{CY}$$

$$\Rightarrow \frac{XA}{XC} = \frac{XD}{XY} \Rightarrow XD = \frac{XA \cdot XY}{XC} \quad ...(iii)$$

From (*ii*) and (*iii*)

$$\frac{XC \cdot XB}{XA} = \frac{XA \cdot XY}{XC}$$

$$\Rightarrow \frac{XB}{XY} = \frac{XA \cdot XA}{XC \cdot XC} \Rightarrow \frac{XB}{XY} = \frac{XA^2}{XC^2}$$

11.

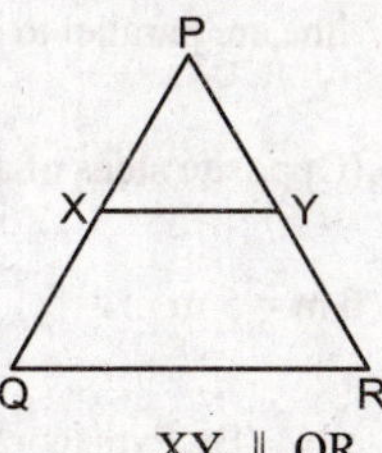

$\because$ XY ∥ QR

$$\therefore \frac{PX}{XQ} = \frac{PY}{YR} \quad \text{(Using B.P.T.)}$$

$$\Rightarrow \frac{1}{3} = \frac{PY}{4.5} \Rightarrow PY = 1.5\text{cm}$$

$\because$ XY ∥ QR

$\therefore$ $\angle PXY = \angle PQR$

$\angle PYX = \angle PRQ$ (corresponding angles)

$\therefore$ $\Delta PXY \sim \Delta PQR$

$$\Rightarrow \frac{PX}{PQ} = \frac{XY}{QR}$$

$$\Rightarrow \frac{1}{PX + XQ} = \frac{XY}{9} \Rightarrow \frac{1}{4} = \frac{XY}{9}$$

$$\Rightarrow XY = \frac{9}{4} = 2.25 \text{ cm}$$

$$\frac{\text{Area } \Delta PXY}{\text{Area } \Delta PQR} = \frac{XY^2}{QR^2} = \frac{(2.25)^2}{9^2} = \frac{1}{16}$$

$$\Rightarrow \frac{A}{\text{Area } \Delta PQR} = \frac{1}{16} \Rightarrow \text{Area of } \Delta PQR = 16A \text{ cm}^2$$

Area of trapezium XYRQ = Area ΔPQR – Area ΔPXY

$= 16A - A = 15A \text{ cm}^2$

12. (*i*) In ΔAGF and ΔADC,

$\angle AGF = \angle ADC$ (Each 90°)

$\angle GAF = \angle DAC$ (Each 45°)

$\therefore$ $\Delta AGF \sim \Delta ADC$

$$\Rightarrow \frac{AG}{AD} = \frac{AF}{AC} \quad ...(i)$$

$$\Rightarrow \frac{AC}{AD} = \frac{AF}{AG}$$

(*ii*) Also, $\angle DAC = \angle GAF$

$\Rightarrow \angle DAC - \angle GAC = \angle GAF - \angle GAC$

$\Rightarrow$ $\angle DAG = \angle CAF$

and $\dfrac{AG}{AD} = \dfrac{AF}{AC}$ [From (*i*)]

$\therefore$ $\Delta ACF \sim \Delta ADG.$

13.

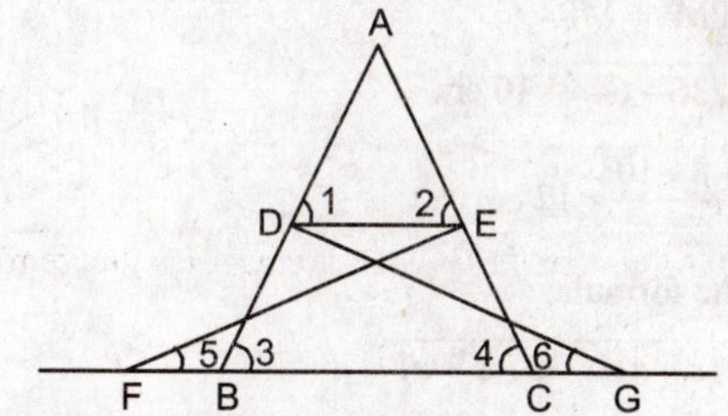

Given : ΔFEC ≅ ΔGBD

and $\angle 1 = \angle 2$

To Prove: ΔADE ~ ΔABC

Proof : In ΔADE,

$\angle 1 = \angle 2$ (Given)

$\Rightarrow$ AE = AD ...(*i*)

(Sides opposite to equal angles are equal)

ΔFEC ≅ ΔGBD (Given)

$\Rightarrow$ BD = EC (CPCT)...(*ii*)

$\angle 3 = \angle 4$

AE + EC = AD + BD

AC = AB ...(*iii*)

Dividing (*i*) by (*iii*),

$$\frac{AD}{AB} = \frac{AE}{AC}$$

and $\angle A = \angle A$ (Common)

$\therefore$ ΔADE ~ ΔABC (SAS similarity)

14. **Given :** A right-angled ΔABC right angled at B.

To Prove : $AC^2 = AB^2 + BC^2$

or $(\text{Hypotenuse})^2 = (\text{Perpendicular})^2 + (\text{base})^2$

Construction : From B draw BD ⊥ AC.

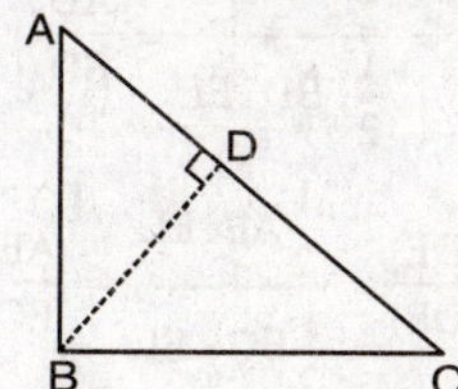

Proof : In right triangles ABC and ABD

$\angle ABC = \angle ADB$ (Each 90°)

$\angle BAC = \angle BAD$ (Common)

$\angle ABD = \angle ACB$ (Remaining angle)

$\therefore$ By AA Similarity, $\Delta ABC \sim \Delta ABD$

$\therefore \quad \frac{AC}{AB} = \frac{AB}{AD} \Rightarrow AB^2 = AC \times AD$ ... (*i*)

Similarly, $\Delta ABC \sim \Delta BCD$

$\therefore \quad \frac{AC}{BC} = \frac{BC}{DC} \Rightarrow BC^2 = AC \times DC$ ... (*ii*)

Adding (*i*) and (*ii*), we get

$AB^2 + BC^2 = AC \times (AD + DC) = AC \times AC$

$\therefore \quad AB^2 + BC^2 = AC^2$ Proved.

IInd Part :

Since PMR be a right triangle

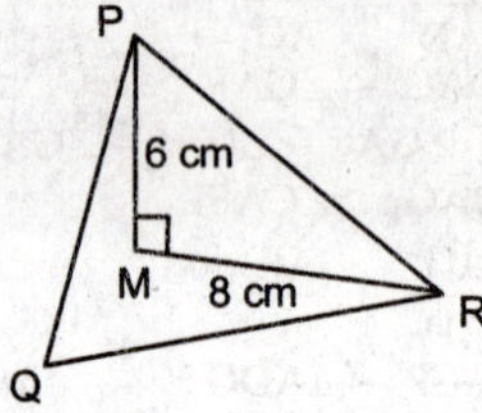

$\therefore \quad PR^2 = PM^2 + MR^2$

$\therefore \quad PR = \sqrt{36+64} = 10$ cm

Now, $s = \frac{6+8+10}{2} = 12$ cm

Now, using the formula,

$$\Delta = \sqrt{s(s-a)(s-b)(s-c)}$$

$$= \sqrt{12(12-10)(12-8)(12-6)}$$

$$= \sqrt{12\times2\times4\times6} = \sqrt{6\times2\times2\times4\times6}$$

$$= \sqrt{6^2\times4^2} = 24 \text{ sq. cm.}$$

15. Given : A triangle ABC, DE || BC, meeting AB at D and AC at E.

To Prove : $\frac{AD}{DB} = \frac{AE}{EC}$

Construction : Join BE, CD and draw EL ⊥ AD.

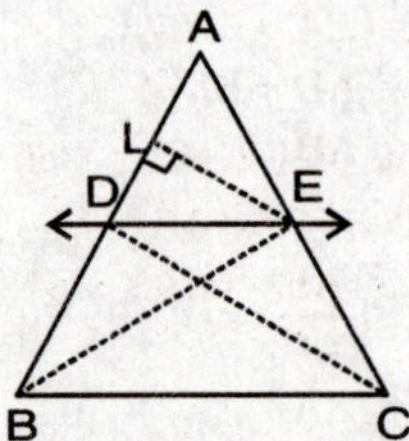

Proof : ΔBDE and ΔCDE are on the same base and between the same parallel BC and DE, hence equal in area, *i.e.*,

$ar(\Delta BDE) = ar(\Delta CDE)$ ... (*i*)

Now, $\frac{ar(\Delta ADE)}{ar(\Delta BDE)} = \frac{\frac{1}{2}.AD.EL}{\frac{1}{2}.BD.EL} = \frac{AD}{BD}$ ... (*ii*)

Similarly, $\frac{ar(\Delta ADE)}{ar(\Delta CDE)} = \frac{\frac{1}{2}AE.EL}{\frac{1}{2}EC\times EL} = \frac{AE}{EC}$ ... (*iii*)

Also, $\frac{ar(\Delta ADE)}{ar(\Delta BDE)} = \frac{ar(\Delta ADE)}{ar(\Delta CDE)}$ [Using (*i*)]

$\Rightarrow \frac{AD}{BD} = \frac{AE}{EC}$ [From (*ii*) and (*iii*)]

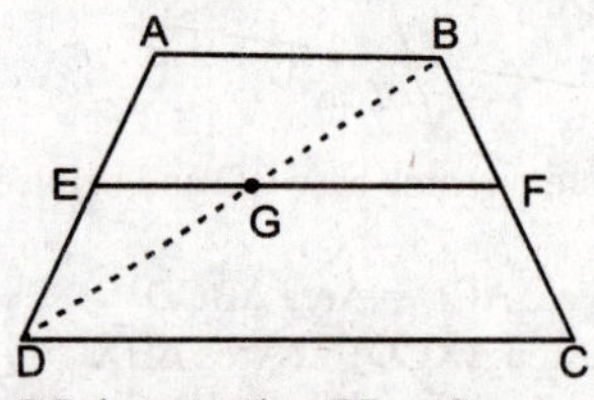

Join BD intersecting EF at G.

In ΔDAB, EG || AB

$\therefore \quad \frac{AE}{DE} = \frac{BG}{GD}$ (Using B.P.T.) ...(*i*)

In ΔDBC, GF || DC

$\therefore \quad \frac{BG}{GD} = \frac{BF}{FC}$ ...(*ii*)

From (*i*) and (*ii*)

$$\frac{AE}{DE} = \frac{BF}{FC}$$

16. IInd Part :

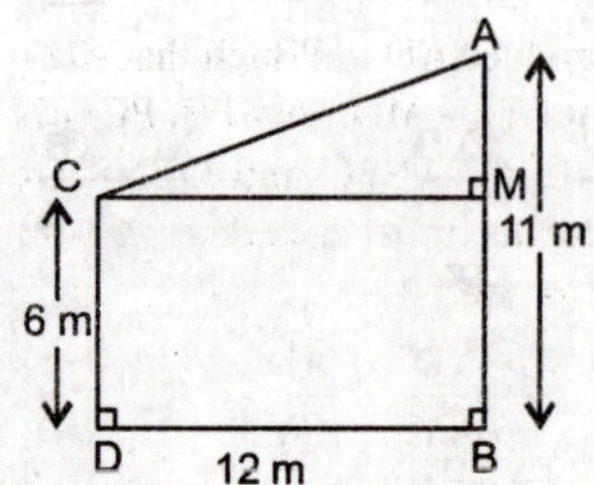

Given : CD ⊥ DB, and AB ⊥ DB

Prove : To find CA.

Construction : Draw CM || DB

Proof : CM || DM (By construction) ...(*i*)

CD ⊥ DB and AB ⊥ DB (Given)

$\Rightarrow$ CD || AB

or CD || MB (Two lines perpendicular to the same line are parallel to each other)

$\Rightarrow$ CDBM is a rectangle

$\Rightarrow$ CM = DB = 12 m (Opposite sides of a rectangle)

MB = CD = 6 m

$\Rightarrow$ AM = AB – BM = 11 m – 6 m = 5 m

In ΔAMC,

$AC^2 = AM^2 + CM^2$ (By Pythagoras theorem)

$= 5^2 + 12^2 = 25 + 144 = 169$

$\Rightarrow$ AC = 13 m

17.

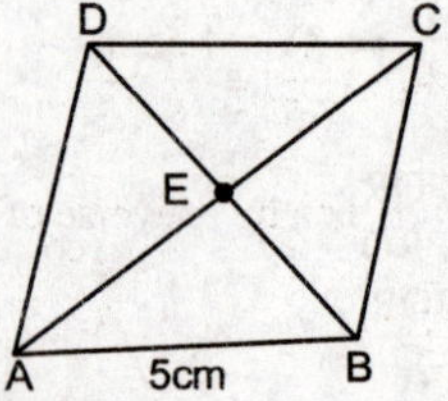

In rhombus ABCD

AB = 5 cm, AC = 6 cm

$\because$ diagonal of a rhombus bisect each other at right angle

∴ In right ΔAEB

$$AB^2 = AE^2 + BE^2 \Rightarrow 5^2 = \left(\frac{1}{2}\times 6\right)^2 + BE^2$$

$\Rightarrow \quad 25 = 9 + BE^2$

$\Rightarrow \quad BE = 4$

$\because \quad BD = 2BE$

$\Rightarrow \quad BD = 2 \times 4 = 8$ cm

18.

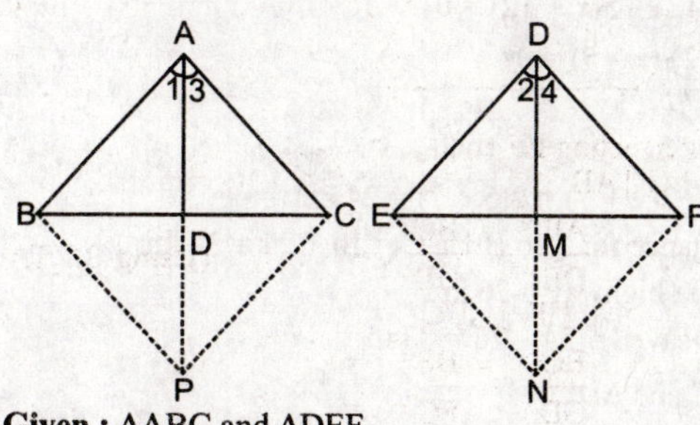

Given : ΔABC and ΔDEF,

$$\frac{AB}{DE} = \frac{AC}{DF} = \frac{AL}{DM} \quad ...(i)$$

To Prove : ΔABC ~ ΔDEF

Construction : Produce AD to P such that AD = DP and produce DM to N such that DM = MN. Join PB, PC, EN and FN.

Proof : Quadrilaterals ABPC and DENF are parallelograms because their diagonals bisect each other at D and M respectively.

$\Rightarrow \quad BP = AC$ (Opp. sides of || gm)

and $\quad EN = DF$

$$\Rightarrow \quad \frac{BP}{EN} = \frac{AC}{DF}$$

$$\Rightarrow \quad \frac{BP}{EN} = \frac{AB}{DE} \quad ...(i)$$

$$i.e., \quad \frac{AB}{DE} = \frac{BP}{EN} \quad ...(ii)$$

$$\text{From } (i) \quad \frac{AB}{DE} = \frac{AD}{DM} = \frac{2AD}{2DM} = \frac{AP}{DN}$$

$$i.e., \quad \frac{AB}{DE} = \frac{AP}{DN} \quad ...(iii)$$

From (*ii*) and (*iii*), we have

$$\frac{AB}{DE} = \frac{BP}{DN} = \frac{AP}{DN}$$

$\Rightarrow \quad \Delta ABE \sim \Delta DEN$

$\Rightarrow \quad \angle 1 = \angle 2 \quad ...(iv)$

Similarly, we can prove

$\Delta ACP \sim \Delta DFN$

$\Rightarrow \quad \angle 3 = \angle 4 \quad ...(v)$

Adding (*iv*) and (*v*) we have

$\angle 1 + \angle 3 = \angle 2 + \angle 4$

$\Rightarrow \quad \angle A = \angle D$

$\Rightarrow \quad \Delta ABC \sim \Delta PQR$ (SAS similarity criterion)

19.

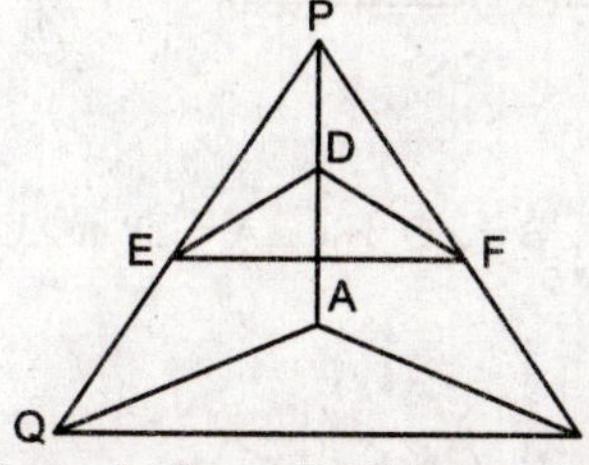

Given : In figure, DE || AQ and DF || AR.

Proof : $\quad EF \parallel QR$

In ΔPAQ, $\quad DE \parallel AQ$ (Given)

$$\therefore \quad \frac{PE}{EQ} = \frac{PD}{DA} \quad \text{(Using BPT) ...}(i)$$

In ΔPAR, $\quad DF \parallel AR$ (Given)

$$\therefore \quad \frac{PD}{DA} = \frac{PF}{FR} \quad ...(ii)$$

From (*i*) and (*ii*),

$$\frac{PE}{EQ} = \frac{PF}{FR} \text{ in } \Delta PQR$$

$\Rightarrow \quad EF \parallel QR$ (Using converse of BPT)

NCERT EXEMPLAR PROBLEMS

Short Answer Questions with Reasoning

1. No, (false) ΔDEF ~ ΔRPQ

$\because \quad \angle D = \angle R$

By order of symmetry

$\angle E = \angle P$

and $\quad \angle F = \angle Q.$

2. Yes,

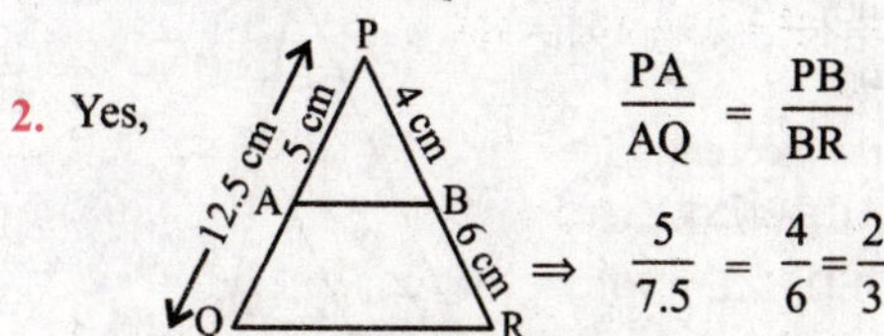

$$\frac{PA}{AQ} = \frac{PB}{BR}$$

$$\Rightarrow \quad \frac{5}{7.5} = \frac{4}{6} = \frac{2}{3}$$

If the sides are in proportion then by converse of BPT AB || QR.

3. No,

ΔQPR ~ ΔSTM

By symmetry

$\angle P = \angle T$

$\angle Q = \angle S$

$\angle R = \angle M$

4. No, (false) ∵ If all the angles of a quadrilateral are equal but all the sides are not in proportion *e.g.,*

In rectangle and square all angles are 90° but sides are not in proportion.

5. Yes,

$$\therefore \quad \frac{a}{b} = \frac{c}{d} = \frac{e}{f} = \frac{a+c+e}{b+d+f}$$

If two sides are in proportion and sum of three sides are in proportion then definitely third side is also in proportion.

6. No, ∵ If two triangles are similar, then ratio of areas of two Δs is equal to ratio of squares of their corresponding sides and squares of their corresponding altitudes.

So, If ΔABC ~ ΔPQR , AD ⊥ BC , PS ⊥ QR

$$\text{So,} \quad \frac{ar(\Delta ABC)}{ar(\Delta PQR)} = \frac{AD^2}{PS^2} = \left(\frac{3}{5}\right)^2 = \frac{9}{25}.$$

7. No, (false) In the given figure Two sides are in proportion then by BPT

DE || BC

So, $\angle D = \angle B$ (Corresponding angles)

$\angle A = \angle A$ (Common angles)

So, $\Delta ADE \sim \Delta ABC$ (AA simlarity)

But in two separate Δs, angle must be included one in between proportional sides such that :

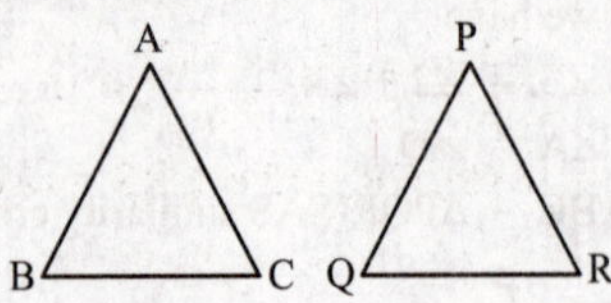

$$\frac{AB}{PQ} = \frac{BC}{QR}$$

(For similarity $\angle B = \angle Q$; Not $\angle A = \angle P$ or $\angle C = \angle R$).

Short Answer Questions

8. PQ || RS

$PQ = 3RS$

$$\Rightarrow \quad \frac{PQ}{RS} = \frac{3}{1}$$

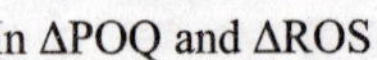

In ΔPOQ and ΔROS

$\angle P = \angle R$ (∵ PQ || RS alternate interior angles)

$\angle Q = \angle S$

∴ $\Delta POQ \sim \Delta ROS$ (AA similarity)

$$\frac{ar(\Delta POQ)}{ar(\Delta ROS)} = \frac{PQ^2}{RS^2}$$

[Ratio of area of two similar triangles is equal to ratio of square of their corresponding sides]

$$\frac{ar(\Delta POQ)}{ar(\Delta ROS)} = \left(\frac{3}{1}\right)^2 = \frac{9}{1}.$$

9. AB || DC

ΔAPO and ΔCQO

$\angle A = \angle C$

$\angle P = \angle Q$ (Alternate interior angles)

$\Delta APO \sim \Delta CQO$ (AA similarity)

$$\frac{OA}{OC} = \frac{AP}{CQ} \Rightarrow OA.CQ = OC.AP.$$

Long Answer Questions

10. A.T.Q.

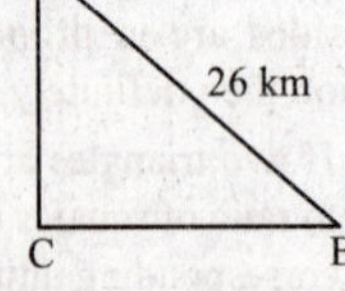

Let $AC = 2x$ km

$CB = 2(x + 7)$ km

In rt. angled ΔACB

$AB^2 = AC^2 + CB^2$

Highway is AB = 26 km

$$(26)^2 = (2x)^2 + \{2(x + 7)\}^2$$

$$\Rightarrow \quad 676 = 4x^2 + 4(x + 7)^2$$

$$\Rightarrow \quad \frac{676}{4} = x^2 + x^2 + 49 + 14x$$

$$\Rightarrow \quad 169 = 2x^2 + 14x + 49$$

$\Rightarrow 2x^2 + 14x + 49 - 169 = 0$

$\Rightarrow 2x^2 + 14x - 120 = 0$

$\Rightarrow x^2 + 7x - 60 = 0$

$\Rightarrow x^2 + 12x - 5x - 60 = 0$

$\Rightarrow x(x + 12) - 5(x + 12) = 0$

$\Rightarrow (x + 12)(x - 5) = 0$

$\Rightarrow x = -12, 5$

So, $AC = 2x = 2 \times 5 = 10$ km

$BC = 2(x + 7) = 2(5 + 7) = 24$ km

Total distance for A to B via C = 10 + 24 = 34 km

A to B via highway = 26 km

Distance saved = 34 – 26 = 8 km.

11. In ΔABC and ΔDEC,

$\angle A = \angle D$ (Each 90°)

$\angle C = \angle C$ (Common angle)

$\Delta ABC \sim \Delta DEC$, (AA similarity)

$$\frac{AB}{DE} = \frac{AC}{DC}$$

$$\frac{6}{1.5} = \frac{AC}{DC}$$

$$\frac{6}{1.5} \times 3 = AC$$

$$12 = AC$$

Distance between woman and pole = AC – DC.

$12 - 3 = 9$ m

12. In ΔPDQ,

$\angle D = 90°$

By Pythagoras theorem

$PQ^2 = QD^2 + PD^2$

$a^2 = c^2 + PD^2$ ... (*i*)

Similarly, In ΔPDR, $\angle D = 90°$

By Pythagoras theorem

$PR^2 = PD^2 + DR^2$

$b^2 = d^2 + PD^2$ ... (*ii*)

Subtracting (*ii*) from (*i*),

$a^2 - b^2 = c^2 - d^2$

$\Rightarrow (a - b)(a + b) = (c - d)(c + d).$

13. $\because \angle A + \angle D = 90°$

In ΔAED∠E = 90°

In ΔAEC∠E = 90°

So, by Pythagoras theorem

$AC^2 = CE^2 + AE^2$ ... (*i*)

Similarly, in ΔBED, ∠E = 90°

$BD^2 = BE^2 + DE^2$ ... (*ii*)

Adding (*i*) and (*ii*),

$AC^2 + BD^2 = CE^2 + AE^2 + BE^2 + DE^2$

$= CE^2 + BE^2 + AE^2 + DE^2$

$= BC^2 + AD^2$ [In ΔAED ∠E = 90° ∵ $AD^2 = AE^2 + DE^2$ or In ΔBEC, $BC^2 = CE^2 + BE^2$]

14. If three lines are perpendicular to one line then they are parallel to each other.

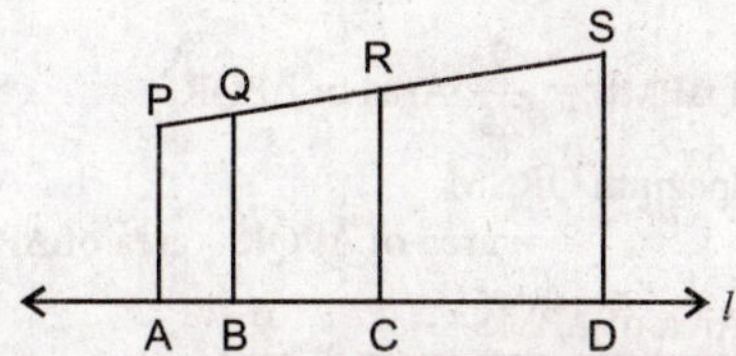

And if three lines (or more than three lines) are parallel to each other then their transversal are in proportion.

So, $\frac{AB}{PQ} = \frac{BC}{QR} = \frac{CD}{RS}$

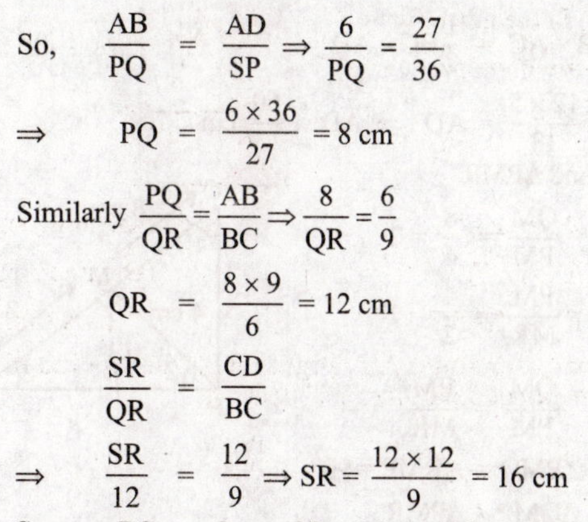

So, $\frac{AB}{PQ} = \frac{AD}{SP} \Rightarrow \frac{6}{PQ} = \frac{27}{36}$

$\Rightarrow \quad PQ = \frac{6 \times 36}{27} = 8 \text{ cm}$

Similarly $\frac{PQ}{QR} = \frac{AB}{BC} \Rightarrow \frac{8}{QR} = \frac{6}{9}$

$QR = \frac{8 \times 9}{6} = 12 \text{ cm}$

$\frac{SR}{QR} = \frac{CD}{BC}$

$\Rightarrow \quad \frac{SR}{12} = \frac{12}{9} \Rightarrow SR = \frac{12 \times 12}{9} = 16 \text{ cm}$

So, $PQ = 8$ cm, $QR = 12$ cm, $RS = 16$ cm.

HOTS QUESTIONS

1. In ΔABC, ΔDEF

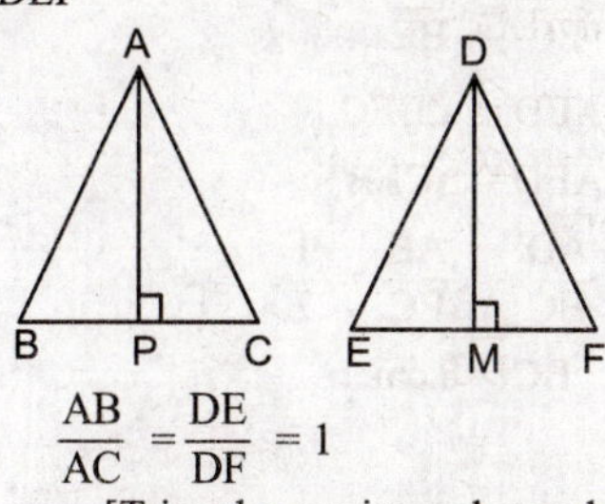

$\frac{AB}{AC} = \frac{DE}{DF} = 1$

[Triangles are isosceles and have equal angles]

As $\angle A = \angle D$

$\therefore \quad \Delta ABC \sim \Delta DEF$

$\Rightarrow \quad \frac{ar\,(\Delta ABC)}{ar\,(\Delta DEF)} = \frac{BC^2}{EF^2}$

$\Rightarrow \quad \frac{81}{25} = \frac{BC^2}{EF^2}$

$\Rightarrow \quad \frac{BC}{EF} = \frac{9}{5}$

Also $\frac{ar\,(\Delta ABC)}{ar\,(\Delta DEF)} = \frac{\frac{1}{2} BC \times AP}{\frac{1}{2} EF \times DM}$

$\Rightarrow \quad \frac{81}{25} = \frac{9}{5} \times \frac{AP}{DM}$

$\Rightarrow \quad \frac{AP}{DM} = 9 : 5.$

2. $\because \quad \Delta ABC \sim \Delta DEF$ (Given)

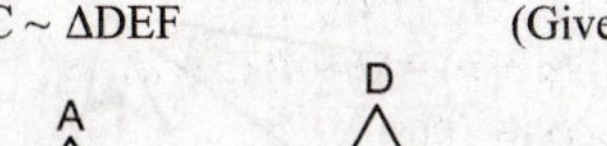

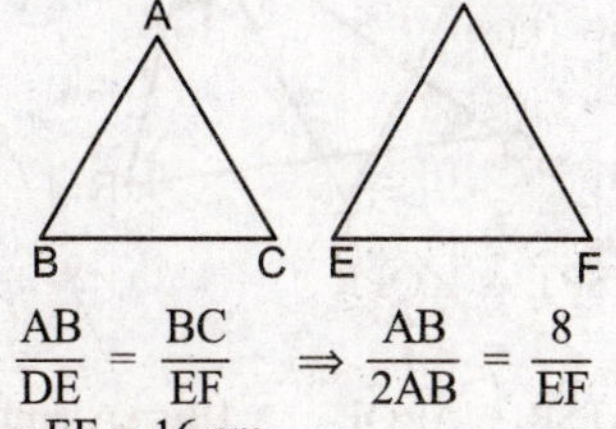

$\therefore \quad \frac{AB}{DE} = \frac{BC}{EF} \Rightarrow \frac{AB}{2AB} = \frac{8}{EF}$

$\Rightarrow \quad EF = 16 \text{ cm}$

3. $\because \quad \Delta ABC \sim \Delta DEF$

$\therefore \quad \frac{AB}{DE} = \frac{BC}{EF} = \frac{AC}{DF}$

$\Rightarrow \quad \frac{3}{DE} = \frac{2}{4} = \frac{2.5}{DF}$

$\Rightarrow \quad DE = 6 \text{ cm}, DF = 5 \text{ cm}$

$\therefore$ Perimeter of ΔDEF = 4 + 6 + 5 = 15 cm

4. $\because$ XY ∥ BC and BY is a transversal

$\therefore \quad \angle 1 = \angle 3$ (Alternate angles) ...(*i*)

Also BY bisects ∠XYC

$\Rightarrow \quad \angle 1 = \angle 2$ ...(*ii*)

From (*i*) and (*ii*) $\angle 2 = \angle 3$

$\therefore$ BC = CY (Sides opp. to equal angles are equal)

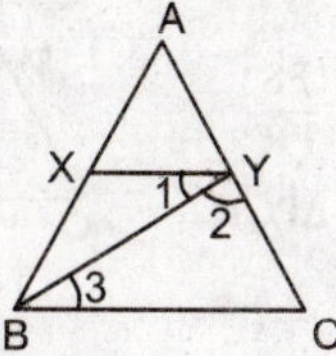

5. In ΔABC and ΔEDF

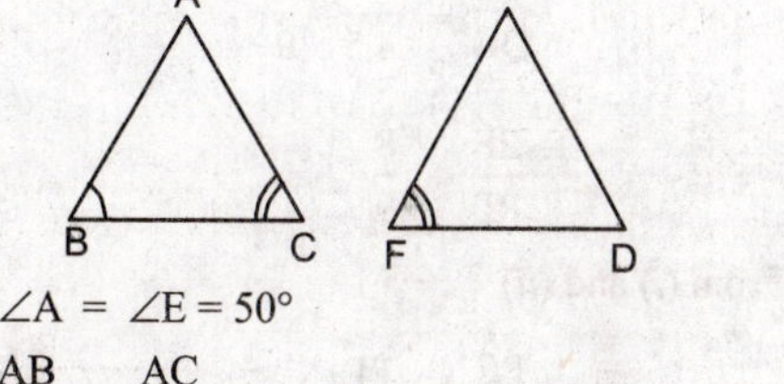

$\angle A = \angle E = 50°$ (Given)

$\frac{AB}{ED} = \frac{AC}{EF}$ (Given)

$\therefore$ ΔABC ~ ΔEDF

$\Rightarrow \quad \angle C = \angle F = 65°$ ($\angle F = 65°$)

$\therefore \quad \angle B = 180° - 50° - 65° = 65°$

$\Rightarrow \quad \angle B = \angle C$

$\Rightarrow \quad AB = AC$ (Sides opp. to equal angles)

6.

$AB^2 + AC^2 = BC^2$ (By Pythagoras theorem)

$\Rightarrow \quad 25 + 144 = BC^2$

$\Rightarrow \quad BC = 13 \text{ cm}$

Area $\Delta ABC = \frac{1}{2} AC \times AB$

Area of $\Delta ABC = \frac{1}{2} BC \times AD$

$\Rightarrow \quad \frac{1}{2}AB \times AC = \frac{1}{2}BC \times AD$

$\frac{12 \times 5}{13} = AD \Rightarrow AD = \frac{60}{13}$ cm

7. In ΔQMP and ΔPMR

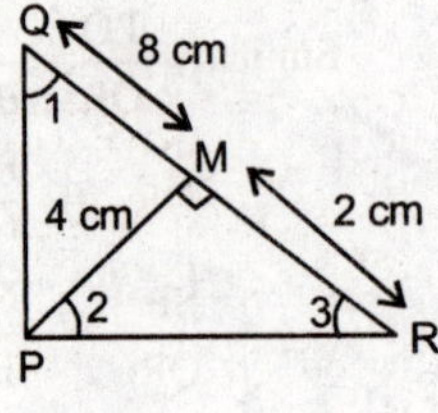

$\frac{QM}{PM} = \frac{8}{4} = 2$

and $\frac{PM}{MR} = \frac{4}{2} = 2$

$\Rightarrow \quad \frac{QM}{PM} = \frac{PM}{MR}$

as $\angle PMQ = \angle RMP = 90°$

$\therefore \quad \Delta QMP \sim \Delta PMR$

$\therefore \quad \angle 1 = \angle 2$

Also $\angle 2 + \angle 3 = 90°$

$\Rightarrow \quad \angle 1 + \angle 3 = 90°$

In ΔQPR,

$\angle P + \angle 1 + \angle 3 = 180°$

$\therefore \quad \angle P = 90°$

8. ∵ QRSM is a trapezium

$\therefore \quad QR \parallel MS$

$\Rightarrow \quad \frac{PM}{PQ} = \frac{PS}{PR}$

$\therefore \quad \Delta PMS \sim \Delta PQR$

$\Rightarrow \quad \frac{ar(\Delta PMS)}{ar(\Delta PQR)} = \frac{MS^2}{QR^2} = \frac{9}{25}$

$\Rightarrow$ Area of $\Delta PMS = \frac{9}{25}$ Area of ΔPQR

Area of trapezium QRSM

= area of ΔPQR – area of ΔPMS

$\therefore \quad \frac{\text{Area of } \Delta PMS}{\text{Area of trapezium QRSM}} = \frac{9}{16}$

9.

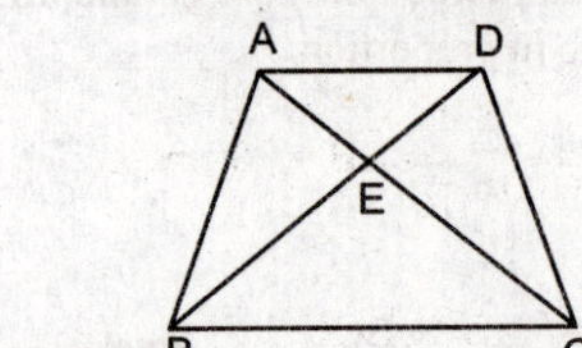

$\because \quad \frac{AE}{EC} = \frac{DE}{BE}$

$\angle AED = \angle BEC$

$\therefore \quad \Delta AED \sim \Delta CEB$

$\therefore \quad \frac{AD}{BC} = \frac{AE}{EC} = \frac{1}{2}$

$\Rightarrow \quad BC = 8$ cm.

VALUE BASED QUESTIONS

1. (*a*) $\frac{PE}{QE} = \frac{4}{4.5} = \frac{8}{9}$ ...(*i*)

$\frac{PF}{RF} = \frac{8}{9}$ ...(*ii*)

From (*i*) and (*ii*)

$\frac{PE}{QE} = \frac{PF}{RF}$

$\therefore$ EF ∥ QR (yes)

(*b*) Value : Environmentally sensitive

2. (*a*)

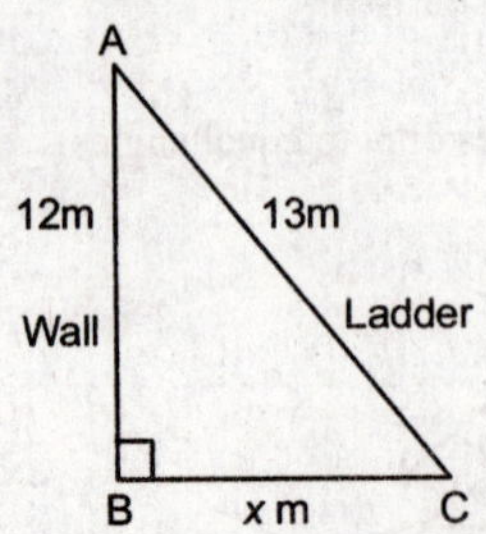

Using Pythagoras theorem

$(13)^2 = (12)^2 + x^2 \Rightarrow x^2 = 25$

$\therefore \quad x = 5$ m.

So, distance of the foot of ladder from the wall = 5m.

(*b*) Values : Respect and care for old persons.

3. (*a*) In ΔABC and ΔBCD

$BC^2 = AC \times CD \Rightarrow \frac{BC}{CD} = \frac{AC}{BC}$

$\angle C = \angle C$ (common)

$\therefore \Delta ABC \sim \Delta BDC$

$\Rightarrow \quad \frac{AB}{BD} = \frac{AC}{BC} \quad \because AB = AC$ (given)

$\therefore BD = BC$

(*b*) Respect for women.

4. (*a*) $\because \quad \angle 1 = \angle 2$ [Given]

$\Rightarrow \angle 1 + \angle QPT = \angle 2 + \angle QPT$

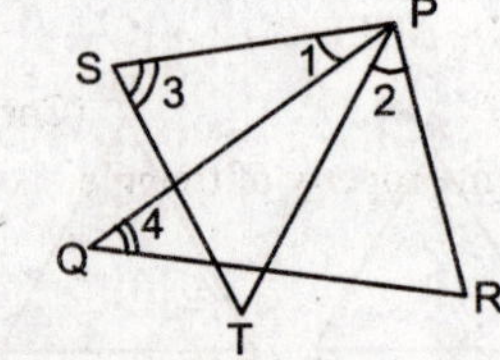

$\Rightarrow \quad \angle SPT = \angle QPR$

Also $\angle 3 = \angle 4$ [Given]

$\Rightarrow \quad \Delta TSP \sim \Delta RQP$ [By AA similarity]

$\Rightarrow \quad \frac{ST}{QR} = \frac{PT}{PR} \Rightarrow ST. PR = PT.QR.$

(*b*) Value : Patriotism

MULTIPLE CHOICE QUESTIONS

1. **(c)** In right ΔATS

$$AS^2 = AT^2 + TS^2$$
$$\Rightarrow 17^2 = 15^2 + TS^2$$
$$\Rightarrow TS^2 = 64 \Rightarrow TS = 8 \text{ cm}$$

∵ ΔATS ~ ΔABC (AA similarity)

$$\Rightarrow \frac{AT}{AB} = \frac{TS}{BC}$$
$$\Rightarrow \frac{15}{AB} = \frac{8}{16} \Rightarrow AB = 30 \text{ cm}$$

In right Δ ABC

$$AC^2 = AB^2 + BC^2$$
$$AC^2 = 900 + 256 = 1156$$
$$AC = \sqrt{1156} = 34 \text{ cm}$$
$$TC = AC - AT = 34 - 15 = 19 \text{ cm}$$

2. **(b)** Here $\frac{PA}{PQ} = \frac{2.4}{6} = 0.4$

$$\frac{PB}{PR} = \frac{3.2}{8} = 0.4$$
$$\Rightarrow \frac{PA}{PQ} = \frac{PB}{PR}, \angle P = \angle P$$

∴ ΔPAB ~ ΔPQR

$$\Rightarrow \frac{AP}{PQ} = \frac{AB}{QR} \Rightarrow \frac{2.4}{6} = \frac{2}{x} \Rightarrow x = 5 \text{ cm}$$

3. **(d)** $\frac{AB}{PQ} = \frac{BC}{QR} = \frac{AC}{PR} \Rightarrow \frac{6}{4.5} = \frac{4}{x} \Rightarrow x = 3$

4. **(c)** Similar figures.

5. **(b)** Principle of similarity of figures.

6. **(d)** Their corresponding sides are in the same ratio.

7. **(c)** Quadrilaterals are similar if their angles are equal and corresponding sides are proportional.

8. **(c)** Thales Theorem.

9. **(b)** Converse of Basic Proportionality Theorem.

10. **(d)** ∵ DE ∥ BC

∴ ∠ABC = 70°. (Corresponding angles)

Using angle sum property of triangle

∠BCA = 60°.

11. **(d)** ∵ MN ∥ QR

$$\frac{PM}{MQ} = \frac{PN}{NR} \Rightarrow \frac{3}{4} = \frac{6}{NR} \Rightarrow NR = 8 \text{ cm}$$

∴ $x = 8 + 6 = 14$ cm.

12. **(a)** DE ∥ BC, if $\frac{AD}{DB} = \frac{AE}{EC} \Rightarrow \frac{1.28}{2.56} = \frac{0.64}{EC}$

⇒ EC = 1.28 cm.

13. **(d)** ΔABC ~ ΔDEF.

14. **(c)** ΔABC ~ ΔEFD.

15. **(d)** ΔQRP ~ ΔSMN.

16. **(d)** $\frac{ar(\Delta ABC)}{ar(\Delta BDE)} = \frac{BC^2}{BD^2}$

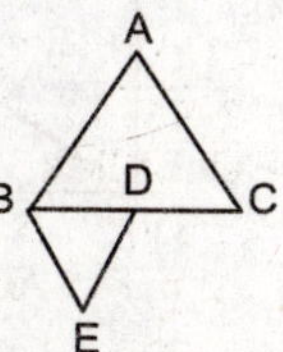

$$= \frac{(2BD)^2}{BD^2} = 4:1.$$

17. **(b)** Ratio of areas of two similar triangles is the square of the ratio of their corresponding sides.

∴ Required ratio = 9 : 49.

18. **(c)** On verification, triangle with sides 7 cm, 24 cm and 25 cm is a right triangle

∵ $(25)^2 = (7)^2 + (24)^2$.

19. **(d)** RHS similarity criterion.

20. **(d)**

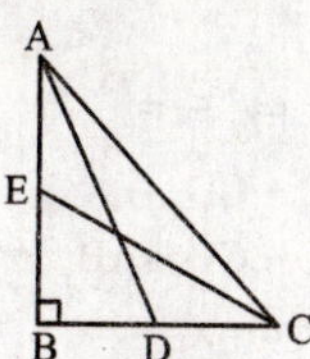

In right angled ΔABC,

$$AD^2 + CE^2 = AB^2 + BD^2 + BC^2 + BE^2$$
$$= AB^2 + \frac{1}{4}BC^2 + BC^2 + \frac{1}{4}AB^2$$
$$\Rightarrow 4(AD^2 + CE^2) = 5(AB^2 + BC^2)$$
$$\Rightarrow 4(25 + 40) = 5(AC^2)$$
$$\Rightarrow AC^2 = \frac{4 \times 65}{5}$$
$$\Rightarrow AC = 2\sqrt{13} \text{ cm.}$$

21. **(b)**

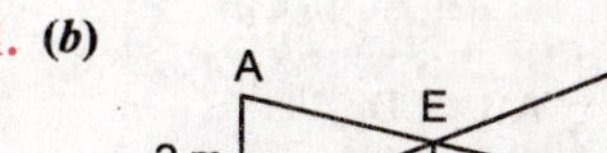

Using the result

$$\frac{1}{AB} + \frac{1}{CD} = \frac{1}{EF}$$

We get $\frac{1}{2} + \frac{1}{8} = \frac{1}{EF}$

$$\therefore \frac{5}{8} = \frac{1}{EF}$$
$$\Rightarrow EF = 1.6 \text{ m.}$$

22. **(b)** ΔBCA ~ ΔBDE

$$\frac{ar(\Delta BCA)}{ar(\Delta BDE)} = \frac{BC^2}{BD^2} = \frac{(20)^2 - (12)^2}{(10)^2}$$
$$\frac{ar(\Delta BCA)}{ar(\Delta BDE)} = \frac{256}{100}$$
$$ar(\Delta BDE) = \frac{100}{256} ar(\Delta BCA)$$
$$= \frac{100}{256} \times \frac{1}{2} \times 12 \times 16 = \frac{75}{2} \text{ cm}^2$$
$$ar(\text{quad. } ADEC) = \frac{1}{2} \times 12 \times 16 - \frac{75}{2} = 96 - \frac{75}{2}$$
$$= \frac{192 - 75}{2} = \frac{117}{2} \text{ cm}^2 = 58\frac{1}{2} \text{ cm}^2$$

23. **(b)**

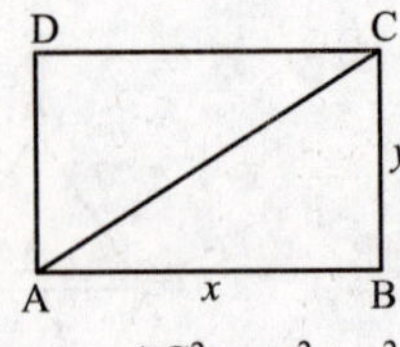

$$AC^2 = x^2 + y^2$$

Also $\quad AC = x + y - \frac{1}{2}x = \frac{1}{2}x + y$

$$\left(\frac{1}{2}x + y\right)^2 = x^2 + y^2$$

$\Rightarrow \quad \frac{1}{4}x^2 + y^2 + xy = x^2 + y^2$

$\Rightarrow \quad xy = \frac{3}{4}x^2$

$\Rightarrow \quad \frac{x}{y} = \frac{4}{3} \Rightarrow \frac{y}{x} = \frac{3}{4}.$

24. **(a)** $\because \quad AB + AD = BC + CD$

$\Rightarrow \quad AD = BC + CD - AB$

$AD = h + d - x$

In right angled ΔACD,

$AD^2 = AC^2 + DC^2$

$(h + d - x)^2 = (x + h)^2 + d^2.$

$(h + d - x)^2 - (h + x)^2 = d^2$

$\Rightarrow (h + d - x - h - x)(h + d - x + h + x) = d^2$

$\Rightarrow \quad (d - 2x)(2h + d) = d^2$

$\Rightarrow \quad 2hd + d^2 - 4hx - 2xd = d^2$

$\Rightarrow \quad 2hd = 4hx + 2xd$

$\Rightarrow \quad x = \frac{hd}{2h + d}.$

25. **(a)** Area of ΔABC, (A) $= \frac{1}{2}$ AB $\times$ CD

$\Rightarrow \quad A = \frac{1}{2} \times c \times CD$

$\Rightarrow \quad CD = \frac{2A}{c}$. (A = Area)

26. **(c)** $\because$ Area of DBCE $= \frac{1}{2}$ Area of ΔABC

$\therefore$ Area of $\Delta ADE = \frac{1}{2}$ Area of ΔABC

$\because \quad DE \parallel BC$

$\Rightarrow \quad \frac{ar(\Delta ADE)}{ar(\Delta ABC)} = \frac{DE^2}{BC^2}$

$\Rightarrow \quad \frac{1}{2} = \frac{DE^2}{4} \Rightarrow DE = \sqrt{2}$ cm.

27. **(d)** We see that $\Delta AB_1C_1 \sim \Delta ABC$

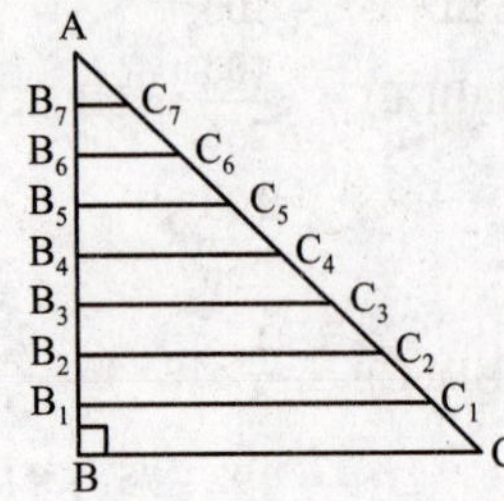

$\therefore \quad \frac{AB}{AB_1} = \frac{BC}{B_1C_1}$

$\Rightarrow \quad \frac{8}{7} = \frac{10}{B_1C_1}$

$\Rightarrow \quad B_1C_1 = \frac{10 \times 7}{8}$

Similarly, $\quad B_2C_2 = \frac{10 \times 6}{8}$

$B_3C_3 = \frac{10 \times 5}{8}$ and so on.

$\therefore$ Sum of lengths of seven line segments

$$= \frac{10 \times 7}{8} + \frac{10 \times 6}{8} + \frac{10 \times 5}{8} + ... + \frac{10 \times 1}{8}$$

$$= \frac{10}{8}[7 + 6 + 5 + ... + 1] = \frac{10}{8} \times 28 = 35 \text{ cm}.$$

28. **(c)**

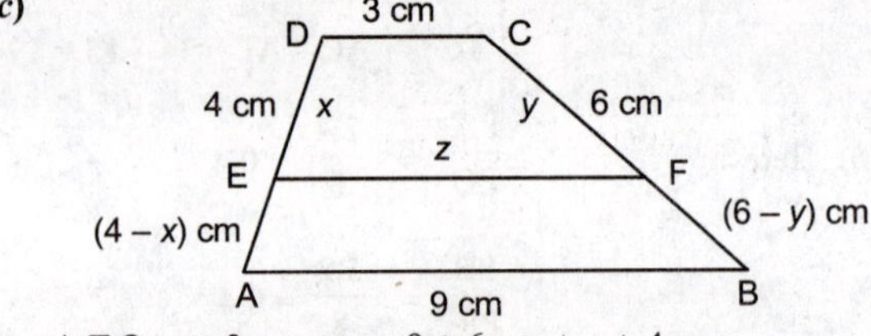

A.T.Q. $x + 3 + y + z = 9 + 6 - y + z + 4 - x$

$\Rightarrow \quad x + 3 + y = 19 - x - y$

$\Rightarrow \quad 2x + 2y = 16$

$\Rightarrow \quad x + y = 8$

$\Rightarrow \quad y = 8 - x$

Also $\quad \frac{x}{4 - x} = \frac{y}{6 - y}$

$\Rightarrow \quad \frac{x}{4 - x} = \frac{8 - x}{6 - 8 + x}$

$\Rightarrow \quad x(x - 2) = (8 - x)(4 - x)$

$\Rightarrow \quad x^2 - 2x = 32 - 12x + x^2$

$\Rightarrow \quad x = 3.2$ cm

$\therefore \quad$ Ratio $= \frac{3.2}{4 - 3.2} = \frac{4}{1}$

29. **(d)**

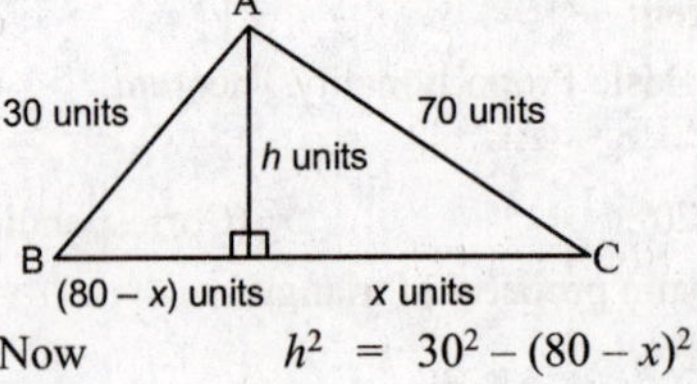

Now $\quad h^2 = 30^2 - (80 - x)^2$

Also $\quad h^2 = 70^2 - x^2$

$\Rightarrow 30^2 - (80 - x)^2 = 70^2 - x^2$

$\Rightarrow \quad 160x = 10400 \Rightarrow x = 65$ units.

30. **(b)**

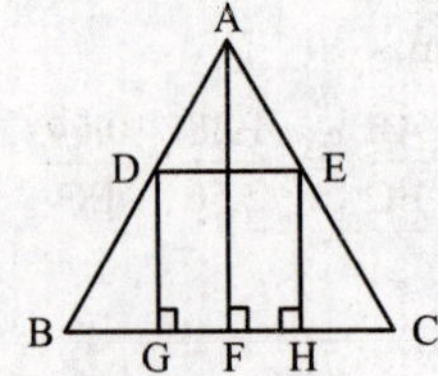

In Δ AFC, AF $\parallel$ HE

$\Rightarrow \quad \Delta AFC \sim \Delta EHC$

$\Rightarrow \quad \frac{AF}{HE} = \frac{FC}{HC}$

$\Rightarrow \quad \frac{24}{6} = \frac{6}{HC} \Rightarrow HC = 3/2$ cm

$\Rightarrow \quad FH = 6 - 3/2 = 9/2$ cm

Similarly $\quad GF = 9/2$

$\therefore \quad GH = \frac{9}{2} + \frac{9}{2} = 9$ cm

Area of rectangle GHED = GH × HE = 9 × 6 = 54 cm^2

31. **(*b*)** $\because$ MN || QR

$\therefore \quad \frac{PM}{MQ} = \frac{PN}{NR}$

$\Rightarrow \quad \frac{x}{10} = \frac{x-2}{6}$

$\Rightarrow \quad 6x = 10x - 20$

$\Rightarrow \quad x = 5$ cm.

32. **(*b*)** $\frac{ar(\Delta ABC)}{ar(\Delta PQR)} = \frac{AD^2}{PM^2}$

$\therefore \quad \frac{81}{121} = \frac{9^2}{PM^2}$

$\Rightarrow \quad PM = 11$ cm.

33. **(*c*)** $\because \quad \frac{CF}{FB} = \frac{CE}{AE}$

$\therefore \quad EF \parallel AB.$

34. **(*b*)** Facts of triangle.

35. **(*c*)** Properties of triangles.

36. **(*b*)** $\Delta PSQ \sim \Delta PSR$ (AA similarity)

$\therefore \quad \frac{PQ}{PR} = \frac{SQ}{SR}$

$\Rightarrow \quad \frac{10}{8} = \frac{5}{x} \Rightarrow x = 4$ cm

37. **(*c*)** $AC^2 = 144\ cm^2$, $AB^2 = 108\ cm^2$

$BC^2 = 36\ cm^2$

Now $AB^2 + BC^2 = 144 = AC^2$

$\therefore \quad \angle B = 90°$

38. **(*c*)** Quadrilateral ABCD ~ Quadrilateral PQRS

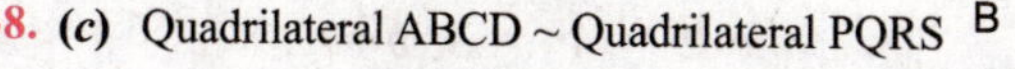

$\therefore \quad \frac{AD}{PS} = \frac{AB}{PS} \Rightarrow \frac{15}{x} = \frac{20}{8} \Rightarrow x = 6$ cm

NCERT EXEMPLAR PROBLEMS

1. **(*b*)** Diagonals of a Rhombus are perpendicular to each other and bisect each other

So AO = 8 cm

BO = 6 cm

$\angle AOB = 90°$

In right angled ΔAOB

$AB^2 = AO^2 + OB^2$

$AB^2 = 8^2 + 6^2$

$= 64 + 36$

$AB = \sqrt{100} = 10$ cm.

2. **(*c*)** $\because \Delta ABC \sim \Delta EDF$

Then. $\frac{AB}{ED} = \frac{BC}{DF} = \frac{AC}{EF}$.

$\Rightarrow \quad AB.DF = ED.BC$

OR $\quad AB.EF = AC.ED$

OR $\quad BC.EF = DF.AC$

$\therefore \quad BC.DE \neq AB.EF$

3. **(*a*)** $\because \frac{AB}{QR} = \frac{BC}{PR} = \frac{CA}{PQ}$

Then, ΔCAB ~ ΔPQR.

4. **(*b*)** Because ΔDEF ~ ΔQRP.

5. **(*d*)** Congruent as well as similar. Congruent by AAS, Similar by AA similarity.

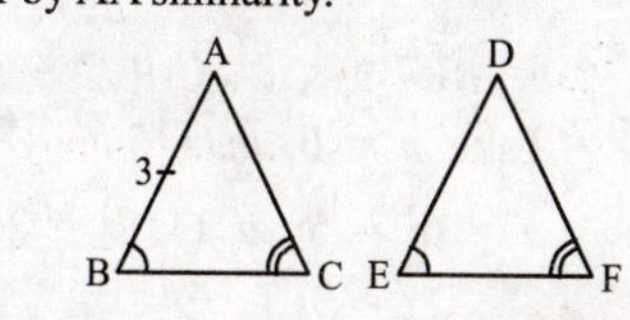

6. **(*c*)**

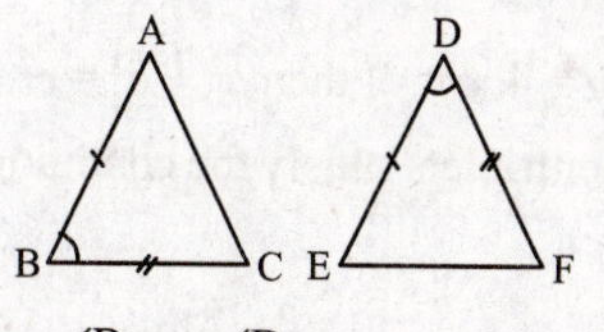

$\angle B = \angle D$

Then the triangles are similar by (SAS similarity)

7. **(*d*)** In ΔPQS, $\quad PS = QS$

$\angle 1 = \angle 2$

In ΔPSR, $\quad PS = SR$

$\angle 3 = \angle 4$

In ΔPQR,

$\angle P + \angle Q + \angle R = 180°$

$\angle 2 + \angle 3 + \angle 1 + \angle 4 = 180°$

$\angle 1 + \angle 3 + \angle 1 + \angle 3 = 180°$

$2\angle 1 + 2\angle 3 = 180°$

$\angle 1 + \angle 3 = 90°$

$\angle P = 90°$

So, In ΔPSR $\quad \angle S = 90°$ $\quad (\because \Delta PQS \sim \Delta PSR)$

So, $\quad PS^2 + RS^2 = PR^2$ (By Pythagoras theorem)

Formative Assessment

Activity/Project 1

- **Topic Covered :** Pythagorean Triplets
- **Objective :** To find Pythagorean triplets.
- **Skills Developed**

  Creativity, analytical thinking.
- **Time Required :** 15 Min.
- **Method**

  Three positive integers a, b, c which satisfy $a^2 + b^2 = c^2$, are called pythagorean triplets.

  Step I. Let n and m be positive integers, where $n > m$.

  Take $a = n^2 - m^2$, $b = 2nm$, $c = n^2 + m^2$

  Step II. Take $n = 2$, $m = 1$

  $a = 2^2 - 1^2 = 3$, $b = 2 \times 2 \times 1 = 4$, $c = 2^2 + 1^2 = 5$

  Take $n = 3$, $m = 2$

  $a = 3^2 - 2^2 = 5$, $b = 2 \times 3 \times 2 = 12$, $c = 3^2 + 2^2 = 13$

  Take $n = 4$, $m = 3$

  $a = 4^2 - 3^2 = 7$, $b = 2 \times 4 \times 3 = 24$, $c = 4^2 + 3^2 = 25$

  Take $n = 5$, $m = 4$

  $a = 5^2 - 4^2 = 9$, $b = 2 \times 5 \times 4 = 40$, $c = 5^2 + 4^2 = 41$

  Take $n = 6$, $m = 5$

  $a = 6^2 - 5^2 = 11$, $b = 2 \times 6 \times 5 = 60$, $c = 6^2 + 5^2 = 61$
- **Observation**

  When $a = 3$, $b = 4$, $c = 5$, then $a^2 + b^2 = c^2$.

  Similarly, other triplets satisfy the condition.

Class Work

Write 10 pythagorean triplets taking different values of n and m.

Homework

Verify that $a = n^2 - m^2$, $b = 2nm$ and $c = n^2 + m^2$ are co-prime numbers.

Quiz Questions

1. 3, 4, 5 are co-prime numbers. State True/False.
2. 10, 12, 14 are pythagorean triplets. State True/False.

Oral Questions

1. Find the area of rectangle, given that its length is 15 cm and the length of its diagonal is 17 cm.
2. Find the area of a square whose diagonal is 10 cm in length.

Suggested Activity/Assignment

Derive a formula for finding pythagorean triplets and verify your result.

Activity/Project 2

- **Topic Covered :** To verify pythagoras theorem by the method of paper cutting and pasting.
- **Pre-requisite Knowledge :**
  1. Area of a square
  2. Construction of right angled triangle
  3. **Pythagoras theorem :** In a right angled triangle, the square of the hypotenuse is equal to the sum of squares of the other two sides containing right angle.
- **Materials Required :**

  Cardboard, coloured pencil, pair of scissors, fevicol, geometry box.
- **Time Required :** 15 Min.
- **Procedure :**
  1. Draw any right triangle ABC, right angled at C, on a poster paper and shade it pink.

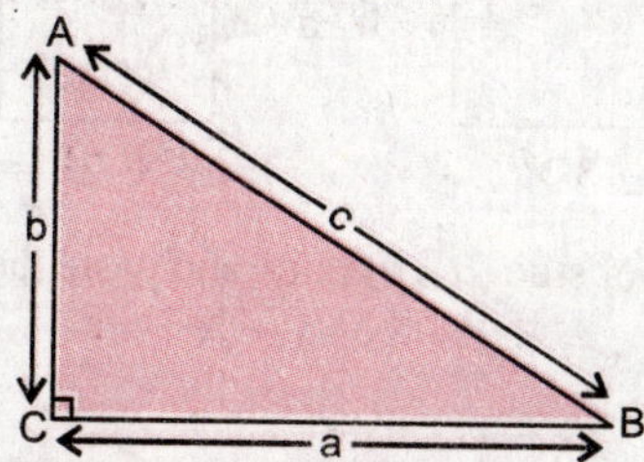

Let the lengths of sides BC, AC and AB be a, b and c units respectively, $a > b$.

2. Draw 8 exact copies of ΔABC on a chart paper. Shade these triangles pink.
3. Cut out these 8 triangles from the chart paper.

4. Draw a square of side c units.

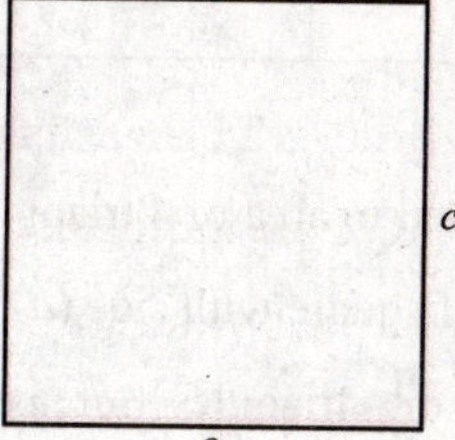

5. Draw another shape as shown below. A square of side 'a' and another square of side 'b' joined together, name this shape 2.

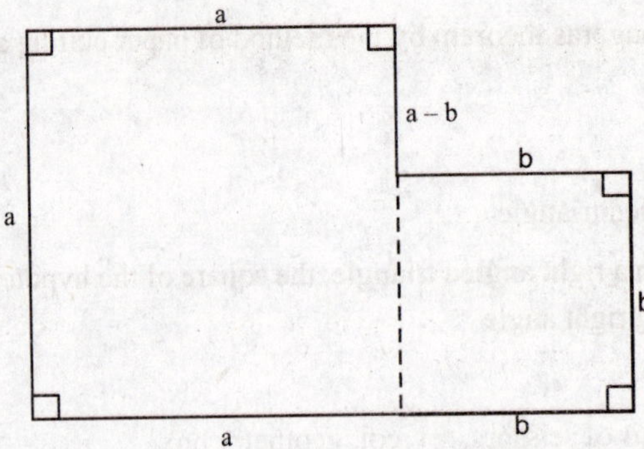

6. Draw two squares of side $(a - b)$ units. Shade it grey and cut it out from the chart paper.

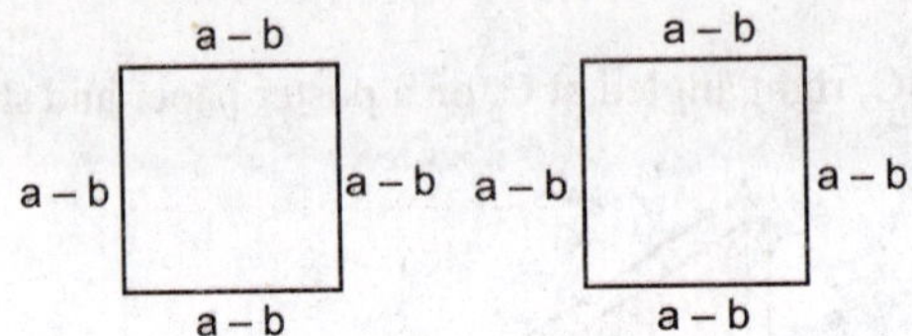

7. Now take 4 triangles and a square of side $(a - b)$ units and paste them on the square with side c units as shown in the figure.

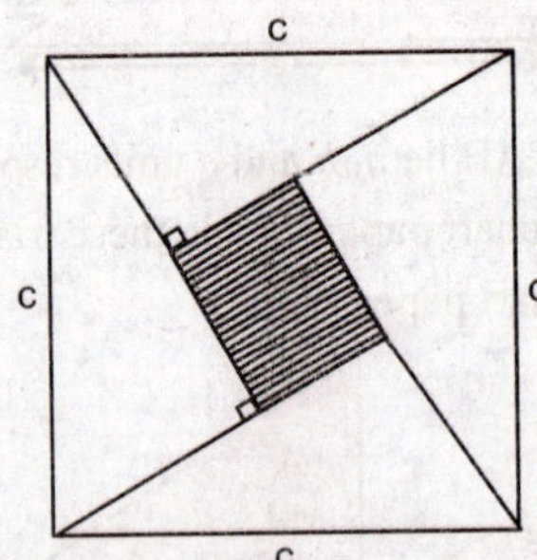

8. Now take remaining 4 triangles and a square with side $(a - b)$ and paste them on the second shape 2 as shown in the figure.

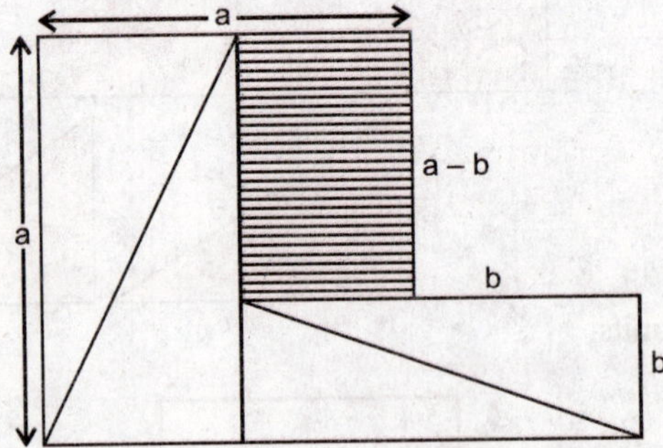

- **Observation**

We find the area of square with side c is equal to area of 4 triangles + area of square with side $(a - b)$.

$\therefore$ $c^2 =$ Area of 4 triangles + area of square with side $(a - b)$

Also area of second shape is equal to area of 4 triangles + area of square with side $(a - b)$ units

$\Rightarrow$ $a^2 + b^2 =$ Area of 4 triangles + area of square with side $(a - b)$ units.

$\therefore$ $c^2 = a^2 + b^2$

Introduction to Trigonometry

CHAPTER 5

• Basic Concepts with Examples

- Trigonometry is the study of relationships between the sides and angles of a right angled triangle.
- **Trigonometric Ratios.** Ratios of sides of right triangle are called Trigonometric ratios.
- **Trigonometric Identities.** Verified relations or an equation involving trigonometric ratios are known as Trigonometric Identities, if it is true for every value of the angle.
-

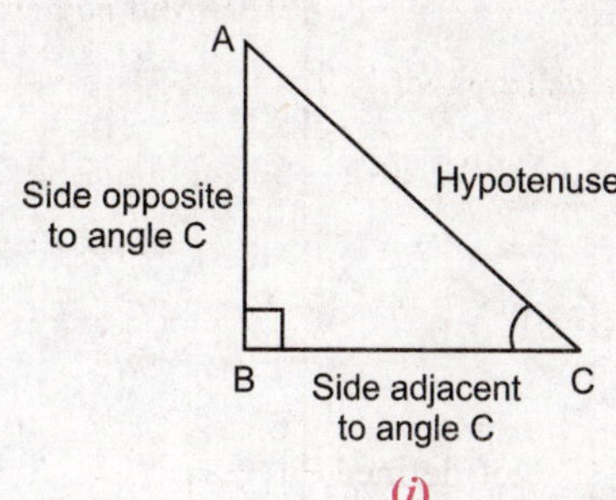

(*i*)

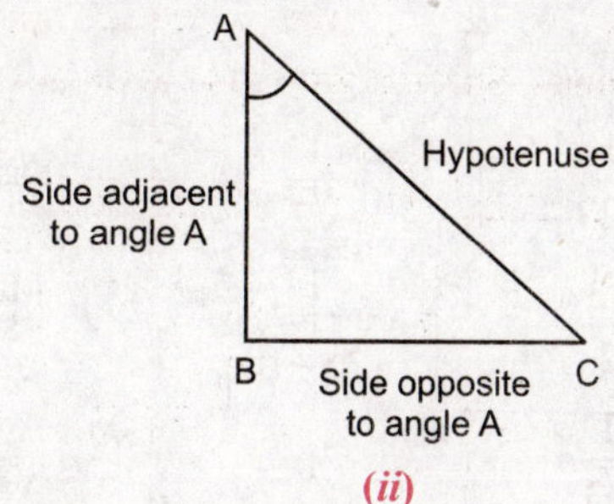

(*ii*)

- **Trigonometric ratios of angle A in right angle triangle ABC right angled at B.**

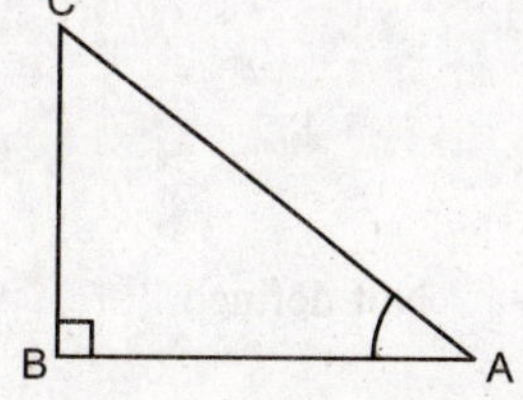

(*i*) sine A $= \dfrac{\text{Side opposite to angle A}}{\text{Hypotenuse}} = \dfrac{BC}{AC}$

(*ii*) cosine A $= \dfrac{\text{Side adjacent to angle A}}{\text{Hypotenuse}} = \dfrac{AB}{AC}$

(*iii*) tangent A $= \dfrac{\text{Side opposite to angle A}}{\text{Side adjacent to angle A}} = \dfrac{BC}{AB}$

(*iv*) cosecant A $= \dfrac{\text{Hypotenuse}}{\text{Side opposite to angle A}} = \dfrac{AC}{BC}$

(*v*) secant A $= \dfrac{\text{Hypotenuse}}{\text{Side adjacent to angle A}} = \dfrac{AC}{AB}$

(*vi*) cotangent A $= \dfrac{\text{Side adjacent to angle A}}{\text{Side opposite to angle A}} = \dfrac{AB}{BC}$

- **Relationship between trigonometric ratios.**

$$\text{cosecant A} = \frac{1}{\text{sine A}}, \ \text{secant A} = \frac{1}{\text{cosine A}}, \ \text{cotangent A} = \frac{1}{\text{tangent A}}.$$

$$\text{tangent A} = \frac{\text{sine A}}{\text{cosine A}}, \ \text{cotangent A} = \frac{\text{cosine A}}{\text{sine A}}$$

- **The abbreviated notations of trigonometric ratios are :**
  - sine A is sin A
  - cosine A is cos A
  - tangent A is tan A
  - secant A is sec A
  - cosecant A is cosec A
  - cotangent A is cot A

- A trigonometric ratio remains same for same angle of different right triangles as

$$\tan A = \frac{BC}{AB} = \frac{DE}{AD} = \frac{FG}{AF}$$

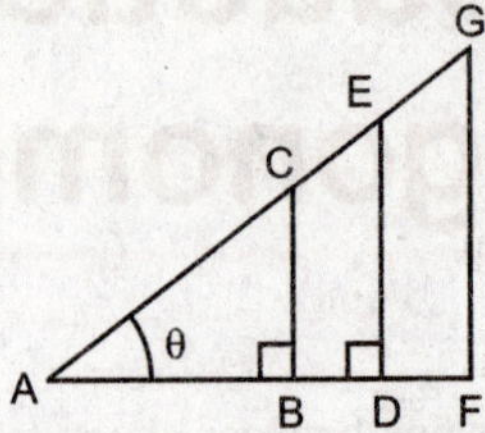

- **Powers of trigonometric ratios.** We may define $(\sin\theta)^2 = \sin^2\theta$ or $\cos^3\theta = (\cos\theta)^3$, etc.
- **Trigonometric ratios of some specific angles :**

A	0°	30°	45°	60°	90°
sin A	0	$\frac{1}{2}$	$\frac{1}{\sqrt{2}}$	$\frac{\sqrt{3}}{2}$	1
cos A	1	$\frac{\sqrt{3}}{2}$	$\frac{1}{\sqrt{2}}$	$\frac{1}{2}$	0
tan A	0	$\frac{1}{\sqrt{3}}$	1	$\sqrt{3}$	Not defined
cosec A	Not defined	2	$\sqrt{2}$	$\frac{2}{\sqrt{3}}$	1
sec A	1	$\frac{2}{\sqrt{3}}$	$\sqrt{2}$	2	Not defined
cot A	Not defined	$\sqrt{3}$	1	$\frac{1}{\sqrt{3}}$	0

- **Trigonometric ratios of complementry angles.**

  (*i*) $\sin(90° - A) = \cos A$ (*ii*) $\cos(90° - A) = \sin A$

  (*iii*) $\tan(90° - A) = \cot A$ (*iv*) $\cot(90° - A) = \tan A$

  (*v*) $\text{cosec}(90° - A) = \sec A$ (*vi*) $\sec(90° - A) = \text{cosec } A$

- **Fundamental trigonometric identities are**

  (*i*) $\cos^2 A + \sin^2 A = 1$ (*ii*) $\sec^2 A = 1 + \tan^2 A$ (*iii*) $\text{cosec}^2 A = 1 + \cot^2 A$

1. If $\sec\theta = \frac{25}{7}$, find the values of $\tan\theta$ and $\text{cosec}\,\theta$.

Sol. In ΔPQO, which is right angle at Q,

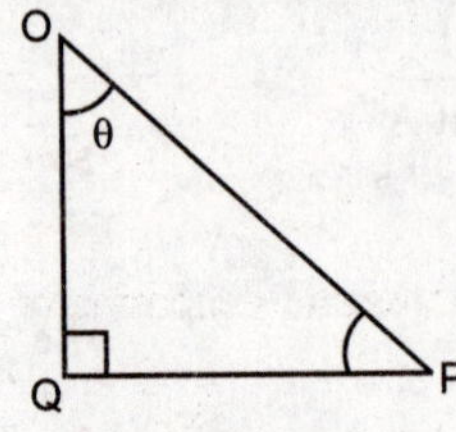

$$\sec\theta = \frac{OP}{OQ} = \frac{25}{7}$$

So, $OP = 25k$ and $OQ = 7k$

$PQ^2 = OP^2 - OQ^2 = (25k)^2 - (7k)^2$

$= 625k^2 - 49k^2 = 576k^2$

$\Rightarrow \quad PQ = \sqrt{576k^2} = 24k$

$\therefore \quad \tan\theta = \frac{PQ}{OQ} = \frac{24}{7}$ and $\text{cosec}\,\theta = \frac{OP}{PQ} = \frac{25}{24}$.

2. In ΔABC, right angle at B, if AB = 12 cm and BC = 5 cm, find

(*i*) sin A and tan A, (*ii*) sin C and cot C.

Sol.

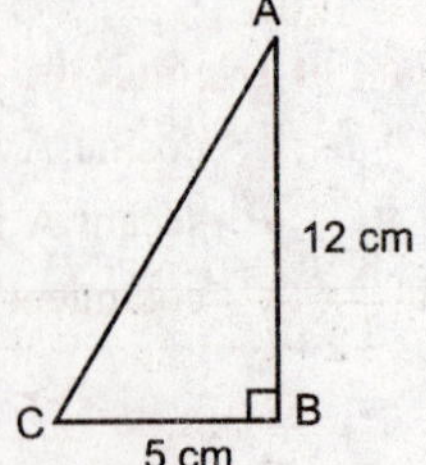

(*i*) In ΔABC, which is right-angled at B, AC is the hypotenuse.

Since, $AC^2 = BC^2 + AB^2$

$= 5^2 + 12^2$

$= 25 + 144 = 169$

$AC = \sqrt{169} = 13$ cm

Now, $\sin A = \dfrac{BC}{AC} = \dfrac{5}{13}$

and $\tan A = \dfrac{BC}{AB} = \dfrac{5}{12}$

(*ii*) $\sin C = \dfrac{AB}{AC} = \dfrac{12}{13}$

and $\cot C = \dfrac{BC}{AB} = \dfrac{5}{12}$

3. If $\cos\theta = \dfrac{3}{5}$, evaluate $\dfrac{\sin\theta - \cot\theta}{2\tan\theta}$

Sol. If ABC is a triangle, right-angled at B, then

$\cos\theta = \dfrac{AB}{AC} = \dfrac{3}{5} \Rightarrow AB : AC = 3 : 5$

Let $AB = 3k$ and $AC = 5k$

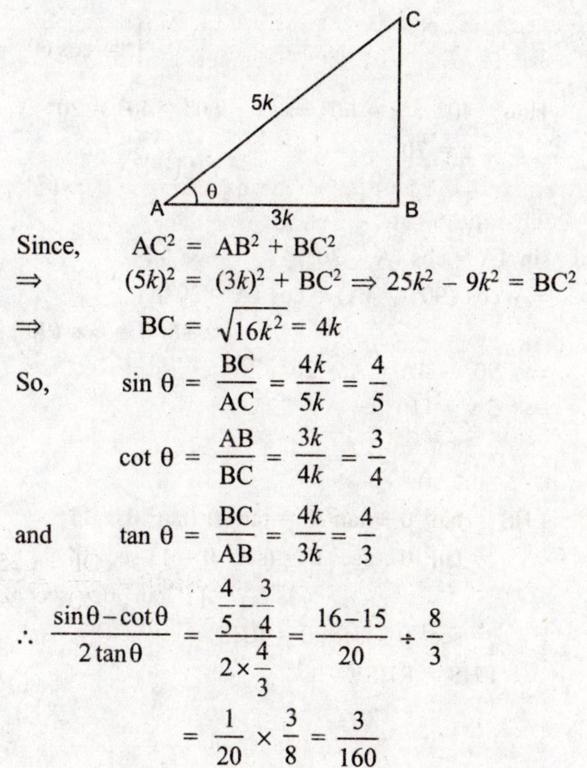

Since, $AC^2 = AB^2 + BC^2$

$\Rightarrow (5k)^2 = (3k)^2 + BC^2 \Rightarrow 25k^2 - 9k^2 = BC^2$

$\Rightarrow BC = \sqrt{16k^2} = 4k$

So, $\sin\theta = \dfrac{BC}{AC} = \dfrac{4k}{5k} = \dfrac{4}{5}$

$\cot\theta = \dfrac{AB}{BC} = \dfrac{3k}{4k} = \dfrac{3}{4}$

and $\tan\theta = \dfrac{BC}{AB} = \dfrac{4k}{3k} = \dfrac{4}{3}$

$\therefore \dfrac{\sin\theta - \cot\theta}{2\tan\theta} = \dfrac{\frac{4}{5} - \frac{3}{4}}{2\times\frac{4}{3}} = \dfrac{16-15}{20} \div \dfrac{8}{3}$

$= \dfrac{1}{20} \times \dfrac{3}{8} = \dfrac{3}{160}$

4. If $\tan A = \sqrt{2} - 1$, show that $\dfrac{\tan A}{1+\tan^2 A} = \dfrac{\sqrt{2}}{4}$.

Sol. $\dfrac{\tan A}{1+\tan^2 A} = \dfrac{\sqrt{2}-1}{1+(\sqrt{2}-1)^2} = \dfrac{\sqrt{2}-1}{1+2+1-2\sqrt{2}} = \dfrac{\sqrt{2}-1}{4-2\sqrt{2}}$

$= \dfrac{\sqrt{2}-1}{4\left(1-\frac{\sqrt{2}}{2}\right)} = \dfrac{\sqrt{2}\left(1-\frac{1}{\sqrt{2}}\right)}{4\left(1-\frac{1}{\sqrt{2}}\right)} = \dfrac{\sqrt{2}}{4}$

Hence Verified.

5. If $\sec\theta = \dfrac{5}{4}$, verify that $\dfrac{\tan\theta}{1+\tan^2\theta} = \dfrac{\sin\theta}{\sec\theta}$.

Sol.

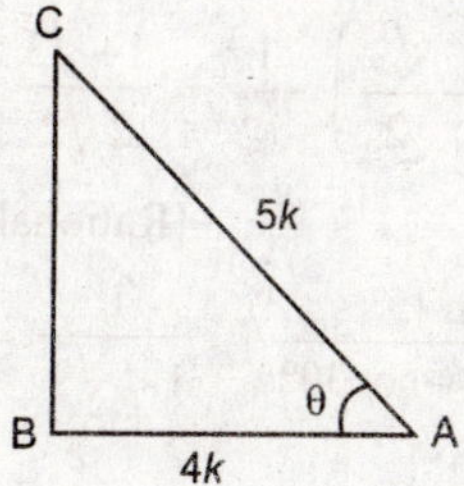

If ABC is a triangle, right-angled at B and $\angle A = \theta$, then

$\sec\theta = \dfrac{AC}{AB} = \dfrac{5}{4}$

$\Rightarrow AC = 5k$ and $AB = 4k$

Since, $AC^2 = AB^2 + BC^2$

$\Rightarrow (5k)^2 = (4k)^2 + BC^2 \Rightarrow BC^2 = 25k^2 - 16k^2 = 9k^2$

$\Rightarrow BC = \sqrt{9k^2} = 3k$

Now, $\dfrac{\tan\theta}{1+\tan^2\theta} = \dfrac{\frac{BC}{AB}}{1+\left(\frac{BC}{AB}\right)^2} = \dfrac{\frac{3k}{4k}}{1+\left(\frac{3k}{4k}\right)^2}$

$= \dfrac{\frac{3}{4}}{\left(1+\frac{9}{16}\right)} = \dfrac{3}{4} \div \dfrac{25}{16}$

$= \dfrac{3}{4} \times \dfrac{16}{25} = \dfrac{12}{25}$

Also, $\dfrac{\sin\theta}{\sec\theta} = \dfrac{\frac{BC}{AC}}{\frac{AC}{AB}} = \dfrac{\frac{3k}{5k}}{\frac{5k}{4k}} = \dfrac{3}{5} \times \dfrac{4}{5} = \dfrac{12}{25}$

Hence verified.

6. If $\tan\theta = \dfrac{4}{3}$, find the value of $\dfrac{3\sin\theta + 2\cos\theta}{3\sin\theta - 2\cos\theta}$.

Sol. Divide the numerator and denominator by $\cos\theta$, we get :

$\dfrac{3\frac{\sin\theta}{\cos\theta}+2}{3\frac{\sin\theta}{\cos\theta}-2} = \dfrac{3\tan\theta+2}{3\tan\theta-2} = \dfrac{3\times\frac{4}{3}+2}{3\times\frac{4}{3}-2} = \dfrac{4+2}{4-2} = \dfrac{6}{2} = 3$

7. Evaluate :

(*i*) $\sin 30° \cos 45° + \cos 30° \sin 45°$

(ii) $\dfrac{\tan 45°}{\sin 30° + \cos 30°}$

(iii) $\dfrac{5\sin^2 30° + \cos^2 45° + 4\tan^2 60°}{2\sin 30° \cos 60° + \tan 45°}$

Sol. (i) $\sin 30° \cos 45° + \cos 30° \sin 45°$

$$= \frac{1}{2} \times \frac{1}{\sqrt{2}} + \frac{\sqrt{3}}{2} \times \frac{1}{\sqrt{2}}$$

$$= \left(\frac{1}{2} + \frac{\sqrt{3}}{2}\right) \frac{1}{\sqrt{2}} = \frac{1+\sqrt{3}}{2\sqrt{2}} = \frac{\sqrt{2}+\sqrt{6}}{4}$$

[Rationalising the denominator]

(ii) $$\frac{\tan 45°}{\sin 30° + \cos 30°} = \frac{1}{\frac{1}{2} + \frac{\sqrt{3}}{2}} = \frac{2}{1+\sqrt{3}}$$

$$= \sqrt{3} - 1$$

[Rationalising the denominator]

(iii) $$\frac{5\sin^2 30° + \cos^2 45° + 4\tan^2 60°}{2\sin 30° \cos 60° + \tan 45°}$$

$$= \frac{5\times\left(\frac{1}{2}\right)^2 + \left(\frac{1}{\sqrt{2}}\right)^2 + 4\times\left(\sqrt{3}\right)^2}{2\times\frac{1}{2}\times\frac{1}{2}+1} = \frac{\frac{5}{4}+\frac{1}{2}+12}{\frac{3}{2}}$$

$$= \frac{7+48}{4\times\frac{3}{2}} = \frac{55}{6} = 9\frac{1}{6}$$

8. Verify the following :

$\cos 60° = 1 - 2\sin^2 30° = 2\cos^2 30° - 1$

Sol. $$1 - 2\sin^2 30° = 1 - 2\times\left(\frac{1}{2}\right)^2 = 1 - 2\times\frac{1}{4}$$

$$= 1 - \frac{1}{2} = \frac{1}{2}$$

$$2\cos^2 30° - 1 = 2\times\left(\frac{\sqrt{3}}{2}\right)^2 - 1$$

$$= 2\times\frac{3}{4} - 1 = \frac{3}{2} - 1$$

$$= \frac{3-2}{2} = \frac{1}{2}$$

But, $\cos 60° = \frac{1}{2}$.

Thus, $\cos 60° = 1 - 2\sin^2 30° = 2\cos^2 30° - 1$.

9. Given A = 30°, verify sin 2A = 2 sin A cos A.

Sol. LHS $= \sin 2A = \sin 2\times 30° = \sin 60° = \frac{\sqrt{3}}{2}$

RHS $= 2\sin A \cos A = 2\sin 30° \cos 30°$

$$= 2\times\frac{1}{2}\times\frac{\sqrt{3}}{2} = \frac{\sqrt{3}}{2}$$

So, LHS = RHS

10. If sin (A + B) = 1 and cos (A – B) = 1, find A and B.

Sol. $\sin(A+B) = 1 \Rightarrow A + B = 90°$, [$\because \sin 90° = 1$]

and $\cos(A-B) = 1 \Rightarrow A - B = 0°$ [$\because \cos 0° = 1$]

$\therefore$ We have $A + B = 90°$ ...(i)

and $A - B = 0°$ ...(ii)

On adding (i) and (ii), we get

$$2A = 90° \Rightarrow A = \frac{90°}{2} = 45°$$

On subtracting (ii) from (i), we get

$$2B = 90° \Rightarrow B = \frac{90°}{2} = 45°$$

$\therefore$ A = 45°, B = 45°

11. If cos (40° + x) = sin 30°, find the value of x.

Sol. $\cos(40° + x) = \sin 30° = \frac{1}{2}$ $\left[\because \sin 30° = \frac{1}{2}\right]$

So, $\cos(40° + x) = \frac{1}{2} \Rightarrow 40° + x = 60°$

$\left[\because \cos 60° = \frac{1}{2}\right]$

Thus, $40° + x = 60° \Rightarrow x = 60° - 40° = 20°$

$\therefore$ $x = 20°$

12. sin 4A = cos (A – 20°), where 4A is an acute angle, find the value of A.

Sol. $\sin 4A = \cos(A - 20°)$

$\Rightarrow \cos(90° - 4A) = \cos(A - 20°)$

[$\because \sin\theta = \cos(90° - \theta)$]

$\Rightarrow 90° - 4A = A - 20°$

$\Rightarrow 5A = 110° \Rightarrow A = 22°$

13. Show that $\tan^4\theta + \tan^2\theta = \sec^4\theta - \sec^2\theta$ for $0° \le \theta \le 90°$.

Sol. LHS $= \tan^4\theta + \tan^2\theta = \tan^2\theta(\tan^2\theta + 1)$

$= \tan^2\theta \cdot \sec^2\theta = (\sec^2\theta - 1)\sec^2\theta$

[$\because \tan^2\theta = \sec^2\theta - 1$]

$= \sec^4\theta - \sec^2\theta =$ RHS

$\therefore$ LHS = RHS

14. Prove that : $\dfrac{\sec\theta + \tan\theta}{\sec\theta - \tan\theta} = \left(\dfrac{1+\sin\theta}{\cos\theta}\right)^2$.

Sol. LHS $= \dfrac{\sec\theta + \tan\theta}{\sec\theta - \tan\theta}$

Multiplying and dividing by $(\sec\theta + \tan\theta)$,

$$= \frac{\sec\theta + \tan\theta}{\sec\theta - \tan\theta} \times \frac{\sec\theta + \tan\theta}{\sec\theta + \tan\theta}$$

$$= \frac{(\sec\theta + \tan\theta)^2}{\sec^2\theta - \tan^2\theta} \qquad [\because (a+b)(a-b) = a^2 - b^2]$$

$$= \frac{(\sec\theta + \tan\theta)^2}{1} \qquad [\because \sec^2\theta - \tan^2\theta = 1]$$

$$= \left(\frac{1}{\cos\theta} + \frac{\sin\theta}{\cos\theta}\right)^2 = \left(\frac{1+\sin\theta}{\cos\theta}\right)^2$$

$\therefore$ LHS = RHS

15. Prove that :

$\sin\theta \,.\, \cos(90° - \theta) + \cos\theta \,.\, \sin(90° - \theta) = 1$

Sol. LHS $= \sin\theta \,.\, \cos(90° - \theta) + \cos\theta \,.\, \sin(90° - \theta)$

$= \sin\theta \,.\, \sin\theta + \cos\theta \,.\, \cos\theta$

$= \sin^2\theta + \cos^2\theta = 1 =$ RHS

16. Prove that :

$\frac{\sin^2 A}{\cos^2 A} + \frac{\cos^2 A}{\sin^2 A} = \frac{1}{\sin^2 A \cos^2 A} - 2$

Sol. LHS $= \frac{\sin^2 A}{\cos^2 A} + \frac{\cos^2 A}{\sin^2 A} = \frac{\sin^4 A + \cos^4 A}{\cos^2 A \sin^2 A}$

By adding and subtracting $2\sin^2 A\cos^2 A$ in the numerator, we get

$$= \frac{(\sin^2 A + \cos^2 A)^2 - 2\sin^2 A\cos^2 A}{\cos^2 A \sin^2 A}$$

$$= \frac{(1)^2 - 2\sin^2 A\cos^2 A}{\cos^2 A\sin^2 A} \qquad [\because \sin^2 A + \cos^2 A = 1]$$

$$= \frac{1 - 2\sin^2 A\cos^2 A}{\cos^2 A\sin^2 A}$$

$$= \frac{1}{\cos^2 A\sin^2 A} - \frac{2\sin^2 A\cos^2 A}{\cos^2 A\sin^2 A} = \frac{1}{\cos^2 A\sin^2 A} - 2$$

$\therefore$ LHS = RHS

17. Prove that :

$(\text{cosec } A - \sin A)(\sec A - \cos A)(\tan A + \cot A) = 1$

Sol. LHS $= (\text{cosec } A - \sin A)(\sec A - \cos A)(\tan A + \cot A)$

$$= \left(\frac{1}{\sin A} - \sin A\right)\left(\frac{1}{\cos A} - \cos A\right)\left(\frac{\sin A}{\cos A} + \frac{\cos A}{\sin A}\right)$$

$$= \left(\frac{1-\sin^2 A}{\sin A}\right)\left(\frac{1-\cos^2 A}{\cos A}\right)\left(\frac{\sin^2 A + \cos^2 A}{\sin A \,.\, \cos A}\right)$$

$$= \frac{\cos^2 A}{\sin A} \cdot \frac{\sin^2 A}{\cos A} \cdot \frac{1}{\sin A \,.\, \cos A} = 1 = \text{RHS}$$

$\therefore$ LHS = RHS

18. Evaluate : $\left(\frac{\sin 27°}{\cos 63°}\right)^2 + \left(\frac{\cos 63°}{\sin 27°}\right)^2$

Sol. $\left(\frac{\sin(90° - 63°)}{\cos 63°}\right)^2 + \left(\frac{\cos(90° - 27°)}{\sin 27°}\right)^2$

$$\left[\because \begin{matrix}\sin(90° - \theta) = \cos\theta \\ \cos(90° - \theta) = \sin\theta\end{matrix}\right]$$

$$= \left(\frac{\cos 63°}{\cos 63°}\right)^2 + \left(\frac{\sin 27°}{\sin 27°}\right)^2 = 1 + 1 = 2$$

19. Given that :

$\cos(A - B) = \cos A\cos B + \sin A\sin B$, find the value of $\cos 15°$.

Sol. $\cos(45° - 30°) = \cos 15°$

$\Rightarrow \cos 15° = \cos 45° \,.\, \cos 30° + \sin 45° \sin 30°$

$$= \frac{1}{\sqrt{2}} \cdot \frac{\sqrt{3}}{2} + \frac{1}{\sqrt{2}} \cdot \frac{1}{2} = \frac{(\sqrt{3}+1)}{2\sqrt{2}}$$

Summative Assessment

Subjective Questions

VERY SHORT ANSWER TYPE QUESTIONS

This category is given for better understanding of the topic and for EXTRA PRACTICE.

PREVIOUS YEARS' QUESTIONS

1. In the given figure, AD = 4 cm, BD = 3 cm and CB = 12 cm find $\cot\theta$. [Delhi 2008]

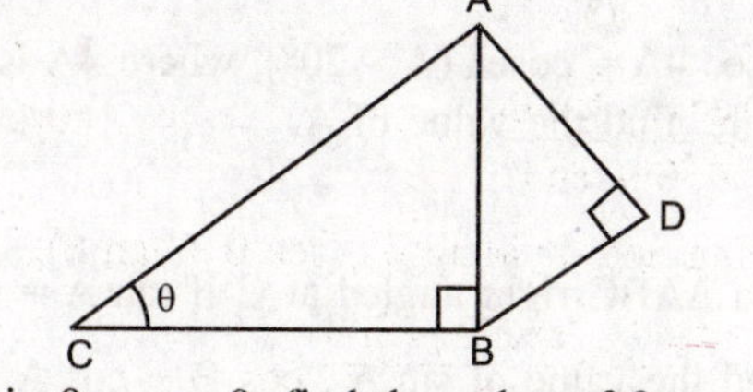

2. If $\sin\theta = \cos\theta$, find the value of θ. [AI 2008]
3. If $\tan\theta = \cot(30° + \theta)$, find the value of θ. [AI 2008]
4. If $\sin 3\theta = \cos(\theta - 6°)$ and 3θ and $(\theta - 6°)$ are acute angles, find the value of θ. [AI 2008]
5. If $\tan A = \frac{5}{12}$, find the value of $(\sin A + \cos A) \,.\, \sec A$. [Foreign 2008]
6. If $\cos A = \frac{7}{25}$, find the value of $\tan A + \cot A$. [Foreign 2008]

7. If $\tan\theta = \frac{1}{\sqrt{3}}$, then evaluate $\left[\frac{\text{cosec}^2\theta - \sec^2\theta}{\text{cosec}^2\theta + \sec^2\theta}\right]$ [Delhi 2008C]

8. If $\tan\theta = \frac{4}{3}$, what is the value of $\left[\frac{\sin\theta + \cos\theta}{\sin\theta - \cos\theta}\right]$ [AI 2008C]

9. If $\cot\theta = \frac{7}{8}$, evaluate $\frac{(1+\sin\theta)(1-\sin\theta)}{(1+\cos\theta)(1-\cos\theta)}$ [AI 2008 C]

10. If $\sin\theta = \frac{1}{3}$, then find the value of $(2\cot^2\theta + 2)$ [Delhi 2009]

11. If $\sec^2\theta\,(1 + \sin\theta)(1 - \sin\theta) = k$, then find the value of k. [AI 2009]

12. (*a*) If $\sec A = \frac{15}{7}$ and $A + B = 90°$, find the value of cosec B. [Foreign 2009]

(*b*) If $3x = \text{cosec}\,\theta$ and $\frac{3}{x} = \cot\theta$, find the value of $3\left(x^2 - \frac{1}{x^2}\right)$. [Delhi 2010]

(*c*) If $6x = \sec\theta$ and $\frac{6}{x} = \tan\theta$, find the value of $9\left(x^2 - \frac{1}{x^2}\right)$. [Foreign 2010]

Practice Questions

13. In a right ΔABC, at right angled at B, if $\tan A = 4/3$, find $\tan C$?

14. If $\cos\theta = \frac{20}{29}$ and $\sin\theta = \frac{21}{29}$, find the value of $\cos^2\theta + \sin^2\theta$.

15. In a ΔABC, right angled at A, if AB = 5, AC = 12 and BC = 13, find sin B and cos C.

16. In the given figure, find tan P – cot R.

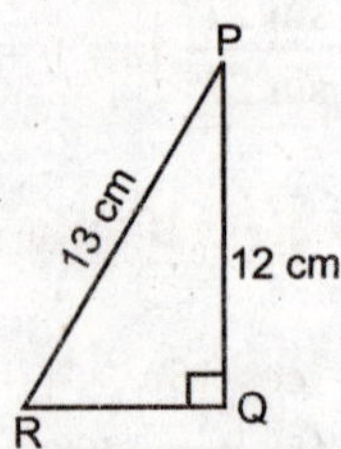

17. In ΔABC, right angled at B, AB = 5 cm and $\angle ACB = 30°$. Determine the length of side AC.

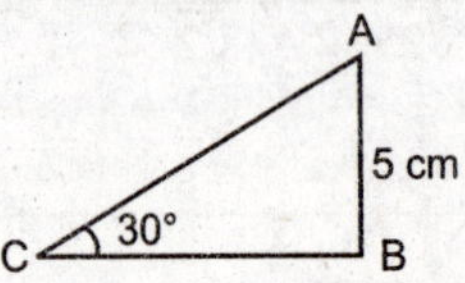

18. In ΔPQR, right angled at Q, PQ = 3 cm and PR = 6 cm, find $\angle QPR$.

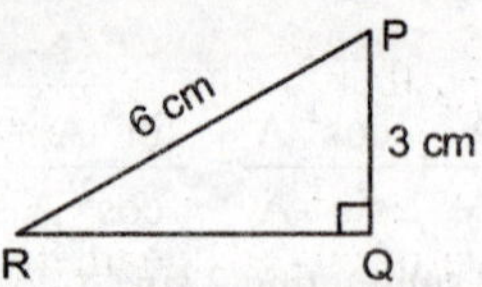

19. If $\sin(A - B) = \frac{1}{2}$, $\cos(A + B) = \frac{1}{2}$, find A and B.

20. If A = 0, then sin 2A = 2 sinA. Is it true ?

21. Evaluate : $\frac{2\tan 30°}{1 + \tan^2 30°}$

22. Evaluate : sin 60°. cos 30° + sin 30°. cos 60°.

23. If tan 2A = cot (A – 18°), where 2A is an acute angle, find the value of A.

24. Express cot 85° + cos 75° in terms of trigonometric ratios of angles between 0° and 45°.

25. Find the value of $9\sec^2 A - 9\tan^2 A$.

SHORT ANSWER TYPE QUESTIONS-I

Previous Years' Questions

1. Without using the trigonometric tables, evaluate

(*i*) $\frac{11\sin 70°}{7\cos 20°} - \frac{4}{7}\frac{\cos 53°\,\text{cosec}\,37°}{\tan 15°\tan 35°\tan 55°.\tan 75°}$ [Delhi 2008]

(*ii*) $\frac{\sin 18°}{\cos 72°} + \sqrt{3}\,.\,[\tan 10°\,.\,\tan 30°.\,\tan 40°\,.\,\tan 50°\,.\,\tan 80°\,]$ [Delhi 2008]

(*iii*) $(\sin^2 25° + \sin^2 65°) + \sqrt{3}\,(\tan 5°\,\tan 15°\,\tan 30°\,\tan 75°\,\tan 85°)$ [AI 2008]

(*iv*) $(\cos^2 25° + \cos^2 65°) + \text{cosec}\,\theta\,.\,\sec(90° - \theta) - \cot\theta\,.\,\tan(90° - \theta)$ [AI 2008]

(*v*) If $7\sin^2\theta + 3\cos^2\theta = 4$, then show that $\tan\theta = \frac{1}{\sqrt{3}}$. [AI 2008]

(*vi*) If sec 4A = cosec (A – 20°), where 4A is an acute angle, find the value of A. [Foreign 2008]

Or

In a ΔABC right angled at C if $\tan A = \frac{1}{\sqrt{3}}$. Find the value of sin A . cos B + cos A sin B.

2. If A, B, C are interior angles of a ΔABC, then show that $\sin\left(\frac{B+C}{2}\right) = \cos\frac{A}{2}$. [Delhi 2008C]

3. Evaluate :

(a) $\frac{\sin 15^\circ \cos 75^\circ + \cos 15^\circ \sin 75^\circ}{\tan 5^\circ.\tan 30^\circ.\tan 35^\circ.\tan 55^\circ \tan 85^\circ}$ [Delhi 2008C]

(b) If $\tan(A + B) = \sqrt{3}$ and $\tan(A - B) = \frac{1}{\sqrt{3}}$, $0 < A + B \le 90^\circ$, $A > B$, find the values of A and B. [Delhi 2008C]

4. Evaluate :

$\cos^2 20^\circ + \cos^2 70^\circ + \sin 48^\circ \sec 42^\circ + \cos 40^\circ \operatorname{cosec} 50^\circ$. [AI 2008C]

5. Simplify : $\frac{\sin^3\theta + \cos^3\theta}{\sin\theta + \cos\theta} + \sin\theta\cos\theta$ [Delhi 2009]

6. If $\cot\theta = \frac{15}{8}$, then evaluate $\frac{(2+2\sin\theta)(1-\sin\theta)}{(1+\cos\theta)(2-2\cos\theta)}$.

Or

Find the value of tan 60° geometrically. [AI 2009]

7. Without using trigonometric tables, evaluate

$\frac{7}{2}\cdot\frac{\cos 70^\circ}{\sin 20^\circ} + \frac{3}{2}\cdot\frac{\cos 55^\circ \operatorname{cosec} 35^\circ}{\tan 5^\circ\cdot\tan 25^\circ.\tan 45^\circ\cdot\tan 85^\circ.\tan 65^\circ}$ [Foreign 2009]

8. Without using trigonometric tables, find the value of the given expression :

$$\frac{\sec(90^\circ-\theta).\operatorname{cosec}\theta - \tan(90^\circ-\theta)\cot\theta + \cos^2 25^\circ + \cos^2 65^\circ}{3\tan 27^\circ.\tan 63^\circ}$$

Or

Find the value of cosec 30° geometrically. [Delhi 2010]

9. Without using trigonometric tables, evaluate the following :

$$\frac{\cos 70^\circ}{3\sin 20^\circ} + \frac{4(\sec^2 59^\circ - \cot^2 31^\circ)}{3} - \frac{2}{3}\sin 90^\circ$$

10. Without using trigonometric tables, find the value of the following :

$\cot\theta . \tan(90^\circ - \theta) - \sec(90^\circ - \theta)\operatorname{cosec}\theta + \sqrt{3} . \tan 12^\circ . \tan 60^\circ . \tan 78^\circ$ [Foreign 2010]

11. Prove that : $\frac{\cos A}{1+\sin A} + \frac{1+\sin A}{\cos A} = 2\sec A$ [CBSE (CCE) 2011]

12. Evaluate :

$$\frac{\tan^2 60^\circ + 4\sin^2 45^\circ + 3\sec^2 30^\circ + 5\cos^2 90^\circ}{\operatorname{cosec} 30^\circ + \sec 60^\circ - \cot^2 30^\circ}$$

[CBSE (CCE) 2011]

13. If $\sqrt{3}\sin\theta - \cos\theta = 0$ and $0^\circ < \theta < 90^\circ$, find the value of θ. [CBSE (CCE) 2012]

14. If $\sin A = \frac{\sqrt{3}}{2}$, find the value of $2\cot^2 A - 1$. [CBSE (CCE) 2012]

Practice Questions

15. Simplify : $(1 + \tan^2\theta)(1 - \sin\theta)(1 + \sin\theta)$.

16. Prove that : $\frac{\sin\theta}{1-\cos\theta} = \operatorname{cosec}\theta + \cot\theta$.

17. Evaluate : $\frac{\tan 35^\circ}{\cot 55^\circ} + \frac{\cot 78^\circ}{\tan 12^\circ} - 1$

18. Find the value of $\sec 50^\circ \sin 40^\circ + \cos 40^\circ \operatorname{cosec} 50^\circ$.

19. If $\sec\theta + \tan\theta = m$ and $\sec\theta - \tan\theta = n$, find the value of $\sqrt{mn}$.

20. Find the value of $\sin^2 20^\circ + \sin^2 70^\circ - \tan^2 45^\circ$.

21. Find the value of θ $(0^\circ < \theta < 90^\circ)$ if $2\cos^2\theta = \frac{1}{2}$.

22. If $\sin\theta + \cos\theta = \sqrt{2}\cos(90^\circ - \theta)$, determine $\cot\theta$.

23. In the given figure, ABCD is a rectangle in which segments AP and AQ are drawn as shown. Find the length of (AP + AQ).

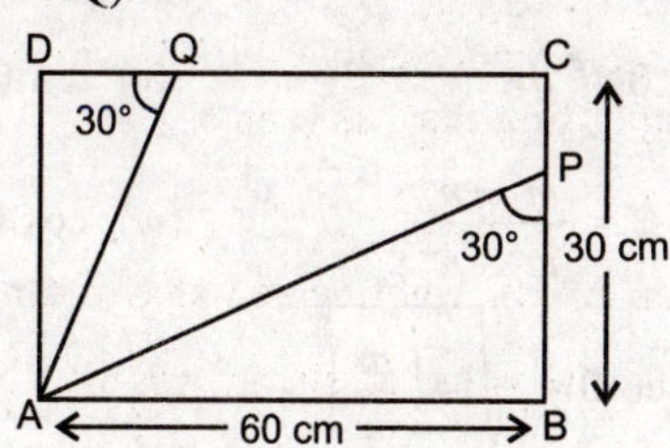

24. Evaluate : $\frac{\cos 80^\circ}{\sin 10^\circ} + \cos 59^\circ \operatorname{cosec} 31^\circ$.

25. In the given figure, ABCD is a rectangle with AD = 8 cm and CD = 12 cm. Line segment CE is drawn, making an angle of 60° with AB, intersecting AB in E. Find the length of CE and BE.

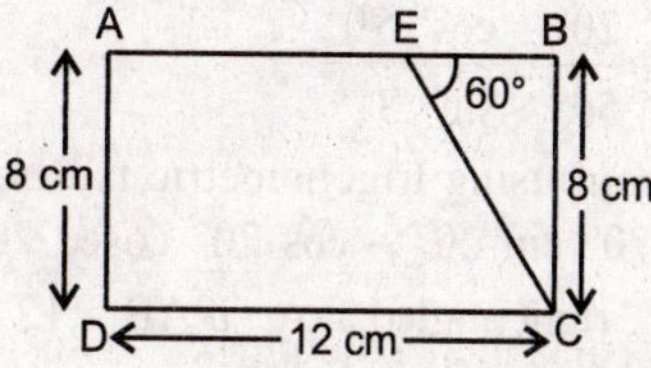

26. Show that $(1 + \cot\theta - \operatorname{cosec}\theta)(1 + \tan\theta + \sec\theta) = 2$.

27. If $\sin\theta + \cos\theta = p$ and $\sec\theta + \operatorname{cosec}\theta = q$, show that $q(p^2 - 1) = 2p$.

28. If $\cos\theta + \sin\theta = \sqrt{2}\cos\theta$, show that $\cos\theta - \sin\theta = \sqrt{2}\sin\theta$.

29. Show that : $\left(1+\frac{1}{\tan^2\theta}\right)\left(1+\frac{1}{\cot^2\theta}\right) = \frac{1}{\sin^2\theta - \sin^4\theta}$

30. In the given figure, ABCD is a rectangle with AD = 12 cm and DC = 20 cm. Line segment DE is drawn making an angle of 30° with AD, intersecting AB in E. Find the length of DE and AE.

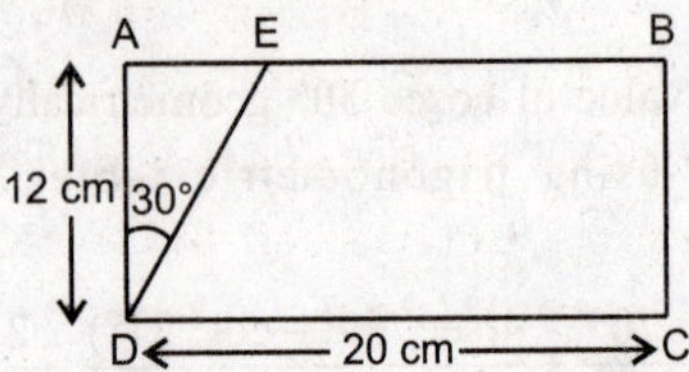

31. If 3 tan θ = 4, find the value of $\dfrac{5\sin\theta - 3\cos\theta}{5\sin\theta + 2\cos\theta}$.

32. Prove that : $\cot\theta - \tan\theta = \dfrac{2\cos^2\theta - 1}{\sin\theta\cos\theta}$.

33. If A = 60° and B = 30°, verify that
cos (A – B) = cos A cos B + sin A sin B.

34. If A = 30°, verify that $\cos 2A = \dfrac{1-\tan^2 A}{1+\tan^2 A}$.

35. *(i)* If A = 60° and B = 30°, verify that
sin (A – B) = sin A cos B – cos A sin B.

(ii) If $\dfrac{\cos\alpha}{\cos\beta} = m$ and $\dfrac{\cos\alpha}{\sin\beta} = n$, show that
$(m^2 + n^2)\cos^2\beta = n^2$. [CBSE (CCE) 2012]

(iii) In the given figure, ABCD is a parallelogram in which DC = 10 cm and BC = $4\sqrt{3}$ cm. AP is perpendicular to DC. If ∠ADC = 60°, find the length of AP.

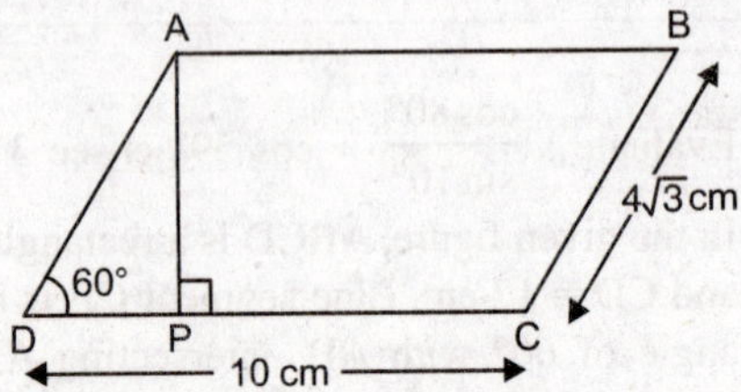

(iv) Without using trigonometric tables, evaluate :
$\dfrac{\cos^2 20° + \cos^2 70°}{\sin^2 59° + \sin^2 31°}$.

(v) Without using trigonometric tables, evaluate :
sec 70° sin 20° – cos 20° cosec 70°.

36. In a ΔABC, right angled at A, if AB = 12 cm, CA = 5 cm and BC = 13 cm, find all the six trigonometric ratios of angle B.

37. In a triangle ABC, right angled at B, the ratio of AB to AC is $1 : \sqrt{2}$. Find the value of $\dfrac{2\tan A}{1+\tan^2 A}$.

38. In a right triangle ABC, right angled at B, the ratio of AB to AC is $1 : \sqrt{2}$. Find the value of $\dfrac{2\tan A}{1-\tan^2 A}$.

39. If $\tan\theta = \dfrac{4}{3}$, show that $\sqrt{\dfrac{1-\sin\theta}{1+\sin\theta}} = \dfrac{1}{3}$.

Find the value of *x*. (Q. 40 to Q. 43)

40. **41.**

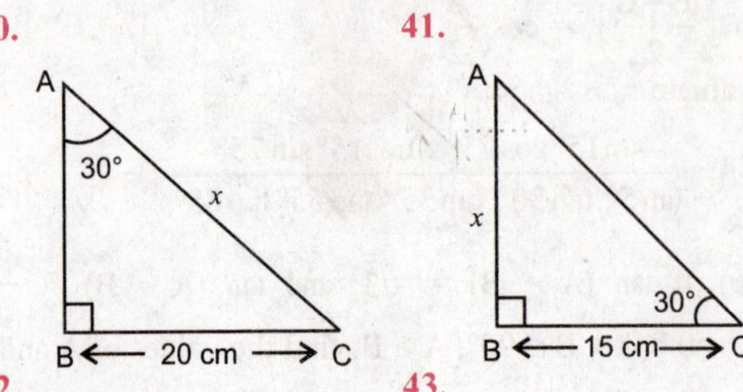

42. **43.**

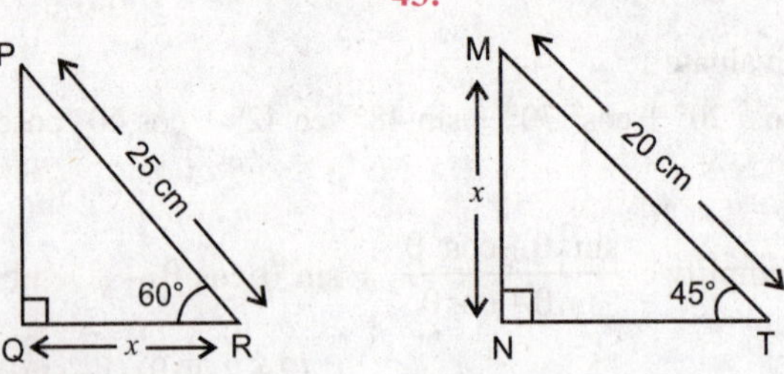

44. Evaluate : 4 cot² 45° – sec² 60° + sin² 60° + cos² 90°.

45. Show that :
2 (cos² 45° + tan² 60°) – 6 (sin² 45° – tan² 30°) = 6.

46. If $\tan\theta + \dfrac{1}{\tan\theta} = 2$, find the value of $\tan^2\theta + \dfrac{1}{\tan^2\theta}$.

47. If the ratio of sine of an angle to its cosine is 3 : 4, then find their actual values.

48. Evaluate in the simplest form :
(i) cos 60° cos 30° + sin 60° sin 30°.
(ii) cos 60° cos 30° – sin 60° sin 30°.

49. Find the value of θ in each of the following :
(i) $2\sin 2\theta = \sqrt{3}$ *(ii)* $2\cos 3\theta = 1$
(iii) $\sqrt{3}\tan 2\theta\ 3 = 0$

50. If A = 30°, verify that :
(i) $\sin 2A = \dfrac{2\tan A}{1+\tan^2 A}$
(ii) cos 3A = 4 cos³ A – 3 cos A.

51. If $\cot\theta = \dfrac{1}{\sqrt{3}}$, show that $\dfrac{\sin^2\theta}{1+\cos^2\theta} = \dfrac{3}{5}$.

52. In a ΔABC right angled at C, if $\tan A = \dfrac{1}{\sqrt{3}}$ and $\tan B = \sqrt{3}$, show that sin A cos B + cos A sin B = 1.

53. ABC is a right triangle, right angled at C. If A = 30° and AB = 40 units, find the remaining two sides of ΔABC.

54. Find the value of other trigonometric ratios, given that
(i) $\cos\theta = \dfrac{5}{13}$ *(ii)* $\tan\theta = \dfrac{1}{\sqrt{3}}$
(iii) $\sec\theta = \sqrt{6}$ *(iv)* $\tan\theta = \dfrac{2mn}{m^2 - n^2}$
(v) $\sin\theta = \dfrac{m^2 - n^2}{m^2 + n^2}$ *(vi)* $\cos\theta = \dfrac{2m}{m^2 + 1}$
(vii) $\mathrm{cosec}\,\theta = \sqrt{1 + \left(\dfrac{m}{n}\right)^4}$

55. If $\tan\theta = \frac{p}{q}$, show that $\frac{p\sin\theta - q\cos\theta}{p\sin\theta + q\cos\theta} = \frac{p^2 - q^2}{p^2 + q^2}$.

56. If $\cot\theta = \sqrt{7}$, show that $\frac{\operatorname{cosec}^2\theta - \sec^2\theta}{\operatorname{cosec}^2\theta + \sec^2\theta} = \frac{3}{4}$.

57. If $\cot\theta = \frac{12}{5}$, show that $\tan^2\theta - \sin^2\theta = \sin^4\theta\sec^2\theta$.

58. Find A and B, if

(*i*) $\sin(A + B) = 1$ and $\cos(A - B) = 1$

(*ii*) $\sin(A + 2B) = \frac{\sqrt{3}}{2}$ and $\cos(A + B) = \frac{1}{2}$

(*iii*) $\tan(A - B) = \frac{1}{\sqrt{3}}$ and $\tan(A + B) = \sqrt{3}$

59. In figure, AD = DB and ∠B is a right angle. Determine $\sin^2\theta + \cos^2\theta$.

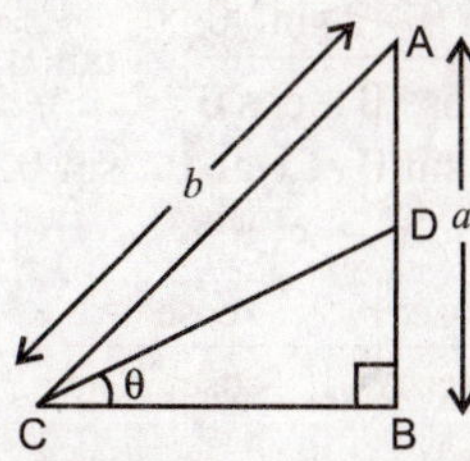

60. Find the value of x if $\tan 3x = \sin 45°. \cos 45° + \sin 30°$.

61. Find the value of x if

$\cos x = \cos 60°. \cos 30° + \sin 60° . \sin 30°$.

62. Find the value of x if

$\cos 2x = \sin 60° . \cos 30° - \cos 60° . \sin 30°$.

63. Given that

$\sin(A + B) = \sin A . \cos B + \cos A . \sin B$, find the value of sin 75°.

64. A rhombus of side 20 cm has two angles of 60° each. Find the length of the diagonals.

65. The altitude AD of a ΔABC in which ∠A is obtuse, is 10 cm. If BD = 10 cm and CD = $10\sqrt{3}$ cm, determine ∠A.

66. An equilateral triangle is inscribed in a circle of radius 6 cm. Find its side.

67. In a rectangle ABCD, AB = 30 cm, ∠BAC = 60°, calculate side BC.

Prove the following identities (68–78) :

68. $(1 + \cot^2\theta)(1 - \cos\theta)(1 + \cos\theta) = 1$.

69. $\frac{\sin\theta}{1 - \cos\theta} + \frac{\tan\theta}{1 + \cos\theta} = \sec\theta . \operatorname{cosec}\theta + \cot\theta$

70. $\frac{\tan\theta - \cot\theta}{\sin\theta . \cos\theta} = \tan^2\theta - \cot^2\theta$

71. $\frac{\sec\theta - \tan\theta}{\sec\theta + \tan\theta} = 1 - 2\sec\theta . \tan\theta + 2\tan^2\theta$

72. $\frac{\sin A + \cos A}{\sin A - \cos A} + \frac{\sin A - \cos A}{\sin A + \cos A} = \frac{2}{2\sin^2 A - 1} = \frac{2}{1 - 2\cos^2 A}$

73. $\frac{\cos A}{1 - \tan A} + \frac{\sin A}{1 - \cot A} = \cos A + \sin A$

74. $\cos^4 A - \cos^2 A = \sin^4 A - \sin^2 A$

75. $\cot^4 A - 1 = \operatorname{cosec}^4 A - 2\operatorname{cosec}^2 A$

76. $\sin^4 A + \cos^4 A = 1 - 2\sin^2 A\cos^2 A$

77. $\sin^6 A + \cos^6 A = 1 - 3\sin^2 A\cos^2 A$

78. $\sin^8\theta - \cos^8\theta = (\sin^2\theta - \cos^2\theta)(1 - 2\sin^2\theta\cos^2\theta)$

79. Show that $\operatorname{cosec}^2\theta + \sin^2\theta$ can never be less than 2.

80. If $\sqrt{3}\cot^2\theta - 4\cot\theta + \sqrt{3} = 0$, then find the value of $\cot^2\theta + \tan^2\theta$.

Prove the following identities (81–98) :

81. $\frac{\cos\theta}{1 - \sin\theta} = \frac{1 + \sin\theta}{\cos\theta}$

82. $\sin^2 A + \frac{1}{1 + \tan^2 A} = 1$

83. $(\sec^2\theta - 1)(\operatorname{cosec}^2\theta - 1) = 1$

84. $(\sec\theta + \cos\theta)(\sec\theta - \cos\theta) = \tan^2\theta + \sin^2\theta$

85. $\tan^2\theta - \sin^2\theta = \tan^2\theta . \sin^2\theta$

86. $\sec^6\theta = \tan^6\theta + 3\tan^2\theta\sec^2\theta + 1$

87. $\frac{(1 + \tan^2\theta)\cot\theta}{\operatorname{cosec}^2\theta} = \tan\theta$

88. $\frac{1 + \cos A}{\sin A} = \frac{\sin A}{1 - \cos A}$

89. $\tan^2 A + \cot^2 A + 2 = \sec^2 A . \operatorname{cosec}^2 A$

90. $\frac{1 + \cos\theta - \sin^2\theta}{\sin\theta(1 + \cos\theta)} = \cot\theta$

91. $\left(\tan\theta + \frac{1}{\cos\theta}\right)^2 + \left(\tan\theta - \frac{1}{\cos\theta}\right)^2 = 2\left(\frac{1 + \sin^2\theta}{1 - \sin^2\theta}\right)$

92. $(\sec A + \tan A - 1)(\sec A - \tan A + 1) = 2\tan A$

93. $(\sec A - \operatorname{cosec} A)(1 + \tan A + \cot A) = \tan A\sec A - \cot A\operatorname{cosec} A$

94. $\sin^2 A\cos^2 B - \cos^2 A\sin^2 B = \sin^2 A - \sin^2 B$

95. $\frac{\tan A + \tan B}{\cot A + \cot B} = \tan A . \tan B$

96. $\tan^2 A\sec^2 B - \sec^2 A\tan^2 B = \tan^2 A - \tan^2 B$

97. $\sec^4 A(1 - \sin^4 A) - 2\tan^2 A = 1$

98. $\cot^2 A\operatorname{cosec}^2 B - \cot^2 B . \operatorname{cosec}^2 A = \cot^2 A - \cot^2 B$

99. If $\cos\theta - \sin\theta = 1$, show that $\cos\theta + \sin\theta = 1$ or -1.

100. If $x = a\sec\theta$, $y = b\tan\theta$, prove that $\frac{a^2}{x^2} - \frac{b^2}{y^2} = 1$.

101. If $\sin 3\theta = \cos(\theta - 6°)$ where 3θ and $(\theta - 6°)$ are acute angles, find the value of θ.

102. Prove that :

$$\frac{1}{\operatorname{cosec} A - \cot A} + \frac{1}{\operatorname{cosec} B - \cot B} + \frac{1}{\operatorname{cosec} C - \cot C} = \operatorname{cosec} A + \operatorname{cosec} B + \operatorname{cosec} C + \cot A + \cot B + \cot C$$

103. If cosec A + cot A = m and cosec A – cot A = n, prove that $mn = 1$.

104. If $\frac{\sin\theta + \cos\theta}{\sin\theta - \cos\theta} = \frac{5}{3}$, find the value of $\frac{7\tan\theta + 2}{2\tan\theta + 7}$.

Prove without using trigonometric tables (105-119) :

105. $\frac{\sin(90° - \theta) . \sin\theta}{\tan\theta} - 1 = -\sin^2\theta$

106. tan 10° . tan 75° . tan 15° . tan 80° = 1

107. tan 1° . tan 2° . tan 3° ...... tan 89° = 1

108. cos 1° . cos 2° . cos 3° ...... cos 180° = 0

109. $\sin^2 5° + \sin^2 10° + ... + \sin^2 85° + \sin^2 90° = 9\frac{1}{2}$

110. $\frac{\sin 36°}{\cos 54°} - \frac{\sin 54°}{\cos 36°} = 0$

111. $\sin^2 25° + \sin^2 65° = 1$

112. $\left(\frac{\tan 20°}{\text{cosec } 70°}\right)^2 + \left(\frac{\cot 20°}{\sec 70°}\right)^2 = 1$

113. (sin 72° + cos 18°) (sin 72° – cos 18°) = 0

114. sin 63° . cos 27° + cos 63° . sin 27° = 1

115. $\frac{\sin 70°}{\cos 20°} + \frac{\text{cosec } 20°}{\sec 70°} - 2\cos 70° . \text{cosec } 20° = 0$

116. $\sin(90° - 0)\cos(90° - \theta) = \frac{\tan\theta}{1 + \cot^2(90° - \theta)}$

117. $\sin^2(90° - \theta)[1 + \cot^2(90° - \theta)] = 1$

118. $\frac{\cos(90° - A) . \sin(90° - A)}{\tan(90° - A)} = \sin^2 A$

119. $\frac{1}{1 + \cos(90° - \theta)} + \frac{1}{1 - \cos(90° - \theta)} = 2\,\text{cosec}^2(90° - \theta)$

120. If A and B are angles of a right angled triangle ABC, right angled at C, prove that $\sin^2 A + \sin^2 B = 1$.

121. Using the formula $\cos 2\theta = 2\cos^2\theta - 1$, find the value of cos 30°, it is being given that $\cos 60° = \frac{1}{2}$.

122. Using the formula $\cos 2\theta = 1 - 2\sin^2\theta$, find the value of sin 30°, if cos 60° = 1/2.

123. Show that $\frac{\cot A + \text{cosec } A - 1}{\cot A - \text{cosec } A + 1} = \frac{1 + \cos A}{\sin A}$.

124. If A and B are acute angles and tan A = 1, sin B = $1/\sqrt{2}$, find the value of cos (A + B).

125. Prove that $\frac{1 + \tan^2\theta}{1 + \cot^2\theta} = \left(\frac{1 - \tan\theta}{1 - \cot\theta}\right)^2$.

126. Prove that $\sqrt{\frac{1 + \sin\theta}{1 - \sin\theta}} = \sec\theta + \tan\theta$.

127. Show that : $(\sin\theta + \text{cosec }\theta)^2 + (\cos\theta + \sec\theta)^2 = 7 + \tan^2\theta + \cot^2\theta$.

128. Prove that $(\text{cosec }\theta - \cot\theta)^2 = \frac{1 - \cos\theta}{1 + \cos\theta}$.

129. Show that $\frac{\sin\theta - 2\sin^3\theta}{2\cos^3\theta - \cos\theta} = \tan\theta$.

130. If $a\cos\theta - b\sin\theta = x$ and $a\sin\theta + b\cos\theta = y$. Prove that $a^2 + b^2 = x^2 + y^2$.

131. Prove that : $\frac{\text{cosec } A}{\text{cosec } A - 1} + \frac{\text{cosec } A}{\text{cosec } A + 1} = 2 + 2\tan^2 A = 2\sec^2 A$

132. If $\text{cosec }\theta = \frac{13}{12}$, find the value of $\frac{2\sin\theta - 3\cos\theta}{4\sin\theta - 9\cos\theta}$.

133. If $\cos\alpha = \frac{1}{2}$ and $\tan\beta = \frac{1}{\sqrt{3}}$, find $\sin(\alpha + \beta)$ where α and β both are acute angles.

134. Find the value of : sin (50° + θ) – cos (40° – θ).

135. If $x = \gamma\cos\alpha\sin\beta$; $y = \gamma\cos\alpha\cos\beta$ and $z = \gamma\sin\alpha$, show that $x^2 + y^2 + z^2 = \gamma^2$.

136. In the figure, ABC is a right angled triangle, D is the mid-point of BC. Show that $\frac{\tan\theta}{\tan\phi} = \frac{1}{2}$

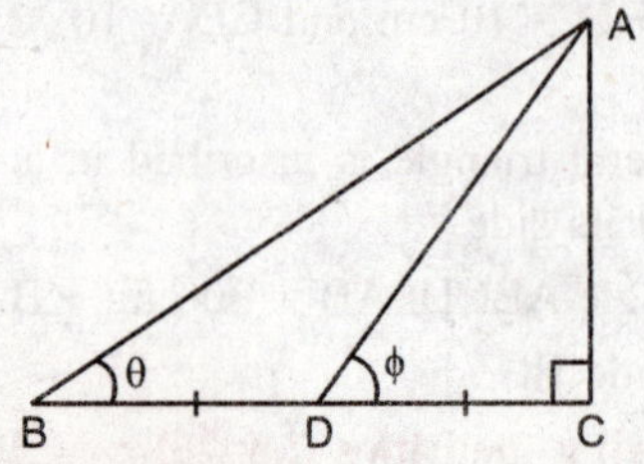

SHORT ANSWER TYPE QUESTIONS-II

PREVIOUS YEARS' QUESTIONS

1. Prove that :

$$\frac{\sin\theta + \cos\theta}{\sin\theta - \cos\theta} + \frac{\sin\theta - \cos\theta}{\sin\theta + \cos\theta} = \frac{2\sec^2\theta}{\tan^2\theta - 1}$$

Or

Evaluate without using trigonometric tables :

$$\frac{\sec^2(90° - \theta) - \cot^2\theta}{2(\sin^2 25° + \sin^2 65°)} + \frac{2\cos^2 60°\tan^2 28°\tan^2 62°}{3(\sec^2 43° - \cot^2 47°)}$$

[Delhi 2006]

2. Prove that :

$$\frac{1}{\text{cosec }\theta - \cot\theta} - \frac{1}{\sin\theta} = \frac{1}{\sin\theta} - \frac{1}{\text{cosec }\theta + \cot\theta}$$

Or

Evaluate without using trigonometric tables :

$$\frac{\text{cosec}^2(90° - \theta) - \tan^2\theta}{4(\cos^2 48° + \cos^2 42°)} - \frac{2\tan^2 30°\sec^2 52°\sin^2 38°}{\text{cosec}^2 70° - \tan^2 20°}$$

[AI 2006]

3. Evaluate without using trigonometric tables :

$$\frac{\sin^2\theta+\sin^2(90-\theta)}{3(\sec^2 61°-\cot^2 29°)}-\frac{3\cot^2 30°\sin^2 54°\sec^2 36°}{2(\operatorname{cosec}^2 65°-\tan^2 25°)}$$

[Foreign 2006]

4. Without using trigonometric tables, evaluate the following:

$$\frac{\cos^2 20°+\cos^2 70°}{\sec^2 50°-\cot^2 40°}+2\operatorname{cosec}^2 58°-2\cot 58°\tan 32°$$
$$-4\tan 13°\tan 37°\tan 45°\tan 53°\tan 77°$$

Or

Prove that : $\sqrt{\frac{\sec\theta-1}{\sec\theta+1}}+\sqrt{\frac{\sec\theta+1}{\sec\theta-1}}=2\operatorname{cosec}\theta$

[Delhi 2006C]

Without using trigonometric table, evaluate the following (5 to 8) :

5. $\frac{\sec 39°}{\operatorname{cosec} 51°}+\frac{2}{\sqrt{3}}\tan 17°\tan 38°\tan 60°\tan 52°\tan 73°$
$-3(\sin^2 31°+\sin^2 59°)$ [AI 2006C]

6. $\frac{3\cos 55°}{7\sin 35°}-\frac{4(\cos 70°.\operatorname{cosec} 20°)}{7(\tan 25°.\tan 45°.\tan 65°.\tan 85°)}$

[Delhi 2007]

7. $\tan 7° . \tan 23° . \tan 60° . \tan 67° . \tan 83° + \frac{\cot 54°}{\tan 36°}+$
$\sin 20° . \sec 70° - 2$ [AI 2007]

8. Prove that : $\frac{\cot A-\cos A}{\cot A+\cos A}=\frac{\operatorname{cosec} A-1}{\operatorname{cosec} A+1}$ [Delhi 2008]

Or

Prove that : $(1 + \cot A - \operatorname{cosec} A)(1 + \tan A + \sec A) = 2$

[Delhi 2008, AI 2010]

9. Prove that : $\sin\theta(1+\tan\theta)+\cos\theta(1+\cot\theta)$
$= \sec\theta + \operatorname{cosec}\theta$. [AI 2008, 2010]

10. Prove that : $(1 + \cot A + \tan A)(\sin A - \cos A)$
$= \sin A \tan A - \cot A . \cos A$. [Foreign 2008]

11. Prove that :

$$(\operatorname{cosec}\theta-\sin\theta).(\sec\theta-\cos\theta)=\frac{1}{\tan\theta+\cot\theta}$$

[AI 2008C, Delhi 2010]

12. Prove that :
$2(\sin^6\theta+\cos^6\theta)-3(\sin^4\theta+\cos^4\theta)+1=0$

[AI 2008C]

13. Without using trigonometric table, evaluate

$$\frac{\cos 58°}{\sin 32°}+\frac{\sin 22°}{\cos 68°}-\frac{\cos 38°\operatorname{cosec} 52°}{\tan 18°.\tan 35°\tan 60°\tan 72°.\tan 55°}$$

[Delhi 2009]

14. Evaluate : $\frac{2}{3}\operatorname{cosec}^2 58°-\frac{2}{3}\cot 58° . \tan 32°-\frac{5}{3}\tan 13°$
$. \tan 37° . \tan 45° . \tan 53° . \tan 77°$

[AI 2009]

15. Prove that : $\sec^2\theta-\left[\frac{\sin^2\theta-2\sin^4\theta}{2\cos^4\theta-\cos^2\theta}\right]=1$ [Foreign 2009]

16. In the figure below, ΔABC is right angled at B, BC = 7 cm and AC – AB = 1 cm. Find the value of cos A + sin A. [CBSE (CCE) 2011]

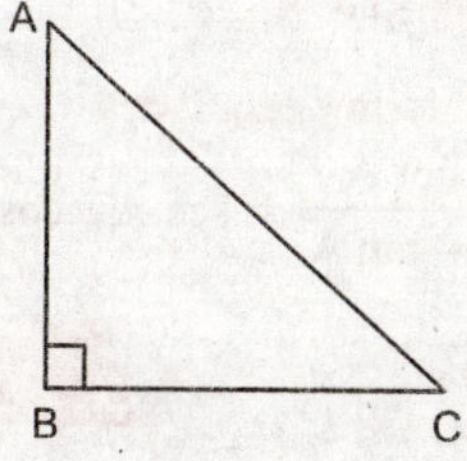

17. If $\cos\theta-\sin\theta=\sqrt{2}\sin\theta$, prove that
$\cos\theta+\sin\theta=\sqrt{2}\cos\theta$. [CBSE (CCE) 2011]

18. If $\operatorname{cosec}(A-B)=2$, $\cot(A+B)=\frac{1}{\sqrt{3}}$,
$0° < (A + B) \le 90°$, $A > B$, then find A and B.

[CBSE (CCE) 2011]

19. Find the value of x if $4\left(\frac{\sec^2 59°-\cot^2 31°}{3}\right)-\frac{2}{3}\sin 90°+$
$3\tan^2 56°\times\tan^2 34°=\frac{x}{3}$ [CBSE (CCE) 2012]

Practice Questions

20. Prove that : $\left(\frac{1+\sin\theta-\cos\theta}{1+\sin\theta+\cos\theta}\right)^2=\frac{1-\cos\theta}{1+\cos\theta}$.

21. If $5\sin\theta+3\cos\theta=4$, find the value of $3\sin\theta-5\cos\theta$.

Prove the following identities (22 to 32) :

22. $(\tan A+\operatorname{cosec} B)^2-(\cot B-\sec A)^2$
$= 2\tan A\cot B(\operatorname{cosec} A+\sec B)$

23. $(\sin A+\sec A)^2+(\cos A+\operatorname{cosec} A)^2$
$= (1+\sec A . \operatorname{cosec} A)^2$

24. $\cot^2 A\left(\frac{\sec A-1}{1+\sin A}\right)+\sec^2 A\left(\frac{\sin A-1}{1+\sec A}\right)=0$

25. $\frac{(1+\cot A+\tan A)(\sin A-\cos A)}{\sec^3 A-\operatorname{cosec}^3 A}=\sin^2 A . \cos^2 A$

26. $\frac{1}{\sec A-1}+\frac{1}{\sec A+1}=2\operatorname{cosec} A . \cot A$

27. $\frac{1}{\cot^2\theta}+\frac{1}{1+\tan^2\theta}=\frac{1}{1-\sin^2\theta}-\frac{1}{\operatorname{cosec}^2\theta}$

28. $\frac{\tan^3\theta}{1+\tan^2\theta}+\frac{\cot^3\theta}{1+\cot^2\theta}=\sec\theta . \operatorname{cosec}\theta-2\sin\theta\cos\theta$

29. $\left(\frac{1}{\sec^2\theta-\cos^2\theta}+\frac{1}{\text{cosec}^2\theta-\sin^2\theta}\right)\sin^2\theta\,.\cos^2\theta$
$$=\frac{1-\sin^2\theta\,.\cos^2\theta}{2+\sin^2\theta\,.\cos^2\theta}$$

30. $\frac{\sin A}{\sec A+\tan A-1}+\frac{\cos A}{\text{cosec} A+\cot A-1}=1$

31. $\frac{\cot^2 A\,(\sec A-1)}{1+\sin A}=\sec^2 A\left(\frac{1-\sin A}{1+\sec A}\right)$

32. $(1+\cot A+\tan A)(\sin A-\cos A)=\frac{\sec A}{\text{cosec}^2 A}-\frac{\text{cosec} A}{\sec^2 A}$

33. Show that $3(\sin\theta-\cos\theta)^4+6(\sin\theta+\cos\theta)^2+4(\sin^6\theta+\cos^6\theta)$ is independent θ.

34. Find value of θ if $\sin^2\theta=1-\cos^2\theta$.

35. Show that $\frac{1}{\sec x-\tan x}-\frac{1}{\cos x}=\frac{1}{\cos x}-\frac{1}{\sec x+\tan x}$.

36. If $\sec\theta=x+\frac{1}{4x}$, prove that $\sec\theta+\tan\theta=2x$ or $\frac{1}{2x}$.
[CBSE (CCE) 2011]

37. If $\sin\theta+\sin^2\theta=1$, prove that $\cos^2\theta+\cos^4\theta=1$.

38. Prove that $\frac{\cos A}{1-\tan A}-\frac{\sin^2 A}{\cos A-\sin A}=\sin A+\cos A$.

39. Prove that $\sec A(1-\sin A)(\sec A+\tan A)=1$.

40. If $\sec\theta+\tan\theta=p$, prove that $\sin\theta=\frac{p^2-1}{p^2+1}$

41. Prove that : $\frac{\tan\theta}{1-\cot\theta}+\frac{\cot\theta}{1-\tan\theta}=1+\tan\theta+\cot\theta$

42. If $\tan\theta+\sin\theta=m$ and $\tan\theta-\sin\theta=n$, show that $m^2-n^2=4\sqrt{mn}$
[CBSE (CCE) 2011]

43. If $\text{cosec}\,\theta-\sin\theta=l$ and $\sec\theta-\cos\theta=m$, show that $l^2m^2(l^2+m^2+3)=1$.

NCERT Questions

Prove the following identities :

44. $\frac{\tan A}{1-\cot A}+\frac{\cot A}{1-\tan A}=\sec A\,.\,\text{cosec} A+1$

45. $\frac{1+\cos\theta-\sin\theta}{\cos\theta-1+\sin\theta}=\text{cosec}\,\theta+\cot\theta$

LONG ANSWER TYPE QUESTIONS

Previous Years' Questions

1. Prove that $\sqrt{\frac{1+\sin\theta}{1-\sin\theta}}+\sqrt{\frac{1-\sin\theta}{1+\sin\theta}}=2\sec\theta$
[CBSE (CCE) 2011]

2. Evaluate :
$$\frac{\sec^2(90°-\theta)-\cot^2\theta}{2(\sin^2 25°+\sin^2 65°)}+\frac{2\sin^2 30°\tan^2 32°.\tan^2 58°}{3(\sec^2 33°-\cot^2 57°)}$$
[CBSE (CCE) 2011]

3. Determine the value of x such that
$$2\,\text{cosec}^2\,30°+x\sin^2 60°-\frac{3}{4}\tan^2 30°=10$$
[CBSE (CCE) 2011]

4. If $A+B=90°$, prove that
$$\sqrt{\frac{\tan A\tan B+\tan A\cot B}{\sec A\sec B}-\frac{\sin^2 B}{\cos^2 A}}=\tan A$$
[CBSE (CCE) 2012]

5. In an acute angled triangle ABC, if $\sin(A+B-C)=\frac{1}{2}$ and $\cos(B+C-A)=\frac{1}{\sqrt{2}}$ find $\angle A$, $\angle B$ and $\angle C$.
[CBSE (CCE) 2012]

NCERT EXEMPLAR PROBLEMS

Short Answer Questions with Reasoning

Write 'True or False' and justify your answer in each of the following :

1. The value of the expression $(\cos^2 23°-\sin^2 67°)$ is positive.
2. The value of the expression $(\sin 80°-\cos 80°)$ is negative.
3. $\sqrt{(1-\cos^2\theta)\sec^2\theta}=\tan\theta$.
4. $(\tan\theta+2)(2\tan\theta+1)=5\tan\theta+2\sec^2\theta$.
5. If a man standing on a platform 3 metres above the surface of a lake observes a cloud and its reflection in the lake, then the angle of elevation of the cloud is equal to the angle of depression of its reflection.
6. The value of $2\sin\theta$ can be $\left(a+\frac{1}{a}\right)$, where a is a positive number, and $a\neq 1$.
7. $\cos\theta=\frac{a^2+b^2}{2ab}$, where a and b are two distinct numbers such that $ab>0$.

SHORT ANSWER QUESTIONS

Prove that (Q8 - Q10) :

8. $(\sqrt{3} + 1)(3 - \cot 30°) = \tan^3 60° - 2 \sin 60°$
9. $\tan \theta + \tan (90° - \theta) = \sec \theta \sec (90° - \theta)$
10. $\dfrac{\cos^2 (45° + \theta) + \cos^2 (45° - \theta)}{\tan (60° + \theta) \tan (30° - \theta)} = 1$

LONG ANSWER QUESTIONS

11. If $1 + \sin^2 \theta = 3 \sin \theta \cos \theta$, then prove that $\tan \theta = 1$ or $\dfrac{1}{2}$.
12. Given that $\sin \theta + 2 \cos \theta = 1$, then prove that $2 \sin \theta - \cos \theta = 2$.
13. If $\tan \theta + \sec \theta = l$, then prove that $\sec \theta = \dfrac{l^2 + 1}{2l}$.

HIGHER ORDER THINKING SKILLS (HOTS) QUESTIONS

1. If $\cot A + \dfrac{1}{\cot A} = 2$, then prove that $\cot^2 A + \dfrac{1}{\cot^2 A} = 2$
2. If $\tan \theta = \dfrac{a}{b}$, prove that $\dfrac{\cos \theta + \sin \theta}{\cos \theta - \sin \theta} = \dfrac{b+a}{b-a}$
3. If $x \sin^3 \theta + y \cos^3 \theta = \sin \theta \cos \theta$ and $x \sin \theta = y \cos \theta$, prove that $x^2 + y^2 = 1$.
4. If $7 \operatorname{cosec} \theta - 3 \cot \theta = 7$, prove that $7 \cot \theta - 3 \operatorname{cosec} \theta = \pm 3$.
5. If $a \cos \theta - b \sin \theta = c$, prove that $a \sin \theta + b \cos \theta = \pm \sqrt{a^2 + b^2 - c^2}$.
6. If $\dfrac{x}{a} \cos \theta + \dfrac{y}{b} \sin \theta = 1$ and $\dfrac{x}{a} \sin \theta - \dfrac{y}{b} \cos \theta = 1$, prove that $\dfrac{x^2}{a^2} + \dfrac{y^2}{b^2} = 2$.
7. If $x = a \cos^3 \theta$, $y = b \sin^3 \theta$, prove that $\left(\dfrac{x}{a}\right)^{2/3} + \left(\dfrac{y}{b}\right)^{2/3} = 1$.
8. If $\tan A = \dfrac{1}{2}$ and $\tan B = \dfrac{1}{3}$, by using $\tan (A + B) = \dfrac{\tan A + \tan B}{1 - \tan A \tan B}$, prove that $A + B = 45°$.
9. If $\sin A + \operatorname{cosec} A = 3$, find the value of $\dfrac{\sin^4 A + 1}{\sin^2 A}$.
10. If $\tan (A + B) = 1$ and $\sin (2A - B) = 1$, find A and B.
11. If $\tan A = n \tan B$ and $\sin A = m \sin B$, prove that $\cos^2 A = \dfrac{m^2 - 1}{n^2 - 1}$.
12. If $\operatorname{cosec} \theta - \sin \theta = m$ and $\sec \theta - \cos \theta = n$, prove that $(m^2 n)^{2/3} + (mn^2)^{2/3} = 1$.
13. If $\cot \theta + \tan \theta = x$ and $\sec \theta - \cos \theta = y$, prove that $(x^2 y)^{2/3} - (xy^2)^{2/3} = 1$.
14. If $\operatorname{cosec} \theta - \sin \theta = a^3$, $\sec \theta - \cos \theta = b^3$, prove that $a^2 b^2 (a^2 + b^2) = 1$.
15. If $a \cos^3 \theta + 3a \cos \theta \sin^2 \theta = m$, $a \sin^3 \theta + 3a \cos^2 \theta \sin \theta = n$, prove that $(m + n)^{2/3} + (m - n)^{2/3} = 2a^{2/3}$.
16. If $2 \cos \theta - \sin \theta = x$ and $\cos \theta - 3 \sin \theta = y$, prove that $2x^2 + y^2 - 2xy = 5$.
17. Compare the areas of the right angled triangles ABC and DEF in which $\angle A = 30°$, $\angle B = 90°$, AC = 4 cm, $\angle D = 60°$, $\angle E = 90°$ and DE = 4 cm.
18. Prove that : $\dfrac{\tan A}{(1 + \tan^2 A)^2} + \dfrac{\cot A}{(1 + \cot^2 A)^2} = \sin A \cos A$.
19. If $(\sec A + \tan A)(\sec B + \tan B)(\sec C + \tan C) = (\sec A - \tan A).(\sec B - \tan B)(\sec C - \tan C)$, prove that each is equal to ± 1.

VALUE BASED QUESTIONS

1. A school organised maths quiz to reward students participating in co-curricular activities group and a group with 100% attendance.
   One of the question asked is as follows.
   If $\sin \theta + \sin^2 \theta = 1$, then find the value of $\cos^2 \theta + \cos^4 \theta$.
   (*a*) Solve the above problem.
   (*b*) Which group should be rewarded ?
2. Who said "There is perhaps nothing which so occupies the middle position of mathematics as trigonometry" ? Swarnim replied "J.F. Herbart".
   (*a*) Do you agree with Swarnim ?
   (*b*) What value is depicted by Swarnim ?

Multiple Choice Questions

PREVIOUS YEARS' QUESTIONS

1. If $\sqrt{2}$ sin (60° – α) = 1, then α is [CBSE (CCE) 2011]

(a) 45° (b) 15°
(c) 60° (d) 30°

2. The maximum value of $\frac{1}{\operatorname{cosec}\theta}$ is [CBSE (CCE) 2011]

(a) 0 (b) 1
(c) $\frac{\sqrt{3}}{2}$ (d) $\frac{1}{\sqrt{2}}$

3. If cos (40° + A) = sin 30°, the value of A is [CBSE (CCE) 2011]

(a) 30° (b) 40°
(c) 60° (d) 20°

4. If cosec θ – cot θ = $\frac{1}{3}$, the value of (cosec θ + cot θ) is

(a) 1 (b) 2 [CBSE (CCE) 2011]
(c) 3 (d) 4

5. The value of $\sin^2 30° - \cos^2 30°$ is [CBSE (CCE) 2011]

(a) $-\frac{1}{2}$ (b) $\frac{\sqrt{3}}{2}$
(c) $\frac{3}{2}$ (d) $\frac{2}{3}$

6. In the given figure, if AB = 14 cm, then the value of tan B is : [CBSE (CCE) 2012]

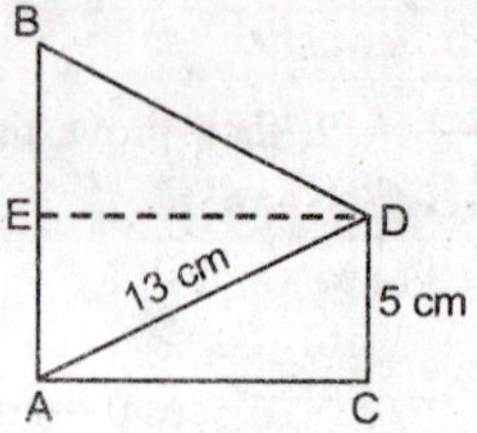

(a) $\frac{4}{3}$ (b) $\frac{14}{3}$
(c) $\frac{5}{3}$ (d) $\frac{13}{3}$

PRACTICE QUESTIONS

7. The study of relationship between the sides and angles of a triangle is known as

(a) geometry (b) coordinate geometry
(c) trigonometry (d) trigometric identities

Read the following line and answer, question 2 to 4.

8. The word trigonometry is derived from the Greek words, tri, gon and metron. In this word 'tri' means

(a) three (b) triangle
(c) try (d) none of these

9. The word 'gon' means

(a) sides (b) gong
(c) triangle (d) right angle

10. 'Metron' means

(a) metro (b) measure
(c) meteor (d) material

11. Ratios of a right triangle with respect to its acute angles are known as

(a) trigonometric identities
(b) trigonometry
(c) trigonometric ratios of the angles
(d) none of these

12. Triangle ABC is a right triangle right angled at B. AC is known as

(a) side adjacent to angle A
(b) side adjacent to angle B
(c) hypotenuse
(d) side opposite to angle C

13. Triangle PQR is a right triangle right angled at Q. PQ is known as

(a) side adjacent to angle P
(b) hypotenuse
(c) side opposite to angle P
(d) side adjacent to angle R

14. Triangle DEF is a right triangle right angled at E. EF is known as

(a) side adjacent to angle D
(b) side opposite to angle D
(c) side adjacent to angle E
(d) hypotenuse

15. Match the Column :

(1) $\dfrac{\text{Side opposite to angle } \theta}{\text{Hypotenuse}}$	(A) $\tan\theta$
(2) $\dfrac{\text{Side adjacent to angle } \theta}{\text{Hypotenuse}}$	(B) $\sin\theta$
(3) $\dfrac{\text{Side opposite to angle } \theta}{\text{Side adjacent to angle } \theta}$	(C) $\cos\theta$
	(D) $\sec\theta$

(*a*) 1 – A, 2 – C, 3 – B
(*b*) 1 – B, 2 – C, 3 – A
(*c*) 1 – B, 2 – C, 3 – D
(*d*) 1 – D, 2 – B, 3 – A

16. Match the Column :

(1) $\dfrac{\text{Hypotenuse}}{\text{Side adjacent to angle } \theta}$	(A) $\text{cosec}\,\theta$
(2) $\dfrac{\text{Side adjacent to angle } \theta}{\text{Side opposite to angle } \theta}$	(B) $\sec\theta$
(3) $\dfrac{\text{Hypotenuse}}{\text{Side opposite to angle } \theta}$	(C) $\tan\theta$
	(D) $\cot\theta$
	(E) $\sin\theta$

(*a*) 1 – A, 2 – C, 3 – B
(*b*) 1 – C, 2 – A, 3 – D
(*c*) 1 – B, 2 – A, 3 – E
(*d*) 1 – B, 2 – D, 3 – A

17. Reciprocal of cos A is
(*a*) sin A (*b*) tan A
(*c*) sec A (*d*) cot A

18. Reciprocal of cosec A is
(*a*) sin A (*b*) cos A
(*c*) sec A (*d*) cot A

19. Reciprocal of cot A is
(*a*) sin A (*b*) tan A
(*c*) sec A (*d*) cosec A

20. Who first used the abbreviated notation 'sin' for sine ratio of an angle ?
(*a*) Aryabhatta (*b*) Edmund Gunter
(*c*) Sir Jonas Moore (*d*) J.F. Herbart

21. If the angle remains the same, the value of the trigonometric ratios of the angle
(*a*) vary with the length of the sides of the triangle
(*b*) do not vary with the length of the sides of the triangle
(*c*) vary with the change in length of hypotenuse only
(*d*) do not vary with the change in length of hypotenuse only

22. If $\sin A = \dfrac{4}{5}$, then $1 - 2\sin^2 A$ =
(*a*) $1 - 2\cos^2 A$ (*b*) $\cos^2 A - \sin^2 A$
(*c*) $2\sin^2 A + 1$ (*d*) $\sin^2 A - \cos^2 A$

23. In ΔABC, right angled at B, AC = 13 cm, AB = 5 cm, then sin A =
(*a*) $\dfrac{5}{13}$ (*b*) $\dfrac{5}{12}$
(*c*) $\dfrac{12}{13}$ (*d*) $\dfrac{13}{12}$

24. $2\cos^2 30° - 1$
(*a*) sin 60° (*b*) cos 60°
(*c*) tan 60° (*d*) sec 60°

25. sin 90° =
(*a*) 2 sin 30° cos 60° (*b*) 2 sin 30° sin 60°
(*c*) 2 sin 45° cos 45° (*d*) 2 cos 30° sin 60°

26. cos 90° =
(*a*) $4\cos^3 30° - 3\cos 30°$ (*b*) sin 60° + sin 30°
(*c*) tan 60° – tan 30° (*d*) 2 cos 60°.cos 30°

27. tan 60° =
(*a*) $\dfrac{2\cos 30°}{1-\cos^2 30°}$ (*b*) $\dfrac{2\tan 30°}{1-\tan^2 30°}$
(*c*) $\dfrac{2\cot 30°}{1-\cot^2 30°}$ (*d*) 2 sin 60°

28. sin A =
(*a*) tan A – cot A (*b*) $\dfrac{\tan A}{\sec A}$
(*c*) tan A – sec A (*d*) cos A – sec A

29. tan A =
(*a*) $\dfrac{\cos A}{\sqrt{1-\cos^2 A}}$ (*b*) $\dfrac{\sec A}{\sqrt{1-\sec^2 A}}$
(*c*) $\dfrac{\sin A}{\sqrt{1-\sin^2 A}}$ (*d*) $\dfrac{1}{\sqrt{1-\sin^2 A}}$

30. sec A =
(*a*) $\dfrac{1}{\cot A}$ (*b*) $\dfrac{1}{\text{cosec}\, A}$
(*c*) $\dfrac{1}{\sqrt{1+\cot^2 A}}$ (*d*) $\dfrac{\sqrt{1+\cot^2 A}}{\cot A}$

31. sin (90° – A) =
(*a*) sin A (*b*) tan A
(*c*) cos A (*d*) cosec A

32. cot (90° – A) =
(*a*) sin A (*b*) tan A
(*c*) cos A (*d*) sec A

33. sec (90° – A) =
(*a*) sin A (*b*) tan A
(*c*) cos A (*d*) cosec A

34. **If tan 2A = cot(A – 18°), where 2A is acute angle and A > 18°, then A =**

(*a*) 54° (*b*) 36°

(*c*) 27° (*d*) 30°

35. **If sec 4A = cosec (A – 20°), where 4A is acute and A > 20°, then 3A =**

(*a*) 63° (*b*) 69°

(*c*) 66° (*d*) 88°

36. **If A, B and C are interior angles of a triangle ABC, then** $\tan\left(\frac{A+B}{2}\right) =$

(*a*) $\sin\frac{C}{2}$ (*b*) $\cos\frac{C}{2}$

(*c*) $\tan\frac{C}{2}$ (*d*) $\cot\frac{C}{2}$

37. $\frac{1+\cot^2 A}{1+\tan^2 A} =$

(*a*) $\tan^2 A$ (*b*) $\sec^2 A$

(*c*) $\text{cosec}^2 A - 1$ (*d*) $1 - \sin^2 A$

38. **If sec A + tan A = *x*, then tan A =**

(*a*) $\frac{2}{x}$ (*b*) $\frac{1}{2x}$

(*c*) $\frac{x^2-1}{2x}$ (*d*) $\frac{2x}{x^2-1}$

39. **If cosec A – cot A =** $\frac{4}{5}$**, then cosec A =**

(*a*) $\frac{47}{40}$ (*b*) $\frac{59}{40}$

(*c*) $\frac{51}{40}$ (*d*) $\frac{41}{40}$

40. **Who said "There is perhaps nothing which so occupies the middle position of mathematics as trigonometry"?**

(*a*) Aryabhata (*b*) Sir Jonas Moore

(*c*) Edmund Gunter (*d*) J.F. Herbart

41. **The value of cos 0° . cos 1° . cos 2° ... cos 90° =**

(*a*) 1 (*b*) 2

(*c*) 3 (*d*) none of these

42. **The value of tan 1° . tan 2° . tan 3° ... tan 89° =**

(*a*) 1 (*b*) 0

(*c*) ∞ (*d*) $\frac{1}{3}$

43. **If sin *x* + cosec *x* = 2, then** $\sin^{19} x + \text{cosec}^{20} x =$

(*a*) 2^{19} (*b*) 2^{20}

(*c*) 2 (*d*) 2^{39}

44. **If *x* = (sec A – tan A) (sec B – tan B) (sec C – tan C) = (sec A + tan A) (sec B + tan B) (sec C + tan C), then *x* =**

(*a*) 0 (*b*) 1

(*c*) – 1 (*d*) ± 1

45. **In a right triangle, the hypotenuse is 2 times as long as its one side. One of the acute angle is**

(*a*) 45° (*b*) 30°

(*c*) 75° (*d*) 25°

46. **If** $\tan^2\theta + \cot^2\theta = 2$**, θ is an acute angle, then** $\tan^3\theta + \cot^3\theta$ **is equal to**

(*a*) 2 (*b*) 3

(*c*) 4 (*d*) 8

47. **If** $\sin x + \sin^2 x = 1$**, then value of** $\cos^2 x + \cos^4 x =$

(*a*) 2 (*b*) 1

(*c*) 0 (*d*) 3

48. **If** $\sin\theta_1 + \sin\theta_2 + \sin\theta_3 = 3$, $0° < \theta_1, \theta_2, \theta_3 \leq 90°$**, then** $\cos\theta_1 + \cos\theta_2 + \cos\theta_3 =$

(*a*) 3 (*b*) 1

(*c*) 0 (*d*) 2

49. **Which of the following is incorrect (θ is acute angle) ?**

(*a*) tan θ = 3 (*b*) sin θ = 3

(*c*) sec θ = 3 (*d*) cot θ = 3

50. **The value of** $2(\sin^6\theta + \cos^6\theta) - 3(\sin^4\theta + \cos^4\theta) + 1 =$

(*a*) 1 (*b*) 2

(*c*) 4 (*d*) 0

51. **The value of** $\sin^2 5° + \sin^2 10° + \sin^2 15° + ... + \sin^2 90°$ **is equal to**

(*a*) 8 (*b*) 8.5

(*c*) 9 (*d*) 9.5

52. **If cosec θ – cot θ =** $\frac{1}{2}$, $0 < \theta < \frac{\pi}{2}$**, then cos θ =**

(*a*) $\frac{5}{3}$ (*b*) $\frac{3}{5}$

(*c*) $\frac{2}{5}$ (*d*) $\frac{4}{5}$

53. **If tan A + cot A = 4, then** $\tan^4 A + \cot^4 A =$

(*a*) 196 (*b*) 194

(*c*) 192 (*d*) 190

54. **If** $x = r\sin\theta.\cos\phi$, $y = r\sin\theta.\sin\phi$ **and** $z = r\cos\theta$**, then the value of** $x^2 + y^2 + z^2$ **is independent of**

(*a*) *r*, θ (*b*) *r*, ϕ

(*c*) θ, ϕ (*d*) *r*

55. **Which of the following is possible ?**

(*a*) $\cos\theta = \frac{7}{5}$ (*b*) $\sin\theta = \frac{13}{12}$

(*c*) $\sec\theta = \frac{4}{5}$ (*d*) tan θ = 41

56. If $\tan\theta = \frac{a}{b}$, then the value of $\frac{a\sin\theta + b\cos\theta}{a\sin\theta - b\cos\theta}$ is

(a) $\frac{a^2-b^2}{a^2+b^2}$ (b) $\frac{a^2+b^2}{a^2-b^2}$

(c) $\frac{a}{a^2+b^2}$ (d) $\frac{b}{a^2+b^2}$

57. If $\sin\theta$ and $\cos\theta$ are the roots of the equation $ax^2 - bx + c = 0$, then a, b, c satisfy the relation

(a) $b^2 - a^2 = 2ac$ (b) $a^2 - b^2 = 2ac$

(c) $a^2 + b^2 = c^2$ (d) $a^2 + b^2 = 2ac$

58. If $0° < \theta < 90°$, then $\sec\theta$ is

(a) > 1 (b) < 1

(c) = 1 (d) 0

59. If sin A = cos A, 0° < A < 90°, then A is equal to

(a) 30° (b) 45°

(c) 60° (d) 90°

60. If $\sin\theta = \sqrt{3}\cos\theta$, $0° < \theta < 90°$, then θ is equal to

(a) 30° (b) 45°

(c) 60° (d) 90°

61. If $3\cot\theta = 2$, then the value of $\tan\theta$

(a) $\frac{2}{3}$ (b) $\frac{3}{2}$

(c) $\frac{3}{\sqrt{13}}$ (d) $\frac{2}{\sqrt{13}}$

62. If $\cos\theta = \frac{7}{8}$ then value of $\sec\theta$ =

(a) $\frac{8}{7}$ (b) $\frac{7}{\sqrt{113}}$

(c) $\frac{8}{\sqrt{113}}$ (d) 1

63. Value of $\frac{\tan 65°}{\cot 25°}$ =

(a) 2 (b) 4

(c) 1 (d) 3

64. Value of (sec A + tan A) (1 – sin A) =

(a) sec A (b) sin A

(c) cosec A (d) cos A

65. Value of (cosec A + cot A) (1 – cos A) =

(a) sec A (b) sin A

(c) cosec A (d) cos A

66. Value of (1 + sin A) (sec A – tan A) =

(a) sin A (b) cos A

(c) tan A (d) sec A

67. cos 1° cos 2° cos 3° ..... cos 90° is equal to

(a) 1 (b) 0

(c) 1/2 (d) –1

NCERT EXEMPLAR PROBLEMS

Multiple Choice Questions

Choose the correct answer from the given four options :

1. The value of the expression
[cosec (75° + θ) – sec (15° – θ) – tan (55° + θ) + cot (35° – θ)] is

(a) –1 (b) 0

(c) 1 (d) $\frac{3}{2}$

2. If cos (α + β) = 0, then sin (α – β) can be reduced to

(a) cos β (b) cos 2β

(c) sin α (d) sin 2α

3. If cos 9α = sin α and 9α < 90°, then the value of tan 5α is

(a) $\frac{1}{\sqrt{3}}$ (b) $\sqrt{3}$

(c) 1 (d) 0

4. If ΔABC is right angled at C, then the value of cos (A + B) is

(a) 0 (b) 1

(c) $\frac{1}{2}$ (d) $\frac{\sqrt{3}}{2}$

5. If sin θ – cos θ = 0, then the value of $(\sin^4\theta + \cos^4\theta)$ is

(a) 1 (b) $\frac{3}{4}$

(c) $\frac{1}{2}$ (d) $\frac{1}{4}$

6. sin (45° + θ) – cos (45° – θ) is equal to

(a) 2 cos θ (b) 0

(c) 2 sin θ (d) 1

ANSWERS

Very Short Answer Type Questions

1. $\frac{12}{5}$ **2.** 45° **3.** 30° **4.** 24° **5.** $\frac{17}{12}$ **6.** $\frac{625}{168}$

7. $\frac{1}{2}$ **8.** 7 **9.** $\frac{49}{64}$ **10.** 18 **11.** 1

12. (*a*) $\frac{15}{7}$ (*b*) $\frac{1}{3}$ (*c*) $\frac{1}{4}$ **13.** $\frac{3}{4}$ **14.** 1 **15.** $\frac{12}{13}, \frac{12}{13}$ **16.** 0

17. 10 cm **18.** 60° **19.** A = 45°, B = 15° **20.** True **21.** $\frac{\sqrt{3}}{2}$ **22.** 1

23. A = 36° **24.** tan 5° + sin 15° **25.** 9

Short Answer Type Questions-I

1. (*i*) 1 (*ii*) 2 (*iii*) 2 (*iv*) 2 (*vi*) A = 22° **Or** 1 **3.** (*a*) $\sqrt{3}$ (*b*) A = 45°, B = 15° **4.** 3

5. 1 **6.** $\frac{225}{64}$ **Or** $\sqrt{3}$ **7.** 5 **8.** $\frac{2}{3}$ or 2 **9.** 1 **10.** 2

11. 2 sec A **12.** 9 **13.** 30° **14.** $-\frac{1}{3}$ **15.** 1 **17.** 1

18. 2 **19.** ±1 **20.** 0 **21.** θ = 60° **22.** $\sqrt{2}-1$ **23.** 180 cm

24. 2 **25.** CE = 9.24 cm, BE = 4.62 cm **30.** DE = $8\sqrt{3}$ cm, AE = $4\sqrt{3}$ cm **31.** $\frac{11}{26}$

35. (*iii*) 6 cm (*iv*) 1 (*v*) 0 **36.** $\sin B = \frac{5}{13}$, $\text{cosec } B = \frac{13}{5}$; $\cos B = \frac{12}{13}$, $\sec B = \frac{13}{12}$; $\tan B = \frac{5}{12}$, $\cot B = \frac{12}{5}$

37. 1 **38.** Not defined **40.** 40 cm **41.** 8.66 cm **42.** 12.5 cm **43.** $10\sqrt{2}$ cm

44. $\frac{3}{4}$ **46.** 2 **47.** sine of angle = $\frac{3}{5}$, cosine of angle = $\frac{4}{5}$ **48.** (*i*) $\frac{\sqrt{3}}{2}$ (*ii*) 0

49. (*i*) 30° (*ii*) 20° (*iii*) 30° **53.** BC = 20 units; AC = $20\sqrt{3}$ units.

54. (*i*) $\sin\theta = \frac{12}{13}$, $\tan\theta = \frac{12}{5}$, $\text{cosec }\theta = \frac{13}{12}$, $\cot\theta = \frac{5}{12}$, $\sec\theta = \frac{13}{5}$.

(*ii*) $\sin\theta = \frac{1}{2}$, $\cos\theta = \frac{\sqrt{3}}{2}$, $\text{cosec }\theta = \frac{2}{1}$, $\sec\theta = \frac{2}{\sqrt{3}}$, $\cot\theta = \sqrt{3}$.

(*iii*) $\sin\theta = \frac{\sqrt{5}}{\sqrt{6}}$, $\tan\theta = \sqrt{5}$, $\cos\theta = \frac{1}{\sqrt{6}}$, $\cot\theta = \frac{1}{\sqrt{5}}$, $\text{cosec }\theta = \frac{\sqrt{6}}{\sqrt{5}}$.

(*iv*) $\sin\theta = \frac{2mn}{m^2+n^2}$, $\cos\theta = \frac{m^2-n^2}{m^2+n^2}$, $\sec\theta = \frac{m^2+n^2}{m^2-n^2}$, $\text{cosec }\theta = \frac{m^2+n^2}{2mn}$, $\cot\theta = \frac{m^2-n^2}{2mn}$

(*v*) $\cos\theta = \frac{2mn}{m^2+n^2}$, $\tan\theta = \frac{m^2-n^2}{2mn}$, $\sec\theta = \frac{m^2+n^2}{2mn}$, $\cot\theta = \frac{2mn}{m^2-n^2}$, $\text{cosec }\theta = \frac{m^2+n^2}{m^2-n^2}$

(*vi*) $\sin\theta = \frac{m^2-1}{m^2+1}$, $\text{cosec }\theta = \frac{m^2+1}{m^2-1}$, $\cos\theta = \frac{2m}{m^2+1}$, $\sec\theta = \frac{m^2+1}{2m}$, $\tan\theta = \frac{m^2-1}{2m}$, $\cot\theta = \frac{2m}{m^2-1}$

(*vii*) $\sin\theta = \frac{n^2}{\sqrt{n^4+m^4}}$, $\cos\theta = \frac{m^2}{\sqrt{n^4+m^4}}$, $\tan\theta = \frac{n^2}{m^2}$, $\text{cosec}\,\theta = \frac{\sqrt{n^4+m^4}}{n^2}$, $\sec\theta = \frac{\sqrt{n^4+m^4}}{m^2}$, $\cot\theta = \frac{m^2}{n^2}$

58. (*i*) A = 45°, B = 45° (*ii*) A = 60°, B = 0° (*iii*) A = 45°, B = 15°

59. 1 **60.** x = 15° **61.** x = 30° **62.** x = 30° **63.** $\frac{\sqrt{3}+1}{2\sqrt{2}}$ **64.** $20\sqrt{3}$, 20 cm

65. ∠A = 105° **66.** $6\sqrt{3}$ cm **67.** $30\sqrt{3}$ **80.** $\frac{10}{3}, \frac{10}{3}$ **101.** θ = 24° **104.** 2

121. $\frac{\sqrt{3}}{2}$ **122.** $\frac{1}{2}$ **124.** 0 **132.** 3 **133.** 1 **134.** 0

Short Answer Type Questions-II

1. Or $\frac{2}{3}$ **2.** Or $\left(\frac{-5}{12}\right)$ **3.** $\frac{-25}{6}$ **4.** – 1 **5.** 0

6. $-\frac{1}{7}$ **7.** $\sqrt{3}$ **13.** $\left(\frac{2\sqrt{3}-1}{\sqrt{3}}\right)$ **14.** – 1

16. $\frac{31}{25}$ **18.** A = 45°, B = 15° **19.** 11 **21.** $\pm 3\sqrt{2}$ **34.** All values of θ

Long Answer Type Questions

2. $\frac{2}{3}$ **3.** 3 **5.** ∠A = 67.5°, ∠B = 37.5°, ∠C = 75°

NCERT Exemplar Problems

1. False **2.** False [sin 80° – sin 10° = Positive, as θ increases, value of sin θ increases]

3. True **4.** True **5.** False **6.** False **7.** False.

Higher Order Thinking Skills (Hots) Questions

9. 7 **10.** A = 45°, B = 0° **17.** $\frac{ar\,(\Delta ABC)}{ar\,(\Delta DEF)} = \frac{1}{4}$

Value Based Questions

1. (*a*) 1 (*b*) both rewarded equally. **2.** (*a*) Yes (*b*) Sincerity

Multiple Choice Questions

1. (*b*) **2.** (*b*) **3.** (*d*) **4.** (*c*) **5.** (*a*) **6.** (*a*)
7. (*c*) **8.** (*a*) **9.** (*a*) **10.** (*b*) **11.** (*c*) **12.** (*c*)
13. (*a*) **14.** (*b*) **15.** (*b*) **16.** (*d*) **17.** (*c*) **18.** (*a*)
19. (*b*) **20.** (*b*) **21.** (*b*) **22.** (*b*) **23.** (*c*) **24.** (*b*)
25. (*c*) **26.** (*a*) **27.** (*b*) **28.** (*b*) **29.** (*c*) **30.** (*d*)
31. (*c*) **32.** (*b*) **33.** (*d*) **34.** (*b*) **35.** (*c*) **36.** (*d*)
37. (*c*) **38.** (*c*) **39.** (*d*) **40.** (*d*) **41.** (*d*) **42.** (*a*)
43. (*c*) **44.** (*d*) **45.** (*b*) **46.** (*a*) **47.** (*b*) **48.** (*c*)
49. (*b*) **50.** (*d*) **51.** (*d*) **52.** (*b*) **53.** (*b*) **54.** (*c*)
55. (*d*) **56.** (*b*) **57.** (*a*) **58.** (*a*) **59.** (*b*) **60.** (*c*)
61. (*b*) **62.** (*a*) **63.** (*c*) **64.** (*d*) **65.** (*b*) **66.** (*b*)
67. (*b*)

NCERT Exemplar Problems

1. (*b*) **2.** (*b*) **3.** (*c*) **4.** (*a*) **5.** (*c*) **6.** (*b*)

SOLUTIONS

VERY SHORT ANSWER TYPE QUESTIONS

1. In right ΔADB,

$$AB^2 = AD^2 + BD^2$$

$\Rightarrow$ $AB^2 = 4^2 + 3^2 \Rightarrow AB = \pm 5$ cm

$\Rightarrow$ AB = 5 cm (Neglecting negative value)

In right ΔABC, $\cot\theta = \dfrac{BC}{AB} = \dfrac{12}{5}$

2. $\sin\theta = \cos\theta$

$\Rightarrow \dfrac{\sin\theta}{\cos\theta} = 1$

$\Rightarrow \tan\theta = 1$ (Also $\tan 45° = 1$)

$\Rightarrow \tan\theta = \tan 45°$

$\Rightarrow \theta = 45°$

3. $\tan\theta = \cot(30° + \theta)$

$\Rightarrow \cot(90° - \theta) = \cot(30° + \theta)$

$\Rightarrow 90° - \theta = 30° + \theta \Rightarrow 2\theta = 60°$

$\Rightarrow \theta = 30°$

4. $\sin 3\theta = \cos(\theta - 6°)$

$\Rightarrow \cos(90° - 3\theta) = \cos(\theta - 6°)$

$\Rightarrow 90° - 3\theta = \theta - 6°$

$\Rightarrow 4\theta = 96° \Rightarrow \theta = 24°$

5. $(\sin A + \cos A).\sec A = \sin A.\sec A + \cos A.\sec A$

$= \sin A \times \dfrac{1}{\cos A} + 1$

$= \tan A + 1$

$= \dfrac{5}{12} + 1 = \dfrac{17}{12}$

6. $\cos A = \dfrac{7}{25}$

We know that $\sin^2 A + \cos^2 A = 1$

$\Rightarrow \sin^2 A = 1 - \cos^2 A$

$\Rightarrow \sin^2 A = 1 - \dfrac{49}{625} = \dfrac{576}{625}$

$\sin A = \sqrt{\dfrac{576}{625}} = \dfrac{24}{25}$

Now $\tan A = \dfrac{\sin A}{\cos A} = \dfrac{24/25}{7/25} = \dfrac{24}{7}$

and $\cot A = \dfrac{1}{\tan A} = \dfrac{7}{24}$

$\therefore \tan A + \cot A = \dfrac{24}{7} + \dfrac{7}{24} = \dfrac{625}{168}$

7. $\tan\theta = \dfrac{1}{\sqrt{3}} \Rightarrow \theta = 30°$

$$\frac{\operatorname{cosec}^2\theta - \sec^2\theta}{\operatorname{cosec}^2\theta + \sec^2\theta} = \frac{\operatorname{cosec}^2 30° - \sec^2 30°}{\operatorname{cosec}^2 30° + \sec^2 30°}$$

$$= \frac{(2)^2 - \left(\frac{2}{\sqrt{3}}\right)^2}{(2)^2 + \left(\frac{2}{\sqrt{3}}\right)^2} = \frac{4 - \frac{4}{3}}{4 + \frac{4}{3}} = \frac{1}{2}$$

8. $$\frac{\sin\theta + \cos\theta}{\sin\theta - \cos\theta} = \frac{\frac{\sin\theta}{\cos\theta} + \frac{\cos\theta}{\cos\theta}}{\frac{\sin\theta}{\cos\theta} - \frac{\cos\theta}{\cos\theta}}$$

$$= \frac{\tan\theta + 1}{\tan\theta - 1} = \frac{\frac{4}{3} + 1}{\frac{4}{3} - 1} = 7$$

9. $$\frac{(1 + \sin\theta)(1 - \sin\theta)}{(1 + \cos\theta)(1 - \cos\theta)} = \frac{1 - \sin^2\theta}{1 - \cos^2\theta} = \frac{\cos^2\theta}{\sin^2\theta} = \cot^2\theta$$

$$= \left(\frac{7}{8}\right)^2 = \frac{49}{64}$$

10. $2(\cot^2\theta + 1) = 2.\operatorname{cosec}^2\theta = \dfrac{2}{\sin^2\theta}$

$= \dfrac{2}{(1/9)} = 18$ $\left(\because \sin^2\theta = \dfrac{1}{9}\right)$

11. $\sec^2\theta(1 + \sin\theta)(1 - \sin\theta) = k$

$\Rightarrow \sec^2\theta(1 - \sin^2\theta) = k$

$\Rightarrow \sec^2\theta.\cos^2\theta = k$

$\Rightarrow k = 1$

12. (*a*) $\sec A = \dfrac{15}{7}$, $\dfrac{H}{B} = \dfrac{AB}{AC} = \dfrac{15}{7}$

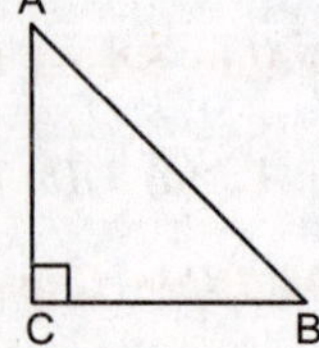

AB = 15*k*, AC = 7*k*

$\operatorname{cosec} B = \dfrac{AB}{AC} = \dfrac{15}{7}$

(*b*) $3x = \operatorname{cosec}\theta$ and $\dfrac{3}{x} = \cot\theta$

$$3\left[x^2 - \frac{1}{x^2}\right] = 3\left[\left(\frac{\operatorname{cosec}\theta}{3}\right)^2 - \left(\frac{\cot\theta}{3}\right)^2\right]$$

$$= 3\left[\frac{\operatorname{cosec}^2\theta - \cot^2\theta}{9}\right] = 3 \times \frac{1}{9} = \frac{1}{3}$$

(*c*) $6x = \sec\theta$ and $\dfrac{6}{x} = \tan\theta$

$\Rightarrow x = \dfrac{\sec\theta}{6}$ and $\dfrac{1}{x} = \dfrac{\tan\theta}{6}$

Consider, $9\left(x^2 - \dfrac{1}{x^2}\right) = 9\left(\dfrac{\sec^2\theta}{36} - \dfrac{\tan^2\theta}{36}\right)$

$$= \frac{9}{36}(\sec^2\theta - \tan^2\theta) = \frac{1}{4} \times 1 = \frac{1}{4}$$

13. $\tan C = \dfrac{3}{4}$

14. $\cos^2\theta + \sin^2\theta = \left(\frac{20}{29}\right)^2 + \left(\frac{21}{29}\right)^2 = 1$

15. $\sin B = \frac{AC}{BC} = \frac{12}{13}$

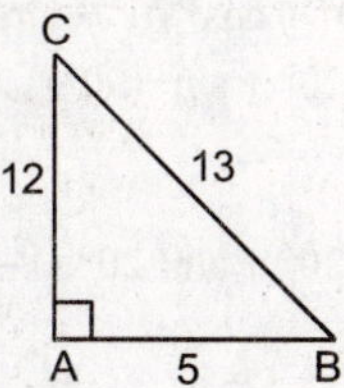

$\cos C = \frac{AC}{BC} = \frac{12}{13}$

16. In right ΔPQR

$$PR^2 = PQ^2 + QR^2$$
$$\Rightarrow (13)^2 = (12)^2 + QR^2$$
$$\Rightarrow QR^2 = 169 - 144 = 25$$
$$\Rightarrow QR = \sqrt{25} = 5 \text{ cm}$$

Now $\tan P = \frac{QR}{PQ} = \frac{5}{12}$

$\cot R = \frac{QR}{PQ} = \frac{5}{12}$

$\therefore \tan P - \cot R = \frac{5}{12} - \frac{5}{12} = 0$

17. $\sin C = \frac{AB}{AC}$

$\Rightarrow \sin 30° = \frac{5}{AC}$

$\Rightarrow \frac{1}{2} = \frac{5}{AC}$

$\Rightarrow AC = 10 \text{ cm}$

18. $\cos(\angle QPR) = \frac{PQ}{PR} = \frac{3}{6} = \frac{1}{2}$

Also, $\cos 60° = \frac{1}{2}$

$\therefore \angle QPR = 60°$

19. $\sin(A - B) = \frac{1}{2}$

$\Rightarrow A - B = 30°$ ...(*i*)

and $\cos(A + B) = \frac{1}{2} \Rightarrow A + B = 60°$ ...(*ii*)

Solving equation (*i*) and (*ii*), we get

$A = 45°$ and $B = 15°$

20. $\sin 2A = \sin 2 \times 0 = \sin 0 = 0$ ...(*i*)

$2 \sin A = 2 \sin 0 = 2 \times 0 = 0$ ...(*ii*)

From (*i*) and (*ii*),

$\sin 2A = 2 \sin A$

It is true.

21. $$\frac{2\tan 30°}{1 + \tan^2 30°} = \frac{2 \times \frac{1}{\sqrt{3}}}{1 + \left(\frac{1}{\sqrt{3}}\right)^2} = \frac{\frac{2}{\sqrt{3}}}{1 + \frac{1}{3}}$$

$$= \frac{\frac{2}{\sqrt{3}}}{\frac{4}{3}} = \frac{2}{\sqrt{3}} \times \frac{3}{4} = \frac{\sqrt{3}}{2}$$

22. $\sin 60° . \cos 30° + \sin 30° . \cos 60°$

$$= \frac{\sqrt{3}}{2} \times \frac{\sqrt{3}}{2} + \frac{1}{2} \times \frac{1}{2} = 1$$

23. $\tan 2A = \cot(A - 18°)$

$\Rightarrow \cot(90° - 2A) = \cot(A - 18°)$

$\Rightarrow 90° - 2A = A - 18°$

$\Rightarrow 3A = 108°$

$\Rightarrow A = 36°$

24. $\cot 85° + \cos 75°$

$= \tan(90° - 85°) + \sin(90° - 75°)$

$= \tan 5° + \sin 15°$

25. $9(\sec^2 A - \tan^2 A) = 9 \times 1 = 9$

SHORT ANSWER TYPE QUESTIONS-I

1. (*i*) $$\frac{11}{7}\frac{\sin 70°}{\cos 20°} - \frac{4}{7}\frac{\cos 53°.\text{cosec } 37°}{\tan 15°.\tan 35°.\tan 55°.\tan 75°}$$

$$= \frac{11}{7}\frac{\sin 70°}{\cos(90° - 70°)} - \frac{4}{7}\frac{\cos 53° \text{cosec}(90° - 53°)}{\tan 15°.\tan 35°.\tan(90° - 35°).\tan(90° - 15°)}$$

$$= \frac{11}{7}\frac{\sin 70°}{\sin 70°} - \frac{4}{7}\frac{\cos 53° \times \sec 53°}{\tan 15° \tan 35°.\cot 35°.\cot 15°}$$

$$= \frac{11}{7} - \frac{4}{7} = 1.$$

(*ii*) $$\frac{\sin 18°}{\cos 72°} + \sqrt{3}(\tan 10° . \tan 30° . \tan 40° . \tan 50° . \tan 80°)$$

$$= \frac{\sin 18°}{\cos(90° - 18°)} + \sqrt{3}\left\{\tan 10° . \frac{1}{\sqrt{3}} . \tan 40° . \tan(90° - 40°) . \tan(90° - 10°)\right\}$$

$$= \frac{\sin 18°}{\sin 18°} + \tan 10° . \tan 40° . \cot 40° . \cot 10°$$

$= 1 + 1 = 2$

(*iii*) $(\sin^2 25° + \sin^2 65°) + \sqrt{3}(\tan 5° \tan 15° \tan 30° . \tan 75° . \tan 85°)$

$$= \sin^2 25° + \cos^2(90° - 65°) + \sqrt{3} \tan 5°.\tan 15° \times \frac{1}{\sqrt{3}} \cot(90° - 75°).\cot(90° - 85°)$$

$= \sin^2 25° + \cos^2 25° + \tan 5°.\tan 15° . \cot 15° . \cot 5°$

$$= 1 + \tan 5°.\tan 15° \times \frac{1}{\tan 15°} \times \frac{1}{\tan 5°}$$

$= 1 + 1 = 2$

(*iv*) $(\cos^2 25° + \cos^2 65°) + \text{cosec}\,\theta . \sec(90° - \theta) - \cot\theta . \tan(90° - \theta)$

$= \cos^2 25° + \sin^2(90° - 65°) + \text{cosec}\,\theta . \text{cosec}\,\theta - \cot\theta . \cot\theta$

$= \cos^2 25° + \sin^2 25° + \text{cosec}^2\,\theta - \cot^2\theta$
$= 1 + 1 = 2$

(*v*) $7\sin^2\theta + 3\cos^2\theta = 4$

$\Rightarrow \quad 3 + 4\sin^2\theta = 4$

$\therefore \quad 3 = 4(1 - \sin^2\theta)$

$\frac{3}{4} = \cos^2\theta$

$\Rightarrow \quad \sec^2\theta = \frac{4}{3}$

$\tan^2\theta = \sec^2\theta - 1 = \left(\frac{4}{3}\right) - 1$

$\tan^2\theta = \frac{1}{3} \Rightarrow \tan\theta = \frac{1}{\sqrt{3}}$

(*vi*) $\sec 4A = \text{cosec}\,(A - 20°)$

$\Rightarrow \quad \text{cosec}\,(90° - 4A) = \text{cosec}\,(A - 20°)$

$\Rightarrow \quad 90° - 4A = A - 20°$

$\Rightarrow \quad 5A = 110°$

$A = 22°.$

Or

$\tan A = \frac{1}{\sqrt{3}} \Rightarrow A = 30°, \ \therefore\ B = 60°$

Putting these values we get.

$\therefore \sin 30° \cos 60° + \cos 30°. \sin 60° = 1$

2. $A + B + C = 180°$

$\Rightarrow \quad B + C = 180° - A$

$\Rightarrow \quad \frac{B+C}{2} = 90° - \frac{A}{2}$

$\Rightarrow \quad \sin\left(\frac{B+C}{2}\right) = \sin\left(90° - \frac{A}{2}\right)$

$\Rightarrow \quad \sin\left(\frac{B+C}{2}\right) = \cos\frac{A}{2}$

3. (*a*) $\frac{\sin 15°.\cos 75° + \cos 15°.\sin 75°}{\tan 5°.\tan 30° \tan 35°.\tan 55°.\tan 85°}$

$= \frac{\sin 15°.\sin(90° - 75°) + \cos 15°.\cos(90° - 75°)}{\tan 5°.\frac{1}{\sqrt{3}}.\tan 35°.\cot(90° - 55°).\cot(90° - 85°)}$

$= \frac{\sin 15°.\sin 15° + \cos 15°.\cos 15°}{\frac{1}{\cot 5°}.\frac{1}{\sqrt{3}}.\frac{1}{\cot 35°}.\cot 35°.\cot 5°}$

$= \frac{\sin^2 15° + \cos^2 15°}{\frac{1}{\sqrt{3}}} = \frac{1}{\frac{1}{\sqrt{3}}} = \sqrt{3}$

(*b*) $\tan(A + B) = \sqrt{3}$

Also, $\tan 60° = \sqrt{3}$

$\Rightarrow \quad A + B = 60°$ ...(*i*)

$\tan(A - B) = \frac{1}{\sqrt{3}}$

Also, $\tan(30°) = \frac{1}{\sqrt{3}} \Rightarrow A - B = 30°$ ...(*ii*)

Solving (*i*) and (*ii*), we get

$A = 45°, B = 15°$

4. $\cos^2 20° + \cos^2 70° + \sin 48°.\sec 42° + \cos 40°.\text{cosec}\,50°$

$= \cos^2 20° + \sin^2(90° - 70°) + \sin 48°.\text{cosec}\,(90° - 42°) + \cos 40°.\sec(90° - 50°)$

$= \cos^2 20° + \sin^2 20° + \frac{1}{\text{cosec}\,48°}.\text{cosec}\,48° + \frac{1}{\sec 40°}\sec 40°$

$= 1 + 1 + 1 = 3$

5. $\frac{\sin^3\theta + \cos^3\theta}{(\sin\theta + \cos\theta)} + \sin\theta\cos\theta$

$= \frac{(\sin\theta + \cos\theta)(\sin^2\theta + \cos^2\theta - \sin\theta\cos\theta)}{(\sin\theta + \cos\theta)} + \sin\theta\cos\theta$

$= 1 - \sin\theta\cos\theta + \sin\theta\cos\theta = 1$

6. Consider, $\frac{(2 + 2\sin\theta)(1 - \sin\theta)}{(1 + \cos\theta)(2 - 2\cos\theta)}$

$= \frac{2(1 + \sin\theta)(1 - \sin\theta)}{(1 + \cos\theta).2(1 - \cos\theta)}$

$= \frac{2(1 - \sin^2\theta)}{2(1 - \cos^2\theta)} = \frac{\cos^2\theta}{\sin^2\theta} = \cot^2\theta = \left(\frac{15}{8}\right)^2 = \frac{225}{64}$

Or

Let ΔABC is an equilateral Δ with each side $= 2a$ units. Draw $AD \perp BC$

$\therefore$ D is mid-point of BC $\Rightarrow BD = a$

In right ΔADB,

$AB^2 = BD^2 + AD^2$

$\Rightarrow \quad (2a)^2 = a^2 + AD^2$

$\Rightarrow \quad 4a^2 - a^2 = AD^2$

$\Rightarrow \quad AD = \sqrt{3a^2} = \sqrt{3}a$

Now in right ΔADB

$\tan B = \frac{AD}{BD}$

$\Rightarrow \quad \tan 60° = \frac{\sqrt{3}a}{a}$ $(\because \angle B = 60°)$

$\Rightarrow \quad \tan 60° = \sqrt{3}$

7. $\frac{7\cos 70°}{2\sin 20°} + \frac{3}{2}\frac{\cos 55°.\text{cosec}\,35°}{\tan 5°.\tan 25°.\tan 45°.\tan 85°.\tan 65°}$

$= \frac{7\sin(90° - 70°)}{2\sin 20°} + \frac{3}{2}$

$+ \frac{\cos 55°.\sec(90° - 35°)}{\tan 5°.\tan 25°.1.\cot(90° - 85°).\cot(90° - 65°)}$

$= \frac{7\sin 20°}{2\sin 20°} + \frac{3}{2}\frac{\frac{1}{\sec 55°}.\sec 55°}{\frac{1}{\cot 5°}.\frac{1}{\cot 25°}.\cot 5°.\cot 25°}$

$= \frac{7}{2} + \frac{3}{2} = \frac{10}{2} = 5$

8. $\dfrac{\sec(90^\circ-\theta).\text{cosec}\,\theta-\tan(90^\circ-\theta)\cot\theta+\cos^2 25^\circ+\cos^2 65^\circ}{3\tan 27^\circ.\tan 63^\circ}$

$=\dfrac{\text{cosec}\,\theta.\text{cosec}\,\theta-\cot\theta.\cot\theta+\cos^2(90^\circ-65^\circ)+\cos^2 65^\circ}{3\tan(90^\circ-63^\circ).\tan 63^\circ}$

$=\dfrac{\text{cosec}^2\theta-\cot^2\theta+\sin^2 65^\circ+\cos^2 65^\circ}{3\cot 63^\circ.\tan 63^\circ}$

$=\dfrac{1+1}{3}=\dfrac{2}{3}$

Or

For the value of cosec 30°, consider an equilateral triangle ABC.

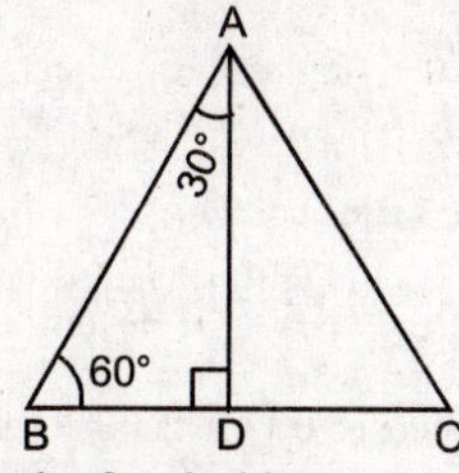

Let $2a$ be the length of each side

$\Rightarrow$ AB = BC = CA = $2a$

Since each angle of equilateral triangle is 60°.

$\therefore$ $\Delta ABD \cong \Delta ACD$

$\therefore$ BD = DC (CPCT)

and $\angle BAD = \angle CAD$

ΔABD is right angled triangle.

BD $=\dfrac{1}{2}$BC $= a$

Now cosec 30° $=\dfrac{AB}{BD}=\dfrac{2a}{a}=2$

Hence, the value of cosec 30° = 2.

9. $\dfrac{\cos 70^\circ}{3\sin 20^\circ}+\dfrac{4(\sec^2 59^\circ-\cot^2 31^\circ)}{3}-\dfrac{2}{3}\sin 90^\circ$

$=\dfrac{\cos(90^\circ-20^\circ)}{3\sin 20^\circ}+\dfrac{4[\sec^2(90^\circ-31^\circ)-\cot^2 31^\circ]}{3}-\dfrac{2}{3}\times 1$

$=\dfrac{\sin 20^\circ}{3\sin 20^\circ}+\dfrac{4[\text{cosec}^2 31^\circ-\cot^2 31^\circ]}{3}-\dfrac{2}{3}$

$=\dfrac{1}{3}+\dfrac{4}{3}-\dfrac{2}{3}=\dfrac{3}{3}=1$

10. Consider $\cot\theta\cdot\tan(90^\circ-\theta)-\sec(90^\circ-\theta)\,\text{cosec}\theta+\sqrt{3}\cdot\tan 12^\circ$ $\tan 60^\circ\tan 78^\circ$

$=\cot\theta\cdot\cot\theta-\text{cosec}\theta\cdot\text{cosec}\theta+\sqrt{3}.\tan 12^\circ.\sqrt{3}\cdot\tan(90^\circ-12^\circ)$

$=(\cot^2\theta-\text{cosec}^2\theta)+3\cdot\tan 12^\circ\cdot\cot 12^\circ$

$=-1+3(1)=2$ $(\because \tan\theta\cot\theta=1)$

11. LHS $=\dfrac{\cos A}{1+\sin A}+\dfrac{1+\sin A}{\cos A}$

$=\dfrac{\cos^2 A+(1+\sin A)^2}{(1+\sin A)\cos A}$

$=\dfrac{\cos^2 A+\sin^2 A+1+2\sin A}{(1+\sin A)\cos A}$

$=\dfrac{2(1+\sin A)}{(1+\sin A)\cos A}$

$=\dfrac{2}{\cos A}=2\sec A=$ RHS

12. $\dfrac{\tan^2 60^\circ+4\sin^2 45^\circ+3\sec^2 30^\circ+5\cos^2 90^\circ}{\text{cosec}\,30^\circ+\sec 60^\circ-\cot^2 30^\circ}$

$=\dfrac{(3)^2+4\times\left(\dfrac{1}{\sqrt{2}}\right)^2+3\times\left(\dfrac{2}{\sqrt{3}}\right)^2+5\times 0}{2+2-(\sqrt{3})^2}=\dfrac{3+2+4}{1}=9$

13. $\sqrt{3}\sin\theta-\cos\theta=0$

$\Rightarrow \sqrt{3}\sin\theta=\cos\theta \Rightarrow \sqrt{3}=\dfrac{\cos\theta}{\sin\theta}$

$\Rightarrow \cot\theta=\sqrt{3}\Rightarrow\theta=30^\circ$

14. $\because \sin A=\dfrac{\sqrt{3}}{2}$ $\therefore A=60^\circ$

Now $2\cot^2 A-1=2.\cot^2 60^\circ-1$

$=2\times\left(\dfrac{1}{\sqrt{3}}\right)^2-1=-\dfrac{1}{3}$

15. $(1+\tan^2\theta)(1-\sin\theta)(1+\sin\theta)$

$=\sec^2\theta(1-\sin^2\theta)$ $(\because 1+\tan^2\theta=\sec^2\theta)$

$=\sec^2\theta\times\cos^2\theta$ $(\because \cos^2\theta+\sin^2\theta=1)$

$=\dfrac{1}{\cos^2\theta}\times\cos^2\theta=1$

16. LHS $=\dfrac{\sin\theta}{1-\cos\theta}=\dfrac{\sin\theta}{1-\cos\theta}\times\dfrac{1+\cos\theta}{1+\cos\theta}$

$=\dfrac{\sin\theta+\sin\theta.\cos\theta}{1-\cos^2\theta}=\dfrac{\sin\theta+\sin\theta.\cos\theta}{\sin^2\theta}$

$=\dfrac{\sin\theta}{\sin^2\theta}+\dfrac{\sin\theta.\cos\theta}{\sin^2\theta}=\dfrac{1}{\sin\theta}+\dfrac{\cos\theta}{\sin\theta}$

$=\text{cosec}\,\theta+\cot\theta$

= RHS

17. $\dfrac{\tan 35^\circ}{\cot 55^\circ}+\dfrac{\cot 78^\circ}{\tan 12^\circ}-1=\dfrac{\tan(90^\circ-35^\circ)}{\cot 55^\circ}+\dfrac{\tan(90^\circ-78^\circ)}{\tan 12^\circ}-1$

$=\dfrac{\cot 55^\circ}{\cot 55^\circ}+\dfrac{\tan 12^\circ}{\tan 12^\circ}-1=1+1-1=1$

18. $\sec 50^\circ\sin 40^\circ+\cos 40^\circ\,\text{cosec}\,50^\circ$

Using $\sin(90^\circ-\theta)=\cos\theta$; we get

$\sin 40^\circ=\sin(90^\circ-50^\circ)=\cos 50^\circ$

and $\cos 40^\circ=\cos(90^\circ-50^\circ)=\sin 50^\circ$

$\therefore$ $\sec 50^\circ\sin 40^\circ+\cos 40^\circ\,\text{cosec}\,50^\circ$

$=\sec 50^\circ\cos 50^\circ+\sin 50^\circ\,\text{cosec}\,50^\circ$

$=\dfrac{1}{\cos 50^\circ}\times\cos 50^\circ+\dfrac{1}{\text{cosec}\,50^\circ}\times\text{cosec}\,50^\circ=1+1=2$

19. $\sec\theta+\tan\theta=m$ ...(*i*), $\sec\theta-\tan\theta=n$ ...(*ii*)

Multiplying (*i*) and (*ii*), we get

$(\sec\theta+\tan\theta)(\sec\theta-\tan\theta)=m.n$

$\Rightarrow \sec^2\theta-\tan^2\theta=mn\Rightarrow mn=1\Rightarrow\sqrt{mn}=\pm 1$

20. $\sin^2 20^\circ+\sin^2 70^\circ-\tan^2 45^\circ$

$=(\sin 20^\circ)^2+(\sin 70^\circ)^2-\tan^2 45^\circ$

Using $\sin(90° - \theta) = \cos\theta$, we get,

$\sin 70° = \sin(90° - 20°) = \cos 20°$

$\therefore \quad \sin^2 20° + \cos^2 20° - \tan^2 45°$

$= 1 - 1 = 0$

21. $2\cos^2\theta = \frac{1}{2} \Rightarrow \cos^2\theta = \frac{1}{4} \Rightarrow \cos\theta = \frac{1}{2}$

Also, $\cos 60° = \frac{1}{2}$ $\qquad (\because 0° < \theta < 90°)$

$\therefore \quad \theta = 60°$

22. $\sin\theta + \cos\theta = \sqrt{2}\,.\cos(90° - \theta)$

$\Rightarrow \sin\theta + \cos\theta = \sqrt{2}\sin\theta$

$\Rightarrow \cos\theta = \sqrt{2}\sin\theta - \sin\theta$

$\Rightarrow \cos\theta = (\sqrt{2} - 1)\sin\theta$

$\Rightarrow \frac{\cos\theta}{\sin\theta} = \sqrt{2} - 1 \Rightarrow \cot\theta = \sqrt{2} - 1$

23. In ΔABP, we have

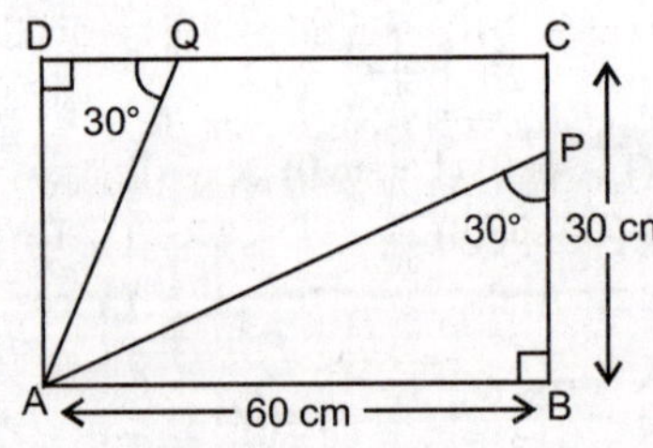

$$\frac{AP}{AB} = \text{cosec } 30°$$

$\Rightarrow \quad AP = 2 \times AB$

$AP = 2 \times 60$ cm

$AP = 120$ cm $\qquad$ ...(*i*)

In ΔADQ, we have

$$\frac{AQ}{AD} = \text{cosec } 30°$$

$\Rightarrow \quad \frac{AQ}{30} = 2 \Rightarrow AQ = 2 \times 30$ cm

$AQ = 60$ cm $\qquad$ ...(*ii*)

Adding (*i*) and (*ii*), we get

$AP + AQ = 120$ cm $+ 60$ cm

$\Rightarrow \quad AP + AQ = 180$ cm

24. $\frac{\cos 80°}{\sin 10°} + \cos 59° \text{ cosec } 31°$

Using $\sin(90° - \theta) = \cos\theta$, we get

$\sin 10° = \sin(90° - 80°) = \cos 80°$

and $\quad \cos 59° = \cos(90° - 31°) = \sin 31°$

$\therefore \quad \frac{\cos 80°}{\sin 80°} + \sin 31° \text{ cosec } 31°$

$$= 1 + \frac{1}{\text{cosec } 31°} \times \text{cosec } 31° = 1 + 1 = 2$$

25. Given. AD = 8 cm, CD = 12 cm, $\angle$CEB = 60°

To find. CE and BE

Sol. In ΔCBE, we have

$$\frac{BE}{CB} = \cot 60°$$

$\Rightarrow \frac{BE}{8} = \frac{1}{\sqrt{3}}$

$\Rightarrow BE = \frac{8}{\sqrt{3}} \times \frac{\sqrt{3}}{\sqrt{3}} = \frac{8\sqrt{3}}{3}$ cm ≈ 4.62 cm

and $\frac{CE}{CB} = \text{cosec } 60°$

$\Rightarrow CE = CB \times \frac{2}{\sqrt{3}}$

$\Rightarrow CE = \frac{8 \times 2}{\sqrt{3}} = \frac{16}{\sqrt{3}} \times \frac{\sqrt{3}}{\sqrt{3}} = \frac{16\sqrt{3}}{3}$ cm ≈ 9.24 cm

26. LHS $= (1 + \cot\theta - \text{cosec }\theta)(1 + \tan\theta + \sec\theta)$

$= 1 + \tan\theta + \sec\theta + \cot\theta + \cot\theta\,.\tan\theta$

$+ \cot\theta\,.\sec\theta - \text{cosec }\theta - \text{cosec }\theta\,.\tan\theta$

$- \text{cosec }\theta\,.\sec\theta$

$$= 1 + \frac{\sin\theta}{\cos\theta} + \frac{1}{\cos\theta} + \frac{\cos\theta}{\sin\theta} + 1 + \frac{\cos\theta}{\sin\theta} \times \frac{1}{\cos\theta} - \frac{1}{\sin\theta}$$

$$- \frac{1}{\sin\theta} \times \frac{\sin\theta}{\cos\theta} - \frac{1}{\sin\theta}\,.\frac{1}{\cos\theta}$$

$$= 2 + \frac{\sin\theta}{\cos\theta} + \frac{\cos\theta}{\sin\theta} - \frac{1}{\sin\theta.\cos\theta}$$

$$= 2 + \frac{\sin^2\theta + \cos^2\theta - 1}{\sin\theta\,.\cos\theta} = 2 + \frac{1-1}{\sin\theta\,.\cos\theta} = 2 = \text{RHS}$$

27. $\sin\theta + \cos\theta = p, \qquad \sec\theta + \text{cosec }\theta = q$

LHS $= q(p^2 - 1) = (\sec\theta + \text{cosec }\theta)\,[(\sin\theta + \cos\theta)^2 - 1]$

$$= \left[\frac{1}{\cos\theta} + \frac{1}{\sin\theta}\right][\sin^2\theta + \cos^2\theta + 2\cos\theta\sin\theta - 1]$$

$$= \left[\frac{\sin\theta + \cos\theta}{\cos\theta\sin\theta}\right][1 + 2\cos\theta\sin\theta - 1] \quad (\because \sin^2\theta + \cos^2\theta = 1)$$

$$= \frac{\sin\theta + \cos\theta}{\cos\theta\sin\theta} \times 2\cos\theta\sin\theta$$

$= 2(\sin\theta + \cos\theta) = 2p \qquad (\because \sin\theta + \cos\theta = p)$

LHS = RHS. $\qquad$ Hence proved.

28. $\cos\theta + \sin\theta = \sqrt{2}\cos\theta$

$\Rightarrow \sin\theta = \sqrt{2}\cos\theta - \cos\theta$

$\Rightarrow \sin\theta = (\sqrt{2} - 1)\cos\theta$

$\Rightarrow \frac{\sin\theta}{\sqrt{2}-1} = \cos\theta \Rightarrow \frac{\sin\theta(\sqrt{2}+1)}{(\sqrt{2}-1)(\sqrt{2}+1)} = \cos\theta$

$\Rightarrow \sqrt{2}\sin\theta + \sin\theta = \cos\theta$

$\Rightarrow \sqrt{2}\sin\theta = \cos\theta - \sin\theta$

29. LHS $= \left(1+\dfrac{1}{\tan^2\theta}\right)\left(1+\dfrac{1}{\cot^2\theta}\right)$

$= (1+\cot^2\theta)(1+\tan^2\theta)$

$= \left(1+\dfrac{\cos^2\theta}{\sin^2\theta}\right)\left(1+\dfrac{\sin^2\theta}{\cos^2\theta}\right)$

$= \left(\dfrac{\sin^2\theta+\cos^2\theta}{\sin^2\theta}\right)\left(\dfrac{\cos^2\theta+\sin^2\theta}{\cos^2\theta}\right) = \dfrac{1}{\sin^2\theta}.\dfrac{1}{\cos^2\theta}$

$= \dfrac{1}{\sin^2\theta\,(1-\sin^2\theta)} = \dfrac{1}{\sin^2\theta-\sin^4\theta}$ = RHS

30. In right ΔDAE,

$\dfrac{AE}{AD} = \tan 30°$

$\Rightarrow \quad \dfrac{AE}{12} = \dfrac{1}{\sqrt{3}}$

$\Rightarrow \quad AE = \dfrac{12}{\sqrt{3}} = 4\sqrt{3}$ cm

Also $\dfrac{DE}{AD} = \sec 30° \Rightarrow \dfrac{DE}{12} = \dfrac{2}{\sqrt{3}}$

$\Rightarrow \quad DE = \dfrac{2}{\sqrt{3}}\times 12 = \dfrac{24}{\sqrt{3}} = 8\sqrt{3}$ cm

31. $3\tan\theta = 4 \Rightarrow \tan\theta = \dfrac{4}{3}$

Now given expression is $\dfrac{5\sin\theta-3\cos\theta}{5\sin\theta+2\cos\theta}$

Dividing numerator and denominator by $\cos\theta$, we get

$$\frac{\dfrac{5\sin\theta}{\cos\theta}-\dfrac{3\cos\theta}{\cos\theta}}{\dfrac{5\sin\theta}{\cos\theta}+\dfrac{2\cos\theta}{\cos\theta}} = \frac{5\tan\theta-3}{5\tan\theta+2}$$

Putting $\tan\theta = \dfrac{4}{3}$, we get, $\dfrac{5\times\frac{4}{3}-3}{5\times\frac{4}{3}+2} = \dfrac{11}{26}$

32. LHS $= \cot\theta - \tan\theta$

$= \dfrac{\cos\theta}{\sin\theta} - \dfrac{\sin\theta}{\cos\theta} = \dfrac{\cos^2\theta-\sin^2\theta}{\sin\theta\cos\theta}$

$= \dfrac{\cos^2\theta-1+\cos^2\theta}{\sin\theta\cos\theta}$ $\quad (\because \sin^2\theta = 1-\cos^2\theta)$

$= \dfrac{2\cos^2\theta-1}{\sin\theta\cos\theta}$

LHS = RHS. Hence proved.

33. A = 60°, B = 30°

LHS $= \cos(A-B) = \cos(60°-30°) = \cos 30° = \dfrac{\sqrt{3}}{2}$

RHS $= \cos A\cos B + \sin A\sin B$

$= \cos 60°\cos 30° + \sin 60°\sin 30°$

$= \dfrac{1}{2}\times\dfrac{\sqrt{3}}{2}+\dfrac{\sqrt{3}}{2}\times\dfrac{1}{2} = \dfrac{\sqrt{3}}{4}+\dfrac{\sqrt{3}}{4} = \dfrac{2\sqrt{3}}{4} = \dfrac{\sqrt{3}}{2}$

LHS = RHS. Hence verified.

34. LHS $= \cos 2A = \cos 2\times 30° = \cos 60° = \dfrac{1}{2}$ ...(*i*)

RHS $= \dfrac{1-\tan^2 A}{1+\tan^2 A} = \dfrac{1-\tan^2 30°}{1+\tan^2 30°}$

$= \dfrac{1-\left(\frac{1}{\sqrt{3}}\right)^2}{1+\left(\frac{1}{\sqrt{3}}\right)^2} = \dfrac{1-\frac{1}{3}}{1+\frac{1}{3}} = \dfrac{\frac{3-1}{3}}{\frac{3+1}{3}}$

$= \dfrac{2}{4} = \dfrac{1}{2}$ ...(*ii*)

From (*i*) and (*ii*),

$\cos 2A = \dfrac{1-\tan^2 A}{1+\tan^2 A}$

35. (*i*) A = 60°, B = 30°

LHS $= \sin(A-B) = \sin(60°-30°) = \sin 30° = \dfrac{1}{2}$ ...(*i*)

RHS $= \sin A.\cos B - \cos A.\sin B$

$= \sin 60°.\cos 30° - \cos 60°.\sin 30°$

$= \dfrac{\sqrt{3}}{2}\times\dfrac{\sqrt{3}}{2}-\dfrac{1}{2}\times\dfrac{1}{2} = \dfrac{3}{4}-\dfrac{1}{4}$

$= \dfrac{2}{4} = \dfrac{1}{2}$ ...(*ii*)

∴ From (*i*) and (*ii*),

$\sin(A-B) = \sin A.\cos B - \cos A.\sin B$

(*ii*) **Given.** $\dfrac{\cos\alpha}{\cos\beta} = m, \quad \dfrac{\cos\alpha}{\sin\beta} = n$

To show. $(m^2+n^2)\cos^2\beta = n^2$

Sol. LHS $= (m^2+n^2)\cos^2\beta$

$= \left(\dfrac{\cos^2\alpha}{\cos^2\beta}+\dfrac{\cos^2\alpha}{\sin^2\beta}\right)\cos^2\beta$

Taking L.C.M, we get

$= \left(\dfrac{\cos^2\alpha\sin^2\beta+\cos^2\alpha\cos^2\beta}{\cos^2\beta\sin^2\beta}\right)\cos^2\beta$

$= \dfrac{\cos^2\alpha(\sin^2\beta+\cos^2\beta)}{\sin^2\beta}$

$= \dfrac{\cos^2\alpha}{\sin^2\beta} = \left(\dfrac{\cos\alpha}{\sin\beta}\right)^2 = n^2$

∴ LHS = RHS. Hence Proved.

(*iii*) $\cos 60° = \dfrac{DP}{AD}$

$\Rightarrow \quad \dfrac{1}{2} = \dfrac{DP}{AD}$ or $\dfrac{1}{2} = \dfrac{DP}{4\sqrt{3}}$

$\Rightarrow \quad DP = 2\sqrt{3}$ cm

∴ $(AD)^2 = (AP)^2 + (DP)^2$ (Using Pythagoras theorem)

$\Rightarrow \quad (4\sqrt{3})^2 = AP^2 + (2\sqrt{3})^2$

$$\Rightarrow \quad 48 = AP^2 + 12$$
$$\Rightarrow \quad 48 - 12 = AP^2$$
$$\Rightarrow \quad 36 = AP^2 \Rightarrow AP = 6 \text{ cm}$$

(*iv*) $\dfrac{\cos^2 20° + \cos^2 70°}{\sin^2 59° + \sin^2 31°} = \dfrac{\cos^2 20° + \sin^2 (90° - 70°)}{\sin^2 59° + \cos^2 (90° - 31°)}$

$$= \frac{\cos^2 20° + \sin^2 20°}{\sin^2 59° + \cos^2 59°} = \frac{1}{1} = 1$$

(*v*) sec 70° . sin 20° – cos 20° . cosec 70°

= cosec (90° – 70°) . sin 20° – cos 20° . sec (90° – 70°)

$$= \text{cosec } 20° \times \frac{1}{\text{cosec } 20°} - \frac{1}{\sec 20°} \times \sec 20°$$

= 1 – 1 = 0

36. $\sin B = \dfrac{AC}{BC} = \dfrac{5}{13}$, $\cos B = \dfrac{AB}{BC} = \dfrac{12}{13}$

$\tan B = \dfrac{AC}{AB} = \dfrac{5}{12}$, $\cot B = \dfrac{AB}{AC} = \dfrac{12}{5}$

$\sec B = \dfrac{BC}{AB} = \dfrac{13}{12}$, $\text{cosec } B = \dfrac{BC}{AC} = \dfrac{13}{5}$

C, 5 cm, 13 cm, A, 12 cm, B

37. $AB : AC = 1 : \sqrt{2} \Rightarrow \dfrac{AB}{AC} = \dfrac{1}{\sqrt{2}}$

∴ $AB = x$ and $AC = \sqrt{2}\,x$, for some x.

By Pythagoras Theorem, we have

$$AC^2 = AB^2 + BC^2$$
$$\Rightarrow (\sqrt{2}\,x)^2 = x^2 + BC^2$$
$$\Rightarrow BC^2 = 2x^2 - x^2$$
$$\Rightarrow BC = x$$

C, A, B

∴ $\tan A = \dfrac{BC}{AB} = \dfrac{x}{x} = 1$,

$$\frac{2\tan A}{1 + \tan^2 A} = \frac{2 \times 1}{1 + (1)^2} = 1$$

38. ∵ $AB : AC = 1 : \sqrt{2}$

Let $AB = x$ and $AC = \sqrt{2}\,x$

In right ΔCBA,

$$AC^2 = BC^2 + AB^2$$
$$\Rightarrow (\sqrt{2}\,x)^2 = BC^2 + x^2$$
$$\Rightarrow 2x^2 - x^2 = BC^2$$
$$\Rightarrow BC^2 = x^2$$
$$\Rightarrow BC = x$$

C, B, A

∴ $\tan A = \dfrac{BC}{AB} = \dfrac{x}{x} = 1$

Now $\dfrac{2\tan A}{1 - \tan^2 A} = \dfrac{2 \times 1}{1 - 1} = \dfrac{2}{0}$ = Not defined

39. $\tan\theta = \dfrac{4}{3}$

Let in right ΔABC, $\tan\theta = \dfrac{4}{3}$

$$\Rightarrow \frac{AB}{BC} = \frac{4}{3}$$

A, B, C, θ

$\Rightarrow AB = 4x$ and $BC = 3x$

Also $AC^2 = AB^2 + BC^2$

$$\Rightarrow AC^2 = (4x)^2 + (3x)^2 = 16x^2 + 9x^2$$

$$AC = \sqrt{25x^2} = 5x$$

$$\therefore \sin\theta = \frac{AB}{AC} = \frac{4x}{5x} = \frac{4}{5}$$

Now, LHS $= \sqrt{\dfrac{1 - \sin\theta}{1 + \sin\theta}} = \sqrt{\dfrac{1 - \frac{4}{5}}{1 + \frac{4}{5}}} = \sqrt{\dfrac{\frac{5-4}{5}}{\frac{5+4}{5}}}$

$= \sqrt{\dfrac{1}{9}} = \dfrac{1}{3}$ = RHS

40. Here, $\dfrac{BC}{AC} = \sin 30° \Rightarrow \dfrac{20}{x} = \dfrac{1}{2} \Rightarrow x = 40$ cm

41. Here, $\dfrac{AB}{BC} = \tan 30° \Rightarrow \dfrac{x}{15} = \dfrac{1}{\sqrt{3}} \Rightarrow x = \dfrac{15}{\sqrt{3}}$

$\Rightarrow x = 5\sqrt{3}$ cm = 8.66 cm

42. Here, $\dfrac{QR}{PR} = \cos 60°$

$\Rightarrow \dfrac{x}{25} = \dfrac{1}{2} \Rightarrow x = \dfrac{25}{2} = 12.5$ cm

43. Here, $\dfrac{MN}{MT} = \sin 45° \Rightarrow \dfrac{x}{20} = \dfrac{1}{\sqrt{2}}$

$\Rightarrow x = \dfrac{20}{\sqrt{2}} = 10\sqrt{2}$ cm

44. $4\cot^2 45° - \sec^2 60° + \sin^2 60° + \cos^2 90°$

$= 4(\cot 45°)^2 - (\sec 60°)^2 + (\sin 60°)^2 + (\cos 90°)^2$

$= 4 \times (1)^2 - (2)^2 + \left(\dfrac{\sqrt{3}}{2}\right)^2 + 0$

$= 4 - 4 + \dfrac{3}{4} + 0 = \dfrac{3}{4}$

45. LHS $= 2(\cos^2 45° + \tan^2 60°) - 6(\sin^2 45° - \tan^2 30°)$

$$= 2\left\{\left(\frac{1}{\sqrt{2}}\right)^2 + \left(\sqrt{3}\right)^2\right\} - 6\left\{\left(\frac{1}{\sqrt{2}}\right)^2 - \left(\frac{1}{\sqrt{3}}\right)^2\right\}$$

$$= 2\left(\frac{1}{2} + 3\right) - 6\left(\frac{1}{2} - \frac{1}{3}\right)$$

$$= 2\left(\frac{1+6}{2}\right) - 6\left(\frac{3-2}{6}\right) = 7 - 1 = 6 = \text{RHS}$$

46. $\tan\theta + \dfrac{1}{\tan\theta} = 2$

$$\Rightarrow \left(\tan\theta + \frac{1}{\tan\theta}\right)^2 = (2)^2$$

$$\Rightarrow \tan^2\theta + \frac{1}{\tan^2\theta} + 2.\tan\theta.\frac{1}{\tan\theta} = 4$$

$$\Rightarrow \qquad \tan^2\theta + \frac{1}{\tan^2\theta} + 2 = 4$$

$$\Rightarrow \qquad \tan^2\theta + \frac{1}{\tan^2\theta} = 2$$

47. Let angle = θ

$$\Rightarrow \qquad \frac{\sin\theta}{\cos\theta} = \frac{3}{4} \Rightarrow \tan\theta = \frac{3}{4}$$

Now, $\qquad \sec^2\theta = 1 + \tan^2\theta$

$$\Rightarrow \qquad \sec^2\theta = 1 + \frac{9}{16} \Rightarrow \sec^2\theta = \frac{25}{16}$$

$$\Rightarrow \qquad \sec\theta = \frac{5}{4}$$

$$\Rightarrow \qquad \cos\theta = \frac{4}{5}$$

$$\Rightarrow \qquad \sin\theta = \sqrt{1-\cos^2\theta} = 1-\left(\frac{4}{5}\right)^2 = \frac{3}{5}$$

48. (*i*) cos 60° . cos 30° + sin 60° . sin 30°

$$= \frac{1}{2}\times\frac{\sqrt{3}}{2} + \frac{\sqrt{3}}{2}\times\frac{1}{2} = \frac{\sqrt{3}}{4} + \frac{\sqrt{3}}{4} = \frac{2\sqrt{3}}{4} = \frac{\sqrt{3}}{2}$$

(*ii*) cos 60° . cos 30° – sin 60° . sin 30°

$$= \frac{1}{2}\times\frac{\sqrt{3}}{2} - \frac{\sqrt{3}}{2}\times\frac{1}{2} = \frac{\sqrt{3}}{4} - \frac{\sqrt{3}}{4} = 0$$

49. (*i*) $2\sin 2\theta = \sqrt{3} \Rightarrow \sin 2\theta = \frac{\sqrt{3}}{2}$

Also, $\sin 60° = \frac{\sqrt{3}}{2}$

$\therefore\ 2\theta = 60° \Rightarrow \theta = 30°$

(*ii*) $2\cos 3\theta = 1 \Rightarrow \cos 3\theta = \frac{1}{2}$

Also, $\cos 60° = \frac{1}{2} \Rightarrow 3\theta = 60° \Rightarrow \theta = 20°$

(*iii*) $\quad \sqrt{3}\tan 2\theta - 3 = 0 \Rightarrow \sqrt{3}\tan 2\theta = 3$

$$\Rightarrow \qquad \tan 2\theta = \frac{3}{\sqrt{3}} \Rightarrow \tan 2\theta = \sqrt{3}$$

Also, $\quad \tan 60° = \sqrt{3}$

$\therefore \qquad 2\theta = 60° \Rightarrow \theta = 30°$

50. (*i*) LHS = $\sin 2A = \sin 60° = \frac{\sqrt{3}}{2}$

$$\text{RHS} = \frac{2\tan A}{1+\tan^2 A} = \frac{2\tan 30°}{1+\tan^2 30°}$$

$$= \frac{2\times\frac{1}{\sqrt{3}}}{1+\left(\frac{1}{\sqrt{3}}\right)^2} = \frac{\frac{2}{\sqrt{3}}}{\frac{4}{3}} = \frac{\sqrt{3}}{2}$$

LHS = RHS.

(*ii*) $\qquad \cos 3\times 30° = 4\cos^3 30° - 3\cos 30°$

$$\cos 90° = 4\left(\frac{\sqrt{3}}{2}\right)^3 - 3\left(\frac{\sqrt{3}}{2}\right)$$

$$0 = \frac{3\sqrt{3}}{2} - \frac{3\sqrt{3}}{2} = 0 \qquad \text{Proved.}$$

51. In right-angled triangle ABC, ∠CAB = θ

$$\cot\theta = \frac{1}{\sqrt{3}} = \frac{BC}{AB}$$

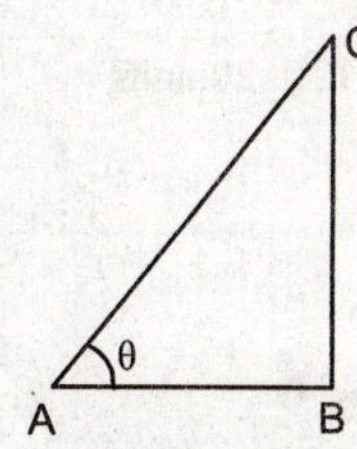

$\therefore \qquad AB = \sqrt{3}$ and BC = 1

$(AC)^2 = (AB)^2 + (BC)^2$ (By Pythagoras theorem)

$= (\sqrt{3})^2 + (1)^2 = 3 + 1 = 4$

$\Rightarrow \qquad AC = 2$

$$\text{LHS} = \frac{\sin^2\theta}{1+\cos^2\theta} = \frac{\frac{3}{4}}{1+\frac{1}{4}} = \frac{3}{5} = \text{RHS.}$$

52. Let us draw a ΔABC, right angled at C in which $\tan B = \sqrt{3}$ and $\tan A = \frac{1}{\sqrt{3}}$

First we consider the trigonometric-ratio for ∠A.

$$\tan A = \frac{1}{\sqrt{3}}$$

Let $\qquad BC = x$ and $AC = \sqrt{3}\,x$ ... (*i*)

and $\qquad \tan B = \sqrt{3}$

$AC = \sqrt{3}\,x$ and $BC = x$

$AB = 2x$ (By Pythagoras theorem)

$$\therefore \sin A = \frac{x}{2x} = \frac{1}{2} \text{ and } \cos A = \frac{\sqrt{3}\,x}{2x} = \frac{\sqrt{3}}{2}$$

Now we consider the trigonometric-ratios of ∠B, we have

Base = x, Perpendicular = $\sqrt{3}\,x$, Hypotenuse = $2x$

$$\therefore \quad \cos B = \frac{x}{2x} = \frac{1}{2} \text{ and}$$

$$\sin B = \frac{AC}{AB} = \frac{\sqrt{3}\,x}{2x} = \frac{\sqrt{3}}{2}$$

Now, LHS = sin A cos B + cos A sin B

$$= \frac{1}{2}\times\frac{1}{2} + \frac{\sqrt{3}}{2}\times\frac{\sqrt{3}}{2}$$

$$= \frac{1}{4} + \frac{3}{4} = \frac{4}{4} = 1 = \text{RHS.}$$

53. Since $\quad \angle A + \angle B + \angle C = 180°$

$30° + \angle B + 90° = 180°$

$\Rightarrow \qquad \angle B = 60°$

Now, $\quad \cos A = \frac{AC}{AB}$

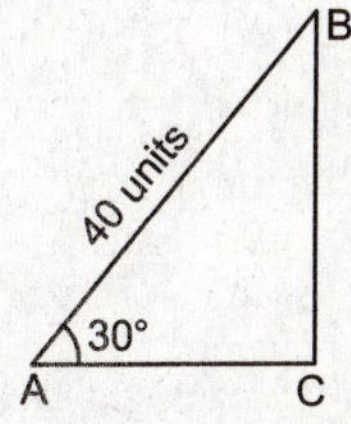

$$\Rightarrow \quad \cos 30° = \frac{AC}{40} \Rightarrow \frac{\sqrt{3}}{2} = \frac{AC}{40}$$

$\Rightarrow \quad AC = 20\sqrt{3}$ units.

and, $\sin A = \dfrac{BC}{AB} \Rightarrow \sin 30° = \dfrac{BC}{40}$

$\Rightarrow \dfrac{1}{2} = \dfrac{BC}{40} \Rightarrow BC = 20$ units

54. (*i*) $\cos\theta = \dfrac{5}{13}$

we know that

$$\sin^2\theta + \cos^2\theta = 1$$

$$\Rightarrow \sin^2\theta = 1 - \cos^2\theta$$

$$\Rightarrow \sin^2\theta = 1 - \frac{25}{169}$$

$$\Rightarrow \sin^2\theta = \frac{144}{169}$$

$$\Rightarrow \sin\theta = \frac{12}{13}$$

$$\Rightarrow \operatorname{cosec}\theta = \frac{1}{\sin\theta} = \frac{13}{12}, \sec\theta = \frac{1}{\cos\theta} = \frac{13}{5}$$

$$\Rightarrow \tan\theta = \frac{\sin\theta}{\cos\theta} = \frac{12/13}{5/13} = \frac{12}{5}$$

$$\Rightarrow \cot\theta = \frac{1}{\tan\theta} = \frac{5}{12}$$

(*ii*) $\tan\theta = \dfrac{1}{\sqrt{3}} \Rightarrow \theta = 30°$

Now, $\sin\theta = \sin 30° = \dfrac{1}{2}$, $\cos\theta = \cos 30° = \dfrac{\sqrt{3}}{2}$

$\cot\theta = \cot 30° = \sqrt{3}$, $\sec\theta = \sec 30° = \dfrac{2}{\sqrt{3}}$

and $\operatorname{cosec}\theta = \operatorname{cosec} 30° = 2$

(*iii*) $\sec\theta = \sqrt{6}$

We know that

$$\tan^2\theta = \sec^2\theta - 1$$

$$\Rightarrow \tan^2\theta = (\sqrt{6})^2 - 1 = 6 - 1 = 5$$

$$\Rightarrow \tan\theta = \sqrt{5}$$

$$\cot\theta = \frac{1}{\tan\theta} = \frac{1}{\sqrt{5}}$$

$$\cos\theta = \frac{1}{\sec\theta} = \frac{1}{\sqrt{6}}$$

$$\tan\theta = \frac{\sin\theta}{\cos\theta}$$

$$\Rightarrow \tan\theta \times \cos\theta = \sin\theta$$

$$\Rightarrow \sqrt{5} \times \frac{1}{\sqrt{6}} = \sin\theta$$

$$\Rightarrow \sin\theta = \frac{\sqrt{5}}{\sqrt{6}}$$

$$\operatorname{cosec}\theta = \frac{1}{\sin\theta} = \frac{\sqrt{6}}{\sqrt{5}}$$

(*iv*) $\tan\theta = \dfrac{2mn}{m^2 - n^2}$

Now $\quad \sec^2\theta = 1 + \tan^2\theta$

$$= 1 + \frac{(2mn)^2}{(m^2 - n^2)^2}$$

$$= \frac{(m^2 - n^2)^2 + 4m^2n^2}{(m^2 - n^2)^2}$$

$$\sec^2\theta = \frac{(m^2 + n^2)^2}{(m^2 - n^2)^2}$$

$$\Rightarrow \sec\theta = \sqrt{\frac{(m^2 + n^2)^2}{(m^2 - n^2)^2}}$$

$$\Rightarrow \sec\theta = \frac{m^2 + n^2}{m^2 - n^2}$$

$$\cos\theta = \frac{1}{\sec\theta} = \frac{m^2 - n^2}{m^2 + n^2}, \cot\theta = \frac{1}{\tan\theta} = \frac{m^2 - n^2}{2mn}$$

$$\tan\theta = \frac{\sin\theta}{\cos\theta} \Rightarrow \tan\theta \,.\, \cos\theta = \sin\theta$$

$$\Rightarrow \frac{2mn}{m^2 - n^2} \times \frac{m^2 - n^2}{m^2 + n^2} = \sin\theta$$

$$\Rightarrow \sin\theta = \frac{2mn}{m^2 + n^2}$$

$$\operatorname{cosec}\theta = \frac{1}{\sin\theta} = \frac{m^2 + n^2}{2mn}$$

(*v*) $\sin\theta = \dfrac{m^2 - n^2}{m^2 + n^2}$

$$\cos^2\theta = 1 - \sin^2\theta \Rightarrow \cos^2\theta = 1 - \frac{(m^2 - n^2)^2}{(m^2 + n^2)^2}$$

$$\Rightarrow \cos\theta = \sqrt{1 - \frac{(m^2 - n^2)^2}{(m^2 + n^2)^2}}$$

$$\cos\theta = \sqrt{\frac{(m^2 + n^2)^2 - (m^2 - n^2)^2}{(m^2 + n^2)^2}} = \sqrt{\frac{4m^2n^2}{(m^2 + n^2)^2}}$$

$$= \frac{2mn}{m^2 + n^2}$$

Now, $\sec\theta = \dfrac{1}{\cos\theta} = \dfrac{m^2 + n^2}{2mn}$

$$\operatorname{cosec}\theta = \frac{1}{\sin\theta} = \frac{m^2 + n^2}{m^2 - n^2}$$

$$\tan\theta = \frac{\sin\theta}{\cos\theta} = \frac{\dfrac{m^2 - n^2}{m^2 + n^2}}{\dfrac{2mn}{m^2 + n^2}} = \frac{m^2 - n^2}{2mn}$$

$$\cot\theta = \frac{1}{\tan\theta} = \frac{2mn}{m^2 - n^2}$$

(*vi*) $\cos\theta = \dfrac{2m}{m^2+1}$

Now, $\sin^2\theta = 1 - \cos^2\theta \Rightarrow \sin^2\theta = 1 - \left(\dfrac{2m}{m^2+1}\right)^2$

$$\Rightarrow \quad \sin\theta = \sqrt{1 - \frac{4m^2}{(m^2+1)^2}} = \sqrt{\frac{(m^2+1)^2 - 4m^2}{(m^2+1)^2}}$$

$$\Rightarrow \quad \sin\theta = \sqrt{\frac{(m^2-1)^2}{(m^2+1)^2}} = \frac{m^2-1}{m^2+1}$$

$$\text{Now, } \sec\theta = \frac{1}{\cos\theta} = \frac{m^2+1}{2m}$$

$$\operatorname{cosec}\theta = \frac{1}{\sin\theta} = \frac{m^2+1}{m^2-1}$$

$$\tan\theta = \frac{\sin\theta}{\cos\theta} = \frac{\dfrac{m^2-1}{m^2+1}}{\dfrac{2m}{m^2+1}} = \frac{m^2-1}{2m}$$

$$\cot\theta = \frac{1}{\tan\theta} = \frac{2m}{m^2-1}$$

(*vii*) $\operatorname{cosec}\theta = \sqrt{1+\left(\dfrac{m}{n}\right)^4} = \sqrt{\dfrac{n^4+m^4}{n^4}} = \dfrac{\sqrt{m^4+n^4}}{n^2}$

$$\cot^2\theta = \operatorname{cosec}^2\theta - 1 \Rightarrow \cot^2\theta = \left(\sqrt{1+\left(\frac{m}{n}\right)^4}\right)^2 - 1$$

$$\Rightarrow \cot^2\theta = 1 + \frac{m^4}{n^4} - 1 = \frac{m^2}{n^2} \Rightarrow \cot\theta = \sqrt{\frac{m^4}{n^4}} = \frac{m^2}{n^2}$$

$$\Rightarrow \tan\theta = \frac{1}{\cot\theta} = \frac{n^2}{m^2}, \ \sin\theta = \frac{1}{\operatorname{cosec}\theta} = \frac{n^2}{\sqrt{m^4+n^4}}$$

$$\Rightarrow \cot\theta = \frac{\cos\theta}{\sin\theta} \Rightarrow \cos\theta = \sin\theta \,.\, \cot\theta$$

$$\Rightarrow \cos\theta = \frac{m^2}{n^2} \times \frac{n^2}{\sqrt{m^4+n^4}} = \frac{m^2}{\sqrt{m^4+n^4}}$$

55. $\tan\theta = \dfrac{p}{q}$

Now, $\quad \text{LHS} = \dfrac{p\sin\theta - q\cos\theta}{p\sin\theta + q\cos\theta}$

Dividing numerator and denominator by $\cos\theta$, we get

$$\frac{p\tan\theta - q}{p\tan\theta + q} = \frac{p\times\dfrac{p}{q} - q}{p\times\dfrac{p}{q} + q} = \frac{p^2-q^2}{p^2+q^2} = \text{RHS}$$

56. $\text{LHS} = \dfrac{\operatorname{cosec}^2\theta - \sec^2\theta}{\operatorname{cosec}^2\theta + \sec^2\theta}$

Dividing the numerator and denominator by $\sec^2\theta$. We get

$$\frac{\cot^2\theta - 1}{\cot^2\theta + 1} = \frac{(\sqrt{7})^2 - 1}{(\sqrt{7})^2 + 1} = \frac{7-1}{7+1} = \frac{6}{8} = \frac{3}{4} = \text{RHS}$$

57. $\cot\theta = \dfrac{12}{5} = \dfrac{\text{Base}}{\text{Perpendicular}} = \dfrac{\text{AB}}{\text{BC}}$

By Pythagoras Theorem,

$$\Rightarrow \quad (\text{AC})^2 = (\text{AB})^2 + (\text{BC})^2$$
$$\Rightarrow \quad (\text{AC})^2 = (12)^2 + (5)^2$$
$$\Rightarrow \quad (\text{AC})^2 = 144 + 25$$
$$\Rightarrow \quad (\text{AC})^2 = 169$$
$$\Rightarrow \quad \text{AC} = 13$$

$$\text{LHS} = \tan^2\theta - \sin^2\theta = \left(\frac{5}{12}\right)^2 - \left(\frac{5}{13}\right)^2 = \frac{25}{144} - \frac{25}{169} = \frac{25\times 25}{144\times 169}$$

$$= \frac{625}{144\times 169}$$

$$\text{RHS} = \sin^4\theta.\sec^2\theta = \left(\frac{5}{13}\right)^4 \times \left(\frac{13}{12}\right)^2 = \frac{25\times 25}{144\times 169} = \frac{625}{144\times 169}$$

LHS = RHS — Hence Proved.

58. (*i*) $\sin(\text{A}+\text{B}) = 1 \quad \therefore \quad \text{A}+\text{B} = 90°$ ... (*i*)

$\cos(\text{A}-\text{B}) = 1 \quad \therefore \quad \text{A}-\text{B} = 0°$ ... (*ii*)

Solving eq. (*i*) and (*ii*), we get

$\text{A} = 45°, \text{B} = 45°$

(*ii*) $\quad \text{A}+2\text{B} = 60°, \text{A}+\text{B} = 60°$

On solving we get, $\text{B} = 0°, \text{A} = 60°$

(*iii*) $\quad \text{A}-\text{B} = 30°, \text{A}+\text{B} = 60°$

On solving $\quad \text{A} = 45°, \text{B} = 15°$

59. In ΔABC, use Pythagoras theorem

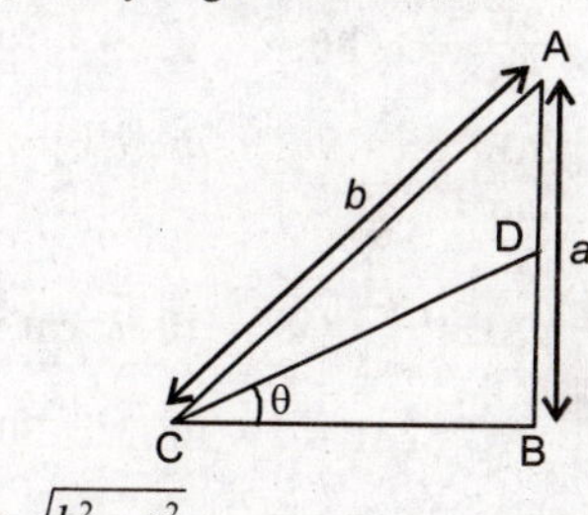

$\text{BC} = \sqrt{b^2 - a^2}$

Now, in ΔDBC, $\text{DB} = \dfrac{1}{2}a$ and $\text{BC} = \sqrt{b^2-a^2}$

$\therefore \quad \text{CD} = \dfrac{\sqrt{4b^2 - 3a^2}}{2}$

On putting values, we get

$\sin^2\theta + \cos^2\theta = 1$

60. $\quad \tan 3x = \sin 45° \,.\, \cos 45° + \sin 30°$

$$\Rightarrow \quad \tan 3x = \frac{1}{\sqrt{2}} \times \frac{1}{\sqrt{2}} + \frac{1}{2} = \frac{1}{2} + \frac{1}{2}$$
$$\Rightarrow \quad \tan 3x = 1$$
$$\Rightarrow \quad 3x = 45° \Rightarrow x = 15°$$

61. $\quad \cos x = \cos 60° \,.\, \cos 30° + \sin 60° \,.\, \sin 30°$

$$\Rightarrow \quad \cos x = \frac{1}{2} \times \frac{\sqrt{3}}{2} + \frac{\sqrt{3}}{2} \times \frac{1}{2}$$

$$\cos x = \frac{\sqrt{3}}{4} + \frac{\sqrt{3}}{4} = \frac{2\sqrt{3}}{4} = \frac{\sqrt{3}}{2}$$

Also, $\cos 30° = \dfrac{\sqrt{3}}{2} \Rightarrow x = 30°$

62. Here, $\cos 2x = \sin 60° \,.\, \cos 30° - \cos 60° \,.\, \sin 30°$

$$\Rightarrow \quad \cos 2x = \frac{\sqrt{3}}{2} \times \frac{\sqrt{3}}{2} - \frac{1}{2} \times \frac{1}{2}$$

$$\Rightarrow \quad \cos 2x = \frac{3}{4} - \frac{1}{4} = \frac{2}{4} = \frac{1}{2}$$

Also, $\cos 60° = \frac{1}{2} \Rightarrow 2x = 60° \Rightarrow x = 30°$

63. Let A = 45°, B = 30°

$\therefore \quad \sin (A + B) = \sin A \,.\, \cos B + \cos A \,.\, \sin B$

$\sin (45° + 30°) = \sin 45° \,.\, \cos 30° + \cos 45° \,.\, \sin 30°$

$$\Rightarrow \quad \sin 75° = \frac{1}{\sqrt{2}} \times \frac{\sqrt{3}}{2} + \frac{1}{\sqrt{2}} \times \frac{1}{2} = \frac{\sqrt{3}}{2\sqrt{2}} + \frac{1}{2\sqrt{2}} = \frac{\sqrt{3}+1}{2\sqrt{2}}$$

64. Let diagonals of the rhombus ABCD intersect at E.

$\therefore \angle BEA = 90°$

$\because \quad \angle EAB = \frac{1}{2} \angle DAB = \frac{1}{2} \times 60° = 30°$

AB = 20 cm

In right ΔAEB

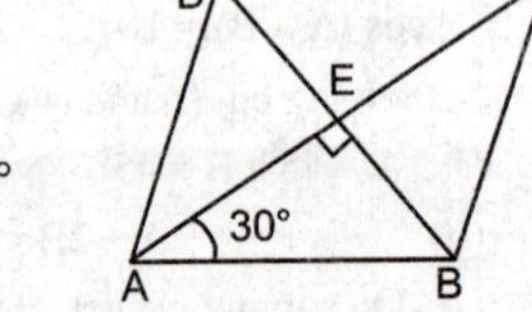

$$\frac{BE}{AB} = \sin 30°$$

$$\Rightarrow \quad \frac{BE}{20} = \frac{1}{2}$$

$\Rightarrow \quad BE = 10$ cm

$BD = 2BE = 2 \times 10 = 20$ cm

Also $\quad \frac{AE}{AB} = \cos 30°$

$$\Rightarrow \quad \frac{AE}{20} = \frac{\sqrt{3}}{2}$$

$$\Rightarrow \quad AE = \frac{\sqrt{3}}{2} \times 20 = 10\sqrt{3} \text{ cm}$$

$$\therefore \quad AC = 2AE = 2 \times 10\sqrt{3} = 20\sqrt{3} \text{ cm}$$

Lengths of diagonals are 20 cm and $20\sqrt{3}$ cm.

65. Now in right ΔADC, $\frac{CD}{AD} = \tan \angle DAC$

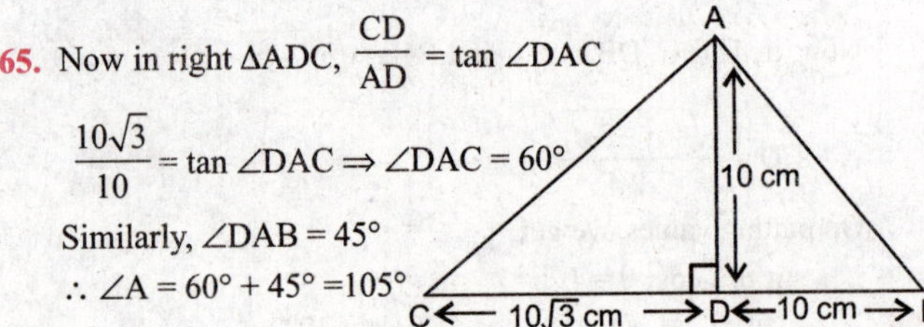

$\frac{10\sqrt{3}}{10} = \tan \angle DAC \Rightarrow \angle DAC = 60°$

Similarly, $\angle DAB = 45°$

$\therefore \angle A = 60° + 45° = 105°$

66.

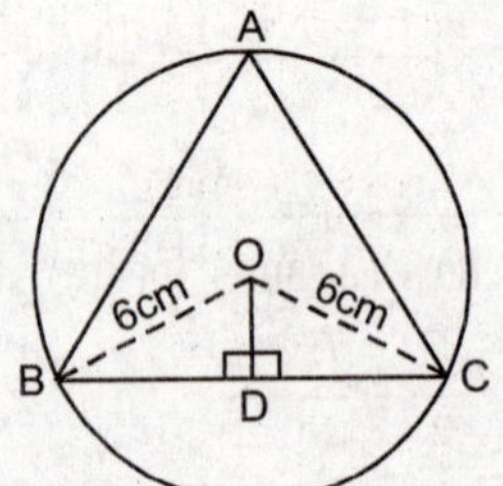

Let O is centre of circle and ΔABC is an equilateral Δ. $OD \perp BC$.

$\because \quad \angle BAC = 60° \quad \therefore \quad \angle BOC = 2\angle BAC = 120°$

$\because \quad OB = OC$ (radii) $\therefore \quad \angle OBD = \angle OCD = 30°$

In right Δ ODB, $\frac{BD}{OB} = \cos 30°$

$$\Rightarrow \quad \frac{BD}{6} = \frac{\sqrt{3}}{2} \Rightarrow BD = \frac{\sqrt{3}}{2} \times 6$$

$\Rightarrow \quad BD = 3\sqrt{3}$ cm

Also $\quad AB = 2BD \quad (\because \perp$ from the centre bisects the chord)

$\therefore \quad AB = 2 \times 3\sqrt{3} = 6\sqrt{3}$ cm

67. In right ΔCBA,

$$\frac{BC}{AB} = \tan 60°$$

$$\Rightarrow \quad \frac{BC}{30} = \sqrt{3}$$

$\Rightarrow \quad BC = 30\sqrt{3}$ cm

68. $LHS = (1 + \cot^2 \theta)(1 - \cos \theta)(1 + \cos \theta)$

$= \text{cosec}^2 \theta (1 - \cos^2 \theta) \qquad (\because 1 + \cot^2 \theta = \text{cosec}^2 \theta)$

$= \frac{1}{\sin^2 \theta} \times \sin^2 \theta = 1 = RHS.$

69. $LHS = \frac{\sin \theta}{1 - \cos \theta} + \frac{\tan \theta}{1 + \cos \theta}$

$$= \frac{\sin \theta}{1 - \cos \theta} + \frac{\frac{\sin \theta}{\cos \theta}}{1 + \cos \theta}$$

$$= \frac{\sin \theta}{1 - \cos \theta} + \frac{\sin \theta}{\cos \theta (1 + \cos \theta)}$$

$$= \frac{\sin \theta \,.\, \cos \theta (1 + \cos \theta) + \sin \theta (1 - \cos \theta)}{(1 - \cos \theta) \cos \theta (1 + \cos \theta)}$$

$$= \frac{\sin \theta \,.\, \cos \theta + \sin \theta \,.\, \cos^2 \theta + \sin \theta - \sin \theta \,.\, \cos \theta}{\cos \theta (1 - \cos^2 \theta)}$$

$$= \frac{\sin \theta \,.\, \cos^2 \theta + \sin \theta}{\cos \theta \,.\, \sin^2 \theta} = \frac{\sin \theta \,.\, \cos^2 \theta}{\cos \theta \,.\, \sin^2 \theta} + \frac{\sin \theta}{\cos \theta \,.\, \sin^2 \theta}$$

$$= \frac{\cos \theta}{\sin \theta} + \frac{1}{\cos \theta \,.\, \sin \theta} = \cot \theta + \sec \theta \,.\, \text{cosec}\, \theta$$

$= \sec \theta \,.\, \text{cosec}\, \theta + \cot \theta = RHS.$

70. $RHS = \tan^2\theta - \cot^2\theta = \frac{\sin^2 \theta}{\cos^2 \theta} - \frac{\cos^2 \theta}{\sin^2 \theta}$

$$= \frac{\sin^4 \theta - \cos^4 \theta}{\sin^2 \theta \cos^2 \theta}$$

$$= \frac{(\sin^2 \theta - \cos^2 \theta)(\sin^2 \theta + \cos^2 \theta)}{\sin^2 \theta \cos^2 \theta}$$

$$= \frac{1}{\sin \theta \cos \theta} \left[\frac{\sin^2 \theta - \cos^2 \theta}{\sin \theta \cos \theta} \right]$$

$$= \frac{1[\tan \theta - \cot \theta]}{\sin \theta \cos \theta} = LHS.$$

71. $LHS = \frac{\sec \theta - \tan \theta}{\sec \theta + \tan \theta} = \frac{\sec \theta - \tan \theta}{\sec \theta + \tan \theta} \times \frac{\sec \theta - \tan \theta}{\sec \theta - \tan \theta}$

$$= \frac{(\sec\theta - \tan\theta)^2}{\sec^2\theta - \tan^2\theta}$$

$$= \frac{\sec^2\theta + \tan^2\theta - 2\sec\theta\tan\theta}{1}$$

$= 1 + \tan^2\theta + \tan^2\theta - 2\sec\theta\tan\theta$

$= 1 + 2\tan^2\theta - 2\sec\theta\tan\theta$ = RHS.

72. $\frac{\sin A + \cos A}{\sin A - \cos A} + \frac{\sin A - \cos A}{\sin A + \cos A} = \frac{2}{2\sin^2 A - 1}$

$$\text{LHS} = \frac{\sin A + \cos A}{\sin A - \cos A} + \frac{\sin A - \cos A}{\sin A + \cos A}$$

$$= \frac{(\sin A + \cos A)^2 + (\sin A - \cos A)^2}{\sin^2 A - \cos^2 A}$$

$$= \frac{\sin^2 A + \cos^2 A + 2\sin A\cos A + \sin^2 A + \cos^2 A - 2\sin A\cos A}{\sin^2 A - \cos^2 A}$$

$$= \frac{1+1}{\sin^2 A - \cos^2 A} \quad (\because \sin^2 A + \cos^2 A = 1)$$

$$= \frac{2}{\sin^2 A - 1 + \sin^2 A} = \frac{2}{2\sin^2 A - 1} = \text{RHS}.$$

$(\because \cos^2 A = 1 - \sin^2 A)$

73. $\text{LHS} = \frac{\cos A}{1 - \tan A} + \frac{\sin A}{1 - \cot A} = \frac{\cos A}{1 - \frac{\sin A}{\cos A}} + \frac{\sin A}{1 - \frac{\cos A}{\sin A}}$

$$= \frac{\cos A}{\frac{\cos A - \sin A}{\cos A}} + \frac{\sin A}{\frac{\sin A - \cos A}{\sin A}}$$

$$= \frac{\cos^2 A}{\cos A - \sin A} - \frac{\sin^2 A}{\cos A - \sin A}$$

$$= \frac{\cos^2 A - \sin^2 A}{\cos A - \sin A} = \frac{(\cos A - \sin A)(\cos A + \sin A)}{(\cos A - \sin A)}$$

$= \cos A + \sin A$ = RHS

74. LHS $= \cos^4 A - \cos^2 A$

$= \cos^2 A(\cos^2 A - 1)$

$= (1 - \sin^2 A)(-\sin^2 A)$

$= \sin^4 A - \sin^2 A$ = RHS.

75. LHS $= \cot^4 A - 1$

$= (\cot^2 A)^2 - 1 = (\cot^2 A + 1)(\cot^2 A - 1)$

$= \text{cosec}^2 A(\cot^2 A - 1)$

$$= \text{cosec}^2 A\left[\frac{\cos^2 A}{\sin^2 A} - 1\right]$$

$$= \text{cosec}^2 A\left[\frac{\cos^2 A - \sin^2 A}{\sin^2 A}\right]$$

$$= \text{cosec}^2 A\left[\frac{1 - 2\sin^2 A}{\sin^2 A}\right]$$

$= \text{cosec}^2 A[\text{cosec}^2 A - 2]$

$= \text{cosec}^4 A - 2\,\text{cosec}^2 A$ = RHS.

76. LHS $= \sin^4 A + \cos^4 A$

$= (\sin^2 A)^2 + (\cos^2 A)^2$

$= (\sin^2 A + \cos^2 A)^2 - 2\sin^2 A \,.\, \cos^2 A$

[Using $a^2 + b^2 = (a + b)^2 - 2ab$]

$= 1 - 2\sin^2 A \,.\, \cos^2 A$ = RHS.

77. LHS $= \sin^6 A + \cos^6 A$

$= (\sin^2 A)^3 + (\cos^2 A)^3$

$= (\sin^2 A + \cos^2 A)^3 - 3\sin^2 A \,.\, \cos^2 A(\sin^2 A + \cos^2 A)$

{Using $a^3 + b^3 = (a + b)^3 - 3ab(a + b)$}

$= 1^3 - 3\sin^2 A \,.\, \cos^2 A \times 1$

$= 1 - 3\sin^2 A \,.\, \cos^2 A$ = RHS.

78. LHS $= (\sin^4\theta)^2 - (\cos^4\theta)^2$

$= (\sin^4\theta + \cos^4\theta)(\sin^4\theta - \cos^4\theta)$

$= (\sin^4\theta + \cos^4\theta)(\sin^2\theta + \cos^2\theta)(\sin^2\theta - \cos^2\theta)$

$= (\sin^2\theta - \cos^2\theta)\{(\sin^2\theta)^2 + (\cos^2\theta)^2$

$+ 2\sin^2\theta\cos^2\theta - 2\sin^2\theta\cos^2\theta\}$

$= (\sin^2\theta - \cos^2\theta)\{(\sin^2\theta + \cos^2\theta)^2 - 2\sin^2\theta\cos^2\theta\}$

$= (\sin^2\theta - \cos^2\theta)(1 - 2\sin^2\theta\cos^2\theta)$ = RHS.

79. $(\text{cosec}\,\theta - \sin\theta)^2 = \text{cosec}^2\theta + \sin^2\theta - 2\,\text{cosec}\,\theta \,.\, \sin\theta$

$\Rightarrow (\text{cosec}\,\theta - \sin\theta)^2 = \text{cosec}^2\theta + \sin^2\theta - 2$

$\Rightarrow \text{cosec}^2\theta + \sin^2\theta = (\text{cosec}\,\theta - \sin\theta)^2 + 2$

$\because (\text{cosec}\,\theta - \sin\theta)^2 \geq 0$

$\therefore \text{cosec}^2\theta + \sin^2\theta \geq 2$

80. $\sqrt{3}\cot^2\theta - 4\cot\theta + \sqrt{3} = 0$

$\Rightarrow \sqrt{3}\cot^2\theta - 3\cot\theta - \cot\theta + \sqrt{3} = 0$

$\Rightarrow \sqrt{3}\cot\theta(\cot\theta - \sqrt{3}) - 1(\cot\theta - \sqrt{3}) = 0$

$\Rightarrow (\cot\theta - \sqrt{3})(\sqrt{3}\cot\theta - 1) = 0$

$\Rightarrow \cot\theta - \sqrt{3} = 0$ or $\sqrt{3}\cot\theta - 1 = 0$

$\Rightarrow \cot\theta = \sqrt{3}$ or $\cot\theta = \frac{1}{\sqrt{3}}$

$\Rightarrow \theta = 30°$ or $\theta = 60°$

When $\theta = 30°$ then

$\cot^2\theta + \tan^2\theta = \cot^2 30° + \tan^2\theta$

$$= (\sqrt{3})^2 + \left(\frac{1}{\sqrt{3}}\right)^2 = 3 + \frac{1}{3} = \frac{10}{3}$$

When $\theta = 60°$ then

$\cot^2\theta + \tan^2\theta = \cot^2 60° + \tan^2 60°$

$$= \left(\frac{1}{\sqrt{3}}\right)^2 + (\sqrt{3})^2 = \frac{1}{3} + 3 = \frac{10}{3}$$

81. $\text{LHS} = \frac{\cos\theta}{1 - \sin\theta} = \frac{\cos\theta(1 + \sin\theta)}{(1 - \sin\theta)(1 + \sin\theta)} = \frac{\cos\theta(1 + \sin\theta)}{1 - \sin^2\theta}$

$$= \frac{\cos\theta(1 + \sin\theta)}{\cos^2\theta} = \frac{1 + \sin\theta}{\cos\theta} = \text{RHS}$$

82. $\text{LHS} = \sin^2 A + \frac{1}{1 + \tan^2 A} = \sin^2 A + \frac{1}{\sec^2 A}$

$= \sin^2 A + \cos^2 A = 1$ = RHS

83. LHS $= (\sec^2\theta - 1)(\text{cosec}^2\theta - 1)$

$= \tan^2\theta \cdot \cot^2\theta = \tan^2\theta \times \dfrac{1}{\tan^2\theta} = 1 = \text{RHS}$

84. LHS $= (\sec\theta + \cos\theta)(\sec\theta - \cos\theta) = \sec^2\theta - \cos^2\theta$

$= (1 + \tan^2\theta) - (1 - \sin^2\theta) = 1 + \tan^2\theta - 1 + \sin^2\theta$

$= \tan^2\theta + \sin^2\theta = \text{RHS}$

85. LHS $= \tan^2\theta - \sin^2\theta = \dfrac{\sin^2\theta}{\cos^2\theta} - \sin^2\theta$

$= \dfrac{\sin^2\theta - \cos^2\theta \cdot \sin^2\theta}{\cos^2\theta} = \dfrac{\sin^2\theta(1-\cos^2\theta)}{\cos^2\theta}$

$= \left(\dfrac{\sin^2\theta}{\cos^2\theta}\right) \cdot \sin^2\theta = \tan^2\theta \cdot \sin^2\theta = \text{RHS}$

86. LHS $= \sec^6\theta = (\sec^2\theta)^3 = (1 + \tan^2\theta)^3$

$= (1)^3 + (\tan^2\theta)^3 + 3 \cdot 1 \cdot \tan^2\theta(1 + \tan^2\theta)$

$= 1 + \tan^6\theta + 3\tan^2\theta \cdot \sec^2\theta = \text{RHS}$

87. LHS $= \dfrac{1+\tan^2\theta}{\text{cosec}^2\theta} \cdot \cot\theta = \dfrac{\sec^2\theta}{\text{cosec}^2\theta} \cdot \cot\theta$

$= \dfrac{\frac{1}{\cos^2\theta}}{\frac{1}{\sin^2\theta}} \times \dfrac{\cos\theta}{\sin\theta} = \dfrac{\sin^2\theta}{\cos^2\theta} \times \dfrac{\cos\theta}{\sin\theta} = \dfrac{\sin\theta}{\cos\theta}$

$= \tan\theta = \text{RHS}$

88. LHS $= \dfrac{1+\cos A}{\sin A} = \dfrac{(1+\cos A)(1-\cos A)}{\sin A(1-\cos A)}$

$= \dfrac{1-\cos^2 A}{\sin A(1-\cos A)} = \dfrac{\sin^2 A}{\sin A(1-\cos A)} = \dfrac{\sin A}{1-\cos A}$

$= \text{RHS}$

89. LHS $= \tan^2 A + \cot^2 A + 2 = \sec^2 A - 1 + \text{cosec}^2 - 1 + 2$

$= \sec^2 A + \text{cosec}^2 A = \dfrac{1}{\cos^2 A} + \dfrac{1}{\sin^2 A}$

$= \dfrac{\sin^2 A + \cos^2 A}{\cos^2 A \cdot \sin^2 A} = \dfrac{1}{\cos^2 A \cdot \sin^2 A}$

$= \dfrac{1}{\cos^2 A} \times \dfrac{1}{\sin^2 A} = \sec^2 A \cdot \text{cosec}^2 A = \text{RHS}$

90. LHS $= \dfrac{1+\cos\theta - \sin^2\theta}{\sin\theta(1+\cos\theta)} = \dfrac{\cos\theta + \cos^2\theta}{\sin\theta(1+\cos\theta)}$

$[\because 1 - \sin^2\theta = \cos^2\theta]$

$= \dfrac{\cos\theta(\cos\theta+1)}{\sin\theta(\cos\theta+1)} = \cot\theta. = \text{RHS}$

91. LHS $= \left(\dfrac{\sin\theta}{\cos\theta} + \dfrac{1}{\cos\theta}\right)^2 + \left(\dfrac{\sin\theta}{\cos\theta} - \dfrac{1}{\cos\theta}\right)^2$

$= \dfrac{(1+\sin\theta)^2 + (\sin\theta - 1)^2}{\cos^2\theta}$

$= \dfrac{1 + 2\sin\theta + \sin^2\theta + \sin^2\theta + 1 - 2\sin\theta}{\cos^2\theta}$

$= \dfrac{2(1+\sin^2\theta)}{\cos^2\theta} = 2\left(\dfrac{1+\sin^2\theta}{1-\sin^2\theta}\right) = \text{RHS.}$

92. LHS $= (\sec A + \tan A - 1)(\sec A - \tan A + 1)$

$= \sec^2 A - \sec A \cdot \tan A + \sec A + \tan A \cdot \sec A - \tan^2 A + \tan A - \sec A + \tan A - 1$

$= \sec^2 A - \tan^2 A + 2\tan A - 1$

$= 1 + 2\tan A - 1 = 2\tan A = \text{RHS}$

93. LHS $= (\sec A - \text{cosec}\, A)(1 + \tan A + \cot A)$

$= \sec A + \sec A \cdot \tan A + \sec A \cdot \cot A - \text{cosec}\, A - \text{cosec}\, A \cdot \tan A - \text{cosec}\, A \cdot \cot A$

$= \dfrac{1}{\cos A} + \dfrac{1}{\cos A} \cdot \dfrac{\sin A}{\cos A} + \dfrac{1}{\cos A} \cdot \dfrac{\cos A}{\sin A} - \dfrac{1}{\sin A} - \dfrac{1}{\sin A} \cdot \dfrac{\sin A}{\cos A} - \dfrac{1}{\sin A} \cdot \dfrac{\cos A}{\sin A}$

$= \dfrac{1}{\cos A} + \dfrac{\sin A}{\cos^2 A} + \dfrac{1}{\sin A} - \dfrac{1}{\sin A} - \dfrac{1}{\cos A} - \dfrac{\cos A}{\sin^2 A}$

$= \dfrac{\sin A}{\cos^2 A} - \dfrac{\cos A}{\sin^2 A} = \dfrac{\sin A}{\cos A \cdot \cos A} - \dfrac{\cos A}{\sin A \cdot \sin A}$

$= \tan A \cdot \sec A - \cot A \cdot \text{cosec}\, A = \text{RHS}$

94. LHS $= \sin^2 A(1 - \sin^2 B) - (1 - \sin^2 A)\sin^2 B$

$= \sin^2 A - \sin^2 A \sin^2 B - (\sin^2 B - \sin^2 A \sin^2 B)$

$= \sin^2 A - \sin^2 A \sin^2 B - \sin^2 B + \sin^2 A \sin^2 B$

$= \sin^2 A - \sin^2 B = \text{RHS.}$

95. LHS $= \dfrac{\tan A + \tan B}{\cot A + \cot B} = \dfrac{\tan A + \tan B}{\frac{1}{\tan A} + \frac{1}{\tan B}}$

$= \dfrac{\tan A + \tan B}{\frac{\tan B + \tan A}{\tan A \cdot \tan B}} = \dfrac{(\tan A + \tan B) \times \tan A \cdot \tan B}{(\tan A + \tan B)}$

$= \tan A \cdot \tan B = \text{RHS}$

96. LHS $= \tan^2 A \cdot \sec^2 B - \sec^2 A \cdot \tan^2 B$

$= \tan^2 A(\tan^2 B + 1) - (1 + \tan^2 A)\tan^2 B$

$= \tan^2 A - \tan^2 B = \text{RHS.}$

97. LHS $= \sec^4 A(1 - \sin^4 A) - 2\tan^2 A$

$= \sec^4 A(1 - \sin^2 A)(1 + \sin^2 A) - 2\tan^2 A$

$= \sec^4 A(\cos^2 A) \cdot (1 + \sin^2 A) - 2\tan^2 A$

$= \sec^4 A \times \dfrac{1}{\sec^2 A}(1 + \sin^2 A) - 2\tan^2 A$

$= \sec^2 A(1 + \sin^2 A) - 2\tan^2 A$

$= \sec^2 A + \sec^2 A \cdot \sin^2 A - 2\tan^2 A$

$= \sec^2 A + \dfrac{1}{\cos^2 A} \cdot \sin^2 A - 2\tan^2 A$

$= \sec^2 A + \tan^2 A - 2\tan^2 A$

$= \sec^2 A - \tan^2 A = 1 = \text{RHS}$

98. LHS $= \cot^2 A \cdot \text{cosec}^2 B - \cot^2 B \cdot \text{cosec}^2 A$

$= \cot^2 A(1 + \cot^2 B) - \cot^2 B(1 + \cot^2 A)$

$= \cot^2 A + \cot^2 A \cdot \cot^2 B - \cot^2 B - \cot^2 B \cdot \cot^2 A$

$= \cot^2 A - \cot^2 B = \text{RHS}$

99. $(\cos\theta - \sin\theta)^2 = (1)^2$

$\Rightarrow \cos^2\theta + \sin^2\theta - 2\sin\theta \cdot \cos\theta = 1$

$\Rightarrow 2\sin\theta \cdot \cos\theta = 0$ ...(*i*)

Now, $(\cos\theta + \sin\theta)^2 = \sin^2\theta + \cos^2\theta + 2\sin\theta \cdot \cos\theta$

$\Rightarrow \quad (\cos\theta + \sin\theta)^2 = 1 + 0$ [Using (*i*)]

$\Rightarrow \quad \cos\theta + \sin\theta = \pm\sqrt{1} = \pm 1$

100. $x = a \sec\theta$

$\Rightarrow \quad \frac{x}{a} = \sec\theta \Rightarrow \frac{x^2}{a^2} = \sec^2\theta$ ...(*i*)

Also $y = b\tan\theta \Rightarrow \frac{y}{b} = \tan\theta$

$\Rightarrow \quad \frac{y^2}{b^2} = \tan^2\theta$ ...(*ii*)

Subtracting equation (*ii*) from (*i*), we get

$\frac{x^2}{a^2} - \frac{y^2}{b^2} = \sec^2\theta - \tan^2\theta \Rightarrow \frac{x^2}{a^2} - \frac{y^2}{b^2} = 1$ Hence proved.

101. $\sin 3\theta = \cos(\theta - 6°)$

$\Rightarrow \quad \cos(90° - 3\theta) = \cos(\theta - 6°)$

$\Rightarrow \quad 90° - 3\theta = \theta - 6°$

$\Rightarrow \quad 90° + 6° = 4\theta$

$\Rightarrow \quad \theta = \frac{96}{4} = 24°$

102. LHS $= \frac{1}{\text{cosec A} - \cot A} + \frac{1}{\text{cosec B} - \cot B} + \frac{1}{\text{cosec C} - \cot C}$

$= \frac{\text{cosec A} + \cot A}{(\text{cosec A} - \cot A)(\text{cosec A} + \cot A)} + \frac{\text{cosec B} + \cot B}{(\text{cosec B} - \cot B)(\text{cosec B} + \cot B)} + \frac{\text{cosec C} + \cot C}{(\text{cosec C} - \cot C)(\text{cosec C} + \cot C)}$

$= \frac{\text{cosec A} + \cot A}{\text{cosec}^2 A - \cot^2 A} + \frac{\text{cosec B} + \cot B}{\text{cosec}^2 B - \cot^2 B} + \frac{\text{cosec C} + \cot C}{\text{cosec}^2 C - \cot^2 C}$

$= \frac{\text{cosec A} + \cot A}{1} + \frac{\text{cosec B} + \cot B}{1} + \frac{\text{cosec C} + \cot C}{1}$

$= \text{cosec A} + \cot A + \text{cosec B} + \cot B + \text{cosec C} + \cot C$

= RHS

103. cosec A + cot A = *m* ...(*i*)

cosec A – cot A = *n* ...(*ii*)

On multiplying equation (*i*) and (*ii*), we get

(cosec A + cot A)(cosec A – cot A) = *m*. *n*

$\Rightarrow \quad \text{cosec}^2 A - \cot^2 A = mn$

$\Rightarrow \quad 1 = mn$ ($\because \text{cosec}^2 A - \cot^2 A = 1$)

Hence proved

104. $\frac{\sin\theta + \cos\theta}{\sin\theta - \cos\theta} = \frac{5}{3}$

$\Rightarrow \quad 3\sin\theta + 3\cos\theta = 5\sin\theta - 5\cos\theta$

$\Rightarrow \quad 8\cos\theta = 2\sin\theta$

$\Rightarrow \quad \frac{\cos\theta}{\sin\theta} = \frac{2}{8} \Rightarrow \cot\theta = \frac{1}{4} \Rightarrow \tan\theta = 4$

Now, $\frac{7\tan\theta + 2}{2\tan\theta + 7} = \frac{7 \times 4 + 2}{2 \times 4 + 7} = \frac{30}{15} = 2$

105. LHS $= \frac{\sin(90° - \theta) . \sin\theta}{\tan\theta} - 1$

$= \frac{\cos\theta . \sin\theta}{\frac{\sin\theta}{\cos\theta}} - 1$

$= \cos\theta . \sin\theta \times \frac{\cos\theta}{\sin\theta} - 1 = \cos^2\theta - 1 = -(1 - \cos^2\theta)$

$= -\sin^2\theta$ = RHS

106. LHS $= \tan 10° . \tan 75° . \tan 15° . \tan 80°$

$= \tan 10° . \tan 75° . \cot(90° - 15°) . \cot(90° - 80°)$

$= \frac{1}{\cot 10°} . \frac{1}{\cot 75°} . \cot 75° . \cot 10° = 1$ = RHS

107. $\tan 1° . \tan 2° . \tan 3° \ldots\ldots \tan 89° = 1$

LHS $= \tan(90° - 89°) \tan(90° - 88°) \tan(90° - 87°) \ldots\ldots \tan 87° \tan 88° \tan 89°$

$= \cot 89° \cot 88° \cot 87° \ldots\ldots \tan 87° \tan 88° \tan 89°$

$= (\cot 89° . \tan 89°)(\cot 88° \tan 88°)(\cot 87° \tan 87°) \ldots\ldots (\cot 44° \tan 44°) . \tan 45°$

$= 1 \times 1 \times 1 \ldots\ldots \times 1 = 1$ = RHS.

($\because \cot\theta \tan\theta = 1$ and $\tan 45° = 1$)

108. LHS $= \cos 1° . \cos 2° . \cos 3° \ldots\ldots \cos 89° . \cos 90° . \cos 91° \ldots\ldots \cos 180°$

$= \cos 1° . \cos 2° . \cos 3° \ldots \cos 89° \times 0 \times \cos 91° \ldots \cos 180°$

= 0 = RHS. ($\cos 90° = 0$)

109. LHS $= \sin^2 5° + \sin^2 10° + \ldots + \sin^2 85° + \sin^2 90°$

$= (\sin^2 5° + \sin^2 85°) + (\sin^2 10° + \sin^2 80°) + \ldots + (\sin^2 40° + \sin^2 50°) + (\sin^2 45° + \sin^2 90°)$

$= \{\sin^2 5° + \cos^2(90° - 85°)\} + \{\sin^2 10° + \cos^2(90° - 80°\} + \ldots + \{\sin^2 40° + \cos^2(90° - 50°)\} + \left\{\left(\frac{1}{\sqrt{2}}\right)^2 + (1)^2\right\}$

$= (1) + (1) + \ldots + 8 \text{ times} + \frac{1}{2} + 1$

$= 8 + \frac{1}{2} + 1 = 9\frac{1}{2}$ = RHS

110. LHS $= \frac{\sin 36°}{\cos 54°} - \frac{\sin 54°}{\cos 36°}$

$= \frac{\cos(90° - 36°)}{\cos 54°} - \frac{\cos(90° - 54°)}{\cos 36°}$

$= \frac{\cos 54°}{\cos 54°} - \frac{\cos 36°}{\cos 36°}$

= 1 – 1 = 0 = RHS

111. LHS $= \sin^2 25° + \sin^2 65°$

$= \sin^2 25° + \cos^2(90° - 65°)$

$= \sin^2 25° + \cos^2 25° = 1$ = RHS

112. LHS $= \left(\frac{\tan 20°}{\text{cosec } 70°}\right)^2 + \left(\frac{\cot 20°}{\sec 70°}\right)^2$

$= (\tan 20° \times \sin 70°)^2 + (\cot 20° \times \cos 70°)^2$

$$= \left(\frac{\sin 20°}{\cos 20°} . \cos(90° - 70°)\right)^2 + \left(\frac{\cos 20°}{\sin 20°} . \sin(90° - 70°)\right)^2$$

$$= \left(\frac{\sin 20°}{\cos 20°} . \cos 20°\right)^2 + \left(\frac{\cos 20°}{\sin 20°} . \sin 20°\right)^2$$

$= (\sin 20°)^2 + (\cos 20°)^2$

$= \sin^2 20° + \cos^2 20° = 1 = \text{RHS}$

113. LHS $= (\sin 72° + \cos 18°)(\sin 72° - \cos 18°)$

$= (\sin 72° + \cos 18°)\{\cos(90° - 72°) - \cos 18°\}$

$= (\sin 72° + \cos 18°)(\cos 18° - \cos 18°)$

$= (\sin 72° + \cos 18°) \times 0 = 0 = \text{RHS}$

114. LHS $= \sin 63° . \cos 27° + \cos 63° . \sin 27°$

$= \sin 63° . \sin(90° - 27°) + \cos 63° . \cos(90° - 27°)$

$= \sin 63° . \sin 63° + \cos 63° . \cos 63°$

$= \sin^2 63° + \cos^2 63° = 1 = \text{RHS}$

115. LHS $= \dfrac{\sin 70°}{\cos 20°} + \dfrac{\operatorname{cosec} 20°}{\sec 70°} - 2\cos 70° . \operatorname{cosec} 20°$

$$= \frac{\cos(90° - 70°)}{\cos 20°} + \frac{\sec(90° - 20°)}{\sec 70°} - 2\cos 70° . \sec(90° - 20°)$$

$$= \frac{\cos 20°}{\cos 20°} + \frac{\sec 70°}{\sec 70°} - 2 \times \frac{1}{\sec 70°} . \sec 70°$$

$= 1 + 1 - 2 = 0 = \text{RHS}$

116. LHS $= \sin(90° - \theta)\cos(90° - \theta)$

$= \cos\theta . \sin\theta$

$$= \cos\theta . \frac{\sin\theta}{\cos\theta} \times \cos\theta$$

$$= \cos^2\theta . \tan\theta = \frac{1}{\sec^2\theta} . \tan\theta$$

$$= \frac{\tan\theta}{1 + \tan^2\theta} = \frac{\tan\theta}{1 + \cot^2(90° - \theta)} = \text{RHS}$$

117. LHS $= \sin^2(90° - \theta)[1 + \cot^2(90° - \theta)]$

$= \cos^2\theta(1 + \tan^2\theta)$

$= \cos^2\theta . \sec^2\theta$

$$= \frac{1}{\sec^2\theta} . \sec^2\theta = 1 = \text{RHS}$$

118. LHS $= \dfrac{\cos(90° - A) . \sin(90° - A)}{\tan(90° - A)}$

$$= \frac{\sin A . \cos A}{\cot A} = \frac{\sin A . \cos A}{\dfrac{\cos A}{\sin A}}$$

$$= \sin A . \cos A \times \frac{\sin A}{\cos A} = \sin^2 A = \text{RHS}$$

119. LHS $= \dfrac{1}{1 + \cos(90° - \theta)} + \dfrac{1}{1 - \cos(90° - \theta)}$

$$= \frac{1}{1 + \sin\theta} + \frac{1}{1 - \sin\theta} = \frac{1 - \sin\theta + 1 + \sin\theta}{(1 + \sin\theta)(1 - \sin\theta)}$$

$$= \frac{2}{1 - \sin^2\theta} = \frac{2}{\cos^2\theta}$$

$= 2\sec^2 = 2\operatorname{cosec}^2(90° - \theta) = \text{RHS}$

120. Since $A + B = 90°$

$\Rightarrow A = 90° - B$

$\Rightarrow \sin A = \sin(90° - B)$

$\Rightarrow \sin A = \cos B$

$\Rightarrow \sin^2 A = \cos^2 B = 1 - \sin^2 B$

$\Rightarrow \sin^2 A + \sin^2 B = 1$

121. $\cos 2\theta = 2\cos^2\theta - 1$

Let $\theta = 30°$

$\Rightarrow \cos(2 \times 30°) = 2\cos^2 30° - 1$

$\Rightarrow \cos 60° = 2\cos^2 30° - 1$

$\Rightarrow \dfrac{1}{2} + 1 = 2\cos^2 30°$

$\Rightarrow 2\cos^2 30° = \dfrac{3}{2}$

$\cos^2 30° = \dfrac{3}{4}$

$\Rightarrow \cos 30° = \sqrt{\dfrac{3}{4}} = \dfrac{\sqrt{3}}{2}$

122. $\cos 2\theta = 1 - 2\sin^2\theta$

Let $\theta = 30°$

$\Rightarrow \cos(2 \times 30°) = 1 - 2\sin^2 30°$

$\Rightarrow \dfrac{1}{2} = 1 - 2\sin^2 30°$

$\Rightarrow 2\sin^2 30° = 1 - \dfrac{1}{2}$

$\Rightarrow 2\sin^2 30° = \dfrac{1}{2}$

$\Rightarrow \sin^2 30° = \dfrac{1}{4}$

$\Rightarrow \sin 30° = \dfrac{1}{2}$.

123. LHS $= \dfrac{\cot A + \operatorname{cosec} A - 1}{\cot A - \operatorname{cosec} A + 1}$

$$= \frac{(\cot A + \operatorname{cosec} A) - (\operatorname{cosec}^2 A - \cot^2 A)}{\cot A - \operatorname{cosec} A + 1}$$

$$= \frac{(\cot A + \operatorname{cosec} A) - (\operatorname{cosec} A + \cot A)(\operatorname{cosec} A - \cot A)}{\cot A - \operatorname{cosec} A + 1}$$

$$= \frac{(\operatorname{cosec} A + \cot A)[1 - (\operatorname{cosec} A - \cot A)]}{\cot A - \operatorname{cosec} A + 1}$$

$$= \frac{(\operatorname{cosec} A + \cot A)(\cot A - \operatorname{cosec} A + 1)}{\cot A - \operatorname{cosec} A + 1}$$

$= \operatorname{cosec} A + \cot A$

$$= \frac{1}{\sin A} + \frac{\cos A}{\sin A} = \frac{1 + \cos A}{\sin A} = \text{RHS}.$$

124. $\tan A = 1 = \tan 45° \Rightarrow A = 45°$

Also, $\sin B = \dfrac{1}{\sqrt{2}} = \sin 45° \Rightarrow B = 45°$

$\therefore \cos(A + B) = \cos(45° + 45°) = \cos 90° = 0$

125. LHS $= \dfrac{1+\tan^2\theta}{1+\cot^2\theta} = \dfrac{1+\tan^2\theta}{1+\dfrac{1}{\tan^2\theta}}$

$= \dfrac{1+\tan^2\theta}{\dfrac{\tan^2\theta+1}{\tan^2\theta}} = \dfrac{\tan^2\theta\,(1+\tan^2\theta)}{(1+\tan^2\theta)}$

$= \tan^2\theta$ ...(*i*)

RHS $= \left(\dfrac{1-\tan\theta}{1-\cot\theta}\right)^2 = \left(\dfrac{1-\tan\theta}{1-\dfrac{1}{\tan\theta}}\right)^2 = \left(\dfrac{1-\tan\theta}{\dfrac{\tan\theta-1}{\tan\theta}}\right)^2$

$= \left(\dfrac{\tan\theta\,(1-\tan\theta)}{\tan\theta-1}\right)^2 = \left(\dfrac{\tan\theta\,(1-\tan\theta)}{-1\,(1-\tan\theta)}\right)^2$

$= \left(\dfrac{\tan\theta}{-1}\right)^2 = \tan^2\theta$ ...(*ii*)

From (*i*) and (*ii*) LHS = RHS

126. LHS $= \sqrt{\dfrac{1+\sin\theta}{1-\sin\theta}}$

$= \sqrt{\dfrac{1+\sin\theta}{1-\sin\theta}\times\dfrac{1+\sin\theta}{1+\sin\theta}} = \sqrt{\dfrac{(1+\sin\theta)^2}{1-\sin^2\theta}}$

$= \dfrac{1+\sin\theta}{\cos\theta} = \dfrac{1}{\cos\theta}+\dfrac{\sin\theta}{\cos\theta} = \sec\theta+\tan\theta$ = RHS.

127. LHS $= (\sin\theta+\mathrm{cosec}\,\theta)^2+(\cos\theta+\sec\theta)^2$

$= \sin^2\theta+\mathrm{cosec}^2\theta+2\sin\theta\,.\,\mathrm{cosec}\,\theta+\cos^2\theta + \sec^2\theta+2\cos\theta\sec\theta$

$= \sin^2\theta+\cos^2\theta+2+2+\mathrm{cosec}^2\theta+\sec^2\theta$

$= 1+4+1+\cot^2\theta+1+\tan^2\theta$

$= 7+\cot^2\theta+\tan^2\theta$ = RHS.

128. LHS $= (\mathrm{cosec}\,\theta-\cot\theta)^2$

$= \left(\dfrac{1}{\sin\theta}-\dfrac{\cos\theta}{\sin\theta}\right)^2$

$= \dfrac{(1-\cos\theta)^2}{\sin^2\theta} = \dfrac{(1-\cos\theta)^2}{1-\cos^2\theta}$

$= \dfrac{(1-\cos\theta)^2}{(1-\cos\theta)(1+\cos\theta)} = \dfrac{1-\cos\theta}{1+\cos\theta}$ = RHS.

129. LHS $= \dfrac{\sin\theta-2\sin^3\theta}{2\cos^3\theta-\cos\theta} = \dfrac{\sin\theta\,(1-2\sin^2\theta)}{\cos\theta\,(2\cos^2\theta-1)}$

$= \dfrac{\sin\theta\,(1-2\sin^2\theta)}{\cos\theta\,\{2(1-\sin^2\theta)-1\}} = \dfrac{\sin\theta\,(1-2\sin^2\theta)}{\cos\theta\,(1-2\sin^2\theta)}$

$= \dfrac{\sin\theta}{\cos\theta} = \tan\theta$ = RHS

130. Given. $a\cos\theta-b\sin\theta=x$

and $a\sin\theta+b\cos\theta=y$

To show. $a^2+b^2=x^2+y^2$

Sol. RHS $= x^2+y^2$

$= (a\cos\theta-b\sin\theta)^2+(a\sin\theta+b\cos\theta)^2$

$= a^2\cos^2\theta+b^2\sin^2\theta-2ab\cos\theta\sin\theta + a^2\sin^2\theta+b^2\cos^2\theta+2ab\cos\theta\sin\theta$

$= a^2(\cos^2\theta+\sin^2\theta)+b^2(\cos^2\theta+\sin^2\theta)$

$= a^2(1)+b^2(1) = a^2+b^2$ = LHS

131. Consider,

LHS $= \dfrac{\mathrm{cosec}\,A}{(\mathrm{cosec}\,A-1)}+\dfrac{\mathrm{cosec}\,A}{(\mathrm{cosec}\,A+1)}$

$= \dfrac{\mathrm{cosec}\,A(\mathrm{cosec}\,A+1)+\mathrm{cosec}\,A(\mathrm{cosec}\,A-1)}{(\mathrm{cosec}\,A-1)(\mathrm{cosec}\,A+1)}$

$= \dfrac{\mathrm{cosec}^2A+\mathrm{cosec}\,A+\mathrm{cosec}^2A-\mathrm{cosec}\,A}{\mathrm{cosec}^2A-1}$

$= \dfrac{2\,\mathrm{cosec}^2A}{\cot^2A} = \dfrac{2(\cot^2A+1)}{\cot^2A} = \dfrac{2\cot^2A+2}{\cot^2A}$

$= \dfrac{2\cot^2A}{\cot^2A}+\dfrac{2}{\cot^2A} = 2+2\tan^2A$ = RHS

$= 2(1+\tan^2A) = 2\sec^2A$ = RHS

132. $\dfrac{2\sin\theta-3\cos\theta}{4\sin\theta-9\cos\theta} = \dfrac{2-3\cot\theta}{4-9\cot\theta}$ ... (*i*)

(Dividing numerator and denominator by $\sin\theta$)

Now $\mathrm{cosec}\;\theta = \dfrac{13}{12}$,

$\cot\theta = \sqrt{\mathrm{cosec}^2\theta-1} = \sqrt{\left(\dfrac{13}{12}\right)^2-1} = \dfrac{5}{12}$

Substituting in (*i*), we have

$\dfrac{2\sin\theta-3\cos\theta}{4\sin\theta-9\cos\theta} = \dfrac{2-3.\dfrac{5}{12}}{4-9.\dfrac{5}{12}} = \dfrac{9}{3} = 3$

133. $\cos\alpha = \dfrac{1}{2} = \cos 60° \Rightarrow \alpha = 60°$

Also $\tan\beta = \dfrac{1}{\sqrt{3}} = \tan 30° \Rightarrow \beta = 30°$

Now $\sin(\alpha+\beta) = \sin(60°+30°) = \sin 90° = 1$

134. $\sin(50°+\theta)-\cos(40°-\theta)$

$= \cos[90°-(50°+\theta)]-\cos(40°-\theta)$

$= \cos(40°-\theta)-\cos(40°-\theta) = 0$

135. $x = \gamma\cos\alpha\,.\,\sin\beta \Rightarrow x^2 = \gamma^2\,.\,\cos^2\alpha\,.\,\sin^2\beta$ ...(*i*)

$y = \gamma\cos\alpha\,.\,\cos\beta \Rightarrow y^2 = \gamma^2\,.\,\cos^2\alpha\,.\,\cos^2\beta$ ...(*ii*)

$z = \gamma\,.\,\sin\alpha \Rightarrow z^2 = \gamma^2\,.\,\sin^2\alpha$ ...(*iii*)

Adding (*i*), (*ii*) and (*iii*), we get

$x^2+y^2+z^2 = \gamma^2\cos^2\alpha\,.\,\sin^2\beta+\gamma^2\cos^2\alpha\,.\,\cos^2\beta+\gamma^2\sin^2\alpha$

$\Rightarrow x^2+y^2+z^2 = \gamma^2\cos^2\alpha\,(\sin^2\beta+\cos^2\beta)+\gamma^2\sin^2\alpha$

$\Rightarrow x^2+y^2+z^2 = \gamma^2\,.\,\cos^2\alpha\,.\,1+\gamma^2\,.\,\sin^2\alpha$

$= \gamma^2\,(\cos^2\alpha+\sin^2\alpha)$

$\Rightarrow x^2+y^2+z^2 = \gamma^2$ Hence proved.

136. In ΔACD, $\tan\phi = \frac{AC}{DC}$

and in ΔACB, $\tan\theta = \frac{AC}{BC}$

Now, $\frac{\tan\theta}{\tan\phi} = \frac{\frac{AC}{BC}}{\frac{AC}{DC}} = \frac{DC}{BC} = \frac{DC}{2DC} = \frac{1}{2}$ Hence proved.

SHORT ANSWER TYPE QUESTIONS-II

1. LHS $= \frac{\sin\theta + \cos\theta}{\sin\theta - \cos\theta} + \frac{\sin\theta - \cos\theta}{\sin\theta + \cos\theta}$

$$= \frac{(\sin\theta + \cos\theta)^2 + (\sin\theta - \cos\theta)^2}{(\sin\theta - \cos\theta)(\sin\theta + \cos\theta)}$$

$$= \frac{\sin^2\theta + \cos^2\theta + 2\sin\theta.\cos\theta + \sin^2\theta + \cos^2\theta - 2\sin\theta.\cos\theta}{\sin^2\theta - \cos^2\theta}$$

$$= \frac{2}{\sin^2\theta - \cos^2\theta} = \frac{\frac{2}{\cos^2\theta}}{\frac{\sin^2\theta}{\cos^2\theta} - \frac{\cos^2\theta}{\cos^2\theta}} = \frac{2\sec^2\theta}{\tan^2\theta} = \text{RHS}$$

Or

$$\frac{\sec^2(90° - \theta) - \cot^2\theta}{2(\sin^2 25° + \sin^2 65°)} + \frac{2\cos^2 60°.\tan^2 28°.\tan^2 62°}{3(\sec^2 43° - \cot^2 47°)}$$

$$= \frac{\text{cosec}^2\theta - \cot^2\theta}{2\{\sin^2 25° + \cos^2(90° - 65°)\}} + \frac{2\times\left(\frac{1}{2}\right)^2.\tan^2 28°.\cot^2(90° - 62°)}{3\{\sec^2 43° - \tan^2(90° - 47°)\}}$$

$$= \frac{1}{2(\sin^2 25° + \cos^2 25°)} + \frac{\frac{1}{2}\times\frac{1}{\cot^2 28°}.\cot^2 28°}{3(\sec^2 43° - \tan^2 43°)}$$

$$= \frac{1}{2\times 1} + \frac{\frac{1}{2}\times 1}{3\times 1} = \frac{1}{2} + \frac{1}{6} = \frac{3+1}{6} = \frac{4}{6} = \frac{2}{3}$$

2. LHS $= \frac{1}{\text{cosec}\theta - \cot\theta} - \frac{1}{\sin\theta}$

$$= \frac{(\text{cosec}\,\theta + \cot\theta)}{(\text{cosec}\,\theta - \cot\theta)(\text{cosec}\theta + \cot\theta)} - \text{cosec}\theta$$

$$= \frac{\text{cosec}\theta + \cot\theta}{\text{cosec}^2\theta - \cot^2\theta} - \text{cosec}\theta$$

$= \text{cosec}\theta + \cot\theta - \text{cosec}\theta = \text{cosec}\theta - (\text{cosec}\theta - \cot\theta)$

$$= \frac{1}{\sin\theta} - \frac{(\text{cosec}\theta - \cot\theta)(\text{cosec}\theta + \cot\theta)}{\text{cosec}\theta + \cot\theta}$$

$$= \frac{1}{\sin\theta} - \frac{\text{cosec}^2\theta - \cot^2\theta}{\text{cosec}\theta + \cot\theta}$$

$$= \frac{1}{\sin\theta} - \frac{1}{\text{cosec}\theta + \cot\theta} = \text{RHS}$$

Or

$$\frac{\text{cosec}^2(90° - \theta) - \tan^2\theta}{4(\cos^2 48° + \cos^2 42°)} - \frac{2\tan^2 30°\sec^2 52°.\sin^2 38°}{\text{cosec}^2 70° - \tan^2 20°}$$

$$= \frac{\sec^2\theta - \tan^2\theta}{4\{\cos^2 48° + \sin^2(90° - 42°)\}} - \frac{2\times\left(\frac{1}{\sqrt{3}}\right)^2.\sec^2 52°.\cos^2(90° - 38°)}{\sec^2(90° - 70°) - \tan^2 20°}$$

$$= \frac{1}{4(\cos^2 48° + \sin^2 48)} - \frac{2\times\frac{1}{3}\times\frac{1}{\cos^2 52°}\times\cos^2 52°}{\sec^2 20° - \tan^2 20°}$$

$$= \frac{1}{4\times 1} - \frac{\frac{2}{3}\times 1}{1} = \frac{1}{4} - \frac{2}{3} = \frac{3-8}{12} = -\frac{5}{12}$$

3. $\frac{\sin^2\theta + \sin^2(90° - \theta)}{3(\sec^2 61° - \cot^2 29°)} - \frac{3\cot^2 30°.\sin^2 54°.\sec^2 36°}{2(\text{cosec}^2 65° - \tan^2 25°)}$

$$= \frac{\sin^2\theta + \cos^2\theta}{3\{\text{cosec}^2(90° - 61°) - \cot^2 29°\}} - \frac{3(\sqrt{3})^2.\sin^2 54°.\text{cosec}^2(90° - 36°)}{2\{\text{cosec}^2(90° - 65°) - \tan^2 25°\}}$$

$$= \frac{1}{3(\text{cosec}^2 29° - \cot^2 29°)} - \frac{3\times 3\frac{1}{\text{cosec}^2 54°}.\text{cosec}^2 54°}{2(\sec^2 25° - \tan^2 25°)}$$

$$= \frac{1}{3\times 1} - \frac{9\times 1}{2\times 1} = \frac{1}{3} - \frac{9}{2} = \frac{2-27}{6} = \frac{-25}{6}$$

4. $\frac{\cos^2(90° - 70°) + \cos^2 70°}{\sec^2(90° - 40°) - \cot^2 40°} + 2(\text{cosec}^2(90° - 32°) - 2\cot(90° - 32°)\tan 32 - 4\tan(90° - 77°)\tan(90° - 53°)\tan 45°$ · $\tan 53°\tan 77°$

$= \frac{\sin^2 70° + \cos^2 70°}{\text{cosec}^2 40° - \cot^2 40°} + 2\sec^2 32° - 2\tan^2 32° - 4\cot 77°\cot 53°\tan 45°\tan 53°\tan 77° = 1 + 2 - 4 = -1$

Or

LHS $= \sqrt{\frac{\sec\theta - 1}{\sec\theta + 1}} + \sqrt{\frac{\sec\theta + 1}{\sec\theta - 1}}$

$$= \sqrt{\frac{\frac{1}{\cos\theta} - 1}{\frac{1}{\cos\theta} + 1}} + \sqrt{\frac{\frac{1}{\cos\theta} + 1}{\frac{1}{\cos\theta} - 1}} = \sqrt{\frac{1 - \cos\theta}{1 + \cos\theta}} + \sqrt{\frac{1 + \cos\theta}{1 - \cos\theta}}$$

$$= \sqrt{\frac{1 - \cos\theta}{1 + \cos\theta}\times\frac{1 - \cos\theta}{1 - \cos\theta}} + \sqrt{\frac{1 + \cos\theta}{1 - \cos\theta}\times\frac{1 + \cos\theta}{1 + \cos\theta}}$$

$$= \sqrt{\frac{(1 - \cos\theta)^2}{1 - \cos^2\theta}} + \sqrt{\frac{(1 + \cos\theta)^2}{1 - \cos^2\theta}}$$

$$= \frac{1 - \cos\theta + 1 + \cos\theta}{\sin\theta} = \frac{2}{\sin\theta}$$

$= 2 \text{ cosec } \theta = \text{RHS}.$

5. $\frac{\sec 39°}{\text{cosec} 51°} + \frac{2}{\sqrt{3}} \tan 17° . \tan 38° . \tan 60° . \tan 52° . \tan 73° - 3(\sin^2 31° + \sin^2 59°)$

$$= \frac{\text{cosec}(90° - 39°)}{\text{cosec} 51°} + \frac{2}{\sqrt{3}} \times \tan 17° . \tan 38° . \sqrt{3} . \cot(90° - 52°) . \cot(90° - 73°) - 3\{\sin^2 31° + \cos^2(90° - 59°)\}$$

$$= \frac{\text{cosec} 51°}{\text{cosec} 51°} + 2 . \frac{1}{\cot 17°} . \frac{1}{\cot 38°} . \cot 38° . \cot 17° - 3(\sin^2 31° + \cos^2 31°)$$

$= 1 + 2 \times 1 - 3 \times 1 = 0$

6. $\frac{3\cos 55°}{7\sin 35°} - \frac{4(\cos 70° . \text{cosec } 20°)}{7(\tan 5° \tan 25° . \tan 45° . \tan 65° . \tan 85°)}$

$$= \frac{3 \sin(90° - 55°)}{7 \sin 35°} - \frac{4 \cos 70° . \sec(90° - 20°)}{7(\tan 5° . \tan 25° . 1 \times \cot(90° - 65°) . \cot(90° - 85°))}$$

$$= \frac{3 \sin 35°}{7 \sin 35°} - \frac{4 \cdot \frac{1}{\sec 70°} . \sec 70°}{7 \times \frac{1}{\cot 5°} . \frac{1}{\cot 25°} \times . 1 \times \cot 25° . \cot 5°}$$

$$= \frac{3}{7} - \frac{4 \times 1}{7 \times 1} = \frac{3}{7} - \frac{4}{7} = \frac{-1}{7}$$

7. $\tan 7° . \tan 23° . \tan 60° . \tan 67° . \tan 83° + \frac{\cot 54°}{\tan 36°} + \sin 20° . \sec 70° - 2$

$$= \tan 7° \tan 23° \times \sqrt{3} \times \cot(90° - 67°) \cot(90° - 83°) + \frac{\tan(90° - 54°)}{\tan 36°} + \sin 20° \text{ cosec } (90° - 70°) - 2$$

$$= \frac{1}{\cot 7°} . \frac{1}{\cot 23°} \times \sqrt{3} \times \cot 23° . \cot 7° + \frac{\tan 36°}{\tan 36°} + \frac{1}{\text{cosec } 20°} \times \text{cosec } 20° - 2$$

$= 1 \times \sqrt{3} + 1 + 1 - 2 = \sqrt{3}$

8. $\text{LHS} = \frac{\cot A - \cos A}{\cot A + \cos A}$

$$= \frac{\frac{\cos A}{\sin A} - \cos A}{\frac{\cos A}{\sin A} + \cos A} = \frac{\frac{\cos A - \sin A \cos A}{\sin A}}{\frac{\cos A + \sin A \cos A}{\sin A}}$$

$$= \frac{\cos A(1 - \sin A)}{\cos A(1 + \sin A)} = \frac{1 - \sin A}{1 + \sin A}$$

$$= \frac{\frac{1}{\sin A} - 1}{\frac{1}{\sin A} + 1} = \frac{\text{cosec } A - 1}{\text{cosec } A + 1} = \text{RHS}.$$

Or

$$= \left(1 + \frac{\cos A}{\sin A} - \frac{1}{\sin A}\right)\left(1 + \frac{\sin A}{\cos A} + \frac{1}{\cos A}\right)$$

$$= \frac{(\sin A + \cos A - 1)(\sin A + \cos A + 1)}{\sin A \cos A}$$

$$= \sin^2 A + \cos^2 A + \frac{2\cos A \sin A - 1}{\sin A \cos A} = 2$$

9. $\text{LHS} = \sin\theta(1 + \tan\theta) + \cos\theta(1 + \cot\theta)$

$= \sin\theta + \sin\theta . \tan\theta + \cos\theta + \cos\theta . \cot\theta$

$$= \sin\theta + \sin\theta . \frac{\sin\theta}{\cos\theta} + \cos\theta + \cos\theta . \frac{\cos\theta}{\sin\theta}$$

$$= \sin\theta + \frac{\cos^2\theta}{\sin\theta} + \frac{\sin^2\theta}{\cos\theta} + \cos\theta$$

$$= \frac{\sin^2\theta + \cos^2\theta}{\sin\theta} + \frac{\sin^2\theta + \cos^2\theta}{\cos\theta} = \frac{1}{\sin\theta} + \frac{1}{\cos\theta}$$

$= \text{cosec}\theta + \sec\theta = \sec\theta + \text{cosec}\theta = \text{RHS}$

10. $\text{LHS} = (1 + \cot A + \tan A)(\sin A - \cos A)$

$= \sin A + \cot A . \sin A + \tan A . \sin A - \cos A - \cot A . \cos A - \tan A . \cos A$

$$= \sin A + \frac{\cos A}{\sin A} . \sin A + \tan A . \sin A - \cos A - \cot A . \cos A - \frac{\sin A}{\cos A} . \cos A$$

$= \tan A . \sin A - \cot A \cos A = \text{RHS}.$

11. $\text{LHS} = (\text{cosec}\theta - \sin\theta)(\sec\theta - \cos\theta)$

$$= \left(\frac{1}{\sin\theta} - \sin\theta\right)\left(\frac{1}{\cos\theta} - \cos\theta\right) = \frac{1 - \sin^2\theta}{\sin\theta} . \frac{1 - \cos^2\theta}{\cos\theta}$$

$$= \frac{\cos^2\theta}{\sin\theta} . \frac{\sin^2\theta}{\cos\theta} = \cos\theta . \sin\theta \quad ...(i)$$

$$\text{RHS} = \frac{1}{\tan\theta + \cot\theta} = \frac{1}{\frac{\sin\theta}{\cos\theta} + \frac{\cos\theta}{\sin\theta}}$$

$$= \frac{1}{\frac{\sin^2\theta + \cos^2\theta}{\cos\theta . \sin\theta}} = \frac{\cos\theta . \sin\theta}{\sin^2\theta + \cos^2\theta} = \frac{\cos\theta . \sin\theta}{1}$$

$= \cos\theta . \sin\theta \quad ...(ii)$

From (*i*) and (*ii*), LHS = RHS

12. $\text{LHS} = 2(\sin^6\theta + \cos^6\theta) - 3(\sin^4\theta + \cos^4\theta) + 1$

$= 2[(\sin^2\theta + \cos^2\theta)^3 - 3\sin^2\theta . \cos^2\theta(\sin^2\theta + \cos^2\theta)] - 3[(\sin^2\theta + \cos^2\theta)^2 - 2\sin^2\theta.\cos^2\theta] + 1$

$= 2[1 - 3\sin^2\theta . \cos^2\theta] - 3[1 - 2\sin^2\theta . \cos^2\theta] + 1$

$= 2 - 6\sin^2\theta.\cos^2\theta - 3 + 6\sin^2\theta.\cos^2\theta + 1 = 0 = \text{RHS}.$

13. $\frac{\cos 58°}{\sin 32°} + \frac{\sin 22°}{\cos 68°} - \frac{\cos 38° . \text{cosec } 52°}{\tan 18° . \tan 35° . \tan 60° . \tan 72° . \tan 55°}$

$$= \frac{\sin(90° - 58°)}{\sin 32°} + \frac{\cos(90° - 22°)}{\cos 68°} - \frac{\cos 38° . \sec(90° - 52°)}{\tan 18° . \tan 35° . \sqrt{3} \times \cot(90° - 72°) \cot(90° - 55°)}$$

$$= \frac{\sin 32°}{\sin 32°} + \frac{\cos 68°}{\cos 68°} - \frac{\frac{1}{\sec 38°} \times \sec 38°}{\frac{1}{\cot 18°} \times \frac{1}{\cot 35°} \times \sqrt{3} \times \cot 18° . \cot 35°}$$

$$= 1 + 1 - \frac{1}{1 \times \sqrt{3}} = 2 - \frac{1}{\sqrt{3}} = \frac{2\sqrt{3} - 1}{\sqrt{3}}$$

14. $\frac{2}{3}\text{cosec}^2 58° - \frac{2}{3}\cot 58° \tan 32° - \frac{5}{3}$ tan 13° tan 37° tan 45° tan 53° tan 77°

$$= \frac{2}{3}\{\text{cosec}^2 58° - \cot 58° \cot(90° - 32°)\} - \frac{5}{3} \tan 13°.\tan 37° \times 1 \times \cot(90° - 53°) \times \cot(90° - 77°)$$

$$= \frac{2}{3}(\text{cosec}^2 58° - \cot 58°.\cot 58°) - \frac{5}{3}\tan 13°.\tan 37°.\cot 37°.\cot 13°$$

$$= \frac{2}{3} \times 1 - \frac{5}{3}\tan 13°.\tan 37° \times \frac{1}{\tan 37°} \times \frac{1}{\tan 13°}$$

$$= \frac{2}{3} - \frac{5}{3} \times 1 = -1$$

15. LHS $= \sec^2\theta - \frac{\sin^2\theta - 2\sin^4\theta}{2\cos^4\theta - \cos^2\theta}$

$$= \sec^2\theta - \frac{\sin^2\theta(1 - 2\sin^2\theta)}{\cos^2\theta(2\cos^2\theta - 1)}$$

$$= \sec^2\theta - \frac{\sin^2\theta(\cos^2\theta - \sin^2\theta)}{\cos^2\theta(\cos^2\theta - \sin^2\theta)} = \frac{1}{\cos^2\theta} - \frac{\sin^2\theta}{\cos^2\theta}$$

$$= \frac{1 - \sin^2\theta}{\cos^2\theta} = \frac{\cos^2\theta}{\cos^2\theta} = 1 = \text{RHS}$$

Hence proved.

16. Let AC = x cm, AB = y cm, $\therefore x - y = 1$ ...(*i*)

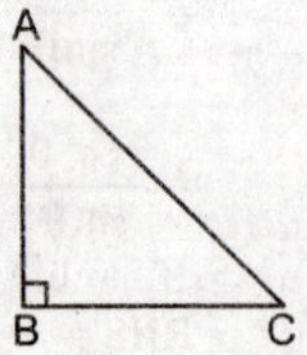

Now, $AC^2 = AB^2 + BC^2$

$\Rightarrow x^2 = y^2 + BC^2$

$\Rightarrow x^2 - y^2 = (7)^2$

$\Rightarrow (x - y)(x + y) = 49$

$\Rightarrow 1 \times (x + y) = 49$ [Using (*i*)]

$\Rightarrow x + y = 49$ ...(*ii*)

Solving (*i*) and (*ii*), we get

$x = 25$ and $y = 24$

Now, cos A + sin A

$$= \frac{AB}{AC} + \frac{BC}{AC} = \frac{24}{25} + \frac{7}{25} = \frac{31}{25}$$

17. Here, $\cos\theta - \sin\theta = \sqrt{2}\sin\theta$

$\Rightarrow \cos\theta = \sqrt{2}\sin\theta + \sin\theta$

$\Rightarrow \cos\theta = (\sqrt{2} + 1)\sin\theta.$

$\Rightarrow \frac{\cos\theta}{\sqrt{2} + 1} = \sin\theta.$

$\Rightarrow \frac{\cos\theta(\sqrt{2} - 1)}{(\sqrt{2} + 1)(\sqrt{2} - 1)} = \sin\theta$

$\Rightarrow \sqrt{2}\cos\theta - \cos\theta = \sin\theta$

$\Rightarrow \sqrt{2}\cos\theta = \sin\theta + \cos\theta$

Hence proved.

18. cosec (A – B) = 2

Also, cosec 30° = 2 $\Rightarrow$ A – B = 30° ...(*i*)

and $\cot(A + B) = \frac{1}{\sqrt{3}}$

Also, $\cot 60° = \frac{1}{\sqrt{3}} \Rightarrow A + B = 60°$ ...(*ii*)

Solving (*i*) and (*ii*), we get A = 45°, B = 15°

19. $4\left(\frac{\sec^2 59° - \cot^2 31°}{3}\right) - \frac{2}{3}\sin 90° + 3\tan^2 56° \tan^2 34° = \frac{x}{3}$

$$\Rightarrow 4\left(\frac{\sec^2 59° - \tan^2(90° - 31°)}{3}\right) - \frac{2}{3} \times 1 + 3\tan^2 56° \cot^2(90° - 34°) = \frac{x}{3}$$

$$\Rightarrow 4\left(\frac{\sec^2 59° - \tan^2 59°}{3}\right) - \frac{2}{3} + 3\tan^2 56° \cot^2 56° = \frac{x}{3}$$

$$\Rightarrow 4 \times \frac{1}{3} - \frac{2}{3} + 3 \times 1 = \frac{x}{3} \Rightarrow \frac{4}{3} - \frac{2}{3} + 3 = \frac{x}{3} \Rightarrow x = 11$$

20. LHS $= \left(\frac{(1 + \sin\theta) - \cos\theta}{(1 + \sin\theta) + \cos\theta}\right)^2$

$$= \frac{(1 + \sin\theta)^2 + (\cos\theta)^2 - 2\cos\theta(1 + \sin\theta)}{(1 + \sin\theta)^2 + (\cos\theta)^2 + 2\cos\theta(1 + \sin\theta)}$$

$$= \frac{1 + \sin^2\theta + 2\sin\theta + \cos^2\theta - 2\cos\theta(1 + \sin\theta)}{1 + \sin^2\theta + 2\sin\theta + \cos^2\theta + 2\cos\theta(1 + \sin\theta)}$$

$$= \frac{1 + 1 + 2\sin\theta - 2\cos\theta(1 + \sin\theta)}{1 + 1 + 2\sin\theta + 2\cos\theta(1 + \sin\theta)} \quad (\because \sin^2\theta + \cos^2\theta = 1)$$

$$= \frac{2 + 2\sin\theta - 2\cos\theta(1 + \sin\theta)}{2 + 2\sin\theta + 2\cos\theta(1 + \sin\theta)}$$

$$= \frac{2(1 + \sin\theta) - 2\cos\theta(1 + \sin\theta)}{2(1 + \sin\theta) + 2\cos\theta(1 + \sin\theta)}$$

$$= \frac{2(1 + \sin\theta)}{2(1 + \sin\theta)}\left[\frac{1 - \cos\theta}{1 + \cos\theta}\right] = \frac{1 - \cos\theta}{1 + \cos\theta}$$

= RHS. Hence proved.

21. $5\sin\theta + 3\cos\theta = 4$

$\Rightarrow (5\sin\theta + 3\cos\theta)^2 = (4)^2$

$\Rightarrow 25\sin^2\theta + 9\cos^2\theta + 30\sin\theta\cos\theta = 16$

$\Rightarrow 25(1 - \cos^2\theta) + 9(1 - \sin^2\theta) + 30\sin\theta\cos\theta = 16$

$\Rightarrow (3\sin\theta - 5\cos\theta)^2 = 18$

$\Rightarrow 3\sin\theta - 5\cos\theta = \pm 3\sqrt{2}$

22. LHS $= (\tan A + \text{cosec } B)^2 - (\cot B - \sec A)^2$

$= \tan^2 A + \text{cosec}^2 B + 2\tan A . \text{cosec } B - (\cot^2 B + \sec^2 A - 2\cot B . \sec A)$

$= \tan^2 A + \text{cosec}^2 B + 2\tan A . \text{cosec } B - \cot^2 B - \sec^2 A + 2\cot B \sec A$

$= (\text{cosec}^2 B - \cot^2 B) - (\sec^2 A - \tan^2 A) + 2\tan A . \text{cosec } B + 2\cot B \sec A$

$= 1 - 1 + 2\tan A . \text{cosec } B + 2\cot B . \sec A$

$= 2\tan A . \text{cosec } B + 2\cot B . \sec A$

$$= 2\left(\frac{\sin A}{\cos A} \times \frac{1}{\sin B} + \frac{\cos B}{\sin B} \times \frac{1}{\cos A}\right)$$

$$= 2\frac{\sin A + \cos B}{\cos A . \sin B} \quad ...(i)$$

RHS $= 2 \tan A . \cot B . (\text{cosec } A + \sec B)$

$= 2 \tan A . \cot B \text{ cosec } A + 2 \tan A \cot B . \sec B$

$$= 2\frac{\sin A}{\cos A} . \frac{\cos B}{\sin B} \times \frac{1}{\sin A} + \frac{2\sin A}{\cos A} . \frac{\cos B}{\sin B} . \frac{1}{\cos B}$$

$$= 2\left[\frac{\cos B}{\cos A . \sin B} + \frac{\sin A}{\cos A . \sin B}\right]$$

$$= 2\left[\frac{\cos B + \sin A}{\cos A . \sin B}\right] \quad ...(ii)$$

From (*i*) and (*ii*),

LHS = RHS Hence proved.

23. Convert sec A.cosec A into $\frac{1}{\cos A}, \frac{1}{\sin A}$ respectively and take LCM.

24. Simplify each term separately

$$\frac{\cos^2 A}{\sin^2 A}\left(\frac{\sec A - 1}{1 + \sin A}\right) + \sec^2 A\left(\frac{\sin A - 1}{1 + \sec A}\right)$$

$$\frac{(1 - \sin A)(\sec A - 1)}{(1 - \cos^2 A)} + \sec^2 A\left(\frac{\sin A - 1}{1 + \sec A}\right)$$

$$= \frac{(1 - \sin A)\sec^2 A}{(1 + \sec A)} + \frac{\sec^2 A(1 - \sin A)}{1 + \sec A} = 0$$

25. LHS $= \frac{(1 + \cot A + \tan A)(\sin A - \cos A)}{\sec^3 A - \text{cosec}^3 A}$

$$= \frac{\left(1 + \frac{\cos A}{\sin A} + \frac{\sin A}{\cos A}\right)(\sin A - \cos A)}{\frac{1}{\cos^3 A} - \frac{1}{\sin^3 A}}$$

$$= \frac{\frac{(\sin A \cos A + \cos^2 A + \sin^2 A)(\sin A - \cos A)}{\sin A . \cos A}}{\frac{\sin^3 A - \cos^3 A}{\sin^3 A . \cos^3 A}}$$

$$= \frac{\sin^3 A - \cos^3 A}{\sin A \cos A} \times \frac{\sin^3 A . \cos^3 A}{\sin^3 A - \cos^3 A}$$

[Using $(a - b)(a^2 + b^2 + ab) = a^3 - b^3$)]

$$= \frac{\sin^3 A . \cos^3 A}{\sin A . \cos A} = \sin^2 A . \cos^2 A = \text{RHS}$$

26. LHS $= \frac{1}{\sec A - 1} + \frac{1}{\sec A + 1}$

$$= \frac{\sec A + 1 + \sec A - 1}{(\sec A - 1)(\sec A + 1)} = \frac{2 \sec A}{\sec^2 A - 1}$$

$$= \frac{2 \sec A}{\tan^2 A} = \frac{2 \times \frac{1}{\cos A}}{\frac{\sin^2 A}{\cos^2 A}} = \frac{2 \times 1}{\cos A} \times \frac{\cos^2 A}{\sin^2 A}$$

$$= 2 \times \cos A \times \frac{1}{\sin A . \sin A}$$

$= 2 \cot A . \text{cosec } A = \text{RHS}$

27. LHS $= \frac{1}{\cot^2 \theta} + \frac{1}{1 + \tan^2 \theta}$

$$= \tan^2\theta + \frac{1}{\sec^2 \theta} = \tan^2\theta + \cos^2\theta$$

$$= \frac{\sin^2 \theta}{\cos^2 \theta} + \cos^2\theta = \frac{\sin^2 \theta + \cos^4 \theta}{\cos^2 \theta} \quad ...(i)$$

RHS $= \frac{1}{1 - \sin^2 \theta} - \frac{1}{\text{cosec}^2\theta} = \frac{1}{\cos^2 \theta} - \sin^2\theta$

$$= \frac{1 - \sin^2 \theta . \cos^2 \theta}{\cos^2 \theta} = \frac{\sin^2 \theta + \cos^2 \theta - \sin^2 \theta . \cos^2 \theta}{\cos^2 \theta}$$

$$= \frac{\sin^2 \theta + \cos^2 \theta (1 - \sin^2 \theta)}{\cos^2 \theta} = \frac{\sin^2 \theta + \cos^2 \theta . \cos^2 \theta}{\cos^2 \theta}$$

$$= \frac{\sin^2 \theta + \cos^4 \theta}{\cos^2 \theta} \quad ...(ii)$$

From (*i*) and (*ii*), LHS = RHS Hence proved

28. LHS $= \frac{\tan^3 \theta}{1 + \tan^2 \theta} + \frac{\cot^3 \theta}{1 + \cot^2 \theta}$

$$= \frac{\sin^3 \theta}{\cos^3 \theta} . \cos^2 \theta + \frac{\cos^3 \theta}{\sin^3 \theta} . \sin^2 \theta$$

$$= \frac{\sin^3 \theta}{\cos \theta} + \frac{\cos^3 \theta}{\sin \theta} = \frac{\sin^4 \theta + \cos^4 \theta}{\cos \theta . \sin \theta}$$

$$= \frac{(\sin^2 \theta + \cos^2 \theta)^2 - 2\sin^2 \theta . \cos^2 \theta}{\sin \theta . \cos \theta} \quad [a^2 + b^2 = (a + b)^2 - 2ab]$$

$$= \frac{1 - 2\sin^2 \theta . \cos^2 \theta}{\sin \theta . \cos \theta}$$

$$= \frac{1}{\sin \theta . \cos \theta} - \frac{2\sin^2 \theta . \cos^2 \theta}{\sin \theta . \cos \theta}$$

$= \sec \theta.\text{cosec } \theta - 2 \sin \theta.\cos \theta = \text{RHS}.$ Hence proved.

29. Simplify each term as

$$\frac{\cos^2 \theta}{\sin^2 \theta(1 + \cos^2 \theta)} + \frac{\sin^2 \theta}{\cos^2 \theta(1 + \sin^2 \theta)}$$

$$= \frac{\cos^4 \theta + \sin^4 \theta + \sin^2 \theta \cos^2 \theta}{(1 + \cos^2 \theta)(1 + \sin^2 \theta)} = \frac{1 - \sin^2 \theta \cos^2 \theta}{2 + \sin^2 \theta \cos^2 \theta}$$

30. LHS $= \frac{\sin A}{\sec A + \tan A - 1} + \frac{\cos A}{\text{cosec } A + \cot A - 1}$

$$= \frac{\sin A}{\frac{1}{\cos A} + \frac{\sin A}{\cos A} - 1} + \frac{\cos A}{\frac{1}{\sin A} + \frac{\cos A}{\sin A} - 1}$$

$$= \frac{\sin A}{\frac{1 + \sin A - \cos A}{\cos A}} + \frac{\cos A}{\frac{1 + \cos A - \sin A}{\sin A}}$$

$$= \frac{\sin A \cos A}{1 + \sin A - \cos A} + \frac{\sin A \cos A}{1 + \cos A - \sin A}$$

$$= \sin A \cos A\left[\frac{1}{1 + \sin A - \cos A} + \frac{1}{1 + \cos A - \sin A}\right]$$

$$= \sin A. \cos A\left[\frac{1 + \cos A - \sin A + 1 + \sin A - \cos A}{(1 + \sin A - \cos A)(1 + \cos A - \sin A)}\right]$$

$$= \sin A.\cos A\left[\frac{2}{(1+\sin A-\cos A)(1+\cos A-\sin A)}\right]$$

$$= \left[\frac{2\sin A.\cos A}{\{1+(\sin A-\cos A)\}\{1-(\sin A-\cos A)\}}\right]$$

$$= \frac{2\sin A.\cos A}{(1)^2-(\sin A-\cos A)^2}$$

$$= \frac{2\sin A.\cos A}{1-(\sin^2 A+\cos^2 A-2\sin A.\cos A)}$$

$$= \frac{2\sin A.\cos A}{1-(1-2\sin A.\cos A)} = \frac{2\sin A.\cos A}{2\sin A.\cos A}$$

= 1 = RHS. Hence Proved.

31. $\text{LHS} = \dfrac{\cot^2 A(\sec A-1)}{1+\sin A}$

$$= \frac{\cot^2 A(\sec A-1)(\sec A+1)(1-\sin A)}{(\sec A+1)\times(1+\sin A)(1-\sin A)}$$

$$= \frac{\cot^2 A(\sec^2 A-1)(1-\sin A)}{(\sec A+1)(1-\sin^2 A)}$$

$$\frac{\frac{1}{\tan^2 A}\times\tan^2 A(1-\sin A)}{(\sec A+1)\times\cos^2 A}$$

$$= \frac{(1-\sin A)}{(\sec A+1)}\times\sec^2 A = \text{RHS}$$

32. LHS = (1 + cot A + tan A) (sin A – cos A)

$$= \left(1+\frac{\cos A}{\sin A}+\frac{\sin A}{\cos A}\right)(\sin A-\cos A)$$

$$= \frac{(\sin A.\cos A+\cos^2 A+\sin^2 A)}{\sin A.\cos A}(\sin A-\cos A)$$

$$= \frac{\sin^3 A-\cos^3 A}{\sin A\cdot\cos A} \qquad [\text{Using } (a-b)(a^2+b^2+ab)=a^3-b^3]$$

$$= \frac{\sin^3 A}{\sin A\cos A}-\frac{\cos^3 A}{\sin A.\cos A}$$

$$= \sin^2 A\times\frac{1}{\cos A}-\cos^2 A\times\frac{1}{\sin A}$$

$$= \frac{1}{\text{cosec}^2 A}\times\sec A-\frac{1}{\sec^2 A}\times\text{cosec } A$$

$$= \frac{\sec A}{\text{cosec}^2 A}-\frac{\text{cosec } A}{\sec^2 A} = \text{RHS}$$

33. $3(\sin\theta-\cos\theta)^4+6(\sin\theta+\cos\theta)^2+4(\sin^6\theta+\cos^6\theta)$

$= 3[(\sin\theta-\cos\theta)^2]^2+6(\sin^2\theta+\cos^2\theta+2\sin\theta.\cos\theta)$
$+4[(\sin^2\theta+\cos^2\theta)^3-3\sin^2\theta.\cos^2\theta(\sin^2\theta+\cos^2\theta)]$

$= 3[\sin^2\theta+\cos^2\theta-2\sin\theta.\cos\theta]^2+6[1+2\sin\theta.\cos\theta]+$
$4(1-3\sin^2\theta.\cos^2\theta)$

$= 3(1+4\sin^2\theta.\cos^2\theta-4\sin\theta\cos\theta)+6+12\sin\theta.$
$\cos\theta+4-12\sin^2\theta\cos^2\theta$

$= 13 \Rightarrow$ independent of θ.

34. $\sin^2\theta = 1-\cos^2\theta$

$\sin^2\theta+\cos^2\theta = 1$

∴ It is an identity which is satisfied for all values of θ.

35. $\text{LHS} = \dfrac{1}{\dfrac{1}{\cos x}-\dfrac{\sin x}{\cos x}}-\dfrac{1}{\cos x} = \dfrac{\cos x}{1-\sin x}-\dfrac{1}{\cos x}$

$$= \frac{\cos x}{1-\sin x}\times\frac{1+\sin x}{1+\sin x}-\frac{1}{\cos x}$$

$$= \frac{\cos x(1+\sin x)}{1-\sin^2 x}-\frac{1}{\cos x}$$

$$= \frac{\cos x(1+\sin x)}{\cos^2 x}-\frac{1}{\cos x}$$

$$= \frac{1+\sin x}{\cos x}-\frac{1}{\cos x} = \frac{1+\sin x-1}{\cos x}$$

$$= \frac{\sin x}{\cos x} = \tan x$$

$\text{RHS} = \dfrac{1}{\cos x}-\dfrac{1}{\dfrac{1}{\cos x}+\dfrac{\sin x}{\cos x}} = \dfrac{1}{\cos x}-\dfrac{\cos x}{1+\sin x}$

$$= \frac{1}{\cos x}-\frac{\cos x(1-\sin x)}{1-\sin^2 x}$$

$$= \frac{1}{\cos x}-\frac{\cos x(1-\sin x)}{\cos^2 x}$$

$$= \frac{1}{\cos x}-\frac{(1-\sin x)}{\cos x} = \frac{1-1+\sin x}{\cos x} = \frac{\sin x}{\cos x} = \tan x$$

∴ LHS = RHS. Hence proved.

36. Since $\tan\theta = \pm\sqrt{\sec^2\theta-1}$

$$= \pm\sqrt{\left(x+\frac{1}{4x}\right)^2-1} = \pm\left(x-\frac{1}{4x}\right)$$

Now,

$$\sec\theta+\tan\theta = \left(x+\frac{1}{4x}\right)\pm\left(x-\frac{1}{4x}\right) = 2x \text{ or } \frac{1}{2x}$$

37. $\sin\theta = 1-\sin^2\theta \Rightarrow \sin\theta = \cos^2\theta$

Now, $\cos^2\theta+\cos^4\theta = \cos^2\theta(1+\cos^2\theta)$

$= \sin\theta(1+\sin\theta) = \sin\theta+\sin^2\theta = 1$

38. $\text{LHS} = \dfrac{\cos A}{1-\dfrac{\sin A}{\cos A}}-\dfrac{\sin^2 A}{\cos A-\sin A}$

$$= \frac{\cos^2 A}{\cos A-\sin A}-\frac{\sin^2 A}{\cos A-\sin A}$$

$$= \frac{\cos^2 A-\sin^2 A}{\cos A-\sin A}$$

$$= \frac{(\cos A+\sin A)(\cos A-\sin A)}{(\cos A-\sin A)}$$

= sin A + cos A = RHS. Hence proved.

39. LHS = sec A (1 – sin A) (sec A + tan A)

$$= \frac{1}{\cos A}(1-\sin A)\left(\frac{1}{\cos A}+\frac{\sin A}{\cos A}\right)$$

$$= \frac{1}{\cos A}(1-\sin A)\left(\frac{1+\sin A}{\cos A}\right)$$

$$= \frac{1-\sin^2 A}{\cos^2 A} = \frac{\cos^2 A}{\cos^2 A} = 1 = \text{RHS}.$$

40. $\sec\theta + \tan\theta = p$

$$\Rightarrow \frac{1}{\cos\theta} + \frac{\sin\theta}{\cos\theta} = p \Rightarrow \frac{1+\sin\theta}{\cos\theta} = p$$

$$\Rightarrow \frac{(1+\sin\theta)^2}{\cos^2\theta} = p^2$$

$$\Rightarrow \frac{(1+\sin\theta)^2}{1-\sin^2\theta} = p^2 \Rightarrow \frac{1+\sin\theta}{1-\sin\theta} = p^2$$

Using componendo and dividendo, we get

$$\frac{1+\sin\theta+1-\sin\theta}{1+\sin\theta-1+\sin\theta} = \frac{p^2+1}{p^2-1}$$

$$\Rightarrow \frac{2}{2\sin\theta} = \frac{p^2+1}{p^2-1} \Rightarrow \sin\theta = \frac{p^2-1}{p^2+1}$$ Hence proved.

41. $\text{LHS} = \frac{\tan\theta}{1-\cot\theta} + \frac{\cot\theta}{1-\tan\theta}$

$$= \frac{\tan\theta}{1-\frac{1}{\tan\theta}} + \frac{\frac{1}{\tan\theta}}{1-\tan\theta} = \frac{\tan^2\theta}{\tan\theta-1} + \frac{1}{\tan\theta(1-\tan\theta)}$$

$$= \frac{\tan^2\theta}{\tan\theta-1} - \frac{1}{\tan\theta(\tan\theta-1)} = \frac{\tan^3\theta-1}{\tan\theta(\tan\theta-1)}$$

$$= \frac{(\tan\theta-1)(\tan^2\theta+\tan\theta+1)}{\tan\theta(\tan\theta-1)} = \frac{\tan^2\theta+\tan\theta+1}{\tan\theta}$$

$$= \frac{\tan^2\theta}{\tan\theta} + \frac{\tan\theta}{\tan\theta} + \frac{1}{\tan\theta} = \tan\theta + 1 + \cot\theta = \text{RHS}.$$

42. $\text{LHS} = m^2 - n^2$

$= (\tan\theta + \sin\theta)^2 - (\tan\theta - \sin\theta)^2$ (Given)

$= \tan^2\theta + \sin^2\theta + 2\tan\theta\sin\theta - \tan^2\theta - \sin^2\theta + 2\tan\theta\sin\theta$

$= 4\tan\theta\sin\theta$

$= 4\sqrt{\tan^2\theta.\sin^2\theta} = 4\sqrt{\tan^2\theta(1-\cos^2\theta)}$

$= 4\sqrt{\tan^2\theta - \sin^2\theta} = 4\sqrt{(\tan\theta+\sin\theta)(\tan\theta-\sin\theta)}$

$= 4\sqrt{mn}$ (From given)

= RHS.

43. $\operatorname{cosec}\theta - \sin\theta = l$

$$\Rightarrow \frac{1}{\sin\theta} - \sin\theta = l$$

$$\Rightarrow \frac{1-\sin^2\theta}{\sin\theta} = l \quad \Rightarrow \frac{\cos^2\theta}{\sin\theta} = l$$

and $\sec\theta - \cos\theta = m$

$$\Rightarrow \frac{1}{\cos\theta} - \cos\theta = m \quad \Rightarrow \frac{1-\cos^2\theta}{\cos\theta} = m$$

$$\Rightarrow \frac{\sin^2\theta}{\cos\theta} = m$$

$\text{LHS} = l^2m^2[l^2 + m^2 + 3]$

$$= \left(\frac{\cos^2\theta}{\sin\theta}\right)^2\left(\frac{\sin^2\theta}{\cos\theta}\right)^2\left[\frac{\cos^4\theta}{\sin^2\theta} + \frac{\sin^4\theta}{\cos^2\theta} + 3\right]$$

$$= \cos^2\theta \,.\, \sin^2\theta\left[\frac{\cos^6\theta + \sin^6\theta + 3\sin^2\theta.\cos^2\theta}{\sin^2\theta.\cos^2\theta}\right]$$

$= (\cos^2\theta + \sin^2\theta)^3 - 3\sin^2\theta \,.\, \cos^2\theta\,[\sin^2\theta + \cos^2\theta]$
$+ [3\cos^2\theta \,.\, \sin^2\theta]$

$= 1 - 3\sin^2\theta \,.\, \cos^2\theta \times 1 + 3\cos^2\theta \,.\, \sin^2\theta = 1 = \text{RHS}.$

44. $\text{LHS} = \frac{\tan A}{1-\cot A} + \frac{\cot A}{1-\tan A}$

$$= \frac{\tan A}{1-\frac{1}{\tan A}} + \frac{\frac{1}{\tan A}}{1-\tan A}$$

$$= \frac{\tan A}{\frac{\tan A-1}{\tan A}} + \frac{\frac{1}{\tan A}}{1-\tan A}$$

$$= \frac{\tan^2 A}{\tan A-1} + \frac{1}{\tan A(1-\tan A)}$$

$$= \frac{\tan^2 A}{(\tan A-1)} - \frac{1}{\tan A(\tan A-1)}$$

$$= \frac{\tan^3 A-1}{\tan A(\tan A-1)} = \frac{(\tan A-1)(\tan^2 A+1+\tan A)}{\tan A(\tan A-1)}$$

$$= \frac{\tan^2 A+1+\tan A}{\tan A} = \frac{\sec^2 A+\tan A}{\tan A} \quad (\because 1+\tan^2 A = \sec^2 A)$$

$$= \frac{\sec^2 A}{\tan A} + \frac{\tan A}{\tan A} = \frac{\frac{1}{\cos^2 A}}{\frac{\sin A}{\cos A}} + 1$$

$$= \frac{1}{\cos^2 A} \,.\, \frac{\cos A}{\sin A} + 1 = \frac{1}{\cos A\sin A} + 1$$

$= \sec A\operatorname{cosec} A + 1 = \text{RHS}.$ Hence proved.

45. $\text{LHS} = \frac{1+\cos\theta-\sin\theta}{\cos\theta-1+\sin\theta}$

$$= \frac{\frac{1}{\sin\theta}+\frac{\cos\theta}{\sin\theta}-\frac{\sin\theta}{\sin\theta}}{\frac{\cos\theta}{\sin\theta}-\frac{1}{\sin\theta}+\frac{\sin\theta}{\sin\theta}} = \frac{\operatorname{cosec}\theta+\cot\theta-1}{\cot\theta-\operatorname{cosec}\theta+1}$$

$$= \frac{(\operatorname{cosec}\theta+\cot\theta)-(\operatorname{cosec}^2\theta-\cot^2\theta)}{(\cot\theta-\operatorname{cosec}\theta+1)}$$

$$= \frac{(\operatorname{cosec}\theta+\cot\theta)-(\operatorname{cosec}\theta+\cot\theta)(\operatorname{cosec}\theta-\cot\theta)}{(\cot\theta-\operatorname{cosec}\theta+1)}$$

$$= \frac{(\operatorname{cosec}\theta+\cot\theta)[1-(\operatorname{cosec}\theta-\cot\theta)]}{\cot\theta-\operatorname{cosec}\theta+1}$$

$$= \frac{(\operatorname{cosec}\theta+\cot\theta)(1-\operatorname{cosec}\theta+\cot\theta)}{(\cot\theta-\operatorname{cosec}\theta+1)}$$

$= \operatorname{cosec}\theta + \cot\theta = \text{RHS}$

LONG ANSWER TYPE QUESTIONS

1. LHS $= \sqrt{\dfrac{1+\sin\theta}{1-\sin\theta}} + \sqrt{\dfrac{1-\sin\theta}{1+\sin\theta}}$

$$= \frac{(\sqrt{1+\sin\theta})^2 + (\sqrt{1-\sin\theta})^2}{\sqrt{1-\sin\theta}\times\sqrt{1+\sin\theta}} = \frac{1+\sin\theta+1-\sin\theta}{\sqrt{1-\sin^2\theta}}$$

$$= \frac{2}{\sqrt{\cos^2\theta}} = \frac{2}{\cos\theta} = 2\sec\theta = \text{RHS}$$

2. LHS $= \dfrac{\sec^2(90^\circ-\theta)-\cot^2\theta}{2(\sin^2 25^\circ+\sin^2 65^\circ)} + \dfrac{2\sin^2 30^\circ\cdot\tan^2 32^\circ\cdot\tan^2 58^\circ}{3(\sec^2 33^\circ-\cot^2 57^\circ)}$

$$= \frac{\text{cosec}^2\theta-\cot^2\theta}{2\{\sin^2 25^\circ+\cos^2(90^\circ-65^\circ)\}} + \frac{2\times\left(\frac{1}{2}\right)^2\times\tan^2 32^\circ\times\cot^2(90^\circ-58^\circ)}{3\{\sec^2 33^\circ-\tan^2(90^\circ-57^\circ)\}}$$

$$= \frac{1}{2(\sin^2 25^\circ+\cos^2 25^\circ)} + \frac{\frac{1}{2}\times\tan^2 32^\circ\times\cot^2 32^\circ}{3(\sec^2 33^\circ-\tan^2 33^\circ)}$$

$$= \frac{1}{2}+\frac{\frac{1}{2}\times 1}{3\times 1} = \frac{1}{2}+\frac{1}{6} = \frac{4}{6} = \frac{2}{3}$$

3. $2\text{cosec}^2 30^\circ + x\sin^2 60^\circ - \dfrac{3}{4}\tan^2 30^\circ = 10$

$$\Rightarrow 2\times(2)^2 + x\times\left(\frac{\sqrt{3}}{2}\right)^2 - \frac{3}{4}\times\left(\frac{1}{\sqrt{3}}\right)^2 = 10$$

$$\Rightarrow \quad 8 + x\times\frac{3}{4} - \frac{1}{4} = 10$$

$$\frac{3x}{4} = 10-8+\frac{1}{4} \quad \Rightarrow \quad \frac{3x}{4} = \frac{9}{4}$$

$$\Rightarrow \quad x = \frac{9}{4}\times\frac{4}{3} \quad \Rightarrow \quad x = 3$$

4. $A + B = 90^\circ \Rightarrow B = 90^\circ - A$

Now, LHS $= \sqrt{\dfrac{\tan A.\tan B+\tan A.\cot B}{\sin A.\sec B} - \dfrac{\sin^2 B}{\cos^2 A}}$

$$= \sqrt{\frac{\tan A.\tan(90^\circ-A)+\tan A.\cot(90^\circ-A)}{\sin A.\sec(90^\circ-A)} - \frac{\sin^2(90^\circ-A)}{\cos^2 A}}$$

$$= \sqrt{\frac{\tan A.\cot A+\tan A.\tan A}{\sin A.\text{cosec} A} - \frac{\cos^2 A}{\cos^2 A}}$$

$$= \sqrt{\frac{1+\tan^2 A}{1} - 1} = \sqrt{\tan^2 A} = \tan A = \text{RHS}$$

5. $\because \angle A + \angle B + \angle C = 180^\circ \Rightarrow \angle A + \angle B = 180^\circ - \angle C$

and $\quad \angle B + \angle C = 180^\circ - \angle A$

Now, $\quad \sin(A+B-C) = \dfrac{1}{2} \Rightarrow A+B-C = 30^\circ$

$\Rightarrow \quad 180^\circ - C - C = 30^\circ \quad \Rightarrow 2C = 150^\circ \Rightarrow C = 75^\circ$

Also $\quad \cos(B+C-A) = \dfrac{1}{\sqrt{2}} \quad \Rightarrow B+C-A = 45^\circ$

$\Rightarrow \quad 180^\circ - A - A = 45^\circ \quad \Rightarrow 2A = 135^\circ \Rightarrow A = 67.5^\circ$

$\therefore \quad \angle B = 37.5^\circ.$

$\angle A = 67.5^\circ, \angle B = 37.5^\circ, \angle C = 75^\circ.$

NCERT EXEMPLAR PROBLEMS

Short Answer Questions with Reasoning

1. False, $\because \cos^2 23^\circ - \sin^2 67^\circ$

$= \cos^2 23^\circ - \cos^2(90^\circ - 67^\circ)$

$= \cos^2 23^\circ - \cos^2 23^\circ = 0$

2. False, $\because \sin 80^\circ - \cos 80^\circ$

$= \sin 80^\circ - \sin(90^\circ - 80^\circ) = \sin 80^\circ - \sin 10^\circ$

$\because \sin 10^\circ$ is less than $\sin 80^\circ$

$\therefore (\sin 80^\circ - \sin 10^\circ)$ is positive.

3. True, $\sqrt{(1-\cos^2\theta)\sec^2\theta} = \tan\theta$

LHS $= \sqrt{\sin^2\theta\times\dfrac{1}{\cos^2\theta}} = \sqrt{\tan^2\theta} = \tan\theta = \text{RHS}.$

4. True, $(\tan\theta + 2)(2\tan\theta + 1)$

$= 2\tan^2\theta + 4\tan\theta + \tan\theta + 2$

$= 2\tan^2\theta + 5\tan\theta + 2$

$= 2(\tan^2\theta + 1) + 5\tan\theta$

$= 2\sec^2\theta + 5\tan\theta \qquad (\because 1+\tan^2\theta = \sec^2\theta)$

5. False, In ΔABC,

$$\frac{h-3}{x} = \tan\theta \quad ...(i)$$

In ΔACF

$$\frac{h+3}{x} = \tan\phi \quad ...(ii)$$

B, A, θ, C, x, φ, $(h-3)$m, h m, 3m, D, E, 3m, h m, h m, F

If angle of elevation equal to the angle of depression.

Then $\quad \theta = \phi$

Then from equation (*i*) and (*ii*)

$$\frac{h-3}{x} = \frac{h+3}{x}$$

h cannot be calculated; so false.

6. False, $\because a + \dfrac{1}{a}$ is more than or equal to 2.

$a \neq 1 \therefore a + \dfrac{1}{a}$ is more than 2.

and 2 sinθ is less than or equal to 2.

7. False, $(a-b)^2 = a^2 + b^2 - 2ab$.

$\because$ a and b are two distinct numbers

$$\therefore \quad (a-b)^2 > 0$$
$$\Rightarrow \quad a^2 + b^2 - 2ab > 0$$
$$\Rightarrow \quad a^2 + b^2 > 2ab$$
$$\Rightarrow \quad \frac{a^2+b^2}{2ab} > 1$$

But cosθ is never more than 1.

$$\therefore \quad \cos\theta \neq \frac{a^2+b^2}{2ab},$$

Short Answer Questions

8. $(\sqrt{3}+1)\ (3-\cot 30°) = \tan^3 60° - 2\sin 60°$

$$\Rightarrow (\sqrt{3}+1)(3-\sqrt{3}) = (\sqrt{3})^3 - 2\left(\frac{\sqrt{3}}{2}\right)$$
$$\Rightarrow \sqrt{3}(\sqrt{3}+1)(\sqrt{3}-1) = 3\sqrt{3}-\sqrt{3}$$
$$\Rightarrow \sqrt{3}\{(\sqrt{3})^2-(1)^2\} = 3\sqrt{3}-\sqrt{3}$$
$$\Rightarrow \sqrt{3}(3-1) = 3\sqrt{3}-\sqrt{3}$$
$$\Rightarrow 2\sqrt{3} = 2\sqrt{3}$$
$$\therefore \text{LHS} = \text{RHS.}$$

9. $\tan\theta + \tan(90°-\theta) = \sec\theta \sec(90°-\theta)$

$$\Rightarrow \tan\theta + \cot\theta = \sec\theta \,.\, \text{cosec}\,\theta$$
$$\Rightarrow \frac{\sin\theta}{\cos\theta}+\frac{\cos\theta}{\sin\theta} = \frac{1}{\cos\theta}\cdot\frac{1}{\sin\theta}$$
$$\Rightarrow \frac{\sin^2\theta+\cos^2\theta}{\sin\theta\cos\theta} = \frac{1}{\sin\theta\cos\theta}$$
$$\Rightarrow \frac{1}{\sin\theta.\cos\theta} = \frac{1}{\sin\theta.\cos\theta}$$
$$\therefore \text{LHS} = \text{RHS} \qquad (\because \sin^2\theta+\cos^2\theta = 1)$$

10. $\dfrac{\cos^2(45°+\theta)+\cos^2(45°-\theta)}{\tan(60°+\theta)\ \tan(30°-\theta)} = 1$

$$\Rightarrow \frac{\cos^2(45°+\theta)+\cos^2\{90°-(45°-\theta)\}}{\tan\{90°-(30°-\theta)\}\ \tan(30°-\theta)} = 1$$
$$\Rightarrow \frac{\cos^2(45°+\theta)+\sin^2(45°+\theta)}{\cot(30°-\theta)\ \tan(30°-\theta)} = \frac{1}{1} = 1$$
$$\Rightarrow \text{LHS} = \text{RHS.}$$

Long Answer Questions

11. $1+\sin^2\theta = 3\sin\theta\cos\theta$

Dividing both sides by $\cos^2\theta$, we get

$$\Rightarrow \frac{1}{\cos^2\theta}+\frac{\sin^2\theta}{\cos^2\theta} = \frac{3\sin\theta\cos\theta}{\cos^2\theta}$$
$$\Rightarrow \sec^2\theta+\tan^2\theta = 3\tan\theta \qquad (\because \sec^2\theta = 1+\tan^2\theta)$$
$$\Rightarrow 1+\tan^2\theta+\tan^2\theta-3\tan\theta = 0$$
$$\Rightarrow 2\tan^2\theta-3\tan\theta+1 = 0$$
$$\Rightarrow 2\tan^2\theta-2\tan\theta-\tan\theta+1 = 0$$
$$\Rightarrow 2\tan\theta(\tan\theta-1)-1(\tan\theta-1) = 0$$
$$\Rightarrow (\tan\theta-1)(2\tan\theta-1) = 0$$
$$\Rightarrow \tan\theta = 1 \text{ or } \frac{1}{2}.$$

Hence proved.

12. $\sin\theta+2\cos\theta = 1$

Squaring both sides, we get

$$(\sin\theta+2\cos\theta)^2 = 1^2$$
$$\Rightarrow \sin^2\theta+4\cos^2\theta+4\sin\theta\cos\theta = 1$$
$$\Rightarrow 1-\cos^2\theta+4(1-\sin^2\theta)+4\sin\theta\cos\theta = 1$$
$$\Rightarrow -\cos^2\theta-4\sin^2\theta+4\sin\theta\cos\theta = 1-5$$
$$\Rightarrow \cos^2\theta+4\sin^2\theta-4\sin\theta\cos\theta = 4$$
$$\Rightarrow (2\sin\theta-\cos\theta)^2 = 4$$

Taking square root of both sides, we get

$$2\sin\theta-\cos\theta = 2$$

Hence proved.

13. $\tan\theta+\sec\theta = l$

$$\Rightarrow \tan\theta = l-\sec\theta$$

Squaring both sides, we get

$$\tan^2\theta = (l-\sec\theta)^2$$
$$\Rightarrow \tan^2\theta = l^2+\sec^2\theta-2l\sec\theta$$
$$\Rightarrow \sec^2\theta-1 = \sec^2\theta+l^2-2l\sec\theta$$
$$\Rightarrow 2l\sec\theta = l^2+1$$
$$\Rightarrow \sec\theta = \frac{l^2+1}{2l}$$

Hence proved.

HOTS QUESTIONS

1. $\cot A+\tan A = 2$

On squaring both sides, we get

$$\cot^2 A+\tan^2 A+2 = 4$$
$$\cot^2 A+\frac{1}{\cot^2 A} = 2$$

2. We have, $\dfrac{\cos\theta+\sin\theta}{\cos\theta-\sin\theta}$

Dividing numerator and denominator by cos θ, we get

$\dfrac{1+\tan\theta}{1-\tan\theta}$, putting values of tan θ, we get

$$\frac{1+\tan\theta}{1-\tan\theta} = \frac{1+\frac{a}{b}}{1-\frac{a}{b}} = \frac{b+a}{b-a}$$

3. $x\sin^3\theta+y\cos^3\theta = \sin\theta\cos\theta \qquad ...(i)$

$x\sin\theta-y\cos\theta = 0 \qquad ...(ii)$

Equation $(ii)\times\sin^2\theta$ and subtracted from (i) we have

$$y(\cos^3\theta+\cos\theta\sin^2\theta) = \sin\theta\cos\theta$$
$$\Rightarrow y\cos\theta = \sin\theta\cos\theta$$
$$\Rightarrow y = \sin\theta$$

Put in equation (ii), we get

$$x\sin\theta = \sin\theta\cos\theta$$
$$\Rightarrow x = \cos\theta$$
$$\Rightarrow x^2+y^2 = \sin^2\theta+\cos^2\theta = 1$$

4. We have $7\,\text{cosec}\,\theta-3\cot\theta = 7$

Squaring both sides, we get

$$49\,\text{cosec}^2\theta+9\cot^2\theta-42\,\text{cosec}\,\theta\cot\theta = 49$$

$\Rightarrow 9\cot^2\theta - 42\,\text{cosec}\,\theta\cot\theta = 49(1 - \text{cosec}^2\theta)$

$\Rightarrow 9\cot^2\theta - 42\,\text{cosec}\,\theta\cot\theta = -49\cot^2\theta$

$\Rightarrow 9\,\text{cosec}^2\theta - 9 - 42\,\text{cosec}\,\theta\cot\theta + 49\cot^2\theta = 0$

$\Rightarrow 49\cot^2\theta + 9\,\text{cosec}^2\theta - 42\,\text{cosec}\,\theta\cot\theta = 9$

$\Rightarrow (7\cot\theta - 3\,\text{cosec}\,\theta)^2 = 9$

$\Rightarrow 7\cot\theta - 3\,\text{cosec}\,\theta = \pm 3$

5. $a\cos\theta - b\sin\theta = c$

$\Rightarrow (a\cos\theta - b\sin\theta)^2 = c^2$

$\Rightarrow a^2\cos^2\theta + b^2\sin^2\theta - 2ab\cos\theta\,.\,\sin\theta = c^2$

$\Rightarrow a^2(1-\sin^2\theta) + b^2(1-\cos^2\theta) - 2ab\cos\theta\,.\,\sin\theta = c^2$

$\Rightarrow a^2 - a^2\sin^2\theta + b^2 - b^2\cos^2\theta - 2ab\cos\theta\,.\,\sin\theta = c^2$

$\Rightarrow a^2\sin^2\theta + b^2\cos^2\theta + 2ab\cos\theta\,.\,\sin\theta = a^2 + b^2 - c^2$

$\Rightarrow (a\sin\theta + b\cos\theta)^2 = a^2 + b^2 - c^2$

$\Rightarrow a\sin\theta + b\cos\theta = \pm\sqrt{a^2 + b^2 - c^2}$

6. $\frac{x}{a}\cos\theta + \frac{y}{b}\sin\theta = 1$

Squaring both sides, we get

$$\frac{x^2}{a^2}\cos^2\theta + \frac{y^2}{b^2}\sin^2\theta + \frac{2xy}{ab}\cos\theta\sin\theta = 1 \quad \ldots(i)$$

and $\frac{x}{a}\sin\theta - \frac{y}{b}\cos\theta = 1$

Squaring both sides, we get

$$\frac{x^2}{a^2}\sin^2\theta + \frac{y^2}{b^2}\cos^2\theta - \frac{2xy}{ab}\cos\theta\sin\theta = 1 \quad \ldots(ii)$$

Adding (*i*) and (*ii*), we get

$$\frac{x^2}{a^2}(\cos^2\theta + \sin^2\theta) + \frac{y^2}{b^2}(\sin^2\theta + \cos^2\theta) = 2$$

$$\Rightarrow \frac{x^2}{a^2} + \frac{y^2}{b^2} = 2 \quad (\because \sin^2\theta + \cos^2\theta = 1)$$

Hence proved.

7. $x = a\cos^3\theta,\ y = b\sin^3\theta$

$$\text{LHS} = \left(\frac{x}{a}\right)^{2/3} + \left(\frac{y}{b}\right)^{2/3}$$

$$= \left(\frac{a\cos^3\theta}{a}\right)^{2/3} + \left(\frac{b\sin^3\theta}{b}\right)^{2/3}$$

$$= (\cos\theta)^{3\times 2/3} + (\sin\theta)^{3\times 2/3} = \cos^2\theta + \sin^2\theta = 1$$

LHS = RHS. Hence proved.

8.
$$\tan(A+B) = \frac{\tan A + \tan B}{1 - \tan A\,.\,\tan B}$$

$$\Rightarrow \tan(A+B) = \frac{\frac{1}{2}+\frac{1}{3}}{1-\frac{1}{2}\times\frac{1}{3}}$$

$$\Rightarrow \tan(A+B) = \frac{\frac{3+2}{6}}{\frac{6-1}{6}} = \frac{5}{5} = 1$$

$$\Rightarrow \tan(A+B) = 1 = \tan 45^\circ$$

$$\Rightarrow A+B = 45^\circ$$

Hence proved.

9. $\sin A + \text{cosec}\,A = 3$

$\Rightarrow (\sin A + \text{cosec}\,A)^2 = 3^2$

$\Rightarrow \sin^2 A + \text{cosec}^2 A + 2\sin A\,.\,\text{cosec}\,A = 9$

$$\Rightarrow \sin^2 A + \frac{1}{\sin^2 A} + 2\sin A\,.\,\frac{1}{\sin A} = 9$$

$$\Rightarrow \frac{\sin^4 A + 1}{\sin^2 A} + 2 = 9$$

$$\Rightarrow \frac{\sin^4 A + 1}{\sin^2 A} = 7$$

10.
$$\tan(A+B) = 1 = \tan 45^\circ$$

$$\Rightarrow A + B = 45^\circ \quad \ldots(i)$$

$$\sin(2A - B) = 1 = \sin 90^\circ$$

$$\Rightarrow 2A - B = 90^\circ \quad \ldots(ii)$$

Adding eq. (*i*) and (*ii*), we get

$$A + B = 45^\circ$$
$$2A - B = 90^\circ$$
$$3A = 135^\circ$$

$$\Rightarrow A = 45^\circ$$

From (*i*),

$$A + B = 45^\circ$$

$$\Rightarrow 45^\circ + B = 45^\circ \Rightarrow B = 0^\circ$$

$\therefore\ A = 45^\circ, B = 0^\circ$

11. $\cot B = \frac{n}{\tan A}$, $\text{cosec}\,B = \frac{m}{\sin A}$

Use $\text{cosec}^2 B - \cot^2 B = 1$.

12. $\text{cosec}\,\theta - \sin\theta = m$

$$\Rightarrow \frac{1}{\sin\theta} - \sin\theta = m$$

$$\Rightarrow \frac{1-\sin^2\theta}{\sin\theta} = m$$

$$\Rightarrow \frac{\cos^2\theta}{\sin\theta} = m \quad \ldots(i)$$

Also, $\sec\theta - \cos\theta = n$

$$\Rightarrow \frac{1}{\cos\theta} - \cos\theta = n$$

$$\Rightarrow \frac{1-\cos^2\theta}{\cos\theta} = n$$

$$\Rightarrow \frac{\sin^2\theta}{\cos\theta} = n \quad \ldots(ii)$$

Now, LHS $= (m^2n)^{2/3} + (mn^2)^{2/3}$

$$= \left\{\left(\frac{\cos^2\theta}{\sin\theta}\right)^2 \frac{\sin^2\theta}{\cos\theta}\right\}^{2/3} + \left\{\frac{\cos^2\theta}{\sin\theta}\times\left(\frac{\sin^2\theta}{\cos\theta}\right)^2\right\}^{2/3}$$

$$= \left(\frac{\cos^4\theta\,.\,\sin^2\theta}{\sin^2\theta\,.\,\cos\theta}\right)^{2/3} + \left(\frac{\cos^2\theta}{\sin\theta}\times\frac{\sin^4\theta}{\cos^2\theta}\right)^{2/3}$$

$$= (\cos^3\theta)^{2/3} + (\sin^3\theta)^{2/3} = \cos^2\theta + \sin^2\theta = 1 = \text{RHS}$$

13. $\cot\theta + \tan\theta = x$

$$\Rightarrow \left(\frac{\cos\theta}{\sin\theta} + \frac{\sin\theta}{\cos\theta}\right) = x$$

$$\Rightarrow \left(\frac{\cos^2\theta + \sin^2\theta}{\sin\theta \,.\, \cos\theta}\right) = x \Rightarrow \frac{1}{\sin\theta \,.\, \cos\theta} = x$$

$$\Rightarrow \quad \sec\theta \,.\, \operatorname{cosec}\theta = x \qquad ...(i)$$

Also $\sec\theta - \cos\theta = y$

$$\Rightarrow \frac{1}{\cos\theta} - \cos\theta = y \Rightarrow \frac{1-\cos^2\theta}{\cos\theta} = y$$

$$\Rightarrow \frac{\sin^2\theta}{\cos\theta} = y \Rightarrow \tan\theta \,.\, \sin\theta = y \qquad ...(ii)$$

$$\text{LHS} = (x^2y)^{2/3} - (xy^2)^{2/3}$$

$$= \{(\sec\theta\ \operatorname{cosec}\theta)^2 \tan\theta \,.\, \sin\theta\}^{2/3} - \{\sec\theta \,.\, \operatorname{cosec}\theta(\tan\theta \,.\, \sin\theta)^2\}^{2/3}$$

$$= \left(\sec^2\theta \,.\, \operatorname{cosec}^2\theta \,.\, \frac{\sin\theta}{\cos\theta} \,.\, \sin\theta\right)^{2/3} - \left(\frac{1}{\cos\theta} \,.\, \frac{1}{\sin\theta} \,.\, \tan^2\theta \,.\, \sin^2\theta\right)^{2/3}$$

$$= (\sec^3\theta)^{2/3} - (\tan^3\theta)^{2/3}$$

$$= \sec^2\theta - \tan^2\theta = 1 = \text{RHS}$$

14. $\operatorname{cosec}\theta - \sin\theta = a^3$

$$\Rightarrow \frac{1}{\sin\theta} - \sin\theta = a^3$$

$$\Rightarrow \frac{1-\sin^2\theta}{\sin\theta} = a^3 \qquad \Rightarrow \frac{\cos^2\theta}{\sin\theta} = a^3$$

$$\Rightarrow \qquad a = \left(\frac{\cos^2\theta}{\sin\theta}\right)^{1/3} \qquad ...(i)$$

Also $\sec\theta - \cos\theta = b^3$

$$\Rightarrow \frac{1}{\cos\theta} - \cos\theta = b^3 \qquad \Rightarrow \frac{1-\cos^2\theta}{\cos\theta} = b^3$$

$$\Rightarrow \frac{\sin^2\theta}{\cos\theta} = b^3 \qquad \Rightarrow b = \left(\frac{\sin^2\theta}{\cos\theta}\right)^{1/3}$$

$$\text{LHS} = a^2b^2(a^2+b^2) = \left(\frac{\cos^2\theta}{\sin\theta}\right)^{2/3}\left(\frac{\sin^2\theta}{\cos\theta}\right)^{2/3}\left[\left(\frac{\cos^2\theta}{\sin\theta}\right)^{2/3} + \left(\frac{\sin^2\theta}{\cos\theta}\right)^{2/3}\right]$$

$$= \left(\frac{\cos^2\theta \,.\, \sin^2\theta}{\sin\theta \,.\, \cos\theta}\right)^{2/3}\left[\left(\frac{\cos^2\theta}{\sin\theta}\right)^{2/3} + \left(\frac{\sin^2\theta}{\cos\theta}\right)^{2/3}\right]$$

$$= (\cos\theta \,.\, \sin\theta)^{2/3}\left[\left(\frac{\cos^2\theta}{\sin\theta}\right)^{2/3} + \left(\frac{\sin^2\theta}{\cos\theta}\right)^{2/3}\right]$$

$$= (\cos\theta \,.\, \sin\theta)^{2/3}\left(\frac{\cos^2\theta}{\sin\theta}\right)^{2/3} + (\cos\theta \,.\, \sin\theta)^{2/3}\left(\frac{\sin^2\theta}{\cos\theta}\right)^{2/3}$$

$$= \left(\cos\theta \,.\, \sin\theta \,.\, \frac{\cos^2\theta}{\sin\theta}\right)^{2/3} + \left(\cos\theta \,.\, \sin\theta \,.\, \frac{\sin^2\theta}{\cos\theta}\right)^{2/3}$$

$$= (\cos^3\theta)^{2/3} + (\sin^3\theta)^{2/3} = (\cos\theta)^2 + (\sin\theta)^2 = 1 = \text{RHS}$$

15. $a\cos^3\theta + 3a\cos\theta\sin^2\theta = m \qquad ...(i)$

$a\sin^3\theta + 3a\cos^2\theta \,.\, \sin\theta = n \qquad ...(ii)$

Adding eq. (*i*) and (*ii*), we get

$a\cos^3\theta + 3a\cos\theta \,.\, \sin^2\theta + a\sin^3\theta + 3a\cos^2\theta \,.\, \sin\theta = m + n$

$\Rightarrow a[\cos^3\theta + 3\cos\theta \,.\, \sin^2\theta + \sin^3\theta + 3\cos^2\theta \,.\, \sin\theta] = m + n$

$\Rightarrow a(\cos\theta + \sin\theta)^3 = (m+n)$

$\Rightarrow [a(\cos\theta + \sin\theta)^3]^{2/3} = (m+n)^{2/3}$

$\Rightarrow \quad a^{2/3}(\sin\theta + \cos\theta)^2 = (m+n)^{2/3} \qquad ...(iii)$

Subtracting eq. (*ii*) from (*i*), we get

$a\cos^3\theta + 3a\cos\theta\sin^2\theta - a\sin^3\theta - 3a\cos^2\sin\theta = m - n$

$\Rightarrow \quad a(\cos\theta - \sin\theta)^3 = m - n$

$[a(\cos\theta - \sin\theta)^3]^{2/3} = (m-n)^{2/3}$

$\Rightarrow \quad a^{2/3}(\cos\theta - \sin\theta)^2 = (m-n)^{2/3} \qquad ...(iv)$

Adding eq. (*iii*) and (*iv*), we get

$a^{2/3}(\cos\theta + \sin\theta)^2 + a^{2/3}(\cos\theta - \sin\theta)^2 = (m+n)^{2/3} + (m-n)^{2/3}$

$\Rightarrow \quad a^{2/3}[(\cos\theta + \sin\theta)^2 + (\cos\theta - \sin\theta)^2] = (m+n)^{2/3} + (m-n)^{2/3}$

$\Rightarrow \quad a^{2/3}(\cos^2\theta + \sin^2\theta + 2\cos\theta \,.\, \sin\theta + \cos^2\theta + \sin^2\theta - 2\sin\theta\cos\theta] = (m+n)^{2/3} + (m-n)^{2/3}$

$\Rightarrow \quad a^{2/3} \times 2 = (m+n)^{2/3} + (m-n)^{2/3}$ Hence proved.

16. $2\cos\theta - \sin\theta = x \qquad ...(i)$

$\cos\theta - 3\sin\theta = y$

$\text{LHS} = 2x^2 + y^2 - 2xy$

$= 2[2\cos\theta - \sin\theta]^2 + (\cos\theta - 3\sin\theta)^2 - 2(2\cos\theta - \sin\theta)(\cos\theta - 3\sin\theta)$

$= 2(4\cos^2\theta + \sin^2\theta - 4\cos\theta \,.\, \sin\theta) + \cos^2\theta + 9\sin^2\theta - 6\cos\theta\sin\theta - 2(2\cos^2\theta - 6\sin\theta\cos\theta - \sin\theta\cos\theta + 3\sin^2\theta)$

$= 8\cos^2\theta + 2\sin^2\theta - 8\cos\theta \,.\, \sin\theta + \cos^2\theta + 9\sin^2\theta - 6\cos\theta \,.\, \sin\theta - 4\cos^2\theta + 14\sin\theta \,.\, \cos\theta - 6\sin^2\theta$

$= 5\cos^2\theta + 5\sin^2\theta - 14\cos\theta\sin\theta + 14\cos\theta \,.\, \sin\theta$

$= 5(\cos^2\theta + \sin^2\theta) = 5 \times 1 = 5 = \text{RHS}$

17.

In right ΔABC,

$$\sin 30^\circ = \frac{BC}{AC}$$

$$\Rightarrow \quad \frac{1}{2} = \frac{BC}{4}$$

$$\Rightarrow \quad BC = 2 \text{ cm}$$

Also $\cos 30^\circ = \dfrac{AB}{AC} \Rightarrow \dfrac{\sqrt{3}}{2} = \dfrac{AB}{4}$

$$\Rightarrow \quad AB = 2\sqrt{3} \text{ cm}$$

In ΔDEF, $\dfrac{FE}{DE} = \tan 60^\circ \Rightarrow \dfrac{FE}{4} = \sqrt{3}$

$$\Rightarrow \quad FE = 4\sqrt{3} \text{ cm}$$

Now $\quad$ area $\Delta ABC = \dfrac{1}{2} \times AB \times BC$

$= \frac{1}{2} \times 2\sqrt{3} \times 2 = 2\sqrt{3}$ sq. cm

and area $\Delta DEF = \frac{1}{2} \times DE \times FE$

$= \frac{1}{2} \times 4 \times 4\sqrt{3} = 8\sqrt{3}$ sq. cm

$\therefore \frac{\text{Area of } \Delta ABC}{\text{Area of } \Delta DEF} = \frac{2\sqrt{3}}{8\sqrt{3}} = \frac{1}{4}$

18. LHS $= \frac{\tan A}{(1+\tan^2 A)^2} + \frac{\cot A}{(1+\cot^2 A)^2}$

$= \frac{\tan A}{(\sec^2 A)^2} + \frac{\cot A}{(\text{cosec}^2 A)^2}$

$= \frac{\sin A}{\cos A}.\cos^4 A + \frac{\cos A}{\sin A}.\sin^4 A$

$= \sin A \cos^3 A + \cos A \sin^3 A$

$= \cos A \sin A (\cos^2 A + \sin^2 A) = \sin A \cos A =$ RHS

19. We have,

$(\sec A + \tan A)(\sec B + \tan B)(\sec C + \tan C)$

$= (\sec A - \tan A)(\sec B - \tan B)(\sec C - \tan C)$ ...(*i*)

Multiply both sides by RHS we get,

$(\sec^2 A - \tan^2 A)(\sec^2 B - \tan^2 B)(\sec^2 C - \tan^2 C)$

$= [(\sec A - \tan A)(\sec B - \tan B)(\sec C - \tan C)]^2$

$1 . 1 . 1 = [(\sec A - \tan A)(\sec B - \tan B)(\sec C - \tan C)]^2$

$\Rightarrow \pm 1 = (\sec A - \tan A)(\sec B - \tan B)(\sec C - \tan C)$

From equation (*i*) we get

$(\sec A + \tan A)(\sec B + \tan B)(\sec C + \tan C) = \pm 1$

VALUE BASED QUESTIONS

1. (*a*) $\sin\theta + \sin^2\theta = 1$

$\Rightarrow \sin\theta = 1 - \sin^2\theta$

Now $\cos^2\theta + \cos^4\theta = \cos^2\theta(1 + \cos^2\theta)$

$\Rightarrow (1 - \sin^2\theta)(1 + 1 - \sin^2\theta)$

$= \sin\theta(1 + \sin\theta)$[using (*i*)] $= \sin\theta + \sin^2\theta = 1$

(*b*) Both should be rewarded equally.

2. (*a*) Yes

(*b*) Sincerity

MULTIPLE CHOICE QUESTIONS

1. (***b***) $\sqrt{2}\sin(60° - \alpha) = 1$

$\Rightarrow \sin(60° - \alpha) = \frac{1}{\sqrt{2}} = \sin 45°$

$\Rightarrow 60° - \alpha = 45° \Rightarrow \alpha = 15°$

2. (***b***) $\frac{1}{\text{cosec}\,\theta} = \sin\theta$

$\therefore$ Maximum value = 1

3. (***d***) $\cos(40° + A) = \sin 30° \Rightarrow \cos(40° + A) = \cos(90° - 30°)$

$\Rightarrow 40° + A = 60° \Rightarrow A = 20°$

4. (***c***) $\text{cosec}\,\theta - \cot\theta = \frac{1}{3}$

Also, $\text{cosec}^2\theta - \cot^2\theta = 1$

$\Rightarrow (\text{cosec}\,\theta + \cot\theta)(\text{cosec}\,\theta - \cot\theta) = 1$

$\Rightarrow (\text{cosec}\,\theta + \cot\theta) \times \frac{1}{3} = 1 \Rightarrow \text{cosec}\,\theta + \cot\theta = 3$

5. (***a***) $\sin^2 30° - \cos^2 30° = \left(\frac{1}{2}\right)^2 - \left(\frac{\sqrt{3}}{2}\right)^2$

$= \frac{1}{4} - \frac{3}{4} = \frac{-2}{4} = \frac{-1}{2}$

6. (***a***) $AC = \sqrt{13^2 - 5^2} = 12$

DE = AC = 12 cm, BE = 14 cm – 5 cm = 9 cm

In ΔBED, $\tan B = \frac{DE}{BE} = \frac{12}{9} = \frac{4}{3}$

7. (***c***) Definition of trigonometry.

8. (***a***) In Greek tri means three.

9. (***a***) In Greek gon means sides.

10. (***b***) In Greek metron means measure.

11. (***c***) Trigonometric ratios of the angles.

12. (***c***) Side opposite to right angle is known as hypotenuse.

13. (***a***) Side adjacent to angle P.

14. (***b***) Side opposite to angle D.

15. (***b***) $\frac{\text{Side opposite to angle } \theta}{\text{Hypotenuse}} = \sin\theta$

$\frac{\text{Side adjacent to angle } \theta}{\text{Hypotenuse}} = \cos\theta$

$\frac{\text{Side opposite to angle } \theta}{\text{Side adjacent to angle } \theta} = \tan\theta$

16. (***d***) Definition of trigonometric ratios.

17. (***c***) $\sec A = \frac{1}{\cos A}$

18. (***a***) $\sin A = \frac{1}{\text{cosec } A}$

19. (***b***) $\tan A = \frac{1}{\cot A}$

20. (***b***) Edmund Gunter

21. (***b***) Do not vary with the length of the sides of the triangle.

22. (***b***) $1 - 2\sin^2 A = \sin^2 A + \cos^2 A - 2\sin^2 A$

$= \cos^2 A - \sin^2 A$

23. **(c)** $AC^2 = AB^2 + BC^2$

$\Rightarrow BC^2 = (13)^2 - (5)^2 = 144 \Rightarrow BC = 12$ cm

$\therefore \sin A = \frac{12}{13}$

24. **(b)** $2\cos^2 30° - 1 = 2 \times \left(\frac{\sqrt{3}}{2}\right)^2 - 1$

$= \frac{3}{2} - 1 = \frac{1}{2} = \cos 60°$

25. **(c)** On verification, $2 \sin 45° . \cos 45°$

$= 2 \times \frac{1}{\sqrt{2}} \times \frac{1}{\sqrt{2}} = 1 = \sin 90°$

26. **(a)** On verification, $4\cos^3 30° - 3\cos 30°$

$= 4 \times \left(\frac{\sqrt{3}}{2}\right)^3 - 3 \times \frac{\sqrt{3}}{2} = 0 = \cos 90°$

27. **(b)** On verification, $\frac{2\tan 30°}{1 - \tan^2 30°} = \frac{2 \times \frac{1}{\sqrt{3}}}{1 - \frac{1}{3}} = \sqrt{3}$

$= \tan 60°$

28. **(b)** $\frac{\tan A}{\sec A} = \frac{\sin A / \cos A}{1/\cos A} = \sin A$

29. **(c)** $\tan A = \frac{\sin A}{\cos A} = \frac{\sin A}{\sqrt{1 - \sin^2 A}}$

30. **(d)** $\sec A = \frac{1}{\cos A} = \frac{\frac{1}{\sin A}}{\frac{\cos A}{\sin A}}$

$= \frac{\text{cosec } A}{\cot A} = \frac{\sqrt{1 + \cot^2 A}}{\cot A}$

31. **(c)** $\sin(90° - A) = \cos A$

32. **(b)** $\cot(90° - A) = \tan A$

33. **(d)** $\sec(90° - A) = \text{cosec } A$

34. **(b)** $\tan 2A = \cot(A - 18°)$

$\Rightarrow \cot(90° - 2A) = \cot(A - 18°)$

$\Rightarrow 90° - 2A = A - 18° \Rightarrow A = 36°$

35. **(c)** Here $\text{cosec}(90° - 4A) = \text{cosec}(A - 20°)$

$\Rightarrow 90° - 4A = A - 20° \Rightarrow A = 22°$

$\Rightarrow 3A = 66°$

36. **(d)** $A + B + C = 180°$

$\Rightarrow A + B = 180° - C$

$\Rightarrow \frac{A+B}{2} = 90° - \frac{C}{2}$

$\tan\left(\frac{A+B}{2}\right) = \tan\left(90° - \frac{C}{2}\right)$

$= \cot \frac{C}{2}$

37. **(c)** $\frac{1 + \cot^2 A}{1 + \tan^2 A} = \frac{1 + \cot^2 A}{1 + \frac{1}{\cot^2 A}}$

$= \frac{\cot^2 A(1 + \cot^2 A)}{\cot^2 A + 1} = \cot^2 A = \text{cosec}^2 A - 1$

38. **(c)** $\sec A + \tan A = x$ ... (*i*)

Also $\sec^2 A - \tan^2 A = 1$

$\Rightarrow (\sec A - \tan A)(\sec A + \tan A) = 1$

$\Rightarrow x(\sec A - \tan A) = 1$

$\therefore \sec A - \tan A = \frac{1}{x}$ ... (*ii*)

From (*i*) and (*ii*),

$\tan A = \frac{x^2 - 1}{2x}$

39. **(d)** $\text{cosec } A - \cot A = \frac{4}{5}$ ... (*i*)

Also $\text{cosec}^2 A - \cot^2 A = 1$

$\Rightarrow (\text{cosec } A - \cot A)(\text{cosec } A + \cot A) = 1$

$\Rightarrow \frac{4}{5}(\text{cosec } A + \cot A) = 1$

$\Rightarrow \text{cosec } A + \cot A = \frac{5}{4}$ ... (*ii*)

From (*i*) and (*ii*), $\text{cosec } A = \frac{41}{40}$.

40. **(d)** J.F. Herbart.

41. **(d)** $\because \cos 90° = 0$

$\therefore$ Product of all terms = 0

42. **(a)** $\tan 1°.\tan 2° ...... \tan 44°.\tan 45°.\tan 46° ...... \tan 89°$

$= \tan 1°.\tan 2° ...... \tan 44° \times 1 \times \cot 44° ...... \cot 1°$

$= 1$

43. **(c)** $\sin x + \text{cosec } x = 2$

$\Rightarrow \sin x + \frac{1}{\sin x} = 2 \Rightarrow \sin^2 x + 1 = 2\sin x$

$\Rightarrow (\sin x - 1)^2 = 0 \Rightarrow \sin x = 1$

$\Rightarrow \text{cosec } x = 1$

$\therefore \sin^{19} x + \text{cosec}^{20} x = 1 + 1 = 2$

44. **(d)** $x = (\sec A - \tan A)(\sec B - \tan B)(\sec C - \tan C)$

$= (\sec A + \tan A)(\sec B + \tan B)(\sec C + \tan C)$

$\Rightarrow (\sec A + \tan A)(\sec B + \tan B)(\sec C + \tan C).$

$(\sec A - \tan A)(\sec B - \tan B)(\sec C - \tan C)$

$= \{(\sec A + \tan A)(\sec B + \tan B)(\sec A + \tan A)\}^2$

$\Rightarrow (\sec^2 A - \tan^2 A)(\sec^2 B - \tan^2 B)(\sec^2 C - \tan^2 C) = x^2$

$\Rightarrow x^2 = 1 \Rightarrow x = \pm 1$

45. **(b)** Let $AB = x \Rightarrow AC = 2x$.

$\therefore \sin C = \frac{AB}{AC} = \frac{x}{2x} = \frac{1}{2}$

$\Rightarrow C = 30°$

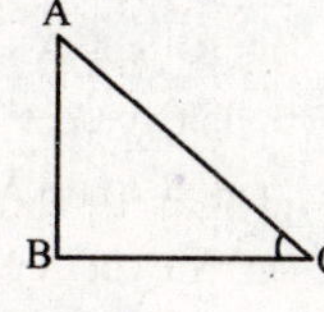

46. **(a)** $\tan^2\theta + \cot^2\theta = 2$

$\Rightarrow \tan^2\theta + \cot^2\theta = 2\tan\theta . \cot\theta$

$\Rightarrow (\tan\theta - \cot\theta)^2 = 0$

$\Rightarrow \tan\theta - \cot\theta = 0 \Rightarrow \tan\theta = \cot\theta$

$\Rightarrow \theta = 45°$

$\therefore \tan^3\theta + \cot^3\theta = \tan^3 45° + \cot^3 45° = (1)^3 + (1)^3 = 2$

47. **(b)** $\sin x + \sin^2 x = 1$

$\Rightarrow \sin x = 1 - \sin^2 x$

$\Rightarrow \sin x = \cos^2 x$

$\cos^2 x + \cos^4 x = \cos^2 x (1 + \cos^2 x)$
$= \sin x (1 + \sin x)$
$= \sin x + \sin^2 x = 1$

48. **(c)** Maximum value of $\sin\theta = 1$
$\therefore \sin\theta_1 + \sin\theta_2 + \sin\theta_3 = 3$
$\Rightarrow \sin\theta_1 = 1, \sin\theta_2 = 1$ and $\sin\theta_3 = 1$
$\Rightarrow \theta_1 = 90°, \theta_2 = 90°$ and $\theta_3 = 90°$
$\therefore \cos\theta_1 + \cos\theta_2 + \cos\theta_3 = 0$

49. **(b)** $\because$ Maximum value of $\sin\theta = 1$
$\therefore \sin\theta = 3$ is incorrect.

50. **(d)** $2(\sin^6\theta + \cos^6\theta) - 3(\sin^4\theta + \cos^4\theta) + 1$
$= 2[(\sin^2\theta + \cos^2\theta)^3 - 3\sin^2\theta . \cos^2\theta (\sin^2\theta + \cos^2\theta)]$
$- 3[(\sin^2\theta + \cos^2\theta)^2 - 2\sin^2\theta . \cos^2\theta] + 1$
$= 2[1 - 3\sin^2\theta . \cos^2\theta] - 3[1 - 2\sin^2\theta . \cos^2\theta] + 1$
$= 2 - 6\sin^2\theta . \cos^2\theta - 3 + 6\sin^2\theta . \cos^2\theta + 1$
$= 0$

51. **(d)** $\sin^2 5° + \sin^2 10° + \sin^2 15° + ... \sin^2 90°$
$= (\sin^2 5° + \sin^2 85°) + (\sin^2 10° + \sin^2 80°)$
$+ ...... + (\sin^2 40° + \sin^2 50°) + \sin^2 45° + \sin^2 90°$
$= (\sin^2 5° + \cos^2 5°) + (\sin^2 10° + \cos^2 10°) + ...$
$+ (\sin^2 40° + \cos^2 40°) + \left(\frac{1}{\sqrt{2}}\right)^2 + 1$
$= 1 + 1 + 1 + ...$ 8 times $+ \frac{1}{2} + 1$
$= 9\frac{1}{2} = 9.5$

52. **(b)** $\operatorname{cosec}\theta - \cot\theta = \frac{1}{2}$ .. (*i*)
Also, $\operatorname{cosec}^2\theta - \cot^2\theta = 1$
$\Rightarrow \operatorname{cosec}\theta + \cot\theta = 2$ ... (*ii*)
From (*i*) and (*ii*),
$\Rightarrow 2\operatorname{cosec}\theta = \frac{5}{2} \Rightarrow \operatorname{cosec}\theta = \frac{5}{4}$
$\Rightarrow \sin\theta = \frac{4}{5} \Rightarrow \cos\theta = \sqrt{1-\sin^2\theta}$
$= \sqrt{1-\left(\frac{4}{5}\right)^2} = \sqrt{1-\frac{16}{25}} = \frac{3}{5}$

53. **(b)** $(\tan A + \cot A)^2 = 4^2$
$\Rightarrow \tan^2 A + \cot^2 A + 2 = 16$
$\Rightarrow \tan^2 A + \cot^2 A = 14$
$\Rightarrow (\tan^2 A + \cot^2 A)^2 = (14)^2$
$\Rightarrow \tan^4 A + \cot^4 A + 2 = 196$
$\Rightarrow \tan^4 A + \cot^4 A = 194$

54. **(c)** $x^2 + y^2 + z^2 = r^2\sin^2\theta\cos^2\phi + r^2\sin^2\theta . \sin^2\phi + r^2\cos^2\theta$
$= r^2\sin^2\theta(\cos^2\phi + \sin^2\phi) + r^2\cos^2\theta$
$= r^2(\sin^2\theta + \cos^2\theta) = r^2$, which is independent of θ, ϕ.

55. **(d)** $\because \cos\theta \le 1, \sin\theta \le 1, \sec\theta \ge 1$
$\therefore$ *a*, *b*, *c* not possible
$\therefore \tan\theta = 41$

56. **(b)** $\frac{a\sin\theta + b\cos\theta}{a\sin\theta - b\cos\theta} = \frac{\frac{a\sin\theta}{\cos\theta} + \frac{b\cos\theta}{\cos\theta}}{\frac{a\sin\theta}{\cos\theta} - \frac{b\cos\theta}{\cos\theta}}$
$= \frac{a\tan\theta + b}{a\tan\theta - b} = \frac{a \times \frac{a}{b} + b}{a \times \frac{a}{b} - b} = \frac{a^2 + b^2}{a^2 - b^2}$

57. **(a)** $\because \sin\theta$ and $\cos\theta$ are the roots,
$\sin\theta + \cos\theta = -\left(\frac{-b}{a}\right)$
and $\sin\theta . \cos\theta = \frac{c}{a}$
$\Rightarrow (\sin\theta + \cos\theta)^2 = \left(\frac{b}{a}\right)^2$
$\Rightarrow \sin^2\theta + \cos^2\theta + 2\sin\theta . \cos\theta = \frac{b^2}{a^2}$
$\Rightarrow 1 + 2\frac{c}{a} = \frac{b^2}{a^2}$
$\Rightarrow a^2 + 2ac = b^2$
$\Rightarrow b^2 - a^2 = 2ac$

58. **(a)** $\because \sec\theta = \frac{1}{\cos\theta}$
$\therefore \sec\theta > 1$

59. **(b)** $\sin A = \cos A$
$\Rightarrow \frac{\sin A}{\cos A} = 1 \Rightarrow \tan A = 1$
$\Rightarrow A = 45°$

60. **(c)** $\sin\theta = \sqrt{3}\cos\theta$
$\Rightarrow \frac{\sin\theta}{\cos\theta} = \sqrt{3}$
$\Rightarrow \tan\theta = \sqrt{3} = \tan 60° \Rightarrow \theta = 60°$

61. **(b)** $3\cot\theta = 2 \Rightarrow \cot\theta = \frac{2}{3} \Rightarrow \tan\theta = \frac{3}{2}$

62. **(a)** $\cos\theta = \frac{7}{8} \Rightarrow \sec\theta = \frac{1}{\cos\theta} = \frac{8}{7}$

63. **(c)** $\frac{\tan 65°}{\cot 25°} = \frac{\cot(90° - 65°)}{\cot 25°} = \frac{\cot 25°}{\cot 25°} = 1$

64. **(d)** $(\sec A + \tan A)(1 - \sin A)$
$= \left(\frac{1}{\cos A} + \frac{\sin A}{\cos A}\right)(1 - \sin A)$
$= \frac{(1 + \sin A)(1 - \sin A)}{\cos A}$
$= \frac{1 - \sin^2 A}{\cos A} = \frac{\cos^2 A}{\cos A} = \cos A$

65. **(b)** $(\operatorname{cosec} A + \cot A)(1 - \cos A)$
$= \left(\frac{1}{\sin A} + \frac{\cos A}{\sin A}\right)(1 - \cos A)$

$$= \left(\frac{1+\cos A}{\sin A}\right)(1-\cos A) = \frac{1-\cos^2 A}{\sin A} = \frac{\sin^2 A}{\sin A} = \sin A$$

66. (*b*) $(1+\sin A)(\sec A - \tan A) = (1+\sin A)\left(\frac{1}{\cos A} - \frac{\sin A}{\cos A}\right)$

$$= (1+\sin A)\frac{(1-\sin A)}{\cos A} = \frac{1-\sin^2 A}{\cos A} = \frac{\cos^2 A}{\cos A} = \cos A$$

67. (*b*) $\cos 1° . \cos 2° . \cos 3°, ..., \cos 90°$
$= \cos 1° . \cos 2° . \cos 3°, ... \times 0 = 0$

NCERT EXEMPLAR PROBLEMS

1. (*b*) $\text{cosec}(75° + \theta) - \sec(15° - \theta) - \tan(55° + \theta) + \cot(35° - \theta)$
$= \text{cosec}\{90° - (15° - \theta)\} - \sec(15° - \theta)$
$- \tan\{90° - (35° - \theta)\} + \cot(35° - \theta)$
$= \sec(15° - \theta) - \sec(15° - \theta) - \cot(35° - \theta) + \cot(35° - \theta)$
$= 0$

2. (*b*) $\cos(\alpha + \beta) = 0 = \cos 90°$
$\Rightarrow \quad \alpha + \beta = 90°$
$\sin(\alpha - \beta) = \sin(\alpha - \beta + \beta - \beta)$
$= \sin(\alpha + \beta - 2\beta)$
$= \sin(90° - 2\beta) = \cos 2\beta$

3. (*c*) $\cos 9\alpha = \sin \alpha$
$\Rightarrow \quad \cos 9\alpha = \cos(90° - \alpha)$
$\Rightarrow \quad 9\alpha = 90° - \alpha$
$\Rightarrow \quad 10\alpha = 90° \Rightarrow \alpha = 9°$
$\therefore \quad \tan 5\alpha = \tan 5 \times 9°$
$= \tan 45° = 1$

4. (*a*) ΔABC is right angled at C,
$\therefore A + B + C = 180°$
$A + B = 180° - 90° = 90° \qquad (\because \angle C = 90°)$
$\cos(A + B) = \cos 90° = 0$

5. (*c*) $\sin\theta - \cos\theta = 0$
$\Rightarrow (\sin\theta - \cos\theta)^2 = 0$
$\Rightarrow \sin^2\theta + \cos^2\theta - 2\sin\theta\cos\theta = 0$
$\Rightarrow -2\sin\theta\cos\theta = -1 \Rightarrow 2\sin\theta\cos\theta = 1$
$\Rightarrow \sin\theta\cos\theta = \frac{1}{2} \quad \Rightarrow \sin^2\theta\cos^2\theta = \frac{1}{4}$
$\sin^4\theta + \cos^4\theta = \sin^4\theta + \cos^4\theta + 2\sin^2\theta\cos^2\theta$
$- 2\sin^2\theta\cos^2\theta$
$= (\sin^2\theta + \cos^2\theta)^2 - 2\sin^2\theta\cos^2\theta$
$= (1)^2 - 2 \times \frac{1}{4} = 1 - \frac{1}{2} = \frac{1}{2}$

6. (*b*) $\sin(45° + \theta) - \cos(45° - \theta)$
$= \sin\{90° - (45° - \theta)\} - \cos(45° - \theta)$
$= \cos(45° - \theta) - \cos(45° - \theta) = 0$

Formative Assessment

Activity/Project 1

- **Topic Covered :** Trigonometry
- **Objective :** Shade all the regions which display correct trigonometric ratios to find the hidden animal.

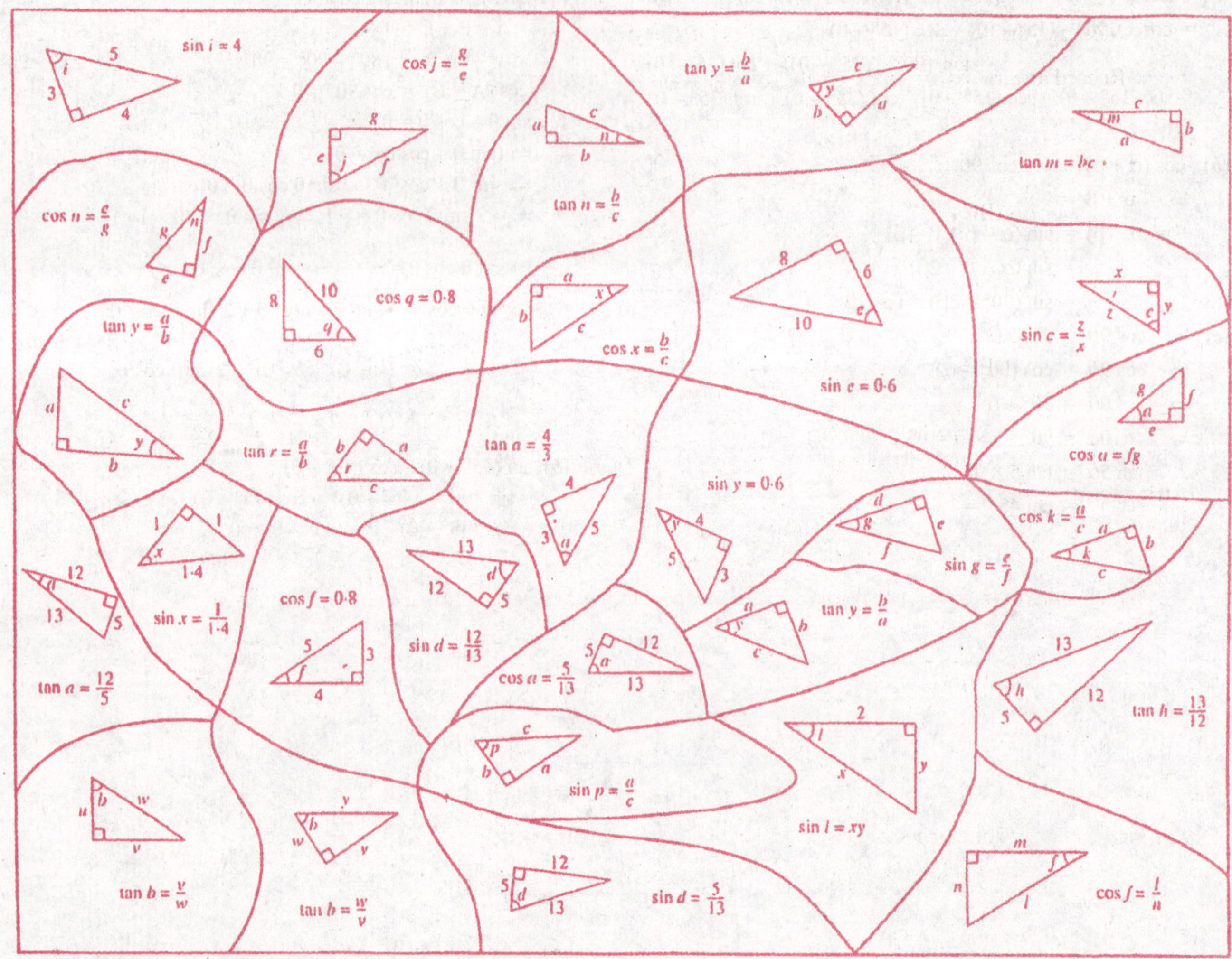

Activity/Project 2

- **Topic Covered :** Trigonometry
- **Objective :** Comparing the ratios of sides of right angled triangle.
- **Skills Developed :**

  Analysis, Creativity, Reasoning.
- **Time Required :** 15 Min.

■ **Method :**

1. Accurately draw three right angled triangles with acute angle of 30° as shown below :

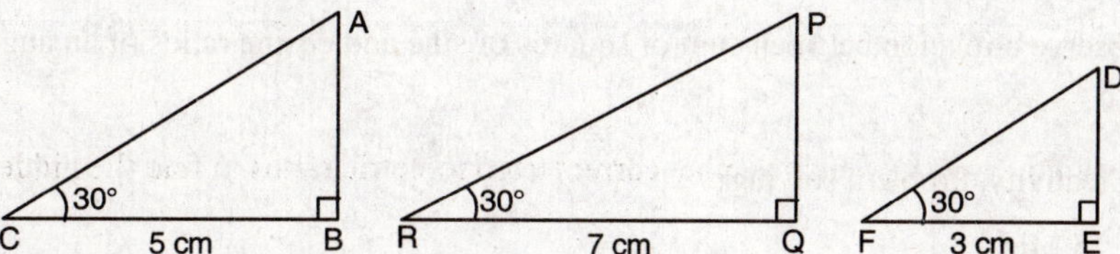

2. Measure the lengths of the other sides of the triangles.

Record the ratios of sides of the triangles in the table given below.

Triangle	Perp/Hyp	Base/Hyp	Perp/Base
Δ ABC	$\frac{AB}{AC}$ =	$\frac{BC}{AC}$ =	$\frac{AB}{BC}$ =
Δ PQR	$\frac{PQ}{PR}$ =	$\frac{QR}{PR}$ =	$\frac{PQ}{QR}$ =
Δ DEF	$\frac{DE}{DF}$ =	$\frac{EF}{DF}$ =	$\frac{DE}{EF}$ =

■ **Observation :**

In a right angled triangle with a given acute angle, ratios of corresponding sides is constant. These ratios are known as trigonometric ratios.

Class Work

1. Which side of the given triangle ABC, with respect to marked angle is adjacent side ?
2. Which side with respect to marked angle is opposite side in Fig. (*i*) ?

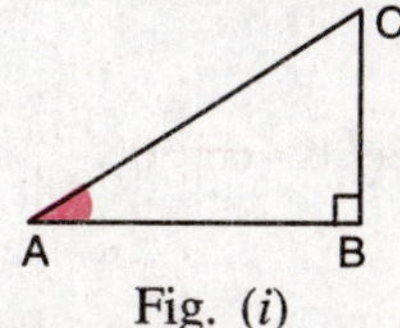

Fig. (*i*)

3. Write the trigonometric name of the ratios of sides in a right triangle.

(*i*) $\frac{\text{Opposite side of an angle}}{\text{Hypotenuse}}$ (*ii*) $\frac{\text{Opposite side of an angle}}{\text{Adjacent side}}$

Homework

Write the trigonometric names of all six possible ratios of the sides of a right angled triangle.

Quiz Questions

1. $\tan A = \frac{\cos A}{\sin A}$. State True/False.
2. cosec A × cos A = 1. State True/False.

Oral Questions

1. Find sin A, if cosec A = 2.
2. Find tan A, if cot A = $\sqrt{3}$

Suggested Activity/Assignment

Make a table showing relationship between sides of right triangle and trigonometric ratios.

Activity/Project 3

- **Topic Covered :** Trigonometry
- **Objective :** To observe a relation between sum of squares of sine and cosine ratios of an angle in a right angled triangle.
- **Skills Developed :**

  Proof analysis, Creativity, Problem solving
- **Time Required :** 15 Min.
- **Method :**

  1. Consider three right angled triangles as shown.

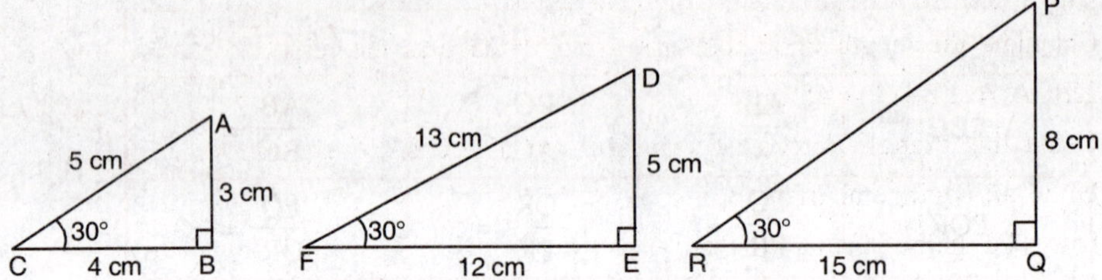

  2. Record the results in the table like the one below :

Δ ABC	sin C =	cos C =	$\sin^2 C + \cos^2 C =$
Δ DEF	sin F =	cos F =	$\sin^2 F + \cos^2 F =$
Δ PQR	sin R =	cos R =	$\sin^2 R + \cos^2 R =$

- **Observation :**

  $\sin^2 \theta + \cos^2 \theta = 1$

Class Work

Record the results for (*i*) $\sec^2 \theta - \tan^2 \theta$ (*ii*) $\text{cosec}^2 \theta - \cot^2 \theta$

Homework

Find the value of (*i*) $\sec^2 \theta (1 - \sin^2 \theta)$ (*ii*) $\sin^2 \theta (1 + \cot^2 \theta)$

Quiz Questions

1. $\sin^2 \theta - \cos^2 \theta = 1$. State True/False.
2. $\cot^2 \theta - \text{cosec}^2 \theta = 1$. State True/False.

Oral Questions

1. Evaluate : $\sin^2 40° + \cos^2 40°$.
2. Evaluate : $\sec^2 50° - \cot^2 40°$.

Suggested Activity/Assignment

List six trigonometric identities.

Activity/Project 4

- **Topic Covered :** Trigonometry
- **Objective :** Finding trigonometric ratios for 30°, 45°, 60°.
- **Skills Developed :**

  Creativity, Analysis, Reasoning
- **Time Required :** 15 Min.
- **Method :**

A. 1. Draw an isosceles right angled triangle as shown.

2. Calculate the length of hypotenuse using Pythagoras theorem.

3. $\angle BCA = \angle BAC = 45°$

4. Using this triangle determine values of sin 45°, cos 45° and tan 45°.

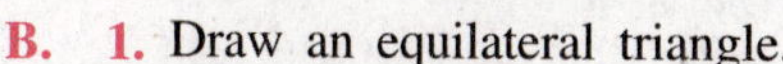

B. 1. Draw an equilateral triangle.

2. Draw PT perpendicular to QR.

3. QT = TR = 2 cm.

4. Find PT using Pythagoras theorem.

5. Calculate $\angle TPQ$.

6. Using triangle PTQ, determine values for sin 60°, cos 60° and tan 60°.

Determine values for sin 30°, cos 30° and tan 30° using the same triangle.

- **Observation :**

$\sin 45° = \frac{1}{\sqrt{2}}$, $\cos 45° = \frac{1}{\sqrt{2}}$, $\tan 45° = 1$ $\quad$ $\sin 60° = \frac{\sqrt{3}}{2}$, $\cos 60° = \frac{1}{2}$, $\tan 60° = \sqrt{3}$

$\sin 30° = \frac{1}{2}$, $\cos 30° = \frac{\sqrt{3}}{2}$, $\tan 30° = \frac{1}{\sqrt{3}}$

Class Work

1. Find sec 45°, cosec 45°, cot 45°, sec 60°, cosec 60°, cot 60°, sec 30°, cosec 30° and cot 30°.
2. Write the pair of ratios for 60° whose product is 1.

Homework

Make a table of values of all trigonometric ratios for angle 0°, 30°, 45°, 60° and 90°.

Quiz Questions

1. $\sin 60° \times \text{cosec } 60° = 1$. State True/False.
2. $\sec 60° = \frac{1}{\cot 60°}$. State True/False.

Oral Questions

1. Evaluate : sin 45° + tan 45°.
2. Evaluate : tan 60° + cot 60°.

Suggested Activity/Assignment

Make a table showing values of all six trigonometric ratios for angles 0°, 30°, 45°, 60° and 90°.

Statistics

• Basic Concepts with Examples

ARITHMETIC MEAN

- The mean of n observations $x_1, x_2, x_3, ... x_n$ is $\bar{x} = \dfrac{x_1 + x_2 + x_3 + x_4 + ... + x_n}{n} = \dfrac{1}{n}\sum_{i=1}^{n} x_i$
- Sum of observations = Number of observations × Mean.
- If n observations in the raw data consist of only k distinct values $x_1, x_2, x_3, ... x_k$ of the variable x occurring with frequencies $f_1, f_2, f_3, ... f_k$ respectively, then

$$\bar{x} = \frac{f_1x_1 + f_2x_2 + ... + f_kx_k}{f_1 + f_2 + ... + f_k} = \frac{\sum_{i=1}^{k} f_ix_i}{\sum_{i=1}^{k} f_i} = \frac{1}{n}\sum_{i=1}^{k} f_ix_i \text{ where, } n = \sum_{i=1}^{k} f_i \text{ denotes total frequency.}$$

- When the raw data are presented in the form of a frequency distribution with equal or unequal, exclusive or inclusive class-intervals, we assume that the frequency in any class is centred at its class marks (or mid-point).

  Class marks of any class-interval = The mean of the upper and the lower limits of the interval

  i.e., $\quad$ Class mark = $\dfrac{\text{upper limit} + \text{lower limit}}{2}$.

- **Assumed mean method for computing mean.** In this method, an arbitrary constant a is chosen as assumed mean somewhere, in the middle of all values of x_i and we find $d_i = x_i - a$.

$$\bar{x} = a + \frac{\Sigma f_i d_i}{\Sigma f_i}$$

- **Step deviation method (Short-cut method) for computing mean.** In this method, an arbitrary constant a is chosen which is called as origin or assumed mean somewhere, in the middle of all values of x_i. If h is the difference of any two consecutive values of x_i, then $u_i = \dfrac{x_i - a}{h}$, $\quad$ Mean $(\bar{x}) = a + \dfrac{h}{n}\sum_{i=1}^{k} f_iu_i = a + h\dfrac{\Sigma f_iu_i}{\Sigma f_i}, i = 1, 2, 3, ... k$.

1. The table below gives the percentage distribution of female teachers in the primary schools of rural areas of various states and union territories of India. Find the mean percentage of female teachers by assumed mean method.

Percentage of female teachers	Number of states/U.T.
15 — 25	6
25 — 35	11
35 — 45	7
45 — 55	4
55 — 65	4
65 — 75	2
75 — 85	1

Sol. Here, $a = 50$

Percentage of female teachers	Number of states /U.T. (f_i)	x_i	$d_i = x_i - 50$	f_id_i
15 — 25	6	20	–30	–180
25 — 35	11	30	–20	–220
35 — 45	7	40	–10	–70
45 — 55	4	50	0	0
55 — 65	4	60	10	40
65 — 75	2	70	20	40
75 — 85	1	80	30	30
Total	$\Sigma f_i = 35$			$\Sigma f_id_i = -360$

$$\bar{x} = a + \frac{\Sigma f_id_i}{\Sigma f_i} = 50 + \frac{-360}{35} = 39.71$$

2. If the mean of the following data is 20.6, find the missing frequency (x).

x	f
10	3
15	10
20	x
25	7
35	5

Sol.

x_i	f_i	$f_i \times x_i$
10	3	30
15	10	150
20	x	$20x$
25	7	175
35	5	175
Total	$\Sigma f_i = 25 + x$	$\Sigma f_i x_i = 530 + 20x$

$$\text{Mean} = \frac{\Sigma f_i x_i}{\Sigma f_i} = \frac{530 + 20x}{25 + x} = 20.6$$

$\therefore$ $530 + 20x = 25 \times 20.6 + 20.6x$

$530 + 20x = 515 + 20.6x$

$$15 = 0.6x \;\Rightarrow x = \frac{15}{0.6} = 25.$$

Hence, missing frequency = 25.

3. The following table gives the enrolment in Higher Secondary Schools in 2012. Find the average enrolment per H.S. School.

Enrolment	No. of schools
20 — 39	526
40 — 59	620
60 — 79	674
80 — 99	717
100 — 119	681
120 — 139	612
140 — 159	540
160 — 179	517
180 — 199	552
Total	5439

Sol. Here, $a = 109.5$, $h = 20$. Then we have the following table for computation of the mean.

Class Interval	Class freq. (f_i)	Class Mid-point (x_i)	$u_i = \frac{x_i - 109.5}{20}$	$f_i u_i$
20— 39	526	29.5	– 4	–2104
40— 59	620	49.5	– 3	–1860
60— 79	674	69.5	– 2	–1348
80— 99	717	89.5	– 1	–717
100—119	681	109.5	0	0
120—139	612	129.5	+ 1	+ 612
140—159	540	149.5	+ 2	+ 1080
160—179	517	169.5	+ 3	+ 1551
180—199	552	189.5	+ 4	+ 2208
Total	5439			$\sum_{i=1}^{9} f_i u_i = -578$

$$\therefore\; \bar{x} = a + h\bar{u}\left[\bar{u} = \frac{\sum f_i u_i}{\sum f_i}\right] \quad [\bar{u} = 0.106]$$

$= 109.5 - 20 \times 0.106 = 109.5 - 2.12 = 107.38$

$\therefore$ Mean enrolment = 107.38.

4. The arithmetic mean of a set of 40 values is 65. If each of the 40 values is increased by 5, what will be the mean of the set of new values.

Sol. Mean of 40 values = 65

Total of 40 values = $65 \times 40 = 2600$

When each value is increased by 5, the total 40 values will be $40 \times 5 = 200$ more than 2600.

So, new total of 40 values = $2600 + 200 = 2800$

$$\text{New mean of 40 values} = \frac{2800}{40} = 70.$$

or Mean is increased by 5. So, new mean = $65 + 5 = 70$.

5. The mean of the following frequency table is 50. But the frequencies f_1 and f_2 in classes 20–40 and 60–80 are missing. Find the missing frequencies.

Class	Frequency
0—20	17
20—40	f_1
40—60	32
60—80	f_2
80—100	19
Total	120

Sol. Here, $a = 50$

Class interval	Class mid-point (x_i)	$d_i = (x_i - a)$	(f_i)	$f_i d_i$
0— 20	10	– 40	17	– 680
20— 40	30	– 20	f_1	$-20f_1$
40— 60	50	0	32	0
60— 80	70	+ 20	f_2	$20f_2$
80— 100	90	+ 40	19	760
Total		$\Sigma d_i = 0$	$\Sigma f_i = 120$	$\Sigma f_i d_i = 0$

$17 + f_1 + 32 + f_2 + 19 = 120$

$f_1 + f_2 = 120 - 68 = 52$ ...(*i*)

$\Sigma f_i d_i = 0$

$\therefore$ $-680 - 20f_1 + 20f_2 + 760 = 0$

$-34 - f_1 + f_2 + 38 = 0$

$\Rightarrow$ $f_1 - f_2 = 4$

$f_1 + f_2 = 52$ [From (*i*)]

$\therefore$ $2f_1 = 52 + 4 = 56$

$$f_1 = \frac{56}{2} = 28,$$

$f_2 = 52 - 28 = 24.$

6. The average score of boys in an examination of a school is 71 and that of girls is 73. The average score of the school in the examination is 71.8, find the ratio of the

number of boys to the number of girls who appeared in the examination.

Sol. Let the number of boys be x and the number of girls be y.

Average score of x boys = 71

$\therefore$ Total score of x boys = $71x$

Average score of y girls = 73

$\therefore$ Total score of y girls = $73y$

Average score of the school = 71.8

$$\therefore \frac{71x + 73y}{x + y} = 71.8$$

$\Rightarrow 71x + 73y = 71.8x + 71.8y$

$\Rightarrow 73y - 71.8y = 71.8x - 71x \Rightarrow 1.2y = 0.8x$

$\Rightarrow \frac{x}{y} = \frac{1.2}{0.8} = \frac{3}{2}$

$\therefore x : y = 3 : 2$

$\Rightarrow$ Number of boys : Number of girls = 2 : 3.

7. The mean weight of 150 students in a class is 60 kg. The mean weight of boys is 70 kg while that of girls is 55 kg. Find the number of boys and girls in the school.

Sol. Let the number of boys be x, their mean weight = 70 kg

Total sum of boys = $70x$

Let the number of girls be y, their mean weight = 55 kg

Total sum of girls = $55y$

Mean weight of 150 students is 60 kg.

$$\therefore \quad 60 = \frac{70x + 55y}{150}$$

$\Rightarrow 70x + 55y = 9000 \Rightarrow 14x + 11y = 1800$ ...(*i*)

Also $x + y = 150$ ...(*ii*)

Solving (*i*) and (*ii*), we get

$x = 50, y = 100$

$\therefore$ Number of boys = 50,

Number of girls = 100.

MODE OF GROUPED DATA

■ **Mode of grouped frequency distribution :**

(*i*) Locate a class with the maximum frequency. This class is known as modal class.

(*ii*) $\text{Mode} = l + \left(\frac{f_1 - f_0}{2f_1 - f_0 - f_2}\right) \times h$,

where, l = lower limit of the modal class.

f_1 = frequency of the modal class.

f_0 = frequency of the class preceding the modal class.

f_2 = frequency of the class succeeding the modal class.

h = size of the class interval (assuming all class sizes to be equal).

8. Find the mode of the following data :

Classes	Frequencies
10 — 15	30
15 — 20	45
20 — 25	75
25 — 30	35
30 — 35	25
35 — 40	15
Total	225

Sol. Maximum frequency is 75.

$\therefore$ Modal class is 20 – 25.

$l = 20, \; f_1 = 75, \; f_0 = 45, \; f_2 = 35, \; h = 5.$

$$\therefore \text{Mode} = l + \left(\frac{f_1 - f_0}{2f_1 - f_0 - f_2}\right) \times h$$

$$= 20 + \frac{75 - 45}{2 \times 75 - 45 - 35} \times 5$$

$$= 20 + \frac{30}{70} \times 5 = 22.14$$

MEDIAN OF GROUPED DATA

Calculation of median for grouped data :

Case (*i*) when the data is discrete.

(*a*) Arrange the data in ascending or descending order of magnitude alongwith their frequencies.

(*b*) Append a column of cumulative frequencies.

(*c*) (*i*) If the total frequency n is odd, then $\left(\frac{n+1}{2}\right)$th observation is the median.

(*ii*) If the total frequency n is even, then the mean of $\frac{n}{2}$th and $\left(\frac{n+2}{2}\right)$th observations is median.

9. The daily wages (in rupees) of 100 workers in a factory are given below :

Daily wages (in ₹)	Number of workers
125	6
130	20
135	24
140	28
145	15
150	4
160	2
180	1

Find the median wage of a worker for the above data.

Sol.

Wages (x_i)	No. of workers (f_i)	Cumulative frequency
125	6	6
130	20	26
135	24	50
140	28	78
145	15	93
150	4	97
160	2	99
180	1	100
	Total = 100	

$n = 100$ (even)

$\therefore$ Median is the mean of $\left(\frac{100}{2}\right)$th and $\left(\frac{100+2}{2}\right)$th observations, *i.e.*, 50th and 51th observations.

$\therefore$ Median $= \frac{135+140}{2} = 137.5$

$\therefore$ Median wage of a worker in the factory is ₹ 137.50

10. Obtain the median for the following frequency distribution :

x_i	f_i
1	8
2	10
3	11
4	16
5	20
6	25
7	15
8	9
9	5

Sol.

x_i	f_i	Cumulative frequency
1	8	8
2	10	18
3	11	29
4	16	45
5	20	65
6	25	90
7	15	105
8	9	114
9	5	119
Total	$\Sigma f_i = 119$	

Here, $n = 119$ (odd)

$\therefore$ Median is $\left(\frac{n+1}{2}\right)$ th observation

$= \left(\frac{119+1}{2}\right)$ th observation, *i.e.*, 60th observation.

From cumulative frequencies, we find 60th observation is 5.

$\therefore$ Median = 5.

Case (*ii*) When the data is continuous and in the form of frequency distribution.

$$\text{Median} = l + \left(\frac{\frac{n}{2} - c}{f}\right) \times h,$$

where, l = lower limit of median class,

f = frequency of median class,

c = cumulative frequency of class preceding the median class,

h = class size,

n = number of observations.

11. Find the median of the following data :

Class interval	Frequency
20 — 30	4
30 — 40	5
40 — 50	6
50 — 60	9
60 — 70	11
70 — 80	12
80 — 90	8
90 — 100	5

Sol.

Classes	Frequency	Cumulative frequency	
20 — 30	4	4	
30 — 40	5	9	
40 — 50	6	15	
50 — 60	9	24	
60 — 70	11	35	⟵ Median class
70 — 80	12	47	
80 — 90	8	55	
90 —100	5	60	
Total	$n = 60$		

$n = \frac{60}{2} = 30$

$\therefore$ 60 – 70 is the median class.

$l = 60, f = 11, c = 24, h = 10$

$$\therefore \text{ Median} = l + \left(\frac{\frac{n}{2} - c}{f}\right) \times h$$

$$= 60 + \left(\frac{30 - 24}{11}\right) \times 10 = 65.45$$

12. Find the median for the following data :

Class interval	Frequency
110 — 119	5
120 — 129	25
130 — 139	40
140 — 149	60
150 — 159	40
160 — 169	25
170 — 179	5

Sol. To compute the median, we have to convert the given distribution to a continuous one.

Class interval	Frequency	Cumulative frequency
109.5 — 119.5	5	5
119.5 — 129.5	25	30
129.5 — 139.5	40	70
139.5 — 149.5	60	130 ←Median
149.5 — 159.5	40	170 class
159.5 — 169.5	25	195
169.5 — 179.5	5	200
Total	$n = 200$	

$\frac{n}{2} = \frac{200}{2} = 100$

$\therefore$ Median class is 139.5 – 149.5.

$l = 139.5, f = 60, c = 70, h = 10$

$$\therefore \text{ Median} = l + \left(\frac{\frac{n}{2} - c}{f}\right) \times h$$

$$= 139.5 + \left(\frac{100 - 70}{60}\right) \times 10 = 144.5.$$

13. If the median of the distribution

Class interval	Frequency
0 — 10	5
10 — 20	x_1
20 — 30	20
30 — 40	15
40 — 50	x_2
50 — 60	5
Total	$n = 60$

is 28.5, find the values of x_1 and x_2.

Sol.

Class interval	Frequency	Cumulative frequency
0 — 10	5	5
10 — 20	x_1	$5 + x_1$
20 — 30	20	$25 + x_1$
30 — 40	15	$40 + x_1$ → Median
40 — 50	x_2	$40 + x_1 + x_2$ class
50 — 60	5	$45 + x_1 + x_2$
Total	$n = 45 + x_1 + x_2$	

It is given $n = 60$

$\Rightarrow 45 + x_1 + x_2 = 60 \quad \Rightarrow x_1 + x_2 = 15$

Median 28.5 lies in the class 20 – 30.

$\therefore$ Median class is 20 – 30.

$l = 20, f = 20, c = 5 + x_1, h = 10$

$$\therefore \quad \text{Median} = l + \left(\frac{\frac{n}{2} - c}{f}\right) \times h$$

$$\Rightarrow \quad 28.5 = 20 + \left(\frac{30 - 5 - x_1}{20}\right) \times 10$$

$\Rightarrow \quad x_1 = 8$

$\therefore \quad x_2 = 7$

Hence, $x_1 = 8$ and $x_2 = 7$.

GRAPHICAL REPRESENTATION OF CUMULATIVE FREQUENCY DISTRIBUTION

(*a*) Cumulative frequency curve or an ogive (of the less than type) : Represent the data in the table graphically.

(*i*) Mark the upper limits of the class intervals on the horizontal axis and their corresponding cumulative frequencies on the vertical axis.

(*ii*) Plot the points corresponding to the ordered pairs given by (upper limit, corresponding cumulative frequency) and join them by a free hand smooth curve.

(*b*) Cumulative frequency curve or an ogive (of the more than type) :

(*i*) Mark the lower limit of the class intervals on the horizontal axis and their corresponding cumulative frequencies on vertical axis.

(*ii*) Plot the points corresponding to the ordered pairs given by (lower limit, corresponding cumulative frequency) and join them by a free hand smooth curve.

■ Median of a grouped data can be obtained graphically as the x-coordinate of the point of intersection of the two ogives (more than type and less than type) for this data.

IMPORTANT : 3 MEDIAN = MODE + 2 MEAN

14. The frequency distribution of marks obtained by 53 students out of 100 in a certain examination is given below :

Marks	No. of students
0 — 10	5
10 — 20	3
20 — 30	4
30 — 40	3
40 — 50	3
50 — 60	4
60 — 70	7
70 — 80	9
80 — 90	7
90 —100	8

(*i*) Draw the cumulative frequency curve or an ogive (of the less than type).

(*ii*) Draw the cumulative frequency curve or an ogive (of the more than type).

Sol. (*i*)

Marks obtained	Number of students (Cumulative frequency)
Less than 10	5
Less than 20	5 + 3 = 8
Less than 30	8 + 4 = 12
Less than 40	12 + 3 = 15
Less than 50	15 + 3 = 18
Less than 60	18 + 4 = 22
Less than 70	22 + 7 = 29
Less than 80	29 + 9 = 38
Less than 90	38 + 7 = 45
Less than 100	45 + 8 = 53

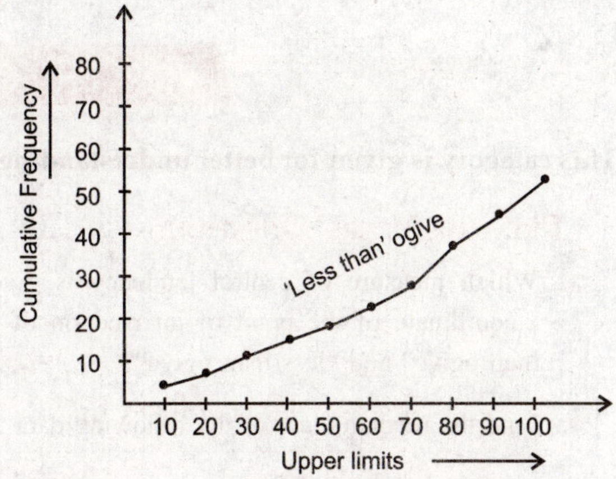

(*ii*)

Marks obtained	Number of students (Cumulative frequency)
More than or equal to 0	53
More than or equal to 10	53 – 5 = 48
More than or equal to 20	48 – 3 = 45
More than or equal to 30	45 – 4 = 41
More than or equal to 40	41 – 3 = 38
More than or equal to 50	38 – 3 = 35
More than or equal to 60	35 – 4 = 31
More than or equal to 70	31 – 7 = 24
More than or equal to 80	24 – 9 = 15
More than or equal to 90	15 – 7 = 8

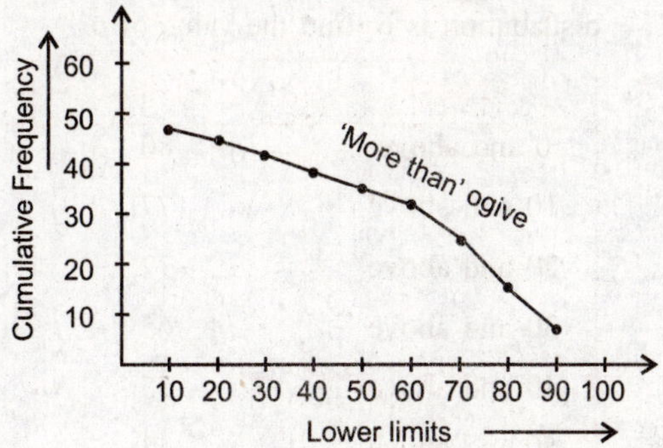

Summative Assessment

Subjective Questions

VERY SHORT ANSWER TYPE QUESTIONS

This category is given for better understanding of the topic and for EXTRA PRACTICE.

Previous Years' Questions

1. Which measure of central tendency is given by the x-coordinate of the point of intersection of the "more than ogive" and "less than ogive"? [Delhi 2008]

2. Find the median class of the following data :[AI 2008]

Marks obtained	Frequency
0 – 10	8
10 – 20	10
20 – 30	12
30 – 40	22
40 – 50	30
50 – 60	18

3. Find the class marks of classes 10 – 25 and 35 – 55. [Foreign 2008]

4. What is the lower limit of the modal class of the following distribution

Age in years	Number of patients
0 – 10	16
10 – 20	13
20 – 30	6
30 – 40	11
40 – 50	27
50 – 60	18

[Foreign 2009]

Practice Questions

5. If the mean of the following distribution is 2.6, then the value of y will be ?

Variable	1	2	3	4	5
Frequency	4	5	y	1	2

6. If $\Sigma f_i = 11$, $\Sigma f_i x_i = 2p + 52$ and the mean of any distribution is 6, find the value of p.

7.

Marks	No. of students
0 and above	80
10 and above	77
20 and above	72
30 and above	65
40 and above	55
50 and above	43
60 and above	28
70 and above	16
80 and above	10
90 and above	8
100 and above	0

Write the proper data (marks) and frequency (No. of students).

8. The following are the marks of 9 students in a class. Find the median of

21, 24, 27, 30, 32, 34, 35, 38, 48.

9. Find the median of the daily wages of ten workers from the following data :

8, 9, 11, 14, 15, 17, 18, 20, 22, 25

10.

x	f	$c.f.$
1	8	8
2	10	18
3	11	29
4	16	45
5	20	65
6	25	90
7	15	105
8	9	114
9	6	120

Find the median for the above frequency distribution.

11.

Marks	No. of students (frequency)	c.f.
0 – 10	5	5
10 – 30	15	F
30 – 60	f	50
60 – 80	8	58
80 – 90	2	60
	N = 60	$N = \Sigma f_i = 60$

Find f and F.

12. Find the mode of the given data : 120, 110, 130, 110, 120, 140, 130, 120, 140, 120

13. Find the mode of the following data :

25, 16, 19, 48, 19, 20, 34, 15, 19, 20, 21, 24, 19, 16, 22, 16, 18, 20, 16, 19.

14. Find the value of x, if the mode of the following data is 25.

15, 20, 25, 18, 14, 15, 25, 15, 18, 16, 20, 25, 20, x, 18

15. The shirt sizes worn by a group of 200 persons, who bought the shirt from a store are as follows:

Shirt size	No. of persons
37	15
38	25
39	39
40	41
41	36
42	17
43	15
44	12

Find the modal shirt size worn by the group.

16. The class marks of a frequency distribution are 6, 10, 14, 18, 22, 26, 30. Find class size.

17. For a particular year, the following is the distribution of the age (in yrs.) of primary school teachers in H.P. :

Age (in yrs)	No. of teachers
16 – 20	11
21 – 25	32
26 – 30	51
31 – 35	49
36 – 40	27
41 – 45	6
46 – 50	4

Find how many teachers are of age less than 31 years.

SHORT ANSWER TYPE QUESTIONS-I

PREVIOUS YEARS' QUESTIONS

1. Convert the following data to a less than type distribution.

C.I.	Frequency
50 – 55	2
55 – 60	8
60 – 65	12
65 – 70	24
70 – 75	38
75 – 80	16

[CBSE (CCE) 2011]

2. Write the frequency distribution table for the following data :

Marks	No. of students
Below 10	0
Below 20	15
Below 30	20
Below 40	30
Below 50	35
Below 60	40

[CBSE (CCE) 2011]

3. Calculate mode of the following data :

Marks Obtained	No. of students
0 – 20	8
20 – 40	10
40 – 60	12
60 – 80	6
80 – 100	3

[CBSE (CCE) 2011]

4. Convert the following frequency distribution to a more then type cumulative frequency distribution

Marks Obtained	No. of students
0 – 20	5
20 – 40	9
40 – 60	12
60 – 80	8
80 – 100	6

[CBSE (CCE) 2012]

5. (*i*) If the mean of the following distribution is 6, find the value of p :

x	2	4	6	10	$p + 5$
y	3	2	3	1	2

(*ii*) In a school 85 boys and 35 girls appeared in a public examination. The mean marks of boys was found to be 40%, whereas the mean marks of girls was 60%. Determine the average marks percentage of the school.

6. The mean score of 25 observations is 80 and the mean score of another 55 observations is 65. Determine the mean score of the whole set of observations.

7. Arun scored 36 marks in English, 44 marks in Hindi, 75 marks in Mathematics and x marks in Science. If he has scored an average of 50 marks, find the value of x.

8. The average height of 30 students is 150 cm. It was detected later that one value of 165 cm was wrongly copied as 135 cm for the computation of mean. Find the correct mean.

9. Find the median wage of a worker engaged at a construction site whose data are given below :

Wages (in ₹)	Number of workers
3500	12
3800	13
4100	25
4500	17
5500	15
6500	12
7000	6

10. Find the median for the following data :

Marks (out of 20)	Number of students
5	4
9	5
10	6
12	12
13	11
16	6
18	4
20	2

11. Find the median marks for the following distribution :

Classes	Number of Students
0 — 10	2
10 — 20	12
20 — 30	22
30 — 40	8
40 — 50	6

12. (*i*) Find the mode of the following data :

25, 16, 19, 48, 19, 20, 34, 15, 19, 20, 21, 24, 19, 16, 22, 16, 18, 20, 16, 19.

(*ii*) If one of the 19's is changed to 16 in the above data, find the new mode.

13. (*i*) Find the value of x, if the mode of the following data is 25 :

15, 20, 25, 18, 14, 15, 25, 15 18, 16, 20, 25, 20, x, 18

(*ii*) If both 20 and 18 are changed to 15 in the above data, find the new mode.

Find the mode of the following distribution (Q.14 to Q. 16)

14.

Class Interval	Frequency
0 — 10	5
10 — 20	8
20 — 30	7
30 — 40	12
40 — 50	28
50 — 60	20
60 — 70	10
70 — 80	10

15.

Class Interval	Frequency
10 — 15	30
15 — 20	45
20 — 25	75
25 — 30	35
30 — 35	25
35 — 40	15

16.

Class Interval	Frequency
25 — 30	25
30 — 35	34
35 — 40	50
40 — 45	42
45 — 50	38
50 — 55	14

17. 20 years ago, when my parents got married, their average age was 23 years, now the average age of my family consisting of myself and my parents is 34 years. What is my present age ?

18. Find the combined mean of a group of 150, if the mean of 50 students is 40 and that of other 100 students is 50.

19. Given that the mean of five numbers is 27. If one of the numbers is excluded, the mean gets reduced by 2. Determine the excluded number.

20. The average marks of A, B and C is 33, while the average marks of B, C and D is 37. If A obtains 30 marks, find the marks obtained by D.

21. The mean monthly salary of the 12 employees of a firm is ₹ 1450. If one more person joins the firm who gets ₹ 1600 per month, what will be the mean monthly salary now ?

22. If the mean of the following data is 12, find p.

x	4	8	p	16	20
y	5	3	12	5	4

23. The mean of a set of 67 values is 35. If each of the 67 values is divided by 4, what will be the mean of new set of values ?

24. The average monthly salary of 15 workers in a factory is ₹ 285. If the salary of the manager is included, the average becomes ₹ 355. What is the manager's salary ?

25. The average weight of A, B, C is 45 kg. If the average weight of A and B be 40 kg. and that of B and C be 43 kg, find the weight of B.

26. Nine persons went to a hotel for taking their meals. Eight of them spent ₹ 12 each on their meals and the ninth spent ₹ 8 more than the average expenditure of all the nine. What was the total money spent by them ?

27. There were 35 students in a hostel. Due to the admission of 7 new students, the expenses of the mess were increased by ₹ 42 per day while the average expenditure per head diminished by ₹ 1. What was the original expenditure of the mess ?

NCERT Questions

28. The following data gives the information on the observed lifetimes (in hours) of 225 electrical components :

Lifetimes (in hours)	Frequency
0 — 20	10
20 — 40	35
40 — 60	52
60 — 80	61
80 — 100	38
100 — 120	29

Determine the modal lifetimes of the components.

29. A student noted the number of cars passing through a spot on a road for 100 periods each of 3 minutes and summarised it in the table given below. Find the mode of the data :

Number of cars	Frequency
0 — 10	7
10 — 20	14
20 — 30	13
30 — 40	12
40 — 50	20
50 — 60	11
60 — 70	15
70 — 80	8

SHORT ANSWER TYPE QUESTIONS-II

Previous Years' Questions

1. The arithmetic mean of the following frequency distribution is 50. Find the value of p.

Class	Frequency
0 — 20	17
20 — 40	p
40 — 60	32
60 — 80	24
80 —100	19

[Delhi 2006]

2. If the mean of the following distribution is 50, find the value of f_1 :

Class	Frequency
0 —20	17
20 — 40	28
40 — 60	32
60 — 80	f_1
80 —100	19

[Delhi 2006C]

3. Find 'p' if the mean of the given data is 15.45.

Class	Frequency
0 — 6	6
6 — 12	8
12 — 18	p
18 — 24	9
24 — 30	7

[CBSE (CCE) 2011]

4. The median of the distribution given below is 14.4. Find the values of x and y, if the sum of frequency is 20.

Class interval	Frequency
0 — 6	4
6 — 12	x
12 — 18	5
18 — 24	y
24 — 30	1

[CBSE (CCE) 2011]

5. Weekly income of 600 families is given below.

Income in ₹	Frequency
0 — 1000	250
1000 — 2000	190
2000 — 3000	100
3000 — 4000	40
4000 — 5000	15
5000 — 6000	5

Find the median. [CBSE (CCE) 2012]

Practice Questions

6. Find the value of p, if the mean of the following distribution is 18 :

x	f
13	8
15	2
17	3
19	4
$20 + p$	$5p$
23	6

7. Find the mean of the following frequency distribution :

Class Interval	Frequency
0 — 10	4
10 — 20	4
20 — 30	7
30 — 40	10
40 — 50	12
50 — 60	8
60 — 70	5
Total	50

8. The data on number of patients attending a hospital in a month are given below. Find the average number of patients attending the hospital in a day.

Numbers of Patients	Number of days attending hospital
0 — 10	2
10 — 20	6
20 — 30	9
30 — 40	7
40 — 50	4
50 — 60	2

Compute mean of the grouped data given in following question nos. (9 to 11) :

9.

Daily wages	No. of workers
12.5 — 17.5	2
17.5 — 22.5	22
22.5 — 27.5	19
27.5 — 32.5	14
32.5 — 37.5	3
37.5 — 42.5	4
42.5 — 47.5	6
47.5 — 52.5	1
52.5 — 57.5	1

10.

Income (in Rupees)	No. of workers
200 — 300	5
300 — 400	36
400 — 500	24
500 — 600	16
600 — 700	9
700 — 800	6
800 — 900	4

11.

Monthly salary	No. of persons
325.5 — 350.5	20
350.5 — 375.5	10
375.5 — 400.5	10
400.5 — 425.5	5
425.5 — 450.5	1
450.5 — 475.5	2
475.5 — 500.5	2

12. Given below is the expenditure (in rupees) on water consumption of 70 houses in a locality. Find the average expenditure (in rupees) per house.

Expenditure on water (in rupees)	Number of houses
15 — 20	7
20 — 25	5
25 — 30	7
30 — 35	8
35 — 40	9
40 — 45	11
45 — 50	7
50 — 55	5
55 — 60	4
60 — 65	4
65 — 70	3

13. The table below gives the frequency distribution of the number of teachers in Higher Secondary Schools in 2012 in India. Find the average number of teachers per Higher Secondary School in India for 2012.

Number of Teachers	Number of H.S. Schools
6 — 10	955
11 — 15	1067
16 — 20	1663
21 — 25	1492
26 — 30	1220
31 — 35	1129
36 — 40	745
41 — 45	637
46 — 50	442

14. The mean of the following frequency table is 50. But the frequencies f_1 and f_2 in classes 20—40 and 60—80 are missing. Find the missing frequencies.

Class interval	Frequency
0 — 20	17
20 — 40	f_1
40 — 60	32
60 — 80	f_2
80 — 100	19
Total	120

15. Following is the distribution of marks obtained by 60 students in Economics test :

Marks	No. of students
More than 0	60
More than 10	56
More than 20	40
More than 30	20
More than 40	10
More than 50	3

Calculate the arithmetic mean.

16. Find the average marks of students from the following table :

Marks	No. of students
Above 0	80
Above 10	77
Above 20	72
Above 30	65
Above 40	55
Above 50	43
Above 60	23
Above 70	16
Above 80	10
Above 90	8
Above 100	0

17. The mean monthly wages of a group of 15 workers in a factory was ₹ 3351. 4 workers whose mean monthly wage was ₹ 3310 left the factory and a new worker at a monthly wage of ₹ 3025 was appointed. Find the mean wage of the remaining group.

18. The following table shows marks secured by 140 students in an examination :

Marks	No. of students
0 — 10	20
10 — 20	24
20 — 30	40
30 — 40	36
40 — 50	20

Calculate mean marks by using all the three methods, *i.e.*, direct method, assumed mean method and step-deviation method.

19. Find the mean age in years from the frequency distribution given below :

Class Interval of age in year	Frequency (f_i)
25 — 29	4
30 — 34	14
35 — 39	22
40 — 44	16
45 — 49	6
50 — 54	5
55 — 59	3

20. Find the mean marks from the following data :

Marks	No. of students
Below 10	5
Below 20	9
Below 30	17
Below 40	29
Below 50	45
Below 60	60
Below 70	70
Below 80	78
Below 90	83
Below 100	85

21. In the following frequency distribution, the frequency of the class-interval (40—50) is missing. It is known that the mean of the distribution is 52. Find the missing frequency.

Wages (in ₹)	No. of Workers
10—20	5
20—30	3
30—40	4
40—50	f
50—60	2
60—70	6
70—80	13

22. Calculate the median for the following data :

Classes	Frequency
20 — 40	12
40 — 60	18
60 — 80	23
80 — 100	15
100 — 120	12
120 — 140	12
140 — 160	8

23. Calculate the median for the following data :

Class interval	Frequency
130 — 140	5
140 — 150	9
150 — 160	17
160 — 170	28
170 — 180	24
180 — 190	10
190 — 200	7

24. Calculate the median for the following data :

Marks obtained	Number of students
Less than 20	0
Less than 30	4
Less than 40	16
Less than 50	30
Less than 60	46
Less than 70	66
Less than 80	82
Less than 90	92
Less than 100	100

Find the median for each of the following data : (Q. 25 to 27)

25.

Class interval	Frequency
130 — 139	4
140 — 149	9
150 — 159	18
160 — 169	28
170 — 179	24
180 — 189	10
190 — 199	7

26.

Class Interval	Frequency
10 — 19	2
20 — 29	4
30 — 39	8
40 — 49	9
50 — 59	4
60 — 69	2
70 — 79	1

27.

Wages (in ₹)	Number of workers
More than 150	Nil
More than 140	12
More than 130	27
More than 120	60
More than 110	105
More than 100	124
More than 90	141
More than 80	150

28. The ages of employees in two factories A and B are given below :

Age of Employees (in Years)	Number of employees in factories	
	A	B
20— 30	5	8
30— 40	26	40
40— 50	78	58
50— 60	104	90
60— 70	98	83

Find the modal age of employees in factory A and factory B.

29. Compare the modal ages of two groups of students A and B appearing for an entrance test :

Age (in years)	Number of students in	
	Group A	Group B
16 — 18	50	54
18 — 20	78	89
20 — 22	46	40
22 — 24	28	25
24 — 26	23	17

NCERT Questions

30. Thirty women were examined in a hospital by a doctor and the number of heart beats per minute were recorded and summarised as follows. Find the mean heart beats per minute for these women, choosing a suitable method.

Number of heart beats per minute	Number of women
65 — 68	2
68 — 71	4
71 — 74	3
74 — 77	8
77 — 80	7
80 — 83	4
83 — 86	2

31. In a retail market, fruit vendors were selling mangoes kept in packing boxes. These boxes contained varying number of mangoes. The following was the distribution of mangoes according to the number of boxes.

Number of mangoes	Number of boxes
50 — 52	15
53 — 55	110
56 — 58	135
59 — 61	115
62 — 64	25

Find the mean number of mangoes kept in a packing box. Which method of finding the mean did you choose ?

32. To find out the concentration of SO_2 in the air (in parts per million, *i.e.*, ppm), the data was collected for 30 localities in a certain city and is presented below :

Concentration of SO_2 (in ppm)	Frequency
0.00 — 0.04	4
0.04 — 0.08	9
0.08 — 0.12	9
0.12 — 0.16	2
0.16 — 0.20	4
0.20 — 0.24	2

Find the mean concentration of SO_2 in the air.

33. The following data gives the distribution of total monthly household expenditure of 200 families of a village. Find the modal monthly expenditure of the families. Also, find the mean monthly expenditure.

Expenditure (in ₹)	Number of families
1000 — 1500	24
1500 — 2000	40
2000 — 2500	33
2500 — 3000	28
3000 — 3500	30
3500 — 4000	22
4000 — 4500	16
4500 — 5000	7

34. The following distribution gives the state-wise teacher - student ratio in higher secondary schools of India. Find the mode and mean of this data. Interpret the two measures.

Number of Students per teacher	Number of States/U.T.
15 — 20	3
20 — 25	8
25 — 30	9
30 — 35	10
35 — 40	3
40 — 45	0
45 — 50	0
50 — 55	2

35. The lengths of 40 leaves of a plant are measured correct to the nearest millimetre, and the data obtained is represented in the following table :

Length (in mm)	Number of leaves
118 — 126	3
127 — 135	5
136 — 144	9
145 — 153	12
154 — 162	5
163 — 171	4
172 — 180	2

Find the median length of the leaves.

LONG ANSWER TYPE QUESTIONS

Previous Years' Questions

1. A survey regarding the heights (in cm) of 50 girls of class X[th] of a school was conducted and the following data was obtained :

Height (in cm)	Number of girls
120 – 130	2
130 – 140	8
140 – 150	12
150 – 160	20
160 – 170	8
Total	50

Find the mean, median and mode of the above data. [Delhi 2008]

2. Find mean, median and mode of the following data :

Classes	Frequency
0 — 20	6
20 — 40	8
40 — 60	10
60 — 80	12
80 —100	6
100 —120	5
120 —140	3

[Foreign 2008]

3. Find the mean, mode and median for the following data:

Marks	Frequency
0 — 10	8
10 — 20	16
20 — 30	36
30 — 40	34
40 — 50	6
Total	100

[Delhi 2008C]

4. The median of the following data is 52.5. Find the values of x and y, if the total frequency is 100.

Class Interval	Frequency
0 — 10	2
10 — 20	5
20 — 30	x
30 — 40	12
40 — 50	17
50 — 60	20
60 — 70	y
70 — 80	9
80 — 90	7
90 — 100	4

[AI 2008C]

5. During the medical check-up of 35 students of a class their weights were recorded as follows :

Weight (in kg)	Number of students
38 — 40	3
40 — 42	2
42 — 44	4
44 — 46	5
46 — 48	14
48 — 50	4
50 — 52	3

Draw a less than type and a more than type ogive from the given data. Hence obtain the median weight from the graph. [AI 2009]

6. Draw 'less than ogive' and 'more than ogive' for the following distribution and hence find its median.

Class	Frequency
20 – 30	8
30 – 40	12
40 – 50	24
50 – 60	6
60 – 70	10
70 – 80	15
80 – 90	25

[AI 2010]

7. Draw 'less than ogive' and 'more than ogive' for the following distribution and hence find its median.

Class	Frequency
20 – 30	25
30 – 40	15
40 – 50	10
50 – 60	6
60 – 70	24
70 – 80	12
80 – 90	8

[AI 2010]

8. If the mean of the following frequency distribution is 65.6, find the missing frequencies (f_1, f_2) :

Class	Frequency
10 – 30	5
30 – 50	8
50 – 70	f_1

70 – 90	20
90 – 110	f_2
110 – 130	2
Total	50

[Foreign 2010]

9. If the mean of the following frequency distribution is 91, find the missing frequencies (f_1, f_2) :

Class	Frequency
0 – 30	12
30 – 60	21
60 – 90	f_1
90 – 120	52
120 – 150	f_2
150 – 180	11
Total	150

[Foreign 2010]

10. Draw a less than type ogive of the following distribution:

Marks	No. of students
0 – 10	5
10 – 20	4
20 – 30	8
30 – 40	10
40 – 50	15
50 – 60	18

Find median from graph. [CBSE (CCE) 2011]

11. Find the median by drawing both types of ogives.

Class interval	Frequency
50 – 60	3
60 – 70	5
70 – 80	9
80 – 90	12
90 – 100	6

[CBSE (CCE) 2011]

12. Find the value of f_1 from the following data, if its mode is 65 :

Class	Frequency
0 — 20	6
20 — 40	8
40 — 60	f_1
60 — 80	12
80 —100	6
100 —120	5

Where frequency 6, 8, f_1 and 12 are in assending order. [CBSE (CCE) 2012]

Practice Questions

13. The median of the following data is 32.5.

Class interval	Frequency
0 — 10	x
10 — 20	5
20 — 30	9
30 — 40	12
40 — 50	y
50 — 60	3
60 — 70	2
Total	40

Find the values of x and y.

14. Compute the median from the following data :

Mid-value	Frequency
115	6
125	25
135	48
145	72
155	116
165	60
175	38
185	22
195	3

15. If the median of the following frequency distribution is 46, find the missing frequencies.

Variable	Frequency
10—20	12
20—30	30
30—40	x
40—50	65
50—60	y
60—70	25
70—80	18
Total	230

16. Find the missing frequencies and the median for the following distribution if the mean is 1.46.

No. of accidents	Frequency (No. of days)
0	46
1	x
2	y
3	25
4	10
5	5
Total	200

17. An incomplete distribution is given below :

Variable	Frequency
10—20	12
20—30	30
30—40	x
40—50	65
50—60	y
60—70	25
70—80	18

You are given that the median value is 46 and the total number of items is 230.

(*i*) Using the median formula and fill up missing frequencies.

(*ii*) Calculate the A.M. of the completed distribution.

NCERT Questions

18. The median of the following data is 525.

Class interval	Frequency
0 — 100	2
100 — 200	5
200 — 300	x
300 — 400	12
400 — 500	17
500 — 600	20
600 — 700	y
700 — 800	9
800 — 900	7
900 — 1000	4

Find the values of x and y, if the total frequency is 100.

19. The following frequency distribution gives the monthly consumption of electricity of 68 consumers of a locality. Find the median, mean and mode of the data and compare them. Also verify that 3 median = mode + 2 mean.

Monthly consumption (in units)	Number of consumers
65 — 85	4
85 — 105	5
105 — 125	13
125 — 145	20
145 — 165	14
165 — 185	8
185 — 205	4

20. 100 surnames were randomly picked up from a local telephone directory and the frequency distribution of the number of letters in the English alphabet in the surnames was obtained as follows :

Number of letters	Number of surnames
1 — 4	6
4 — 7	30
7 — 10	40
10 — 13	16
13 — 16	4
16 — 19	4

Determine the median number of letters in the surnames. Find the mean number of letters in the surnames ? Also, find the modal size of the surnames.

21. The following distribution gives the daily income of 50 workers of a factory.

Daily income (in ₹)	Number of workers
100 — 120	12
120 — 140	14
140 — 160	8
160 — 180	6
180 — 200	10

Convert the distribution above to a less than type cumulative frequency distribution, and draw its ogive.

22. During the medical check up of 35 students of a class, their weights were recorded as follows :

Weight(in kg)	Number of students
Less than 38	0
Less than 40	3
Less than 42	5
Less than 44	9
Less than 46	14
Less than 48	28
Less than 50	32
Less than 52	35

Draw a less than type ogive for the given data. Hence obtain the median weight from the graph and verify the result by using the formula.

23. The following table gives production yield per hectare of wheat of 100 farms of a village.

Production yield (in kg/ha)	Number of farms
50 — 55	2
55 — 60	8
60 — 65	12
65 — 70	24
70 — 75	38
75 — 80	16

Change the distribution to a more than type distribution, and draw its ogive.

Short Answer Questions with Reasoning

1. The median of an ungrouped data and the median calculated when the same data is grouped are always the same. Do you think that this is a correct statement ? Give reason.

2. In calculating the mean of grouped data, grouped in classes of equal width, we may use the formula

$$\bar{x} = a + \frac{\sum f_i d_i}{\sum f_i}$$

Where a is the assumed mean a must be one of the mid-points of the classes. Is the last statement correct ? Justify your answer.

3. Is it true to say that the mean, mode and median of grouped data will always be different ? Justify your answer.

4. Will the median class and modal class of grouped data always be different ? Justify your answer.

Short Answer Questions

5. The following table shows the cumulative frequency distribution of marks of 800 students in an examination:

Marks	Number of students
Below 10	10
Below 20	50
Below 30	130
Below 40	270
Below 50	440
Below 60	570
Below 70	670
Below 80	740
Below 90	780
Below 100	800

Construct a frequency distribution table for the data above.

6. Form the frequency distribution table from the following data :

Marks (out of 90)	Number of candidates
More than or equal to 80	4
More than or equal to 70	6
More than or equal to 60	11
More than or equal to 50	17
More than or equal to 40	23
More than or equal to 30	27
More than or equal to 20	30
More than or equal to 10	32
More than or equal to 0	34

7. Find the unknown entries a, b, c, d, e, f in the following distribution of heights of students in a class :

Height (in cm)	Frequency	Cumulative frequency
150 — 155	12	a
155 — 160	b	25
160 — 165	10	c
165 — 170	d	43
170 — 175	e	48
175 — 180	2	f
Total	50	

Long Answer Questions

8. Find the mean marks of students for the following distribution :

Marks	Number of students
0 and above	80
10 and above	77
20 and above	72
30 and above	65
40 and above	55
50 and above	43
60 and above	28
70 and above	16
80 and above	10
90 and above	8
100 and above	0

9. Determine the mean of the following distribution :

Marks	Number of students
Below 10	5
Below 20	9
Below 30	17
Below 40	29
Below 50	45
Below 60	60
Below 70	70
Below 80	78
Below 90	83
Below 100	85

10. The median of the following data is 50. Find the values of p and q, if the sum of all the frequencies is 90.

Marks	Frequency
20 — 30	p
30 — 40	15
40 — 50	25
50 — 60	20
60 — 70	q
70 — 80	8
80 — 90	10

HIGHER ORDER THINKING SKILLS (HOTS) QUESTIONS

1. The interest paid on each of the three different sums of money yielding 1%, 2% and 4% simple interest per annum respectively, is the same. Find the average yield percent on the total sum invested.

2. The mean weight of 100 students in a class is 46 kg. The mean weight of the boys is 50 kg and of girls 40 kg. Find the number of boys in the class.

VALUE BASED QUESTIONS

1. To highlight child labour problem, some students organised a *javelin* throw competition. 50 students participated in this competition. The distance (in metres) thrown are recorded below :

Distance (in m)	Number of students
0 — 20	6
20 — 40	11
40 — 60	17
60 — 80	12
80 — 100	4

(*a*) Construct a cumulative frequency table.

(*b*) Draw cumulative frequency curve (less than type) and calculate the median distance thrown.

(*c*) Calculate the median distance by using the formula for median.

(*d*) Are the median distance calculated in (*a*) and (*b*) same ?

(*e*) Which value is depicted by students ?

2. Students of a locality decided to create an awareness about 'Save Electricity Campaign'. They recorded the monthly consumption of electricity of 68 consumers of that locality. The following frequency distribution gives the monthly consumption of electricity of these consumers of that locality.

Monthly consumption (in units)	Number of consumers
65 — 85	4
85 — 105	5
105 — 125	13
125 — 145	20
145 — 165	14
165 — 185	8
185 — 205	4

(*a*) Find the median, mean and mode of the data.

(*b*) Which value is depicted by students ?

3. Students of class X of a school decided to donate their pocket money to purchase food packets for flood victims in a near by village. They packed the food packets in different boxes. These boxes contained varying number of food packets. The following was the distribution of food packets according to the number of boxes.

Number of food packets	Number of boxes
50 — 52	15
53 — 55	110
56 — 58	135
59 — 61	115
62 — 64	25

(*a*) Find the mean number of food packets kept in a packing box.

(*b*) What value is depicted by the students ?

Multiple Choice Questions

Previous Years' Questions

1. **The middle most observation of a statistical data has value which is called**

(*a*) mean of the data
(*b*) mode of the data
(*c*) median of the data
(*d*) none of these

2. **The mean and median of same data are 24 and 26 respectively. The value of mode is : [CBSE (CCE) 2011]**

(*a*) 23 (*b*) 26
(*c*) 25 (*d*) 30

3. **Which of the following is not a measure of central tendency : [CBSE (CCE) 2011]**
(*a*) Mean
(*b*) Median
(*c*) Class interval
(*d*) Mode

4. **Mode is the value of the variable which has : [CBSE (CCE) 2012]**
(*a*) maximum frequency
(*b*) minimum frequency
(*c*) mean frequency
(*d*) middle most frequency

Practice Questions

5. **The central value of the set of observation is called**
(*a*) mean (*b*) median
(*c*) mode (*d*) A.M.

6. **The relationship between mean, median and mode for a moderately skewed distribution is**
(*a*) mode = median – 2 mean
(*b*) mode = 3 median – 2 mean
(*c*) mode = 2 median – 3 mean
(*d*) mode = median – mean

7. **The mean of 25 observations is 9. If each observation is increased by 4, then the new mean is**
(*a*) 10 (*b*) 11
(*c*) 12 (*d*) 13

8. **In an arranged series of an even number of 2*n* terms the median is**
(*a*) $(n)^{th}$ term
(*b*) $(n+1)^{th}$ term
(*c*) mean of $(n)^{th}$ term and $(n+1)^{th}$ term
(*d*) none of these

9. **A set of numbers consists of four 5's, six 7's, ten 9's, eleven 12's, three 13's, two 14's. The mode of this set of numbers is**
(*a*) 9 (*b*) 12
(*c*) 7 (*d*) 14

10. **Mode and mean of a data are 12*k* and 15*k*. Median of the data is**
(*a*) 12*k* (*b*) 14*k*
(*c*) 15*k* (*d*) 16*k*

11. **d_i is the deviation of x_i from assumed mean *a*.**
If mean = $x + \dfrac{\sum f_i d_i}{\sum f_i}$, then *x* is
(*a*) class size (*b*) number of observations
(*c*) assumed mean (*d*) none of these

12. **If $\bar{x}$ is the mean of a distribution, then $\sum f_i(x_i - \bar{x})$ is equal to**
(*a*) 0 (*b*) mean
(*c*) sum of observations (*d*) none of these

13. **The mean of first *n* natural number is**
(*a*) $\dfrac{n(n+1)}{2}$ (*b*) $n(n+1)$
(*c*) $\dfrac{n+1}{2}$ (*d*) $n+1$

14. **Mean of 100 items is 49. It was discovered that three items which should have been 60, 70, 80 were wrongly read as 40, 20, 50 respectively. The correct mean is**
(*a*) 48 (*b*) 49
(*c*) 50 (*d*) 60

15. **If the mode of a data is 18 and the mean is 24, then median is**
(*a*) 18 (*b*) 24
(*c*) 22 (*d*) 21

16. **The mean of a set of numbers is $\bar{x}$. If each number is multiplied by *k*, then the mean of the new set is**
(*a*) $\bar{x} + k$ (*b*) $\dfrac{\bar{x}}{k}$
(*c*) $\bar{x} - k$ (*d*) $k.\bar{x}$

17. **If mean = (3 median – mode) . *k*, then the value of *k* is**
(*a*) 1 (*b*) 2
(*c*) 1/2 (*d*) 3/2

18. **A car travels from city A to city B, 120 km apart at an average speed of 50 km/h. It then makes a return trip at an average speed of 60 km/h. It covers another 120 km distance at an average speed of 40 km/h. The average speed over the entire 360 km will be**
(*a*) $\dfrac{50+60+40}{3}$ km/h (*b*) $\left(\dfrac{3}{\frac{1}{50}+\frac{1}{60}+\frac{1}{40}}\right)$ km/h
(*c*) $\dfrac{300}{50+60+40}$ km/h (*d*) none of these

19. **Mean of *n* numbers $x_1, x_2, \ldots x_n$ is *m*. If x_n is replaced by *x*, then new mean is**
(*a*) $m - x_n + x$ (*b*) $\dfrac{nm - x_n + x}{n}$
(*c*) $\dfrac{(n-1)m + x}{n}$ (*d*) $\dfrac{m - x_n + x}{n}$

20. **If mean of 1, 2, 3, ... *n* is $\dfrac{6n}{11}$, then *n* is**
(*a*) 10 (*b*) 12
(*c*) 11 (*d*) 13

21. **The mean of n terms is $\overline{x}$. If first term is increased by 1, second term by 2 and so on, then new mean is**

(a) $\overline{x}+\frac{n+1}{2}$ (b) $\overline{x}+\frac{n-1}{2}$

(c) $\overline{x}+\frac{n}{2}$ (d) $\overline{x}+n$

22. **Which of the following is most effected by extreme observations**

(a) mode (b) median

(c) mean (d) none of these

23. **The mean age of combined group of men and women is 30 years. If the mean of the age of men and women are respectively 32 and 27, then the percentage of women in the group is**

(a) 30 (b) 20

(c) 50 (d) 40

24. **The mean of 5 numbers is 18. One number is excluded their mean becomes 16. Then the excluded number is**

(a) 18 (b) 25

(c) 30 (d) 26

25. **The mean and median of 100 items are 50 and 52 respectively. The value of the largest item is 100. It was later found that it is 110 not 100. The true mean and median are**

(a) 50.10, 52 (b) 50, 52

(c) 50.20, 52 (d) none of these

26. **The median of set of 9 distinct observations is 20.5. If each of the largest 4 observations of the set is increased by 2, then the median of the new set**

(a) is increased by 2

(b) is decreased by 2

(c) is two times of the original number

(d) Remains the same as that of the original set.

27. **The median from the table is**

Value	7	8	9	10	11	12	13
Frequency	2	1	4	5	6	1	3

(a) 11 (b) 10

(c) 12 (d) 11.5

28. **The A.M. of a set of 50 numbers is 38. If two numbers of the set namely 55 and 45 are discarded, the A.M. of the remaining set of numbers is**

(a) 37.5 (b) 38.5

(c) 36 (d) 37

29. **Find the missing frequencies in the following frequency distribution if it is given that the mean is 1.46**

Variables	0	1	2	3	4	5	Total
Frequency	46	?	?	25	10	5	200

(a) 76, 36 (b) 37, 38

(c) 76, 38 (d) 70, 38

30. **Construction of cumulative frequency table is useful in determining the**

(a) mean (b) median

(c) mode (d) none of these

NCERT EXEMPLAR PROBLEMS

MULTIPLE CHOICE QUESTIONS

1. Choose the correct answer from the given four options : In the formula

$$\overline{x} = a+\frac{\sum f_i d_i}{\sum f_i}$$

for finding the mean of grouped data d_i's are deviation from a of

(a) lower limits of the classes

(b) upper limits of the classes

(c) mid-points of the classes

(d) frequencies of the class marks

2. While computing mean of grouped data, we assume that the frequencies are

(a) evenly distributed over all the classes

(b) centred at the classmarks of the classes

(c) centred at the upper limits of the classes

(d) centred at the lower limits of the classes

3. In the formula $\overline{x} = a + h\left(\frac{\sum f_i u_i}{\sum f_i}\right)$, for finding the mean of grouped frequency distribution, u_i =

(a) $\frac{x_i+a}{h}$ (b) $h(x_i-a)$

(c) $\frac{x_i-a}{h}$ (d) $\frac{a-x_i}{h}$

4. The abscissa of the point of intersection of the less than type and of the more than type cumulative frequency curves of a grouped data gives its

(a) mean

(b) median

(c) mode

(d) all the three above

5. For the following distribution :

Marks	Number of students
Below 10	3
Below 20	12
Below 30	27
Below 40	57
Below 50	75
Below 60	80

the modal class is

(*a*) 10 – 20
(*b*) 20 – 30
(*c*) 30 – 40
(*d*) 50 – 60

6. The times, in seconds, taken by 150 atheletes to run a 110 m hurdle race are tabulated below :

Class	Frequency
13.8 – 14.0	2
14.0 – 14.2	4
14.2 – 14.4	5
14.4 – 14.6	71
14.6 – 14.8	48
14.8 – 15.0	20

The number of atheletes who completed the race in less then 14.6 seconds is :

(*a*) 11 (*b*) 71
(*c*) 82 (*d*) 130

ANSWERS

Very Short Answer Type Questions

1. Median **2.** 30 — 40 **3.** 17.5, 45 **4.** 40 **5.** $y = 8$ **6.** $p = 7$

7.

Marks	0 – 10	10 – 20	20 – 30	30 – 40	40 – 50	50 – 60	60 – 70	70 – 80	80 – 90	90 – 100
No. of students	3	5	7	10	12	15	12	6	2	8

8. 32 **9.** 16 **10.** 5 **11.** $f = 30$, F = 20 **12.** 120 **13.** 19
14. $x = 25$ **15.** 40 **16.** 4 **17.** 94

Short Answer Type Questions-I

3. 45 **5.** (*i*) 7 (*ii*) 45.83 **6.** 69.69 **7.** 45 **8.** 151 cm **9.** ₹ 4300
10. 12 **11.** 25 **12.** (*i*)19 (*ii*) 16 **13.** (*i*) $x = 25$, (*ii*)15 **14.** 46.67 **15.** 22.14
16. 38.33 **17.** 16 years **18.** 46.67 **19.** 35 **20.** 42 marks **21.** ₹ 1461.54
22. $p = 12$ **23.** 8.75 **24.** ₹ 1405 **25.** 31 kg **26.** ₹ 117 **27.** ₹ 420
28. 65.625 hours **29.** 44.7 cars

Short Answer Type Questions-II

1. $p = 28$ **2.** $f_1 = 24$ **3.** $p = 10$ **4.** $x = 4, y = 6$ **5.** 1263.158 **6.** $p = 1$
7. 38.2 **8.** 28.67 **9.** 27.85 **10.** ₹ 472 **11.** ₹ 373.5 **12.** ₹ 39.57
13. 27.75 **14.** $f_1 = 28, f_2 = 24$ **15.** 26.5 **16.** 51.125 **17.** ₹ 3337.52 **18.** 25.86 marks
19. 39.36 years **20.** 48.41 marks **21.** 7 **22.** 77.39 **23.** 166.79 **24.** 62
25. 166.3 **26.** 40.61 **27.** 116.67 **28.** Factory A : 58.125; Factory B : 58.205
29. A = 18.93, B = 18.83 **30.** 75.9 **31.** 57.19 **32.** 0.099 ppm
33. ₹ 1847.83, ₹ 2662.5
34. 30.6, 29.2; Most states/U.T. have a student teacher ratio of 30.6 and on an average, this ratio is 29.2. **35.** 146.75 mm

Long Answer Type Questions

1. Mean = 150.25, Median = 151.5, Mode = 154 **2.** Mean = 59.9, Median = 61.6, Mode = 65
3. Mean = 26.4, Mode = 29.09, Median = 27.22 **4.** $x = 9, y = 15$
6. Median = 60 **7.** Median = 50 **8.** $f_1 = 12$, $f_2 = 3$ **9.** $f_1 = 34$, $f_2 = 20$ **10.** median = 42 **11.** median = 80.4
12. $f_1 = 10$ **13.** $x = 3; y = 6$ **14.** 153.8 **15.** $x = 34, y = 46$ **16.** $x = 76, y = 38$; median = 1
17. (*i*) $x = 34$ and $y = 46$ (*ii*) mean = 45.87 **18.** $x = 9, y = 15$
19. Median = 137 units; Mean = 137.05 units; Mode = 135.77 units. The three measures are approximately the same in this case.
20. Median = 8.05; Mean = 8.32; Model size = 7.88 **22.** Median = 46.5 kg

NCERT Exemplar Problems

1. (False) Not always
2. (False) Not necessary
3. No
4. No

5.

Marks	No. of students
0 – 10	10
10 – 20	40
20 – 30	80
30 – 40	140
40 – 50	170
50 – 60	130
60 – 70	100
70 – 80	70
80 – 90	40
90 – 100	20

6.

Marks	No. of candidates
0 – 10	2
10 – 20	2
20 – 30	3
30 – 40	4
40 – 50	6
50 – 60	6
60 – 70	5
70 – 80	2
80 – 90	4

7. $a = 12, b = 13, c = 35, d = 8, e = 5, f = 50$
8. 51.75
9. 48.41
10. $p = 5, q = 7$

Higher Order Thinking Skills (Hots) Questions

1. 1 : 71%
2. 60

Value Based Questions

1. (*a*)

Distance (in m)	No. of students		Cumulative frequency
0 – 20	6	Less than 20	6
20 – 40	11	Less than 40	17
40 – 60	17	Less than 60	34
60 – 80	12	Less than 80	46
80 – 100	4	Less than 100	50

(*b*) 49.4 m

(*c*) 49.41 m

(*d*) yes

(*e*) Concern for others

2. (*a*) Median = 137 units, Mean = 137.07 units, Mode = 135.76 units

   (*b*) Social service

3. (*a*) 57.19

   (*b*) Concern for others

Multiple Choice Questions

1. (*c*) 2. (*d*) 3. (*c*) 4. (*a*) 5. (*b*) 6. (*b*) 7. (*d*)
8. (*c*) 9. (*b*) 10. (*b*) 11. (*c*) 12. (*a*) 13. (*c*) 14. (*c*)
15. (*c*) 16. (*d*) 17. (*c*) 18. (*b*) 19. (*b*) 20. (*c*) 21. (*a*)
22. (*c*) 23. (*d*) 24. (*d*) 25. (*a*) 26. (*d*) 27. (*b*) 28. (*a*)
29. (*c*) 30. (*b*)

NCERT Exemplar Problems

1. (*c*) 2. (*b*) 3. (*c*) 4. (*b*) 5. (*c*) 6. (*c*)

SOLUTIONS

VERY SHORT ANSWER TYPE QUESTIONS

1. Median.

2.

Marks obtained	f	$c.f.$	
0 – 10	8	8	
10 – 20	10	18	
20 – 30	12	30	
30 – 40	22	52	← Median class
40 – 50	30	82	
50 – 60	18	100	

$n = 100 \Rightarrow \frac{n}{2} = 50$

∴ Median class is 30–40.

3. Class marks of class 10 – 25 = $\frac{10+25}{2} = 17.5$

Class marks of class 35 – 55 = $\frac{35+55}{2} = 45$

4. Modal class = 40–50

Lower limit = 40

5.

x_i	f_i	$f_i x_i$
1	4	4
2	5	10
3	y	$3y$
4	1	4
5	2	10
	$\Sigma f_i = 12 + y$	$\Sigma f_i x_i = 28 + 3y$

$\overline{x} = 2.6$

$\overline{x} = \frac{\Sigma f_i x_i}{\Sigma f_i} \Rightarrow 2.6 = \frac{28+3y}{12+y}$

$\Rightarrow \quad 2.6\,(12+y) = 28 + 3y$

$\Rightarrow \quad 31.2 + 2.6y = 28 + 3y$

$\Rightarrow \quad 31.2 - 28 = 3y - 2.6y$

$\Rightarrow \quad 3.2 = 0.4y$

$\Rightarrow \quad y = \frac{3.2}{0.4} = 8$

6. $\overline{x} = \frac{\Sigma f_i x_i}{\Sigma f_i} \Rightarrow 6 = \frac{2p+52}{11}$

$\Rightarrow \; 2p = 66 - 52 \;\Rightarrow\; 2p = 14 \;\Rightarrow\; p = 7$

7.

Class Interval	Frequency
0 – 10	3
10 – 20	5
20 – 30	7
30 – 40	10
40 – 50	12
50 – 60	15
60 – 70	12
70 – 80	6
80 – 90	2
90 – 100	8

8. $n = 9$

Median $= \left(\frac{n+1}{2}\right)$ th term

$= \left(\frac{9+1}{2}\right)$ th term

$= 5^{\text{th}}$ term $= 32$

9. $n = 10$

Median $= \frac{1}{2}\left[\frac{n}{2}\text{th} + \left(\frac{n}{2}+1\right)\text{th}\right]$ term

$= \frac{1}{2}\,[5^{\text{th}} + 6^{\text{th}}]$ term

$= \frac{1}{2}[15+17] = \frac{1}{2} \times 32 = 16$

10. $n = 120 \Rightarrow \frac{n}{2} = 60$

∴ Median is average of 60th and 61st observation

$\Rightarrow$ Median $= \frac{5+5}{2} = 5$

11.

Marks	No. of students	$c.f.$
0 – 10	5	5
10 – 30	15	15 + 5 = 20 = F
30 – 60	50 – 20 = 30 = f	50
60 – 80	8	58
80 – 90	2	60
	N = 60	N = Σfi = 60

∴ f = 30, F = 20.

12.

x	Tally Marks	f
110	||	2
120	||||	4
130	||	2
140	||	2

∴ 120 has maximum frequency *i.e.*, 4

∴ Mode is 120

13.

x	Tally Marks	f
15	|	1
16	||||	4
18	|	1
19	~~||||~~	5
20	|||	3
21	|	1
22	|	1
24	|	1
25	|	1
34	|	1
48	|	1

Since 19 has maximum frequency *i.e.*, 5, so mode is 19.

14.

Observations	Tally Marks	f
14	|	1
15	|||	3
16	|	1
18	|||	3
20	|||	3
25	|||	3

Observations 15, 18, 20, and 25 all have the same frequency *i.e.*, 3. For 25 be the mode of the data it should have the maximum frequency.

$\therefore$ 25 should repeat itself at least once more *i.e.*, x should be 25.

15. Since size 40 is worn by maximum persons *i.e.*, 41

$\therefore$ Modal shirt size will be 40.

16. Class size = difference between two consecutive class marks *i.e.*, 10 – 6 = 4

17. Number of teachers of age less than 31 years = Number of teachers in the age group of

(16 – 20) + (21 – 25) + (26 – 30) = 11 + 32 + 51= 94

SHORT ANSWER TYPE QUESTIONS-I

1.

Less than	50	55	60	65	70	75	80
Frequency	0	2	10	22	46	84	100

2.

Class Interval	10 - 20	20 - 30	30 - 40	40 - 50	50 - 60
No. of Students	15	5	10	5	5

3.

Marks obtained	No. of students
0 – 20	8
20 – 40	10
40 – 60	12
60 – 80	6
80 – 100	3

$\because$ Maximum frequency = 12

$\therefore$ Modal class = 40 – 60

Now, $l = 40, f_0 = 10, f_1 = 12, f_2 = 6, h = 20$

$$\text{Mode} = l + \left(\frac{f_1 - f_0}{2f_1 - f_0 - f_2}\right) \times h$$

$$= 40 + \frac{12 - 10}{2 \times 12 - 10 - 6} \times 20 = 40 + \frac{2}{8} \times 20 = 45$$

4.

Marks obtained	Number of students
More than or equal to 0	40
More than or equal to 20	35
More than or equal to 40	26
More than or equal to 60	14
More than or equal to 80	6
More than or equal to 100	0

5. (*i*)

x	y	xy
2	3	6
4	2	8
6	3	18
10	1	10
$p + 5$	2	$2p + 10$
	$\Sigma y = 11$	$\Sigma xy = 52 + 2p$

$$\text{Mean} = \frac{\Sigma xy}{\Sigma y} \Rightarrow \frac{52 + 2p}{11} = 6$$

$\Rightarrow$ $2p = 66 - 52$

$\Rightarrow$ $2p = 14 \Rightarrow p = 7$

(*ii*) $$x_{\bar{c}} = \frac{n_1.\bar{x}_1 + n_2.\bar{x}_2}{n_1 + n_2}$$

$$\Rightarrow x_{\bar{c}} = \frac{85 \times 40 + 35 \times 60}{85 + 35} = \frac{3400 + 2100}{120}$$

$$= \frac{5500}{120} = 45.83$$

6. Mean score of 25 observations = 80

$\therefore$ Total score of 25 observations = 25 × 80 = 2000

Mean score of 55 observations = 65

$\therefore$ Total score of 55 observations = 55 × 65 = 3575

Total score of (25 + 55 = 80) observations = 2000 + 3575 = 5575

$\therefore$ Mean score of all the observation $= \frac{5575}{80} = 69.68 \simeq 69.69$

7. Average marks is 50

$$\Rightarrow \quad 50 = \frac{36 + 44 + 75 + x}{4}$$

$\Rightarrow$ $200 = 155 + x \quad \Rightarrow x = 45$

8. Average height of 30 students = 150 cm

$\therefore$ Total height of 30 students = 150 × 30 = 4500 cm

Total height (correct) of 30 students

= 4500 – 135 + 165 = 4530 cm

$\therefore$ Correct mean height $= \frac{4530}{30} = 151$ cm.

9.

Wages (in ₹)	Number of workers	*c.f.*
3500	12	12
3800	13	25
4100	25	50
4500	17	67
5500	15	82
6500	12	94
7000	6	100

$n = 100 \text{ (even)} \Rightarrow \frac{n}{2} = 50$

$$\text{Median} = \frac{1}{2}\,(50^{\text{th}} + 51^{\text{st}})$$

$$= \frac{1}{2}\,(4100 + 4500) = \frac{8600}{2} = 4300$$

10.

Marks (out of 20)	Number of students (f_i)	c.f.
5	4	4
9	5	9
10	6	15
12	12	27
13	11	38
16	6	44
18	4	48
20	2	50

$N = 50 \Rightarrow \frac{n}{2} = 25$

$\therefore$ Median is 12.

11.

Classes	Number of students	c. f.
0 – 10	2	2
10 – 20	12	14
20 – 30	22	36
30 – 40	8	44
40 – 50	6	50

$n = 50, \frac{n}{2} = \frac{50}{2} = 25$, Median Class = 20 – 30

$l = 20, f = 22, c.f. = 14, h = 10$

$$\text{Median} = l + \frac{\left(\frac{n}{2} - c.f.\right)}{f} \times h$$

$$= 20 + \frac{(25-14)}{22} \times 10 = 20 + \frac{11}{22} \times 10$$

$$= 20 + 5 = 25$$

12. (*i*) Mode is 19.

(*ii*) If 19 is changed to 16, then frequency of 16 is 5 and frequency of 19 becomes 4 *i.e.*, 16 has maximum frequency $\Rightarrow$ 16 is the mode.

13. (*i*) Mode of given data is 25. $\therefore$ 25 has maximum frequency. Hence, the value of x is 25.

(*ii*) If 20 and 18 both are changed to 15. Then frequency of 15 becomes the maximum *i.e.*, 9 and new mode is 15.

14.

Class Interval	Frequency
0 – 10	5
10 – 20	8
20 – 30	7
30 – 40	12
40 – 50	28
50 – 60	20
60 – 70	10
70 – 80	10

Modal Class 40 – 50, $\text{Mode} = l + \left(\frac{f_1 - f_0}{2f_1 - f_0 - f_2}\right) \times h$

l = lower limit of the modal class

f_1 = frequency of the modal class

f_0 = frequency of the class preceding the modal class

f_2 = frequency of the class succeeding the modal class.

h = Class size

$$\text{Mode} = 40 + \left(\frac{28-12}{2 \times 28 - 12 - 20}\right) \times 10 = 40 + \left(\frac{16}{56-32}\right) \times 10$$

$$= 40 + \left(\frac{16}{24}\right) \times 10 = 40 + \frac{20}{3} = 46.666 = 46.67$$

15.

Class Interval	Frequency
10 – 15	30
15 – 20	45
20 – 25	75
25 – 30	35
30 – 35	25
35 – 40	15

Modal Class is 20 – 25,

$l = 20, f_1 = 75, f_0 = 45, \quad f_2 = 35, h = 5$

$$\text{Mode} = 20 + \left(\frac{75-45}{2 \times 75 - 45 - 35}\right) \times 5$$

$$= 20 + \frac{30}{70} \times 5 = 20 + \frac{15}{7} = 22.142 = 22.14$$

16.

Class Interval	Frequency
25 – 30	25
30 – 35	34
35 – 40	50
40 – 45	42
45 – 50	38
50 – 55	14

Maximum frequency = 50

$\therefore$ Modal class = 35 – 40

$\therefore \quad l = 35, f_0 = 34, f_1 = 50, f_2 = 42, h = 5$

$$\text{Mode} = l + \frac{f_1 - f_0}{2f_1 - f_0 - f_2} \times h$$

$$= 35 + \frac{50-34}{2 \times 50 - 34 - 42} \times 5$$

$$= 35 + \frac{16}{24} \times 5 = 38.33$$

17. Average age of parents 20 years ago = 23 years

$\therefore$ Their total ages 20 years ago = 23 × 2 = 46 years

Let my parents age = x years

Sum of present age of parents = 46 + 20 × 2 = 86 years

$$\text{Average age of family} = \frac{86 + x}{3}$$

$$\Rightarrow 34 = \frac{86+x}{3} \Rightarrow 102 = 86 + x \Rightarrow x = 16 \text{ years}$$

18. $n_1 + n_2 = 150, n_1 = 50, \bar{x}_1 = 40, n_2 = 100, \bar{x}_2 = 50$

$$\bar{x} = \frac{n_1\bar{x}_1 + n_2\bar{x}_2}{n_1 + n_2} = \frac{50 \times 40 + 100 \times 50}{150}$$

$$= \frac{2000 + 5000}{150} = \frac{7000}{150} = \frac{700}{15} = 46.666 = 46.67 \text{ app.}$$

19. Mean of 5 numbers = 27

Sum of 5 numbers = 27 × 5 = 135

If one number is excluded, then mean of remaining 4 numbers = 27 – 2 = 25

Sum of remaining 4 numbers = 25 × 4 = 100

$\therefore$ Excluded number = 135 – 100 = 35.

20. $\frac{A+B+C}{3} = 33 \Rightarrow B+C = 99 - A$ ...(*i*)

$\frac{B+C+D}{3} = 37 \Rightarrow B+C = 111 - D$ ...(*ii*)

$B+C = 99 - 30 = 69 \quad \because A = 30$

Putting in (*ii*)

$69 = 111 - D$

$D = 111 - 69 = 42$

21. Sum of salary of 12 employees $= n \times \text{mean} =$ ₹ 12×1450

Salary of 13th person = ₹ 1600

Total salary of 13 employees $= 12 \times 1450 + 1600$

$\text{Mean} = \frac{12 \times 1450 + 1600}{13} = 1461.538 =$ ₹ 1461.54

22.

x	y	xy
4	5	20
8	3	24
p	12	$12p$
16	5	80
20	4	80
	$\Sigma y = 29$	$\Sigma xy = 204 + 12p$

$\text{Mean} = \frac{\Sigma xy}{\Sigma y}$

$12 = \frac{204 + 12p}{29}$

$348 - 204 = 12p$

$144 = 12p$

$p = 12$

23. Sum of 67 observations $= 35 \times 67$

i.e., $x_1 + x_2 + x_3 + ... + x_4 = 35 \times 67 = 2345$

$\Rightarrow \frac{x_1}{4} + \frac{x_2}{4} + \frac{x_3}{4} + ... + \frac{x_4}{4}$

$= \frac{1}{4}(x_1 + x_2 + x_3 + ... + x_4) = \frac{1}{4} \times 2345$

$\Rightarrow \text{Mean} = \frac{2345}{4 \times 67} = 8.75$

24. Sum of salaries of 15 workers $= n \times \bar{x}$

$= 15 \times 285 = 4275$

Including Manager's salary average of 16 workers = 355

Let manager's salary be ₹ x

$\Rightarrow \frac{4275 + x}{16} = 355$

$\Rightarrow 4275 + x = 5680$

$\Rightarrow x = 5680 - 4275$

$\Rightarrow x = 1405$

Manager's salary is ₹ 1405.

25. Average weight of A, B, C is 45 kg.

$\Rightarrow \frac{A+B+C}{3} = 45$

$\Rightarrow A + B + C = 135$ ...(*i*)

Average weight of A and B is 40 kg.

$\Rightarrow \frac{A+B}{2} = 40$

$\Rightarrow A + B = 80$ ...(*ii*)

Average weight of B and C is 43 kg.

$\Rightarrow \frac{B+C}{2} = 43$

$\Rightarrow B + C = 86$ ...(*iii*)

Subtracting (*ii*) and (*i*)

$A + B + C = 135$

$-A - B = -80$

$\Rightarrow C = 55$

Putting in (*iii*) we get

$\Rightarrow B + 55 = 86 \Rightarrow B = 31$

Weight of B is 31 kg.

26. Expenditure of eight = ₹ 12×8 = ₹ 96

Let average expenditure of all 9 is ₹ x

$\Rightarrow \frac{96 + x + 8}{9} = x \Rightarrow 104 + x = 9x$

$\Rightarrow 8x = 104 \Rightarrow x = \frac{104}{8} = 13$

Total money spent by all $= 104 + x = 104 + 13 =$ ₹ 117.

27. Let original total expenditure of mess was ₹ x

$\Rightarrow$ Average expenditure per student = ₹ $\frac{x}{35}$

New expenditure of mess = ₹ $x + 42$

Average expenditure per student

$= \frac{x+42}{42} = \frac{x}{42} + 1$

$\Rightarrow \frac{x}{42} + 1 = \frac{x}{35} - 1$

$\Rightarrow 1 + 1 = \frac{x}{35} - \frac{x}{42}$

$\Rightarrow 2 = \frac{6x - 5x}{210}$

$\Rightarrow x = 420$

Original expenditure of mess was ₹ 420.

28. Modal class of the given data is 60–80

Here, $l = 60, f_1 = 61, f_0 = 52, f_2 = 38$ and $h = 20$

$$\text{Mode} = l + \left[\frac{f_1 - f_0}{2f_1 - f_0 - f_2}\right] \times h$$

$$= 60 + \left[\frac{61 - 52}{122 - 52 - 38}\right] \times 20 = 60 + \frac{9 \times 20}{32}$$

$$= 60 + \frac{45}{8} = 60 + 5.625 = 65.625 \text{ hrs.}$$

29. Modal class is (40–50) and its frequency is 20

Here, $l = 40, \; f_1 = 20, \; f_0 = 12, \; f_2 = 11, \; h = 10$

$$\text{Mode} = l + \left[\frac{f_1 - f_0}{2f_1 - f_0 - f_2}\right] \times h$$

$$= 40 + \left[\frac{20 - 12}{40 - 12 - 11}\right] \times 10$$

$$= 40 + \frac{80}{17} = 40 + 4.7 = 44.7 \text{ cars}$$

SHORT ANSWER TYPE QUESTIONS-II

1.

Class	x_i	Frequency f_i	$f_i x_i$
0 – 20	10	17	170
20 – 40	30	p	$30p$
40 – 60	50	32	1600
60 – 80	70	24	1680
80 – 100	90	19	1710
Total		$\Sigma f_i = 92 + p$	$\Sigma f_i x_i = 5160 + 30p$

$$\text{Mean} = \frac{\Sigma f_i x_i}{\Sigma f_i}$$

$$\Rightarrow \quad 50 = \frac{5160 + 30p}{92 + p}$$

$$\Rightarrow \quad 50 \times 92 + 50p = 5160 + 30p$$

$$\Rightarrow \quad 50p - 30p = 5160 - 4600$$

$$\Rightarrow \quad 20p = 560$$

$$\Rightarrow \quad p = \frac{560}{20} = 28$$

2.

Class	Mid-value x_i	Frequency f_i	$f_i x_i$
0 – 20	10	17	170
20 – 40	30	28	840
40 – 60	50	32	1600
60 – 80	70	f_1	$70f_1$
80 – 100	90	19	1710
Total		$\Sigma f_i = 96 + f_1$	$\Sigma f_i x_i = 4320+70f_1$

$$\text{Mean} = \frac{\Sigma f_i x_i}{\Sigma f_i}$$

$$\Rightarrow \quad 50 = \frac{4320 + 70f_1}{96 + f_1}$$

$$\Rightarrow \quad 50 \times 96 + 50f_1 = 4320 + 70f_1$$

$$\Rightarrow \quad 4800 + 50f_1 = 4320 + 70f_1$$

$$\Rightarrow \quad 4800 - 4320 = 70f_1 - 50f_1$$

$$\Rightarrow \quad 480 = 20f_1$$

$$\Rightarrow \quad f_1 = \frac{480}{20} = 24$$

3.

Class	x_i	f_i	$f x_i$
0 – 6	3	6	18
6 – 12	9	8	72
12 – 18	15	p	$15p$
18 – 24	21	9	189
24 – 30	27	7	189
Total		$\Sigma f_i = 30 + p$	$\Sigma f_i x_i = 468 + 15p$

$$\text{Mean} = \frac{468 + 15p}{30 + p}$$

$$\Rightarrow \quad \frac{468 + 15p}{30 + p} = 15.45$$

$$468 + 15p = 463.5 + 15.45p$$

$$468 - 463.5 = 15.45p - 15p$$

$$4.5 = 0.45p$$

$$p = \frac{4.5}{0.45}$$

$$\Rightarrow \quad p = 10$$

4.

Class Interval	f	cf
0 – 6	4	4
6 – 12	x	$4 + x$
12 – 18	5	$9 + x$
18 – 24	y	$9 + x + y$
24 – 30	1	$10 + x + y$
Total	$10 + x + y$	

$n = 20 \Rightarrow \frac{n}{2} = 10$

Median = 14.4

$\therefore$ Median class = 12 – 18

$\therefore$ $l = 12$, $c = 4 + x$, $f = 5$, $h = 6$

$$\text{Median} = l + \frac{\frac{n}{2} - c}{f} \times h$$

$$14.4 = 12 + \frac{10 - (4 + x)}{5} \times 6$$

$$2.4 = \frac{6 - x}{5} \times 6 \quad \Rightarrow \quad \frac{2.4 \times 5}{6} = 6 - x$$

$$\Rightarrow \quad x = 4, \therefore y = 6$$

5.

Income in ₹	Number of families (f)	c.f.
0 – 1000	250	250
1000 – 2000	190	440
2000 – 3000	100	540
3000 – 4000	40	580
4000 – 5000	15	595
5000 – 6000	5	600

$n = 600 \Rightarrow \frac{n}{2} = 300$

$\therefore$ Median class = 1000 – 2000

$l = 1000$, $c = 250$, $f = 190$, $h = 1000$

$$\text{Median} = l + \frac{\frac{n}{2} - c}{f} \times h$$

$$= 1000 + \frac{300 - 250}{190} \times 1000$$

$$= 1000 + \frac{5000}{19} = 1263.158$$

6.

x	f	fx
13	8	104
15	2	30
17	3	51
19	4	76
$20 + p$	$5p$	$100p + 5p^2$
23	6	138
Total	$\Sigma f_i = 23+5p$	$\Sigma f_i x_i = 399 + 100p + 5p^2$

$$\text{Mean} = \frac{\Sigma f_i x_i}{\Sigma f_i}$$

$$\Rightarrow \quad 18 = \frac{399 + 100p + 5p^2}{23 + 5p}$$

$\Rightarrow$ $18 \times 23 + 90p = 399 + 100p + 5p^2$

$\Rightarrow$ $5p^2 + 100p - 90p + 399 - 414 = 0$

$\Rightarrow$ $5p^2 + 10p - 15 = 0$

$\Rightarrow$ $p^2 + 2p - 3 = 0$

$\Rightarrow$ $p^2 + 3p - p - 3 = 0$

$\Rightarrow$ $p(p + 3) - 1(p + 3) = 0$

$\Rightarrow$ $(p + 3)(p - 1) = 0$

$\Rightarrow$ either $p + 3 = 0$ or $p - 1 = 0$

$\Rightarrow$ $p = -3$ or $p = 1$

Frequency can't be negative, so, $p = 1$

7. Let $a = 35$, $h = 10$

C. I.	Class mark (x_i)	f_i	$u_i = \frac{x_i - a}{h}$	$f_i u_i$
0 – 10	5	4	–3	–12
10 – 20	15	4	–2	–8
20 – 30	25	7	–1	–7
30 – 40	35	10	0	0
40 – 50	45	12	1	12
50 – 60	55	8	2	16
60 – 70	65	5	3	15
Total		$\Sigma f_i = 50$		$\Sigma f_i u_i = 16$

$$\text{Mean} = a + \frac{\Sigma f_i u_i}{\Sigma f_i} \times h$$

$$= 35 + \frac{16}{50} \times 10$$

$$= 35 + 3.2 = 38.2$$

8. Let $a = 25$

Number of patients	x_i	Number of days attending hospital f_i	$d_i = x_1 - a$	$f_i d_i$
0 – 10	5	2	–20	–40
10 – 20	15	6	–10	–60
20 – 30	25	9	0	0
30 – 40	35	7	10	70
40 – 50	45	4	20	80
50 – 60	55	2	30	60
Total		$\Sigma f_i = 30$		$\Sigma f_i d_i = 110$

$$\text{Mean} = a + \frac{\Sigma f_i d_i}{\Sigma f_i}$$

$$= 25 + \frac{110}{30} = 25 + 3.666$$

$$= 25 + 3.67 = 28.67$$

9. Let $a = 35$, $h = 5$

Daily wages	x_i	f_i	$u_i = \frac{x_i - a}{h}$	$f_i u_i$
12.5 – 17.5	15	2	–4	–8
17.5 – 22.5	20	22	–3	–66
22.5 – 27.5	25	19	–2	–38
27.5 – 32.5	30	14	–1	–14
32.5 – 37.5	35	3	0	0
37.5 – 42.5	40	4	1	4
42.5 – 47.5	45	6	2	12
47.5 – 52.5	50	1	3	3
52.5 – 57.5	55	1	4	4
Total		$\Sigma f_i = 72$		$\Sigma f_i u_i = -103$

$$\bar{x} = a + \frac{\Sigma f_i u_i}{\Sigma f_i} \times h = 35 + \frac{(-103) \times 5}{72}$$

$$= 35 - 7.152 = 27.847 = 27.85$$

10. Let $a = 550$, $h = 100$

Income (in ₹)	x_i	No. of workers (f_i)	$u_i = \frac{x_i - a}{h} = \frac{x_i - 550}{100}$	$f_i u_i$
200 – 300	250	5	–3	–15
300 – 400	350	36	–2	–72
400 – 500	450	24	–1	–24
500 – 600	550	16	0	0
600 – 700	650	9	1	9
700 – 800	750	6	2	12
800 – 900	850	4	3	12
Total		$\Sigma f_i = 100$		$\Sigma f_i u_i = -78$

$$\text{Mean} = a + \frac{\sum f_i u_i}{\sum f_i} \times h = 550 + \left(\frac{-78}{100}\right) \times 100 = ₹\,472$$

11. Let $a = 413$, $h = 25$

Monthly Salary	x_i	No. of person (f_i)	$u_i = \frac{x_i - a}{h}$	$f_i u_i$
325.5 – 350.5	338	20	–3	–60
350.5 – 375.5	363	10	–2	–20
375.5 – 400.5	388	10	–1	–10
400.5 – 425.5	413	5	0	0
425.5 – 450.5	438	1	1	1
450.5 – 475.5	463	2	2	4
475.5 – 500.5	488	2	3	6
Total		$\Sigma f_i = 50$		$\Sigma f_i u_i = -79$

$$\bar{x} = a + \frac{\Sigma f_i u_i}{\Sigma f_i} \times h = 413 + \frac{(-79)}{50} \times 25$$

$$= 413 - \frac{79}{2} = 413 - 39.5 = 373.5$$

12. Let $a = 42.5$, $h = 5$

Expenditure of water (in ₹)	x_i	No. of houses f_i	$u_i = \frac{x_i - a}{h} = \frac{x_i - 42.5}{5}$	$f_i u_i$
15 – 20	17.5	7	–5	–35
20 – 25	22.5	5	–4	–20
25 – 30	27.5	7	–3	–21
30 – 35	32.5	8	–2	–16
35 – 40	37.5	9	–1	–9
40 – 45	42.5	11	0	0
45 – 50	47.5	7	1	7
50 – 55	52.5	5	2	10
55 – 60	57.5	4	3	12
60 – 65	62.5	4	4	16
65 – 70	67.5	3	5	15
Total		$\Sigma f_i = 70$		$\Sigma f u_i = -41$

$$\text{Mean} = a + \frac{\sum f_i u_i}{\sum f_i} \times h = 42.5 + \frac{-41}{70} \times 5 = ₹\,39.57$$

13. Let $a = 28$, $h = 5$

Number of teachers	x_i	Number of H.S. Schools	$u_i = \frac{x_i - a}{h}$	$f_i u_i$
5.5 – 10.5	8	955	–4	–3820
10.5 – 15.5	13	1067	–3	–3201
15.5 – 20.5	18	1663	–2	–3326
20.5 – 25.5	23	1492	–1	–1492
25.5 – 30.5	28	1220	0	0
30.5 – 35.5	33	1129	1	1129
35.5 – 40.5	38	745	2	1490
40.5 – 45.5	43	637	3	1911
45.5 – 50.5	48	442	4	1768
Total		$\Sigma f_i = 9350$		$\Sigma f u_i = -2340$

$\overline{x} = 28 - 0.25 = 27.75$

14.

C.I.	x_i	f_i	$f_i x_i$
0–20	10	17	170
20–40	30	f_1	$30 f_1$
40–60	50	32	1600
60–80	70	f_2	$70 f_2$
80–100	90	19	1710
Total		$\Sigma f_i = 120$	$\Sigma f x_i = 3480 + 30 f_1 + 70 f_2$

Now, $\quad 17 + f_1 + 32 + f_2 + 19 = 120$

$\Rightarrow \quad 68 + f_1 + f_2 = 120$

$\Rightarrow \quad f_1 + f_2 = 52$

$\Rightarrow \quad f_2 = 52 - f_1$

Now, $\quad \overline{x} = \frac{3480 + 30 f_1 + 70 f_2}{120} = 50$

$\Rightarrow \quad 30 f_1 + 70 f_2 = 2520$

$\Rightarrow \quad 3 f_1 + 7 f_2 = 252$

$\Rightarrow \quad 3 f_1 + 7(52 - f_1) = 252$

$\Rightarrow \quad -4 f_1 + 364 = 252$

$\Rightarrow \quad -4 f_1 = -112$

$f_1 = 28$

and $\quad f_2 = 52 - f_1 = 52 - 28 = 24.$

15.

Class interval	No. of students f_i	mid value x_i	$f_i x_i$
0 – 10	60 – 56 = 4	5	20
10 – 20	56 – 40 = 16	15	240
20 – 30	40 – 20 = 20	25	500
30 – 40	20 – 10 = 10	35	350
40 – 50	10 – 3 = 7	45	315
50 – 60	3	55	165
Total	$\Sigma f_i = 60$		$\Sigma f x_i = 1590$

$$\text{Mean} = \frac{\sum f_i x_i}{\sum f_i} = \frac{1590}{60} = 26.5.$$

16. Construct frequency table we have $a = 55$, $h = 10$

C.I.	f_i	x_i	$u_i = \frac{x_i - a}{h}$	$f_i u_i$
0 – 10	3	5	–5	–15
10 – 20	5	15	–4	–20
20 – 30	7	25	–3	–21
30 – 40	10	35	–2	–20
40 – 50	12	45	–1	–12
50 – 60	20	55	0	0
60 – 70	7	65	1	7
70 – 80	6	75	2	12
80 – 90	2	85	3	6
90 – 100	8	95	4	32
Total	$\Sigma f_i = 80$			$\Sigma f u_i = -31$

$$\overline{x} = 55 + \frac{(-31) \times 10}{80} = 55 - \frac{31}{8}$$

$= 55 - 3.875 = 51.125$

17. Total monthly wages of 15 workers = 15 × 3351 = ₹ 50265

Total monthly wages of 4 worker who left the factory
= 4 × 3310 = 13240

Total monthly wages of remaining 11 workers
= 50265 – 13240 = 37025

When a new worker joins total wages of 12 workers
= 37025 + 3025 = 40050

$$\text{Mean monthly wages} = \frac{40050}{12} = ₹\,3337.50$$

18. Let $a = 25$, $h = 10$

Marks	Mid-value x_i	No. of students f_i	f_ix_i	$d_i = x_i - a$	f_id_i	$u_i = \frac{x_i - a}{h}$	f_iu_i
0 – 10	5	20	100	–20	–400	–2	–40
10 – 20	15	24	360	–10	–240	–1	–24
20 – 30	25	40	1000	0	0	0	0
30 – 40	35	36	1260	10	360	1	36
40 – 50	45	20	900	20	400	2	40
Total		$\Sigma f_i = 140$	$\Sigma f_ix_i = 3620$		$\Sigma f_id_i = 120$		$\Sigma f_iu_i = 12$

Mean (direct method) $= \frac{\Sigma f_ix_i}{\Sigma f_i} = \frac{3620}{140} = 25.857 = 25.86$

Mean (short-cut method) $= a + \frac{\Sigma f_id_i}{\Sigma f_i} = 25 + \frac{120}{140} = 25.86$

Mean (step deviation method) $= a + \frac{\Sigma f_iu_i}{\Sigma f_i} \times h = 25 + \frac{12}{140} \times 10 = 25.86$

19. Given distribution is not continuons. Making it continuous we have gap $30 - 29 = 1$. Subtracting $\frac{1}{2} = 0.5$ from lower limit and adding 0.5 to upper limit, we get the continuous distribution as follows. Let $a = 42$, $h = 5$

Class Interval	x_i	f_i	$u_i = \frac{x_i - 42}{5}$	f_iu_i
24.5 – 29.5	27	4	–3	–12
29.5 – 34.5	32	14	–2	–28
34.5 – 39.5	37	22	–1	–22
39.5 – 44.5	42	16	0	0
44.5 – 49.5	47	6	1	6
49.5 – 54.5	52	5	2	10
54.5 – 59.5	57	3	3	9
Total		$\Sigma f_i = 70$		$\Sigma f_iu_i = -37$

$$\bar{x} = a + \frac{\sum f_iu_i}{\sum f_i} \times h = 42 + \frac{(-37)}{70} \times 5 = 42 - 2.6428 = 39.35$$

20. Extracting frequency distribution we have $a = 45$, $h = 10$

C.I.	f_i	x_i	$u_i = \frac{x_i - a}{h}$	f_iu_i
0 – 10	5	5	–4	–20
10 – 20	4	15	–3	–12
20 – 30	8	25	–2	–16
30 – 40	12	35	–1	–12
40 – 50	16	45	0	0
50 – 60	15	55	1	15
60 – 70	10	65	2	20
70 – 80	8	75	3	24
80 – 90	5	85	4	20
90 –100	2	95	5	10
Total	$\Sigma f_i = 85$			$\Sigma f_iu_i = 29$

$$\bar{x} = a + \frac{\sum f_iu_i}{\sum f_i} \times h = 45 + \frac{29}{85} \times 10 = 45 + 3.4117 = 48.41$$

21.

Wages (in ₹)	x_i	No. of workers (f_i)	f_ix_i
10 – 20	15	5	75
20 – 30	25	3	75
30 – 40	35	4	140
40 – 50	45	f	$45f$
50 – 60	55	2	110
60 – 70	65	6	390
70 – 80	75	13	975
Total		$\Sigma f_i = 33 + f$	$\Sigma f_ix_i = 1765 + 45f$

Let the missing frequency be f.

$$\text{Mean} = \frac{\Sigma f_ix_i}{\Sigma f_i}$$

$\Rightarrow \quad 52 = \frac{1765 + 45f}{33 + f}$

$\Rightarrow \quad 52 \times (33 + f) = 1765 + 45f$

$\Rightarrow \quad 1716 + 52f = 1765 + 45f$

$\Rightarrow \quad 52f - 45f = 1765 - 1716$

$\Rightarrow \quad 7f = 49$

$\Rightarrow \quad f = 7$

22.

Classes	f	$c.f.$
20 – 40	12	12
40 – 60	18	30
60 – 80	23	53
80 – 100	15	68
100 – 120	12	80
120 – 140	12	92
140 – 160	8	100
Total	100	

$n = 100 \Rightarrow \frac{n}{2} = 50 \;\therefore\;$ Median class = 60 – 80

$l = 60, c.f. = 30, f = 23, h = 20$

$$\text{Median} = l + \frac{\frac{n}{2} - c.f.}{f} \times h = 60 + \frac{50-30}{23} \times 20$$
$$= 77.39$$

23.

Class Interval	Frequency	c.f.
130 – 140	5	5
140 – 150	9	14
150 – 160	17	31
160 – 170	28	59
170 – 180	24	83
180 – 190	10	93
190 – 200	7	100
Total	100	

$$n = 100 \Rightarrow \frac{n}{2} = 50$$

Median class = 160 – 170, $l = 160, c.f. = 31, f = 28, h = 10$

$$\text{Median} = 160 + \left[\frac{50-31}{28}\right] \times 10 = 160 + \frac{19}{28} \times 10$$
$$= 160 + 6.785 = 166.785 = 166.79$$

24. Extracting frequency distribution

Class interval	frequency (f)	c.f.
20 – 30	4	4
30 – 40	12	16
40 – 50	14	30
50 – 60	16	46
60 – 70	20	66
70 – 80	16	82
80 – 90	10	92
90 – 100	8	100

$$n = 100 \Rightarrow \frac{n}{2} = 50$$

Median class is 60 – 70, $l = 60, c.f. = 46, f = 20, h = 10$

$$\text{Median} = 60 + \left(\frac{50-46}{20}\right) \times 10 = 60 + \frac{4}{2} = 60 + 2 = 62$$

25. Making given distribution continuous by subtracting $\frac{1}{2} = 0.5$ from lower limit and adding $\frac{1}{2} = 0.5$ to upper limit.

Class interval	f	c.f.
129.5 – 139.5	4	4
139.5 – 149.5	9	13
149.5 – 159.5	18	31
159.5 – 169.5	28	59
169.5 – 179.5	24	83
179.5 – 189.5	10	93
189.5 – 199.5	7	100

$$n = 100 \Rightarrow \frac{n}{2} = 50$$

Median class = 159.5 – 169.5, $l = 159.5, c.f. = 31, f = 28, h = 10$

$$\text{Median} = 159.5 + \left(\frac{50-31}{28}\right) \times 10 = 159.5 + \frac{19}{28} \times 10$$
$$= 166.285 = 166.3$$

26. Changing it to continuous distribution.

Class interval	Frequency	c.f.
9.5 – 19.5	2	2
19.5 – 29.5	4	6
29.5 – 39.5	8	14
39.5 – 49.5	9	23
49.5 – 59.5	4	27
59.5 – 69.5	2	29
69.5 – 79.5	1	30

$n = 30, \frac{n}{2} = 15$, Median class = 39.5 – 49.5

$l = 39.5, c.f. = 14, f = 9, h = 10$

$$\text{Median} = 39.5 + \left(\frac{15-14}{9}\right) \times 10 = 39.5 + \frac{1}{9} \times 10$$
$$= 39.5 + 1.11 = 40.61$$

27.

Wages (in ₹)	f	c.f.
80 – 90	9	9
90 – 100	17	26
100 – 110	19	45
110 – 120	45	90
120 – 130	33	123
130 – 140	15	138
140 – 150	12	150

$$n = 150 \Rightarrow \frac{n}{2} = 75$$

$\therefore$ Median class = 110 – 120

$l = 110, f = 45, c.f. = 45, h = 10$

$$\text{Median} = l + \frac{\frac{n}{2} - cf}{f} \times h = 110 + \frac{75-45}{45} \times 10 = 116.67$$

28.

Age of employees in years	Number of employees in factories	
C.I.	A	B
20 – 30	5	8
30 – 40	26	40
40 – 50	78	58
50 – 60	104	90
60 – 70	98	83

For factory A, max. frequency = 104.

$\therefore$ Modal class = 50 – 60.

$$\text{Mode} = l + \left[\frac{f_1 - f_o}{2f_1 - f_o - f_2}\right] \times h$$

f_1 = frequency of modal class = 100

f_0 = frequency of preeceding class = 78

f_2 = frequency of succeding class = 98

h = class size = 10

l = lower limit of the modal class = 50

$$\text{Mode} = 50 + \left[\frac{104-78}{2 \times 104 - 78 - 98}\right] \times 10 = 50 + \frac{26}{32} \times 10$$
$$= 50 + 8.125 = 58.125$$

For factory B, max. frequency = 90.

$\therefore$ Modal class = 50 – 60.

$$\text{Mode} = 50 + \left(\frac{90-58}{2\times 90-58-83}\right)\times 10 = 50 + \frac{32\times 10}{39}$$

$$= 50 + 8.205 = 58.205$$

29.

Age in years	Number of students in	
	Group A	Group B
16 – 18	50	54
18 – 20	78	89
20 – 22	46	40
22 – 24	28	25
24 – 26	23	17

Maximum frequency for groups.

A = 78 B = 89

Modal class = 18 – 20, Modal class = 18 – 20

$$\text{Mode for group A} = 18 + \left(\frac{78-50}{2\times 78-50-46}\right)\times 2$$

$$= 18 + \frac{28}{156-96}\times 2 = 18 + \frac{28}{60}\times 2 = 18 + \frac{28}{30}$$

$$= 18 + 0.933 = 18.933$$

$$\text{Mode for group B} = 18 + \frac{(89-54)\times 2}{2\times 89-54-40}$$

$$= 18 + \frac{35\times 2}{178-94} = 18 + \frac{70}{84} = 18 + \frac{10}{12} = 18 + \frac{5}{6}$$

$$= 18 + 0.833 = 18.833$$

30. Let $a = 75.5$, $h = 3$

Number of heart beats per min	Number of women f_i	Class mark x_i	$u_i = \frac{x_i - 75.5}{3}$	$f_i u_i$
65 – 68	2	66.5	–3	–6
68 – 71	4	69.5	–2	–8
71 – 74	3	72.5	–1	–3
74 – 77	8	75.5	0	0
77 – 80	7	78.5	1	7
80 – 83	4	81.5	2	8
83 – 86	2	84.5	3	6
Total	$n = 30$			$\Sigma f_i u_i = 4$

By step deviation method

$$\text{mean} = a + h\times\frac{1}{n}\Sigma f_i u_i = 75.5 + 3\times\frac{1}{30}\times 4$$

$$= 75.5 + 0.4 = 75.9$$

31.

Number of mangoes	Number of boxes f_i	Class mark x_i	$u_i = \frac{x_i - 57}{3}$	$f_i u_i$
49.5 – 52.5	15	51	–2	–30
52.5 – 55.5	110	54	–1	–110
55.5 – 58.5	135	57	0	0
58.5 – 61.5	115	60	1	115
61.5 – 64.5	25	63	2	50
Total	$\Sigma f_i = 400$			$\Sigma f_i u_i = 25$

$a = 57$, $h = 3$, $\Sigma f_i = 400$ and $\Sigma f_i u_i = 25$

By step deviation method,

$$\text{Mean} = a + h\times\frac{1}{\Sigma f_i}\times\Sigma f_i u_i = 57 + 3\times\frac{1}{400}\times 25 = 57.19$$

32.

Concentration of SO_2 (in ppm)	frequency f_i	x_i	$f_i u_i$
0.00 – 0.04	4	0.02	0.08
0.04 – 0.08	9	0.06	0.54
0.08 – 0.12	9	0.1	0.90
0.12 – 0.16	2	0.14	0.28
0.16 – 0.20	4	0.18	0.72
0.20 – 0.24	2	0.22	0.44
Total	$\Sigma f_i = 30$		$\Sigma f_i u_i = 2.96$

$$\text{Mean} = \frac{\Sigma f_i u_i}{\Sigma f_i} \Rightarrow \frac{2.96}{30} = 0.986 \simeq 0.99 \text{ ppm.}$$

33. Modal class of the data is (1500 – 2000)

$l = 1500, f_m = 40, f_1 = 24, f_2 = 33, h = 500$

$$\text{Mode} = 1500 + \left\{\frac{40-24}{2\times 40-24-33}\right\}\times 500 = 1500 + \frac{16\times 500}{23}$$

$$= 1500 + \frac{8000}{23} = 1500 + 347.83 = 1847.83$$

Let $a = 2750$, $h = 500$

Expenditure (in ₹)	Number of families f_i	Class mark x_i	$u_i = \frac{x_i - 2750}{500}$	$f_i u_i$
1000 – 1500	24	1250	–3	–72
1500 – 2000	40	1750	–2	–80
2000 – 2500	33	2250	–1	–33
2500 – 3000	28	2750	0	0
3000 – 3500	30	3250	1	30
3500 – 4000	22	3750	2	44
4000 – 4500	16	4250	3	48
4500 – 5000	7	4750	4	28
	$n = 200$			$\Sigma f_i u_i = -35$

$$\text{Mean} = a + h\times\frac{1}{n}\Sigma f_i u_i = 2750 + 500\times\frac{1}{200}\times(-35)$$

$$= 2750 - \frac{5\times 35}{2}$$

$$= 2750 - 87.50 = 2662.50$$

Thus, the modal monthly expenditure is ₹ 1847.83 and the mean monthly expenditure is ₹ 2662.50.

34. Modal class is (30 – 35) as its frequency is 10.

$$\therefore \quad \text{Mode} = 30 + \left\{\frac{10-9}{2\times 10-9-3}\right\}\times 5$$

Now, let $a = 32.5$, $h = 5$

$$= 30 + \frac{5}{8} = 30.6$$

Number of students per teacher	Number of states/U.T. f_i	Class mark x_i	$u_i = \frac{x_i - 32.5}{5}$	$f_i u_i$
15 – 20	3	17.5	–3	–9
20 – 25	8	22.5	–2	–16
25 – 30	9	27.5	–1	–9
30 – 35	10	32.5	0	0
35 – 40	3	37.5	1	3
40 – 45	0	42.5	2	0
45 – 50	0	47.5	3	0
50 – 55	2	52.5	4	8
Total	$n = 35$			$\Sigma f_i u_i = -23$

$$\text{Mean} = 32.5 + 5 \times \frac{1}{35} \times (-23) = 32.5$$

$$= \frac{23}{7} = 32.5 - 3.3 = 29.2$$

Hence, mode = 30.6, and mean = 29.2, we conclude that most states/U.T. have a student teacher ratio of 30.06 and on an average, the ratio is 29.2.

35.

Length (in mm)	Number of leaves f_i	c.f.
117.5–126.5	3	3
126.5–135.5	5	8
135.5–144.5	9	17
144.5–153.5	12	29
153.5–162.5	5	34
162.5–171.5	4	38
171.5–180.5	2	40

$n = 40 \Rightarrow \frac{n}{2} = 20$

$\therefore$ Median class = 144.5 – 153.5

$l = 144.5, c.f. = 17, f = 12, h = 9$

$$\text{Median} = l + \frac{\frac{n}{2} - c.f.}{f} \times h$$

$$= 144.5 + \frac{20 - 17}{12} \times 9 = 144.5 + \frac{9}{4}$$

$= 146.75$ mm.

LONG ANSWER TYPE QUESTIONS

1.

Height (in cm)	Number of girls	Cumulative frequency
120 – 130	2	2
130 – 140	8	10
140 – 150	$12 = f_0$	$22 = c.f.$
150 – 160	$20 = f_1$	42
160 – 170	$8 = f_2$	50
Total	50	

$n = 50 \Rightarrow \frac{n}{2} = 25$

$\therefore$ Median class = 150 – 160

$l = 150, c.f. = 22, f = 20, h = 10$

$$\therefore \quad \text{Median} = l + \frac{\frac{n}{2} - c.f.}{f} \times h = 150 + \frac{25 - 22}{20} \times 10$$

$= 150 + 1.5 = 151.5$

Modal class = 150 – 160

$l = 150, h = 10, f_1 = 20, f_0 = 12, f_2 = 8$

$$\text{Mode} = l + \left(\frac{f_1 - f_0}{2f_1 - f_0 - f_2}\right) \times h$$

$$= 150 + \frac{20 - 12}{2 \times 20 - 12 - 8} \times 10 = 150 + 4 = 154$$

Now, Mode = 3 Median – 2 Mean

$\Rightarrow$ 154 = 3 × 151.5 – 2 Mean $\Rightarrow$ 154 – 454.5 = – 2 Mean

$\Rightarrow$ 300.5 = 2 Mean $\Rightarrow$ Mean = $\frac{300.5}{2}$ = 150.25

2.

Classes	Frequency	Cumulative frequency	
0 – 20	6	6	
20 – 40	8	14	
40 – 60	10	24	
60 – 80	12	36	← Median Class
80 – 100	6	42	
100 – 120	5	47	
120 – 140	3	50	
	$n = 50$		

$\therefore \quad \frac{n}{2} = 25$

Median class = (60 – 80)

$l = 60, f = 12, c.f. = 24, h = 20.$

$$\text{Median} = l + \frac{\frac{n}{2} - c.f.}{f} \times h$$

$$= 60 + \frac{25 - 24}{12} \times 20 = 60 + \frac{1 \times 5}{3}$$

$$= \frac{180 + 5}{3} = \frac{185}{3} = 61.6$$

Modal class = (60 – 80) as its frequency is 12

$h = 20, l = 60, f_1 = 12, f_0 = 10, f_2 = 6.$

$$\text{Mode} = l + \frac{f_1 - f_0}{2f_1 - f_0 - f_2} \times h$$

$$= 60 + \frac{12 - 10}{2 \times 12 - 10 - 6} \times 20$$

$$= 60 + \frac{2}{8} \times 20 = 65$$

Now,

$$\text{Mode} = 3\text{ Median} - 2\text{ Mean}$$
$$65 = 3(61.6) - 2\text{ Mean}$$
$$2\text{ Mean} = 184.8 - 65$$
$$2\text{ Mean} = 119.8$$

$$\text{Mean} = \frac{119.8}{2} = 59.9$$

$\therefore$ Mean = 59.9

Median = 61.6

Mode = 65

3. Let $a = 25$, $h = 10$

Marks	Frequency f_i	c.f.	Mid value (x_i)	$u_i = \frac{x_i - a}{h}$	$f_i u_i$
0 – 10	8	8	5	–2	–16
10 – 20	16	24	15	–1	–16
20 – 30	36	60	25	0	0
30 – 40	34	94	35	1	34
40 – 50	6	100	45	2	12
Total	$\Sigma f_i = 100$				$\Sigma f_i u_i = 14$

$$\text{Mean} = a + \frac{\Sigma f_i u_i}{\Sigma f_i} \times h = 25 + \frac{14}{100} \times 10 = 25 + 1.4 = 26.4$$

$n = 100 \Rightarrow \frac{n}{2} = 50$, So, Median class = 20 – 30.

$$\text{Median} = 20 + \frac{50 - 24}{36} \times 10 = 20 + \frac{26}{36} \times 10 = 20 + 7.22 = 27.22$$

Modal class = 20 – 30.

$$\text{Mode} = 20 + \frac{(36 - 16)}{2 \times 36 - 16 - 34} \times 10 = 20 + \frac{20 \times 10}{22} = 20 + 9.09 = 29.09$$

4.

C.I.	f	c.f.
0 – 10	2	2
10 – 20	5	7
20 – 30	x	$7 + x$
30 – 40	12	$19 + x$
40 – 50	17	$36 + x$
50 – 60	20	$56 + x$
60 – 70	y	$56 + x + y$
70 – 80	9	$65 + x + y$
80 – 90	7	$72 + x + y$
90 – 100	4	$76 + x + y$
	$\Sigma f_i = 76 + x + y$	

As given $\Sigma f_i = 100$

$\Rightarrow \quad 76 + x + y = 100$

$\Rightarrow \quad x + y = 24 \quad$ ...(i)

Median = 52.5, $n = 100 \Rightarrow \frac{n}{2} = 50$

Median class is 50 – 60 (as given median is 52.5.)

$\Rightarrow$ Using formula for the median.

$$52.5 = 50 + \frac{[50 - (36 + x)]}{20} \times 10$$
$$= 50 + \frac{14 - x}{2}$$

$$52.5 - 50 = \frac{14 - x}{2} \Rightarrow 2.5 \times 2 = 14 - x$$

$\Rightarrow \quad 5 = 14 - x \Rightarrow x = 14 - 5 = 9$

Putting in equation (i) we get $9 + y = 24$

$\Rightarrow \quad y = 24 - 9 = 15$

5.

Weight (in kg)	No. of students
Less than 38	0
Less than 40	3
Less than 42	5
Less than 44	9
Less than 46	14
Less than 48	28
Less than 50	32
Less than 52	35

Weight (in kg)	No. of students
More than or equal to 38	35
More than or equal to 40	32
More than or equal to 42	30
More than or equal to 44	26
More than or equal to 46	21
More than or equal to 48	7
More than or equal to 50	3
More than or equal to 52	0

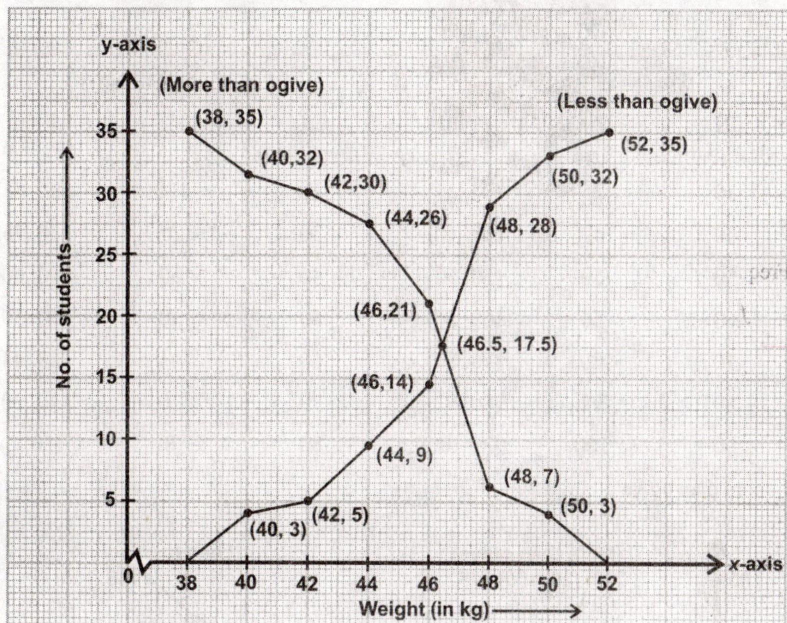

Curves intersect at (46.5, 17.5) $\therefore$ Median weight = 46.5

6.

C.I.	f	For 'less than ogive'			For 'more than ogive'		
		C.I. (less than)	$c.f$	Point	C.I. (more than)	$c.f$	Point
20–30	8	30	8	(30, 8)	20	100	(20, 100)
30–40	12	40	20	(40, 20)	30	92	(30, 92)
40–50	24	50	44	(50, 44)	40	80	(40, 80)
50–60	6	60	50	(60, 50)	50	56	(50, 56)
60–70	10	70	60	(70, 60)	60	50	(60, 50)
70–80	15	80	75	(80, 75)	70	40	(70, 40)
80–90	25	90	100	(90, 100)	80	25	(80, 25)

'Less than ogive' & 'more than ogive' curves

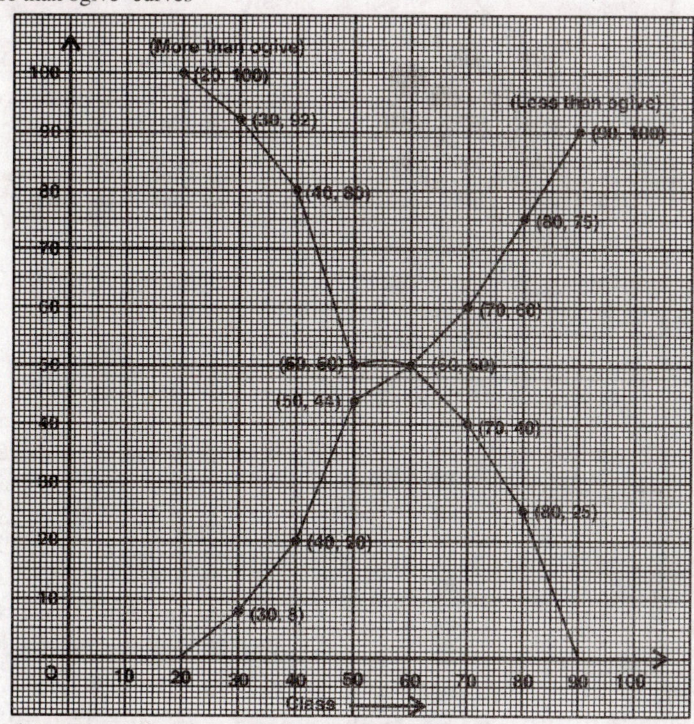

We notice both the curves intersect at (60, 50) $\therefore$ Median = 60.

7. Table for 'less than ogive' and 'more than ogive'

C.I.	f	For 'less than ogive'			For 'more than ogive'		
		C.I. (less than)	c.f	Point	C.I. (more than)	c.f	Point
20–30	25	30	25	(30, 25)	20	100	(20, 100)
30–40	15	40	40	(40, 40)	30	75	(30, 75)
40–50	10	50	50	(50, 50)	40	60	(40, 60)
50–60	6	60	56	(60, 56)	50	50	(50, 50)
60–70	24	70	80	(70, 80)	60	44	(60, 44)
70–80	12	80	92	(80, 92)	70	20	(70, 20)
80–90	8	90	100	(90, 100)	80	8	(80, 8)

'Less than ogive' and 'more than ogive'.

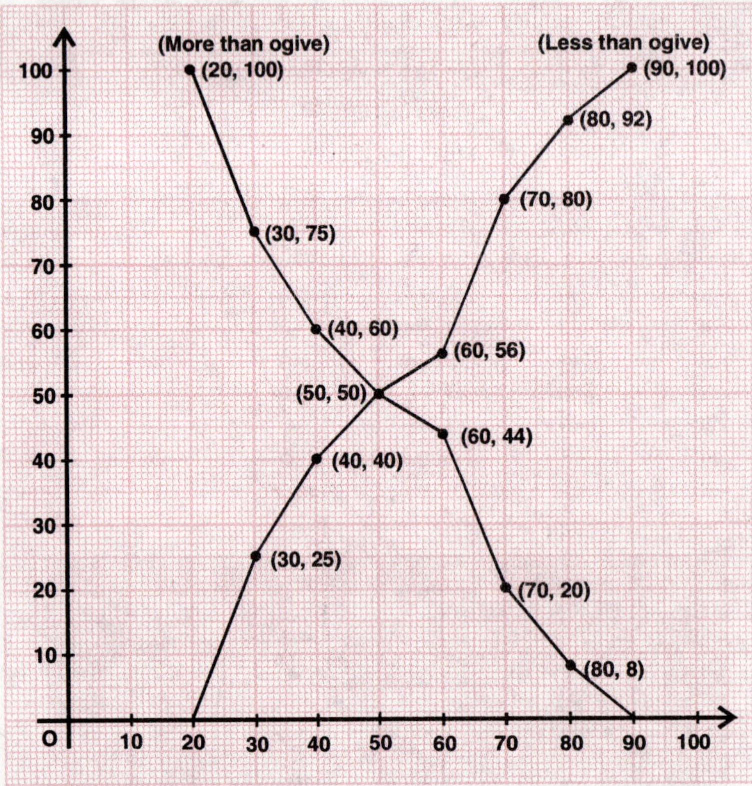

We notice both the curves intersect at (50, 50)

$\therefore$ Median = 50.

8. $a = 60, h = 20$

C.I.	x	f_i	$d_i = \frac{x-60}{20}$	f_id_i
10 – 30	20	5	– 2	– 10
30 – 50	40	8	– 1	– 8
50 – 70	60	f_1	0	0
70 – 90	80	20	1	20
90 – 110	100	f_2	2	$2f_2$
110 – 130	120	2	3	6
Total		$\Sigma f_i = 35 + f_1 + f_2$		$\Sigma f_id_i = 2f_2 + 8$

$$\text{Mean} = a + \frac{\Sigma f_id_i}{\Sigma f_i} \times h$$

$$\Rightarrow \quad 65.6 = 60 + \frac{2f_2 + 8}{50} \times 20$$

$$\Rightarrow \quad 5.6 = \frac{(2f_2 + 8)2}{5} \Rightarrow 28 = 4f_2 + 16$$

$$\Rightarrow \quad 12 = 4f_2 \Rightarrow f_2 = 3$$

As given $[\Sigma f_i = 50]$

Also $35 + f_1 + f_2 = 50$

$\Rightarrow 35 + f_1 + 3 = 50 \Rightarrow f_1 = 12$

$\therefore \quad f_1 = 12, f_2 = 3.$

9. $a = 75, h = 30.$

C.I.	x_i	f_i	$d_i = \frac{x-75}{30}$	$f_i d_i$
0 – 30	15	12	– 2	– 24
30 – 60	45	21	– 1	– 21
60 – 90	75	f_1	0	0
90 – 120	105	52	1	52
120 – 150	135	f_2	2	$2f_2$
150 – 180	165	11	3	33
Total		$\Sigma f_i = 96 + f_1 + f_2$		$\Sigma f_i d_i = 2f_2 + 40$

$$\text{Mean} = a + \frac{\Sigma f_i d_i}{\Sigma f_i} \times h$$

$$91 = 75 + \frac{2f_2 + 40}{150} \times 30 \Rightarrow 16 = \frac{2f_2 + 40}{5}$$

$$\Rightarrow \quad 80 = 2f_2 + 40 \Rightarrow 2f_2 = 40 \Rightarrow f_2 = 20.$$

As given $\Sigma f_i = 150$

$$\Rightarrow \quad 96 + f_1 + f_2 = 150$$

$$\Rightarrow \quad 96 + f_1 + 20 = 150$$

$$\Rightarrow \quad f_1 = 150 - 116 = 34$$

$$\therefore \quad f_1 = 34, f_2 = 20$$

10.

Class Interval	f (No. of students)	Marks	$c.f.$
0 – 10	5	Marks less than 10	5
10 – 20	4	Marks less than 20	9
20 – 30	8	Marks less than 30	17
30 – 40	10	Marks less than 40	27
40 – 50	15	Marks less than 50	42
50 – 60	18	Marks less than 60	60

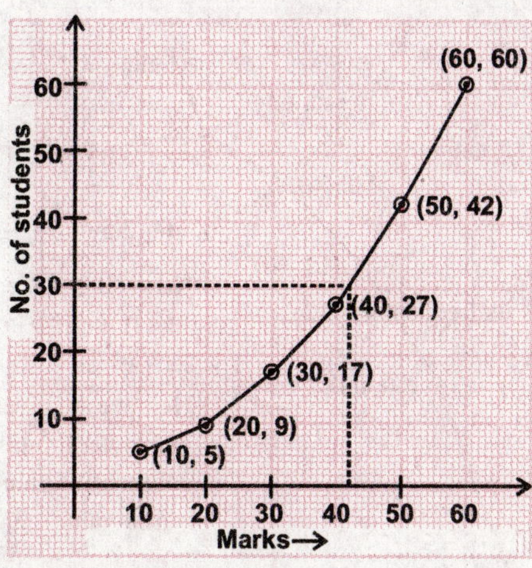

Median = 42

11.

Less than 60	3
Less than 70	8
Less than 80	17
Less than 90	29
Less than 100	35

More than or equal to 50	35
More than or equal to 60	32
More than or equal to 70	27
More than or equal to 80	18
More than or equal to 90	6
More than or equal to 100	0

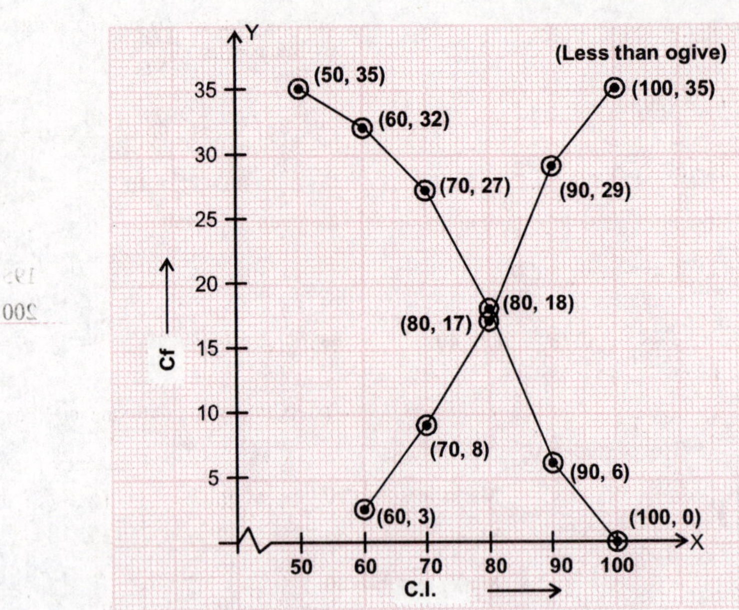

Curve intersect at (80.4, 17.5)

$\therefore$ Median = 80.4

12.

Class	Frequency
0 – 20	6
20 – 40	8
40 – 60	f_1
60 – 80	12
80 – 100	6
100 – 120	5

$\because$ Mode = 65 $\therefore$ Modal class = 60 – 80 as its frequency is 12.

$l = 60,\ f = 12,\ f_0 = f_1,\ f_2 = 6,\ h = 20.$

$$\text{Mode} = l + \frac{f - f_0}{2f - f_0 - f_2} \times h$$

$$\Rightarrow \quad 65 = 60 + \frac{12 - f_1}{2 \times 12 - f_1 - 6} \times 20$$

$$5 = \frac{12 - f_1}{18 - f_1} \times 20$$

$$90 - 5f_1 = 240 - 20f_1$$

$$15f_1 = 150 \Rightarrow f_1 = 10.$$

13.

C.I	f	$c.f.$
0 – 10	x	x
10 – 20	5	$x + 5$
20 – 30	9	$x + 14$
30 – 40	12	$x + 26$
40 – 50	y	$x + y + 26$
50 – 60	3	$x + y + 29$
60 – 70	2	$x + y + 31$
Total	40	

$\because$ Median = 32.5

$\therefore$ Median class = 30 – 40

$\therefore$ $l = 30,\ f = 12,\ n = 40,\ c.f. = 9 + 5 + x = 14 + x.,$ $h = 10.$

$$\text{Median} = l + \left(\frac{\frac{n}{2} - c.f}{f}\right) \times h$$

$$\Rightarrow \quad 32.5 = 30 + \frac{20 - (14 + x)}{12} \times 10$$

$$\Rightarrow \quad 2.5 = \frac{6 - x}{12} \times 10 \Rightarrow x = 3.$$

Now $x + 5 + 9 + 12 + y + 3 + 2 = 40$

$\Rightarrow \quad 3 + 5 + 9 + 12 + y + 3 + 2 = 40$

$\Rightarrow \quad y = 6$

14. Here we are given the mid-values. So, we should first find the upper and lower limits of the various classes. The difference between two consecutive values is $h = 125 - 115 = 10$. So, lower limit of a class = mid-value – $h/2$, and upper limit = mid-value + $h/2$.

Mid-value	Class groups	Frequency	Cumulative Frequency
115	110 — 120	6	6
125	120 — 130	25	31
135	130 — 140	48	79
145	140 — 150	72	151
155	150 — 160	116	267
165	160 — 170	60	327
175	170 — 180	38	365
185	180 — 190	22	387
195	190 — 200	3	390
Total		$n = \Sigma f_i = 390$	

We have $n = 390$. $\therefore \ \frac{n}{2} = \frac{390}{2} = 195$

The cumulative frequency just greater than $n/2 = 195$ is 267 and the corresponding class is 150 – 160. So, 150 – 160 is the median class.

$\therefore \quad l = 150, f = 116, h = 10, c.f. = 151.$

$$\therefore \quad \text{Median} = l + \frac{\frac{n}{2} - c.f.}{f} \times h$$

$$\Rightarrow \text{Median} = 150 + \frac{195 - 151}{116} \times 10 = 153.8$$

15. Let the missing frequencies be x and y.

C.I.	Frequency	Cumulative frequency
10 – 20	12	12
20 – 30	30	42
30 – 40	x	$42 + x$
40 – 50	65	$107 + x$
50 – 60	y	$107 + x + y$
60 – 70	25	$132 + x + y$
70 – 80	18	$150 + x + y$
Total	230	

Given $150 + x + y = 230 \Rightarrow x + y = 80$ ...(*i*)

Median =46 ⇒ Median class is 40 – 50.

∴ Using formula

$$\text{Median} = l + \left(\frac{\frac{n}{2} - c.f.}{f}\right) \times h$$

$$46 = 40 + \left\{\frac{\frac{230}{2} - (42 + x)}{65}\right\} \times 10$$

$$\Rightarrow \quad 6 = \frac{115 - 42 - x}{65} \times 10$$

$$\Rightarrow \quad \frac{6 \times 65}{10} = 73 - x \Rightarrow \frac{390}{10} = 73 - x$$

$$\Rightarrow \quad 39 = 73 - x$$

$$\Rightarrow \quad x = 73 - 39 = 34$$

$$x = 34$$

Putting in (*i*) we get

$$x + y = 80 \Rightarrow 34 + y = 80$$

$$\Rightarrow \quad y = 80 - 34 = 46$$

16.

No. of accidents x_i	Frequency f_i	$f_i x_i$
0	46	0
1	x	x
2	y	$2y$
3	25	75
4	10	40
5	5	25
	$\Sigma f_i = 86 + x + y = 200$	$\Sigma f_i x_i = 140 + x + 2y$

$$\text{Mean} = \frac{\Sigma f_i x_i}{\Sigma f_i}$$

$$\Rightarrow \quad 1.46 = \frac{140 + x + 2y}{200} \Rightarrow 292 = 140 + x + 2y.$$

$\Rightarrow \quad x + 2y = 152$ ...(*i*)

Also $86 + x + y = 200$

$\Rightarrow \; x + y = 114$ ...(*ii*)

Solving equation (*i*) and (*ii*), we get $x = 76, y = 38$.

x_i	f_i	c.f.
0	46	46
1	76	122
2	38	160
3	25	185
4	10	195
5	5	200

N = 200 ⇒ N/2 = 100

∴ Median = 1

17. (*i*) Let the missing frequencies be x and y.

C.I.	Frequency	Cumulative frequency
10 – 20	12	12
20 – 30	30	42
30 – 40	x	$42 + x$
40 – 50	65	$107 + x$
50 – 60	y	$107 + x + y$
60 – 70	25	$132 + x + y$
70 – 80	18	$150 + x + y$
	230	

Given $150 + x + y = 230 \Rightarrow x + y = 80$ ...(*i*)

Median =46 ⇒ Median class is 40 – 50.

∴ Using formula

$$\text{Median} = l + \frac{\frac{n}{2} - c.f.}{f} \times h$$

Here, $l = 40$, $c.f. = 42 + x$, $f = 65$, $h = 10$, $n = 230$

$$46 = 40 + \left\{\frac{\frac{230}{2} - (42 + x)}{65}\right\} \times 10$$

$$\Rightarrow \quad 6 = \frac{115 - 42 - x}{65} \times 10$$

$$\Rightarrow \quad \frac{6 \times 65}{10} = 73 - x \Rightarrow \frac{390}{10} = 73 - x$$

$$\Rightarrow \quad 39 = 73 - x \Rightarrow x = 73 - 39 = 34$$

$$x = 34$$

Putting in (*i*), we get

$$x + y = 80 \Rightarrow 34 + y = 80$$

$$\Rightarrow \quad y = 80 - 34 = 46$$

(*ii*) For mean, $a = 45$, $h = 10$

Class interval	Frequency f_i	x_i	$u_i = \frac{x_i - a}{h}$	$f_i u_i$
10 – 20	12	15	–3	–36
20 – 30	30	25	–2	–60
30 – 40	34	35	–1	–34
40 – 50	65	45	0	0
50 – 60	46	55	1	46
60 – 70	25	65	2	50
70 – 80	18	75	3	54
	$\Sigma f_i = 230$			$\Sigma f_i u_i = 20$

$$\bar{x} = a + \frac{1}{\Sigma f_i} \times (\Sigma f_i u_i) \times h = 45 + \frac{1}{230} \times 20 \times 10$$

$= 45.8695 = 45.87$

18. Median class is 500 – 600 as given that median is 525.

Class interval	Frequency	Cumulative frequency
0 – 100	2	2
100 – 200	5	7
200 – 300	x	$7 + x$
300 – 400	12	$19 + x$
400 – 500	17	$36 + x$
500 – 600	20	$56 + x$
600 – 700	y	$56 + x + y$
700 – 800	9	$65 + x + y$
800 – 900	7	$72 + x + y$
900 – 1000	4	$76 + x + y$
Total	$n = 100$	

Given $n = 100 \Rightarrow 76 + x + y = 100$

$\Rightarrow \quad x + y = 24 \quad$ ...(*i*)

Using formula for median, we have

$$\text{Median} = l + \frac{\frac{n}{2} - c.f.}{f} \times h$$

Here, $l = 500, n = 100, c.f. = 36 + x, f = 20, h = 100$

$$525 = 500 + \left\{\frac{50 - (36 + x)}{20}\right\} \times 100$$

$$\Rightarrow \quad 25 = \frac{14 - x}{20} \times 100 \quad \Rightarrow \quad 25 = (14 - x) \times 5$$

$$\Rightarrow \quad 5 = 14 - x \quad \Rightarrow \quad x = 14 - 5 = 9$$

Now from (*i*), we get

$$9 + y = 24 \quad \Rightarrow \quad y = 24 - 9 = 15$$

Hence $\quad x = 9, y = 15$

19. Here, $a = 135$, $h = 20$

Monthly consumption (in units)	x_i	Number of consumers f_i	Cumulative frequency	$d_i = \frac{x_i - 135}{20}$	$f_i d_i$
65 – 85	75	4	4	–3	–12
85 – 105	95	5	9	–2	–10
105 – 125	115	13	22	–1	–13
125 – 145	135	20	42	0	0
145 – 165	155	14	56	1	14
165 – 185	175	8	64	2	16
185 – 205	195	4	68	3	12
Total		$\Sigma f_i = 68$			$\Sigma f_i d_i = 7$

$$\bar{x} = a + h \,.\, \bar{u} \quad \Rightarrow \quad \bar{x} = 135 + 20 \times \frac{7}{68} = 137.05 \quad (n = 68)$$

$\therefore \quad n = 68 \quad \Rightarrow \quad \frac{n}{2} = 34$, so, Median class = 125 – 145, $l = 125, n = 68, f = 20, c = 22, h = 20$.

$$\text{Median} = l + \frac{\left(\frac{n}{2} - c.f.\right)}{f} \times h = 125 + \frac{\left(\frac{68}{2} - 22\right)}{20} \times 20$$

$= 125 + (34 - 22) = 125 + 12 = 137$

Modal class = 125 – 145 as it frequency is 20, $f_1 = 20, f_0 = 13, f_2 = 14, h = 20$.

$$\text{Mode} = l + \left(\frac{f_1 - f_0}{2f_1 - f_0 - f_2}\right) \times h = 125 + \left(\frac{20 - 13}{40 - 13 - 14}\right) \times 20$$

$$= 125 + \frac{7 \times 20}{13} = 125 + \frac{140}{13} = 125 + 10.77 = 135.77$$

Verification: $3 \times \text{Median} - 2 \text{ mean} = 3 \times 137 - 2 \times 137.05 = 411 - 274.10 = 136.90 \approx \text{Mode}$

20. Here, $a = 8.5$, $h = 3$

No. of letters	Number of surnames (f_i)	c.f.	x_i	$u_i = \frac{x_i - a}{h} = \frac{x_i - 8.5}{3}$	$f_i u_i$
1 – 4	6	6	2.5	–2	–12
4 – 7	30	36	5.5	–1	–30
7 – 10	40	76	8.5	0	0
10 – 13	16	92	11.5	1	16
13 – 16	4	96	14.5	2	8
16 – 19	4	100	17.5	3	12
Total	$\Sigma f_i = 100$				$\Sigma f_i u_i = -6$

$$n = 100, \frac{n}{2} = 50$$

$\therefore$ Median class = 7 – 10

$l = 7$, $c.f. = 36$, $f = 40$, $h = 3$

$$\text{Median} = l + \frac{n/2 - c.f.}{f} \times h$$

$$= 7 + \frac{50 - 36}{40} \times 3 = 7 + 1.05 = 8.05$$

$$\text{Mean} = a + \frac{\Sigma f_i u_i}{\Sigma f_i} \times h$$

$$= 8.5 + \frac{-6}{100} \times 3 = 8.5 - 0.18 = 8.32$$

Maximum frequency = 40

$\therefore$ Modal class = 7 – 10

$\therefore l = 7, f_0 = 30, f_1 = 40, f_2 = 16, h = 3$

$$\text{Mode} = l + \frac{f_1 - f_0}{2f_1 - f_0 - f_2} \times h$$

$$= 7 + \frac{40 - 30}{2 \times 40 - 30 - 16} \times 3 = 7 + \frac{30}{34} = 7.88$$

21.

Daily income (in ₹)	Number of workers (frequency) f_i	Daily income (in ₹) (less than type)	Cumulative frequency
100 – 120	12	Less than 120	12
120 – 140	14	Less than 140	26
140 – 160	8	Less than 160	34
160 – 180	6	Less than 180	40
180 – 200	10	Less than 200	50
Total	$n = 50$		

$n = 50$ given, $\frac{n}{2} = 25$. On the graph, we plot the points (120, 12), (140, 26), (160, 34), (180, 40), (200, 50)

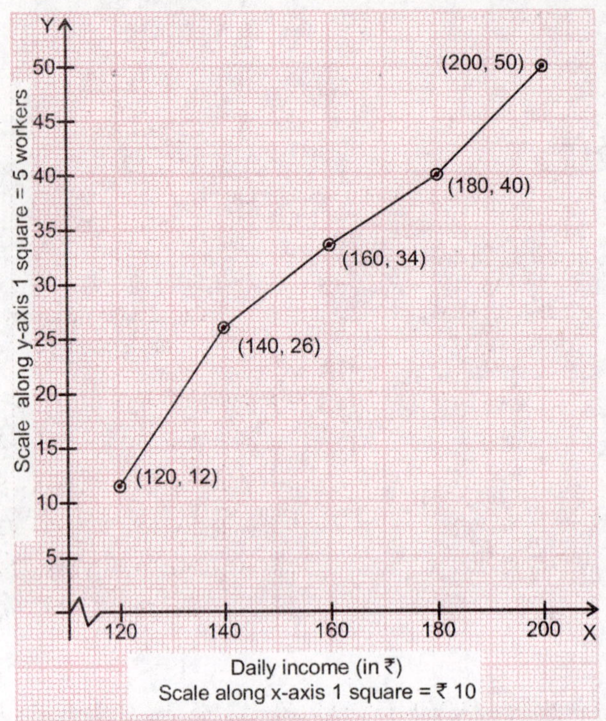

22.

Weight (in kg.)	Number of students (frequency) f_i	Cumulative frequency less than type	
36 – 38	0	Less than 38	0
38 – 40	(3 – 0) = 3	Less than 40	3
40 – 42	(5 – 3) = 2	Less than 42	5
42 – 44	(9 – 5) = 4	Less than 44	9
44 – 46	(14 – 9) = 5	Less than 46	14
46 – 48	(28 – 14) = 14	Less than 48	28
48 – 50	(32 – 28) = 4	Less than 50	32
50 – 52	(35 – 32) = 3	Less than 52	35
Total	n = 35		

To draw less than type ogive, we plot the points (38, 0), (40, 3),(42, 5), (44, 9), (46, 14), (48, 28), (50, 32) and (52, 35) on the graph.

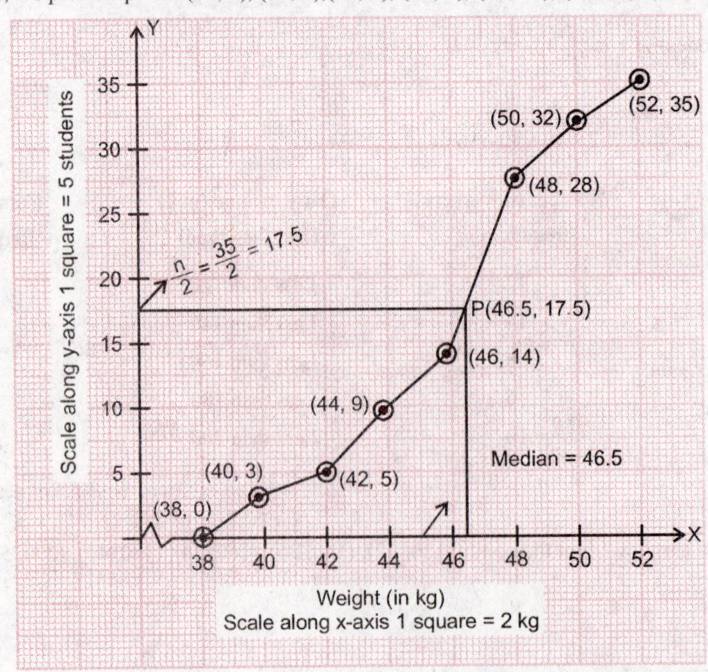

Median from the graph = 46.5 kg.

Median class is (46 – 48) see in the table.

We have $l = 46, f = 14, c.f = 14, n = 35, h = 2$

$$\text{Median} = l + \left\{\frac{\frac{n}{2} - c.f}{f}\right\} \times h = 46 + \left\{\frac{\frac{35}{2} - 14}{14}\right\} \times 2 = 46 + \frac{1}{2} = 46.5 \text{ kg.}$$

Hence, the median is the same as we have noticed from the graph.

23.

Production yield in (kg / ha)	Number of farms (frequency) f_i	Cumulative frequency more than type	
50 – 55	2	50 or more than 50	100 = 100
55 – 60	8	55 or more than 55	(100 – 2) = 98
60 – 65	12	60 or more than 60	(98 – 8) = 90
65 – 70	24	65 or more than 65	(90 – 12) = 78
70 – 75	38	70 or more than 70	(78 – 24) = 54
75 – 80	16	75 or more than 75	(54 – 38) = 16
Total	$n = 100$		

Now, we will draw ogive by plotting the points (50, 100), (55, 98), (60, 90), (65, 78), (70, 54), (75, 16) on the graph.

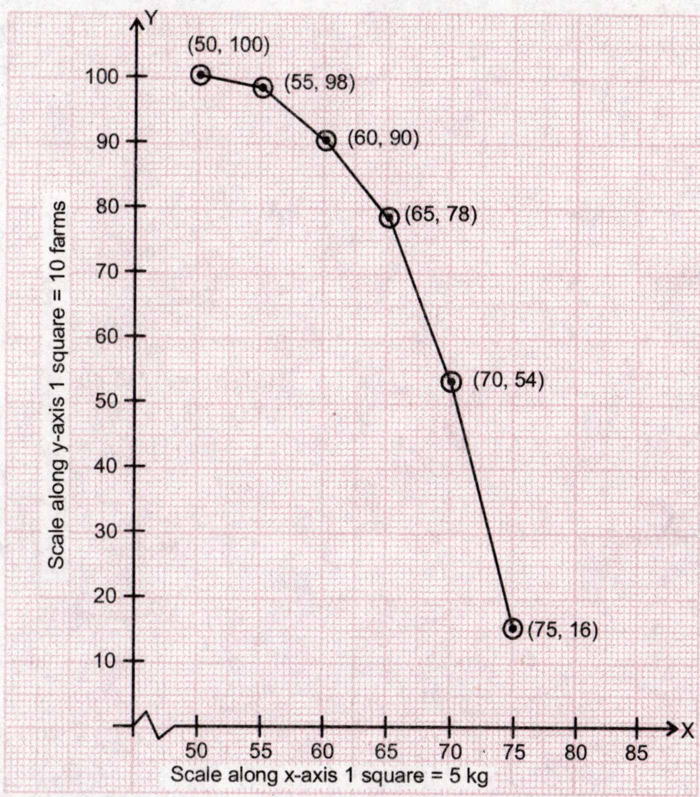

NCERT EXEMPLAR PROBLEMS

1. False.
2. False.
3. No. The values of these three measures can be the same. It depends on the type of data.
4. No. It may be same or different.

5. Frequency distribution table :

Marks	No. of students
0 – 10	10
10 – 20	40
20 – 30	80
30 – 40	140
40 – 50	170
50 – 60	130
60 – 70	100
70 – 80	70
80 – 90	40
90 – 100	20

6.

Marks	No. of candidates
0 – 10	2
10 – 20	2
20 – 30	3
30 – 40	4
40 – 50	6
50 – 60	6
60 – 70	5
70 – 80	2
80 – 90	4

7.

$$a = 12$$
$$b = 25 - 12 = 13$$
$$c = 25 + 10 = 35$$
$$d = 43 - 35 = 8$$
$$e = 48 - 43 = 5$$
$$f = 48 + 2 = 50$$

8. By step deviation method

Let $a = 55$ and $h = 10$

Class	Frequency f_i	Mid-value x_i	$u_i = \frac{x_i - a}{h}$	$u_i f_i$
0 – 10	3	5	– 5	– 15
10 – 20	5	15	– 4	– 20
20 – 30	7	25	– 3	– 21
30 – 40	10	35	– 2	– 20
40 – 50	12	45	– 1	– 12
50 – 60	15	55	0	0
60 – 70	12	65	1	12
70 – 80	6	75	2	12
80 – 90	2	85	3	6
90 – 100	8	95	4	32
Total	$\Sigma f_i = 80$			$\Sigma f u_i = -26$

Mean $\bar{x} = a + \left(\frac{\Sigma f_i u_i}{\Sigma f_i}\right) \times h = 55 + \left(\frac{-26}{80}\right) \times 10$

$= 55 - 3.25 = 51.75$

9. $a = 45, h = 10$

Class	Frequency f_i	Mid-value x_i	$u_i = \frac{x_i - a}{h}$	$f_i u_i$
0 – 10	5	5	–4	–20
10 – 20	4	15	–3	–12
20 – 30	8	25	–2	–16
30 – 40	12	35	–1	–12
40 – 50	16	45	0	0
50 – 60	15	55	1	15
60 – 70	10	65	2	20
70 – 80	8	75	3	24
80 – 90	5	85	4	20
90 – 100	2	95	5	10
Total	$\Sigma f_i = 85$			$\Sigma f_i u_i = 29$

$$\bar{x} = a + \left(\frac{\Sigma f_i u_i}{\Sigma f_i}\right) \times h = 45 + \left(\frac{29}{85}\right) \times 10 = 45 + 3.41 = 48.41$$

10.

Marks	Frequency	c.f
20 – 30	p	p
30 – 40	15	$15 + p$
40 – 50	25	$40 + p$
50 – 60	20	$60 + p$
60 – 70	q	$60 + p + q$
70 – 80	8	$68 + p + q$
80 – 90	10	$78 + p + q$
Total	$\Sigma f_i = 78 + p + q = 90$	

Because median is 50

Median class is 50–60.

A.T.Q. $78 + p + q = 90$

$p + q = 12$ ... (i)

$$\text{Median} = l + \left(\frac{\frac{n}{2} - c.f.}{f}\right) \times h$$

Median $= 50, l = 50$

$$\frac{n}{2} = \frac{90}{2} = 45$$

$c.f. = 40 + p, f = 20, h = 10$

$$50 = 50 + \left[\frac{45 - (40 + p)}{20}\right] \times 10$$

$$50 - 50 = \frac{5 - p}{2}$$

$$0 \times 2 = 5 - p \Rightarrow p = 5.$$

Putting the value of p in (i)

$$p + q = 12$$
$$q = 12 - 5 = 7$$
$$p = 5, q = 7$$

HOTS QUESTIONS

1. Let amount invested = ₹ x, y and z

$\therefore \quad \frac{1}{100}x = \frac{2}{100}y = \frac{4}{100}y = k.$

$\Rightarrow \quad x = 100k, y = 50k, y = 25k$

$\therefore$ Total investment = $100k + 50k + 25k = 175k.$

Total interest = $k + k + k = 3k.$

$\therefore$ average yield = $\frac{3k}{175k} \times 100 = 1.71\ \%$

2. Total weight of 100 students = 46 × 100 = 4600 kg.

let Number of boys = x

$\therefore$ Total weight of boys = $50x$

Number of girls = $100 - x$

Total weight of girls = $(100 - x)$ 40 kg. = $(4000 - 40x)$ kg

$\Rightarrow \quad 4000 - 40x + 50x = 4600 \Rightarrow 10x = 600$

$x = 60$

$\therefore$ Number of boys = 60

VALUE BASED QUESTIONS

1. Cumulative frequency table.

(*a*) C.I.	f		c.f.
0 – 20	6	Less than 20	6
20 – 40	11	Less than 40	17
40 – 60	17	Less than 60	34
60 – 80	12	Less than 80	46
80 – 100	4	Less than 100	50

(*b*)

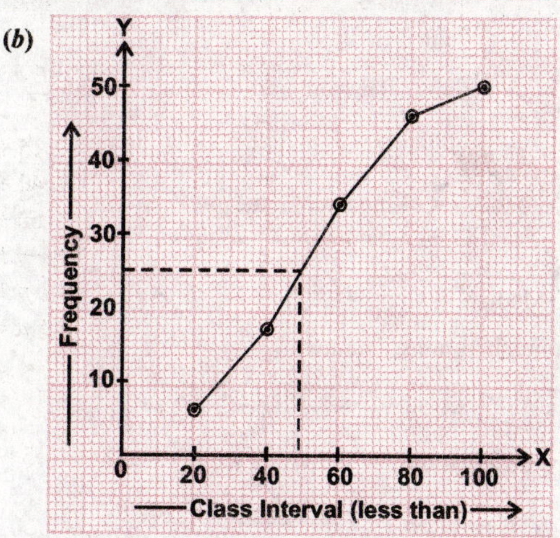

(*c*) Median class $\frac{N}{2} = 25$

From graph Median = 49 m

$$\text{Median} = l + \left[\frac{\frac{n}{2} - c.f.}{f}\right] \times h$$

$$l = 40, \ \frac{n}{2} = 25,$$

$$c.f. = 17, \ f = 17, \ h = 20.$$

$$\text{Median} = 40 + \left[\frac{25-17}{17}\right] \times 20$$

$$= 40 + \frac{8}{17} \times 20 = 40 + \frac{160}{17}$$

$$= 40 + 9.411 = 49.41 \text{ m.}$$

(*d*) Yes, median is approximately same.

(*e*) Concern for others.

2. (*a*)

Monthly consumption (in units)	x_i	Number of consumers f_i	Cumulative frequency	$\mu_i = \frac{x_i - 135}{20}$	$f_i\mu_i$
65 – 85	75	4	4	–3	–12
85 – 105	95	5	9	–2	–10
105 – 125	115	13	22	–1	–13
125 – 145	135	20	42	0	0
145 – 165	155	14	56	1	14
165 – 185	175	8	64	2	16
185 – 205	195	4	68	3	12
Total		$\Sigma f_i = 68$			$\Sigma f_i\mu_i = 7$

$\bar{x} = a + h\,.\,\bar{u} \Rightarrow \bar{x} = 135 + 20 \times \frac{7}{68} = 137.05$ units $(n = 68)$

$\therefore$ Median class = 125 – 145, $l = 125$, $n = 68$, $f = 20$, $c = 22$, $h = 20$.

$$\text{Median} = l + \frac{\left(\frac{n}{2} - c.f.\right)}{f} \times h = 125 + \frac{\left(\frac{68}{2} - 22\right)}{20} \times 20$$

$$= 125 + (34 - 22) = 125 + 12 = 137$$

Modal class = 125 – 145, $f_1 = 20, f_0 = 13, f_2 = 14, h = 20$.

$$\text{Mode} = l + \left(\frac{f_1 - f_0}{2f_1 - f_0 - f_2}\right) \times h = 125 + \left(\frac{20-13}{40-13-14}\right) \times 20$$

$$= 125 + \frac{7 \times 20}{13} = 125 + \frac{140}{13} = 125 + 10.77 = 135.77$$

Verification: 3 × Median – 2 mean = 3 × 137 – 2 × 137.05 = 411 – 274.10 = 136.90 ≈ Mode

(*b*) Social service

3. (*a*)

Number of Food packets	Number of boxes f_i	Class mark x_i	$u_i = \frac{x_i - 57}{3}$	$f_i u_i$
49.5 – 52.5	15	51	–2	–30
52.5 – 55.5	110	54	–1	–110
55.5 – 58.5	135	57	0	0
58.5 – 61.5	115	60	1	115
61.5 – 64.5	25	63	2	50
Total	$n = 400$			$\Sigma f_i u_i = 25$

Let $a = 57$, $h = 3$, $n = 400$ and $\Sigma f_i u_i = 25$

By step deviation method,

$$\text{Mean} = a + h \times \frac{1}{n} \times \Sigma f_i u_i = 57 + 3 \times \frac{1}{400} \times 25 = 57.19$$

(*b*) Concern for others.

MULTIPLE CHOICE QUESTIONS

1. (*c*) Definition of median
2. (*d*) Mode = 3 median – 2 mean = 3×26 – 2 × 24 = 78 – 48 = 30
3. (*c*) Class Interval.
4. (*a*) Maximum frequency.
5. (*b*) Median
6. (*b*) Mode = 3 median – 2 mean
7. (*d*) $\bar{x} = 9, n = 25 \Rightarrow$ sum of observations = 225

   Total increase = 100

   $\text{New mean} = \frac{325}{25} = 13.$
8. (*c*) Mean of nth term and $(n + 1)$th term
9. (*b*) 12, ∵ it has maximum values.
10. (*b*) ∵ Mode = 3 median – 2 mean

    ⇒ $12k$ = 3 median – 2 × $15k$

    ⇒ $42k$ = 3 median

    ⇒ Median = $14k$.
11. (*c*) ∵ $\text{Mean} = \text{assumed mean} + \frac{\Sigma f_i d_i}{\Sigma f_i}$

    ∴ x = assumed mean.
12. (*a*) Sum of deviations from mean = 0.
13. (*c*) $\text{Mean} = \frac{1+2+3+...+n}{n} = \frac{\frac{n(n+1)}{2}}{n} = \frac{n+1}{2}$
14. (*c*) Sum of 100 observations = 100 × 49 = 4900

    Correct sum = 4900 – [40 + 20 + 50] + [60 + 70 + 80] = 5000

    ∴ $\text{Correct mean} = \frac{5000}{100} = 50.$
15. (*c*) ∵ Mode = 3 median – 2 mean

    ⇒ 18 = 3 median – 2 × 24

    ⇒ $\text{Median} = \frac{66}{3} = 22.$
16. (*d*) Let numbers are $x_1, x_2, ... x_n$.

    ∵ $\text{Mean} = \frac{x_1 + x_2 + ... + x_n}{n} = \bar{x}$

    New observations = $kx_1, kx_2 ... kx_n$

    $\text{New mean} = \frac{kx_1 + kx_2 + ... kx_n}{n} = \frac{k(x_1 + x_2 + ... + x_n)}{n} = k\bar{x}$
17. (*c*) ∵ Mode = 3 median – 2 mean

    ⇒ 2 mean = 3 median – mode

    ⇒ $\text{mean} = \frac{1}{2}(\text{3 median – mode})$

    ⇒ $k = \frac{1}{2}$
18. (*b*) $T_1 = \frac{120}{50}$ hrs, $T_2 = \frac{120}{60}$ and $T_3 = \frac{120}{40}$

    $\text{Total time} = \left(\frac{120}{50} + \frac{120}{60} + \frac{120}{40}\right)$ hrs

    Total distance = 3 × 120 km

    $$\text{Average speed} = \frac{3 \times 120}{\frac{120}{50} + \frac{120}{60} + \frac{120}{40}} = \left(\frac{3}{\frac{1}{50} + \frac{1}{60} + \frac{1}{40}}\right) \text{km/h}$$

19. (*b*) Mean $= \dfrac{x_1 + x_2 + x_3 + ... + x_n}{n} = m$

$\Rightarrow x_1 + x_2 + ... + x_n = nm$

$\Rightarrow x_1 + x_2 + ... x_{n-1} + x_n = nm$

$\Rightarrow x_1 + x_2 + ... + x_{n-1} = nm - x_n$

New sum $= x_1 + x_2 + ... + x_{n-1} + x$

New mean $= \dfrac{x_1 + x_2 + ... + x_{n-1} + x}{n} = \dfrac{nm - x_n + x}{n}$

20. (*c*) Mean $= \dfrac{1 + 2 + 3 + ... + n}{n} = \dfrac{6n}{11}$

$$\Rightarrow \quad \dfrac{\frac{n(n+1)}{2}}{n} = \dfrac{6n}{11}$$

$$\Rightarrow \quad \dfrac{n+1}{2} = \dfrac{6n}{11}$$

$$\Rightarrow \quad 11n + 11 = 12n \Rightarrow n = 11.$$

21. (*a*) Let terms are $x_1, x_2, ... x_n$

Mean $= \dfrac{x_1 + x_2 + x_3 + ... + x_n}{n} = \bar{x}$

$\Rightarrow x_1 + x_2 + x_3 + ... + x_n = n \cdot \bar{x}$

New observations

$x_1 + 1, x_2 + 2, x_3 + 3, ..., x_n + n$

New sum $= x_1 + 1 + x_2 + 2 + x_3 + 3 + ... x_n + n$

$= x_1 + x_2 + ... x_n + 1 + 2 + 3 + ... n$

$= n\bar{x} + \dfrac{n(n+1)}{2}$

New mean $= \dfrac{n\bar{x} + \frac{n(n+1)}{2}}{n}$

$= \bar{x} + \dfrac{n+1}{2}$

22. (*c*) Mean

23. (*d*) Let no. of men = x, women = y.

Total age of the group = $30(x + y)$

Total age of men = $32x$,

Total age of women = $27y$

$$\Rightarrow \quad 30(x + y) = 32x + 27y$$

$$\Rightarrow \quad 30x + 30y = 32x + 27y \Rightarrow x = \dfrac{3}{2}y$$

$$\therefore \quad \text{\% of women} = \dfrac{y}{x + y} \times 100$$

$$\Rightarrow \dfrac{y}{\frac{3}{2}y + y} \times 100 = 40\%$$

24. (*d*) Sum of 5 numbers = 5 × 18 = 90

Sum of 4 numbers = 16 × 4 = 64

Excluded number = 90 – 64 = 26

25. (*a*) ∵ Value of the largest item has changed, so median will remain same.

∴ Median = 52.

Now sum of observations

= 50 × 100 – 100 + 110 = 5010

∴ New mean $= \dfrac{5010}{100} = 50.1$

26. (*d*) No. of observations = 9

∴ median = 5th observation

∵ The largest four observations are increased

∴ 5th observation remains unchanged.

27. (**b**)

Value	7	8	9	10	11	12	13
f	2	1	4	5	6	1	3
$c.f.$	2	3	7	12	18	19	22

$n = 22 \quad \Rightarrow \quad \dfrac{n}{2} = 11$, so, median is 10

Median = mean of 11th and 12th observations = 10.

28. (*a*) Sum of 50 observations = 50 × 38 = 1900

Sum of remaining 48 observations

= 1900 – 55 – 45 = 1800

∴ Mean $= \dfrac{1800}{48} = 37.5$

29. (*c*) Let missing frequencies are x and y.

x_i	0	1	2	3	4	5
f_i	46	x	y	25	10	5
f_ix_i	0	x	$2y$	75	40	25

$$\therefore \quad \text{Mean} = \dfrac{\sum f_ix_i}{\sum f_i} = \dfrac{x + 2y + 140}{200} = 1.46$$

$$\Rightarrow \quad x + 2y + 140 = 292$$

$$\Rightarrow \quad x + 2y = 152 \qquad ...(i)$$

$$\text{Also} \quad x + y + 86 = 200$$

$$\Rightarrow \quad x + y = 114 \qquad ...(ii)$$

Solving (*i*) and (*ii*), we get

$x = 76, y = 38.$

30. (*b*) Median is calculated with the help of cumulative frequency table.

NCERT EXEMPLAR PROBLEMS

1. (*c*) mid-points of the classes.

2. (*b*) centred at the classmarks of the classes.

3. (*c*) $\dfrac{x_i - a}{h}$.

4. (*b*) median.

5. (*c*) (30 – 40).

6. (*c*) 82.

Formative Assessment

Activity/Project 1

- **Topic Covered :** Statistics.
- **Objective :** To find set of numbers when mean, median and mode is given.
- **Skills Developed**

Visualising averages, working systematically.

- **Time Required :** 20 Min.
- **Method**

There are five positive whole numbers with the following properties :

- Mean = 4 • Median = 3 • Mode = 3 • If the range is 10, can you identify numbers ?

How many different sets of five numbers can you find that satisfy these conditions ?

The problem will be discussed with the students in the following manner. First of all, because the median is 3, we know that if we write the numbers in numerical order, the middle number has to be 3.

Because the mode is 3, there must be at least one other 3 and since we are writing this in numerical order, a 3 must at least occupy a place on one side of the middle 3.

1st number : Has to be equal to, or less than 3

2nd number : Has to be equal to, or less than 3

3rd number : Has to be 3

4th number : Has to be equal to, or greater than 3

5th number : Has to be greater than 3 (if the largest number was 3, the mean could not be 4)

Since the mean is 4, we know that all the numbers must add up to 20 (20 divided by the number 5 is 4). We can guarantee two 3s in the group, so the remaining numbers must total 14.

Now we need to make a list of all the possible groups whose starting number is less than 3, and apart from 3 no number is repeated twice. The only time that this repetition would work, would be with three 3*s*, and two of another number, but no whole number multiplied by 2 equals 11 (20 minus three 3*s*).

Here is our list :

1, 2, 11

1, 3, 10

1, 4, 9

1, 5, 8

1, 6, 7 (1, 7, 6 = 1, 6, 7 because ultimately the numbers will be written numerically)

2, 3, 9

2, 4, 8

2, 5, 7

3, 3, 8

3, 4, 7

3, 5, 6

Written with the other numbers, the full sets are :

1, 2, 3, 3, 11

1, 3, 3, 3, 10

1, 3, 3, 4, 9

1, 3, 3, 5, 8

1, 3, 3, 6, 7

2, 3, 3, 3, 9

2, 3, 3, 4, 8

2, 3, 3, 5, 7

3, 3, 3, 3, 8

3, 3, 3, 4, 7

3, 3, 3, 5, 6

If the range is 10 the numbers must be 1, 2, 3, 3, 11, because this is the only group with a range of 10.

Class Work

Another five positive whole numbers have the following properties :

- Mean = 31
- Median = 33
- Mode = 34
- Range = 8

How many different sets of five numbers can you find that fit these criteria ?

Homework

Take ten positive numbers. Find mean, median and mode.

Quiz Questions

3 Mode = 2 Median – Mean. State True/False.

Oral Questions

If mean = 5, median = 7, then find mode.

Suggested Activity/Assignment

Find set of all five positive numbers with mean = 5, mode = 4, median = 4.

Activity/Project 2

- **Topic Covered :** Statistics.
- **Objective :** Is the mean of the squares of two numbers greater than, or less than, the square of their means ?
- **Skills Developed**

  Visualising averages, working systematically.
- **Time Required :** 20 Min.
- **Method**

  Let the two numbers be p and q. Mean = $(p + q)/2$. Square of mean = $(p + q)^2/4 = (p^2 + 2pq + q^2)/4$

  Mean of squares = $(p^2 + q^2)/2$

  $(p^2 + q^2)/2 - (p^2 + 2pq + q^2)/4 = (p - q)^2/4 > 0$

  Note that this difference, $(p - q)^2/4$, is zero if $p = q$ and positive for all other choices of p and q. So, if the numbers are equal then the mean of the squares is equal to the square of the mean. Otherwise the mean of the square is greater than the square of mean.

Class Work

Check the result for numbers 24 and 45 in above activity.

Homework

Check the result for 50 and 50 in above activity.

Quiz Questions

Mean of five numbers is always greater than their median. T/F.

Oral Questions

Find median of first five natural numbers.

Suggested Activity/Assignment

Find the relationship between mean of cubes of three numbers and cube of their mean.

PART - II

- **DESIGN OF QUESTION PAPER**
- **PRACTICE PAPERS**

QUESTION PAPER DESIGN

Mathematics – X (SA – 1)							
Time – 3 Hours							**Marks – 90**
S. No.	**Typology of Questions**	**Very Short Answer (VSA) (1 Mark)**	**Short Answer-I (SA) (2 Marks)**	**Short Answer-II (SA) (3 Marks)**	**Long Answer (LA) (4 Marks)**	**Total Marks**	**% Weightage**
1	**Remembering- (Knowledge based** Simple recall questions, to know specific facts, terms, concepts, principles, or theories, identify, define, or recite, information)	1	2	2	3	23	26%
2	**Understanding- (Comprehension-** to be familiar with meaning and to understand conceptually, interpret, compare, contrast, explain, paraphrase, or interpret information)	1	1	1	2	14	16%
3	**Application** (Use abstract information in concrete situation, to apply knowledge to new situations; Use given content to interpret a situation, provide an example, or solve a problem)	1	2	3	2	22	24%
4	**High Order Thinking Skills (Analysis & Synthesis-** Classify, compare, contrast, or differentiate between different pieces of information; Organize and/or integrate unique pieces of information from a variety of sources)	1	1	4	1	19	21%
5	**Creating, Evaluation and Multi-Disciplinary-** (Generating new ideas, product or ways of viewing things Appraise, judge, and/or justify the value or worth of a decision or outcome, or to predict outcomes based on values)	–	–	–	3*	12	13%
	TOTAL	4×1=4	6×2=12	10×3=30	11×4=44	90	100%

* One of the LA (4 marks) will assess the values inherent in the texts.

Time Management for the Practice Papers

Note: 15 minutes will be allotted to each examinee for reading the question paper before commencement of the examination.

Marks Per Question	Expected Time per Question	Number of Questions	Total Expected Time	Average expected time for each category
1	1 — 2 min	4	4 — 8 min	7 min
2	3 — 5 min	6	18 — 30 min	26 min
3	5 — 6 min	10	50 — 60 min	57 min
4	6 — 7 min	11	66 — 77 min	75 min
	Revision Time for the whole paper			15 min
Total		**31**		**180 min**

Assess yourself on the Grading Chart

Time limit	Sum of time limit	Grade	Remarks
Lower time limit	138 minutes	**Brillant** as he has finished the paper before the expected time	Very well-prepared but needs to revise thoroughly so as to score 100% marks.
Upper time limit	175 minutes	**Average** as there is no time left for revision.	Needs more practice and increase speed so as to keep time for revision.
Expected time limit	165 minutes	**Excellent** as he is left with 15 minutes for revision.	Well-prepared but remove mistakes, if any, during revision so as to get 100% marks.

Practice Paper 1

Time Allowed : *3 Hours* | *Maximum Marks* : *90*

Questions 1-4 (1 Mark), 5-10 (2 Marks), 11-20 (3 Marks), 21-31 (4 Marks)

SECTION A

1. Find HCF of 3.5, 6.5 and 8.5.

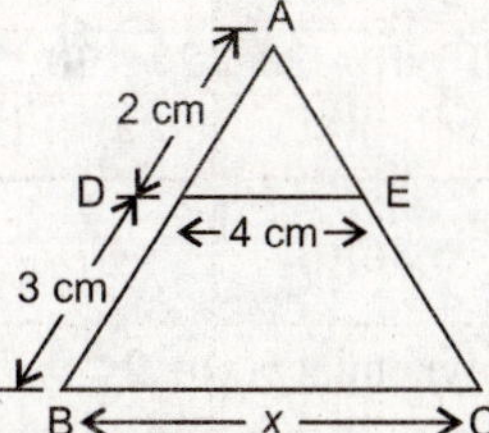

2. In the given fig., if DE || BC, then find the value of x.

3. A set of numbers consists of four 5s, six 7s, ten 9s, eleven 12s, three 13s and two 14s. Find the mode of this set of numbers.

4. Two lines are given to be parallel. The equation of one of the lines is $4x + 3y = 14$. Find the equation of the second line.

SECTION B

5. If $7\sin^2\theta + 3\cos^2\theta = 4$, show that $\tan\theta = \frac{1}{\sqrt{3}}$

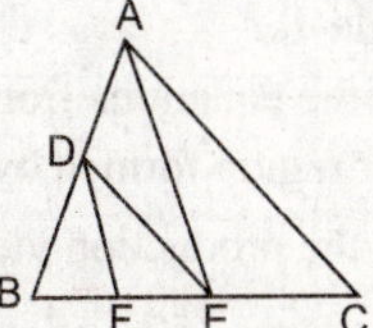

6. In the given fig, DE || AC and DF || AE. Prove that $\frac{FE}{BF} = \frac{EC}{BE}$

7. The following distribution gives the daily income of 50 workers of a factory.

Daily Income (in ₹)	100 – 120	120 – 140	140 – 160	160 – 180	180 – 200
Number of Workers	12	14	8	6	10

Write the above distribution as less than type cumulative frequency distribution.

8. If $\cos\alpha = \frac{1}{2}$ and $\tan\beta = \frac{1}{\sqrt{3}}$, find $\sin(\alpha + \beta)$ where α and β both are acute angles.

9. If 1 is a zero of polynomial $p(x) = ax^2 - 3(a - 1) - 1$, then find the value of a.

10. The coach of a cricket team buys 3 bats and 6 balls for ₹ 3900. Later she buys another bat and 2 more balls of the same kind for ₹ 1300. Represent this situation algebraically.

SECTION C

11. Show that any positive odd integer is of the form $4q + 1$ or $4q + 3$, where q is a positive integer.

12. Prove that $\frac{2\sqrt{3}}{5}$ is an irrational number.

13. In a competitive examination, one mark is awarded for each correct answer while half mark is deducted for each wrong answer. Jayanti answered 120 questions and got 90 marks. How many questions did she answer correctly?

14. If α, β are zeroes of the polynomial $x^2 - 2x - 15$, then form a quadratic polynomial whose zeroes are (2α) and (2β).

15. Prove that $(\text{cosec}\,\theta - \sin\theta)(\sec\theta - \cos\theta) = \frac{1}{\tan\theta + \cot\theta}$.

16. If $\cos\theta + \sin\theta = \sqrt{2}\cos\theta$, show that $\cos\theta - \sin\theta = \sqrt{2}\sin\theta$.

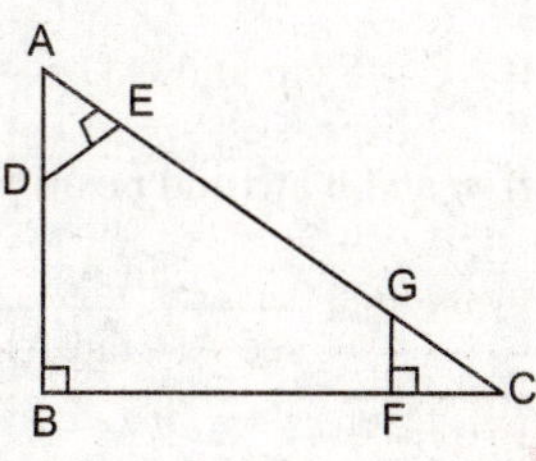

17. In the given fig., $AB \perp BC$, $FG \perp BC$ and $DE \perp AC$. Prove that $\Delta ADE \sim \Delta GCF$

18. ΔABC and ΔDBC are on the same base BC and on opposite sides of BC. O is the point of intersection of AD and BC. Prove that $\dfrac{\text{area }(\Delta ABC)}{\text{area }(\Delta DBC)} = \dfrac{AO}{DO}$.

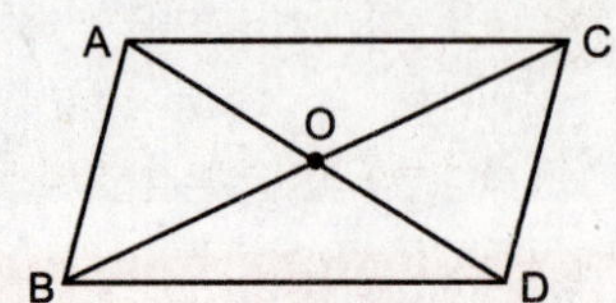

19. Find mean of the following frequency distribution, using step-deviation method:

Class	0 – 10	10 – 20	20 – 30	30 – 40	40 – 50
Frequency	7	12	13	10	8

20. Varun decided to teach maths to weak students in a slum school. He recorded the marks obtained by the students in their last term. The following frequency distribution gives the marks obtained by the students.

Marks	0 – 10	10 – 20	20 – 30	30 – 40	40 – 50	50 – 60	60 – 70	70 – 80	80 – 90	90 – 100
No. of students	5	3	4	3	3	4	7	9	7	8

Find the median of the data.

SECTION D

21. Find other zeroes of the polynomial $p(x) = 2x^4 + 7x^3 - 19x^2 - 14x + 30$ if two of its zeroes are $-\sqrt{2}$ and $\sqrt{2}$.

22. Prove that the ratio of the areas of two similar triangles is equal to the ratio of the squares of their corresponding sides.

23. Evaluate $\dfrac{\sec\theta\operatorname{cosec}(90° - \theta) - \tan\theta\cot(90° - \theta) + \sin^2 55° + \sin^2 35°}{\tan 10°\tan 20°\tan 60°\tan 70°\tan 80°}$

24. If $\sec\theta + \tan\theta = p$, prove that $\sin\theta = \dfrac{p^2 - 1}{p^2 + 1}$

25. Draw the graphs of following equations:
$2x - y = 1$, $x + 2y = 13$ and
(*i*) find the solution of the equations from the graph.
(*ii*) shade the triangular region formed by the lines and the y-axis.

26. The following table gives the production yield per hectare of wheat of 100 farms of a village.

Production yield in kg/hectare	50 – 55	55 – 60	60 – 65	65 – 70	70 – 75	75 – 80
Number of farms	2	8	12	24	38	16

Change the above distribution to more than type distribution and draw its ogive.

***27.** 3 chairs and 2 tables cost ₹ 700 and 5 chairs and 3 tables cost ₹ 1100. A man purchased 2 chairs and 2 tables and paid ₹ 700 to the shopkeeper. Shopkeeper said "It is not the correct amount" and returned him the balance. Find the correct amount and amount paid back by the shopkeeper. Which value is depicted by the shopkeeper?

28. The median of the following data is 32.5. Find the value of x and y.

Class interval	Frequency
0 – 10	x
10 – 20	5
20 – 30	9
30 – 40	12
40 – 50	y
50 – 60	3
60 –70	2
	40

29. In the given figure, ΔABC is a right triangle, right angled at B. AD and CE are the two medians drawn from A and C respectively.

If AC = 5 cm and AD = $\dfrac{3\sqrt{5}}{2}$ cm, find the length of CE.

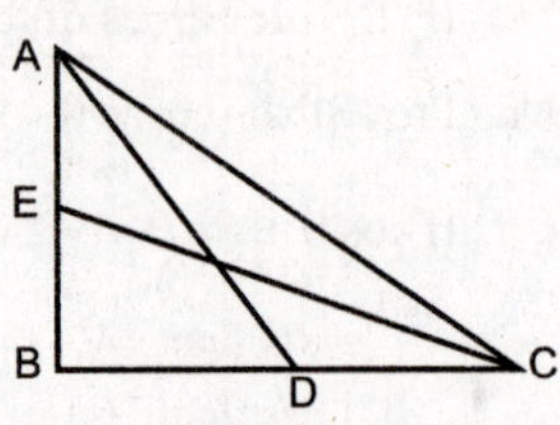

30. If m and n are odd positive integers, then $m^2 + n^2$ is even, but not divisible by 4. Justify.

31. Prove that 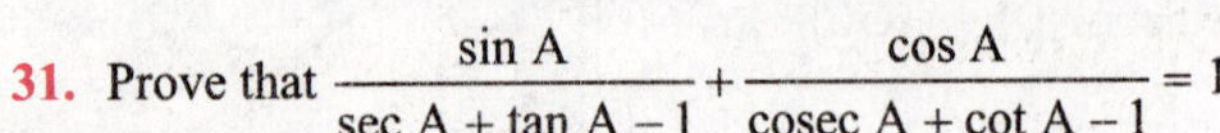 $\dfrac{\sin A}{\sec A + \tan A - 1} + \dfrac{\cos A}{\operatorname{cosec} A + \cot A - 1} = 1$

* Value Based Question

Solutions

1. $3.5 = 7 \times 0.5, 6.5 = 13 \times 0.5, 8.5 = 17 \times 0.5$

HCF = 0.5

2. $\Delta ADE \sim \Delta ABC$

$\therefore \frac{AD}{AB} = \frac{DE}{BC} \Rightarrow \frac{2}{5} = \frac{4}{x} \Rightarrow 2x = 20 \Rightarrow x = 10$ cm

3. 12 (Maximum frequency is of 12).

4. $4x + 3y = k$ where k is any real number and $k \neq 14$.

5. $7 \sin^2 \theta + 3 \cos^2 \theta = 4$ [Given]

$\Rightarrow 7 \sin^2 \theta + 3(1 - \sin^2 \theta) = 4$ $[\because \sin^2 \theta + \cos^2 \theta = 1]$

$\Rightarrow 7 \sin^2 \theta + 3 - 3 \sin^2 \theta = 4$

$\Rightarrow 4 \sin^2 \theta = 4 - 3$

$\Rightarrow \sin^2 \theta = \frac{1}{4} \Rightarrow \sin \theta = \frac{1}{2}$

$\Rightarrow \sin \theta = \sin 30°$

$\Rightarrow \theta = 30°$

$\therefore \tan \theta = \tan 30° = \frac{1}{\sqrt{3}}$ proved.

6. **Given:** ΔABC in which DE || AC and DF || AE.

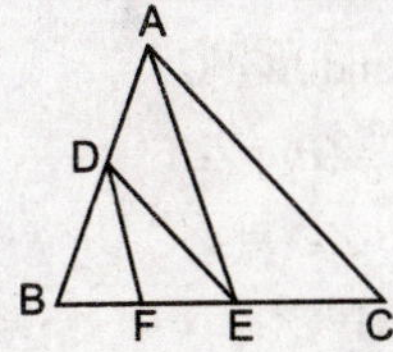

To Prove: $\frac{FE}{BF} = \frac{EC}{BE}$

Proof: In ΔABE, DF || AE

$\therefore \frac{AD}{DB} = \frac{FE}{BF}$ [Thales Theorem] ...(*i*)

Again in ΔABC, DE || AC

$\therefore \frac{AD}{DB} = \frac{EC}{BE}$ [Same reason] ...(*ii*)

From (*i*) and (*ii*), we have

$\frac{FE}{BF} = \frac{EC}{BE}$ Proved.

7.

Daily Income (in ₹)	No. of Workers (*c.f.*)
Less than 120	12
Less than 140	26
Less than 160	34
Less than 180	40
Less than 200	50

8. $\cos \alpha = \frac{1}{2} = \cos 60° \Rightarrow \alpha = 60°$

$\tan \beta = \frac{1}{\sqrt{3}} = \tan 30° \Rightarrow \beta = 30°$

$\therefore \sin(\alpha + \beta) = \sin(60° + 30°) = \sin 90° = 1.$

9. $\because$ 1 is a zero of $p(x)$.

$\Rightarrow p(1) = 0$

$\Rightarrow a(1)^2 - 3(a - 1) - 1 = 0$

$\Rightarrow a - 3a + 3 - 1 = 0 \Rightarrow 2a = 2$

$\Rightarrow a = 1$

10. Let cost of 1 bat be ₹ x and cost of 1 ball be ₹ y.

A.T.Q., $3x + 6y = 3900$

and $x + 2y = 1300$

11. To show $4q + 1$ or $4q + 3$ are positive odd integers.

By Euclid Division algorithm

$a = bq + r$, where q and r are positive integers, such that $0 \leq r < b$.

Taking $a = 4q + r$, taking $b = 4$, such that $0 \leq r < 4$

$\therefore$ Possible remainders are 0, 1, 2, 3

$\therefore$ Integer a is in the form of $4q$, $4q + 1$, $4q + 2$ or $4q + 3$

Integer a is in the form of odd integers, $\therefore$ a cannot be $4q$ and $4q + 2$ because these are positive even integers, therefore, any positive odd integers are in the form $4q + 1$ or $4q + 3$. Proved.

12. Let $\frac{2\sqrt{3}}{5} = \frac{a}{b}$ be a rational number where $b \neq 0$

$\therefore \sqrt{3} = \frac{5a}{2b}$

$\because \frac{5a}{2b}$ is a rational number.

$\Rightarrow \sqrt{3}$ is also a rational number.

But actually $\sqrt{3}$ is not a rational number, hence it gives us contradiction.

$\therefore$ Our assumption is wrong.

Hence $\frac{2\sqrt{3}}{5}$ is an irrational number.

13. Total questions attempted = 120

Total marks obtained = 90

Let the number of correct answers be x and number of wrong answers be y.

$\therefore$ A.T.Q., $x + y = 120$ ... (*i*)

$\because$ 1 mark is given for correct answer and $\frac{1}{2}$ mark is deducted for wrong answer.

$$\therefore \quad x - \frac{1}{2}.y = 90 \Rightarrow \frac{2x-y}{2} = 90$$

or $\quad 2x - y = 180 \quad$ ... (*ii*)

Adding (*i*) and (*ii*), we get $x + y = 120$

and $\quad 2x - y = 180$

$$\Rightarrow 3x = 300 \Rightarrow x = \frac{300}{3} = 100$$

Putting the value of x in equation (*i*), we get

$$100 + y = 120$$

$$\Rightarrow \quad y = 120 - 100 = 20$$

$\therefore$ Correct answered questions = 100

14. Polynomial $p(x) = x^2 - 2x - 15$

Here, $a = 1, \; b = -2$ and $c = -15$

$\therefore$ α and β are the zeroes of $p(x)$

$\therefore$ Sum of zeroes $(\alpha + \beta) = -\frac{b}{a} \quad \Rightarrow \alpha + \beta = 2$

Product of zeroes $(\alpha . \beta) = \frac{c}{a} \quad \Rightarrow \alpha . \beta = -15$

Given zeroes are 2α and 2β.

Sum of zeroes $= 2\alpha + 2\beta = 2(\alpha + \beta) = 2 \times 2 = 4$

$[\because \alpha + \beta = 2]$

Product of zeroes $= 2\alpha \cdot 2\beta = 4\alpha\beta = 4 \times (-15) = -60$

$[\because \alpha \cdot \beta = -15]$

$\therefore$ Required polynomial

$= x^2 -$ (sum of zeroes) $. x +$ product of zeroes

$= x^2 - 4x - 60$

15. L.H.S. = $(\text{cosec } \theta - \sin\theta)(\sec\theta - \cos\theta)$

$$= \left(\frac{1}{\sin\theta} - \sin\theta\right)\left(\frac{1}{\cos\theta} - \cos\theta\right)$$

$$= \left(\frac{1-\sin^2\theta}{\sin\theta}\right)\left(\frac{1-\cos^2\theta}{\cos\theta}\right) = \frac{\cos^2\theta}{\sin\theta} \cdot \frac{\sin^2\theta}{\cos\theta}$$

$$= \sin\theta . \cos\theta \qquad \left[\begin{array}{l}\because 1 - \sin^2\theta = \cos^2\theta \\ \text{and } 1 - \cos^2\theta = \sin^2\theta\end{array}\right]$$

$$\text{R.H.S.} = \frac{1}{\tan\theta + \cot\theta} = \frac{1}{\frac{\sin\theta}{\cos\theta} + \frac{\cos\theta}{\sin\theta}} = \frac{1}{\frac{\sin^2\theta + \cos^2\theta}{\sin\theta . \cos\theta}}$$

$$= \frac{\sin\theta . \cos\theta}{\sin^2\theta + \cos^2\theta} = \sin\theta . \cos\theta \quad [\because \sin^2\theta + \cos^2\theta = 1]$$

$\therefore$ L.H.S. = R.H.S. Proved.

16. $\cos\theta + \sin\theta = \sqrt{2}\cos\theta$

$$\Rightarrow \quad \sin\theta = \sqrt{2}\cos\theta - \cos\theta$$

$$\Rightarrow \quad \sin\theta = (\sqrt{2} - 1)\cos\theta$$

$$\Rightarrow \quad \frac{\sin\theta}{\sqrt{2}-1} = \cos\theta$$

$$\Rightarrow \quad \frac{\sin\theta(\sqrt{2}+1)}{2-1} = \cos\theta$$

$$\Rightarrow \quad \sqrt{2}\sin\theta + \sin\theta = \cos\theta$$

$$\Rightarrow \quad \cos\theta - \sin\theta = \sqrt{2}\sin\theta$$

17.

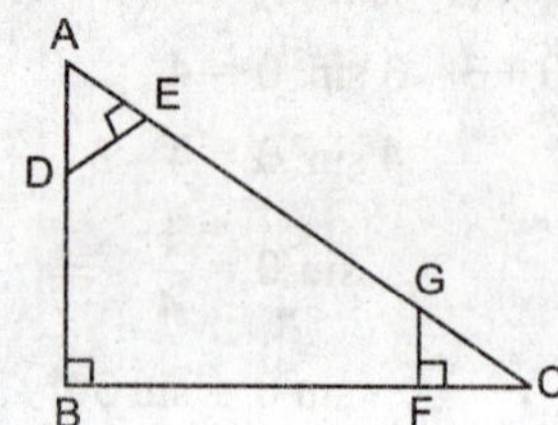

Given: ΔABC in which $\angle B = 90°$

$GF \perp BC$

and $DE \perp AC$.

To prove: $\Delta ADE \sim \Delta GCF$

Proof: In ΔABC and ΔGFC

$\angle B = \angle F$ [Each 90° given]

$\angle C = \angle C$ [Common angle]

$\therefore$ $\Delta ABC \sim \Delta GFC$ [AA similarity]

$\therefore$ $\angle A = \angle FGC$

[IIIrd corresponding angle of the Δ]

Now, in ΔADE and ΔGCF,

$\angle E = \angle F$ [Each 90°]

$\angle DAE = \angle FGC$ [Proved above]

$\therefore$ $\Delta ADE \sim \Delta GCF$ (AA similarity)

18. Given: ΔABC and ΔDBC are on the same base BC and O is the point of intersection of AD and BC.

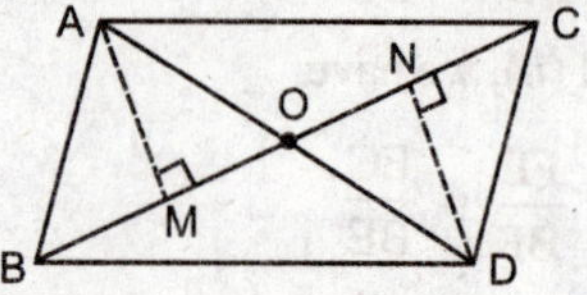

To prove: $\frac{\text{ar } \Delta ABC}{\text{ar } \Delta DBC} = \frac{AO}{DO}$

Construction: Draw $AM \perp BC$ and $DN \perp BC$.

Proof: Now in ΔAMO and ΔDNO,

$\angle AMO = \angle DNO$ [Each 90°]

$\angle AOM = \angle DON$ [V.O.angles]

$\therefore$ $\Delta MOA \sim \Delta NOD$ [A.A. similarity]

$\therefore$ $\frac{AM}{DN} = \frac{AO}{DO}$ ...(*i*)

[Ratio of corresponding sides of similar triangles]

Again, $\frac{\text{ar } \Delta ABC}{\text{ar } \Delta DBC} = \frac{\frac{1}{2} \times BC \times AM}{\frac{1}{2} \times BC \times DN}$

$\Rightarrow$ $\frac{\text{ar } \Delta ABC}{\text{ar } \Delta DBC} = \frac{AM}{DN}$ ...(*ii*)

From (*i*) and (*ii*)

$\frac{\text{ar } \Delta ABC}{\text{ar } \Delta DBC} = \frac{AO}{DO}$ Proved. $\left[\because \frac{AM}{DN} = \frac{AO}{DO}\right]$

19.

C.I	frequency (f_i)	mid value (x_i)	$u_i = \frac{x_i - a}{h}$	$f_i u_i$
0 – 10	7	5	– 2	– 14
10 – 20	12	15	– 1	– 12
20 – 30	13	25	0	0
30 – 40	10	35	1	10
40 – 50	8	45	2	16
	$\Sigma f_i = 50$			$\Sigma f_i u_i = -26 + 26 = 0$

Let assumed mean $(a) = 25$

Mean $(\bar{x}) = a + \frac{\Sigma f_i u_i}{\Sigma f_i} \times h = 25 + \frac{0}{50} \times 10$

$\therefore$ Mean $(\bar{x}) = 25$

20.

C.I	frequency	*c.f.*
0 – 10	5	5
10 – 20	3	8
20 – 30	4	12
30 – 40	3	15
40 – 50	3	18
50 – 60	4	22 $\to$ *c.f.*
60 – 70	7 $\to f$	29
70 – 80	9	38
80 – 90	7	45
90 – 100	8	53

$\frac{n}{2} = \frac{53}{2} = 26.5$ which lies in class with of 29.

$\therefore$ Median class $= 60 - 70$

Median $= l + \frac{\frac{n}{2} - c.f.}{f} \times h$

where *c.f.* = Cumulative frequency of class preceding the median class.

l = Lower limit of median class.

f = Frequency of the median class.

$\therefore$ Median $= 60 + \frac{26.5 - 22}{7} \times 10$

$= 60 + \frac{4.5}{7} \times 10 = 60 + \frac{45}{7}$

$= 60 + 6.43 = 66.43$

21. Polynomial $p(x) = 2x^4 + 7x^3 - 19x^2 - 14x + 30$ has two zeroes $-\sqrt{2}$ and $\sqrt{2}$.

$\therefore (x + \sqrt{2})(x - \sqrt{2})$ is a factor of $p(x)$ or $(x^2 - 2)$ is a factor of $p(x)$

For other factor.

$$
\begin{array}{r|l}
 & 2x^2 + 7x - 15 \\
x^2 - 2 & 2x^4 + 7x^3 - 19x^2 - 14x + 30 \\
 & \underline{\overset{-}{2x^4} \quad \overset{+}{-4x^2}} \\
 & 7x^3 - 15x^2 - 14x + 30 \\
 & \underline{\overset{-}{7x^3} \quad \overset{+}{-14x}} \\
 & -15x^2 + 30 \\
 & \underline{\overset{+}{-15x^2} \overset{-}{+30}} \\
 & \times
\end{array}
$$

$\therefore$ $2x^4 + 7x^3 - 19x^2 - 14x + 30$

$= (x^2 - 2)(2x^2 + 7x - 15)$

$= (x^2 - 2)(2x^2 + 10x - 3x - 15)$

$= (x^2 - 2)[2x(x + 5) - 3(x + 5)]$

$= (x^2 - 2)(x + 5)(2x - 3)$

For other two zeroes, $x + 5 = 0 \Rightarrow x = -5$

and $2x - 3 = 0 \Rightarrow 2x = 3 \Rightarrow x = \frac{3}{2}$

$\therefore$ Other two zeroes are -5 and $\frac{3}{2}$.

22.

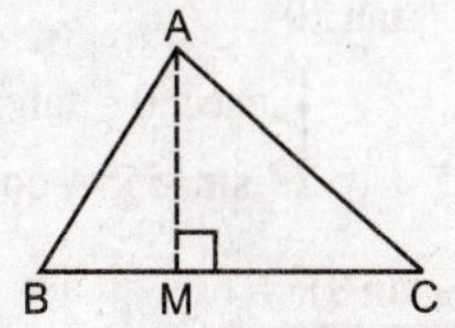

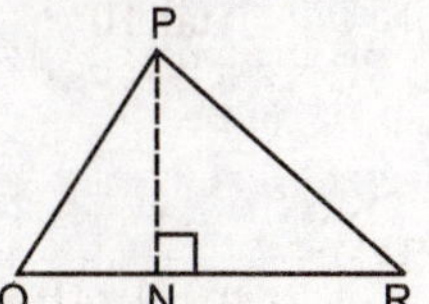

Given: $\Delta ABC \sim \Delta PQR$

To Prove: $\frac{\text{ar } \Delta ABC}{\text{ar } \Delta PQR} = \frac{AB^2}{PQ^2} = \frac{BC^2}{QR^2} = \frac{AC^2}{PR^2}$

Construction: Draw $AM \perp BC$ and $PN \perp QR$.

Proof: $\dfrac{\text{ar } \Delta ABC}{\text{ar } \Delta PQR} = \dfrac{\frac{1}{2} BC \times AM}{\frac{1}{2} QR \times PN}$

$\Rightarrow \quad \dfrac{\text{ar } \Delta ABC}{\text{ar } \Delta PQR} = \dfrac{BC}{QR} \times \dfrac{AM}{PN}$ ...(*i*)

$\because \quad \Delta ABC \sim \Delta PQR$ [Given]

$\therefore \quad \dfrac{AB}{PQ} = \dfrac{BC}{QR} = \dfrac{AC}{PR}$ ...(*ii*)

[Ratio of corresponding sides of similar triangles]

In ΔABM and ΔPQN,

$\angle B = \angle Q$

[Corresponding angles of similar triangles]

$\angle AMB = \angle PNQ$ [Each 90°]

$\therefore \quad \Delta BMA \sim \Delta QNP$ [AA similarity]

$\therefore \quad \dfrac{AM}{PN} = \dfrac{AB}{PQ}$

But $\quad \dfrac{AB}{PQ} = \dfrac{BC}{QR}$ $\left[\text{From } (ii)\ \dfrac{AB}{PQ} = \dfrac{BC}{QR}\right]$

$\dfrac{AM}{PN} = \dfrac{BC}{QR}$ ...(*iii*)

From equation (*i*) and (*iii*),

$\dfrac{\text{ar } \Delta ABC}{\text{ar } \Delta PQR} = \dfrac{BC}{QR} \times \dfrac{BC}{QR} = \dfrac{BC^2}{QR^2}$

Similarly, $\dfrac{\text{ar } \Delta ABC}{\text{ar } \Delta PQR} = \dfrac{AB^2}{PQ^2} = \dfrac{BC^2}{QR^2} = \dfrac{AC^2}{PR^2}$

23.

$$\frac{\sec\theta . \text{cosec}(90° - \theta) - \tan\theta . \cot(90° - \theta) + \sin^2 55° + \sin^2 35°}{\tan 10° . \tan 20° . \tan 60° . \tan 70° . \tan 80°}$$

$$= \frac{\sec\theta . \sec\theta - \tan\theta . \tan\theta + \sin^2 55 + \sin^2(90° - 55°)}{\tan 10° \cdot \tan(90° - 10°) . \tan 20° . \tan(90° - 20°) . \sqrt{3}}$$

$$= \frac{\sec^2\theta - \tan^2\theta + \sin^2 55° + \cos^2 55°}{\tan 10° . \cot 10° . \tan 20° . \cot 20° \times \sqrt{3}}$$

$$= \frac{1 + 1}{\tan 10° \times \frac{1}{\tan 10°} . \tan 20° . \frac{1}{\tan 20°} \times \sqrt{3}} = \frac{2}{\sqrt{3}}$$

$\left[\because \sec^2\theta - \tan^2\theta = 1, \ \sin^2 55° + \cos^2 55° = 1\right]$

24. R.H.S. $= \dfrac{p^2 - 1}{p^2 + 1} = \dfrac{(\sec\theta + \tan\theta)^2 - 1}{(\sec\theta + \tan\theta)^2 + 1}$

$$= \frac{\sec^2\theta + \tan^2\theta + 2\sec\theta . \tan\theta - 1}{\sec^2\theta + \tan^2\theta + 2\sec\theta . \tan\theta + 1}$$

$$= \frac{1 + \tan^2\theta + \tan^2\theta + 2\sec\theta . \tan\theta - 1}{1 + \tan^2\theta + \tan^2\theta + 2\sec\theta . \tan\theta + 1}$$

$$= \frac{2\tan^2\theta + 2\sec\theta . \tan\theta}{2 + 2\tan^2\theta + 2\sec\theta \tan\theta}$$

$$= \frac{2\left[\frac{\sin^2\theta}{\cos^2\theta} + \frac{1}{\cos\theta} . \frac{\sin\theta}{\cos\theta}\right]}{2\left[1 + \frac{\sin^2\theta}{\cos^2\theta} + \frac{1}{\cos\theta} . \frac{\sin\theta}{\cos\theta}\right]}$$

$$= \frac{\frac{\sin^2\theta + \sin\theta}{\cos^2\theta}}{\frac{\cos^2\theta + \sin^2\theta + \sin\theta}{\cos^2\theta}} = \frac{\sin\theta(1 + \sin\theta)}{(1 + \sin\theta)}$$

$= \sin\theta = \text{L.H.S.}$

$\therefore \quad$ L.H.S. = R.H.S. Proved

25. 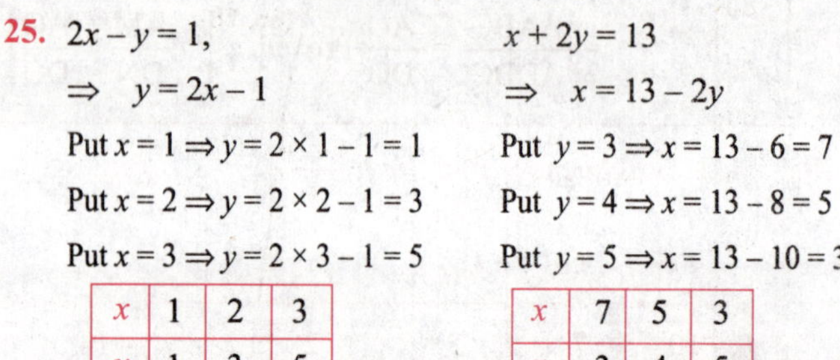

$2x - y = 1$, $\qquad x + 2y = 13$

$\Rightarrow \quad y = 2x - 1$ $\qquad \Rightarrow \quad x = 13 - 2y$

Put $x = 1 \Rightarrow y = 2 \times 1 - 1 = 1$ $\qquad$ Put $y = 3 \Rightarrow x = 13 - 6 = 7$

Put $x = 2 \Rightarrow y = 2 \times 2 - 1 = 3$ $\qquad$ Put $y = 4 \Rightarrow x = 13 - 8 = 5$

Put $x = 3 \Rightarrow y = 2 \times 3 - 1 = 5$ $\qquad$ Put $y = 5 \Rightarrow x = 13 - 10 = 3$

x	1	2	3
y	1	3	5

x	7	5	3
y	3	4	5

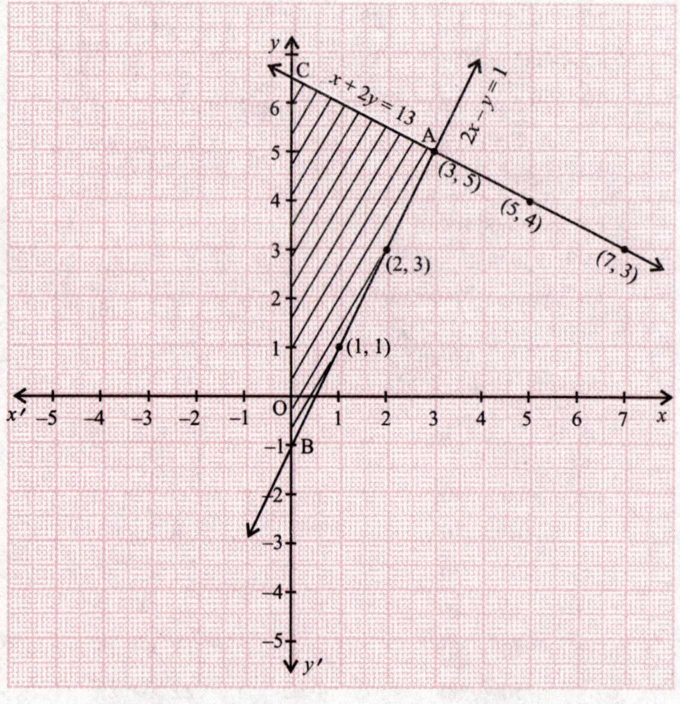

(*i*) Points of intersection of these graph lines is A.

Coordinates of A are $x = 3$ and $y = 5$

(*ii*) Triangular shaded region between lines and *y*-axis is ΔABC.

26.

More than	c.f.
≥ 50	100
≥ 55	98
≥ 60	90
≥ 65	78
≥ 70	54
≥ 75	16

More than ogive

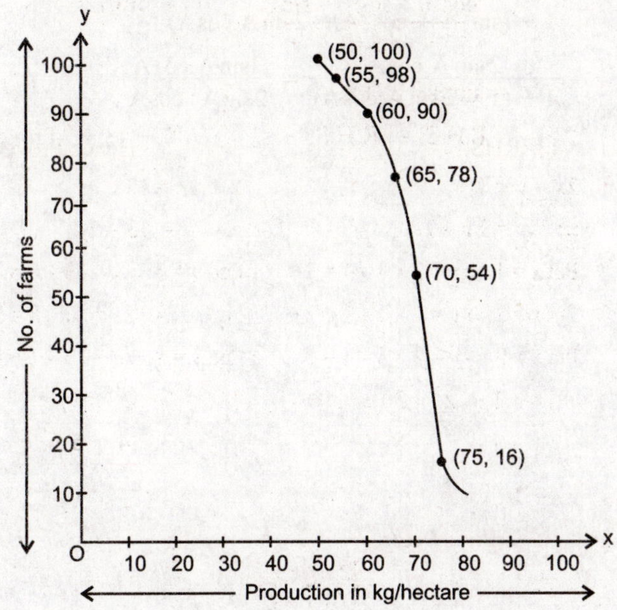

27. Let cost of one chair be ₹ x and one table be ₹ y.

ATQ $\quad 3x + 2y = 700 \quad$ ...(*i*)

and $\quad 5x + 3y = 1100 \quad$ ...(*ii*)

On solving (*i*) and (*ii*), we get

Cost of 2 chairs and 2 tables = ₹ 600.

Amount paid back by shopkeeper

= ₹ 700 – ₹ 600 = ₹ 100

Value: Honesty

28.

Class interval	Frequency	c.f.
0 – 10	x	x
10 – 20	5	$x+5$
20 – 30	9	$x+14$
30 – 40	12	$x+26$
40 – 50	y	$x+y+26$
50 – 60	3	$x+y+29$
60 – 70	2	$x+y+31$
Total	40	

$\because$ Median = 32.5

$\therefore$ Median class = 30 – 40

Also $\quad x + y + 31 = 40$

$\Rightarrow \quad x + y = 9 \quad$ ...(*i*)

Now, $\quad \text{median} = l + \dfrac{\frac{n}{2} - c.f.}{f} \times h$

$\Rightarrow \quad 32.5 = 30 + \dfrac{20 - (x+14)}{12} \times 10$

$\Rightarrow \quad 32.5 = 30 + \dfrac{(6-x)}{12} \times 10$

$\Rightarrow \quad 390.0 = 360 + 60 - 10x$

$\Rightarrow \quad x = 3$

$\therefore$ From (*i*)

$3 + y = 9 \Rightarrow y = 6$

29. In right ΔABD,

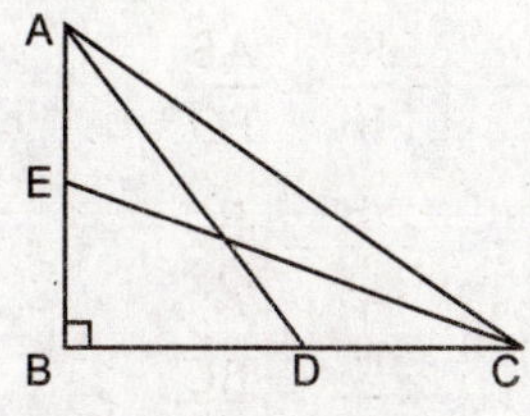

$AD^2 = AB^2 + BD^2$

$\Rightarrow \quad AD^2 = AB^2 + \frac{1}{4}BC^2 \quad$ ...(*i*)

Similarly, in ΔEBC, $CE^2 = BC^2 + \frac{1}{4}AB^2 \quad$ ...(*ii*)

Also, in ΔABC, $\quad AC^2 = AB^2 + BC^2 \quad$ ...(*iii*)

Adding (*i*) and (*ii*) we get

$AD^2 + CE^2 = AB^2 + \frac{1}{4}BC^2 + BC^2 + \frac{1}{4}AB^2$

$\Rightarrow 4AD^2 + 4CE^2 = 5(AB^2 + BC^2) \quad$ ...(*iv*)

Using (*iii*); eq (*iv*) becomes

$4AD^2 + 4CE^2 = 5AC^2 \Rightarrow 4 \times \frac{45}{4} + 4CE^2 = 5 \times 25$

$\Rightarrow \quad 4CE^2 = 80 \Rightarrow CE = 2\sqrt{5}$ cm

30. Let $m = 2x + 1$ and $n = 2y + 1$ for some positive integers x and y.

Then $\quad m^2 + n^2 = (2x+1)^2 + (2y+1)^2$

$= 4x^2 + 1 + 4x + 4y^2 + 1 + 4y$

$= 4x^2 + 4y^2 + 4x + 4y + 2$

$= 2[2x^2 + 2y^2 + 2x + 2y + 1]$

= even integer.

again $\quad m^2 + n^2 = 4[(x^2 + y^2) + (x + y)] + 2$

Let $\quad q = (x^2 + y^2) + (x + y)$

then $\quad m^2 + n^2 = 4q + 2$

Since $m^2 + n^2$ is even and leaves remainder 2 on dividing by 4.

$\therefore$ $m^2 + n^2$ is not divisible by 4.

31. LHS $= \dfrac{\sin A}{\sec A + \tan A - 1} + \dfrac{\cos A}{\operatorname{cosec} A + \cot A - 1}$

$$= \frac{\sin A}{\frac{1}{\cos A} + \frac{\sin A}{\cos A} - 1} + \frac{\cos A}{\frac{1}{\sin A} + \frac{\cos A}{\sin A} - 1}$$

$$= \frac{\sin A}{\frac{1 + \sin A - \cos A}{\cos A}} + \frac{\cos A}{\frac{1 + \cos A - \sin A}{\sin A}}$$

$$= \frac{\sin A \cos A}{1 + \sin A - \cos A} + \frac{\sin A \cos A}{1 + \cos A - \sin A}$$

$$= \sin A \cos A\left[\frac{1}{1 + \sin A - \cos A} + \frac{1}{1 + \cos A - \sin A}\right]$$

$$= \sin A.\cos A\left[\frac{1 + \cos A - \sin A + 1 + \sin A - \cos A}{(1 + \sin A - \cos A)(1 + \cos A - \sin A)}\right]$$

$$= \sin A.\cos A\left[\frac{2}{(1 + \sin A - \cos A)(1 + \cos A - \sin A)}\right]$$

$$= \left[\frac{2 \sin A.\cos A}{\{1 + (\sin A - \cos A)\}\{1 - (\sin A - \cos A)\}}\right]$$

$$= \frac{2 \sin A.\cos A}{(1)^2 - (\sin A - \cos A)^2}$$

$$= \frac{2 \sin A.\cos A}{1 - (\sin^2 A + \cos^2 A - 2 \sin A.\cos A)}$$

$$= \frac{2 \sin A.\cos A}{1 - (1 - 2 \sin A.\cos A)} = \frac{2 \sin A.\cos A}{2 \sin A.\cos A}$$

$= 1 =$ RHS. Hence Proved.

Practice Paper 2

Time Allowed : 3 Hours | *Maximum Marks : 90*

General Instructions: As given in Practice Paper 1

SECTION A

1. $\frac{195}{650}$ will have a terminating decimal expansion. State True or False. Justify your answer.

2. Find the median of the observation:

x_i	5	6	7	8	9	10
f_i	4	5	7	9	7	6

3. In the given figure, MN || PR, then

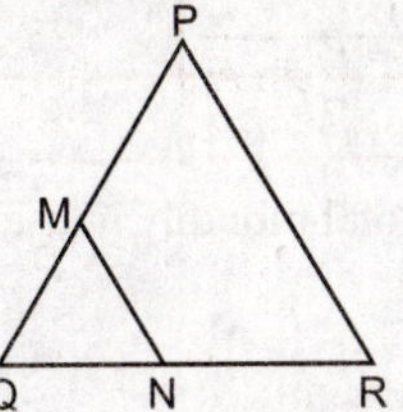

Show that PM × NQ = NR × MQ

4. Find the value of k so that the following system of equations has no solution:
$3x - y - 5 = 0$, $6x - 2y + k = 0$

SECTION B

5. Prove that $\frac{\cos\theta}{1-\sin\theta} = \sec\theta + \tan\theta$.

6. A cricketer has a mean score of 58 runs in nine innings. Find out how many runs are to be scored in the tenth inning to raise the mean score to 61.

7. In the given figure, ∠A = ∠B and AD = BE, show that DE || AB.

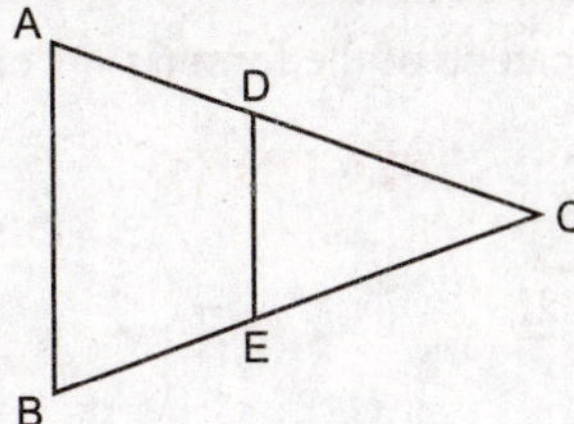

8. Evaluate: $\frac{2\tan 30^\circ}{1+\tan^2 30^\circ}$

9. Find a quadratic polynomial whose zeroes are $3+\sqrt{5}$ and $3-\sqrt{5}$.

10. For what value of k, the system of equations $kx - 3y + 6 = 0$, $4x - 6y + 12 = 0$ represents parallel lines?

SECTION C

11. Using Euclid's division algorithm, find HCF of 216 and 1176.

12. Draw the graphs of the equation $x - y + 1 = 0$ and $3x + 2y - 12 = 0$. Determine the coordinates of the vertices of the triangle formed by these lines and the x-axis.

13. Find the value of $\frac{\cos^2 20^\circ + \cos^2 70^\circ}{\sin^2 31^\circ + \sin^2 59^\circ} + \sin^2 64^\circ + \cos 64^\circ \sin 26^\circ$

14. Two isosceles triangles have equal vertical angles and their areas are in the ratio 81 : 25. Find the ratio of their corresponding heights.

15. In the adjoining figure, E is a point on side CB produced of an isosceles ΔABC with $AB = AC$.

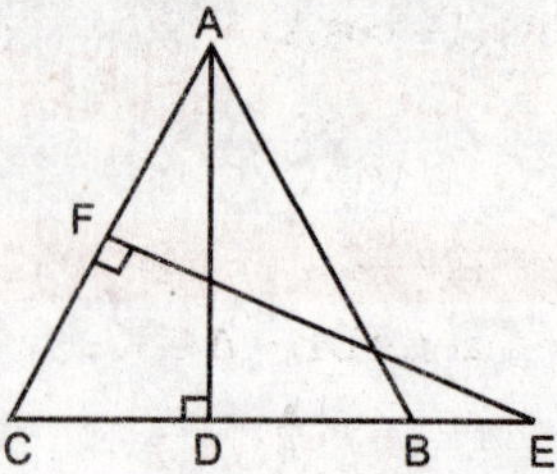

If $AD \perp BC$ and $EF \perp AC$, prove that $\Delta ABD \sim \Delta ECF$.

16. Show that $\dfrac{\sin\theta - 2\sin^3\theta}{2\cos^3\theta - \cos\theta} = \tan\theta$.

17. If two zeroes of the polynomial $x^4 + 3x^3 - 20x^2 - 6x + 36$ are $\sqrt{2}$ and $-\sqrt{2}$, find the other zeroes of the polynomial.

18. If the mean of the following data is 18.75, find the value of p:

x_i	10	15	p	25	30
f_i	5	10	7	8	2

19. The following data gives the distribution of total monthly household expenditure of 200 families of a village. Find the modal monthly expenditure of the families.

Expenditure (in ₹)	Number of families
1000 — 1500	24
1500 — 2000	40
2000 — 2500	33
2500 — 3000	28
3000 — 3500	30
3500 — 4000	22
4000 — 4500	16
4500 — 5000	7

20. Show that square of an odd positive integer can be of the form $6q + 1$ or $6q + 3$ for some integer q.

SECTION D

21. If $\tan\theta + \sec\theta = l$, then prove that $\sec\theta = \dfrac{l^2+1}{2l}$.

22. State and prove pythagoras theorem.

23. It can take 12 hours to fill a swimming pool using two pipes. If the pipe of larger diameter is used for 4 hours and the pipe of smaller diameter is used for 9 hours, only half of the pool can be filled. How long would it take for each pipe to fill the pool separately?

***24.** Students of a locality decided to create an awareness about 'save electricity compaign'. They recorded the weekly consumption of electricity of 56 families as given in the table:

Weekly consumption	0 — 10	10 — 20	20 — 30	30 — 40	40 — 50
No. of families	16	12	18	6	4

(*a*) Find the mean.

(*b*) Which value is depicted by the students?

25. α, β are zeroes of the polynomial $x^2 - 6x + a$. Find the value of a if $3\alpha + 2\beta = 20$.

---

*Value Based Question

26. Prove that $\dfrac{\cos A}{1-\sin A}+\dfrac{\sin A}{1-\cos A}+1=\dfrac{\sin A\cos A}{(1-\sin A)(1-\cos A)}$.

27. Solve for x and y: $6x + 3y = 8x + 9y - 5 = 10x + 12y - 8$

28. In figure, $\Delta FEC \cong \Delta GBD$ and $\angle 1 = \angle 2$. Prove that $\Delta ADE \sim \Delta ABC$.

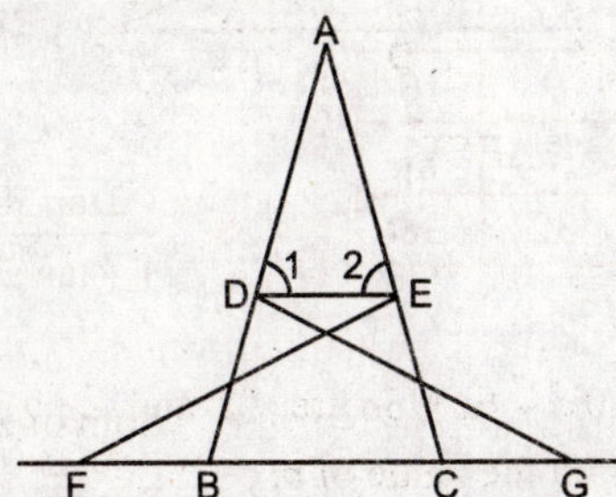

29. If the mean of the following frequency distribution is 65.6, find the missing frequencies (f_1, f_2) :

Class	Frequency
10 – 30	5
30 – 50	8
50 – 70	f_1
70 – 90	20
90 – 110	f_2
110 – 130	2
Total	50

30. Show that p^2 will leave a remainder 1 when divided by 8, if p is an odd positive integer.

31. Prove that $\dfrac{\tan^3\theta}{1+\tan^2\theta}+\dfrac{\cot^3\theta}{1+\cot^2\theta}=\sec\theta\,.\,\text{cosec}\,\theta-2\sin\theta\cos\theta$

Solutions

1. $\frac{195}{650} = \frac{3 \times 13 \times 5}{2^1 \times 13 \times 5^2} = \frac{3}{2 \times 5}$

It has terminating decimal expansion because denominator is of the form $2^n \times 5^m$.

2. (*d*)

x	5	6	7	8	9	10
f	4	5	7	9	7	6
cf	4	9	16	25	32	38

$\frac{N}{2} = \frac{38}{2} = 19$

$\therefore$ Median = 8.

3. In ΔPQR, MN $\parallel$ PR

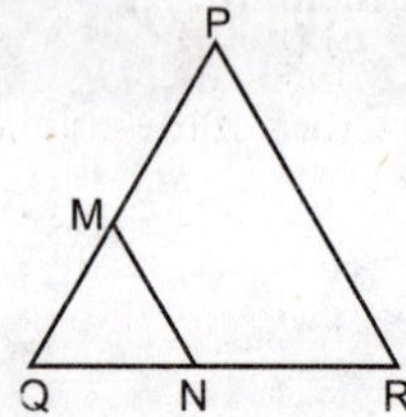

$\therefore \quad \frac{QM}{PM} = \frac{QN}{NR}$ [Using Thales Theorem]

$\Rightarrow$ QM.NR = QN.PM

$\Rightarrow$ PM.NQ = NR.MQ

4. Here $a_1 = 3, b_1 = -1, c_1 = -5$,

and $a_2 = 6, b_2 = -2, c_2 = k$.

For no solution,

$\frac{a_1}{a_2} = \frac{b_1}{b_2} \neq \frac{c_1}{c_2} \Rightarrow \frac{3}{6} = \frac{-1}{-2} \neq \frac{-5}{k}$

$\Rightarrow \frac{1}{2} \neq \frac{-5}{k} \Rightarrow k \neq -10$

5. L.H.S. $= \frac{\cos\theta}{1 - \sin\theta} = \frac{\cos\theta(1 + \sin\theta)}{1 - \sin^2\theta}$

$= \frac{\cos\theta(1 + \sin\theta)}{\cos^2\theta} = \frac{1 + \sin\theta}{\cos\theta} = \sec\theta + \tan\theta$ = R.H.S.

6. Mean score in nine innings = 58

$\therefore$ Total score in nine innings = 9 × 58 = 522

Mean score in ten innings = 61

$\therefore$ Total score in ten innings = 61 × 10 = 610

Runs to be scored in tenth inning = 610 – 522 = 88 runs

7. In ΔACB, $\because \angle A = \angle B$ (Given)

$\therefore$ AC = BC ... (*i*)

Also AD = BE ... (*ii*)

Dividing (*i*) by (*ii*) we get,

$\frac{AC}{AD} = \frac{BC}{BE}$

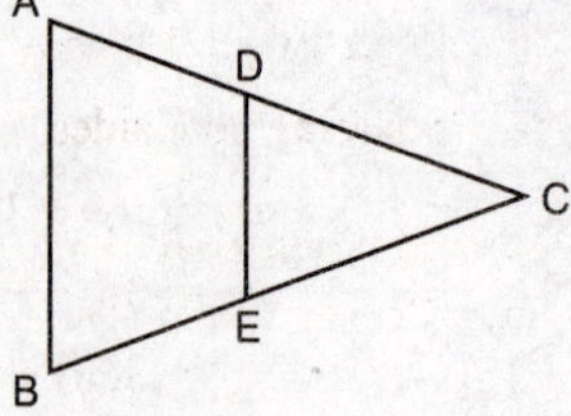

$\Rightarrow \quad \frac{AC}{AD} - 1 = \frac{BC}{BE} - 1$

$\Rightarrow \quad \frac{AC - AD}{AD} = \frac{BC - BE}{BE} \Rightarrow \frac{CD}{AD} = \frac{CE}{BE}$

$\therefore$ DE $\parallel$ AB (using converse of (B.P.T.)

8. $\frac{2\tan 30^\circ}{1 + \tan^2 30^\circ} = \frac{2 \times \frac{1}{\sqrt{3}}}{1 + \left(\frac{1}{\sqrt{3}}\right)^2} = \frac{2}{\sqrt{3}} \times \frac{3}{4} = \frac{\sqrt{3}}{2}$

9. Sum of zeroes, $S = (3 + \sqrt{5}) + (3 - \sqrt{5}) = 6$

Product of zeroes, $P = (3 + \sqrt{5}).(3 - \sqrt{5}) = 9 - 5 = 4$

Polynomial $p(x) = x^2 - (S)x + P = x^2 - 6x + 4$

10. The given system of equations will represent parallel lines if it has no solution.

For no solution, $\frac{a_1}{a_2} = \frac{b_1}{b_2} \neq \frac{c_1}{c_2}$

i.e. $\frac{k}{4} = \frac{-3}{-6} \neq \frac{6}{12} \Rightarrow \frac{k}{4} = \frac{1}{2}$ and $\frac{k}{4} \neq \frac{1}{2}$

$\Rightarrow \quad \frac{k}{4} = \frac{1}{2} \Rightarrow k = \frac{4}{2} = 2$ and $k \neq 2$

$\therefore$ no value of k.

11. 1176, 216

$1176 = 216 \times 5 + 96$

$216 = 96 \times 2 + 24$

$96 = 24 \times 4 + 0$

$\therefore$ HCF is 24.

12.

$x - y = -1$

x	0	–1	2
y	1	0	3

$3x + 2y = 12$

x	0	4	2
y	6	0	3

Required vertices are A (–1, 0), B (2, 3) and C (4, 0).

13. $\dfrac{\cos^2 20° + \cos^2 70°}{\sin^2 31° + \sin^2 59°} + \sin^2 64° + \cos 64° \sin 26°$

$= \dfrac{\sin^2(90° - 20°) + \cos^2 70°}{\sin^2(90° - 59°) + \sin^2 59°} + \sin^2 64° + \cos 64° \cos(90° - 26°)$

$= \dfrac{\sin^2 70° + \cos^2 70°}{\cos^2 59° + \sin^2 59°} + \sin^2 64° + \cos^2 64°$

$= \dfrac{1}{1} + 1 = 2$

14. In ΔABC and ΔDEF

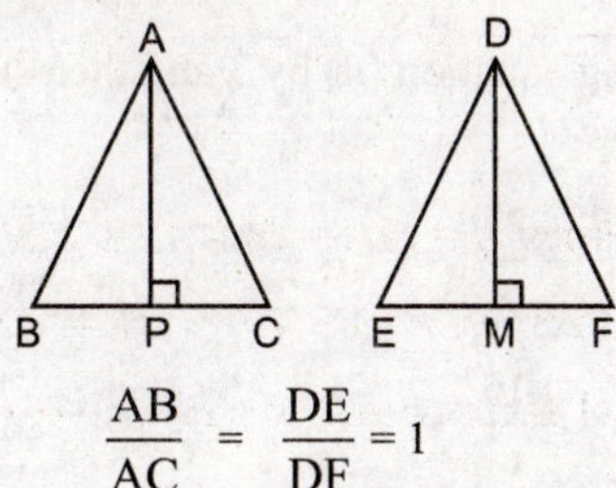

$$\frac{AB}{AC} = \frac{DE}{DF} = 1$$

and $\angle A = \angle D$ [Triangles are isosceles and have equal vertical angles]

$\therefore$ $\Delta ABC \sim \Delta DEF$

$\Rightarrow$ $\dfrac{\text{Area}(\Delta ABC)}{\text{Area}(\Delta DEF)} = \dfrac{BC^2}{EF^2} \Rightarrow \dfrac{81}{25} = \dfrac{BC^2}{EF^2}$

$\Rightarrow$ $\dfrac{BC}{EF} = \dfrac{9}{5}$

Also, $\dfrac{\text{Area}(\Delta ABC)}{\text{Area}(\Delta DEF)} = \dfrac{\frac{1}{2}BC \times AP}{\frac{1}{2}EF \times DM}$

$\Rightarrow$ $\dfrac{81}{25} = \dfrac{9}{5} \times \dfrac{AP}{DM}$

$\Rightarrow$ $\dfrac{AP}{DM} = 9 : 5.$

15. ΔABC is an isosceles Δ.

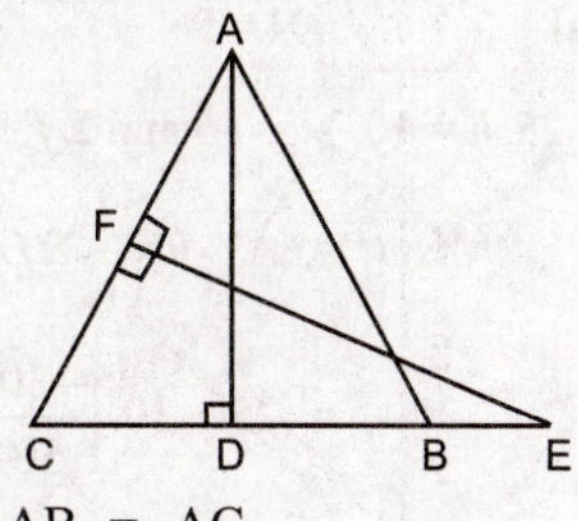

AB = AC [Given]

$\therefore$ $\angle B = \angle C$ [Angles opp to equal sides]

Now, in Δ ABD and ΔECF, we have

$\angle ABD = \angle ECF$ [$\because \angle B = \angle C$]

$\angle ADB = \angle EFC = 90°$

because AD ⊥ BC and EF ⊥ AC

$\therefore$ $\Delta ABD \sim \Delta ECF$ [AA similarity]

Proved.

16. LHS $= \dfrac{\sin\theta - 2\sin^3\theta}{2\cos^3\theta - \cos\theta}$

$= \dfrac{\sin\theta(1 - 2\sin^2\theta)}{\cos\theta(2\cos^2\theta - 1)}$

$= \dfrac{\sin\theta(\sin^2\theta + \cos^2\theta - 2\sin^2\theta)}{\cos\theta(2\cos^2\theta - \sin^2\theta - \cos^2\theta)}$

$= \dfrac{\sin\theta}{\cos\theta}\left[\dfrac{\cos^2\theta - \sin^2\theta}{\cos^2\theta - \sin^2\theta}\right] = \tan\theta = \text{RHS}.$

17. Let $f(x) = x^4 + 3x^3 - 20x^2 - 6x + 36$

Since $\sqrt{2}$ and $-\sqrt{2}$ are zeroes of $f(x)$.

$\therefore (x - \sqrt{2})(x + \sqrt{2}) = x^2 - 2$ is a factor of $f(x)$.

By division algorithm, we have

$x^4 + 3x^3 - 20x^2 - 6x + 36$

$= (x^2 - 2)(x^2 + 3x - 18)$

$= (x^2 - 2)(x + 6)(x - 3)$

$\therefore$ The other zeroes of the polynomial are –6 and 3.

18. Mean = 18.75

x_i	f_i	$f_i x_i$
10	5	50
15	10	150
p	7	$7p$
25	8	200
30	2	60
Total	$\Sigma f_i = 32$	$\Sigma f_i x_i = 460 + 7p$

Using direct method:

$$\text{Mean} = \frac{\Sigma f_i x_i}{\Sigma f_i} = \frac{460 + 7p}{32}$$

$\Rightarrow$ $18.75 = \dfrac{460 + 7p}{32}$

$\Rightarrow$ $18.75 \times 32 = 460 + 7p$

$\Rightarrow$ $140 = 7p$

$\Rightarrow$ $p = 20$

19. Modal class = 1500 – 2000

[$\because$ this class has the highest frequency]

$\therefore$ $l = 1500,\ f_0 = 24,\ f_1 = 40,\ f_2 = 33,\ h = 500$

$\therefore$ Mode $= l + \left(\dfrac{f_1 - f_0}{2f_1 - f_0 - f_2}\right) \times h$

$= 1500 + \left(\dfrac{40 - 24}{2 \times 40 - 24 - 33}\right) \times 500$

$= 1500 + \dfrac{16}{23} \times 500 = ₹1847.82$

20. Let $a = 6q + 1$ be any odd positive integer.

Squaring both sides, we get

$a^2 = (6q + 1)^2$

$\Rightarrow$ $a^2 = 36q^2 + 1 + 12q$

$= 36q^2 + 12q + 1$

$$= 6(6q^2 + 2q) + 1$$
$$= 6m + 1 \text{ where } m = 6q^2 + 2q$$

Again let $a = 6q + 3 \Rightarrow a^2 = (6q + 3)^2$

$$\Rightarrow \quad a^2 = 36q^2 + 36q + 9$$
$$= 36q^2 + 36q + 6 + 3$$
$$= 6(6q^2 + 6q + 1) + 3$$
$$= 6n + 3$$

where $n = 6q^2 + 6q + 1$

Also, let $a = 6q + 5$

$$\Rightarrow \quad a^2 = (6q + 5)^2$$
$$\Rightarrow \quad a^2 = 36q^2 + 60q + 25$$
$$= 36q^2 + 60q + 24 + 1$$
$$= 6(6q^2 + 10q + 4) + 1$$
$$= 6p + 1$$

where $p = 6q^2 + 10q + 4$.

∴ Any square of odd positive integer is of the form $6q + 1$ or $6q + 3$.

21. $\tan\theta + \sec\theta = l$

Squaring both sides, we get

$(\tan\theta + \sec\theta)^2 = l^2$

$\Rightarrow \tan^2\theta + \sec^2\theta + 2\tan\theta\sec\theta = l^2$

$\Rightarrow \sec^2\theta - 1 + \sec^2\theta + 2\tan\theta\sec\theta = l^2$

$\Rightarrow 2\sec^2\theta + 2\tan\theta\sec\theta = l^2 + 1$

$\Rightarrow \sec\theta = \dfrac{l^2 + 1}{2(\tan\theta + \sec\theta)} = \dfrac{l^2 + 1}{2l}$. Hence proved.

22.

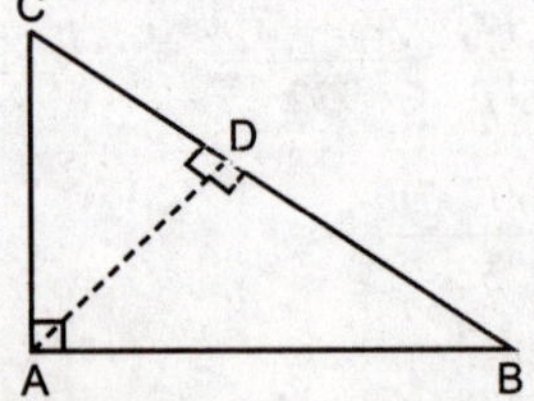

Given: In ΔABC, ∠A = 90°

To prove: $BC^2 = AC^2 + AB^2$

Construction: AD ⊥ BC

Proof: ΔADB ~ ΔCAB

$$\Rightarrow \frac{AD}{AC} = \frac{BD}{AB} = \frac{AB}{BC} \Rightarrow \frac{BD}{AB} = \frac{AB}{BC}$$

$$\Rightarrow AB^2 = BD.BC \quad ...(i)$$

Also ΔADC ~ ΔBAC

$$\therefore \frac{AD}{AB} = \frac{CD}{AC} = \frac{AC}{BC}$$

$$\Rightarrow \frac{CD}{AC} = \frac{AC}{BC} \Rightarrow AC^2 = BC.CD \quad ...(ii)$$

(*i*) + (*ii*)

$$AB^2 + AC^2 = BC.CD + BC.BD$$
$$\Rightarrow AB^2 + AC^2 = BC(CD + BD)$$
$$AB^2 + AC^2 = BC.BC = BC^2$$

23. Let the diameter of larger pipe be y and smaller pipe be x.

According to 1st condition

$$\frac{1}{x} + \frac{1}{y} = \frac{1}{12}$$

$$\Rightarrow \frac{12}{x} + \frac{12}{y} = 1 \quad ...(i)$$

2nd condition

$$\frac{4}{y} + \frac{9}{x} = \frac{1}{2} \quad ...(ii)$$

Multiplying equation (*ii*) by 3 and then subtracting from (*i*), we have.

$$\frac{12}{x} - \frac{27}{x} = 1 - \frac{3}{2}$$

$$\frac{-15}{x} = \frac{-1}{2}$$

$$x = 30 \text{ h.}$$

∴ From equation (*i*)

$$\frac{12}{30} + \frac{12}{y} = 1 \quad \Rightarrow \frac{12}{y} = 1 - \frac{2}{5}$$

$$\Rightarrow \frac{12}{y} = \frac{3}{5} \quad \Rightarrow y = 20 \text{ h.}$$

24. (*a*)

C.I.	Mid value (x_i)	f_i	$u_i = \dfrac{x_i - 25}{10}$	$f_i u_i$
0 – 10	5	16	–2	–32
10 – 20	15	12	–1	–12
20 – 30	25	18	0	0
30 – 40	35	6	1	6
40 – 50	45	4	2	8
Total		N = 56		$\Sigma f_i u_i = -30$

Let A = 25, $h = 10$, N = 56 and $\Sigma f_i u_i = -30$

$$\therefore \quad \text{Mean} = A + h\left\{\frac{1}{N}\Sigma f_i u_i\right\}$$

$$= 25 + 10\left\{\frac{-30}{56}\right\} = 25 - \frac{300}{56}$$

$$= \frac{1400 - 300}{56} = \frac{1100}{56}$$

$$\therefore \quad \text{Mean} = 19.6$$

(*b*) Social service

25. ∵ α, β are the roots of the polynomial $x^2 - 6x + a$

$$\therefore \quad \alpha + \beta = \frac{-b}{a} = -(-6) = 6$$

$$\Rightarrow \quad \alpha + \beta = 6 \quad ...(i)$$

Also, $3\alpha + 2\beta = 20$ (given) ...(*ii*)

Solving (*i*) and (*ii*), we get

$$\beta = -2$$

Hence, –2 is a root of $x^2 - 6x + a$.

$\Rightarrow \quad (-2)^2 - 6(-2) + a = 0$

$\Rightarrow \quad a = -16$

26. LHS $= \dfrac{\cos A}{1-\sin A} + \dfrac{\sin A}{1-\cos A} + 1$

$$= \frac{\cos A(1-\cos A) + \sin A(1-\sin A) + (1-\sin A)(1-\cos A)}{(1-\sin A)(1-\cos A)}$$

$$= \frac{\cos A - \cos^2 A + \sin A - \sin^2 A + 1 - \cos A - \sin A + \sin A \cos A}{(1-\sin A)(1-\cos A)}$$

$$= \frac{-(\cos^2 A + \sin^2 A) + 1 + \sin A \cos A}{(1-\sin A)(1-\cos A)}$$

$$= \frac{\sin A \cos A}{(1-\sin A)(1-\cos A)} = \text{RHS}$$

Hence proved.

27. $6x + 3y = 8x + 9y - 5$

$\Rightarrow \quad 2x + 6y = 5 \quad \text{... (i)}$

Also, $8x + 9y - 5 = 10x + 12y - 8$

$\Rightarrow \quad 2x + 3y = 3 \quad \text{... (ii)}$

On solving (*i*) and (*ii*),

$$3y = 2 \Rightarrow y = \frac{2}{3}$$

$\therefore$ From (*i*), $\quad 2x = 5 - 6 \times \dfrac{2}{3} = 5 - 4 = 1$

$\Rightarrow \quad x = \dfrac{1}{2}$

28.

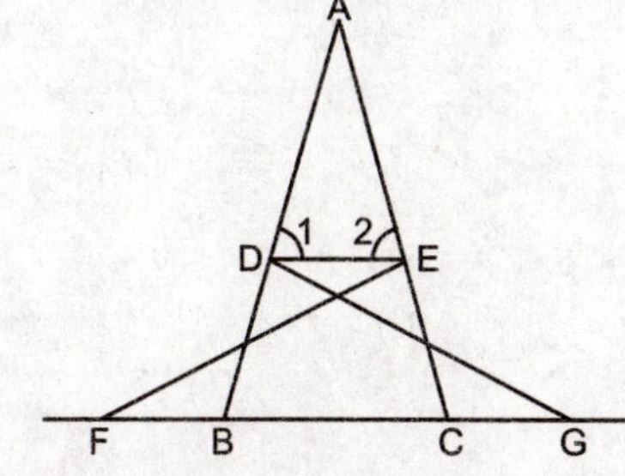

Given: $\Delta FEC \cong \Delta GBD$

and $\angle 1 = \angle 2$

To Prove: $\Delta ADE \sim \Delta ABC$

Proof: In ΔADE,

$\angle 1 = \angle 2$ (Given)

$\Rightarrow \quad AD = AE \quad \text{...(i)}$

[Sides opposite to equal angles are equal]

$\Delta FEC \cong \Delta GBD$ (Given)

$\Rightarrow \quad BD = EC$ (CPCT)...(*ii*)

Adding (*i*) and (*ii*) we have,

$\Rightarrow \quad AD + BD = AE + EC$

$AB = AC \quad \text{...(iii)}$

Dividing (*i*) by (*iii*),

$$\frac{AD}{AB} = \frac{AE}{AC}$$

and $\angle A = \angle A$ (Common)

$\therefore \quad \Delta ADE \sim \Delta ABC$ (SAS similarity)

29. Let A = 60, $h = 20$

C.I.	x_i	f_i	$u_i = \dfrac{x_i - 60}{20}$	$f_i u_i$
10 – 30	20	5	– 2	– 10
30 – 50	40	8	– 1	– 8
50 – 70	60	f_1	0	0
70 – 90	80	20	1	20
90 – 110	100	f_2	2	$2f_2$
110 – 130	120	2	3	6
Total		$\Sigma f_i = 50$		$\Sigma f_i u_i = 2f_2 + 8$

$$\text{Mean} = A + \frac{\Sigma f_i u_i}{\Sigma f_i} \times h$$

$\Rightarrow \quad 65.6 = 60 + \dfrac{(2f_2 + 8)}{50} \times 20$

$\Rightarrow \quad 5.6 = \dfrac{(2f_2 + 8) \times 2}{5}$

$\Rightarrow \quad 28 = 4f_2 + 16$

$\Rightarrow \quad 12 = 4f_2$

$\Rightarrow \quad f_2 = 3$

Also, $\quad 35 + f_1 + f_2 = 50$

$\Rightarrow \quad 35 + f_1 + 3 = 50$

$\Rightarrow \quad f_1 = 12$

$\therefore \quad f_1 = 12, f_2 = 3.$

30. $\because$ p is an odd positive integer.

$\therefore \quad p = 4q + 1$ or $4q + 3$ for some integer q.

When $\quad p = 4q + 1$

then $\quad p^2 = 16q^2 + 8q + 1$

$= 8(2q^2 + q) + 1$

$\Rightarrow \quad p^2 = 8m + 1$

$\Rightarrow$ p^2 leaves remainder 1 when divided by 8.

When $\quad p = 4q + 3$

$\Rightarrow \quad p^2 = 16q^2 + 24q + 9$

$\Rightarrow \quad p^2 = 16q^2 + 24q + 8 + 1$

$\Rightarrow \quad p^2 = 16q^2 + 24q + 8 + 1$

$= 8(2q^2 + 3q + 1) + 1$

$= 8m + 1$

$\therefore$ p^2 leaves a remainder 1 when divided by 8.

31. LHS $= \dfrac{\tan^3\theta}{1+\tan^2\theta} + \dfrac{\cot^3\theta}{1+\cot^2\theta}$

$= \dfrac{\sin^3\theta}{\cos^3\theta}.\cos^2\theta + \dfrac{\cos^3\theta}{\sin^3\theta}.\sin^2\theta$

$= \dfrac{\sin^3\theta}{\cos\theta} + \dfrac{\cos^3\theta}{\sin\theta} = \dfrac{\sin^4\theta + \cos^4\theta}{\cos\theta.\sin\theta}$

$= \dfrac{(\sin^2\theta+\cos^2\theta)^2 - 2\sin^2\theta.\cos^2\theta}{\sin\theta.\cos\theta}$

$[\because a^2 + b^2 = (a+b)^2 - 2ab]$

$= \dfrac{1 - 2\sin^2\theta.\cos^2\theta}{\sin\theta.\cos\theta} = \dfrac{1}{\sin\theta.\cos\theta} - \dfrac{2\sin^2\theta.\cos^2\theta}{\sin\theta.\cos\theta}$

$= \sec\theta.\text{cosec}\,\theta - 2\sin\theta.\cos\theta$ = RHS. Hence proved.

Practice Paper 3

Time Allowed : *3 Hours* | *Maximum Marks* : *90*

General Instructions: As given in Practice Paper 1

SECTION A

1. D and E are points on the sides AB and AC of a ΔABC such that DE||BC. If AD = 6 cm, BD = 9 cm and AE = 8 cm, then find EC.

2. Is $7 \times 5 \times 3 \times 11 + 11$ a composite number? Justify your answer.

3. Find the median of the following observations

x_i	2	3	4	5	6	7	8
f_i	3	5	7	9	3	2	2

4. Find the value of K so that the following system of equation has unique solution: $2x + 4y - 7 = 0$, $8x + Ky - 5 = 0$

SECTION B

5. Find the zeroes of the polynomial $5x^2 + 17x + 12$.

6. Solve for x and y: $11x - 7y = 1$; $9x - 4y = 6$.

7. Find A and B, if $\sin(A + B) = 1$ and $\cos(A - B) = 1$.

8. In the given figure, $\angle A = \angle B$ and AD = BE, $\therefore$ DE || AB. State True or False.

Justify your answer.

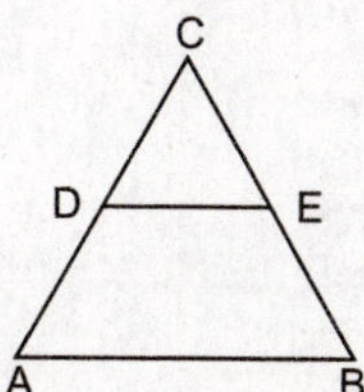

9. Find the median wage of a worker engaged at a construction site whose data are given below:

Wages (in ₹)	Number of workers
3500	12
3800	13
4100	25
4500	17
5500	15
6500	12
7000	6

10. If $\sec A = \frac{15}{7}$ and $A + B = 90°$, find the value of cosec B.

SECTION C

11. Find the LCM and HCF of 312 and 27 and verify; LCM × HCF = Product of the numbers.

12. Prove that: $\cos\theta \sin\theta - \frac{\sin\theta\cos(90° - \theta)\cos\theta}{\sec(90° - \theta)} - \frac{\cos\theta\sin(90° - \theta)\sin\theta}{\operatorname{cosec}(90° - \theta)} + \operatorname{cosec}(90° - \theta) = \frac{1}{\cos\theta}$.

13. In a ΔABC, $AB = AC$ and D is a point on side AC such that $BC^2 = AC \times CD$.

    Prove that $BD = BC$.

14. Prove that the line segments joining the mid-points of the sides of a triangle, form four triangles each of which is similar to the original triangle.

15. A and B are two points 150 km apart on a highway. Two cars start from A and B at the same time. If they move in the same direction, they meet in 15 hours. But if they move in the opposite directions, they meet in one hour. Find their speeds.

16. Two zeroes of the polynomial $P(x) = 2x^4 - 3x^3 - 3x^2 + 6x - 2$ are $\sqrt{2}$ and $-\sqrt{2}$ respectively. Find the remaining zeroes of $P(x)$, if any.

17. Show that: $\tan^4\theta + \tan^2\theta = \sec^4\theta - \sec^2\theta$.

18. The following data gives the distribution of total monthly household expenditure of 200 families of a village. Find the modal monthly expenditure of the families.

Expenditure (in ₹)	Number of families
1000 — 1500	24
1500 — 2000	40
2000 — 2500	33
2500 — 3000	28
3000 — 3500	30
3500 — 4000	22
4000 — 4500	16
4500 — 5000	7

19. Find the median of the following distribution:

Class	12.5 – 17.5	17.5 – 22.5	22.5 – 27.5	27.5 – 32.5
Frequency	2	22	19	14

20. Show that square of any positive integer is either of the form $3m$ or $3m + 1$. Where m is a positive integer.

SECTION D

21. The ratio of the areas of similar triangles is equal to the ratio of the squares of their corresponding sides. Prove it.

22. If $\frac{x}{a}\cos\theta + \frac{y}{b}\sin\theta = 1$ and $\frac{x}{a}\sin\theta - \frac{y}{b}\cos\theta = 1$, prove that $\frac{x^2}{a^2} + \frac{y^2}{b^2} = 2$.

23. Find the mean marks of students from the following cumulative frequency table:

Marks	Number of Students
0 and above	80
10 and above	77
20 and above	72
30 and above	65
40 and above	55
50 and above	43
60 and above	28
70 and above	16
80 and above	10
90 and above	8
100 and above	0

24. On dividing the polynomial $4x^4 - 5x^3 - 39x^2 - 46x - 2$ by the polynomial $g(x)$, the quotient is $x^2 - 3x - 5$ and the remainder is $-5x + 8$. Find the polynomial $g(x)$.

25. Solve graphically the system of linear equations:

$2x + y = 8 \; ; x + 1 = 2y.$

Also, find the co-ordinates of the points where the lines meet the y-axis.

26. Find the value of sin 30° geometrically.

*27. 3 chairs and 2 tables cost ₹ 700 and 5 chairs and 3 tables cost ₹ 1100. A man purchased 2 chairs and 2 tables and paid ₹ 700 to the shopkeeper. Shopkeeper said, "It is not the correct amount" and returned him the balance. Find the correct amount and amount paid back by the shopkeeper. Which value is depicted by the shopkeeper?

28. In the given figure, M is the mid-point of the side CD of parallelogram ABCD. BM when joined meets AC at L and AD produced to E. Prove that EL = 2BL.

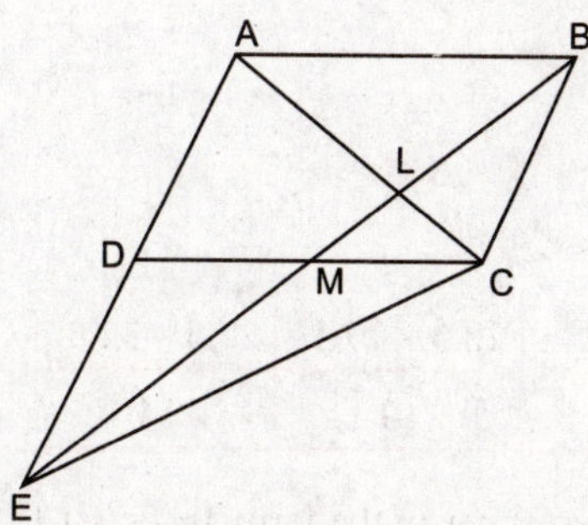

29. Draw 'less than ogive' and 'more than ogive' for the following distribution and hence find its median.

Class	20 – 30	30 – 40	40 – 50	50 – 60	60 – 70	70 – 80	80 – 90
Frequency	10	8	12	24	6	25	15

30. If $\sec A = x + \frac{1}{4x}$, then prove that $\sec A + \tan A = 2x$ or $\frac{1}{2x}$.

31. Use Euclid's division algorithm to show that the cube of any positive integer is of the form $9m$, $9m + 1$ or $9m + 8$.

* Value Based Question

Answers

1. 12 cm
2. Yes
3. 5
4. $k \neq 16$
5. $-1, \frac{-12}{5}$
6. $x = 2, y = 3$
7. A = 45°, B = 45°
8. True
9. ₹ 4300
10. $\frac{15}{7}$
11. HCF = 3, LCM = 2808
15. Speed of car at A = 80 km/h.; speed of car at B = 70 km/h.
16. $\frac{1}{2}, 1$
18. ₹ 1847.826
19. 23.68
23. 51.75.
24. $4x^2 + 7x + 2$
25. $x = 3, y = 2, (0, 8), (0, \frac{1}{2})$
26. $\frac{1}{2}$
27. ₹ 600, ₹ 100, Honesty
29. Median = 58.3

Practice Paper 4

Time Allowed : 3 Hours | **Maximum Marks : 90**

General Instructions: As given in Practice Paper 1

SECTION A

1. Find the value of k for which the pair of linear equations $2x - 3y + 7 = 0$ and $4x - ky - 14 = 0$ will have infinite solutions.

2. Find mode, if mean of 10 observations is 5 and median is 4.

3. If Δ AMB ~ Δ CMD,

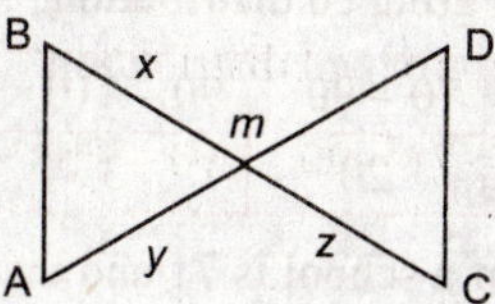

and BM = x, AM = y, CM = z, then find DM.

4. HCF of two numbers are 9 and 360 respectively. If one number is 45, write the other number.

SECTION B

5. Solve for x and y:

$15x + 22y = 5$; $\quad 40x + 55y = 13$

6. Write a quadratic polynomial whose one zero is 5 and sum of zeroes is –12.

7. In figure, $\angle 1 = \angle 2$, $\angle 3 = \angle 4$,

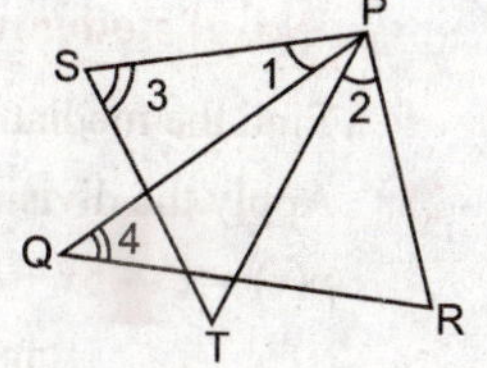

show that: PT . QR = PR.ST.

8. If A = 30°, then verify: $\tan 2A = \dfrac{2\tan A}{1 - \tan^2 A}$.

9. If the mean of the following distribution is 2.6, then find the value of y.

Variable	1	2	3	4	5
Frequency	4	5	y	1	2

10. If $\cot\theta = \dfrac{7}{8}$, evaluate $\dfrac{(1+\sin\theta)(1-\sin\theta)}{(1+\cos\theta)(1-\cos\theta)}$.

SECTION C

11. State whether 10.064 is rational or not. If rational, express in $\dfrac{p}{q}$ form.

12. Using division algorithm, find the quotient and remainder on dividing $8x^4 + 14x^3 - 2x^2 + 8x - 12$ by $4x^2 + 3x - 2$.

13. Show that: $\dfrac{1}{\sec x - \tan x} - \dfrac{1}{\cos x} = \dfrac{1}{\cos x} - \dfrac{1}{\sec x + \tan x}$

14. Evaluate: $\sin\theta\cos\theta - \dfrac{\sin\theta\cos(90° - \theta)\cos\theta}{\sec(90° - \theta)} - \dfrac{\cos\theta\sin(90° - \theta)\sin\theta}{\operatorname{cosec}(90° - \theta)}$.

15. In figure, DE || AQ and DF || AR. Prove that EF || QR.

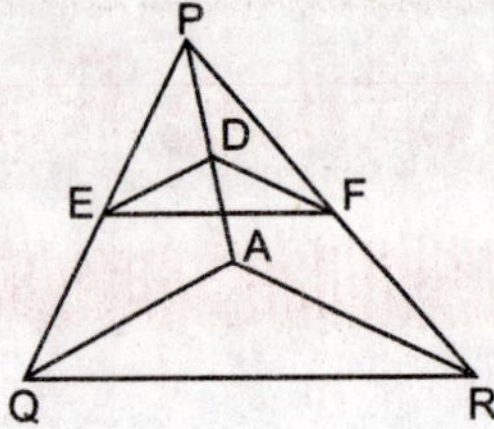

16. Draw the graphs of the following equations:

$$3x - 4y + 6 = 0;\ 3x + y - 9 = 0$$

Also, determine the co-ordinates of the vertices of the triangle formed by these lines and the *x*-axis.

17. L and M are the mid-points of AB and BC respectively of ΔABC, right-angled at B. Prove that $4LC^2 = AB^2 + 4BC^2$.

18. Calculate the arithmetic mean of the following grouped distribution.

Class	10 – 30	30 – 50	50 – 70	70 – 90	90 – 110	110 –130	Total
Frequency	5	8	12	20	3	2	50

19. The average score of boys in an examination of a school is 71 and that of girls is 73. The average score of the school in the examination is 71.8. Find the ratio of the number of boys to the number of girls who appeared in the examination.

20. Show that any positive odd integer is of the form $4q + 1$ or $4q + 3$, where q is some integer.

SECTION D

21. In a right triangle, prove that the square on the hypotenuse is equal to the sum of the squares of the other two sides.

22. Prove that: $\dfrac{\tan\theta + \sec\theta - 1}{\tan\theta - \sec\theta + 1} = \dfrac{1 + \sin\theta}{\cos\theta}$

23. In a ΔABC, $\angle A = x°$, $\angle B = (3x - 2)°$, $\angle C = y°$. Also $\angle C - \angle B = 9°$. Find the three angles.

24. The following table gives the marks obtained by 80 students in a selection test:

Marks	Below 10	Below 20	Below 30	Below 40	Below 50	Below 60
No. of students	3	12	27	57	75	80

Find the median marks.

25. Apply the division algorithm to find the quotient and remainder on dividing $p(x)$ by $g(x)$, where

$p(x) = x^4 - 5x + 6,\ g(x) = 2 - x^2$

26. Prove that: $\dfrac{\tan\theta}{1 - \cot\theta} + \dfrac{\cot\theta}{1 - \tan\theta} = 1 + \tan\theta + \cot\theta$

27. Without using trigonometric tables, evaluate:

$$\frac{3\tan 25°\tan 40°\tan 50°\tan 65° - \frac{1}{2}\tan^2 60°}{4(\cos^2 29° + \cos^2 61°)}$$

*28. A group consists of honest and extremely kind people. If 8 honest people and 12 extremely kind people can finish a piece of work in 10 days while 6 honest and 8 extremely kind people can finish it in 14 days, then find the time taken by one honest people alone and that by extremely kind people alone to finish the work. Which of the above values you prefer more?

29. ABCD is a trapezium such that BC || AD and AB = 4cm. If the diagonal AC and BD intersects at E such that $\dfrac{AE}{EC} = \dfrac{DE}{BE} = \dfrac{1}{2}$, then find DC.

30. Find the mean, mode and median of the following frequency distribution:

Class	0 – 10	10 – 20	20 – 30	30 – 40	40 – 50	50 – 60	60 – 70
Frequency	8	7	15	20	12	8	10

31. Show that $\dfrac{2 + \sqrt{5}}{3}$ is an irrational number.

*Value Based Question

Answers

1. No value of k
2. 2
3. $\frac{xz}{y}$
4. 72
5. $x = \frac{1}{5}, y = \frac{1}{11}$
6. $x^2 + 12x - 85$
9. 8
10. $\frac{49}{64}$
11. yes it is rational = $\frac{1258}{125}$
12. Quotient = $2x^2 + 2x - 1$; Remainder = $15x - 14$
14. 0
16. (–2, 0), (3, 0), (2, 3)
18. 65.6
19. 3 : 2
23. 25°, 73°, 82°
24. 34.33
25. Quotient = $-x^2 - 2$; Remainder = $-5x + 10$
27. $\frac{3}{8}$
28. Honest people = 140 days, Kind people = 280 days, Preferably honesty.
29. 8 cm
30. Mean = 35.63, Median = 35, mode = 33.85

Practice Paper 5

Time Allowed : 3 Hours *Maximum Marks : 90*

General Instructions: As given in Practice Paper 1

SECTION A

1. If mean of n observations is 30 and their mode is 45, then find the median.

2. In Δ ABC, AB = $6\sqrt{3}$ cm, AC = 12 cm and BC = 6 cm. Find $\angle$B.

3. Express $\left(\frac{17}{8}+\frac{7}{80}\right)$ as a decimal representation.

4. For what value of k, the graphs of the equations, $3x + 7y + 5 = 0$ and $9x + ky + 5 = 0$ will represent two parallel lines?

5. If k is a zero of $4x^2 - 7x + 3$, find the value of k.

6. The sum of the digits of a two digit number is 15. The number is decreased by 27, if the digits are reversed. Represent this situation algebraically in two variables.

SECTION B

7. If $\sin \alpha = \frac{1}{\sqrt{2}}$ and $\tan \beta = 1$, find the value of $\sin(\alpha + \beta)$ where α and β are both acute angles.

8. In the given figure, $\angle ACB = 90°$, $CD \perp AB$.

   Prove that $\frac{CB^2}{CA^2} = \frac{BD}{AD}$.

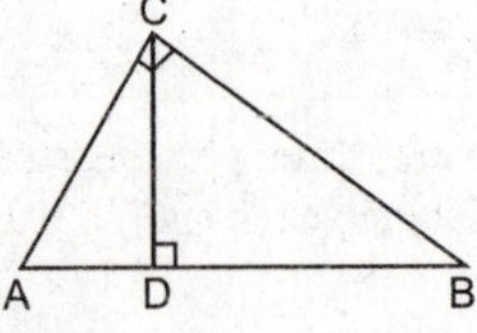

9. Find the median.

x_i	5	7	9	11	13	15	17
f_i	4	6	7	8	3	2	1

10. If $\sin \theta = \frac{1}{3}$, then find the value of $(2 \cot^2 \theta + 2)$.

SECTION C

11. Find k so that $x^2 + 2x + k$ is a factor of $2x^4 + x^3 - 14x^2 + 5x + 6$.

12. Solve the following by the method of cross multiplication:

    $(a - b)x + (a + b)y = a^2 - 2ab - b^2, (a + b)(x + y) = a^2 + b^2$.

13. Show that $5 - \sqrt{3}$ is an irrational number.

14. Prove that the ratio of the areas of two similar triangles is equal to the squares of their corresponding sides.

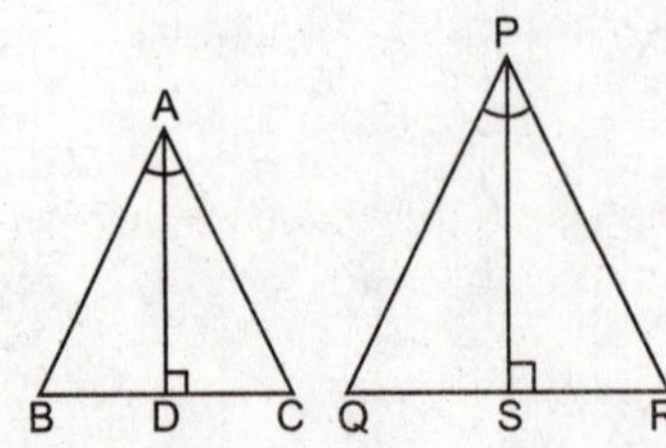

Using the above, do the following :

In fig., ΔABC and ΔPQR are isosceles triangles in which $\angle A = \angle P$.

If $\frac{\text{area}(\Delta ABC)}{\text{area}(\Delta PQR)} = \frac{9}{16}$, find $\frac{AD}{PS}$.

15. In the given figure, if ABCD is a trapezium in which AB || DC || EF, then show that $\frac{AE}{ED} = \frac{BF}{FC}$.

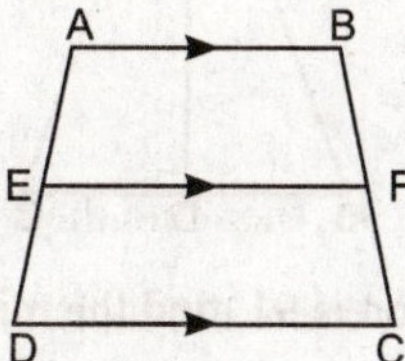

16. Prove that: $\sin^4\theta - \cos^4\theta = \sin^2\theta - \cos^2\theta$.

17. If a cosec A = p and b cot A = q, then prove that $\frac{p^2}{a^2} - \frac{q^2}{b^2} = 1$.

18. The following distribution gives the state wise teacher and student ratio in higher secondary school of India.

Number of Students	15–20	20–25	25–30	30–35	35–40	40–45	45–50	50–55
Number of teachers	3	8	9	10	3	0	0	2

Find the mode.

19. Find the missing frequencies and the median for the following distribution, if the mean is 1.46.

No. of accidents	0	1	2	3	4	5	Total
Frequency (No. of days)	46	?	?	25	10	5	200

20. On a morning walk, three persons step off together and their steps measure 40 cm, 42 cm and 45 cm respectively. What is the minimum distance each should walk so that each can cover the same distance in complete steps?

SECTION D

21. Show that: $\frac{\tan^3\theta}{1+\tan^2\theta} + \frac{\cot^3\theta}{1+\cot^2\theta} = \sec\theta \operatorname{cosec}\theta - 2\sin\theta\cos\theta$.

22. State and prove converse of Pythagoras theorem.

23. Draw the graph of the equations $x - y + 1 = 0$ and $3x + 2y - 12 = 0$. Determine the coordinates of the vertices of the triangle formed by these lines and the x-axis and shade the triangular region.

24. The following table gives production yield per hectare of wheat of 100 farms of a village.

Production yield (in kg/ha)	50 – 55	55 – 60	60 – 65	65 – 70	70 – 75	75 – 80
Number of farms	2	8	12	24	38	16

Change the distribution to a more than type distribution and draw its ogive.

***25.** A group consist of 19 honest people and 17 punctual people.

(*a*) Write a quadratic polynomial whose roots are equal to number of honest people and number of punctual people.

(*b*) Which value do you prefer ?

(*c*) Also, find the zeroes of this polynomial.

26. Prove that: $(\sin\theta + \operatorname{cosec}\theta)^2 + (\cos\theta + \sec\theta)^2 = 7 + \tan^2\theta + \cot^2\theta$

27. Prove that: $\sin\theta(1 + \tan\theta) + \cos\theta(1 + \cot\theta) = \sec\theta + \operatorname{cosec}\theta$.

28. A train covered a certain distance at a uniform speed. If the train would have been 6 km/hr. faster, it would have taken 4 hours less than the scheduled time. And if the train were slower by 6 km/hr., it would have taken 6 hours more than the scheduled time. Find the length of the journey.

*Value Based Question

29. In the given figure, perpendiculars OD, OE and OF are drawn to sides BC, CA and AB respectively from a point O in the interior of a ΔABC. Prove that: $AF^2 + BD^2 + CE^2 = OA^2 + OB^2 + OC^2 - OD^2 - OE^2 - OF^2$.

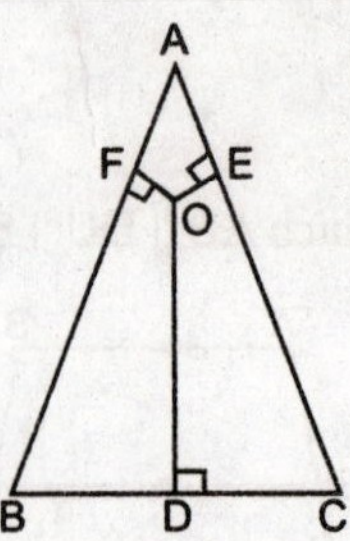

30. If the mean of the following frequency distribution is 91, find the missing frequencies (f_1, f_2) :

Class	Frequency
0 – 30	12
30 – 60	21
60 – 90	f_1
90 – 120	52
120 – 150	f_2
150 – 180	11
Total	150

31. Show that a number and its cube leaves the same remainder when divided by 6.

Answers

1. 35
2. 90°
3. 2.2125
4. $k = 21$
5. $k = 1, \frac{3}{4}$
6. $x + y = 15 \quad x - y = -3$
7. 1
9. 9
10. 18
11. $k = -3, -1,$
12. $x = a + b, y = \frac{-2ab}{a+b}$
14. 3 : 4
18. 30.62
19. Frequencies = 76, 38; median = 1
20. 2520 cm
23. (2, 3), (4, 0) and (–1, 0)
25. (*a*) $k(x^2 - 36x + 323)$ (*b*) Honesty (*c*) 17, 19
28. Distance = 720 km
30. $f_1 = 34, f_2 = 20$